Physicochemical Treatment Processes

VOLUME 3
HANDBOOK OF ENVIRONMENTAL ENGINEERING

Physicochemical Treatment Processes

Edited by

Lawrence K. Wang, PhD, PE, DEE
Zorex Corporation, Newtonville, NY
Lenox Institute of Water Technology, Lenox, MA
Krofta Engineering Corporation, Lenox, MA

Yung-Tse Hung, PhD, PE, DEE
Department of Civil and Environmental Engineering
Cleveland State University, Cleveland, OH

Nazih K. Shammas, PhD
Lenox Institute of Water Technology, Lenox, MA

HUMANA PRESS TOTOWA, NEW JERSEY

999 Riverview Drive, Suite 208
Totowa, New Jersey 07512

humanapress.com

For additional copies, pricing for bulk purchases, and/or information about other Humana titles, contact Humana at the above address or at any of the following numbers: Tel.: 973-256-1699; Fax: 973-256-8341; E-mail: humana@humanapr.com

This publication is printed on acid-free paper. ∞
ANSI Z39.48-1984 (American Standards Institute)
Permanence of Paper for Printed Library Materials.

Cover design by Patricia F. Cleary.

eISBN 1-59259-820-x

Printed in the United States of America. 10 9 8 7 6 5 4 3 2 1

Library of Congress Cataloging-in-Publication Data

Physicochemical treatment processes / edited by Lawrence K. Wang, Yung-Tse Hung, Nazih K. Shammas.
p. cm. — (Handbook of environmental engineering)
Includes bibliographical references and index.
ISBN 1-58829-165-0 (v. 3 : alk. paper)
1. Water—Purification. 2. Sewerage—Purification. I. Wang, Lawrence K. II. Hung, Yung-Tse. III. Shammas, Nazih K. IV Series: Handbook of environmental engineering (2004) ; v. 3.
TD170 .H37 2004 vol. 3
[TD430]
628 s—dc22 [628.1/ 2004002102

Preface

The past 30 years have seen the emergence of a growing desire worldwide to take positive actions to restore and protect the environment from the degrading effects of all forms of pollution: air, noise, solid waste, and water. Because pollution is a direct or indirect consequence of waste, the seemingly idealistic demand for "zero discharge" can be construed as an unrealistic demand for zero waste. However, as long as waste exists, we can only attempt to abate the subsequent pollution by converting it to a less noxious form. Three major questions usually arise when a particular type of pollution has been identified: (1) How serious is the pollution? (2) Is the technology to abate it available? and (3) Do the costs of abatement justify the degree of abatement achieved? The principal intention of the *Handbook of Environmental Engineering* series is to help readers formulate answers to the last two questions.

The traditional approach of applying tried-and-true solutions to specific pollution problems has been a major contributing factor to the success of environmental engineering, and has accounted in large measure for the establishment of a "methodology of pollution control." However, realization of the ever-increasing complexity and interrelated nature of current environmental problems makes it imperative that intelligent planning of pollution abatement systems be undertaken. Prerequisite to such planning is an understanding of the performance, potential, and limitations of the various methods of pollution abatement available for environmental engineering. In this series of handbooks, we will review at a tutorial level a broad spectrum of engineering systems (processes, operations, and methods) currently being utilized, or of potential utility, for pollution abatement. We believe that the unified interdisciplinary approach in these handbooks is a logical step in the evolution of environmental engineering.

The treatment of the various engineering systems presented in *Physicochemical Treatment Process* shows how an engineering formulation of the subject flows naturally from the fundamental principles and theories of chemistry, physics, and mathematics. This emphasis on fundamental science recognizes that engineering practice has in recent years become more firmly based on scientific principles rather than its earlier dependency on empirical accumulation of facts. It is not intended, though, to neglect empiricism when such data lead quickly to the most economic design; certain engineering systems are not readily amenable to fundamental scientific analysis, and in these instances we have resorted to less science in favor of more art and empiricism.

Because an environmental engineer must understand science within the context of application, we first present the development of the scientific basis of a particular subject, followed by exposition of the pertinent design concepts and operations, and detailed explanations of their applications to environmental quality control or improvement. Throughout this series, methods of practical design calculation are illustrated by numerical examples. These examples clearly demonstrate how organized, analytical reasoning leads to the most direct and clear solutions. Wherever possible, pertinent cost data have been provided.

Our treatment of pollution-abatement engineering is offered in the belief that the trained engineer should more firmly understand fundamental principles, be more aware of the similarities and/or differences among many of the engineering systems, and exhibit greater flexibility and originality in the definition and innovative solution of environmental pollution problems. In short, environmental engineers should by conviction and practice be more readily adaptable to change and progress.

Coverage of the unusually broad field of environmental engineering has demanded an expertise that could only be provided through multiple authorships. Each author (or group of authors) was permitted to employ, within reasonable limits, the customary personal style in organizing and presenting a particular subject area, and, consequently, it has been difficult to treat all subject material in a homogeneous manner. Moreover, owing to limitations of space, some of the authors' favored topics could not be treated in great detail, and many less important topics had to be merely mentioned or commented on briefly. All of the authors have provided an excellent list of references at the end of each chapter for the benefit of the interested reader. Because each of the chapters is meant to be self-contained, some mild repetition among the various texts was unavoidable. In each case, all errors of omission or repetition are the responsibility of the editors and not the individual authors. With the current trend toward metrication, the question of using a consistent system of units has been a problem. Wherever possible the authors have used the British system along with the metric equivalent or vice versa. The authors sincerely hope that this doubled system of unit notation will prove helpful rather than disruptive to the readers.

The goals of the *Handbook of Environmental Engineering* series are: (1) to cover the entire range of environmental fields, including air and noise pollution control, solid waste processing and resource recovery, biological treatment processes, water resources, natural control processes, radioactive waste disposal, thermal pollution control, and physicochemical treatment processes; and (2) to employ a multithematic approach to environmental pollution control because air, water, land, and energy are all interrelated. The organization of the series is mainly based on the three basic forms in which pollutants and waste are manifested: gas, solid, and liquid. In addition, noise pollution control is included in one of the handbooks in the series.

This volume, *Physicochemical Treatment Processes*, has been designed to serve as a basic physicochemical treatment text as well as a comprehensive reference book. We hope and expect it will prove to be of□high value to advanced undergraduate or graduate students, to designers of water and wastewater treatment systems, and to research workers. The editors welcome comments from readers in all these categories. It is our hope that this book will not only provide information on the physical, chemical, and mechanical treatment technologies, but will also serve as a basis for advanced study or specialized investigation of the theory and practice of the individual physicochemical systems covered.

The editors are pleased to acknowledge the encouragement and support received from their colleagues and the publisher during the conceptual stages of this endeavor. We wish to thank the contributing authors for their time and effort, and for having patiently borne our reviews and numerous queries and comments. We are very grateful to our respective families for their patience and understanding during some rather trying times.

Lawrence K. Wang
Yung-Tse Hung
Nazih K. Shammas

Contents

Contributors

E. Robert Baumann, PhD • *Department of Civil Engineering, Iowa State University of Science and Technology, Ames, IA*

Chein-Chi Chang, PhD, PE • *District of Columbia Water and Sewer Authority, Washington, DC*

Shoou-Yuh Chang, PhD, PE • *Department of Civil and Environmental Engineering, North Carolina A&T State University, Greensboro, NC*

Durgananda Singh Chaudhary, PhD • *Faculty of Engineering, University of Technology Sydney (UTS), New South Wales, Australia*

J. Paul Chen, PhD • *Department of Chemical and Biomolecular Engineering, National University of Singapore, Singapore*

Frank DeLuise, ME, PE • *Emeritus Professor, Department of Mechanical Engineering, University of Rhode Island, Kingston, RI*

Edward M. Fahey, ME • *DAF Environmental, LLC, Hinsdale, MA*

Joseph R. V. Flora, PhD • *Department of Civil & Environmental Engineering, University of South Carolina, Columbia, SC*

Ramesh K. Goel, PhD • *Department of Civil and Environmental Engineering, University of Wisconsin, Madison, WI*

Pin Jing He, PhD • *School of Environmental Science and Engineering, Tongji University, Shanghai, China*

Frederick B. Higgins, PhD • *Civil and Environmental Engineering Department, Temple University, Philadelphia, PA*

Yung-Tse Hung, PhD, PE, DEE • *Department of Civil and Environmental Engineering, Cleveland State University, Cleveland, OH*

Jerry Y. C. Huang, PhD • *Department of Civil Engineering, University of Wisconsin–Milwaukee, Milwaukee, WI*

Inder Jit Kumar, PhD • *Eustance & Horowitz, P.C., Consulting Engineers, Circleville, NY*

Duu-Jong Lee, PhD • *Department of Chemical Engineering, National Taiwan University, Taipei, Taiwan*

Kathleen Hung Li, MS • *NEC Business Network Solutions, Irving, TX*

Yan Li, PE, MS • *Department of Environmental Management, State of Rhode Island, Providence, RI*

Howard Lo, PhD • *Department of Biological, Geological and Environmental Sciences, Cleveland State University, Cleveland, OH*

Huu Hao Ngo, PhD • *Faculty of Engineering, University of Technology Sydney (UTS), New South Wales, Australia*

Nazih K. Shammas, PhD • *Graduate Environmental Engineering Program, Lenox Institute of Water Technology, Lenox, MA*

Jerry R. Taricska, PhD, PE • *Hole Montes Inc., Naples, FL*

JOO-HWA TAY, PhD, PE • *Division of Environmental and Water Resource Engineering, Nanyang Technological University, Singapore*

DAVID A. VACCARI, PhD, PE, DEE • *Department of Civil, Environmental and Ocean Engineering, Stevens Institute of Technology, Hoboken, NJ*

SARAVANAMUTHU VIGNESWARAN, PhD, DSc, CPEng • *Faculty of Engineering, University of Technology Sydney (UTS), New South Wales, Australia*

LAWRENCE K. WANG, PhD, PE, DEE • *Zorex Corporation, Newtonville, NY; Lenox Institute of Water Technology, Lenox, MA; and Krofta Engineering Corporation, Lenox, MA*

JY S. WU, PhD • *Department of Civil Engineering, University of North Carolina at Charlotte, Charlotte, NC*

ZUCHENG WU, PhD • *Department of Environmental Science and Engineering, Zhejiang University, Hangzhou, People's Republic of China*

JOHN Y. YANG, PhD • *Niagara Technology Inc., Williamsville, NY*

PAO-CHIANG YUAN, PhD • *Technology Department, Jackson State University, Jackson, MS*

1
Screening and Comminution

Frank Deluise, Lawrence K. Wang, Shoou-Yuh Chang, and Yung-Tse Hung

CONTENTS

1. FUNCTION OF SCREENS AND COMMINUTORS

In order for water and wastewater treatment plants to operate effectively, it is necessary to remove or reduce early in the treatment process large suspended solid material that might interfere with operations or damage equipment. Removal of solids may be accomplished through the use of various size screens placed in the flow channel. Any material removed may then be ground to a smaller size and returned to the process stream or disposed of in an appropriate manner such as burying or incineration. An alternative to actual removal of the solids by screening is to reduce the size of the solids by grinding them while still in the waste stream; this grinding process is called comminution (1–8). Coarse screens (bar racks) and comminutors are usually located at the very beginning of a treatment process, immediately preceding the grit chambers (Fig. 1). To ensure continuous operation in a flow process, it is desirable to have the screens or comminutors installed in parallel in the event of a breakdown or to provide for overhaul of a unit. With this arrangement, flow is primarily through the comminutor and diverted to the coarse (bar) screens only when necessary to shut down the comminutor. Fine screens are usually placed after the coarse (bar) screens.

From: *Handbook of Environmental Engineering, Volume 3: Physicochemical Treatment Processes*
Edited by: L. K. Wang, Y.-T. Hung, and N. K. Shammas © The Humana Press Inc., Totowa, NJ

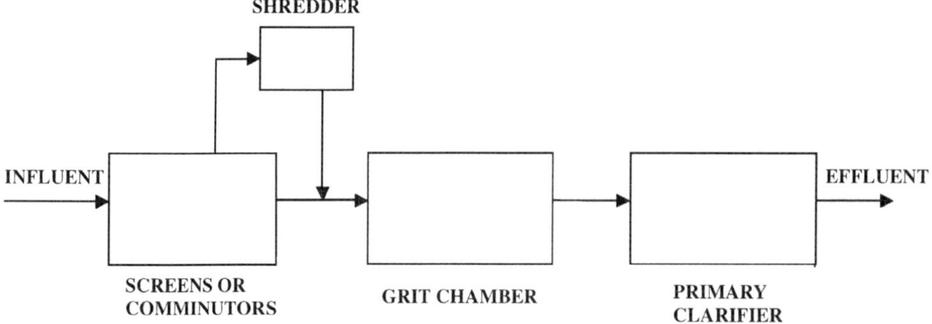

Fig. 1. Location of screens and comminutors in a wastewater treatment plant.

2. TYPES OF SCREENS

2.1. Coarse Screens

Screens may be classified as coarse or fine. Coarse screens are usually called bar screens or racks and are used where the wastewater contains large quantities of coarse solids that might disrupt plant operations. These bar screens consist of parallel bars spaced anywhere from 1.27 cm (1/2 in.) to 10.16 cm (4 in.) apart with no cross-members other than those required for support. The size of the spacing depends on the type of waste being treated (size and quantity of solids) and the type of equipment being protected downstream in the plant. These screens are placed either vertically or at an angle in the flow channel. Installing screens at an angle allows easier cleaning (particularly if by hand) and more screen area per channel depth, but obviously requires more space.

2.2. Fine Screens

Fine screens have openings of less than 0.25 in. and are used to remove solids smaller than those retained on bar racks. They are used primarily in water or wastewater containing little or no coarse solids. In many instances, fine screens are used for the recovery of valuable materials that exist as finely divided solids in industrial waste streams. Most fine screens use a relatively fine mesh screen cloth (openings anywhere from 0.005 to 0.126 in.) rather than bars to intercept the solids. A screen cloth covers discs or drums, which rotate through the wastewater. The disc-type screen (Fig. 2) is a vertical hoop with a screen cloth covering the area within the hoop, and mounted on a horizontal shaft that is positioned slightly above the surface of the water. Water flows through the screen parallel to the horizontal shaft and the solids are retained on the screen, which carries them out of the water as it rotates. Solids may then be removed from the upper part of the screen by water sprays or mechanical brushing.

The drum-type screen (Fig. 3) consists of a cylinder covered by a screen cloth with the drum rotating on a horizontal axis, slightly less than half submerged. Wastewater enters the inside of the drum at one end and flows outward through the screen cloth. Solids collect inside the drum on the screen cloth and are carried out of the water as the drum rotates. Once out of the water, the solids may be removed by backwater sprays, forcing the solids off the screen into collecting troughs.

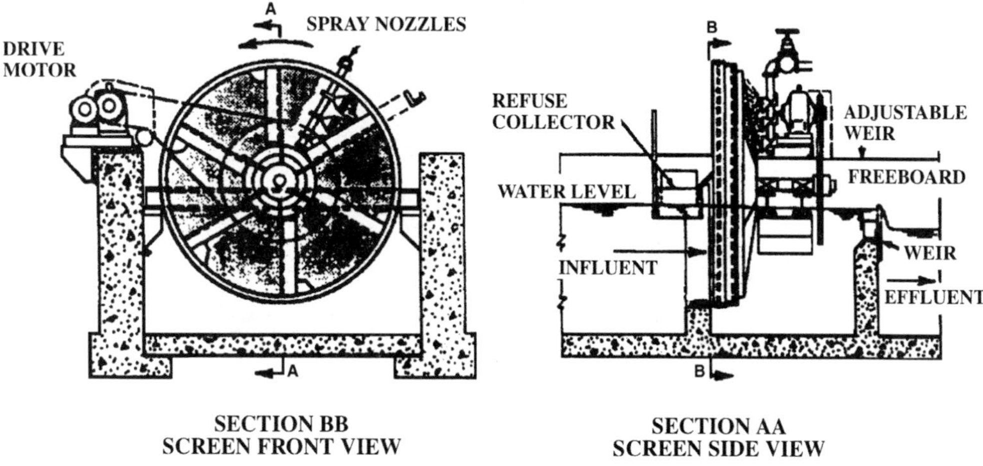

Fig. 2. Revolving disc screen: (a) screen front (inlet side) view and (b) screen side view section.

3. PHYSICAL CHARACTERISTICS AND HYDRAULIC CONSIDERATIONS OF SCREENS

The physical characteristics of bar racks and screens depend on the use for which the unit is intended. Coarse bar racks, sometimes called trash racks, with 7.62 or 10.16 cm (3 or 4 in.) spacing are used to intercept unusually large solids and therefore must be of rugged construction to withstand possible large impacts. Bar screens with smaller spacing may be of less rugged construction. As previously mentioned, the spacing between bars depends on the size and quantity of solids being intercepted. Although a screen's primary purpose is to protect equipment in a sewage-treatment plant, spacings smaller than 2.54 cm (1 in.) are usually not necessary because today's sewage sludge pumps can handle solids passing through the screen. Typical bar screens are shown in Fig. 4.

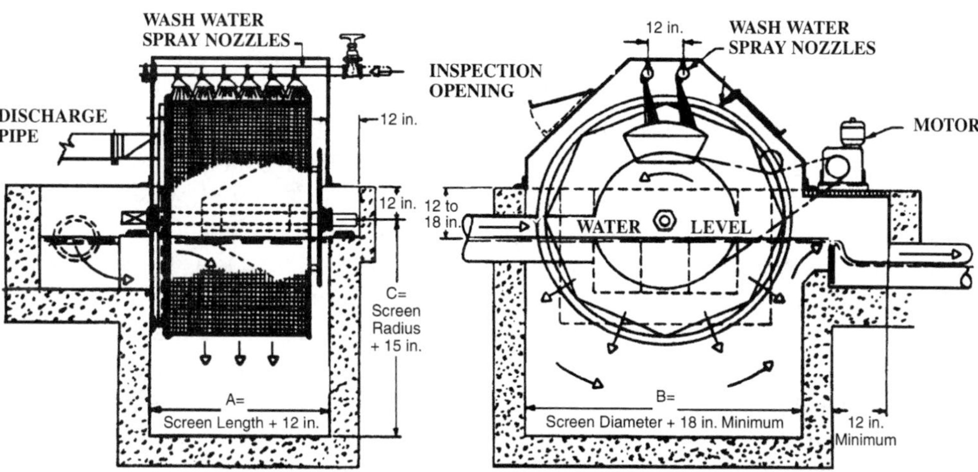

Fig. 3. Revolving drum screen.

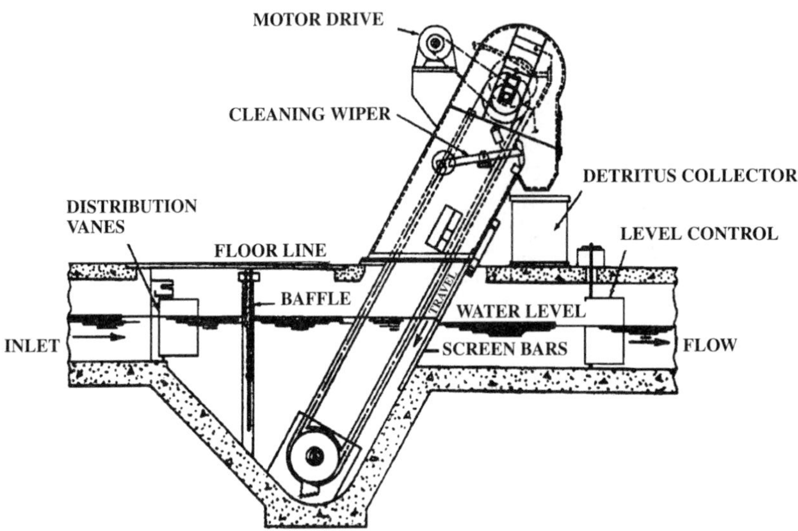

Fig. 4. Elements of a mechanical bar screen and grit collector.

The screen bars are usually rectangular in cross-section and their size depends on the size (width and depth) of the screen channel as well as the conditions under which the screen will be operating. The longer the unsupported length of the bar, the larger is the required cross-section. Bars up to 1.83 m (6 ft) in length are usually no smaller than 0.635 × 5.08 cm (1/4 × 2 in.), while bars up to 3.66 m (12 ft) long might be 0.952 × 6.35 cm (3/8 × 2.5 in.). Longer bars or bars used for operating conditions causing unusual stress might be as large as 1.59 × 7.62 cm (5/8 × 3 in.). The bars must be designed to withstand bending as well as impact stresses due to the accumulation of solids on the screen.

Many screens, particularly those that are hand-cleaned, are installed with bars at an angle between 60° and 90° with the horizontal. With the bars placed at an angle, the screenings will tend to accumulate near the top of the screen. In addition, the velocity through the screen will be low enough to prevent objects from being forced through the screen. Optimum horizontal velocity through the bars is approx 0.610 m/s (2 ft/s). If velocities get too low, sedimentation will take place in the screen channel. In the design of the screen channel, it is desirable to have the flow evenly distributed across the screen by having several feet of straight channel preceding the screen. Flow entering at an angle to the screen would tend to create uneven distribution of solids across the screen and prevent the proper operation of the equipment.

The required size of the screen channel depends on the volume flow rate and the free space available between the bars. If a net area ratio is defined as the free area between bars divided by the total area occupied by the screen, then a table such as Table 1 may be set up showing the net area ratio for various combinations of bar size openings.

The bar spacing should be kept as large as practical and the bar thickness as small as practical in order to obtain the highest net area ratio possible. Once the volume flow rates are known and the net area ratio is determined, the screen channel size may be determined. The maximum volume flow rate in cubic meters per second divided by the optimum velocity of 0.610 m/s will yield the net area required. This net area divided by

Table 1
Net Area Ratios for Bar Size and Openings

Bar size		Opening		
cm	in.	cm	in.	Net area ratio
0.635	1/4	1.27	1/2	0.667
0.635	1/4	2.54	1	0.800
0.635	1/4	3.81	1 1/2	0.856
0.952	3/8	1.27	1/2	0.572
0.952	3/8	2.54	1	0.728
0.952	3/8	3.81	1 1/2	0.800
1.270	1/2	1.27	1/2	0.500
1.270	1/2	2.54	1	0.667
1.270	1/2	3.81	1 1/2	0.750

the net area ratio selected will give the total wet area required for the channel. With this known area, the width and depth of the channel may be determined. Usually the maximum width or depth of the channel is limited by considerations other than the actual screening process. Too wide a screen could present problems in cleaning, and therefore the maximum practical width for a channel is about 4.27 m (14 ft); the minimum width is about 0.610 m (2 ft). The depth of liquid in the channel is usually kept as shallow as possible so that the head loss through the plant will be a minimum. The wet area divided by the known limiting width or depth will thus provide the dimensions of the channel.

From Bernoulli's equation, the theoretical head loss for frictionless, adiabatic flow through the bar screen is

$$h = \frac{V_2^2 - V_1^2}{2g} \tag{1}$$

where h = head loss, m (ft), V_2 = velocity through bar screen, m/s (ft/s), V_1 = velocity ahead of bar screen, m/s (ft/s), and $g = 9.806$ m/s^2 (32.17 ft/s^2).

To determine the actual head loss, the above expression may be modified by a discharge coefficient, C_D, to account for deviation from theoretical conditions. Values of C_D should be determined experimentally, but a typical average value is 0.7. The equation then becomes

$$h = \frac{V_2^2 - V_1^2}{C_D 2g} \tag{2}$$

$$h = 0.0728\left(V_2^2 - V_1^2\right) \text{ with SI units} \tag{2a}$$

$$h = 0.0222\left(V_2^2 - V_1^2\right) \text{ with English units} \tag{2b}$$

4. CLEANING METHODS FOR SCREENS

Bar screens or racks may be cleaned by hand or by machine. Hand-cleaning limits the length of screen that may be used to that which may be conveniently raked by hand. The cleaning is accomplished using a specially designed rake with teeth that fit between the bars of the rack. The rake is pulled up toward the top of the screen carrying the screenings with it. At the top of the screen, the screenings are deposited on a grid or perforated plate for drainage and then removed for shredding and return to the channel or for incineration or burial. Hand-cleaning requires a great deal of manual labor and is an unpleasant job. Because hand-cleaning is not continuous, plant operations may be materially affected by undue plugging of the screens before cleaning as well as by large surges of flow when the screens are finally cleaned. Plugging of the screens could cause troublesome deposits in the lines leading to the bar screens, and surges after cleaning could disrupt the normally smooth operations of units following the screens.

Mechanical cleaning overcomes many of the problems associated with hand-cleaning. Although the initial cost of a mechanically cleaned screen will be much greater than for a hand-cleaned screen, the improvement in plant efficiency, particularly in large installations, usually justifies the higher cost. The ability to operate the cleaning mechanism on an automatically controlled schedule avoids the flooding and surging through the plant associated with plugging and unplugging of the screens. After a short while, a preset automatic cleaning cycle may be easily established to keep the bars relatively clear at all times.

Mechanically cleaned screens use moving rakes attached to either chains or cables to carry the screenings to the top of the screen. At the top of the screen, rake wiper blades sweep the screenings into containers or onto conveyor belts for disposal. The teeth on the rakes project between the screen bars either from the front or the back of the rack. Both methods have their advantages and disadvantages. The front-cleaned models have the rakes passing down through the wastewater in front of the rack and then up the face of the rack. This method provides excellent cleaning efficiency, but the rakes may potentially become jammed as they pass through any accumulation of solids at the base of the screen on the downward travel. A modification of the front-cleaned model has the rakes traveling down behind the screen and through a boot under the screen, and then moving up the front of the screen. The back-cleaned models eliminate the jamming problem by having the rakes travel down through the water behind the screen and then travel up behind the screen with teeth projecting through the bars far enough to pick up solids deposited on the front of the screen. In models where the rake travels up the back of the screen, the bars are fixed only at the bottom of the screen because the rake must project all the way through the bars. It is thus possible for the bars to move as they are supported only by the traveling rake teeth. With movement of the bars, it is possible for solids substantially larger than those designed for to pass through the screen. Another drawback of the back-cleaned screen is that any solids not removed from the rakes because of faulty wiper blades are returned to the flow behind the screen. Several manufacturers have modified both the front- and back-cleaned screens to help reduce some of these problems.

5. QUANTITY AND DISPOSAL OF SCREENINGS

The quantity of screenings is obviously greatly affected by the type and size of screen openings and the nature of the waste stream being screened. The curves in Fig. 5 show the average and maximum quantities of screenings in cubic feet per 10^6 gallons (ft^3/MG) that might be obtained from sewage for different sized openings between bars. Data for these curves were obtained from 133 installations of hand-cleaned and mechanically cleaned bar screens in the United States. It can be seen that the average

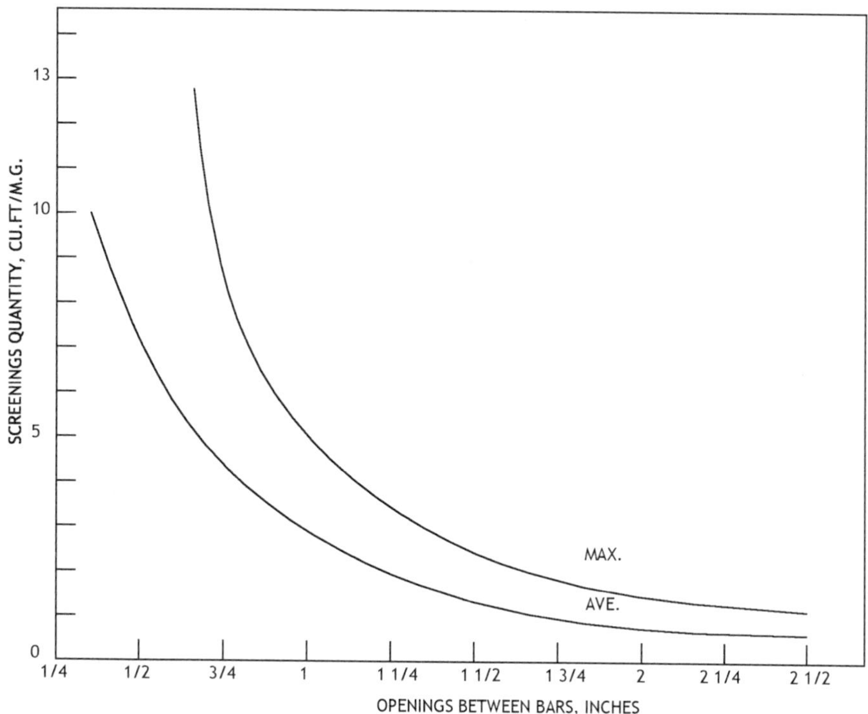

Fig. 5. Quantity of screenings from wastewater as a function of openings between bars. (*Source*: US EPA)

screenings vary from 71.1 $m^3/10^6$ m^3 (9.5 ft^3/MG) for a 0.952 cm (3/8 in.) opening to 3.74 $m^3/10^6$ m^3 (0.5 ft^3/MG) for a 6.35 cm (2.5 in.) opening. Taking a common opening of 2.54 cm (1 in.), the average quantity of screenings expected would be about 22.4 $m^3/10^6$ m^3 (3 ft^3/MG), and the maximum quantity expected would be 37.4 $m^3/10^6$ m^3. Fine screens with openings from 0.119 to 0.318 cm (3/64 to 1/8 in.) have typical screenings of 224.4 to 37.4 $m^3/10^6$ m^3 (30 to 5 ft^3/MG) of sewage flow. The density of all screenings from a typical municipal sewage treatment plant is approx 800–960 kg/m^3 (50–60 lb/ft^3).

Screenings may be disposed of by grinding and returning them to the flow, by burial in landfill areas or at the plant site, or by incineration. Incineration usually requires partial dewatering of the screenings by some type of pressing and therefore is not usually practical except for large installations with large volumes of screenings.

6. COMMINUTORS

The handling and disposal of screenings is at best a disagreeable and expensive procedure unless the product has some recovery value. To overcome this problem, devices were developed to cut up large screened material into small, relatively uniform size solids, without removal from the line of flow. These devices are generally referred to as comminutors (8–14). Figure 6 shows the essential elements of a comminutor, and Fig. 7 shows a crosssection of a typical comminutor. Various methods are used to accomplish the cutting of the solids.

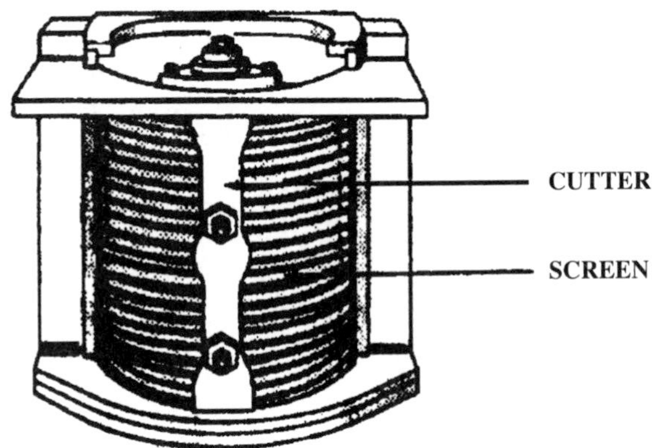

Fig. 6. Essential elements of a comminutor.

One type of comminuting device uses a slotted, rotating drum mounted vertically in the flow channel. Liquid passes through the slots down through the bottom of the drum and into the downstream channel. The solids are retained on the outside of the drum and carried by the drum to stationary comb bars mounted against the main casing of the comminutor. Mounted on the drum are hardened cutting teeth and shear bars (usually removable for sharpening or replacement) that pass through the comb bars, thereby cutting the solids. The small particles that result from the cutting operation then pass through the slots of the drum with the liquid flow.

Another type of device uses a stationary vertical semicircular screen grid (installed convex to the flow), with rotating circular discs on whose edges are mounted the cutting teeth. The grid intercepts the larger solids, while smaller solids pass through the clearing space between the grid and cutter discs. The rotating cutter teeth move the intercepted solids around to a stationary cutter comb where the solids are sheared as the teeth pass through the comb.

A third type of comminutor also uses a stationary vertical semicircular screen grid with horizontal slots, but is installed concave rather than convex to the flow. Ahead of the screen, a vertical arm with a cutter bar attached oscillates back and forth so the teeth on the cutting bar pass between the horizontal slots. The oscillating cutter bar carries the trapped solids to a stationary cutter bar mounted on the screen grid where the teeth of the cutters mesh and thereby shear the solids.

Various size comminutors are commercially available. For low flows, units as small as 10.16 cm (4 in.) in diameter are available, while units with 137.16 cm (54 in.) diameter can handle flows up to 3.15 m^3/s (72 million gallons / d [MGD]). Most of the units use slot widths of either 0.635 cm (1/4 in.) or 0.952 cm (3/8 in.). Power requirements vary from 186 W (1/4 hp) for the smaller units to 1491 W (2 hp) for the larger units.

7. ENGINEERING SPECIFICATIONS AND EXPERIENCE

7.1. Professional Association Specifications

The Water Pollution Control Federation (WPCF) Technical Practice Committee explains the screening process and equipment (1), as well as the types of bar screens and bar racks and the differences between them.

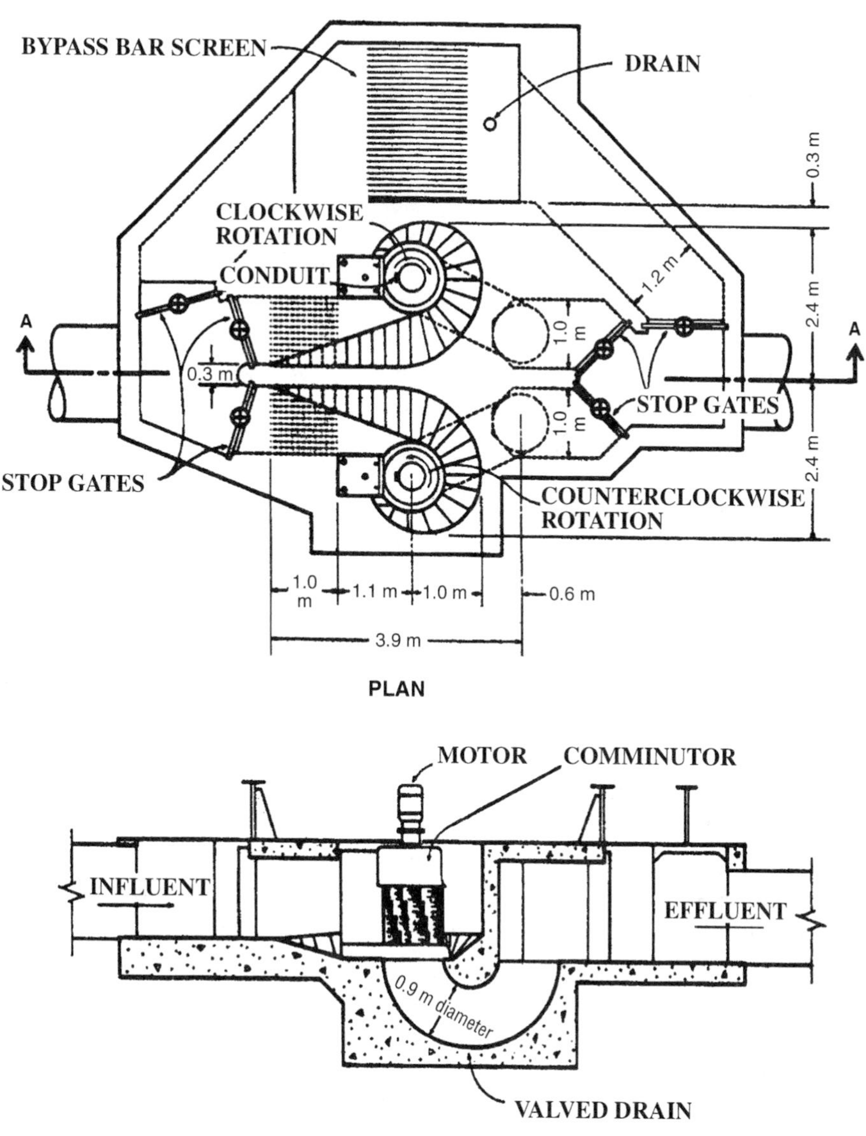

Fig. 7. Crosssection of a comminutor.

Detailed information is also given by the WPCF on screening equipment operation. Equipment should be checked frequently to ensure that it runs correctly. Screen overflow should be prevented and cleanliness maintained in order to prevent or eliminate (a) decay of organic matter, (b) offensive odors, and (c) pathogens. On dry days, daily removal of debris is sufficient. However, on rainy days, debris should be removed more frequently because leaves and other matter from combined sewer overflow (CSO) may be transported to the plant (1).

Screening equipment may require troubleshooting for several reasons: abnormal operational circumstances (unexpected loads of debris that clog or jam the screening equipment), equipment failure, and control failure. If a mechanically cleaned screen lacks blubber-control systems, it could suddenly receive huge loads of debris that jam its raking mechanisms.

Proper maintenance of screening equipment includes performing routine checks of components for obstructions, proper alignment, constant speed, and unusual vibrations and sounds. Screeches may result from a lack of lubrications, while thumps may mean the components are loose or broken. Proper lubrication is an important preventive maintenance procedure. Chain-driven bar screens require frequent replacement of chains, sprockets, and other parts that appear to be badly worn. Periodic removal of a link may be required to make certain that a chain rides smoothly on the sprockets.

A description of comminutors, grinders, and various bar screens, such as trash racks, manually cleaned screens, and mechanically cleaned screens is provided by the Water Environment Federation (WEF) Manual of Practice (2). The types of mechanically cleaned screens include chain- or cable-driven screens, reciprocating rake screens, centenary screens, and continuously self-cleaning screens. Trash racks, which are usually used in combined systems that have very large debris, are bar screens with large openings of 38 to 150 mm. The oldest mechanical-screening device is the chain- or cable-driven screen, which uses a chain or cable to move the rake teeth through the screen openings. They are produced as front clean/front return, front clean/rear return, and back clean/rear return. The front clean/front return has proven to be the most efficient. The up and down motion of the reciprocating rake screen reduces the risk of jamming and, because their parts are not submerged, they permit simple inspection and maintenance. The reciprocating screen is at a disadvantage because the single rake limits the ability to handle excessive loads and requires high overhead clearance. Cantenary screens have heavy tooth rakes, secured against the screen by the weight of its chain and a curved transition piece at the base that provides for effective removal of solids confined at the bottom. Continuous self-cleaning screens are comprised of a belt of plastic or stainless- steel elements that are pulled through the wastewater to provide screening along the entire length of the screen and are designed with vertical and horizontal limiting devices. The size of openings may range from 1 to more than 76 mm. The continuous screening motion provides effective removal of a large number of solids, but has the disadvantage of possible carryover of solids due to its front clean/back return design.

When designing mechanical bar screens, the following parameters should be considered: (a) bar spacing, construction materials, and dimensions; (b) depth of channel, width, and approach velocity; (c) discharge height; (d) angle of screen; (e) screen cover to obstruct wind and improve appearance; (f) coatings for overall unit; (g) drive unit service factor; (h) drive motor sized and enclosure; (i) spare parts; (j) stipulation of unneeded screen or bypass manual screen; and (k) head loss through unit.

The designer must consider the effects of the backwater caused by the head loss through the screen when considering a screen location. Many installations comprise an overflow weir to a bypass channel to avoid upstream surcharging if the screen becomes affected by power failure or mechanical problems.

In the past, most screening devices were placed downstream from grit chambers to prevent grit damage of comminutor teeth and combs. However, screening devices are presently placed upstream because they are more cost effective and cause fewer problems than downstream placement. A structural enclosure for screening devices is most favorable under windy and freezing climate conditions. An enclosure also reduces the amount of maintenance required and improves aesthetics.

7.2. Engineering Experience

Liu and Liptak (3) stated that the combined mechanical screen and grit collector can be used for small- and medium-sized plants. It is similar to the front cleaned mechanical screen, but rakes are connected to one or more perforated buckets and a steep hopper to collect the grit precedes the screen. The disadvantage of the system is that screenings and grit are mixed (3).

Some plants use coarse-mesh screens instead of screens and comminutors. Wastewater travels through a basket of wires or rods with a mesh size 1 in. or more. Coarse suspended matter is left in the basket.

Revolving drum screens may be characterized as having either outward or inward flow. With outward flow, the wastewater can move toward the drum from a direction parallel to its axis. Solids are captured on the inside of the screen. With inward flow, wastewater travels perpendicular to the drums axis and solids are captured on the outside of the drum. In both systems, the captured solids are lifted above the water level as the drum slowly rotates. Solids are usually removed by water spray, which is the disadvantage of these systems because solids are then mixed with great amount of spray water (3).

The revolving vertical disk screen is another screening device that employs the same principles as the revolving drum but uses a slowly revolving disc screen. The screen is positioned in the approach channel totally blocking the flow so that it travels through the screen. Solids are raised above the liquid level and washed by water spray. The screen consists of a 2–60-mesh stainless-steel wire cloth and is not suited for handling very large objects, large amounts of suspended objects, or greasy, gummy or sticky solids (3).

The inclined revolving disk screen consists of a round flat plate revolving on an axis inclined 10° to 25°, and the disk is comprised of bronze plates with slots 1/6 to 1/2 in. wide. As the liquid passes through the lower two-thirds of the plates, solids are captured, elevated above the water, and removed by brushes.

The traveling water screen, which has limited use in sewage treatment, consists of several inclined screen trays on two strands of steel chain. The head wheel is powered by a motor that moves screen trays through the sewage for disposal of solids by jets of water. The trays then return to the wastewater. Vibrating screens are used in the food packing industry to capture grease and meat particles, remove manure, catch animal hair, remove feathers from poultry, and retain vegetable and fruit particles from canning wastes. Vibration reduces the clogging of screens, which are flat and covered by stainless-steel cloth of 20 to 200 mesh.

Microscreens have openings as small as 20 μm and are used to remove fine suspended solids from effluent in tertiary treatment units. Hydrasieves is used for industrial effluent in treatment in plants that require an efficiency of 20–35% suspended solids and biochemical oxygen demand (BOD) removal. No power is needed to operate except to lift the water to the headbox of the screen. The microscreens are self-cleaning and require little maintenance. Wastewater is supplied by gravity or pumped into the headbox of the microscreen consisting of three slopes of 25°, 35°, and 45°.

Table 2
General Characteristics of Bar Screens

Item	Hand cleaned	Mechanically cleaned
Bar screen size		
Width, in.	¼ to ⅝	¼ to ⅝
Depth, in.	1 to 3	1 to 3
Spacing, in.	1 to 2	⅝ to 3
Slope from vertical, deg	30 to 45	0 to 30
Approach velocity, fps	1 to 2	2 to 3
Allowable head loss, in.	6	6

(*Source*: US Army).

8. ENGINEERING DESIGN

8.1. Summary of Screening Design Considerations

Screening devices are designed to remove large floating objects that may otherwise damage pumps and other equipment, obstruct pipelines, and interfere with the normal operation of the treatment facilities. As discussed in previous sections, screens used in water and wastewater treatment facilities or in pumping stations are generally classified as fine screens or bar screens.

Fine screens are those with openings of less than 0.25 in. These screens have been used as a substitute for sedimentation tanks to remove suspended solids prior to biological treatment. However, few plants today use this concept of solids removal. Fine screens may be of the disc, drum, or bar type. Bar-type screens are available with openings of 0.005 to 0.126 in.

Bar screens are used mainly to protect pumps, valves, pipelines, and other devices from being damaged or clogged by large floating objects. These screens are sometimes used in conjunction with comminuting devices. Bar screens consist of vertical or inclined bars spaced at equal intervals (usually 0.5–4 in.) across the channel where water or wastewater flows. These devices may be cleaned manually or mechanically. Bar screens with openings exceeding 2.5 in. are also termed trash racks.

The quantity of screenings removed by bar screens usually depends on the size of the bar spacing. Because handling and disposal of screenings is one of the most disagreeable jobs in wastewater treatment, it is usually recommended that the quantity of screenings be kept to a minimum. Amounts of screenings from normal domestic wastes have been reported from 0.5 to 5 ft^3/MG of wastewater treated. Screenings may be disposed of by burial, incineration, grinding, and digestion.

Bar screen designs are based mainly on average and peak wastewater flow. Normal design and operating parameters are usually presented in the manufacturer's specifications. The literature (1–7) presents a thorough discussion of the design, operation, and maintenance of screening devices. General characteristics of bar and fine screens are presented in Tables 2 and 3, respectively. Figure 4 shows a mechanically cleaned bar rack.

8.1.1. Screen Design Input Data

The following input data are required for the design of screens:

Table 3
General Characteristics of Fine Screens

Item	Disc	Drum
Fine screen		
Openings, in.	0.126 to 0.009 (6 to 60 mesh)	0.126 to 0.009 (6 to 60 mesh)
Diameter, ft	4 to 18	3 to 5
Length, ft		4 to 12
rpm		4

(*Source*: US Army).

1. Wastewater Flow
    - Average daily flow, MGD
    - Maximum daily flow, MGD
    - Peak wet weather flow, MGD
2. Wastewater Characteristics
    - Alkalinity and acidity (pH adjustment may be required)
    - pH (pH adjustment may be required)

8.1.2. Screen Design Parameters

The screen's design parameters are summarized below:

1. Type of bar screen
    - Manually cleaned
    - Mechanically cleaned
2. Velocity through bar screen, ft/s (Table 2)
3. Approach velocity, ft/s (Table 2)
4. Maximum head loss through screen, in. (Table 2)
5. Bar spacing, in. (Table 2)
6. Slope of bars, degree (Table 2)
7. Channel width, ft
8. Width of bar, in.
9. Shape factor

8.1.3. Screen Design Procedures

The procedures for screen design are:

Step 1: Consult equipment manufacturer's specifications and select a bar screen that meets design requirements.
Step 2: Calculate head loss through the screen. It should be noted that when screens start to become clogged between cleanings in manually cleaned screens, head loss will increase.

$$H_e = B(W/b)^{4/3}\left[\left(v^2 \sin^2 A\right)/2g\right] \tag{3}$$

where H_e = head loss through the screen, ft, B = bar shape factor:

B = 2.42 for sharp edged rectangular bars
= 1.83 for rectangular bars with semicircular upstream faces
= 1.79 for circular bars

= 1.67 for rectangular bars with semicircular upstream and downstream faces
= 0.76 for rectangular bars with semicircular upstream faces and tapering in a symmetrical curve to a small circular downstream face (teardrop)

W = maximum width of bars facing the flow, in., b = minimum width of the clear spacing between pairs of bars, in., v = longitudinal approach velocity, ft/s, A = angle of the rack with horizontal, degree, g = gravitational acceleration.

Step 3: Calculate average water depth.

$$D_a = (Q_a)(1.54)/[(W_c)(V)] \tag{4}$$

where D_a = average water depth, ft, Q_a = average flow, MGD, W_c = channel width, ft, V = average velocity, ft/s.

Step 4: Calculate maximum water depth.

$$D_m = D_a(Q_p/Q_a) \tag{5}$$

where D_m = maximum water depth, ft, D_a = average water depth, ft, Q_p = peak flow, MGD, Q_a = average flow, MGD.

8.1.4. Screen Design Output Data

Output data for screen design include:

1. Bar size, in.
2. Bar spacing, in.
3. Slope of bars from horizontal, degree
4. Head loss through screen, ft
5. Approach velocity, ft/s
6. Average flow-through velocity, ft/s
7. Maximum flow-through velocity, ft/s
8. Screen channel width, ft
9. Channel depth, ft

8.2. Summary of Comminution Design Considerations

Comminution is defined as (a) the act of reducing to a fine powder or to small particles, (b) the state of being comminuted, or (c) fracture into a number of pieces (10). Readers are referred to another book, entitled *Comminution Practices* (11) and other references (12–14) for more information on recent extensive research, innovative comminution devices, new process control strategies, and modeling and simulation of conventional comminution devices to improve their energy efficiencies. Additional simple design considerations of the comminution process equipment (9) are summarized below.

Comminutors are screens equipped with a device that cuts and shreds the screenings without removing them from the waste stream. Thus, comminuting devices eliminate odors, flies, and other nuisances associated with other screening devices. A variety of comminuting devices are available commercially.

Comminutors are usually located behind grit removal facilities in order to reduce wear on the cutting surfaces. They are frequently installed in front of pumping stations to protect the pumps against clogging by large floating objects.

The comminutor size is based usually on the volume of waste to be treated. Treatment plants with a wastewater flow below 1 MGD normally use one comminutor. Table 4 summarizes design characteristics of comminutors (9).

Table 4
Comminutor Size Selection

Drum diameter (in.)	Drum (rpm)	Avg slot width (in.)	Horse power	Standard sizes: Height	Standard sizes: Net weight (lb)	Rates of flow Avg 12-hr day time (MGD)	Maximum hourly rates of flow (MGD)
4	56	1/4	1/4	2 ft 3.25 in.	175	0 to 0.035	0.09
7	56	1/4	1/4	4 ft 3 in.	450	0.03 to 0.113	0.24
7	56	1/4	1/4	4 ft 3 in.	450	0.06 to 0.200	0.36
10	45	1/4	1/2	4 ft 5 in.	650	0.17 to 0.720	1.08
15	37	1/4	3/4	4 ft 11.5 in.	1100	0.25 to 1.820	2.40
25	25	3/8	1.5	5 ft 9.5 in.	2100	0.97 to 5.100	6.10
25	25	3/8	1.5	6 ft 11.5 in.	3500	1.00 to 9.400	11.10
36	15	3/8	2	9 ft 4.5 in.	8500	1.30 to 20.00	24.00

(*Source*: US Army).

In wastewater treatment facilities for recreation areas, a comminutor may be installed in the wet well to protect the pump from large floating objects. In the treatment of vault waste, a comminutor may be included as an integral part of a vault waste holding station. Figure 4 illustrates such a comminutor.

8.2.1. Comminutor Design Input Data

The following input data is required for the design of comminutors:

1. Wastewater flow
   - Average daily flow, MGD
   - Maximum daily flow, MGD
   - Peak wet weather flow, MGD
2. Wastewater characteristics (13)
   - Alkalinity and acidity (pH adjustment may be required)
   - pH (pH adjustment may be required)

8.2.2. Comminutor Design Procedures

Comminutor should be selected from equipment manufacturer's catalogs to correspond to maximum wastewater flows.

8.2.3. Comminutor Design Output Data

Output data for comminutor design include:

1. Comminutor specifications
2. Number of comminutors

9. DESIGN EXAMPLES

9.1. *Example 1: Bar Screen Design*

Bar screens are frequently used for catch basin screening (15), stormwater pretreatment (15,16), raw water inlet screening, and raw sewage screening (17). The following is an example showing how the bar screens are designed for raw sewage screening.

A sewage treatment plant has a maximum daily flow of 0.131 m^3/s (3 MGD). Design a typical bar screen system assuming the velocity through the screen is 0.610 m/s or (2 ft/s), and the design data in Table 1 are to be used.

Solution

1. Selection of the net area ratio (R) = 0.728 from Table 1
2. Selection of bar size = 0.952 cm (3/8 in.)
3. Selection of bar opening = 2.54 cm (1 in.)
4. Determination of the required net flow area (A_f)

$$A_f = Q_p / V_2 \tag{6}$$

where A_f = required net flow area, ft^2, Q_p = peak influent flow, m^3/s or ft^3/s, V_2 = velocity through bar screen, m/s or ft/s, then

$$A_f = Q_p / V_2 = \left(0.131 \text{ m}^3/\text{s}\right)/\left(0.610 \text{ m/s}\right) = 0.215 \text{ m}^2$$

$$A_f = Q_p / V_2 = \left(4.641 \text{ ft}^3/\text{s}\right)/\left(2 \text{ ft/s}\right) = 2.32 \text{ ft}^2$$

5. Determination of the required total wet flow area (A_{wf})

$$A_{wf} = A_f / R \tag{7}$$

where A_{wf} = required total wet flow area, ft^2, R = net area ratio, then

$$A_{wf} = A_f / R$$

$$A_{wf} = 0.215 \text{ m}^2/0.728 = 0.295 \text{ m}^2$$

$$A_{wf} = 2.32 \text{ ft}^2/0.728 = 3.18 \text{ ft}^2$$

6. Determination of the maximum depth of water (D_m)

$$D_m = A_{wf} / W_c \tag{8}$$

where, D_m = maximum depth of water, m or ft, W_c = channel width, m or ft. If the channel width is set at 0.915 m (3 ft), then the depth of liquid would be

$$D_m = A_{wf} / W_c = 0.295 \text{ m}^2/0.915 \text{ m} = 0.323 \text{ m}$$

$$D_m = 3.18 \text{ ft}^2/3 \text{ ft} = 1.06 \text{ ft}$$

This would be the depth of liquid in the channel assuming there were no effects from other parts of the plant following the bar screen. The depth may actually be greater or less than the calculated value if units subsequent to screening increase or decrease the resistance to flow.

9.2. *Example 2: Bar Screen Head Loss*

Calculate the head loss of the bar screen system designed in Example 1.

Solution

1. Determination of the velocity ahead of bar screen (V_1)

$$V_1 = Q_p / A_{wf} \tag{9}$$

$$V_1 = Q_p / A_{wf} = \left(0.131 \text{ m}^3/\text{s}\right)/\left(0.295 \text{ m}^2\right) = 0.444 \text{ m/s}$$

$$V_1 = (4.641 \text{ ft}^3/\text{s})/(3.18 \text{ ft}^2) = 1.46 \text{ ft/s}$$

2. Selection of the velocity through screen $V_2 = 0.610$ m/s = 2 ft/s
3. Determination of the head loss (h)

$$h = 0.0728(V_2^2 - V_1^2) \text{ with SI units} \quad (2a)$$

$$h = 0.0728(0.610^2 - 0.444^2) = 0.0127 \text{ m}$$

$$h = 0.0222(V_2^2 - V_1^2) \text{ with English units} \quad (2b)$$

$$h = 0.0222(2^2 - 1.46^2) = 0.0415 \text{ ft}$$

9.3. *Example 3: Plugged Bar Screen Head Loss*

To demonstrate the effect of plugging, assume the screen area is cut in half by the screenings. Determine the head loss under this plugging situation.

Solution

In this case since $Q = AV$, the velocity through the screen would double. Therefore the head loss would be

$$h = 0.0728(1.220^2 - 0.444^2) = 0.094 \text{ m}$$

$$h = 0.0222(4^2 - 1.46^2) = 0.308 \text{ ft}$$

This is a sevenfold increase in head loss when the screen becomes half plugged. This demonstrates the necessity for regular cleaning of the screen.

9.4. *Example 4: Screen System Design*

(a) *Step 1:* Select a mechanically cleaned bar screen from Table 2 with bar screen size of width = 1/14 in., depth = 1 in., spacing = 5/8 in., slope = 10°, approach velocity = 2 ft/s, and allowable head loss = 6 in.

(b) *Step 2:* Calculate head loss through screen:

$$H_e = B(W/b)^{4/3}[(v^2 \sin^2 A)/2g] \quad (3)$$

where H_e = head loss through the screen, ft, B = bar shape factor = 1.83 for rectangular bars with semicircular upstream faces, W = maximum width of bars facing the flow = 1/4 in., b = minimum width of the clear spacing between pairs of bars = 5/8 in., v = longitudinal approach velocity = 2 ft/s, A = angle of the rack with horizonta = 10° g = gravitational acceleration = 32.2 ft/s^2:

$$H_e = 1.83[(1/4)/(5/8)]^{4/3}[(2^2 \sin^2 10)/2(32.2)] = 0.001 \text{ ft}$$

(c) *Step 3:* Calculate the average water depth:

$$D_a = (Q_a)(1.54)/[(W_c)(V)] \quad (4)$$

where D_a = average water depth, ft, Q_a = average flow = 1 MGD, W_c = channel width = 1.23 ft, V = average velocity = 2 ft/s:

$$D_a = (1)(1.54)/[(1.23)(2)] = 0.63 \text{ ft}$$

(d) *Step 4:* Calculate maximum water depth:

$$D_m = D_a(Q_p/Q_a) \quad (5)$$

where D_m = maximum water depth, ft, D_a = average water depth = 0.63 ft, Q_p = peak flow = 2 MGD, Q_a = average flow = 1 MGD, then

$$D_m = D_a(Q_p/Q_a) = 0.63(2/1) = 1.26 \text{ ft}$$

NOMENCLATURE

A	angle of the rack with horizontal, degree
A_f	required net flow area, m^2 or ft^2
A_{wf}	required total wet flow area, m^2 or ft^2
b	minimum width of the clear spacing between pairs of bars, m or in.
B	bar shape factor 2.42 for sharp edged rectangular bars 1.83 for rectangular bars with semicircular upstream faces 1.79 for circular bars 1.67 for rectangular bars with semicircular upstream and downstream faces 0.76 for rectangular bars with semicircular upstream faces and tapering in a symmetrical curve to a small circular downstream face (teardrop)
C_D	discharge coefficient
D_a	average water depth, m or ft
D_m	maximum water depth, m or ft
g	gravitational acceleration = 9.806 m/s^2 = 32.17 ft/s^2
h	head loss, m or ft
H_e	head loss through the screen, m or ft
Q_a	average influent flow, m^3/s, or ft^3/s, or MGD
Q_p	peak influent flow, m^3/s, or ft^3/s, or MGD
R	net area ratio
v	longitudinal approach velocity, m/s or ft/s
V	average velocity, m/s or ft/s
V_1	velocity ahead of bar screen, m/s or ft/s
V_2	velocity through bar screen, m/s or ft/s
W	maximum width of bars facing the flow, m or in.
W_c	channel width, m or ft

REFERENCES

1. WPCF, *Operation of Municipal Wastewater Treatment Plants*, Water Pollution Control Federation, Washington, DC, WPCF Operations and Maintenance Subcommittee, Vol. 11, pp. 444–451 (1990).
2. WEF, *Design of Municipal Wastewater Treatment Plants*, Water Environment Federation, Washington, DC, WEF Manual of Practice No.8, pp. 390–405 (1992).
3. D. H. F. Liu and B. G. Liptak, *Wastewater Treatment*, Lewis Publishers, Boca Raton, FL, pp. 136–139 (2000).

4. WPCF and ASCE, *Sewage Treatment Plant Design*, WPCF Manual of Practice No. 8, 1959, 1961, 1967, 1968, Water Pollution Control Federation, Washington, DC, American Society of Civil Engineers, New York, NY (1968).
5. G. M. Fair, J. C. Geyer, and D. A. Okun, *Water Purification and Wastewater Treatment and Disposal: Water and Wastewater Engineering*, Vol. 2, Wiley, New York, NY (1968).
6. B. L. Goodman, *Design Handbook of Wastewater Systems: Domestic, Industrial, and Commercial*, Technomic, Westport, CT (1971).
7. Metcalf and Eddy, Inc., *Wastewater Engineering: Collection, Treatment, and Disposal*, McGraw-Hill, New York, NY (1972).
8. Great Lakes–Upper Mississippi River Board of State Sanitary Engineers, *Recommended Standards for Sewage Works (Ten States Standards)*, Health Education Service, Albany, NY (2002).
9. US Army, *Engineering and Design—Design of Wastewater Treatment Facilities Major Systems*, Engineering Manual No. 1110-2-501. US Army, Washington, DC (1978).
10. Editor, Comminution, *Brainy Dictionary*, www.brainydictionary.com/words/co/comminution 145997.html (2003).
11. K. S. Kawatra, Comminution practices, Society of Mechanical Engineering, www.min-eng.com/commin_store.html (1997).
12. H. Cho and L. G. Austin, An equation for the breakage of particles under impact, *Power Technology*, **132**, No. 2 (2003).
13. T. W. Chenje, D. J. Simbi, and E. Navara, The role of corrosive wear during laboratory milling, *Minerals Engineering*, **16**, No. 7 (2003).
14. Editor, A quantification of the benefits of high pressure rolls crushing in an operating environment, *Minerals Engineering*, **16**, No. 9 (2003).
15. Labosky, L., Stormwater management—catch basin. Environmental Protection, **15**, No. 5, pp. 26 (2004).
16. C. Yapijakis, R. L. Trotta, C. C. Chang, and L. K. Wang. Stormwater management and treatment. In: *Handbook of Industrial and Hazardous Wastes Treatment* (L. K. Wang, Y. T. Hung, H. H. Lo, and C. Yapijakis, eds.). Marcel Dekker, Inc., NY, NY. pp. 873–921 (2004).
17. L. K. Wang, Y. T. Hung, and N. K. Shammas (eds.), *Advanced Physicochemical Treatment Process*. Humana Press, Totowa, NJ (2005).

2
Flow Equalization and Neutralization

Ramesh K. Goel, Joseph R.V. Flora, and J. Paul Chen

CONTENTS

1. INTRODUCTION

Flow equalization and chemical neutralization and are two important components of water and wastewater treatment. Chemical neutralization is employed to balance the excess acidity or alkalinity in water, whereas flow equalization is a process of controlling flow velocity and flow composition. In a practical sense, chemical neutralization is the adjustment of pH to achieve the desired treatment objective. Flow equalization is necessary in many municipal and industrial treatment processes to dampen severe variations in flow and water quality. Both these processes have been practiced in the water and wastewater treatment field for several decades. Thtis chapter will present an overview of these two processes, the chemistry behind neutralization, design considerations, and their industrial application.

2. FLOW EQUALIZATION

Flow equalization is used to minimize the variability of water and wastewater flow rates and composition. Each unit operation in a treatment train is designed for specific wastewater characteristics. Improved efficiency and control are possible when all unit operations are carried out at uniform flow conditions. If there exists a wide variation in flow composition over time, the treatment efficiency of the overall process performance may degrade severely. These variations in flow composition

From: *Handbook of Environmental Engineering, Volume 3: Physicochemical Treatment Processes*
Edited by: L. K. Wang, Y.-T. Hung, and N. K. Shammas © The Humana Press Inc., Totowa, NJ

could be due to many reasons, including the cyclic nature of industrial processes, the sudden occurrence of storm water events, and seasonal variations. To dampen these variations, equalization basins are provided at the beginning of the treatment train. The influent water with varying flow composition enters this basin first before it is allowed to go through the rest of the treatment process. Equalization tanks serve many purposes. Many processes use equalization basins to accumulate and consolidate smaller volumes of wastewater such that full scale batch reactors can be operated. Other processes incorporate equalization basins in continuous treatment systems to equalize the waste flow so that the effluent at the downstream end can be discharged at a uniform rate.

Various benefits are ascribed by different investigators to the use of flow equalization in wastewater treatment systems. Some of the most important benefits are listed as follows (1–6):

1. Equalization improves sedimentation efficiency by improving hydraulic detention time.
2. The efficiency of a biological process can be increased because of uniform flow characteristics and minimization of the impact of shock loads and toxins during operation.
3. Manual and automated control of flow-rate-dependent operations, such as chemical feeding, disinfection, and sludge pumping, are simplified.
4. Treatability of the wastewater is improved and some BOD reduction and odor removal is provided if aeration is used for mixing in the equalization basin.
5. A point of return for recycling concentrated waste streams is provided, thereby mitigating shock loads to primary settlers or aeration basin.

Sometimes it is thought that equalization tanks also serve the purpose of dilution. However, the United States Environmental Protection Agency (US EPA) does not consider the use of equalization tanks as an alternative to achieve dilution. The US EPA's viewpoint is that dilution is mixing of more concentrated waste with greater volumes of less concentrated waste such that the resulting wastewater does not need any further treatment.

Equalization basins in a treatment system can be located in-line or off-line. Figures 1a,b depict the typical layouts of both types of equalization practice with respect to the rest of the unit operations. In in-line equalization, 100% incoming raw wastewater directly enters into the equalization basin, which is then pumped directly to other treatment units (e.g., primary treatment units). However, for side-line or off-line equalization, the basin does not directly receive the incoming wastewater. Rather, an overflow structure diverts excess flow from the incoming raw wastewater into the basin. Water is pumped from the basin into the treatment stream to augment the flow as required.

Two basic configurations are recommended for an equalization basin: variable volume and constant volume. In a variable volume configuration, the basin is designed to provide a constant effluent flow to the downstream treatment units. However, in the case of a constant volume basin, the outflow to other treatment units changes with changes in the influent. Both configurations have their uses in different applications. For example, variable volume type basins are used in industrial applications where a low daily volume is expected. Variable volume equalization basins can also be used for municipal wastewater treatment applications.

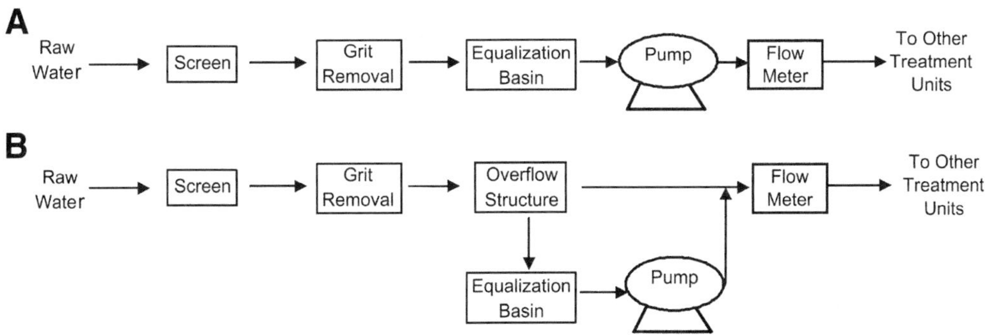

Fig. 1. (a) In-line and (b) off-line flow equalization.

2.1. *Flow Equalization Basin Calculations*

Computation of the volume of an equalization basin is the key design requirement and is based on inflow variation over time. There are two methods used to compute equalization volume. One procedure is based on the characteristic diurnal flow pattern, whereas the other is based on the mass loading pattern of a particular constituent.

The first method relies on computing the equalization volume based on the excess daily average flow storage. The required volume is determined graphically by constructing a hydrograph. The function of the basin is to store flows in excess of the average daily flow and to divert this flow during times when the inflow is less that the average daily flow. The second method computes the volume based on mass loading variations within an acceptable range.

In general, the first method is regarded as a flow balance approach and the second method is regarded as a composition balance approach. Flow balance is the most common method for computing equalization basin volume. The selection of a particular method depends on the type of flow, flow variations, and overall composition of the flow. Flow balance is used when the composition of incoming water is relatively constant but the flow varies over time. The composition balance method is used when the rate of inflow is fairly constant and the composition varies with time.

In the flow balance method, a plot of cumulative volume versus time is developed, which is the well-known Rippl diagram (7). The steps required to create a Rippl diagram and to use this diagram to calculate the equalization volume are outlined as follows:

- The first step is to draw a cumulative volume curve based on the wastewater flow. The volume that flows within a specified periodic time period is calculated based on the flow. The cumulative volume is obtained by adding the volume at the start of a preselected time period to the volume in the next time period. The resulting volume is then added to the volume in the subsequent time period. This process is continued until a cycle of low-flow and high-flow is completed (typically 24 h).
- The second step is to determine the required equalization volume by drawing a line parallel to the average flow rate and tangent to the cumulative influent flow diagram. The equalization volume is calculated by the vertical distance from the point of tangency to the straight line. There could be several points of tangency on the cumulative influent flow curve. However, care should be taken in selecting the points to be taken into consideration for equalization volume calculations.

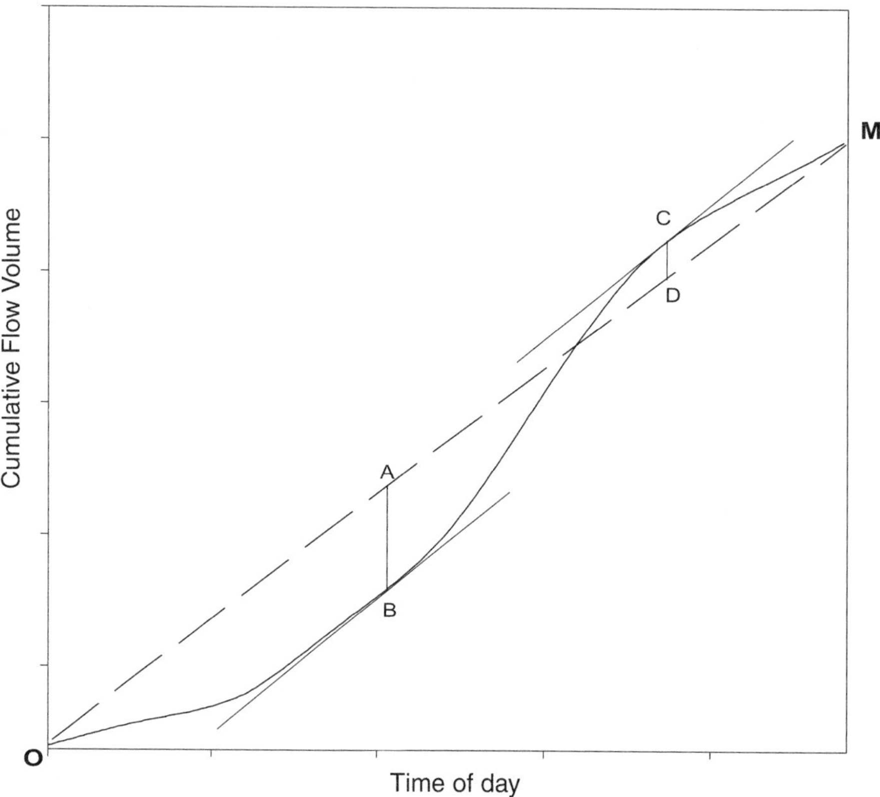

Fig. 2. Volume calculation of equalization basin.

The theory behind the method is explained with Fig. 2, which shows a typical cumulative influent volume curve for the average daily flows. In this figure, the cumulative volume is plotted on the *y*-axis against the time of day. The resulting graph is shown by an irregularly shaped curve. If the curve is linear, then the flow is constant. When the tail end (O) of this cumulative influent volume curve is joined with the top end (M), the average flow curve (shown by dotted line) is obtained. Lines parallel to average daily flow line (dotted line) and tangent to mass flow curve are then drawn. The points of tangency are (B) and (C). From these points of tangency, vertical straight lines are drawn until these vertical lines intersect the average daily flow line. The points of intersection are given at (A) and (D). The required equalization volume will be equal to sum of the vertical distances AB and CD. At the first point (B) of tangency, the storage basin is empty and beyond this point, the basin begins to fill and continues until the basin becomes full at upper point (C) of tangency.

The volume calculated based on the hydrograph method is the theoretical volume. In practice, the volume will be always greater than the theoretical because of the following reasons:

- A minimum volume of water is always required in an equalization basin for mixing and aeration equipment inside the basin to operate.

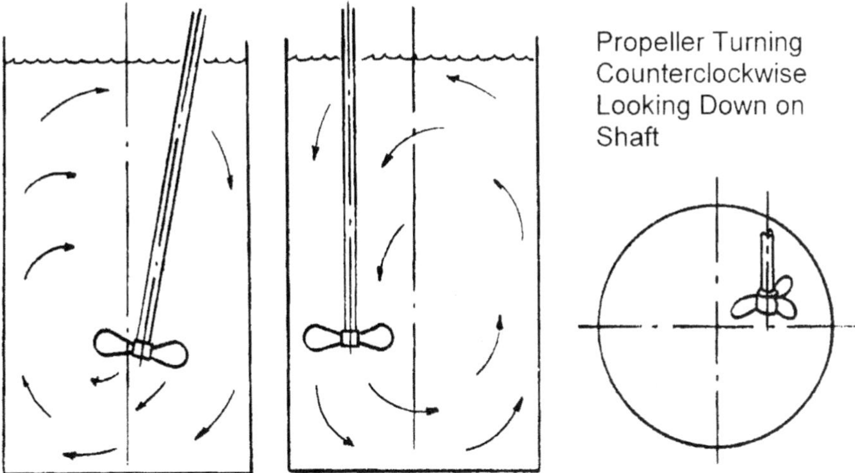

Fig. 3. Illustration of mixer (9).

- Sometimes, concentrated waste downstream in the treatment plant is returned to the equalization basin. To avoid odor problems, dilution of such returned waste is needed and the diluted water is stored in the equalization basin.
- Some free board is always provided to accommodate unforeseen changes in diurnal flow.

Flow equalization is more routinely employed in industry than at municipal facilities because many industries use batch production processes (2,3). However, there are now also a large number of municipal equalization basin installations.

2.2. *Mixing and Aeration Requirements*

Mixers are often employed in equalization basins to achieve homogeneity in and to aerate the wastewater. Various types of mixers are available. The classification of mixers depends on the flow pattern the mixers produce. The commonly used mixers have either axial or radial patterns, with axial mixers most prevalently used in industries (8).

Axial mixers can further be subdivided into other categories, the most common of which are propeller mixers and turbine mixers. Propeller mixers are used primarily when rapid mixing is needed. The axial propeller mixer can be either fixed or portable, depending on the mixer size and application. The size of top-entering propeller mixers range from 0.37 to 2.24 kW, although many industrial designs limit the size to 0.75 kW and a maximum shaft length of 1.83 m (8). Propeller mixers are usually mounted angularly off center. The advantage with this type of arrangement is that complete top to bottom mixing can be achieved. Typically the maximum water volume that is recommended for a propeller mixer is 3.785 m^3 (1000 gal). As shown in Fig. 3, the mixer shaft should enter at 15° from vertical and at a point off the centerline.

The speed ranges for both portable and fixed mounted propeller mixers are 1750 rpm and 350–420 rpm, respectively. The high speed provides a high degree of shear with low draft velocity, causing instant mixing. Low speeds provide less shear force and may allow selective setting of larger and heavier particles.

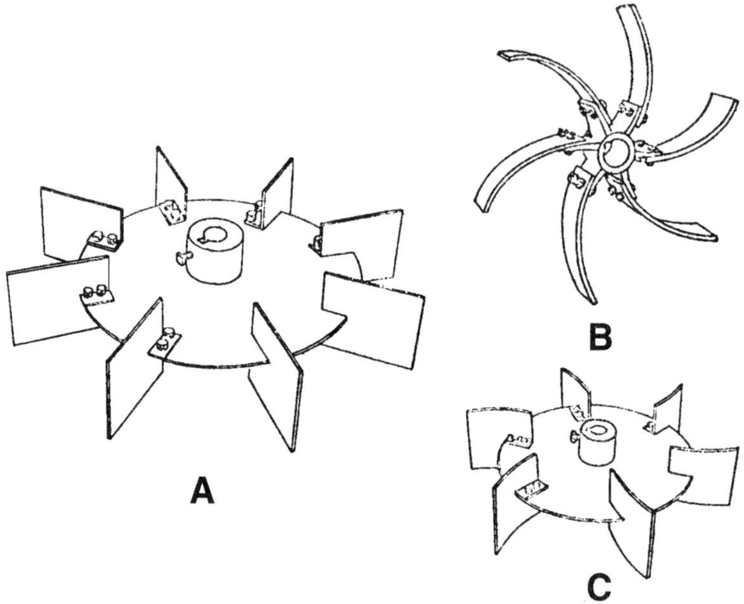

Fig. 4. Typical Radial Turbine Impeller: (a) flat blade; (b) spiral backswept; (c) curved Blade (9).

Other classes of axial mixers include turbine mixers. They can induce both axial as well as radial flow. Axial turbine impellers are pitched blade or fan turbines, whereas radial turbine impellers are flat blade, curved blade, or with a spiral backswept blade (shown in Fig. 4). The curved and spiral backswept impellers are used in high viscous applications such as sodium hydroxide or soda ash neutralization. Axial turbines are used for large scale mixing involving liquid solid suspensions. Turbines mixers are usually fixed mounted, vertically in fully baffled tanks. Turbine impeller diameters are generally one third of the tank diameter.

2.3. *Mixer Unit*

The design of an economically feasible mixer unit requires an assessment of power requirements, laboratory scale up studies, and the selection of either a batch or continuous system, hydraulic retention time, vessel geometry, and type of mixing unit (6). The following sections discuss some of these design considerations.

2.3.1. *Power Requirements*

Under turbulent hydraulic conditions (i.e., when the Reynolds number is greater than 10^5), the following formula can be used to determine the power requirements of an impeller mixer (6),

$$P = \rho K_T n^3 D^5 \tag{1}$$

where P = power requirement, N-m/s, ρ = density of the fluid, kg/m^3, K_T = constant dependent on impeller size and shape, n = impeller revolutions per second, s^{-1}, D = diameter of impeller, m.

Table 1
Values of K_T for Impeller Design

Impeller type	K_T
Propeller (square pitch, three blades)	0.32
Propeller (pitch of two, three blades)	1.0
Turbine (six flat blades)	6.30
Turbine (six curved blades)	4.80
Turbine (six arrowhead blades)	4.00
Fan Turbine (six blades)	1.65
Flat paddle (two blades)	1.70
Shrouded turbine (six curved blades)	1.08
Shrouded turbine (with stator, no baffles)	1.12

Source: Ref. 6.

Some of the typical K_T values for design purposes are given in Table 1 (6). These K_T values are for mixing impellers rotating at the center of cylindrical tanks with a flat bottom, four baffles at the tank wall, baffle width of 10% of the tank diameter, and impeller diameter equal to one-third of the tank diameter.

The Camp and Stein mean velocity gradient, G, is used to describe the intensity of mixing in the tank. G is related to the amount of power dissipated in the tank and typically ranges from 500 to 1500 s^{-1} for rapid mixing (2). G can be calculated as follows:

$$G = \left(\frac{P}{V\mu}\right)^{\frac{1}{2}} \tag{2}$$

where V = mixing tank volume, m^3, μ = absolute viscocity of the fluid, $N\text{-}s/m^2$.

To ensure adequate mixing, the tank is sized to obtain a detention time, t_d, in the range of 5–30 s for rapid mixing. This results in $G \times t_d$ values of at least 2500, where

$$G \times t_d = \left(\frac{P}{V\mu}\right)^{\frac{1}{2}} \times \frac{V}{Q} = \frac{1}{Q}\left(\frac{PV}{\mu}\right)^{\frac{1}{2}} \tag{3}$$

where Q = flow rate, m^3/s.

2.3.2. *Laboratory Scale Up*

The usual practice involves the determination of design parameters in laboratory scale experiments and then generalizing these parameters for full-scale applications. Problems are often encountered during the scaling up of laboratory parameters for full-scale applications. Careful considerations should be given while selecting design parameters from laboratory experiments. The selection should be based on experience, similarity, and testing accuracy. If budget permits, it is always beneficial to test the design parameters found in laboratory experiments and in pilot-scale experiments. Once these parameters prove their suitability in pilot-scale experiments, they can further be used for full-scale operations.

2.3.3. *Vessel Geometry*

Vessel geometry plays a significant role in achieving overall mixing efficiency. However, the selection of vessel geometry is dictated by process considerations. As a general rule, circular tanks are more efficient in achieving proper mixing than square or rectangular tanks. For circular tanks, a liquid depth equal to tank diameter is generally employed. For tanks less than 4000 L, compact turbine mixers are the most practical.

3. NEUTRALIZATION

Neutralization is a common practice in wastewater treatment and waste stabilization. If a waste stream is found to be hazardous because of corrosivity, neutralization is the primary treatment used. Moreover, neutralization is used as a pretreatment system before a variety of biological, chemical, and physical treatment processes. Since many chemical treatment processes, such as metal precipitation, coagulation, phosphorus precipitation, and water softening are pH dependent, the pH of these processes is adjusted to achieve maximum process efficiency. Furthermore, the pH of the effluent wastewater from different industrial activities also requires adjustment prior to its discharge into receiving water bodies. The US EPA has set pH standards for different types of water; for example, the pH range required to protect marine aquatic life is 5–9 (10).

Neutralization is the process of adjusting the pH of water through the addition of an acid or a base, depending on the target pH and process requirements. Some processes such as boiler operations and drinking water standards need neutral water at a pH of 7. Water or wastewater is generally considered adequately neutralized if (1) its damage to metals, concrete, or other materials is minimal; (2) it has little effect on fish and aquatic life; (3) it has no effect on biological matter (i.e., biological treatment systems).

In chemical industrial treatment, neutralization of excess alkalinity or acidity is often required. One of the critical items in neutralizing the water is to determine the nature of the substances that cause acidity and alkalinity. This is generally achieved in laboratory-scale experiments by preparing titration curves showing the quantity of alkaline or acidic material necessary to adjust the pH of the target wastewater. The nature of titration curves obtained in these experiments is critical in determining the proper chemical type and dose. Methods used for pH adjustment should be selected on the basis of costs associated with the neutralizing agent and equipment requirements for dispensing the agent.

In neutralization, several parameters need to be assessed and evaluated before the actual pH adjustment is carried out. These parameters are discussed in the following sections.

3.1. *pH*

pH is the reference indicator for neutralization. Many chemical processes, such as metal precipitation and water softening, which are involved in neutralization, are pH dependent. pH is the negative logarithm of the H^+ ion activity in solution

$$\mathrm{pH} = -\log\{\mathrm{H}^+\} \tag{4}$$

If the ionic strength of the waters is not very high (less than 0.01 *M*), the activity of hydrogen ions can be replaced with the molar concentration of hydrogen ions,

If the ionic strength is high, correction factors using the Debye–Hückel equation or Davies equation can be commonly used (10).

In most practical applications, the pH scale ranges from 1 to 14. In pure water and in the absence of materials other than H^+ and OH^-, water behaves ideally and activity equals molar concentration. Under these conditions, $[H^+]$ equals $[OH^-]$ as required by electroneutrality. At 25°C, the ion product of water ($K_w = [H^+][OH^-]$) is 10^{-14}.

The process of neutralization is not only limited to bringing the pH to 7; it is invariably used in the processes, where pH adjustment to other than 7 is required depending on the chemical process in question. For example, some processes like biological wastewater treatment require pH to be near neutral, whereas other processes like metal precipitation require pH to be in the alkaline range. Some of the important chemical processes, where pH plays a significant role and where pH adjustment through neutralization is often required, are metal adsorption and biosorption, chemical precipitation, water softening, coagulation, water fluoridation, and water oxidation (11–14).

3.2. *Acidity and Alkalinity*

Alkalinity is the capacity of water to neutralize acids, whereas acidity is the capacity of water to neutralize bases. The amount of acid or base to be used in the neutralization process depends upon the respective amount of acidity and alkalinity.

The most important source of both alkalinity and acidity in natural waters is from the carbonate system. However, if the wastewater comes from industrial sources, OH^- or H^+ is also a major contributory factor to alkalinity or acidity, respectively. For example, water from acid mine drainage contains a large amount of acidity because of the presence of sulfuric acid produced from the oxidation of pyrite. Both acidity and alkalinity are expressed in terms of acid/base equivalents. In water and wastewaters where the predominant ions controlling pH are $[H^+]$, $[OH^-]$, $[HCO_3^-]$, and $[CO_3^{2-}]$, the forms of alkalinity encountered are hydroxide, carbonate, and bicarbonate. These three forms of alkalinity altogether constitute total alkalinity.

Alkalinity and acidity are determined by titration. For wastewater samples whose pH is above 8.3, titration is made in two steps. In the first step, the pH is brought down to 8.3; in the second step, the pH is brought down to about 4.5. When the pH of wastewater is below 8.3, a single titration curve is made. When the pH of wastewater reaches 8.3, all carbonate present in wastewater converts to bicarbonate according to the following reaction;

$$CO_3^{2-} + H^+ \rightarrow HCO_3^- \tag{5a}$$

As titration proceeds, bicarbonate goes to carbon dioxide when the pH reaches at 4.5. Carbon dioxide and water together form weak carbonic acid:

$$HCO_3^- + H^+ \rightarrow H_2CO_3 \tag{5b}$$

If it is assumed that carbonate species and OH^- are the only chemical constituents causing alkalinity, the three forms of alkalinity can be defined based on pH. When pH of water is above 8.3, all three forms of alkalinities are present. As a rule of thumb, caustic alkalinity is absent if the pH of the water is below 10, and carbonate alkalinity is absent if the pH is below 8.3.

Mathematically, alkalinity can be expressed by considering the volume of acid required to drop the pH from or above 10 to 8.3 and then to 4.5. If the initial water composition requires V_p mL of acid to reach 8.3 and V_c is the volume of acid required to reach pH 4.5, then following holds true (30):

If $V_c = 0$, alkalinity is due to $[OH^-]$ only
If $V_c = V_p$, alkalinity is only due to carbonate
If $V_p > V_c$, major alkalinity specie are hydroxide and carbonate
If $V_p < V_c$, major alkalinity species are bicarbonate and carbonates.

In general mathematical terms, the total alkalinity can be expressed using the following equation:

$$\text{Total alkalinity in eq/L} = 2\left[CO_3^{2-}\right]+\left[HCO_3^-\right]+\left[OH^-\right]-\left[H^+\right] \tag{6}$$

The terms on the right-hand side of Eq. (6) are in mol/L. Alternatively, alkalinity can be expressed in terms of mg/L as $CaCO_3$ (13). Alkalinity of individual species is calculated by

$$\text{Alkalinity of species}_i\left(\text{mg/L as CaCO}_3\right) = \text{species}_i\left(\text{mg/L}\right)\times\frac{EW_{\text{CaCO}_3}}{EW_{\text{species}_i}} \tag{7}$$

where *EW* is the equivalent weight. *EW* values of $CaCO_3$, CO_3^{2-}, HCO_3^-, OH^-, and H^+ are 50, 30, 61, 17 and 1, respectively. Therefore Eq. (6) is revised to:

$$\text{Total alkalinity in mg/L as CaCO}_3 = \left(CO_3^{2-}\right)+\left(HCO_3^-\right)+\left(OH^-\right)-\left(H^+\right) \tag{8}$$

The terms on the right-hand side of Eq. (8) are in mg/L as $CaCO_3$.

The acidity of water is defined in a similar fashion. In the case of acidity also, there are two equivalence points, one at pH 4.5 and the other at pH 8.3. Depending on the pH, the water can have mineral acidity, CO_2 acidity, and total acidity. When pH of the water sample lies below 4.5, the amount of base added to raise the pH to 4.5 is the mineral acidity. In the same way, the amount of base required to raise the solution pH to 8.3 is called CO_2 acidity. Total acidity corresponds to the amount of base added to raise the pH to the carbonate equivalence point (above 8.3). Mathematically, the total acidity can be expressed as follows:

$$\text{Total acidity in eq/L} = 2\left[H_2CO_3\right]+\left[HCO_3^-\right]+\left[H^+\right]-\left[OH^-\right] \tag{9}$$

The terms on the right-hand side of Eq. (9) are in mol/L. The ranges of acidity and alkalinity are shown in Fig. 5.

3.3. *Buffer Capacity*

The word "buffer" stands for the stubbornness against any change. In environmental chemistry, buffers are always defined in the context of pH. pH buffers are those that resist any changes in solution pH when an acid or a base is added into the solution. They are very important in chemical neutralization processes. Buffers generally contain a mixture of weak acid and their salts (conjugate base) or weak bases and their conjugate acid. A solution buffered at a particular pH will contain an acid that can react with an externally added base and vice versa. The overall efficiency and

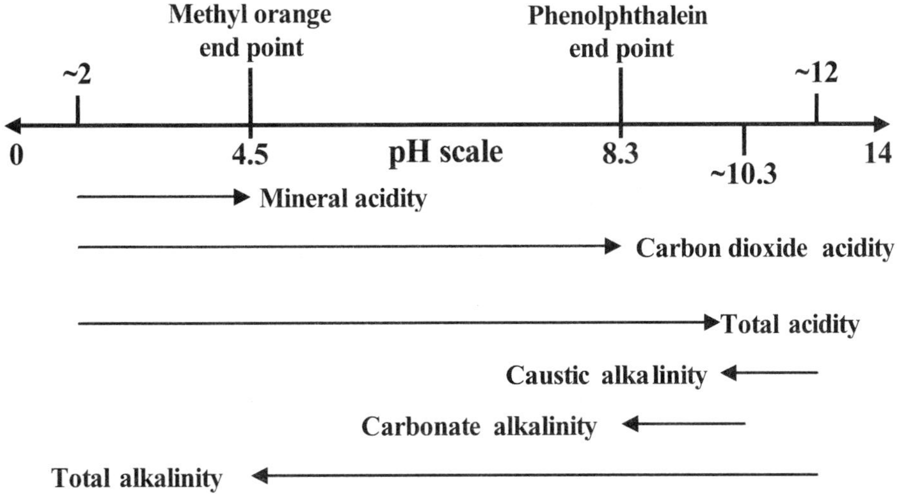

Fig. 5. Ranges of acidity and alkalinity.

chemical cost of the neutralization process depend on the presence of pH buffers in wastewaters.

To define the theory behind how pH buffers act, let us take an example. Consider a solution containing 0.06 *M* acetic acid and 0.06 *M* sodium acetate. When a small amount of hydroxide is added in form of sodium hydroxide, the acetic acid present in the solution ionizes to produce H^+, which reacts with the hydroxide added. In similar fashion, if an acid is added to the solution, the acetate takes up the added H^+ to form acetic acid.

In natural waters and wastewaters, the buffering capacity arises due to the presence of phosphates, carbonates, and other weak organic acids. The mineral composition of natural waters is regulated by a buffer system involving natural clay minerals such illite and kaolinite. Careful consideration should be given while neutralizing such waters. If the buffering capacity of the water or wastewater to be neutralized is not taken into account, the actual amount of neutralizing chemical required may vary widely and causes operational problems.

3.4. Hardness

Hardness in waters arises from the presence of multivalent metallic cations (30). The principal hardness-causing cations are calcium, magnesium, ferrous iron, and manganous ions. This parameter is important in water-softening processes. The part of the total hardness that is chemically equivalent to the bicarbonate plus carbonate alkalinities is called carbonate hardness. When both hardness and alkalinity are expressed in mg/L as $CaCO_3$, these two are be related as follows:
When alkalinity < total hardness,

$$\text{Carbonate hardness (in mg/L)} = \text{alkalinity (in mg/L)}$$

When alkalinity > total hardness,

$$\text{Carbonate hardness (in mg/L)} = \text{total hardness (in mg/L)}$$

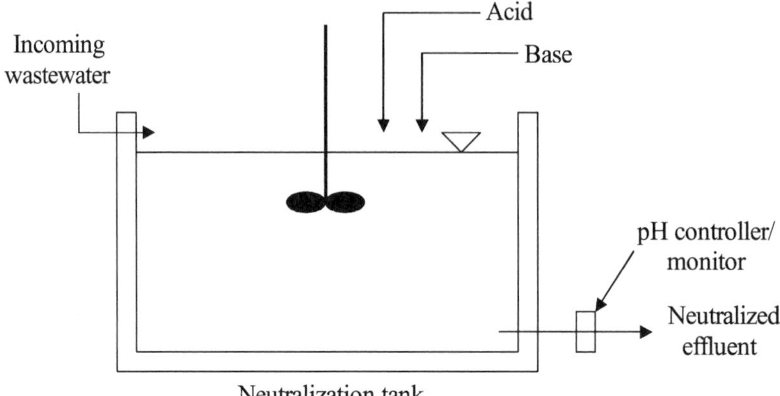

Fig. 6. Continuously operated neutralization tank.

4. NEUTRALIZATION PRACTICES

Neutralization can be carried out in either batch or continuous mode. In batch mode, the effluent is retained until its quality meets specifications before release. Several processes can be simultaneously carried out when the process is performed batchwise. Batch processes are good for small scale treatment plants or small waste volume. For large volumes, a continuous neutralization process is typically used. Figure 6 shows a typical schematic of a continuous neutralization reactor. The use of a batch neutralizing system or continuous flow system depends upon several considerations. In general, continuous flow-through systems are used when

- Influent flow is relatively constant and sudden variations are not expected.
- The influent flow characteristics are essentially constant.
- Effluent chemistry is not very critical. An example is when the process is a part of multistage neutralization process.

Batch neutralization systems are used when:

- There are large fluctuations in influent properties (i.e., flow and pH).
- The influent wastewater contains concentrated acids or bases.
- The effluent quality has stringent discharge limits.

Neutralization tanks should be constructed with a corrosion-resistant material or should be lined to prevent corrosion. Addition of an acid or an alkali should be controlled by continuous pH measurement, either by withdrawing samples periodically and measuring the pH or by installing an online pH meter that gives continuous pH readings.

4.1. Neutralization of Acidity

The most widely used methods to balance acidity by adding a proper alkaline solution are outlined below (6):

- Mixing alkaline and acidic wastes such that the net effect is nearly neutral pH.
- Passing the acidic water through a limestone bed. This water should not contain limestone-coating substances such as metal salts or sulfuric or hydrofluoric acids.

- Mixing acid waste with lime slurries or dolomitic slurries.
- Supplementing acidic wastewater with proper amounts of caustic soda or soda ash (Na_2CO_3).

Acidic wastes are neutralized either by adding lime alkalis or by adding sodium alkalis. The most commonly used lime alkalis are quicklime (CaO) and hydrated or slaked lime ($Ca(OH)_2$) (13–15). Sodium alkalis involve the use of caustic soda (NaOH) or soda ash (Na_2CO_3). Calcium and magnesium oxides are considerably less expensive than sodium alkalis and are used more widely (6). Because these oxides are moderately soluble in water, they are typically slurried. Calcium or magnesium alkalis produce more sludge than do sodium alkalis.

Sodium alkali rapidly reacts with acidic wastes and produces soluble neutral salts when combined with most acidic wastewaters. Between the two types of sodium alkalis, caustic soda is a stronger alkali than soda ash. Caustic soda is available in anhydrous form at various concentrations. Soda ash can be purchased as dry granular material. Liquid caustic soda is produced and supplied in a concentration range of 50–73%. Most industries use a 50% caustic soda solution. The specific gravity ranges from 1.47 to 1.53 depending on the temperature. Caustic soda is very corrosive in nature. Hence all containers and lines that come in to contact with caustic soda during use or shipment should be carefully selected.

Soda ash, when used as sodium carbonate monohydrate, contains 85.48% sodium carbonate and 14.52% water of crystallization. Hydrated soda ash loses water of crystallization when heated. Heptahydrated and decahydrated are other forms of soda ash used in neutralization practices. Dissolving monohydrated soda ash in water generates heat while heptahydrate and decahydrate absorbs heat in contact with water. Bagged soda ash should not be stored in humid places. Furthermore, excessive air circulation should be avoided. Soda ash contains 99.2% sodium carbonate when shipped.

4.2. Neutralization of Alkalinity

Lowering the pH of a solution is sometimes necessary in some treatment processes or when wastewater is to be discharged in open streams. Discharge of effluent with a pH greater than 8.5 is undesirable and lowering the pH is generally achieved either by adding an acid or by adding carbon dioxide. The process of adding carbon dioxide is called recarbonation and is often practiced in industrial wastewater neutralization. The commonly used acids for pH adjustment of alkaline wastewaters are sulfuric acid (H_2SO_4), hydrochloric acid (HCl), and nitric acid (HNO_3). Among them, sulfuric acid is the most widely used neutralizing agent. Use of nitric acid is restricted because of more stringent nutrient effluent limitations. There is no direct relationship between pH and alkalinity. Hence, titration curves should be established in laboratories before the design of an alkaline wastewater neutralization system. Sulfuric acid used in wastewater treatment could be 77.7% concentration or 97% concentration with an approximate specific gravity of 1.83 (1,8,9). Sulfuric acid releases a significant amount of heat when added to water. Precautionary measures must be taken to avoid any chemical accident due to the heat generated when practicing neutralization with sulfuric acid. Hydrochloric acid has an average specific gravity of 1.17 and an acid content of 33% by weight. Properly lined tanks should be used to store this classification of hydrochloric acid. Generally polyvinyl chloride tanks or lined steel tanks are used.

4.3. Common Neutralization Treatments

The application of neutralization varies from industry to industry. The most common application includes neutralization of acidic waste from mining industries, in chemical precipitation, water softening, wastewater coming out from electronic manufacturing plants, and coagulation and flocculation in wastewater-treatment plants. Neutralization is also required for treated wastewater if the pH of such water is found to be higher or lower than the permissible discharge limits. Some of the applications of neutralization are discussed in the following sections.

4.3.1. *Water Softening*

As explained earlier, hardness of water is caused by the presence of polyvalent metal cations. The major disadvantages of using this type of water are the increased consumption of soap required to produce lather when bathing or washing clothes and the formation of scales in boilers if this hard water is used for generating steam. Chemical precipitation is commonly employed to soften the water, where alkalis are added to the water to raise the pH and precipitate the metal ions in the forms of hydroxides and carbonates.

The softened waters usually have high pH values in the range of 10.5 and are supersaturated with calcium carbonate and magnesium hydroxide. For further use of such high pH waters, acid neutralization is applied. Adjustment of pH toward neutrality is accomplished either by recarbonation or by adding sulfuric acid (30).

pH adjustment by recarbonation can proceed in two different ways: one-stage recarbonation or two-stage recarbonation. In one-stage recarbonation, enough CO_2 is passed only one time to drop the pH to the desired level. When sulfuric acid is used in place of CO_2 in one-stage recarbonation, the process is simply called one-stage neutralization. In two-stage recarbonation, CO_2 is added to water at two different points after excess lime treatment. At the first point of addition, the CO_2 is passed to precipitate calcium carbonate. In the next step, CO_2 is added to adjust pH to acceptable levels. Figure 7 shows a schematic of one-stage and two-stage recarbonation.

4.3.2. *Metal Precipitation*

Metal precipitation through formation of metal hydroxide is one of the common methods of metal removal in industries. At high pH, most of the metal hydroxides are insoluble and come out of the solution in the form of metal hydroxide precipitates. Metals are precipitated as the hydroxides through the addition of lime or another base to raise pH to an optimum value (10–12,30–32). Metal carbonate precipitates can also be formed once soluble carbonate solutions such as sodium carbonate are added into metal solutions. Because pH is the most important parameter in precipitation, control of pH is crucial to the success of the process.

4.3.3. *Mine Drainage*

The wastewater coming out of mining industries is highly acidic due to the presence of sulfuric acid in appreciable quantities. Acid water coming out of mining industries is one of the common problems prevalent in United States and around the world. Sulfide minerals, mainly pyrite (FeS_2), which are often present in mine waste, can generate acid mine drainage when the waste comes in contact with water and air. Pyrite oxidizes to

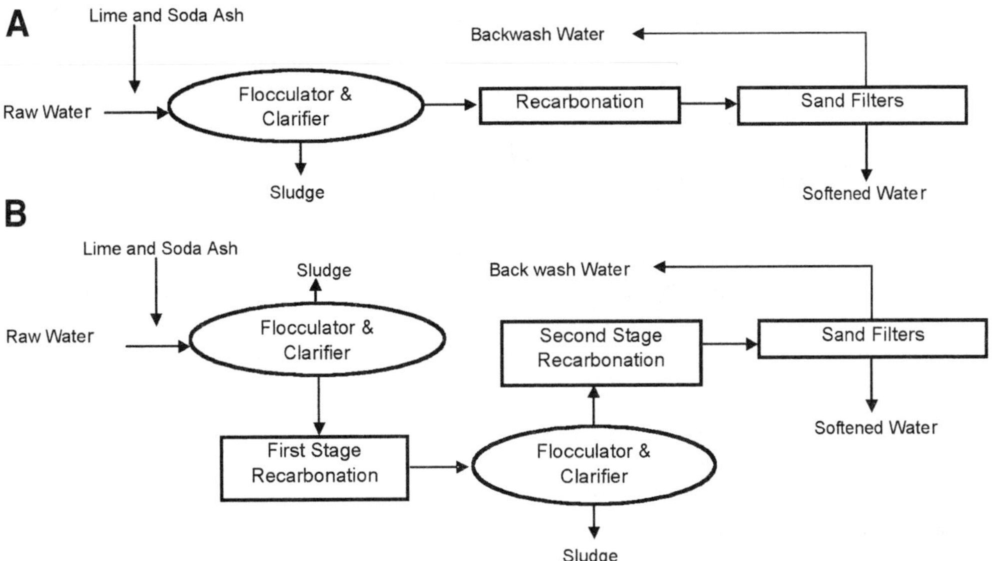

Fig. 7. Recarbonation in water treatment: (a) one-stage recarbonation; (b) two-stage recarbonation.

release sulfuric acid into water, resulting in a pH decrease that can be lower than 2. Most of the metals coming in the mine waste dissolve at this pH, resulting in water that is toxic to aquatic life. Chemical treatment by neutralization and subsequent precipitation is often applied to acid mine drainage. The pH range for point source discharge set by the US EPA is in the range of 6–9. The alkali comparison for acid mine drainage is given in Table 2 (8).

4.3.4. Metal Sorption

Activated carbons have successfully been used for metal removal (16–18). They are normally used for filtration of suspended solids, as well as adsorption of both organic and metal substances. Removal of heavy metal ions from waste streams by inexpensive recyclable biosorbents has emerged as an innovative technique in the last two decades (11,19). The major advantage is the high removal efficiencies for metal ions. This

Table 2
Alkali Comparison for Treatment of Acid Mine Drainage

Alkali	Formula	Molecular weight	Equivalent weight	Factor[a]
Ca Neutralizers				
Hydrated lime	$Ca(OH)_2$	74.10	37.05	1.35
Quicklime	CaO	56.08	28.04	1.78
Limestone	$CaCO_3$	100.08	50.04	1.00
Mg neutralizers				
Dolomitic lime	$Mg(OH)_2$	58.03	29.15	1.72
Na neutralizers				
Caustic soda	$NaOH$	39.99	39.99	1.25
Soda ash	Na_2CO_3	105.99	53	0.94

[a]Factor to convert CaO to $CaCO_3$ equivalence.

process is normally termed as biosorption. Normally biodegradation isn't involved as most biosorbents are inactive. The term "biosorption" is used simply because the biosorbents are made from organisms, such as bacteria and seaweed.

Numerous studies have shown that the sorption of metal ions from aqueous solutions is strongly pH dependent. An increase of the solution pH results in a decrease of positive surface charge and an increase of negatively charged sites and, eventually, an increase of metal ion binding. Normally the pH effect becomes less important when the pH is above 4–6. The metal ion adsorption onto activated carbon increases from 5% to 99% from pH 2.0 to 5.5 (16–18). Sorption experiments using calcium alginate beads (a biosorbent) demonstrated that the metal removal percentages increased from 0 to almost 100% (for metal concentrations < 0.1 m*M*) from pH 1.2 to 4 and a plateau was established at a pH > 4 (11). Therefore, neutralization pretreatment must be performed if the initial pH value of metal waste stream is less than 6.

5. pH NEUTRALIZATION PRACTICES

5.1. *Passive Neutralization*

In most cases, wastewater equalization is used to dampen out short-term extreme pH variations and allow excess acid to neutralize excess base materials, and vice versa, wherever possible. The equalization can be either on-line or off-line depending on the magnitude of the flows involved.

Off-line equalization is frequently practiced for small flow batch-wise release such as those associated with the regeneration of a plant's process water ion exchange columns. The isolation and off-line blending of the acid and caustic regeneration streams allow an industry to minimize the amounts of neutralization agents required to produce a wastewater that is suitable for downstream processes. The schematic of a typical industrial neutralization process is shown in Fig. 8.

5.2. *In-Plant Neutralization*

Industrial facilities that generate a continuous wastewater stream that is consistently acidic or basic can practice in-plant neutralization by metering a known quantity of the opposite neutralization agent into the sewer system. The combination of mixing that occurs in the pipe lines and in on-site equalization tanks can be sufficient to avoid costly pH adjustment systems.

5.3. *Influent pH Neutralization*

Industrial wastewaters produced by non-continuous processes that are characterized to be outside the allowable range for either direct discharge to a treatment plant or on-site treatment must be collected and the pH adjusted with a neutralization system. The most common influent pH adjustment chemicals are sulfuric acid, carbon dioxide, sodium hydroxide, calcium hydroxide, and magnesium hydroxide.

Carbon dioxide is frequently used as an in-pipe neutralization agent because of its rapid dissolution rate. The addition rate of the carbon dioxide is controlled by an in-line pH sensor in combination with a proportional pH controller and a wastewater flow meter.

The remaining neutralization agents are normally applied using a flowthrough neutralization tank containing a mechanical agitator capable of providing vigorous

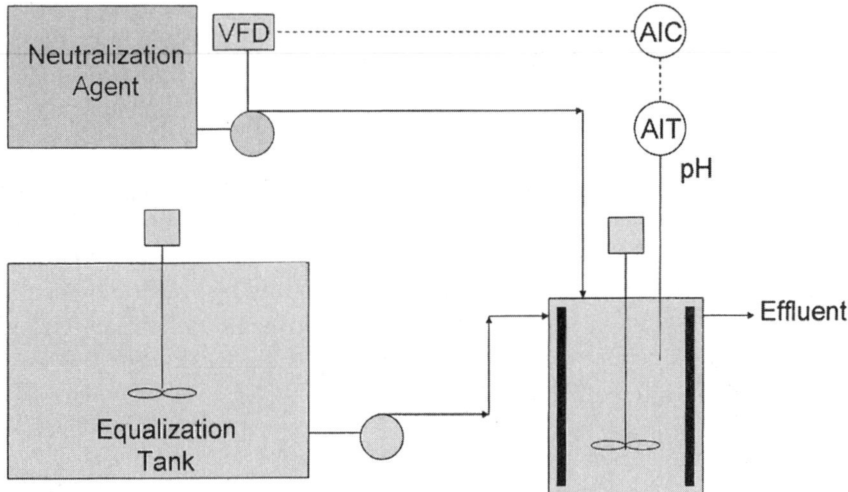

Fig. 8. Industrial neutralization practice.

agitation and instantaneous blending of wastewater and the neutralization agent. The neutralization tank contains a pH sensor that is connected to a proportional pH controller that sends a control signal to either a neutralization metering pump or control valve that is used to precisely meter the required amount of chemical to meet the instantaneous demand. These systems operate best when there is a constant (i.e., pumped) flow rate entering the neutralization tank and upstream equalization is practiced. Variable influent flow rate and a dynamic influent pH range present difficult neutralization problems that increase the complexity and cost of the neutralization equipment.

If the pH of the influent wastewater is typically more than 2 pH units away from the desired set point, the system is normally designed with two pH tanks in series. The first tank is designed to provide a rough pH adjustment and the second provides the fine tuning of the wastewater's pH.

Typical neutralization tank design provides for a hydraulic residence time between 10 and 20 min. The 20-min design factor is normally for systems that use either calcium or magnesium hydroxide slurries for base addition. The additional time is required to allow complete dissolution of the solids and to avoid downstream pH creep associated with post-neutralization tank reactions.

In some cases, the influent neutralization system is designed to raise the pH high enough to provide the needed alkalinity for a downstream process such as biological nitrification.

5.4. In-Process Neutralization

In well-mixed and buffered biological systems, the designer may elect to practice pH adjustment within the biological system's aeration basin. Under these circumstances, the designer must be careful to design redundant pH sensors and a control system that will protect the bacteria from malfunctioning mechanical or instrumentation systems.

Such systems are normally applied where the neutralization agent demand is low and continuous in nature such that the addition system can meet the demand but cannot rapidly shift the pH of the system.

5.5. *Effluent Neutralization*

Effluent neutralization is not normally required for a biological treatment system discharging to either a sewage treatment plant or an outfall. Normally, effluent neutralization would only be required for a physicochemical treatment system discharging to an NPDES outfall. Effluent neutralization maybe required by a sewage treatment plant if the receiving plant has an excess of alkalinity in the influent wastewater. Negotiations can frequently result in a pretreatment permit that allows discharge of treated wastewater with a pH as high as 10 (20,21).

5.6. *Chemicals for Neutralization*

It may be difficult to hold a pH of 6–8 as a slight change in hydrogen concentration can bring about wide swings in pH. Acidic industrial water can be neutralized by slaked lime [also called hydrated lime, calcium hydroxide, $Ca(OH)_2$], caustic soda (sodium hydroxide, NaOH), and soda ash (sodium carbonate, Na_2CO_3) (15).

As calcium hydroxide is less expensive than others, it is commonly applied for pH neutralization. High calcium quicklime known as calcium oxide (CaO) and dolomitic quicklime (a mixture of CaO and MgO) are typical commercial limes. The composition is very much dependent on their sources and manufacturing procedures. High calcium quicklime produces a high calcium hydrated lime that contains around 70% CaO; a dolomitic hydrate from a dolomitic quicklime has around 45% CaO and 34% MgO.

Selection of the above neutralizing agents depends on a series of factors, including their cost, expense of transportation, handling in plant, preparation for usage, and investment in facility, storage, safety, and labor costs. Caustic soda is poisonous, thus must be carefully handled. Emergency eyewashes and showers must be provided close to the chemical storage and operation area in case of an accident.

Lime may lose its efficiency as the solution pH approaches 7. In addition, the presence of organics can cause a significant amount of sludge, which is classified as a hazardous waste and must be treated. Thus, caustic soda as the principal neutralizing agent can be used in order to reduce the sludge production rate (15). However, the operational cost will be increased due to its high purchase expense. Lime can be used first to bring up the water to a slightly higher pH and subsequently caustic soda can be applied, resulting in the reduction of total operational cost.

Titration experiments are highly recommended to obtain the optimal dosage of neutralizing agents. However, if the composition of wastewater is known, one can use commercially available computer stimulation programs to get the dosage. MINEQL is one of the programs and has been widely used (22). In the program, chemical reactions, including solution reactions, precipitation and sorption reactions, in conjunction with the mass balances of different species are considered and solved numerically. It has been successfully used for many cases, including adsorption of heavy metals and metal pollution of groundwater (16,23).

5.7. *Encapsulated Phosphate Buffers for* In Situ *Bioremediation*

During *in situ* bioremediation of subsurface sediments and groundwater, changes in pH could be neutralized by the environmentally controlled release into the subsurface of phosphate buffers encapsulated in a polymer coating (24,32). The capsules are not designed or expected to move through the aquifer to specific contaminated areas. During *in situ* applications, it is anticipated that the encapsulated buffers would be added through a series of monitoring wells or drive points at specific locations. As groundwater flows through those points, the pH of the groundwater would be modified. This system would be analogous to that of *in situ* treatment walls in which a reactive barrier is created through which the groundwater flows. The reactive barrier is not mobile. Once the capsules have been used up, more could be added as necessary and any management of the introduction system, such as de-fouling, could be accomplished during that time.

The capsules are designed to release buffer (KH_2PO_4 or K_2HPO_4) into sediment pore water as a function of the polymer material used as the outer coating. Polymer coatings can be designed to dissolve at specific pH levels, releasing the buffer only when necessary and mediating not only processes that increase pH, but those that decrease pH as well.

The KH_2PO_4 microcapsules designed for application have an average diameter of 1 mm and are coated with a polymer that dissolves at pH levels above 7.0. It was shown that the encapsulated KH_2PO_4 buffer controlled pH under denitrification conditions in activated sludge suspended culture. The pH rise from 7.0 to 8.6 after 2 d of incubation was mediated to 7.0 ± 0.2 pH units in microcosms containing the encapsulated buffer (25).

Encapsulation technology has been examined for *in situ* bioremediation of subsurface environments. Vesper et al. encapsulated sodium percarbonate as 0.25–2.0 mm grains in order to provide a source of oxygen (from hydrogen peroxide) to enhance aerobic biodegradation of propylene glycol in soil (26). Encapsulated bacteria added to 0.2-μm dialysis bags and lowered into contaminated subsurface sediment have been used to enhance remediation of atrazine (27). dos Santos et al. reported the use of co-immobilized nitrifiers and denitrifiers to remove nitrogen from wastewater systems (28). Lin et al. co-immobilized fungal cells, cellulose co-substrate, and activated carbon in alginate beads in order to concentrate pentachlorophenol for microbial degradation (29). Encapsulation in environmental systems usually entails applications such as these, in which bacteria or slow release compounds are used to directly enhance biodegradation.

6. DESIGN OF A NEUTRALIZATION SYSTEM

The engineering design of a successful neutralization system involves several steps. Engineering design should be based on several factors such as optimum process parameters, laboratory-scale tests and their results, and, finally, cost analysis. Practical aspects such as availability of neutralizing agent in the near vicinity and thus reduced transportation costs play an important role in process design. The important steps involved in neutralization process design are outlined below.

All neutralization process, irrespective of type of waste, share several basic features and operate on the principle of acid–base reaction. Successful design of a neutralization process should consider the following;

- Influent wastewater parameters
- Type of neutralizing agent used
- Availability of land
- Laboratory scale experimental results

The overall design of neutralization process involves the design of the following features:

1. Neutralization basin
2. Neutralization agent requirements based on theoretical and treatability studies
3. Neutralization agent storage (e.g., silo, silo side valve, dust collector, and foundation design)
4. Neutralization agent feeding system
5. Flash mixer design

7. DESIGN EXAMPLES

7.1. Example 1

The flow rate at different time levels is given in Table 3. Calculate the volume of an equalization basin based on the characteristic diurnal flow.

Solution

The example asks for equalization basin volume based on diurnal flow. Hence, the hydrograph method described above is used for the calculation. It is assumed that the rate of inflow between any two consecutive time events is constant. The first step is to calculate the total volume entering the basin on an hourly basis by multiplying the flow rate in gal/min with 60 min. Then the cumulative flow is calculated as shown in Table 3 with the corresponding hydrograph shown in Fig. 9.

As explained in the previous section, the equalization volume will be sum of the vertical distances between the points of tangency of cumulative volume curve and the average daily flow. In this example, there are three such vertical distances. In Fig. 9, these distances are shown by AB, CD, and EF. However, close observation reveals that the equalization basin starts filling up at point B and continues until the cumulative volume curve reaches at point E. The equalization basin fills up to point C also, and continues beyond this point until it reaches point E. Hence the equalization volume is given by summation of AB and EF.

$$\text{Equalization volume} = AB + EF = 6000 + 40{,}000 = 46{,}000 \text{ gal.}$$

7.2. Example 2

Design a neutralization basin with 20 min detention time and a complete neutralization system for an industrial effluent with the following characteristics: flow rate = 0.792 MGD, pH = 3.5, acidity as mg/L $CaCO_3$ = 605, sulfate = 1300 mg/L, suspended solids = 65 mg/L.

Solution

Neutralization basin
Assume water depth = 5 ft and detention time period = 20 min:

Table 3
Wastewater Flow Variation with Time

Time	Flow rate (gpm)	Total volume (gal)	Cumulative volume (gal)
8 AM	70	4200	4200
9 AM	90	5400	9600
10 AM	235	14100	23700
11 AM	315	18900	42600
12 PM	279	16740	59340
1 PM	142	8520	67860
2 PM	85	5100	72960
3 PM	110	6600	79560
4 PM	78	4680	84240
5 PM	148	8880	93120
6 PM	234	14040	107160
7 PM	300	18000	125160
8 PM	382	22920	148080
9 PM	202	12120	160200
10 PM	78	4680	164880
11 PM	60	3600	168480
12 PM	68	4080	172560
1 AM	57	3420	175980
2 AM	42	2520	178500
3 AM	72	4320	182820
4 AM	77	4620	187440
5 AM	47	2820	190260
6 AM	57	3420	193680
7 AM	30	1800	195480

$$\text{Required volume} = 0.792\times10^{6}\,\frac{\text{gal}}{\text{d}}\times\frac{1\ \text{d}}{24\ \text{h}}\times\frac{1\ \text{h}}{60\ \text{min}}\times 20\ \text{min}$$

$$= 11\times10^{3}\,\text{gal} = 1470\ \text{ft}^{3}$$

$$\text{Surface area required} = \frac{1470\ \text{ft}^{3}}{5\ \text{ft}} = 294\ \text{ft}^{2}$$

The neutralization basin can be of square, rectangular, or circular cross section. For a square basin, each side should be 17.1 ft.

Lime Requirement (Theoretical)
Lime requirement will be calculated based on the amount of acidity present in water. Assuming 70% lime efficiency, theoretical lime required is

$$\text{Theoretical daily lime requirement} = 605\,\frac{\text{mg}}{\text{L}}\times\frac{1\ \text{mol Ca(OH)}_2}{1.35\ \text{mol CaCO}_3}\times\frac{1}{0.7}$$

$$= 640\,\frac{\text{mg Ca(OH)}_2}{\text{L}}$$

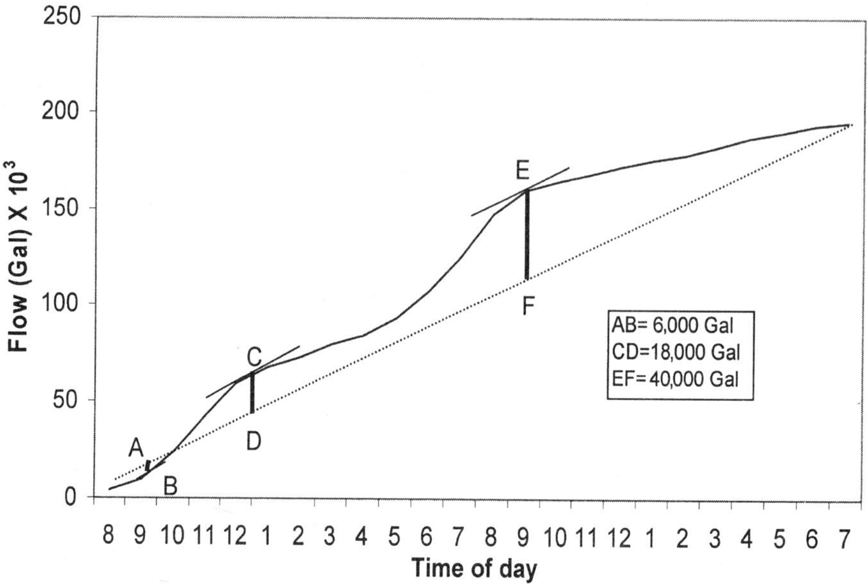

Fig. 9. Hydrograph for volume calculation of equalization basin.

$$\text{In terms of lb/d, the amount required} = 640\frac{\text{mg}}{\text{L}} \times 0.792\times 10^6\frac{\text{gal}}{\text{d}} \times 3.79\frac{\text{L}}{\text{gal}} \times \frac{1\text{ kg}}{10^6\text{ mg}} \times \frac{\text{lb}}{0.45\text{ kg}} = 4270\frac{\text{lb}}{\text{d}}$$

To assess the actual lime requirement, laboratory-scale titration experiments need to be performed. In general, the actual lime requirements are always higher than the theoretical requirement, because of other chemicals present. In this particular example, 15% extra is added to fulfill that requirement:

$$\text{Actual lime requirement} = 4270\frac{\text{lb}}{\text{d}} \times 1.15 = 4910\frac{\text{lb}}{\text{d}}$$

The above calculation gives a preliminary estimate of the amount of lime to be used. Although it is a good estimate, a treatability study must be performed when designing a large plant. A firmer estimate of the amount to be purchased on a regular basis will depend on actual usage. Once the amount of lime required is calculated, further design requires the selection of the type of lime used. The most common forms of lime used in industries are quicklime, limestone, and hydrated lime. In this example, quicklime is used, because this is the most widely used form of lime:

$$\text{Quicklime required} = 4910\frac{\text{lb}}{\text{d}} \times \frac{56\text{ g quick lime}}{74\text{ g hydrated lime}} = 3720\frac{\text{lb}}{\text{d}}$$

The actual total quicklime requirement will depend on the average efficiency of the slaker. Let us assume a 90% slaker efficiency:

$$\text{Actual CaO requirement} = 3720\frac{\text{lb}}{\text{d}} \times \frac{1}{0.9} = 4130\frac{\text{lb}}{\text{d}}$$

Design of Lime Silo
Silo is designed for a storage capacity for 7 d:

$$\text{Silo capacity} = 4130\frac{\text{lb}}{\text{d}} \times 7\ \text{d} = 28{,}900\ \text{lb}$$

Using the density of quick lime of 30 lb/ft^3,

$$\text{the silo volume required} = \frac{28{,}900\ \text{lb}}{30\dfrac{\text{lb}}{\text{ft}^3}} = 960\ \text{ft}^3$$

Assuming a diameter of 8 ft, the required side wall height is 19.1 ft. Provide a side wall height of 20 ft with a 60° hopper angle.

Other Silo Equipment
Design of other silo parts such as bin activator, dust collector, and bin level indicators is based on personal judgment and experience of designer. In general, bin activators are sized one half of the silo diameter, and the dust collector size varies with the module.

Lime Feeding, Slurry, and Tank
The hourly rate of lime required is

$$4130\frac{\text{lb}}{\text{d}} \times \frac{1\text{d}}{24\ \text{h}} = 172\frac{\text{lb}}{\text{h}}$$

In practice, a 10–15% slurry is desired. Let us assume a 15% slurry. To calculate the slurry volume and water requirements, the specific gravity of the slurry can be determined experimentally in the laboratory or can be obtained from the supplier. If the slurry specific gravity is 1.05, then the slurry has a unit weight of 8.8 lb/gal, of which 15% is due to the lime. Thus, the slurry flowrate required is

$$\text{Slurry flowrate} = 172\frac{\text{lb CaO}}{\text{h}} \times \frac{\text{gal}}{8.81\text{b slurry}} \times \frac{1\ \text{lb slurry}}{0.15\ \text{lb CaO}} = 130\frac{\text{gal}}{\text{h}}$$

If a 30-min detention time is provided, the required tank volume is

$$130\frac{\text{gal}}{\text{h}} \times 0.5\text{h} = 65\ \text{gal} = 8.7\ \text{ft}^3$$

If we use a cylindrical tank with diameter equal to height, we obtain $D = H = 2.2$ ft.

Flash Mix Tank
Assume a detention time of 3 min:

$$\text{Volume} = \frac{3}{60}\text{h} \times \left(\left(0.792 \times 10^6\frac{\text{gal}}{\text{d}} \times \frac{1\ \text{d}}{24\ \text{h}}\right) + 130\frac{\text{gal}}{\text{h}}\right) = 1660\ \text{gal} = 222\ \text{ft}^3$$

If we use a cylindrical tank with diameter equal to height, we obtain $D = H = 6.6$ ft.

NOMENCLATURE

D diameter of impeller, m
G mean velocity gradient, s^{-1}
K_T constant dependent on impeller size and shape

n impeller revolutions per second, s^{-1}
P power requirement, N-m/s
Q flow rate, m^3/s
V mixing tank volume, m^3
t_d detention time, s
V_p volume of acid added to a solution to reach a pH of 8.3 during titration, mL
V_c volume of acid added to a solution to reach a pH of 4.5 during titration, mL
ρ density of the fluid, kg/m^3
μ absolute viscocity of the fluid, $N\text{-}s/m^2$

REFERENCES

1. US EPA, *An Appraisal of Neutralization Processes to Treat Coal Mine Drainage.* EPA-670/2-73-093, U.S. Environmental Protection Agency, Washington, DC, 1973.
2. Metcalf & Eddy Inc., *Wastewater Engineering: Treatment Disposal Reuse*, 4th ed., McGraw-Hill, New York, 2002.
3. R. A. Corbitt, *Wastewater Disposal*, McGraw-Hill, New York, 1989.
4. E. R. Alley, *Water Quality Control Handbook*, McGraw-Hill, New York, 2000.
5. W. W. J. Eckenfelder, *Industrial Water Pollution Control*, 3rd ed., McGraw-Hill, New York, 2000.
6. WEF/ASCE, *Design of Municipal Wastewater Treatment Plants*, 4th ed., Water Environment Federation and American Society of Civil Engineers, 1998.
7. US Army Corps of Engineers, *Engineering and Design—Hydrologic Engineering Requirements for Reservoirs*, CECW-EH-Y, Washington, DC, 1997.
8. US EPA, *Design Manual—Neutralization of Acid Mine Drainage*, U.S. Environmental Protection Agency, Municipal Environmental Research Laboratory, EPA-600/2-83-001, U.S. Environmental Protection Agency Technology, Cincinnati, OH, 1983.
9. US EPA, *Evaluation of Flow Equalization at a Small Wastewater Treatment Plant*, US Environmental Protection Agency, Municipal Environmental Research Laboratory, EPA-600/2-76-181, U.S. Environmental Protection Agency, Cincinnati, OH, 1976.
10. W. Stumm and J. J. Morgan, *Aquatic Chemistry*, John Wiley and Sons, New York, 1981.
11. J. P. Chen and L. Wang, Characterization of a Ca-alginate based ion exchange resin and its applications in lead, copper and zinc removal. *Separation Science and Technology*, **36**(16), 3617–3637 (2001).
12. J. P. Chen and H. Yu, Lead removal from synthetic wastewater by crystallization in a fluidized-bed reactor, *Journal of Environmental Science and Health, Part A-Toxic/Hazardous Substances & Environmental Engineering*, **A35**(6), 817–835 (2000).
13. M. L. Davis and D. A. Cornwell, *Introduction to Environmental Engineering*, 3rd ed., McGraw-Hill, New York, 1998.
14. F. N. Kemmer, *The Nalco Water Handbook*, McGraw-Hill, New York, 1988.
15. C. A. Hazen and J. I. Myers, Neutralization tactics for acidic industrial wastewater. In: *Process Engineering for Pollution Control and Waste Minimization* (D. L. Wise, ed.), Marcel Dekker, New York, 1994.
16. J. P. Chen and S. N. Wu, Acid/base treated activated carbons: characterization of functional group and metal adsorptive properties, *Langmuir*, **20**(6), 2233–2242 (2004).
17. J. P. Chen and M. S. Lin, Equilibrium and kinetics of metal ion adsorption onto a commercial H-type granular activated carbon: Experimental and modeling Studies, *Water Research*, **35**(10), 2385–2394 (2001).
18. J. P. Chen and S. N. Wu, Study on EDTA-chelated copper adsorption by granular activated carbon, *Journal of Chemical Technology and Biotechnology*, **75**(9), 791–797 (2000).

19. J. P. Chen, L. Hong, S. N. Wu, and L. Wang, Elucidation of interactions between metal ions and Ca-alginate based ion exchange resin by spectroscopic analysis and modeling simulation, *Langmuir*, **18**(24), 9413–9421 (2002).
20. US EPA, *Flow Equalization*, EPA 625/4-74-006, US Environmental Protection Agency, Washington DC, 1974.
21. US EPA, Process Design Manual for Upgrading Existing Treatment Plants, EPA 625/1-71-004a, U.S. Environmental Protection Agency, Washington DC 1974.
22. W. D. Schecher and D. C. McAvoy, *MINEQL+ Chemical Equilibrium Modeling System*, version 4.5 for Windows. Environmental Research Software, Hallowell, ME, 2001.
23. J. P. Chen and S. Yiacoumi, Transport modeling of depleted uranium (DU) in subsurface systems, *Water, Air, and Soil Pollution*, **140**(1–4), 173–201 (2002).
24. C. M. Rust, C. M. Aelion, and J. R. V. Flora, Control of pH during denitrification in subsurface sediment microcosms using an encapsulated phosphate buffer, *Water Research*, **34**(5), 1447–1454 (2000).
25. B. Vanukuru, J. R. V. Flora, M. F. Petrou, and C. M. Aelion, Control of pH during denitrification using an encapsulated phosphate buffer. *Water Research*, **32**(9), 2735–2745 (1998).
26. S. J. Vesper, L. C. Murdoch, S. Hayes, and W. J. Davis-Hoover, Solid oxygen source for bioremediation in subsurface soils. *Journal of Hazardous Materials* **36**(3), 265–274 (1994).
27. M. R. Shati, D. Ronen, and R. Mandelbaum, Method for *in situ* study of bacterial activity in aquifers. *Environmental Science and Technology*, **30**(8), 2646–2653 (1996).
28. V. A. P. M. dos Santos, M. Bruijnse, J. Tramper, and R. H. Wijffels, The magic-bead concept: an integrated approach to nitrogen removal with co-immobilized micro-organisms. *Applied Microbiology and Biotechnology*, **45**(4), 447–453 (1996).
29. J. Lin, H. Y. Wang, and R. F. Hickey, Use of coimmobilized biological systems to degrade toxic organic compounds, *Biotechnology and Bioengineering*, **38**(3), 273–279 (1991).
30. L. K. Wang, Y. T. Hung, and N. S. Shammas (eds.), *Physicochemical Treatment Processes*. Humana Press, Totowa, NJ (2005).
31. L. K. Wang, N. S. Shammas, and Y. T. Hung (eds.), *Advanced Physicochemical Treatment Processes*. Humana Press, Totowa, NJ (2005).
32. L. K. Wang, Y. T. Hung, H. H. Lo, and C. Yapijakis (eds.), *Handbook of Industrial and Hazardous Wastes Treatment*. Marcel Dekker, Inc., NY, NY. (2004).

3
Mixing

J. Paul Chen, Frederick B. Higgins, Shoou-Yuh Chang, and Yung-Tse Hung

CONTENTS

1. INTRODUCTION

Mixing is an important operation in many types of facilities utilized in various industries, including chemical production and environmental pollution control (1–10). Solids may be shredded and blended to promote uniform and complete combustion in modern incinerators. In water and wastewater treatment operations, mixing may be involved in equalization, dispersion of chemicals, enhancement of reaction kinetics, and prevention of solids deposits. Prior to the selection of equipment or the design of specific facilities, it is first necessary to consider the various reasons for mixing and the underlying principles of each.

Dispersion, for the purpose of distributing reactants in a bulk medium or achieving uniformity of concentration within a mixture, is primarily a turbulent diffusion phenomenon. Equally efficient dispersion may be obtained through high turbulence for a short period of time or low turbulence for a prolonged period. Design values can be expressed in terms of energy input per unit volume of the bulk material mixed. For a flash mixer associated with chemical feed into water or wastewater, a frequently cited design parameter is horsepower per unit mixing chamber volume for a specified detention period (10–15).

Mixing to obtain solids suspension or transport is characterized by time maintenance of eddy velocities well in excess of the settling velocities being handled and simultaneously obtaining boundary velocities sufficient to resuspend any solids that may reach the bottom of the conduit or chamber. Traditional design approaches for solids transport have centered on the specification of average velocities known to be

From: *Handbook of Environmental Engineering, Volume 3: Physicochemical Treatment Processes*
Edited by: L. K. Wang, Y.-T. Hung, and N. K. Shammas © The Humana Press Inc., Totowa, NJ

adequate for the conveyance of certain materials. Thus, the commonly quoted figures of 0.61 m/s minimum velocity for domestic sewage in pipes, 0.76 m/s for storm sewage (where sand is encountered), and 10–15 m/s average velocity for gas-borne dusts in ventilation ducting (16–19).

The promotion of chemical and biological reactions within a bulk medium involves the transport of reactants by eddies within the bulk medium, and the molecular diffusion of reactants to the reaction surface. The reaction rate is enhanced by increased turbulence; however, in environmental engineering applications, upper limits on eddy velocities are imposed by the need to prevent the disruption of solid particles by excessive shear forces associated with extreme turbulence. This problem is especially critical in reactions involving chemical flocculation and activated sludge. Within the environmental engineering field, design approaches have been largely empirical. Chemical engineers have, however, been somewhat more successful in developing at least semiquantitative approaches to the design of mixed reactors.

Mixing to maintain temperature is similar in principle to the dispersion mixing case presented above (12,13). However, where heat transfer across a solid boundary is desired, the rate of transfer is enhanced by large temperature gradients adjacent to the boundary. This is achieved through high wall shear and the resultant decrease in thickness of the boundary layer. Proper design involves a balance between the increasing cost of achieving higher degrees of turbulence and the decreasing cost of providing the correspondingly smaller heat exchange surface.

In many applications, mixing may be required to meet several objectives simultaneously. In an aerobic activated sludge treatment of organic wastewater, the air provides oxygen to the mixed liquor (16–20). It provides mixing to disperse the oxygen, enhance the biological uptake of the oxygen, and prevent deposition of solids in the bioreactor. An upper limit on velocity gradient is imposed by the desire to prevent excessive breakdown of the biological floc. In a case of this type, optimization is impossible and design becomes a compromise of the various functions.

In a conventional activated sludge basin, the volume of air required for oxygen supply, applied to a basin of proper geometry, will provide acceptable solids suspension without excessive breakdown and an adequate oxygen uptake rate for the biochemical processes.

In the following sections of this chapter, mixing will be treated from a fundamental standpoint. The basic principles and equipment will be presented in a comprehensive manner with examples to illustrate design computations.

2. BASIC CONCEPTS

The term "mixing" is applied to operations that tend to eliminate nonuniformities in chemical and/or physical properties of materials. Mixing is accomplished by movement of matter between various parts of a mass. For fluids, the movement results from the combined effects of bulk flow and both eddy and molecular diffusion. The mixture produced by application of the fluid motions noted above is a completely random distribution. This randomness may be apparent in dry solids mixing and in solids suspensions because of the relatively few particles in a sample. However, in fluid blends the random character of the mixture cannot be discerned because the particles are molecules, and therefore the number of particles in any sample is several orders of magnitude larger than mixtures that include solid particles (1–6).

Bulk diffusion involves the distribution of materials through relatively massive movements to remote parts of an unmixed system. Injection of a high-velocity liquid stream into a tank of a second liquid will create a jet motion that may carry a considerable distance before dissipation into turbulent motion is complete. In mechanically mixed systems, the pumped flow created by impellers of many types results in bulk flow. Bulk flow in liquids is almost always accompanied by eddy and molecular diffusion, although the latter may become negligible in high-viscosity systems (1,2,21–24).

Molecular diffusion is a product of relative molecular motion. In any gaseous or liquid system where there are two or more kinds of molecules, if we wait long enough, the molecules will intermingle and form a uniform mixture on a submicroscopic scale (by submicroscopic, we mean larger than molecular, but smaller than visual by the best microscope). This view is consistent with the definition of a mixture, for we know that, if we were to use a molecular scale, we would still observe individual molecules of the two kinds, and these would always retain their separate identities. The ultimate goal in any mixing process would be this submicroscopic homogeneity, where molecules are uniformly distributed over the field; however, the molecular diffusion process alone is generally not fast enough for practical mixing needs.

If turbulence can be generated, eddy-diffusion effects can be used to enhance the mixing process. The turbulent process can be used to break up fluid elements to some limiting point; however, because of the macroscopic nature of turbulence, one would not expect the ultimate level of breakup to be anywhere near molecular size. Because energy is required for this reduction in scale, the limiting scale should be associated with the smallest of the energy-containing eddies. This might be considered as the eddy size, which characterizes the dissipation range. One might also use the microscale as a measure. In any case, this size will be large when compared with molecular dimensions. No matter how far we reduce the scale, we still have pure components. Depending on the size observed, any one of these levels in scale might be considered mixed; however, from a view of submicroscopic homogeneity, where molecules are uniformly distributed over the field, none is mixed. Without molecular diffusion, this ultimate mixing cannot be obtained (2,9,13,24–30).

Molecular diffusion promotes the movement of the different molecules across the boundaries of fluid elements, thus reducing the difference in properties between elements. This reduction in degree of segregation will occur with or without turbulence; however, turbulence can help speed the process by breaking the fluid into many small clumps, thus allowing more area for molecular diffusion. When diffusion has reduced the intensity of segregation to zero, the system is mixed. The molecules are distributed uniformly over the field. In systems where the reaction is to occur, the need for submicroscopic mixing is apparent, for without it, the only chemical reaction that could occur would be on the surface of the fluid clumps (12,17–19,30–35).

Each of the bulk-diffusion phenomena tends to reduce the scale of segregation by spreading a contaminant over a wider area. The molecular diffusion is enhanced because of the larger area. It is important to note that if the molecular diffusion is rapid enough, the system may be almost submicroscopically mixed by the time the bulk diffusion has spread the contaminant over the field (36–41).

2.1. *Criteria for Mixing*

The mixing process can be visualized as a breakdown of the larger eddies to smaller, and finally to the smallest eddies, at which point the mixing scale becomes small enough for turbulence no longer to act. Eddies become so small that viscous shear forces prevent turbulent motion, and molecular diffusion becomes the controlling factor. Molecular diffusion completes the mixing process by eventually providing molecular homogeneity. In reality, the two processes of breakdown and diffusion occur at the same time. However, the assumption of a stepwise process will aid in the discussion to follow. If the fluids to be mixed are gases, the molecular diffusion is very high and the diffusion time extremely short. But if the fluids are liquids, the molecular diffusion is slow, and becomes very important. The slow diffusion time, in the case of liquids, requires knowledge of the turbulence, so that an estimate of the size of the smallest eddy and the time for molecular diffusion can be made (1,12,13).

In order to approach this problem, it is necessary to quantify the degree of mixing of the system under study (1). In addition, it must be recognized that two processes are occurring, the breakup of eddies and diffusion. Mixing parameters can be defined in terms of scale of segregation and intensity of segregation. These parameters describe the mixing process and can be estimated from measurable statistical values. The only major restriction on the parameters is that they cannot be applied to cases where gross segregation occurs as in the initial moments of mixing.

The scale of segregation of a mixture is a measure of the size of regions of segregation within the mixture. The smaller the scale of segregation would cause a better mixture to occur. Consider the example of dispersion of the pigment in the bulk. The scale of segregation in this case is an area on the inspected surface that does not have mean composition of the bulk of the mixture. The divergence from mean composition may vary within all areas of segregation. The intensity of segregation is the measure of this divergence. The lower the intensity of segregation is, the better the mixture is. Both the "scale of segregation" and the "intensity of segregation" are the measure of quality of mixture.

The scale of segregation is analogous to the scale of turbulence used in the treatment of fluid motion. However, because the concentration term is a scalar, there is one term instead of nine:

$$C(r)=\frac{\overline{a(x)a(x+r)}}{a'^2} \tag{1}$$

where $C(r)$ is the Eulerian concentration correlation, a is the deviation $A–A'$ (A is concentration fraction of liquid A, A' is the average), and a' is the root-mean-square (rms) fluctuation. For two points separated in space by a distance r, the deviations from the mean values of concentration of liquid A will be large compared to the rms deviation squared. As the liquid is mixed, the numerator and denominator of Eq. (1) will approach the same values yielding a concentration correlation of unity. $C(r)$ may be integrated over the distance r to produce a linear value termed "scale of segregation" defined as

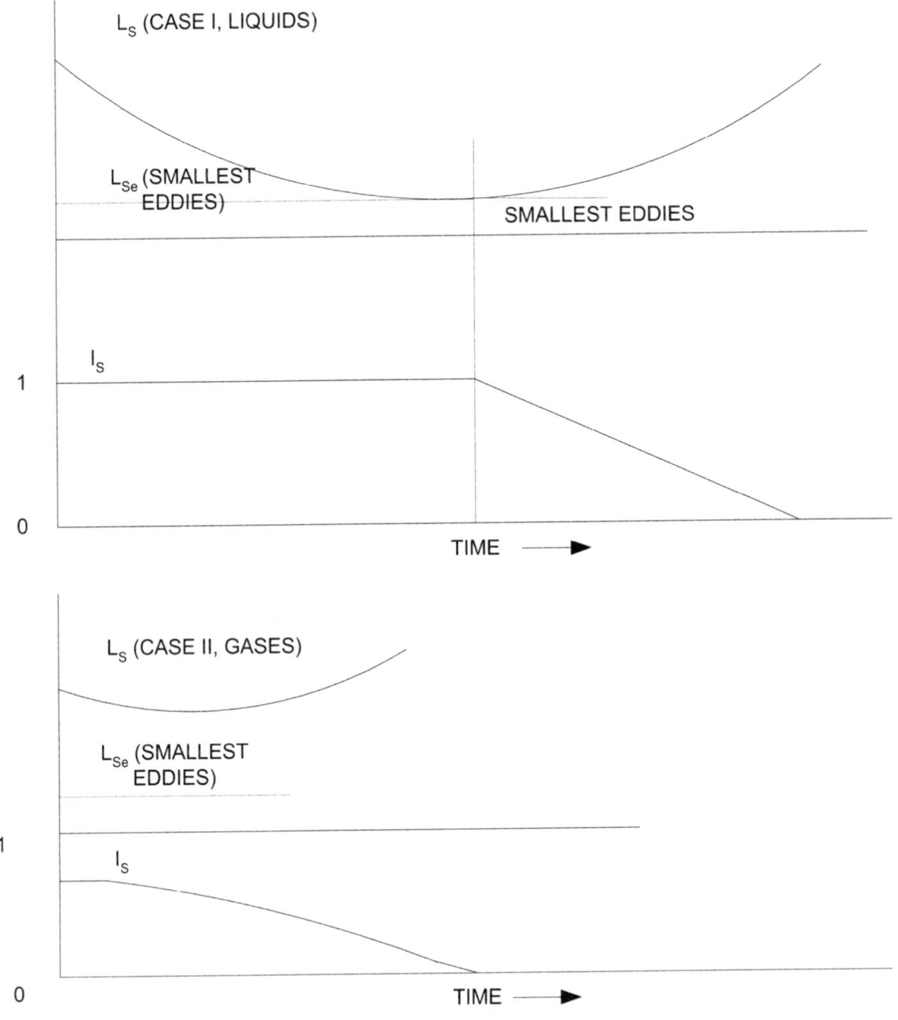

Fig. 1. Intensity and scale of segregation.

$$L_s = \int C(r)dr \tag{2}$$

As illustrated in Fig. 1, the initial value of L_s will be large, but will decrease as $C(r)$ approaches unity. Because $C(r)$ remains constant once the system is mixed as far as practical by eddy-diffusion, the value of L_s will increase with outward diffusion and consequent increase in the magnitude of r.

The scale of segregation is an average over relatively wide values of r, and thus is a good measure of the large-scale process (breakup of the eddies) but not of the small-scale diffusional process. In the liquid system with slow molecular diffusion, the scale would decrease rapidly to some small value (smallest eddy size) and then increase slowly as molecular diffusion completes the mixing (case I of Fig. 1). The increase in scale is due to an apparent increase in eddy size because of outward diffusion. The value would increase indefinitely with r, since $C(r)$ is unity everywhere in the uniform medium. In a gas, where the molecular diffusion is very rapid, the scale may not be reduced appreciably before diffusional effects become controlling (case II of Fig. 1). For the rapid diffusion gas system, and in the latter part of the liquid mixing process when the scale is small, the intensity of segregation gives a better description of the degree of mixing.

The intensity of segregation (also called the degree of segregation) is defined as

$$I_s = a'^2 / a_0'^2 \tag{3}$$

It is measured at a point in the fluid for a long enough period of time to obtain a true average. The subscript zero refers to the initial value. The intensity of segregation is unity for initial time and for complete segregation. The intensity of segregation drops to zero when the mixture is uniform ($a'^2 = 0$). If there were no diffusion, and only the smallest possible eddies were present, the value of I_s would still be one; thus, the intensity of segregation is a good measure of the degree of completion of the diffusional process (see Fig. 1).

Equation (3) gives the simplest form of the intensity of segregation and is defined as a function of time-averaged variables at a point. For a complete definition of a given system, one would have to specify the variation over the entire volume. As a simple example, let us consider a plug flow in which two fluids are to be mixed. It will be assumed that each fluid is initially uniformly distributed across the pipe cross section on a macroscopic scale under the condition of complete segregation (I_s is unity). As the fluid moves down the tube in plug flow, mixing will occur as a result of the turbulent field and diffusion, and the value of I_s will decrease to zero in the limit of molecular uniformity. I_s must be measured over some small but finite volume. If this volume is too small, submicroscopic variations will be detected (statistical fluctuations in the number of molecules present), and if the volume is too large, the measurement would become insensitive and approach the average value of the system. For many problems (such as non-ideal mixers used for reactors), a detailed study of the variation in I_s over the entire reactor is not desirable, and some space average of the entire system is used.

2.2. *Mixing Efficiency*

The concept of "complete" or "ideal" mixing is often encountered in the literature related to equalization and chemical reactions occurring in continuous-flow reactors. Many of the formulations presented in the chapter on equalization (chapter 2) are predicated on this assumption. In the equalization context, ideal mixing connotes a basin within which the concentration of a tracer substance is everywhere equal to that in the effluent. Short-circuiting and time lags in distribution of incoming materials will always result in departures from ideal performance; however, the ideal may be closely approximated in real systems where relatively high degrees of agitation are used. In chemical reaction applications, reactor performance is affected by the macroscale mixing described above but may also be influenced on the microscale by the effect of mixing on the rate of chemical reaction. For a complete discussion of reaction kinetics, one of the many textbooks on the subject should be consulted (3,41).

On the macroscale, the test of mixing performance in continuous-flow systems consists of whether the effluent characteristics match those predicted by mathematical models in which complete mixing is assumed. Deviation from expected performance may indicate either that the tank contents are inadequately mixed or that short-circuiting is occurring from the inlet to the outlet. From the designers' standpoint, however, the selection of specific mixing criteria for a given application is an uncertain process. Generalizations are possible in terms of analogies with batch mixers or through empirical approaches related to power input, shear gradients, or basin turnover time. In addition, a few studies have been performed on mixing efficiency as a function of mixing effort. The relationship between mixing intensity can be determined by measuring the ratio of inflow to circulation with the vessel and the approach to theoretical performance. Pulse tracers are used and performance is evaluated comparing the slope of the concentration gradient in the effluent to that expected based on a mathematical model assuming ideal mixing. Marr and Johnson pointed out that, in all cases, the slope of the curve relating concentration and time was less than theoretical, indicating slower than expected flushing of the tracer from the vessel (4). The final equation is

$$\text{Slope ratio} = 2\left(\frac{q-q_f}{q}\right)\left(1-\sqrt{1-\frac{q_f}{q}}\right) \quad (4)$$

where q_f is inlet flow, m^3/s and q is circulation flow within the vessel due to the impeller, m^3/s.

For an inflow rate equal to one-tenth the circulation rate, the slope ratio is 92%, while at two-tenths the value is 85%. For design purposes, a design circulation rate of five times the inflow is frequently taken as a sufficiently close approximation to the ideal.

Where chemical reactions occur, the approach to mixing design goes beyond the concepts of approximating ideal performance by tracer behavior. In real systems, influent–effluent relationships may be determined for the ideal case and then modifying either empirically or using relationships such as Eq. (4) to develop residence-time distribution functions that may be used with kinetic models to predict performance.

Zweitering defined two extremes of mixing that may occur within a reactor. Each may be uniquely defined in terms of the residence-time distribution function (5). The least amount of mixing occurs in what is termed a completely segregated reactor. The term "complete segregation" means that the fluid elements entering the reactor remain essentially intact in passing through the vessel. Under this condition, the fluid elements act as batch reactors subject to reaction times determined by the residence-time distribution. Expected performance is determined by

$$w_i^N = w_i^{N-1} + \int_0^\infty f^N(t) \int_0^\infty r_i^N(w_i, \rho, T)\,dt'dt \quad (5)$$

where w_i^N is the mass fraction of the ith component in reactor N; w_i^{N-1} is same as above in the preceding reactor of a series; $w_i^N(w_i, \rho, T)$ = mass rate of production of the

ith component in the Nth tank per unit volume, g/cm^{3}s (w_i,ρ, and T are mass, density, and temperatures which prevail and affect r_i^N); t = time, s; and $f^N(t)dt$ = fraction of the mass flow through the reactor with a residence time in the interval t to $t + \Delta t$.

Zweitering used the term "maximum mixedness" in conjunction with the largest possible fluid interaction consistent with the residence-time distribution (5). In an ideal stirred reactor each fluid element has a uniform probability of mixing with any other element. This probability is consistent with the residence-time distribution of an ideal reactor but is not when the residence-time distribution is altered by non-ideal mixing. Zweitering determined that the conversion in a maximum mixedness reactor would be predicted by the solution of the following differential equation:

$$\frac{dw_i^N}{dh} = \frac{r_i^N}{\rho}(w_i, \rho, T) + \left(w_i^{N-1} - w_i^N\right)\frac{f^N(-h)}{\int_{-h}^{\infty} f^N(t)dt} \tag{6}$$

The term h in Eq. (6) is a dummy variable, which is termed the "mixing history parameter". Depending on the shape of the residence-time distribution function, the function of h $\left[f^N(-h)\Big/\int_{-h}^{\infty} f^N(t)dt\right]$ may equal zero, a positive constant, or infinity. The zero value is obtained for a residence-time frequency that decreases as t^{-n}. A positive constant is obtained when the residence-time frequency function approaches zero at large t. Where the residence-time frequency function approaches zero in a stepwise fashion at finite t, the infinite value is obtained.

The significance of Eqs. (5) and (6) is that limits may be established on the conversion to be expected in a reactor subject to the two extremes of mixing. If the predicted conversions are close to the same value, mixing is of relatively little importance and may be adequately handled by empirical rules. However, where significant deviation is found, mixing may warrant more attention in design. Normally, microscale mixing is not a significant factor in zero- and first-order reactions, but can play an extremely important role in reactions of increasing complexity.

2.3. *Fluid Shear*

In water and wastewater applications, shear forces in mixed fluids have been considered important owing to the necessity of preventing the rupture of delicate floc particles in the coagulation process. Shear is related to velocity gradient by the relationship:

$$\tau = \mu' G \tag{7}$$

where μ' is fluid viscosity (dynamic), kg·s/m^2 (lb·s/ft^2) and G = velocity gradient, m/s m (ft/s ft).

Camp proposed that flocculation basins be designed on the basis of controlling the rms velocity gradient in flocculation basins according to the formula (6,13,16–19,41):

$$\overline{G} = \sqrt{\frac{W}{\mu'}} \tag{8}$$

where W is the rate of power dissipation in a unit volume of fluid. The concept may be applied to various types of mixing systems by determining the appropriate power dissipation. For baffled mixing basins and conduits:

$$W = \frac{\rho h_f}{t} \tag{9}$$

where ρ is unit weight of fluid, kg/m³(lb/ft³); h_f = head loss, m (ft); and t = detention time, s.

For diffused air mixing systems, we have

$$W^* = \frac{62.4Q_a}{V} \times \frac{34H}{\frac{11}{2}+34} \tag{10}$$

where Q_a is air flow, ft³/s (cfs); V is volume of liquid, m³; and H = depth of diffusers, m.

For a given velocity gradient, we have

$$Q_a^* = \frac{\overline{G}^2 \mu' V}{4890 \log \frac{H+34}{34}} \tag{11}$$

In the case of rotating paddles, Camp derived the equation (6):

$$W^* = \frac{239 C_D (1-k)^3 S_s^3}{V} \sum A r_b^3 \tag{12}$$

where C_D is drag coefficient of blades; k is water velocity relative to blade velocity (0.24–0.32); S_s is speed of rotation of blades in revolutions, s; A is area of a given blade, ft², and r_b = distance to the centroid of a given blade, ft.

The above equation was generalized by Hudson and Wolfner to apply to any mechanically mixed system (7):

$$\overline{G}^* = 425\left(\frac{\text{hp}_\text{w}}{t'}\right) \tag{13}$$

where hp_w is horsepower applied to the water in hp/l0⁶ gpd flow rate, and t' = detention time, mins.

Because rapid mix systems are associated with flocculators and since G' is a measure of turbulence in the liquid, the concept of using the product of C and time as a measure of mixing intensity–duration and therefore completeness, naturally followed. Recommended values for the product $\overline{G}t$ are presented in the subsequent design sections for various processes.

3. MIXING PROCESSES AND EQUIPMENT

3.1. Mixing in Turbulent Fields

Many attempts have been made to characterize the nature of turbulence in fluid flow. These have been previously discussed as they relate to mixing (20). One of better approaches, developed by Beek and Miller, is amenable to calculation of mixing length requirements without measurement of actual parameters in the system to be considered (8). With slight modification of the original equations, Brodkey showed acceptable correlation with at least limited experimental data (2).

Beek and Miller's approach to quantify the relationship between turbulence and mixing length is founded on considerations of energy spectra as defined by the wave number (reciprocal eddy size) approach.

The basic equation below contains terms related to energy spectra [$E_s(k)$] in terms of wave numbers (k^X), mass diffusivity (D_m) and time (t):

$$\frac{\partial E_s(k)}{\partial t} = -2\left[D_m + \beta\int_k^{\infty}\sqrt{E(k')/(k')^3\,dk'}\right]k^2 E_s(k) + 2\beta\sqrt{E(k)/k^3}\int_0^k k'' E_s(k'')dk'' \tag{14}$$

Owing to space considerations, the reader may refer to the original authors for a detailed discussion of the basis of the equation. It will suffice at this point to state that the equation has been numerically integrated for various values of k. Integration was accomplished by assuming isotropic turbulence (uniform in three dimensions) enclosed in a pipe and conveyed by a uniform mean velocity. The isotropic assumption does not hold in real pipes and channels because turbulent intensity is larger at the walls, but, if the subsequent relationships are considered to apply along the centerline of flow, the approach is subject to only minor error.

The largest velocity eddy defined as one-fourth the pipe diameter:

$$\frac{1}{k_0} = \frac{d}{4} \tag{15}$$

where k_0 = reciprocal of the largest eddy size, 1/m (l/ft) and d = pipe diameter, m (ft).

A second parameter is defined as:

$$\alpha' = \left(\frac{1}{N_{Sc}}\right)\left(\frac{1}{N_{\mathrm{Re}}}\right)\left(\frac{8}{u'/\bar{u}}\right) \tag{16}$$

where N_{Sc} is the Schmidt number = $\mu/\rho D_v$; μ is viscosity, kg/s m (lb/s ft); ρ is density, kg/m^3 (lb/ft^3); D_v is diffusion coefficient, m^2/s; N_{Re} is Reynolds number = $\bar{u}d/\upsilon$; $\bar{u}$ *is* mean velocity, m/s; d = pipe diameter, m; υ is kinematic viscosity, m^2/s; and u' is rms velocity, m/s.

The relationship between rms velocity and mean velocity is defined by

$$\left(\frac{u'}{u}\right) = 0.56 f^{1/3} g_s^{1/2} \tag{17}$$

where f is the friction factor associated with the flow and g_s is a number determined graphically from Fig. 2 using the relationship:

$$X_s = 0.164 f^{1/4} \mathrm{N}_{\mathrm{Re}}^{3/4} \tag{18}$$

A dimensionless time of mixing may be calculated by the relationship:

$$\sigma = k_0 u' t \tag{19}$$

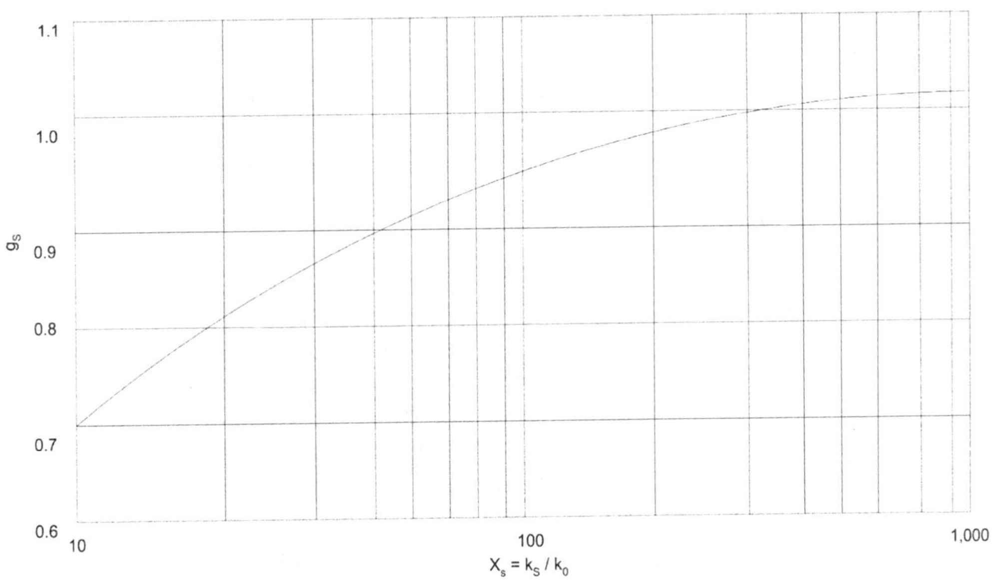

Fig. 2. Dependence of g_S on the reduced wave number.

The results of the approach may be summarized in graphical form for at least representative values. Figure 3 demonstrates Beet and Millers' results for a typical gas ($N_{Sc} = 1$) and a typical liquid (N_{Sc}) = 2300 (8). The time parameter is defined as a function of Reynolds number. Although it appears from the figure that gas mixing is independent of Reynolds number, this is not actually true because the term u′ affects the Reynolds number and also appears in the time parameter. The approach to linear

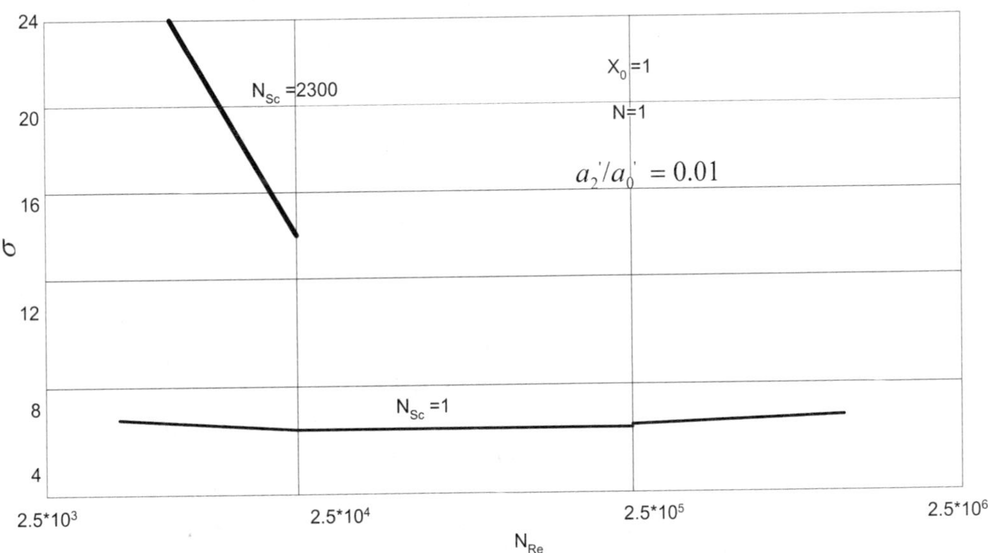

Fig. 3. Dimensionless mixing time parameter vs Reynolds number for 99% complete mixing.

mixing described above has not been widely developed, although correlation is possible with experimental data (2).

Insufficient evidence is available to definitely establish the validity of the approach and the calculations tend to be unwieldy. In practical applications, empirical approaches are used and these are presented in subsequent sections related to design.

3.2. *Mechanical Mixing Equipment*

In environmental engineering applications, mechanical mixing equipment consist of shaft-mounted impellers, which may be classified into groups consisting of propellers (screw type), turbine (flat blades, high speed), and paddles (flat blade, large size, low speed). Propeller and turbine types are used interchangeable in chemical mixing operations. Partially submerged turbine units are also common in mechanical surface aeration. Paddles are most commonly used in applications where fragile solids (alum floc) are handled or frequently with relatively high viscosity materials such as sludge (20,27).

In this section, the various types of impellers in common use will be described and relationships developed for power consumption and impeller discharge. This information will be applied to mixing applications in subsequent sections.

3.2.1. *Impeller Characteristics*

Propellers are of the axial flow type (discharge flow parallel to the agitator shaft) and may be used in low viscosity liquids almost without restriction as to the size and shape of the vessel. The circulating capacity is high and, as with a jet, entrainment of surrounding liquid occurs. Circulation rate is very sensitive to an imposed head and care must be exercised when applying propellers to a draft tube or circulating pump system.

The modified marine-type propeller is in almost universal use today in the three-blade style. Older literature treats the two- and four-blade style and a few current applications still use special designs. Total blade area is usually stated as the ratio of developed or projected area to disk area and typical values range from 0.45 to 0.55. It should be noted that the driving or operating face of a blade is flat or concave while the back side is convex.

Individual blade slope varies continuously from root to tip but specification of pitch of a propeller is on the basis of its being a segment of a screw. Pitch is the theoretical advance per revolution. In general, industry has standardized on a "square" pitch, i.e., a pitch value equal to the diameter. When an odd pitch is used, it is stated as the second term, such as 8×12 in. for a 1.5:1 pitch.

Definition of the way the blades are pitched is related to a viewpoint and direction of rotation. Marine practice derives from screw thread nomenclature and defines a left-hand propeller as one which thrusts the fluid downward when rotating clockwise viewed from above. Conversely, a right-hand propeller would thrust upward under the same operating conditions. This definition will be used throughout this text where pitched impellers are described. Unfortunately, not all manufacturers of propeller agitators have standardized on this designation.

The term "turbine" has been applied to a wide variety of impellers without regard to design, direction of discharge, or character of flow (9,41). A turbine can be defined as "an impeller with essentially constant blade angle with respect to a vertical-plane, over its entire

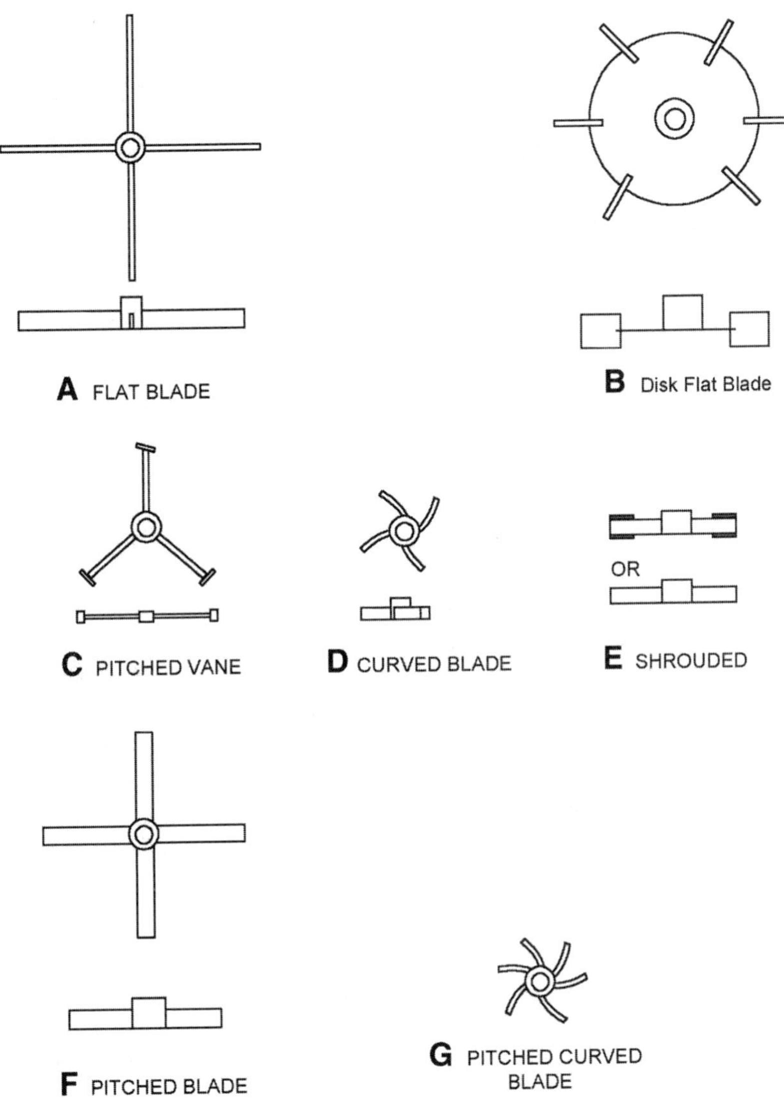

Fig. 4. Turbine impeller designs.

length or over finite sections, having blades either vertical or set at an angle less than 90 with the vertical." Blades may be curved or flat. The number of blades varies widely.

There are two basic physical forms of the turbine, the flat-blade radial discharging style and the pitched-blade axial discharging type. Pitched-blade type differs from propellers only in having a fixed blade angle. All others are modifications of these and, in most cases, performance is affected in only a minor way. Serious alteration of performance must come from changes in geometry. Figure 4 illustrates the more popular types in commercial use.

The flat blade or straight blade turbine discharges radially, deriving suction from both top and bottom. Customary operation is in a peripheral speed range from 600 to 900 ft/min. Blade widths are generally one-fifth to one-eighth of the diameter.

The disk flat blade type is widely used industrially and has been employed in many investigations. While it has essentially the same performance characteristic as the flat-blade turbine, the difference in power consumption is marked providing greater efficiency in energy use.

The pitched vane turbine is simply an adaptation of the disk type with the area reduced by pitching the blades to the vertical plane. Its advantage is the ability to support a large operating diameter and speed without high power consumption. Very little quantitative power or performance data have been published on this impeller.

The curved blade turbine, also termed the "backswept" or "retreating blade" turbine, has blades that curve away from the direction of rotation. This modification of the flat-blade style is commonly thought to reduce the mechanical shear effect at the impeller periphery. Industrial usage in suspensions of friable solids is widespread.

Addition of a plate, full or partial, to the top or bottom planes of a radial flow turbine will control the suction and discharge pattern. In Fig. 4C, the upper unit has annular rings on top and bottom. The lower design is fully shrouded on top to restrict suction to the lower side. Flow restriction may be useful in multiple impellers mounted on a single shaft or where unusual circulation patterns are required. A full shroud on the lower surface of an impeller that is located near the liquid surface will increase the vortex considerably, e.g., for gas re-entrainment.

The pitched blade impeller has a constant blade angle over its entire blade length. Its flow characteristic is primarily axial but a radial component exists and can predominate if the impeller is located close to the tank bottom or the blade angle is high. The blade angle can be anywhere up to 90°, but 45° is the commercial standard.

Sloping the blades of a curved-blade style to combine the effects of Fig. 4D is possible and has been practiced occasionally. No performance or power data are available and the high cost of construction of this impeller would eliminate it from consideration in all but with special applications.

The paddle in its basic form holds a fundamental place in industrial mixing practice because it has been used so long, although considerable conflict of nomenclature exists. In its basic form a common description would consist of usually two blades, horizontal or vertical, with a relatively large diameter compared to the tank in which it operates. Actually, by both physical form and power correlation the basic paddle is simply a turbine-type impeller, but it is worthwhile to retain the distinction for two reasons. First, the bulk of the technical literature treating the basic paddle is based on operation in the laminar range, or in the transition and turbulent range without baffles. Turbine impellers are not normally considered for either of these conditions. To avoid added confusion, the term "paddle" will be retained in referencing the applicable literature. Second, an impeller of the basic paddle design is not particularly effective for many process operations involving high viscosities. For this reason numerous other impeller configurations have evolved from it. It is thus convenient to consider these designs as a group, as shown in Fig. 5.

The simplest form of paddle is a single horizontal flat beam. The ratio of impeller diameter to tank diameter is usually in the range of 0.5–0.9 with a peripheral speed of 250–450 ft/min. Paddles used in the United States have generally had ratios of width to diameter from one-sixth to one-twelfth, but European practice is in the neighborhood of one-fourth to one-sixth.

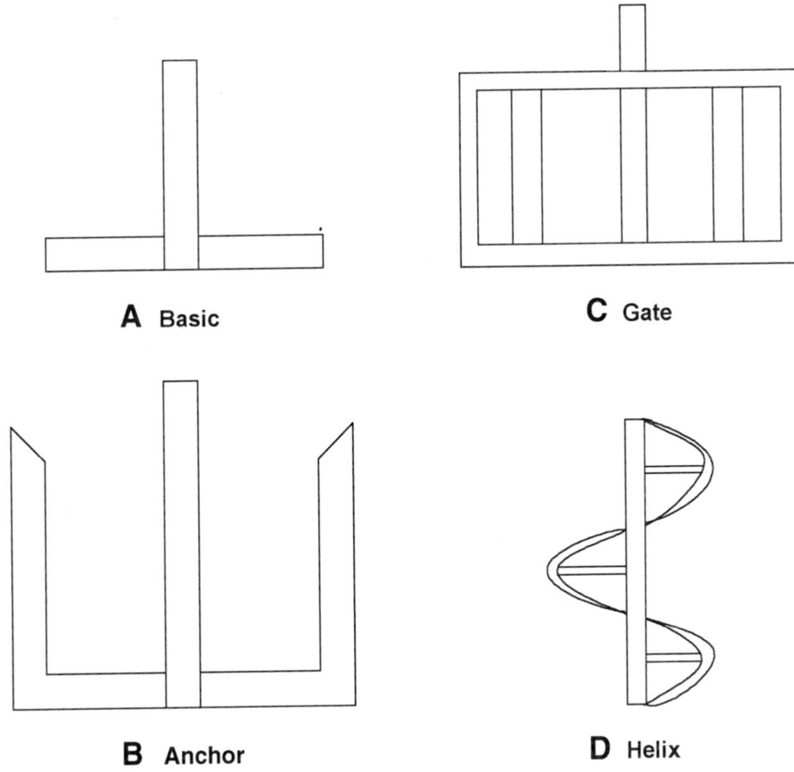

Fig. 5. Paddle impeller designs.

Contouring a simple paddle to the shape of a tank bottom gives the anchor or "horseshoe" style. Extent of the blade may be limited to the lower vessel tangent line or the blades may continue upward along the straight side. Clearance between blade and vessel shell is kept small depending on tank diameter and possible heat transfer needs.

The gate paddle consists of multiple arms with connecting vertical members. This design is often adopted for structural reasons in large tanks. In horizontal configuration, this type of unit is widely used in flocculation and coagulation applications.

In physical form the helical configuration least resembles the basic paddle. It does, however, operate in the laminar range at normally large diameters and is an important member of the paddle group. One traditional use of a helix or screw is in a vertical draft located within a larger tank. Draft tube impellers usually occupy one-third to one-half of the tank diameter and may pump downward. A helical ribbon with a diameter nearly equal to the vessel diameter was occasionally used in the past for blending solids. In recent years this type of impeller has been adapted for many types of applications.

The addition of scrapers to paddle impellers can be used to eliminate the stagnant film adjacent to the vessel wall and often results in a marked improvement in heat transfer capacity. Scrapers are usually hinged, with spring or hydraulic loading.

3.2.2. *Power Consumption*

3.2.2.1. Dimensionless Relationship

The general dimensionless equation for agitator power was derived by the early investigators using dimensional analysis. In keeping with other fluid motion concepts, it was observed that impeller power should be a function of the geometry of the impeller and the tank, the properties of the fluid (viscosity and density), the rotational speed of the impeller, and gravitational force. The Buckingham pi theorem gives the following general dimensionless equation for the relationship of the variables (9,13,41):

$$f\left(\frac{D^2N\rho}{\mu},\frac{DN^2}{g},\frac{Pg_c}{\rho N^3D^5},\frac{D}{T},\frac{D}{Z},\frac{D}{C},\frac{D}{p},\frac{D}{W},\frac{D}{l},\frac{n_2}{n_1}\right)=0 \quad (20)$$

where D is impeller diameter, in. (ft), T is tank diameter, m (ft); Z is liquid depth, m (ft); C is clearance of impeller off vessel bottom, m (ft); W is blade width, m (ft); p is pitch of blades, m/rev (ft/rev); n is number of blades; l is blade length, m (ft); ρ is density, kg/m^3(lb/ft^3); μ is viscosity, kg/s m (lb/s ft); P is power, kg m/s (ft lb/s); N is impeller rotational speed, rev/s; g is gravitational acceleration, m/s^2 (ft/s^2); and g_c is Newton's law conversion factor, 9.81 m/s^2 (32.2 ft/s^2).

Equality of all individual groups in Eq. (20) ensures similarity between systems of different size. The various groupings are generally classified into geometric, kinematic, and dynamic factors. The last seven terms in Eq.(20) represent the condition of geometric similarity, which requires that all corresponding dimensions in systems of different size bear the same ratio to each other. The reference dimension used is the impeller diameter. The last term in Eq.(20) is not a linear dimension relationship but may be inserted to accommodate differing numbers of impeller blades.

Equation (20) assumes the simple case of a single impeller centered on the axis of a vertical cylindrical flat bottom tank. Where required, additional terms could be inserted to accommodate an off-center impeller, baffling, or other geometric variables.

Given geometric similarity, two systems are dynamically similar when the ratios of all corresponding forces are equal. Kinematic similarity requires that velocity vectors at corresponding points be in the same ratio and directed along similar lines of action. These two similarity criteria are presented together because they are interrelated in a fluid system. For strictly geometrically similar systems, Eq. (20) may be stated as

$$f\left(\frac{D^2N\rho}{\mu},\frac{DN^2}{g},\frac{Pg_c}{\rho N^3D^5}\right)=0 \quad (21)$$

Thus, we have

$$f\left(\frac{\rho\bar{u}L}{\mu},\frac{\bar{u}^2}{L_g},\frac{\Delta p}{\rho\bar{u}^2}\right)=0 \quad (22)$$

where $\bar{u}$ is velocity, L is characteristic length, and Δp is pressure difference.

The groups in this equation are the same as those of Eq. (21), as will be demonstrated below, and a definite physical significance may be attributed to each group.

The first group in Equation (22), $\frac{\rho\bar{u}L}{\mu}$ is the Reynolds number and represents the ratio of inertial forces to viscous forces. Because this ratio determines whether the flow is laminar or turbulent, Reynolds number is a critical group in correlating power. In similar systems, any convenient velocity and length may be used in the Reynolds number. For agitation, the impeller diameter is generally accepted as the characteristic length, while velocity is equated with speed of rotation times diameter (ND). Substitution gives

$$N_{\mathrm{Re}} = \frac{\rho(ND)(D)}{\mu} = \frac{D^2 N\rho}{\mu} \tag{23}$$

which is identical to the group derived by dimensional analysis.

The group $\frac{\bar{u}^2}{Lg}$ in Eq. (22) is known as the Froude number and represents the ratio of inertial to gravitational forces (9,13,41). Substituting the characteristic terms into this group gives for an agitator:

$$N_{\mathrm{Fr}} = \frac{(ND)^2}{Dg} = \frac{DN^2}{g} \tag{24}$$

In enclosed flow problems, gravitational effects are unimportant and the Froude number is not a significant variable. However, most agitation operations are carried out with a free liquid surface in the tank. The surface profile and, therefore, the flow pattern are affected by the influence of gravity. This is particularly noticeable in unbaffled tanks. Where vortexing occurs, the shape of the free surface represents a balancing of gravitational and inertial forces.

The term $\frac{\Delta p}{\rho\bar{u}^2}$ is equivalent to the Euler number in enclosed flow and represents the ratio of the pressure difference (force) producing flow to inertial forces. For mixers, ND is used as a reference velocity; Δp is related to power consumption as the pressure distribution over the surface of the impeller blades could, in theory, be integrated to give torque acting on the impeller. Power could then be calculated directly from the total torque and the rpm of the impeller. In practice, the pressure distribution is not known, but in dynamically similar systems it can be shown that Δp and power are related by

$$k\frac{P}{ND^3} = \Delta p \tag{25}$$

Making this substitution into the Euler number together with the reference velocity $\bar{u} = ND$ gives

$$\frac{\Delta p}{\rho\bar{u}^2} = \frac{kP/ND^3}{\rho(ND)^2} = \frac{kP}{\rho N^3 D^5} \tag{26}$$

To make the power number dimensionless, pound-force is reduced to units of pound-mass. The unknown constant k serves no purpose and is omitted so that the Euler number for agitators is expressed as

$$N_P = \frac{Pg_c}{\rho N^3 D^5} \tag{27}$$

The resulting dimensionless ratio is termed the power number and is extremely important in impeller correlations. An understanding of the physical significance of the power number is enhanced by considering it as a drag coefficient or friction factor. The drag coefficient of a solid body immersed in a flowing stream is usually defined as

$$C_D = \frac{F_D g_c}{(\rho \bar{u}^2/2)A} \tag{28}$$

where C_D is drag coefficient; F_D is drag force on the body, kg (lb); $\bar{u}$ is velocity of flowing stream, m/s (ft/s); and A = cross sectional area of the body, $m^2(ft^2)$.

For geometrically similar impellors:

$$\bar{u} \,\alpha\, ND$$
$$A \,\alpha\, D^2$$
$$P \,\alpha\, NF_D D$$

Introduction of these relationships into Eq. 28 gives

$$C_D \alpha \frac{(P/ND)g_c}{\rho(ND)^2 D^2} \tag{29}$$

Simplifying,

$$C_D \alpha \frac{Pg_c}{\rho N^3 D^5} \tag{30}$$

or

$$C_D \alpha N_P \tag{31}$$

The analogy of C_D to N_P is a useful observation because correlations of drag coefficients and power number bear many relationships to each other. For pressure drop in pipes, the use of friction factor is analogous to N_P for impellers and for immersed bodies.

3.2.2.2. Power Prediction

Equation (20) may be written in the following form:

$$N_P = k(N_{Re})^a (N_{Fr})^b \left(\frac{T}{D}\right)^c \left(\frac{Z}{D}\right)^d \left(\frac{C}{D}\right)^e \left(\frac{p}{D}\right)^f \left(\frac{w}{D}\right)^g \left(\frac{l}{D}\right)^h \left(\frac{n_1}{n_2}\right)^i \tag{32}$$

For geometric similarity, the terms relating to length may be dropped, yielding

$$N_P = K_1 (N_{Re})^a (N_{Fr})^b \tag{33}$$

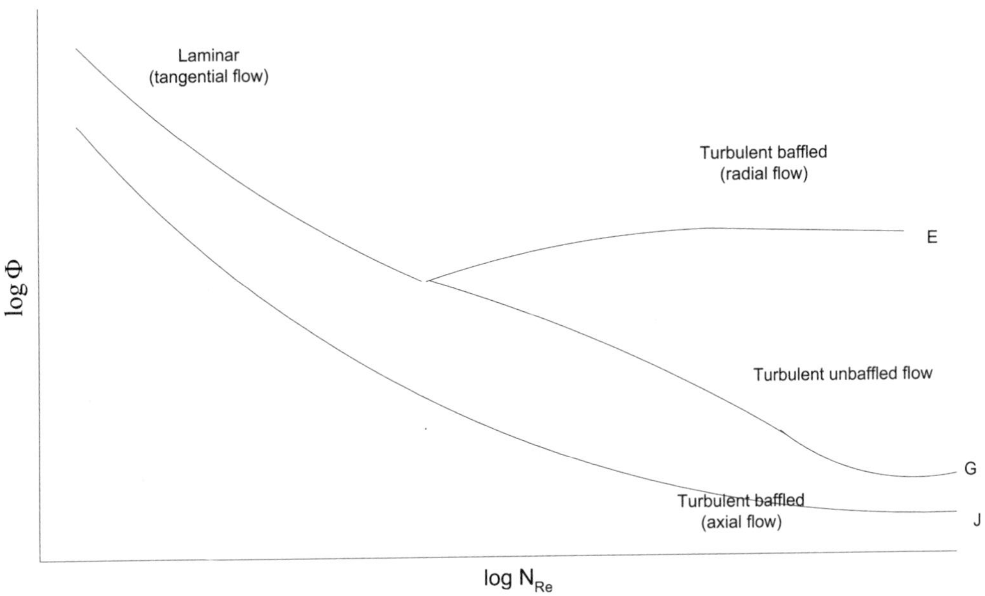

Fig. 6. Typical impeller power curves.

In presenting data graphically, the usual technique in fluid flow is to use the Reynolds number as abscissa in a logarithmic plot. To facilitate this, Eq. (33) can be modified to:

$$\phi = \frac{N_{\mathrm{P}}}{\left(N_{\mathrm{Fr}}\right)^{b}} = K_1\left(N_{\mathrm{Re}}\right)^{a} \tag{34}$$

For a fully baffled tank (no vortex), the exponent b on the Froude number generally equals 0 and $\phi = N_{\mathrm{P}}$.

Typical curves of ϕ vs N_{Re} are shown in Fig. 6 for configurations often used in practice. The similarity to the Moody diagram used in enclosed flow is obvious. For fully baffled conditions, and in the laminar range, ϕ can be assumed to be N_{P}.

In highly turbulent flow and for high Reynolds numbers in fully baffled tanks, N_{P} approaches a constant value which may be called K'. Substituting in Eq. (27) and solving for P:

$$P = \frac{K'}{g_c}\rho N^3 D^5 \tag{35}$$

Thus in the turbulent range with geometric similarity, power can be stated to be proportional to density, to impeller speed cubed, to diameter to the fifth power, and to be independent of viscosity.

In the laminar regime, the initial portions of the curves in Fig. 6 represent the viscous range of flow and the slope shown is constant for all types of impellers. Evidence for a slope of −1 is plentiful and, because Froude effects are unimportant in this range,

$$N_{\mathrm{P}} = K''\left(N_{\mathrm{Re}}\right)^{-1} \tag{36}$$

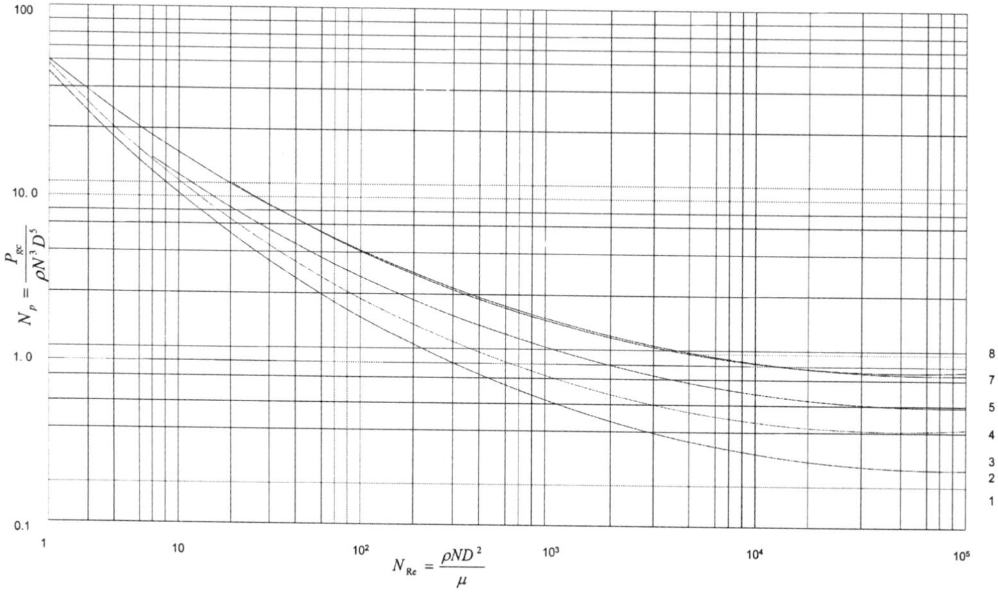

Fig. 7. Propeller power correlation.

Substituting into Eq. (27) and solving for power as above:

$$P = \frac{K''}{g_c} \mu N^2 D^3 \tag{37}$$

Recognizing certain limitations, the above relationships can be used to predict the power required to turn an impeller of a standard design at any speed in any fluid media.

Figure 7 shows the power correlation developed by Bates, Fondy, and Fenic for modified marine propellers of the three-blade style (10). The data are for a single impeller and were taken from Rushton et al. (11). Blade shape and area ratio for the propellers

Table 1
Values of ϕ for Three-Blade Propellers

			ϕ at N_{Re} of		
Curve	p/D	D/T	5	300	10^5
1*	1.0	0.33	8.3	0.60	0.22
2*	1.0	0.31	8.3	0.60	0.25
3	1.0	0.40	9.7	0.75	0.30
4	1.0	0.33	9.7	0.82	0.35
5	1.4	0.33	9.7	1.04	0.54
6*	2.0	0.31	8.7	1.00	0.52
7	1.8	0.30	9.7	1.27	0.86
8	2.0	0.31	8.7	1.10	1.0

*No baffles.

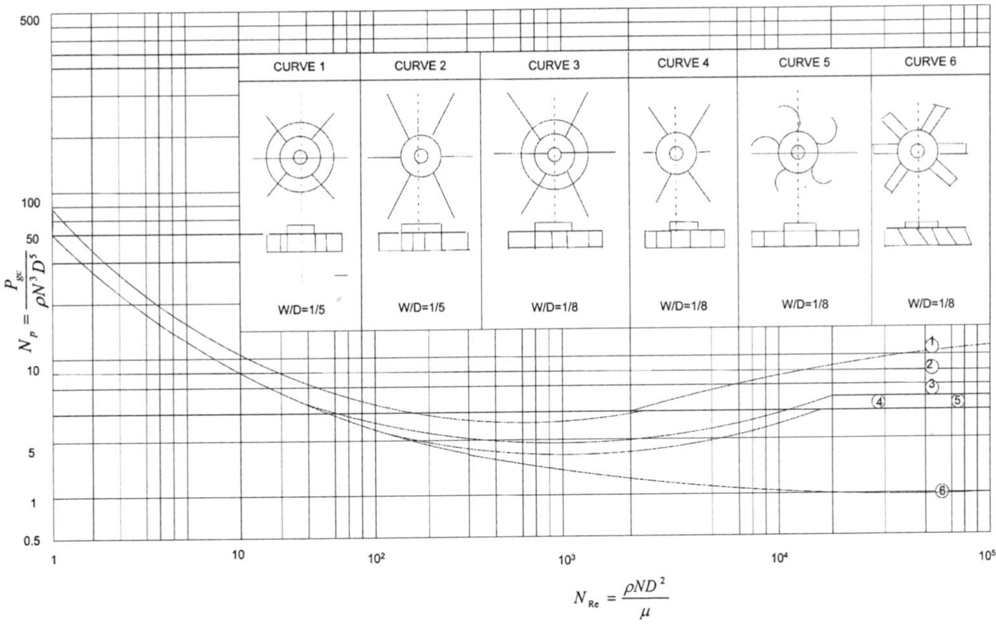

Fig. 8. Turbine power correlation.

from these two sources differ and thus there is an expected difference in power level. Where baffling is used, it was assumed that the angle mount used to eliminate swirl was an equivalent design.

For convenience in evaluation, the power function ϕ is reproduced in Table 1 for three representative values of impeller Reynolds number. For the impellers used, fully developed baffled turbulence was achieved at $N_{Re} = 10^5$. While the unbaffled power function continues to decrease past this point, it is a reasonable limit for useful operation. Turbine impellers in baffled vessels have been widely used in trial mixing in recent years. Much of the research on mixing has been devoted to this application.

In many cases, however, the published reports failed to describe the geometry of the impeller and the vessel in sufficient detail to allow comparison of absolute power consumption. The useable data are shown in Fig. 8.

The correlations for the several types of turbines have characteristically different curve forms in the transition and turbulent range, although all impellers reach a constant value of Reynolds numbers slightly in excess of 10,000. In the laminar range, the nominal slope of –1 found for propellers also applies to turbines. The flat-blade turbine, curves 2 and 4, exhibits a dip below the fully turbulent value, but the transition range extends only from $N_{Re} = 15$ to 1500. The disk and curved-blade styles, curves 1, 3, and 5, extend the transition range to about l0 and also show a similar dip below the fully turbulent range. Curve 6, for pitched-blade turbines, has a shape similar to that of the propellers in Fig. 7, both are axial flow type units.

It should be noted that a difference in power requirement exists between the disk style of construction and the flat-blade turbine at high Reynolds number. Curve 2 applies to an open flat-blade style with a full blade originating at the hub. Although it

Table 2
Values of ϕ for Turbine Agitators Propellers

Type	D^b/w	n	D/T	C/D	Baffles (no) T/w_b	ϕ at N_{Re} of 5	300	10^5
Flat blade (4)[a]	8.0	6	0.33	1	(4) 12	10.0	2.1	2.6
Flat blade (2)[a]	5.0	6	0.33	1	(4) 12	14.0	3.4	4.0
Disk $D/l = 4(1)$[a]	5.0	6	0.33	1	(4) 10	14.0	3.4	5.0
Disk $D/l = 2$ (3)[a]	8.0	6	0.33	1	(4) 12	10.0	2.0	3.0
Curved blade (5)[a]	8.0	6	0.33	1	(4) 12	10.0	1.9	2.6
Curved blade disk	5.0	6	0.22–0.31	1	(4) 10	14.0	3.4	4.8
Pitched blade, 45° (6)[a]	8.0	6	0.33	1	(4) 12	10.0	1.5	1.3
Arrowhead	5.0	6	0.31–0.47	1	(4) 10	14.2	3.4	3.9
Arrowhead	—	6	0.44	—	(4) 8.7	—	—	2.4
Flat blade	4.1	2	0.23–0.37	0.35–0.25	(4) 8 & 10	—	—	1.83
Flat blade	3.6	2	0.31–0.52	0.55	(4) 11	—	—	2.32
Flat blade	5.0	2	0.36–0.63	0.5–1.2	(4) 21	9.7	—	1.94
Flat blade	5.0	6	0.36–0.63	0.5–1.2	(4) 21	17.4	—	4.1
Flat blade	8.0	6	0.36–0.63	0.5–1.2	(4) 21	14.5	—	2.5
Flat blade	1.5	2	0.3	—	(4) 15	18	7.0	8.8
Flat blade	1.5	2	0.5	—	(4) 15	10	4.2	6.0

*Refers to curves on Fig. 8.
[a]For pitched turbines, based on horizontally projected w.

has a longer blade than the disk style, it consumes approx 25% less power. With reduced height to diameter ratio (curves 3 and 4), the difference is approx 15%.

Table 2 is a compilation of power functions at three representative Reynolds numbers prepared by Bates, Fondy and Fenic (10).

Paddle power data will be presented in a different manner from that used for propellers and turbines. The typical N_P–N_{Re} plot could be applicable, but in the laminar range, equations will suffice. Data for turbulent mixing is quite incomplete and varies sufficiently with paddle type to make graphical representation meaningless at the present time.

For a multiple bladed paddle, the general correlation equation is of the form:

$$\frac{Pg_c}{\rho N^3 D^5} = K\left(\frac{w}{D}\right)^g \left(\frac{D^2 N_P}{\mu}\right) \tag{38}$$

This may be rewritten in the form

$$\frac{Pg_c}{\rho N^3 D^5} = K\left(\frac{w}{D}\right)^g \tag{39}$$

where the exponent g is related to the number of blades. For the two-blade paddle, O'Connell and Mack evaluated the terms K and g based on experimental data and found (14)

$$P = \frac{113}{g_c} \mu N^2 D^{2.48} w^{0.52} \tag{40}$$

Note that the exponent g equals 0.52. Hirsekorn and Miller independently obtained a value for the exponent of 0.5 (15). Their value of K was close to 113 at $N_{\text{Re}} = 1$, but decreased with smaller N_{Re} to a constant value of 95. The equation above applies to Reynolds numbers up to at least 10 and for some configurations above this value.

For efficient utilization of mixing power applied with any type of impeller, swirl of the tank contents must be avoided to maintain high relative velocities at the impeller blades. This is normally accomplished by baffling placed adjacent to the tank walls, but may also be achieved by mounting the impeller shaft off-center in the tank. Eccentric mounting is most common with propeller or axial-flow turbines. With radial flow types, lateral loads imposed on the impeller shaft tend to reduce bearing life substantially.

Baffling practice for cylindrical tanks in industrial applications has essentially been standardized. Four baffles are commonly used with a width equal to one-twelfth the tank diameter and a height essentially equal to the liquid depth. Baffles are usually set a few inches away from the tank walls and bottom to avoid stagnant zones or solids deposits.

3.3. *Impeller Discharge*

The relationships between impeller geometry, rotational speed, and other variables have been investigated in considerable detail for centrifugal pumps. The basic principles relating head, fluid velocity, flow, power, and geometry may also be applied to rotating impellers in agitated vessels. In this section are described the theoretical and empirical relationships that have been obtained by various investigators between discharge velocities and flow rates, impeller geometry, and rotational speed. This will ultimately be linked to power consumption.

In order to develop the relationship between flow and power, it is convenient to start with a curved blade turbine, which is analogous to the centrifugal pump. Figure 9 shows the pertinent characteristics and velocity vectors associated with the discharge from the periphery of the impeller. For a flow outward from the center of the impeller, Q, the absolute velocity, u, may be resolved into components consisting of the tangential velocity u_t created by the rotation of the impeller, and the velocity directed outward along the blade u_b. For any impeller of known blade angle β and flow, the radial velocity u_r may be calculated. By application of basic geometry all components of the vector diagram may be calculated from the information above.

If it is assumed that only the forces exerted at the periphery of the blade are significant, the force exerted by the impeller on the fluid is equal to the change in angular momentum of the fluid, or the flow times mass density times the change in velocity:

$$F = \frac{Q\rho}{g_c}(u \cos \alpha) \tag{41}$$

Because torque is force times distance, the torque may be written

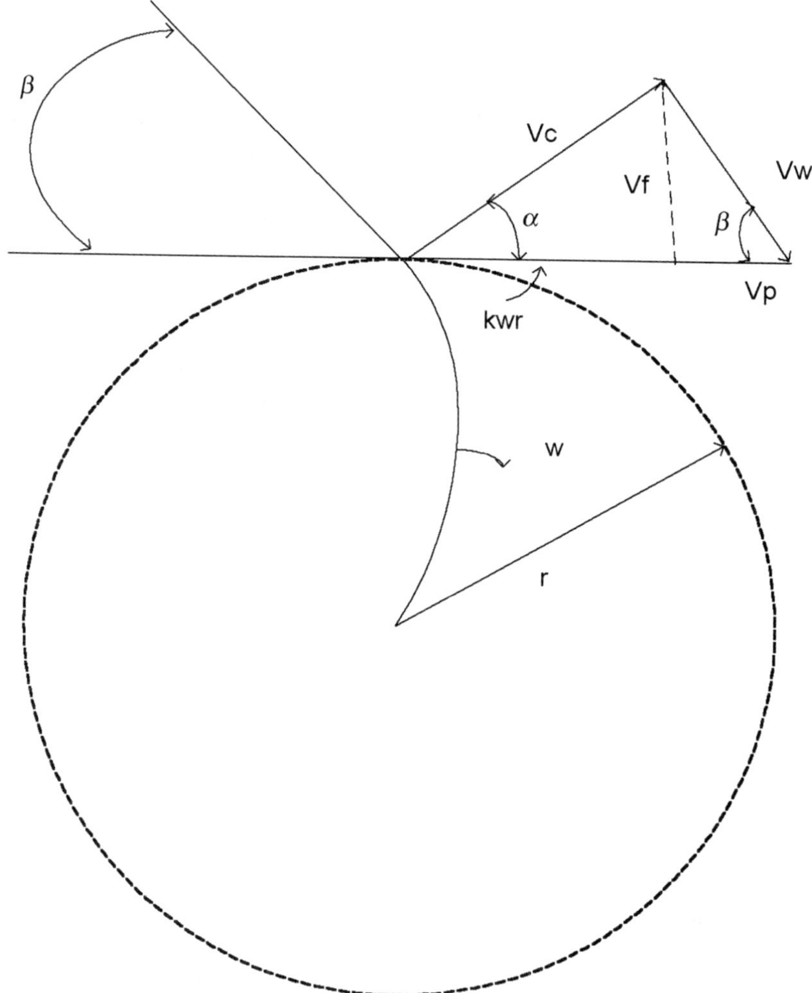

Fig. 9. Discharge velocity vectors for a curved-blade turbine.

$$T = Q\frac{\rho}{g_c}(u\cos\alpha)r \tag{42}$$

For reasons to be noted later, Gray introduced the following notation (19):

$$u\cos\alpha = kwr \tag{43}$$

where k is a decimal fraction relating the tangential component of u with the tangential velocity u_t. Using this terminology, Eq. (42) may be rewritten in the form

$$T = Q\frac{\rho}{g_c}kw^2r^2 \tag{44}$$

Because power is torque times speed of rotation,

$$P = Tw = Q\frac{\rho}{g_c}kw^2r^2 \tag{45}$$

Substituting the more conventional units:

$$w = \frac{2\pi N}{60} \tag{46}$$

where N = revolutions per minute

$$r = D/2 \tag{47}$$

$$P = \pi^2 Q\frac{\rho}{g_c}kN^2D^2 \tag{48}$$

Gray draws on the concepts of theoretical head to derive from the vector diagram the expression (19)

$$u_r = wr\sqrt{2k(1-k)} \tag{49}$$

This is inserted into the equation for flow from the impeller

$$Q = 2\pi bru_r \tag{50}$$

to yield,

$$Q = 2\pi bwr^2\sqrt{2k(1-k)} \tag{51}$$

which may be converted to the form

$$Q = \pi^2 bND^2\sqrt{2k(1-k)} \tag{52}$$

Defining the discharge number of an impeller as

$$N_Q = \frac{Q}{ND^3} \tag{53}$$

Equation (48) may be converted to the form:

$$N_Q = \frac{Q}{ND^3} = \frac{g_cP}{\pi^2 k\rho N^3D^5} \tag{54}$$

Typical discharge numbers were compiled by Gray and are shown in Table 3 (19). The power number has previously been defined as

$$N_P = \frac{g_cP}{\rho N^3D^5} \tag{55}$$

Inserting this expression into Eq. (54) above:

$$N_Q = \frac{N_P}{\pi^2 k} \tag{56}$$

Table 3
Comparison of Discharge Rates of Impellers

Impeller type	No. of blades	D/T[a]	b/D[b]	Z_1/T[c]	Z_L/T	No. of baffles	D (in.)	Speed (rps)	$N_{Re} \times 10^{-5}$	N_Q
Propeller (p^e)	3						6		1.6	0.40
Propeller ($p = 1$)	3	0.2–0.6		0.4–0.8	0.7–1.1	3	2–5	5–15	0.4–2	0.61
Propeller	3	0.17–0.6				4	3.6–12.3	2–15		0.42
Propeller ($p = 1$)	3									0.47
Propeller ($p = 1$)	3									0.60
Turbine	4	1/6	1/5	0.45	1.33	4	6	5–7	1–1.7	0.59
Turbine	3	1/6	1/5	0.45	1.33	4	6	6–10	1.5–2	0.50
Turbine	4	0.35	1/5	0.35	1.04	4	4	1.7–3.3	0.17–0.34	0.47
Turbine	6	0.53	1/5	0.35	2.8		6			2.9
Turbine	6	0.50	1/5	0.50	1	4			10	1.9
Turbine	8	0.51	1/5	0.50	1	8	11.8	1.45	1.3	1.34
Turbine			1/5							1.25
Turbine	8	0.31	1/3	0.50	1	8	7.1	2.3	0.76	2.9
Turbine			1/5							2.3
Pitched blade	8	0.51	0.14	0.50	1	8	11.8		1.3	0.87
Pitched blade			0.14							1
Turbine	8	0.51	1/5	0.50	1	8	11.8		1.3	1.2
Turbine			1/5							1.26

[a] D/T = ratio of impeller diameter to vessel diameter.
[b] b/D = ratio of blade width to impeller diameter.
[c] Z_1/T = ratio of "axial distance from impeller center to bottom of vessel" to vessel diameter.

For the assumptions of negligible entrance losses and no friction losses through the impeller, the relationship between flow and discharge for a given impeller should be based strictly on geometric considerations. However, in practice the term k is modified to incorporate loss factors and thus becomes an empirical coefficient. Insufficient data are available to accurately determine the relationship of power and discharge number; Gray recommended the values of 0.5 for marine propellers and $0.93 \times T/D$ for six-bladed turbines ($b/D = 1/5$), both in baffled vessels (13,19).

3.4. *Motionless Mixers*

In an effort to avoid the capital and operating costs associated with agitators, mixers with no moving parts have been developed. In the most common form, convoluted surface vanes are inserted into a pipe section. For viscous mixtures in particular, mechanical bulk mixing across the pipe cross-section, combined with flow splitting on the leading edge of the vanes, can accomplish mixing within a surprisingly short distance. Velocity through the vaned section may be well below that necessary for turbulence. The advantages for viscous mixtures are obvious. The cost is relatively low and maintenance is avoided with non-clogging mixtures. Higher costs are required for pumps and pump head, but this is offset to some degree by the reliability of a pump compared to agitation equipment.

Motionless mixers have occasionally been used in the air pollution field for many years. Multiple layers of screens or short sections of straightening vanes can be inserted into ducts or stacks to enhance mixing of gaseous pollutants. This is usually done to obtain greater pollutant concentration uniformity at a sampling section that must be located in a short length of ducting where adequate mixing cannot be obtained from natural turbulence alone. An additional advantage is that greater uniformity of velocity distribution is also obtained. This will enhance the accuracy of flow measurements.

Interest has been shown in the use of packed or tray towers as a biological reactor in small activated sludge plants. Hsu et al. found that the oxygen transfer efficiency in such towers with convoluted surface mixers in bulk form. Attached growth was a significant factor in reducing the effect of cell washout, but no plugging was encountered.

3.5. *Mixing in Batch and Continuous Flow Systems*

If two miscible fluids or a single fluid and a readily soluble substance are mixed in a vessel, initial differences in properties are reduced quickly at first but with a decreasing rate. For batch systems, the time required to reach a desired degree of uniformity will depend on the fluid flow pattern and velocity distribution, which is itself dependent on fluid properties, vessel geometry, and agitator characteristics (29,30,32).

Van de Vusse developed a correlation of the variables that affect batch mixing times, using a Schieren method to determine when refractive index differences disappeared (1,21). Differences in refractive index between various points in the fluid result in bending of the light beam and shadows were projected on to a screen, which were related to the refractive index pattern. The mixing time was defined as the period required for elimination of shadows. Water and dilute acetic acid or water and dilute solutions of glycerol in water were used in the experiments. Initially, the two liquids formed two clearly defined layers and upon agitation the fluid was blended. In some cases, a small quantity of liquid was injected while the agitator was running.

Van de Vusse developed correlations of mixing times in terms of the following dimensionless groups:

$$\frac{\phi Q}{V} = \text{number of circulations of fluid to obtain a desired uniformity} \quad (57)$$

$$\frac{D^2 N\rho}{\mu} = N_{\text{Re}} \quad (58)$$

$$\frac{D^2N^2\rho}{gZ\Delta p} = \text{ratio of kinetic head to static head (a modified Froude number)} \quad (59)$$

$$D/T = \text{ratio of impller diameter to vessel diameter.} \quad (60)$$

$$Z/T = \text{ratio of fluid depth to vessel diameter.} \quad (61)$$

Experimental mixing-time data were obtained using several types of agitators in an unbaffled vessel. For each of the types of impellers, an appropriate expression for Q was formulated as follows:

$$\text{Two blade paddle: } Q\alpha ND^2 b\pi^2 \quad (62)$$

$$\text{Propeller: } Q\alpha ND^2\rho \quad (63)$$

$$\text{Pitched-blade paddle: } Q\alpha ND^2 b\pi^2 \sin^m \gamma \quad (64)$$

$$\text{Curved– blade turbine: } Q\alpha ND^2 b\pi^2 \sin\beta \quad (65)$$

The above functions of flow may be substituted into the expression $\frac{\phi Q}{V}$ in Eq. (57) to provide a basis for correlation of mixing data. The basic equation is as follows:

$$\frac{\phi Q}{V}\alpha\left(\frac{\rho D^2 N^2}{g_c Z\Delta p}\right)^a \quad (66)$$

where the exponent "a" was found to be -0.25 for propellers, -0.30 for turbines, and -0.35 for a pitched blade paddle. Van de Vusse also found Reynolds number to be an essential variable and plotted his correlations with Reynolds number as ordinate and the function

$$\frac{\phi Q}{V}\left(\frac{\rho D^2 N^2}{g_c Z\Delta p}\right)^a$$

as abscissa. The specific correlation curves are not reproduced herein, but, as might be expected, the value of the function above is large at low Reynolds number and decreases to approach a constant value of approximately 10^5.

The mixing of acidic and basic solutions in water was studied using turbine agitators in baffled vessels (22,23). In the studies of Norwood and Metzner, each of the turbines had six flat blades with $b/D = 1/5$ and were located 35% of the distance from the vessel bottom to the upper liquid surface. The horizontal dimension of the baffles was one-tenth the tank diameter. The vessels varied from 5.67 to 15.5 in. diameter, the turbines from 2 to 6 in. diameter, and fluid depths from 6 to 12 in.

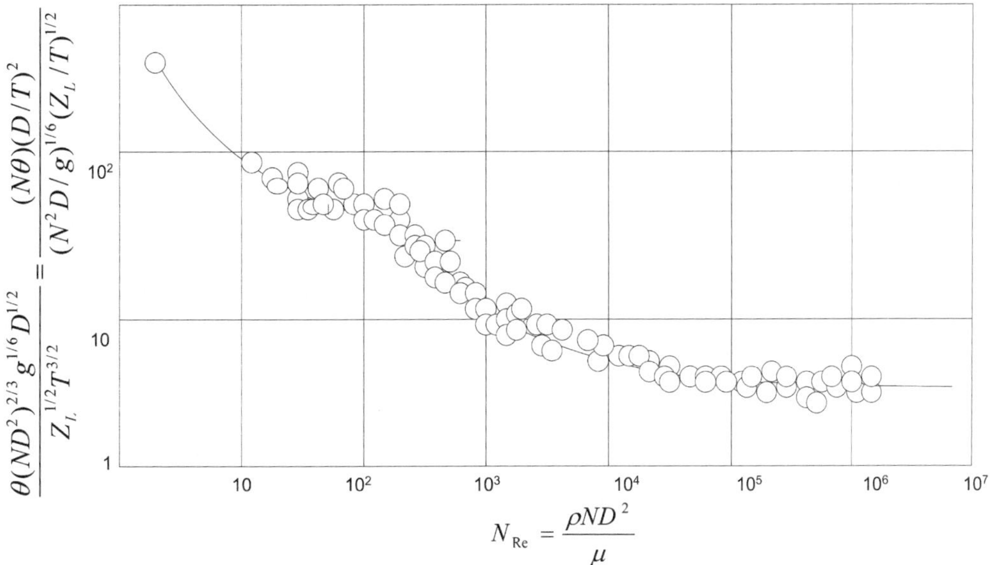

Fig. 10. Mixing time correlation for turbines.

Mixing times were determined by adding a basic solution containing a methyl red indicator to the vessel (34). Subsequently, an equivalent amount of acid solution was added to the vessel at a point near the rotating agitator. Initially the region at the impeller appeared red in color while the solution throughout the rest of the vessel was yellow. The time required for the disappearance of the red color was taken as the mixing time. The red color was found to persist longest at the impeller.

The data were correlated on the basis of Reynolds number versus a mixing function similar but not identical to that of van de Vusse (21). The correlation was excellent over the range of variable tested and is shown in Fig. 10. Significant breaks in slope were noted at Reynolds numbers of 400 and 1200. These are apparently real and relate to circulatory patterns as full turbulence is approached. Full turbulence is reached at a Reynolds number of approx 105. It should be noted that English units should be used in conjunction with Fig. 10.

Fox and Gex determined the mixing requirements for propeller agitation of vessels in a manner similar to Norwood and Metzner (22,24). Cylindrical vessels were used with diameters of 1/2 to 14 ft. The smaller tanks were glass and the larger tanks were steel. Propeller diameters ranged from 1 to 22 in. and the pitch was equal to the diameter. The locations of the agitators were not specified but were stated to be positioned so that no general swirl or rotation was produced.

In the 14-ft-diameter vessel, a small amount of hardened oil was added to a batch of unhardened oil, then the propeller was started. Samples were taken from each of three sample ports and analyzed for iodine value. A plot of iodine value versus time was used to determine the time required for the variation in iodine value to approach zero.

In the smaller vessels, the time for neutralization of HCl by an equivalent amount of NaOH was measured by visual observation of the disappearance of the red color

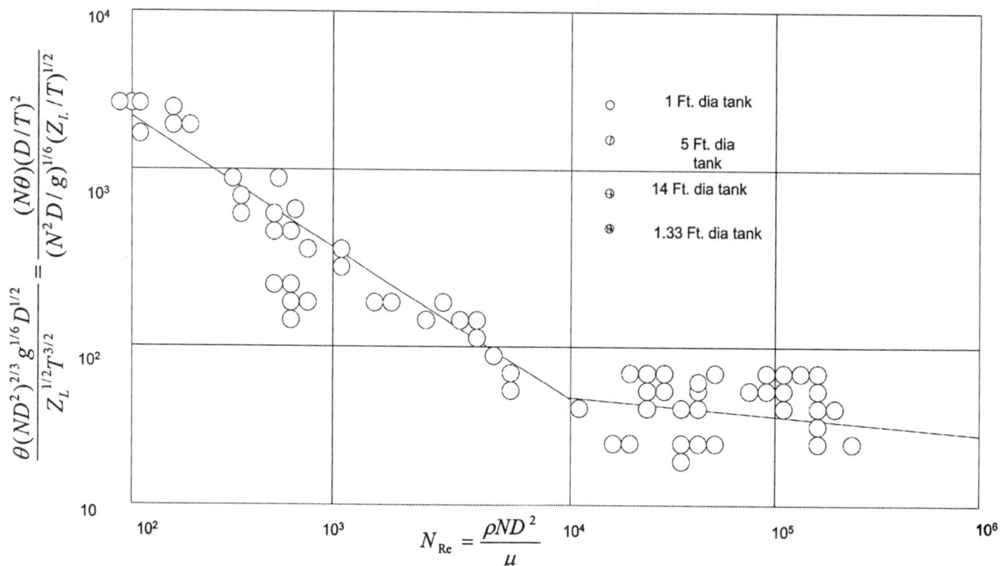

Fig. 11. Mixing time correlation for propellers.

due to a phenolphthalein indicator. In these tests, agitation was started prior to adding the reactants. The phenolphthalein indicator was mixed with the alkaline solution initially present in the vessel, and acid added subsequently. In further tests, a range of fluid viscosities from 0.5 to 400 CP was obtained by using water and water solutions of glycerol or carboxymethyl cellulose.

Fox and Gex determined experimentally the separate effects on mixing time of propeller diameter, rotational speed, depth of liquid in the tank, and liquid viscosity (24). They obtained the following equations for turbulent and laminar flow, respectively:

$$\phi = f\left(Z^{1/2}, T, N^{-5/6}, D^{-10/6}, \mu^a, \rho^b, g^c\right) \quad \text{(Turbulent)} \tag{67}$$

$$\phi = f\left(Z^{1/2}, T, N^{-5/6}, D^{-20/6}, \mu^a, \rho^b, g^c\right) \quad \text{(Laminar)} \tag{68}$$

The values of a, b, and c can be found by dimensional analysis and the variables are rearranged to obtain the following equations:

$$\phi = \frac{C_1 Z^{1/2} T}{N_{Re}^{1/6}\left(ND^2\right)^{4/6} g^{1/6}} \quad \text{(Turbulent)} \tag{69}$$

$$\phi = \frac{C_2 Z^{1/2} T}{N_{Re}^{1/6}\left(ND^2\right)^{4/6} g^{1/6}} \quad \text{(Laminar)} \tag{70}$$

The above equations were plotted in dimensionless form by Fox and Gas and the resulting relationship is reproduced in Fig. 11 to show the degree of correlation (24). English units should be used in the equations.

In continuous flow systems, it may be desired to determine the degree of mixing required to ensure that the basin performs as a "completely" or "ideally" mixed system. As previously discussed, this implies that the mixing is of sufficient intensity and duration to ensure that the effect of the basin on a concentration difference between inlet and outlet is in accordance with that predicted mathematically. There have been many attempts to develop design relationships, but, unfortunately, these have not been generally applicable to real systems. However, the following general relationships may be stated:

1. Impeller discharge rate must exceed the inflow rate by a minimum of a factor of 5.
2. The mixing time (or detention period) should be substantially in excess of that required for complete mixing of an additive in a batch-mixed system using the same agitator.
3. The basin volume should be sufficiently large relative to the inflow to allow the contents to be recalculated at least 10 times during the theoretical detention period.

The generalizations above are obviously crude and, for this reason, empirical approaches based on power applied for a given time have been widely used in chemical and environmental engineering practice. This approach is presented subsequently in the section related to facilities design.

3.6. *Suspension of Solids*

The first experiments designed to evaluate the variables affecting solids suspension were reported in 1930s (1,27,28,35,41). In an unbaffled tank, sand concentration (at any point above the bottom) decreased with an increase in sand particle size and increased with impeller speed up to a "critical point," which they suggested as a criterion for effectiveness of agitation. At this point the sand concentration reached a maximum. At higher agitator speeds, lower suspended sand concentrations were obtained because of centrifugal separation of the sand. The use of baffles could have decreased or eliminated this separation.

Another criterion for effectiveness of solids suspension was formulated and reported in early 1930s (1,27,36,37). Using a grab sampling tube to measure sand concentrations, a "mixing index" was calculated as the arithmetic average of the "percentage mixed" values for a series of sample locations. The percentage mixed values were determined for each sample location as the ratio of the actual concentration of sand present to the theoretical average in the vessel as a whole assuming uniform mixing. For a four-bladed turbine in an unbaffled tank, the mixing index was found to increase with impeller speed to approx 90 % at which point it leveled off. The mixing index also increased with the viscosity of the liquid.

In subsequent studies of solids suspension in viscous liquids, Hirsekorn and Miller showed "complete" suspension was affected by vessel dimensions, viscosity, particle size, and settling rate (15). It was also noted that the portion of liquid nearest the surface was still free of solids when complete suspension was achieved. A slurry–liquid interface was evident and below this interface the slurry was essentially uniform. Zwietering found a similar interface near the free surface under conditions of complete suspension (28). By increasing the agitator speed well above that required to lift the solids from the bottom, solids could eventually be dispersed throughout the entire liquid volume. The ultimate degree of uniformity would, however, be limited by the centrifugal effects of swirl in an unbaffled tank.

Vessel geometry, impeller construction, baffling, and impeller speed have been mentioned as the critical factors in the design of an effective mixer for suspending a slurry. These factors may generally be adjusted freely by the designer. Certain additional properties of solid–liquid systems may intuitively be recognized as important, although in real mixing applications, these are often fixed. Included in this category are particle density; solids concentration; density and viscosity of liquid phase; size, size range, and shape of solid particles; and hindered settling at extremely high solids concentrations.

Hindered settling influences can assist in maintaining a homogeneous suspension. Impellers for solids suspension fall into two basic categories. The first is represented by the various types of paddles such as the single or multiple straight-blade designs rotating on either vertical or horizontal axes. Turbulent motion is induced by physically pushing the material in its path and entraining adjacent solids in the vortices trailing the blades. Little or no directed velocity is imparted to the fluid. The second, more commonly used type include propellers and turbines, both of which develop a directional velocity. The fluid, along with adjacent material, flows in a definite pattern through the vessel and returns to the suction side or point of the impeller. Most slurries are of a non-Newtonian nature and are dependent on velocities in excess of a "critical" value to ensure flow of the slurry. In propellers the stream is axial, but in turbines the flow is centrifugally developed and radial. The pitched blade turbine uses a combination of both axial and radial forces to produce a diagonal flow with an angle dependent on the degree of pitch, number of blades, blade width, and rotational speed.

Selection of the preferred type of impeller will depend on the concentration of solids in the slurry, the flow characteristics of the mixture, and the vessel shape. A mixture which cannot readily be pumped centrifugally must be transported by a paddle-type element. Pumpable slurries will be mixed most efficiently by propellers or turbines. In environmental engineering applications, paddles are used to mix fragile solids (such as alum floc) because of their lack of high velocity gradients. Turbines are more common where solids disintegration is unimportant or cannot occur (19,39,40).

Liquid and air jets or bubbles may also be used for conveying slurries in vessels and tanks. Air bubbles can induce mass velocities through the creation of apparent differences in liquid density. Where the velocities are directed so as to produce rotation of the liquid, solids may be suspended and dispersed uniformly throughout the basin. For low density solids in semi-viscous slurries (such as sewage sludge), air lift pump principles may also be used to produce mixing by bulk flow. Air lifts, however, are poorly suited for the suspension of freely settling particles.

Circulation within a vessel for the purpose of solids suspension may be "upwardly directed," to physically lift the dispersed phase, or "universally directed," to distribute the solids into the fluid. The patterns developed by various impellers are illustrated in Figs. 12–18. The types shown in Figs. 12–18 produce an upward lifting stream. The turbine in Fig. 14 is the most effective type for the suspension of fast settling, granular solids in a low-viscosity fluid such as activated carbon in water. The turbine placed directly on the vessel bottom provides the maximum scouring velocity at the location where critical suspension velocity is required. Energy efficiency is high because only one directional change at the wall is required to produce upward flow. Stator blades may be placed outside the impeller to produce a straight, radial discharge in all directions

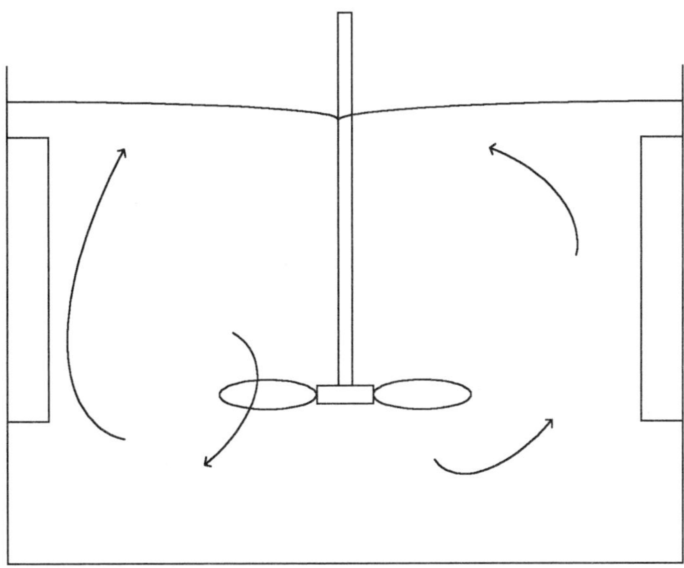

Fig. 12. Propeller in baffled vessel.

from the center. This further increases efficiency by creating an essentially vertical pattern without a rotational component. Loss of velocity in the stators is less than 5% when properly designed. The raised turbine (Fig. 15) produces a figure-eight circulation pattern and depends on the lower recycle stream to lift slurry up to the bottom eye of the impeller for redistribution. This pattern is commonly used and is adequate for many industrial applications.

When located above the bottom of a vessel, a turbine may be tilted to distort the circulation pattern of a viscous slurry. The depth of the figure-eight pattern is increased and the vertical rate of interchange of the mixture through the plane of the impeller is multiplied. The horizontal discharge of the conventional turbine makes it most effective on slurries, particularly fibrous, in shallow, large diameter tanks, relative to other impeller types.

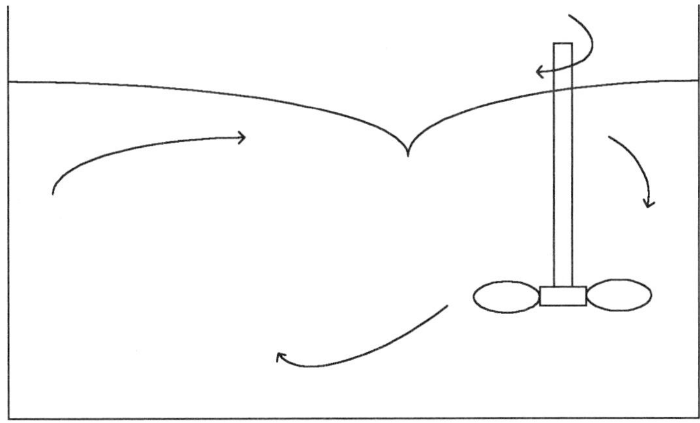

Fig. 13. Off-center propeller.

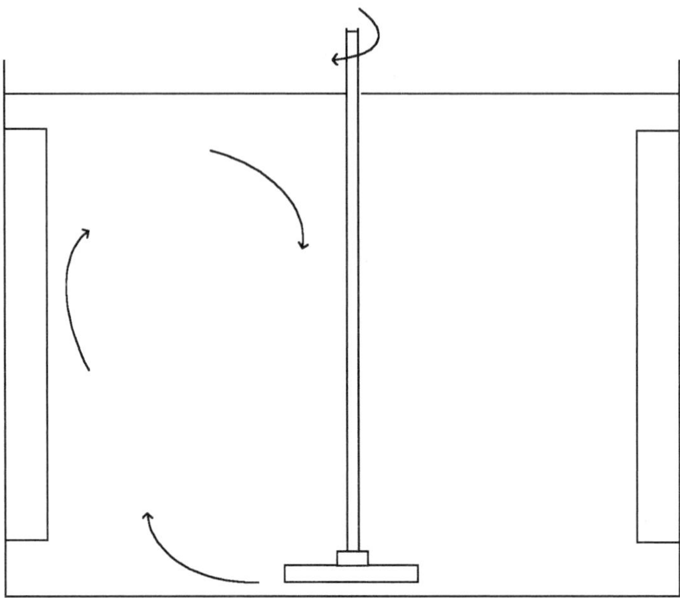

Fig. 14. Bottom-mounted turbine in baffled vessel.

Figure 16 shows a specialized turbine used primarily in the dispersion of sludge in waste treatment service in which sludge concentration and density are low and the volume is large.

The high-speed disk is a circular, horizontal plate with small peripheral vanes or circumferential convolutions simulating blades, which is operated in the 1800–3500 rpm

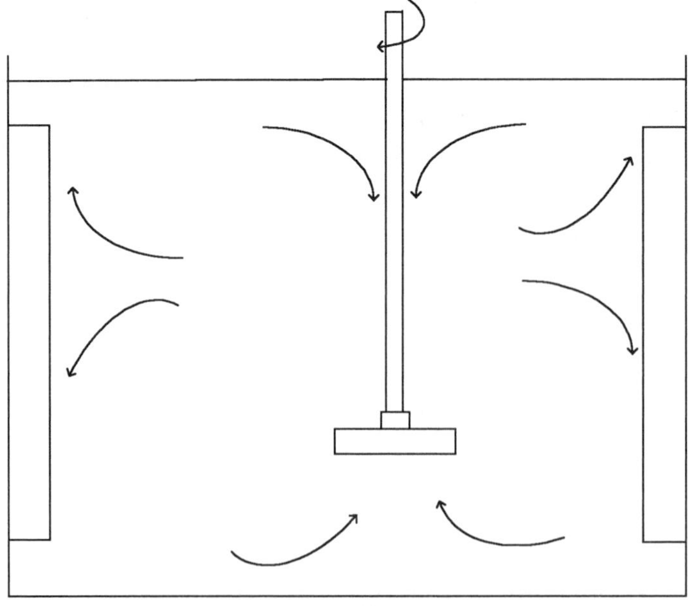

Fig. 15. Raised turbine in baffled vessel.

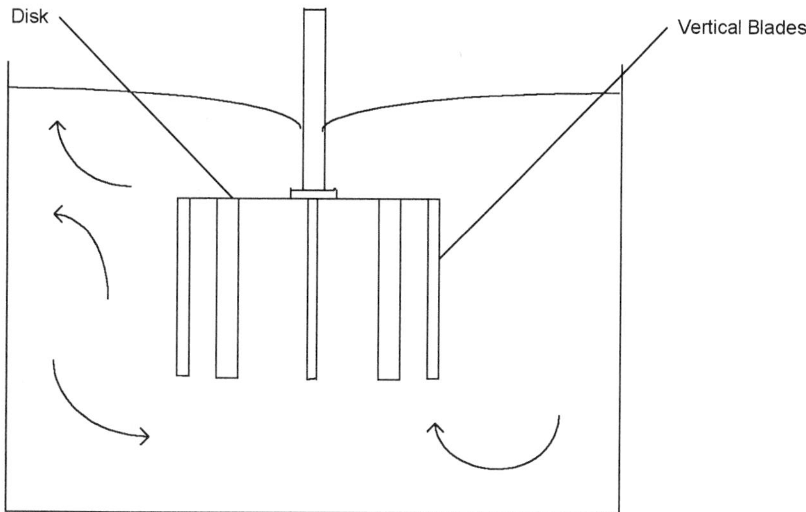

Fig. 16. Vorti-mix turbine.

range. It is dependent on an unusually high-velocity radial discharge to achieve suspension. Owing to its size and volumetric capacity, the disk is particularly applicable to non- or semi-Newtonian mixtures of fibrous materials such as wood or paper pulp or perhaps sewage sludge in small-scale equipment. Imposed stream velocities are sufficiently above the critical velocity of such slurries to momentarily convert them into free-flowing fluids. The propeller, Figs. 12 and 13, creates a downward flow which scours the vessel bottom and rises in the outer, annular space to the surface. To be effective, the upward flow induced must be significantly above the settling velocity of the particles. By using a large-diameter (relative to the tank) three- or four-blade propeller with a wide blade face, large flows may be induced at relatively low impeller speeds. The ratio of tank diameter to propeller diameter is often taken as

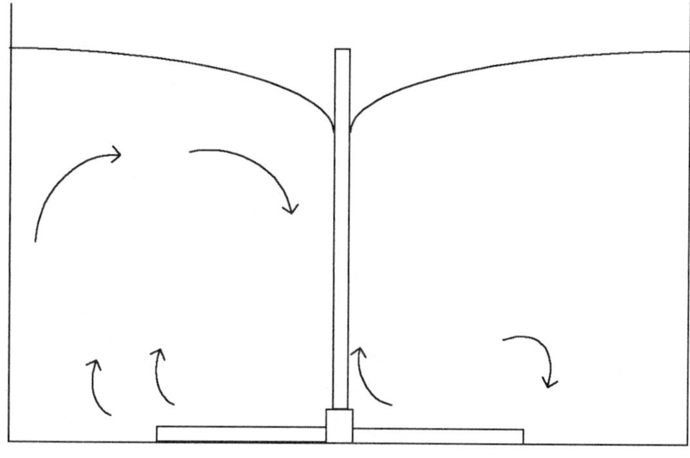

Fig. 17. Paddle mixer.

Fig. 18. Paddle-wheel flocculator.

four. The propeller is most effective in slurries in which hindrance or apparent viscosity of the mixture assists in the support of the solid particles, or for relatively high solids concentration mixtures.

The paddle type mixer shown in Fig. 17 cannot produce stream velocities sufficient for suspension of dense solids but has been applied to highly concentrated or viscous solids. Multiblade paddles have been used in high-viscosity mixtures, usually non-Newtonian (such as polymers), and with simple suspensions in large tanks in which homogeneity is not required but the solids are sufficiently light and flocculent to be easily dispersed. The horizontal shaft flocculator (Fig. 18) is often used for the latter type of application. Homogeneity is required, but low liquid turbulence is needed to preserve the floc. The horizontally mounted paddles physically move from the bottom to the top of the channel providing gentle lifting and circulation throughout the vertical cross section.

A free liquid layer between the upper boundary of solids and the liquid surface is often seen where insufficient mixing power is applied. As the impeller speed is increased, the elevation of the solids surface rises until the tank contents are completely mixed. In Fig. 19, the height of the suspension, h, represents the solids at the critical suspension speed where the tank bottom is just swept clean. The increment h' represents the clear layer of liquid between slurry interface and liquid surface. The slurry interface can be raised to equal $h + h'$ by increasing the impeller and therefore, the recirculating stream velocity. An alternative design technique would be to reduce the depth of slurry to h, if all other factors remained constant. A third alternative would be to change the tank and impeller dimensions at the same slurry volume so that the suspension—height to impeller-diameter ratio, h/D, and the impeller-diameter to tank-diameter ratio, D/T—remain the same.

The latter procedure is normally followed in commercial designs. Since power increases roughly as the cube of the speed in the turbulent regime, maintaining the speed at the minimum required for suspension and changing the vessel and impeller geometry will result in minimizing the power required for complete suspension. Critical speed would be selected to simultaneously produce a uniform dispersion and complete suspension.

A series of experimental studies of the factors affecting uniformity of particle suspension have been carried out in the past. A variety of solids, liquids, and impellers have

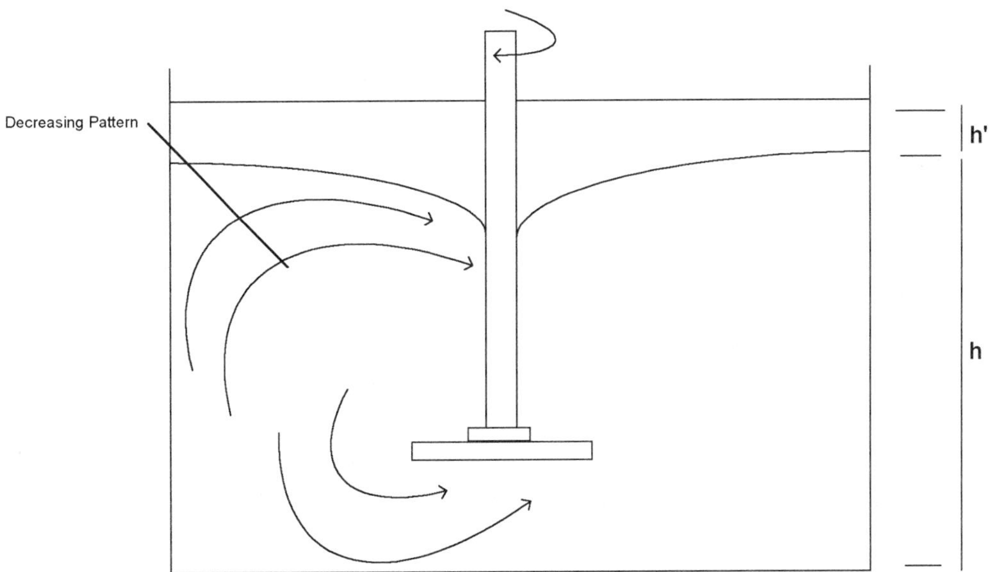

Fig. 19. Effect of slurry viscosity on solids suspension height for a raised turbine.

been used in baffled and unbaffled vessels. A critical point was determined at which local particle concentration leveled off at a point below complete mixing as impeller speed was increased. It was concluded that swirl in the unbaffled vessel inhibited production of a uniform concentration. Hixson and Tenney's mixing index similarly leveled off at 90 (percentage of theoretical concentration) with increasing turbine speed when sand and water were suspended in an unbaffled tank.

Zwietering studied the mixing of sand and sodium chloride with a variety of nonsolvent liquids in baffled tanks used several impeller types (28). He defined a critical speed (N_c) at which no solids remained on the vessel bottom although noting that this did not necessarily provide complete uniformity throughout the tank. Correlation of his data yielded the equation:

$$N_c = \psi\left(\frac{\tau}{D}\right)^t \frac{g^{0.45}\left(\rho_p - \rho\right)^{0.45} \mu^{0.1} D_p^{\ 0.2} (100R)^{0.13}}{D^{0.85}\,\rho^{0.55}} \tag{71}$$

where R is the weight ratio of solid to liquid, and ψ and τ are constants applicable to various impeller types, and relative blade heights ranged from 1.0 to 2.0 (averaging 1.5) and the exponent, t, was found to be approx 1.4.

A similar relationship was derived from experiments using sand and iron ore in a variety of liquids at a 1 to 4 weight ratio. Tests were run in an unbaffled vessel with three-blade square-pitch propellers. The resulting equation for critical speed N_c is as follows:

$$N_c = 0.105\frac{g^{0.6}\,\rho_p^{\ 0.8} D_p^{\ 0.4} T^{1.9}}{\mu^{0.2}\rho^{0.6} D^{2.5}} \tag{72}$$

The values of ρ and μ are for the liquid medium.

Single and multiple turbine impellers were used with thorium oxide and glass bead suspensions to study the effect of solids concentration, particle size, impeller elevation, and system geometry on height of the slurry interface. A definition of "suspension criterion" *(S)* for description of a condition similar to the critical speed of Zwietering was correlated by (28)

$$S = \frac{ND^{0.85}}{\upsilon^{0.1} D_p^{0.2} \left(g\Delta\rho/\rho\right)^{0.45} (100R)^{0.13}} \tag{73}$$

An impeller power per mass of liquid relationship was developed for the impeller speed at which "complete" suspension was attained (clearing of the bottom):

$$\left(\frac{P_c g_c}{M}\right)^{2/3} = \frac{gD_p^{1.3}\left(\rho_p - \rho\right)}{16\rho} \tag{74}$$

The above equation applies to the suspension of low concentration slurries of sand and resins.

All of the equations above apply to both laminar and turbulent conditions. In each case, it was noted that additional speed above critical was essential for homogeneous distribution of the solid phase throughout the fluid mass. Although each of the correlating relationships is valid for the conditions used in its derivation, all of the studies were conducted on a relatively small scale. Extrapolation to full scale might require adjustment of the applicable constants, but full-scale data are not available. Velocity loss due to entrainment of fluid by the high velocity discharge stream and viscous drag do not scale as a straight-line function of size in geometrically similar larger vessels. The velocity decreases with distance traveled by the slurry from the point of maximum velocity at the point of discharge, and the circulation pattern becomes smaller relative to the impeller diameter as equipment size is increased. A higher discharge velocity or a larger impeller to-tank diameter ratio is required to maintain a similar flow pattern in the larger vessel.

Since the above equations are empirical in nature and are not dimensionally homologous, it is recommended that English units be employed. Definitions and appropriate units are as follows:

D = Impeller diameter, ft.
D_p = Particle diameter, ft.
g = Acceleration of gravity, ft/s^2.
g_c = Gravitational conversion factor, lb/lb-(force) × ft/s^3.
M = Slurry weight, lb.
N = Impeller rotational speed, rev/min.
N_c = Impeller speed at which no particles remain on the tank volume, rev/min.
P = Impeller power at which no particles remain on the tank bottom in horsepower.
R = Weight ratio of solid to liquid, lb/lb.
S = Suspension criterion of Eq.(73) in units of 0 to 100.
t = Coefficient in equation.
T = Tank diameter, ft.
μ_t = Kinematic viscosity of liquid, lb/ft-s.
υ = Dynamic viscosity of liquid, ft^2/s.
ψ = Constant in Eq.(71).

ρ = Unit weight of liquid, lb/ft^3.
ρ_p = Unit weight of particle, lb/ft^3.
$\Delta\rho$ = Difference in unit weights of solid and liquid, lb/ ft^3.

3.7. *Static Mixer*

Static mixers or motionless mixers are an efficient and reliable mixing device for the application of homogeneous distribution of several liquid and/or gaseous media. They are fins, obstructions, or channels mounted in pipes, designed to promote mixing as fluid flows through the mixer. Most static mixers use some method of first dividing the flow, then rotating, channeling, or diverting the flow, before recombining it. Other static mixers create additional turbulence to enhance mixing. The power input to the mixing process is a result of pressure loss through the mixer.

Static mixers serve to put liquid into motion in order to achieve homogeneity of composition and avoid the sedimentation process. They are driven by auxiliary equipment, such as a shaft, speed reducer, or electric motor, to provide mixing action. The mixer consists of a mixing tube with inside welded mixing elements, which are specially designed for highest mixing degree at an extremely low pressure drop. Form and dimension of the mixing elements effects an intensive mixing of the media, which have to be mixed. The mixing degree is determinated by operating pressure and flow velocity. For example: the mixing degree for the media water/ozonized water is 99%.

The elements in the static mixer are rectangular plates twisted at 180°. According to the direction of the twist, one can distinguish a right element and a left element. All elements have a standard length of 1.5 times their diameter. The static mixer effectively mixes fluids, executing the operations of division of flow, radial mixing, and flow reversal. As fluids flow along the curves of each element, they are rotated radially toward the pipe wall, or rotated back to the center. Fluids are bisected as they pass each element. The partition number can be $N = 2n$, where n is the number of elements. As fluids pass each element, they change their direction to the right or to the left, and the force of inertia that suddenly occurs creates a strong flow reversal motion that results in stirring and mixing of the fluids. Low viscous substances that are mutually soluble are mixed using the process of flow reversal. It should be noted that such substances can be sufficiently mixed using a relatively small number of elements. Even in the case of low viscous substances that are mutually insoluble, such as water and oil, dispersion is possible through the process of radial mixing. Noted that as they pass through the elements, the diameters of the dispersion particles become smaller, which makes the process of dispersion possible. Of the three mixing principles, the mixing of high viscous substances uses mainly the operations of division of flow and radial mixing. An increase in the number of stripes can be noted with each passing through the elements.

They function by forcing sediment to flow in one direction and overcome the resistance during a liquid circulation flow in open reservoirs, ditches, and canals. Static mixers are used to intensify physical and chemical processes in liquids, particularly the processes of gas and solid dissolution. Gas dissolution is usually used in sediment /waste water/anaerobic process. The intensified mixing operation is applied in order to lengthen the distance covered by gas bubbles and to prevent smaller bubbles from joining into bigger ones. Direct drive, fast rotating mixers may also be used to prevent surface scum from coming into existence and to destroy any surface scum that has already appeared. Its applications include dissolving of gases in liquids, oxygen in water, wastewater or sludge, ozone in water or wastewater, carbon dioxide in water, mixing of liquids, neutralization by acids or alkaline solutions, precipitants and flocculants in water, mixing of gases, or cooling fumes by means of fresh air (33,43–47).

When determining which of the many available static mixers would be best for the given application, there are two main factors to consider, the material to be processed and the rate at which it must be processed. If the process media or the finished product is acidic, a high-purity chemical, a pharmaceutical, of a specific polymer type, or if the mixer itself will have multiuse applications, it is best to select a lined or coated static mixer, which provides a nonstick, chemically resistant mixing area. The speed at which a mixer can process materials, also known as its flow rate, is a rated measurement based on the volume of product the mixer can process during given period of time. If the needed flow rate is known, it is easy to find a static mixer that may operate within this range.

The static mixer can be operated in either laminar or turbulent flows. In the case of laminar flows, the components are distributed in thinner layers, whereas eddies are generated in the case of turbulent flows. Owing to space limitation, the detailed design procedure is not given here; however, one can find it in the literature (44,47).

There are several new developments in static mixers. For example, Sulzer Chemtech has brought a new type of mixer onto the market (44). It comprises a narrow ring (without flanges) that is clamped between two pipe flanges. A single mixing element, together with a dosing port for additives, is integrated within this ring. The mixing element comprises three blades with specially shaped edges to break up the flow. The additive is fed behind the central blade. This development reduces the installation length significantly compared with traditional mixers that comprise a housing normally welded together from a pipe and two flanges. It does not incur a greater pressure drop. The other advantages are simple design, mounting, and cleaning, and lower cost in purchase, installation, operation and maintenance.

4. DESIGN OF FACILITIES

In the following sections, applications of the preceding materials along with newly introduced empirical techniques are considered. The approach will be basic to the various mixing functions in so far as is possible, although examples will be drawn largely from water and wastewater practice. For process factors that will also influence the overall design of most systems in which mixing is employed, the reader is referred to the appropriate chapter related to the process design in question.

4.1. *Pipes, Ducts, and Channels*

In liquid systems, theoretical approaches to defining the length required for complete mixing are rudimentary. The following procedure may be used to produce an approximate result for water (or for a typical gas flowing in a duct).

1. Compute the Reynolds number for the flow by:

$$N_{\mathrm{Re}} = \frac{\bar{u}D}{\upsilon}$$

where $\bar{u}$ is velocity, m/s (ft/s), D is pipe diameter, m (ft); and υ is kinematic viscosity, m^2/s (ft^2/s).

2. Using Fig. 3, determine the appropriate value of σ (extrapolating, if required).
3. Compute:

$$k_0 = \frac{4}{D}$$

$$X_s = 0.164 f^{1/4} N_{\text{Re}}^{\ 3/4}$$

for friction factor f.

4. Determine g_s from Fig. 2 and calculate:

$$u' = 0.56 f^{1/3} g_s^{\ 1/2}\, \bar{u}$$

where u′ is root mean square velocity, m/s (ft/s) and $\bar{u}$ is mean velocity, m/s (ft/s).

5. Compute the time and length of flow for mixing by:

$$t(\text{in s}) = \frac{\sigma}{k_0 u'}$$

$$L(\text{m or ft}) = \bar{u}t$$

For non-circular conduits or open channels , the diameter of a hydraulically equivalent circular pipe could be computed for substitution in the above procedure by the equation:

$$D_e = 1.53 A^{0.38} R^{0.24} \tag{75}$$

where A is cross-section area of conduit, $m^2(ft^2)$ and R is hydraulic radius = A divided by the wetted perimeter, m (ft).

The concept of power times time may also be applied to mixing in any type of conduit. In water treatment practice, rapid-mix units having a velocity gradient of 300 m/s/m and a detention time of 10–30 s have been found to be adequate for dispersion of chemicals. This produces a product $\overline{Gt}$ ranging from 3000 to 9000. From Eq. (8), one will have:

$$\bar{G} = \sqrt{\frac{Wg}{\mu}}$$

and

$$W = \frac{\rho h_f}{t}$$

where μ is absolute viscosity, ρ is unit weight of water, kg/m^3 (lb/ft^3); h_f is head loss, m (ft); and t is contact time, s.

Assuming that any reasonable combination of C and t producing a desired product will achieve adequate mixing, the time may be calculated as:

$$\bar{G}^2 = \frac{\rho h_f g}{t\mu} = \frac{\rho f \dfrac{L}{D}\dfrac{\bar{u}^2}{2g} g}{t\mu}$$

$$\text{Substituting } \bar{u} = \frac{L}{t}$$

$$\overline{G}^2 = \rho f \frac{L}{D}\frac{L^2/t^2}{2}\frac{1}{t\mu} = \frac{\rho f \overline{u} L^2}{2Dt^2\mu}$$

solving for L:

$$L = \left(\frac{2D\mu(\overline{G}t)^2}{\rho f \overline{u}}\right)^{1/2} \tag{76}$$

The prediction of the mixing length required for gases may be approached by the procedure of Beek and Miller in a manner analogous to that outlined for water (8,41). The variation in mixing length with velocity, however, will be much less than that for water. For many applications, an empirical rule of 10 pipe diameters is taken for flows in the moderate to high Reynolds number range.

The question of suspension of particles in flowing fluids is omitted from consideration in this chapter, because conduits are rarely used for particle mixing in environmental engineering practice. Generally, the concern is with minimum velocities to prevent deposition. This aspect is adequately approached on a design basis through application of the widely recognized (but empirical) carrying velocities such as 0.6 m/s (2.0 ft/s) for domestic wastewater; 0.75 m/s (2.5 ft/s) for storm wastewater; 610 to 915 m/min (2000 to 3000 ft/min) for dusts in air ducts.

Example 1

Estimate the length of 0.30 m (12 in.) steel pipe required to ensure essentially complete mixing of an injected, soluble chemical. The velocity is 0.5 m/s (1.64 fps). How long should the same pipe be for flocculation?

Solution

Using the method described by Beek and Miller, (8) and following the steps outlined above.

1. $N_{\text{Re}} = \frac{\overline{u}D}{\upsilon} = \frac{(0.5\,\text{m/s})\times(0.30\ \text{m})}{1\times10^{-6}\text{m/s}} = 1.5\times10^5$

2. $\sigma \cong 6$

3. $k_0 = \frac{4}{D} = \frac{4}{0.3\ \text{m}} = 13.3\times\frac{1}{\text{m}}$

$$\begin{aligned} X_s &= 0.164 f^{1/4} N_{\text{Re}}^{3/4} \\ &= 0.164(0.018)^{1/4}(1.5\times10^5)^{3/4} \\ &= 0.164(0.366)(7622) \\ &= 4.58. \end{aligned}$$

4. From Fig. 2, $g_s = 1.01$

$$\begin{aligned} \overline{u} &= 0.56(f)^{1/3}(g_s)^{1/2}\overline{u} \\ &= 0.56(0.018)^{1/3}(1.01)^{1/2}(0.5\ \text{m/s}) = 0.0734\ \text{m/s} \end{aligned}$$

5. $t = \frac{\sigma}{\text{k}_0\text{u}'} = \frac{6.0}{(13.3/\text{m})\times(0.073\ \text{m/s})} = 6.15\ \text{s}$

$$\text{Pipe length} = 6.15\ \text{s} \times 0.5\ \text{m/s} = 3.07\ \text{m}(10.1\ \text{ft})$$

This is approximately equal to 10 pipe diameters. By use of the equation adapted from Camp, the length may also be calculated (6):

$$L = \left(\frac{2D\mu(\overline{G}t)^2}{\rho f u} \right)^{1/2}$$

Selecting a *Gt* value of 3000:

$$L = \left(\frac{2 \times 0.3\ \text{m} \times 1.005 \times 10^{-3}\ \frac{\text{kg}}{\text{s} \cdot \text{m}} (3 \times 10^3)^2}{1000\ \frac{\text{kg}}{\text{m}^3} \times f \times 0.5\ \text{m/s}} \right)^{1/2} = \left(\frac{10.85}{f} \right)^{1/2}$$

From a Moody diagram in any hydraulics text using an absolute roughness of 4.57×10^{-5} m (0.00015 ft) and $N_{Re} = 1.5 \times 10^5$.
Using $f = 0.018$ from the preceding calculation,

$$L = \left(\frac{10.85}{0.018} \right)^{1/2} = 24.5\,\text{m} \quad (\text{or } 81\ \text{ft})$$

This result differs from the preceding answer by a factor of eight. The discrepancy is largely due to the extreme conservative approach with which $\overline{G}t$ values have been chosen for rapid mix units in water treatment plants. Although the first computation cannot be regarded as precise, a length calculated on the basis of $\overline{G}t$ values of 500 to 1000 should be sufficiently conservative for design practice when linear mixing in pipes or channels is contemplated.

4.2. Self-Induced and Baffled Basins

The velocity gradient time approach as proposed by Camp is the only available design technique for rapid-mix basins (6,17). The self-induced principle involves discharging a high velocity inlet stream tangentially along the wall of a small mixing basin. The resulting turbulence induced by the jet promotes mixing. Baffled basins are used to some extent in flocculation of water, although newer plants generally use mechanical mixing systems. Design of either system may be based on $(\overline{G}t)$ concepts with 3000–9000 used for rapid mix units and 10^4 and 10^5 for flocculation basins. Owing to the fragile floc, the velocity gradient, $\overline{G}t$, is also kept below 100 m/s/m. The procedure outlined above for conduits may be applied to both self-induced and baffled basins by redefining the head loss term.

Self-induced: h_f = velocity head of the entering flow, m.
Baffled: h_f = total head loss in basin, m.

Example 2

A baffled flocculation basin is to be designed for a velocity gradient of 60 m/s/m, a $\overline{G}t$ value of 60 m/s and an allowable head loss of 0.5 m (1.6 ft). What detention period is required ?

Solution:

Combining Eq. (8) and (9):

$$\bar{G} = \sqrt{\frac{\rho h_f g}{t\mu}}$$

Solving for t:

$$\bar{G}^2 = \frac{\rho h_f g}{t\mu}$$

$$t = \frac{\rho h_f g}{\bar{G}^2 \mu} = \frac{(1000\ \text{kg/m}^3) \times 0.5\ \text{m/s} \times 9.81\ \text{m/s}}{(60/\text{s})^2 \times 1.005 \times 10^{-3}\ \text{kg/s}\cdot\text{m}} = 1356\ \text{s} = 23\ \text{min}$$

4.3. Mechanically Mixed Systems

The design of mechanically mixed systems will be divided from a functional standpoint into four basic types of system. The most simple, the dispersion unit, is typified by the mechanically agitated, rapid-mix basin. Its purpose is simply to spread an additive through a carrier solution in a reasonably uniform manner. In equalization facilities, the basin volume relative to the influent flow is fixed by considerations of a desired degree of smoothing of influent flow characteristics. The function of mixing is to ensure that the desired degree of averaging is obtained, or in other words, that "ideal mixing" is achieved. Reactors may be of either the batch type, typified by the sludge digester, or of the continuous-flow type, characterized by flocculation basins and activated sludge units. In long-detention period activated sludge units with mechanical aerators, the critical mixing requirement is the suspension of solids.

4.3.1. Dispersion

The classical environmental engineering approach to mechanically assisted mixing has involved the familiar technique of selecting a desired product of velocity gradient and time. For flash mixers in water treatment, the typical values lie between 3000 and 9000 based on a $\bar{G}$ of approx 300/s and 10–30 seconds. The design procedure is identical to that employed for conduits and baffled basins above except that the velocity gradient is defined by

$$\bar{G} = \sqrt{\frac{102Pg}{V\mu}}$$

where P is power delivered by the agitator to the water in kW (0.745 × horsepower); V is reactor volume in m^3(0.0283 × ft^3); and μ is absolute viscosity in kg/s-m.

Refinement of the above procedure is possible. The necessary correlations of impeller characteristics have not been achieved for continuous flow systems. For batch mixing in vessels, the product of the impeller discharge and the mixing time must equal 1 to 1.5 times the vessel volume to achieve complete mixing. It appears that rapid-mixing could be correlated in the same manner to give a design basis more consistent with the basic principles of mixing.

4.3.2. Equalization

As previously discussed, mixing in equalization basins should be sufficiently complete to ensure that the desired averaging effect is achieved between inlet and outlet flow properties. Unfortunately, efforts to quantify the mixing requirements in terms of basin size and shape have been unsuccessful except for the establishment of empirical guidelines.

By definition, as equalization facility will have a relatively long detention period if averaging of flow properties is to be attained. Complete mixing cannot be achieved unless the pump discharge rate substantially exceeds the inflow rate and unless circulation within the basin is such as to convey the fluid from and to all parts of the basin. The ability of a mechanical mixer to circulate the contents of a basin will depend on its liquid pumping capacity and on the geometry of the basin. Deep basins will facilitate the establishment of large-scale circulation patterns, while shallow basins will tend to produce flow reversal at a short distance from the impeller.

In order for complete mixing to be achieved, the entire contents of the basin must be circulated through the impeller a number of times during the detention period. Available data suggest that the circulation ratio should be at least 10 for a single basin impeller, calculated on the basis of actual impeller discharge rate (which is less than the actual circulation rate). For multiple impellers, the circulation ratio should be substantially in excess of 10. In long detention facilities such as aerated lagoons, the ability to maintain solids in suspension will commonly override the requirements for complete mixing and the circulation rate may exceed 100. Empirically determined power requirements for aerated lagoons range from 7.9×10^{-3} kW/m^3 (0.3 hp/1000 ft^3) in deep basins to 1.3×10^{-2} kW/m^3(0.5 hp/1000 ft^3) for shallow basins.

Where detention periods are less than 24 h in activated sludge or aerated lagoon applications, the mixing associated with oxygen transfer (either mechanical or diffused air) will generally provide complete mixing.

Until basic information becomes available, the design of mixing facilities in equalization basins will necessarily require the exercise of considerable engineering judgment. The following guidelines are offered:

1. The proportions of equalization basins should be established to provide equal, approximately square areas for each aerator used. The depth should be as large as possible consistent with overall economy. Steep side walls are preferred to minimize solids deposits and facilitate mixing.
2. The general recommendations of 8×10^{-3} kW/m^3(0.3 hp per thousand gallons) for deep basins to 1.3×10^{-2} kW/m^3 (0.5 hp per thousand gallons) for aerated lagoons may be decreased slightly if solids are not a factor, but complete mixing will not be achieved much below these values.
3. The manufacturers of mechanical agitators will furnish information on impeller flow rate and recommended maximum mixing circle diameter for each of their units.
4. Impeller discharge times detention time should be checked to ensure that the product provides for recirculation of the basin volume at least 10 times within the detention period for single impellers, and a substantially greater value for multiple impellers.
5. If mechanical aerators are used for mixing, care should be taken to select a type that permits a large pumped flow per unit horsepower input. Surface aerators vary widely in this regard.
6. Impeller discharge per unit horsepower decreases with total horsepower of a single unit.

Therefore, several small units may provide greater mixing efficiency than a single large unit for the same power costs.

Example 3

Determine the mixing requirements and basin proportions for an earthen equalization basin designed to provide a one-day detention period for a flow of 0.7 m^3/min (185 gpm). Assume that the basin will be lined, that the maximum side slope is 45°, and that surface aerators will be used although oxygen supply is not a critical factor.

Solution

The total volume of the aerated lagoon is

$$V = 0.7\,\text{m}^3/\text{min} \times 1440\,\text{min/d} \times 1\text{d} = 1008\ \text{m}^3 \left(3.56 \times 10^4\,\text{ft}^3\right)$$

Assuming a relatively shallow basin, the total power requirement will be approximately:

$$P = 1.3 \times 10^{-3}\ \text{kW/m}^3 \times 1008\ \text{m}^3 = 13.1\ \text{kW}(17.6\ \text{hp}).$$

Based on manufacturers' catalogs, information on commercial units' characteristics may be obtained. Subsequent calculations are performed for a series of equipment from a company. Data on appropriate units is summarized in following table.

Model No.	Horsepower	Pumping rate (gpm)	Complete mixing diameter (ft)	Minimum operating depth (ft)	Normal operating depth (ft)
OX 825	5	2885	50	3	6–12
OX 1160	10	4650	55	3.5	6.5–14
OX 1260	20	7750	75	4	7.5–16

From the information above, it may be seen that for a given horsepower input, pumping rate (and O_2 transfer) is higher for several small units than for a single large unit. For this application, two 10 hp units would provide good mixing at reasonable economy with at least some safety in the event of failure of a single unit. Four aerators would provide greater reliability, but at a higher cost.

Assuming this choice of a two 10 hp units, the basin configuration will be taken as a 2:1 ratio rectangle to provide equal, square mixing zones for each impeller. This will produce the most uniform mixing intensity throughout the basin. If practical, the depth will be established within the "normal" range to avoid the use of antierosion assemblies for shallow depths or draft-tubes for large depths. The volume of a frustrum of a pyramid is:

$$V = \frac{D}{3}\left(A_1, A_2, \sqrt{A_1, A_2}\right)$$

Setting the bottom dimensions as L_1 and $W_1 = 0.5L_1$

$$A_1 = L_1 W_1 = 0.5\ L_1^{\ 2}$$

For 45° side slope:

$$L_2 = L_1 + 2D$$

$$W_2 = \text{W}_1 + 2D = 0.5L_1 + 2D$$

$$A_2 = (L_1 + 2D)\ (0.5L_1 + 2D)$$

Assuming a liquid depth of 3 m:

$$1008 = \frac{3}{3}\left[0.5L_1^{\,2} + \left(L_1 + 2D\right)\left(0.5L_1 + 2D\right)\sqrt{0.5L_1^{\,2}\left(L_1 + 2D\right)\left(0.5L_1 + 2D\right)}\right]$$

By trial and error, the bottom length required is 21.5 m (70.5 ft). This would give bottom dimensions of 21.5 m (70.5 ft) by 10.75 m (35.3 ft) and water surface dimensions of 27.5 m (90 ft) by 16.75 m (55 ft). Allowing 0.5 m (1.6 ft) freeboard, the dimensions of the actual top surface would be 28.5 m (93.5 ft) by 17.75 m (58 ft). Checking the pumping rate/volume/inflow relationships:

$$Q = 2 \times 4650 \text{ gpm} = 9300 \text{ gpm} = 35.2 \text{ m}^3 / \text{min}$$

$$\frac{Q_{\text{Re circ}}}{Q_{\text{inflow}}} = \frac{35.2 \text{ m}^3/\text{min}}{0.7 \text{ m}^3/\text{min}} = 50$$

Both of the above criteria are sufficiently large to insure that the basin will be completely mixed.

4.3.3. Batch Mixers and Reactors

Batch mixing applications are to be found in the preparation of chemical solutions or the addition of conditioning agents to sludge. Mixing will generally occur in cylindrical or rectangular tanks with the aid of turbine or propeller type mechanical mixers. Applications of this type are common in chemical engineering practice and the correlations may be applied in a rational manner. Two different approaches may be used to the selection of an appropriate mixer. These are outlined below.

Method 1

(a) Select an impeller type, diameter, and speed suitable to the tank and commercial applications.
For simple mixing, the propeller type will consume less power than a turbine. Suitable diameters are in the range of 0.1 to 0.25 times the tank diameter for turbines or smaller for propellers.

(b) Calculate the impeller Reynolds number and determine the mixing time correlation factor from Figs. 10 or 11 as appropriate. Solve for the mixing time.

(c) If the mixing time is shorter or longer than desired, alter the impeller diameter or speed and repeat the process above.

(d) Determine the impeller power number from Figs. 7 or 8 and solve for the required motor horsepower.

Method 2

(a) Select a design value for the number of circulations of the tank contents in the range of 1.0 to 1.5.

(b) Select a desired mixing time and calculate the impeller discharge from the relationship:

$$\frac{\phi Q}{V} = \text{No. of times circulated}$$

where ϕ is mixing time, min; Q is impeller discharge, m^3/min (ft^3/min); and V is vessel volume, in (ft^3).

(c) Determine the desired type of impeller and its discharge number from Table 3 and power numbers from Figs. 7 or 8.

(d) Size the impeller using the discharge number relationship and determine the power required from the power number.

It should be noted that both of the procedures above are approximate and therefore may yield somewhat different values. Where possible, manufacturer's data concerning discharge and power numbers should be substituted for the generalized values. For mixing of sludge, the procedures remain valid except that a kinematic viscosity of 5×10^{-4} ft^2/s (4.6×10^{-5} m^2/s) (31) should be used and power should be increased to provide sufficient starting torque.

Example 4

A facility is to be provided to dilute 92% sulfuric acid to 20% for use in a neutralization feed unit. If the tank is to contain 3800 L (1000 gal), determine the dimensions, the mixing requirements, and the time required for preparation of an acid batch.

Solution

Assume a cylindrical tank of "square" design with the liquid depth equal to approx 85% of the tank depth to provide freeboard.
Compute the tank dimensions:

$$V = 3800 \text{ L (approx 1000 gal)} 3.8 = 3.8 \text{ m}^3 (133.7 \text{ ft}^3)$$

$$= \frac{\pi D^2}{4} \times 0.85D = 3.8 \text{ m}^3$$

$$D^3 = \frac{3.8 \text{ m}^3 \times 4}{\pi \times 0.85} = 5.69 \text{ m}^3 (200.3 \text{ ft}^3)$$

$$D = 1.79 \text{ m } (5.85 \text{ ft})$$

Adjust the dimensions to a more standard value by choosing the diameter to be 2 m (6.56 ft).

$$\text{Liquid depth} = \frac{3.8 \text{ m}^3}{\frac{\pi}{4}(2 \text{ m})^2} = 1.21 \text{ m}, (3.97 \text{ ft})$$

$$\text{Tank depth} = \frac{1.21 \text{ m}}{0.85} = 1.42 \text{ m} (4.67 \text{ ft})$$

Use a 2 m diameter × 1.5 m high tank with liquid depth of 1.2 m.

Four baffles will be installed in the tank, equally spaced around the periphery. The width will be taken as 1/12 the tank diameter or 0.17 m (0.5 ft). The baffles will be positioned on brackets approx 0.075 m (3 in.) away from the tank walls and bottom. The height of the baffles will be taken as 1.0 m (3.3 ft) to ensure submergence of the upper end protecting against splash. Using method 2 above, mixing may be assumed complete after 1.5 circulations of the tank contents. Because the power requirement will be low, the desired mixing time will be set at 5 min.

$$\frac{\phi Q}{V} = \frac{5 \text{ min} \times Q}{3800 \text{L}} = 1.5$$

$$Q = 1140 \text{ L/min} (301 \text{ gpm})$$

Because high shear is not required for acid dilution, a propeller type mixer will suffice and will give optimum efficiency. From Table 3 N_Q is found to vary from 0.40 to 0.61. Using a typical value:

$$N_Q = 0.5 = \frac{Q}{ND^3}$$

For a rotational speed of 1725 rpm,

$$D^3 = \frac{Q}{0.5N} = \frac{1.14 \text{ m}^3/\text{min}}{0.5 \times 1725 \text{ rev/min}} = 0.00132 \text{ m}^3$$

$$\text{D} = 0.110 \text{ m (4.5 in.)}$$

$$N_{\text{Re}} = \frac{D^2 N\rho}{\mu} = \frac{(11.0 \text{ cm})^2 (1785 \text{ rpm} \times 1 \text{ min / 60 s})(1 \text{ g/cm}^3)}{0.01002 \text{ P}} = 3.5 \times 10^5$$

From Fig. 7, $N_p \cong 0.3$ for square pitch propellers.

$$N_p = \frac{Pg_c}{\rho N^3 D^5} = 0.3$$

$$P = \frac{0.3(1000 \text{ kg/m}^3)(1785/60)^3 (0.110 \text{ m})^5}{9.81 \text{ m/s}} = 11.6 \text{ kg} \cdot \text{m/s}$$
$$= 0.113 \text{ kw}(0.152 \text{ hp})$$

A 0.25 hp propeller mixer with 0.11 m (4.5 in.) diameter and 1725 rpm speed will be used. Total batch preparation time will consist of the time required for acid addition plus 5 min to complete mixing. For a 9.2:2 dilution ratio, the volume of acid to be added is:

$$\text{Volume acid} = \frac{3800\text{L}}{4.6} = 206\text{L}$$

At 10 liters/min, the time required for batch preparation is:

$$\text{time} = \frac{206\text{L}}{10\text{L/min}} + 5 \text{ min} = 26 \text{ min}$$

4.3.4. *Continuous Flow Reactors*

Continuous-flow reaction vessels encountered in environmental engineering applications generally consist of flocculation and coagulation basins and activated sludge units of the various types. In activated sludge and related systems, the introduction of oxygen by either diffused air or mechanical aerators is sufficient to provide complete mixing if the detention period is less than 24 h. For longer aeration periods, the requirements for mixing relate to solids suspension as previously discussed. Activated sludge units, therefore, need not be discussed further under this heading.

The classical approach to the design of flocculation and coagulation facilities has involved the concepts of velocity gradient and time considerations outlined in Sections 4.1 and 4.3 and previously applied to linear, self-induced, and rapid mixing. Velocity gradients must be sufficient to ensure particle contact for reaction, but not so great as to disrupt the reacted particles. For water treatment units using iron or aluminum as coagulating agents, design values of $\bar{G}$ have ranged up to 100 fps/ft and G_t values of 10^4 and 10^5. Care must be taken to keep paddle speeds low (0.5–1 fps) to avoid high velocity gradients at the tip. For wastewater treatment, $\bar{G}$ and $\bar{G}_t$ values should be kept below those required for water because relatively massive chemical doses are used. As the process parameters related to flocculation are beyond the scope of this chapter, and ample illustrations of the application of the velocity gradient concept to the selection of mixers have been presented, examples of the determination of power requirements will be omitted.

4.3.5. Solids Suspension

The principles of mixing as they relate to solids suspension have been presented earlier. Equations (71) and (72) were presented that relate fluid, tank, and particle properties to mixing requirements in terms of either impeller speed or power input required to lift the particles off the bottom of the tank. All three equations were derived from laboratory tests and indicate the degree of mixing necessary to suspend the particles, but without necessarily producing a uniform slurry. The equations may be used to determine the mixing requirements for small slurry applications such as activated charcoal feed tanks, but the predicted power requirements should be considered as minimum requirements because full-size units will almost always require greater power input.

The mixing of high concentration slurries will tend to contract the circulation patterns in a mixing tank closer to the impeller than would be expected for pure liquid systems. For high particle densities, in particular, uniform suspension will be difficult to obtain unless the impeller is relatively large compared to the tank. However, if the impeller approaches the side walls closely, the circulation will be restricted and improperly mixed zones will be present. Optimum impeller diameters will range from 20% to 40% of the tank diameter with the high values representing high solids density and size applications and the lower values representing low solids density, fine particles, and solids concentrations rising above 40 % by weight.

Cylindrical or square tanks are normally used for chemical feed slurries. The liquid depth should not greatly exceed the tank diameter if uniform slurries are to be obtained. Where taller tanks are necessary for conservation of space, multiple impellers may be used on the same drive shaft. Either propeller or turbine type impellers may be used in shallow tanks. However, turbine impellers are preferred for multiple impeller designs in tall tanks. Impeller spacing on a common shaft is usually three to four impeller diameters.

Baffles are essential in cylindrical tanks if centerline agitation is used. Typical baffling consists of four vertical plates one-tenth to one-twelfth the tank diameter and offset from the tank wall a distance of one-fourth to one-half their own width (Fig. 20). Baffling is not generally necessary in square tanks and may be avoided in cylindrical tanks by offsetting the impeller a distance of one-fourth to one-half the tank radius from the centerline.

Solids suspension in unusually shaped basins such as the truncated trapezoid commonly used for earthen lagoons or in large shallow tanks is not amenable to mathematical analysis based on the present state of knowledge. The reader is cautioned not to attempt

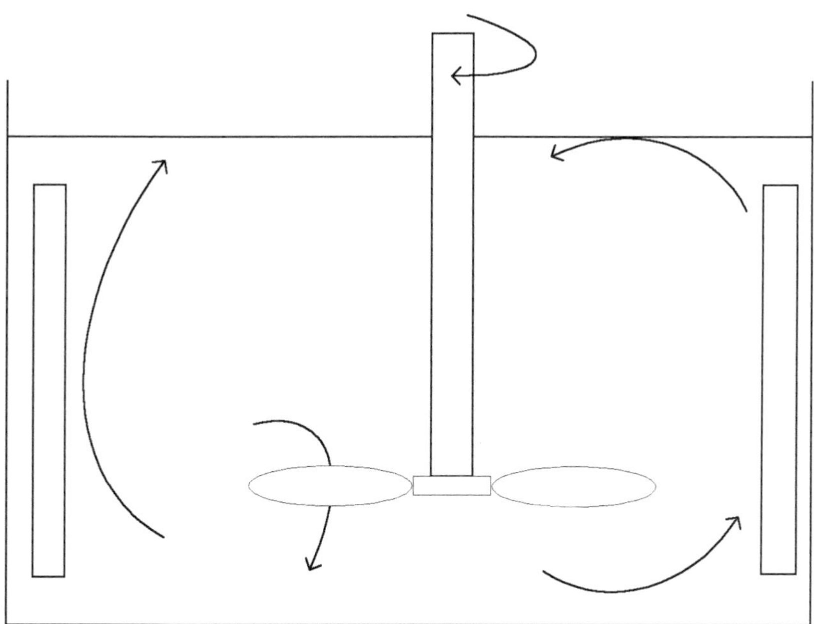

Fig. 20. Cyclinderical tank bafflin.

to apply the design relationships cited above for applications other than cylindrical or square tanks.

Example 6

A 2.0-m-diameter cylindrical vessel, 3.0 m high, is to be used to prepare a bentonite slurry in the proportions of 18 kg clay per 100 kg water. Determine the power required to suspend the slurry if the specific gravity of the clay is 2.7 and the size is such as to pass a 100 mesh sieve. A six-blade turbine mixer will be used.

Solution

Using Zwitering's technique as shown in Eq. (71)

$$N_c = \frac{1.5\left(\frac{T}{D}\right)^{1.4} g^{0.45}\left(\frac{\rho_p - \rho}{\rho}\right)^{0.45} \upsilon^{0.1} D_p{}^{0.2}(100R)^{0.13}}{D^{0.85}}$$

The units are:

N_c = Critical suspension speed (rev/sec).
T = Tank diameter = 2.0 m (6.56 ft).
D = Impeller diameter, m.
g = Acceleration of gravity = 9.81 m/s^2(32.2 ft/sec^2).
ρ_p = Unit weight. of clay = 2700 kg/m^3(168 lb/ft).
ρ = Unit weight. of water = 1000 kg/m^3 (62.4 lb/ft).
υ = Kinematic viscosity = 10^{-6} m^2/s (1.08×10^{-5} ft^2/s).
D_p = Particle diameter = 1.49×10 m (4.92×10 ft) for a 100 mesh sieve.
R = Ratio of weight. of clay to that of water = 0.18

Assuming a desired impeller speed, the required diameter may be from the formula above. For 600 rpm, $N = 10$ rps:

$$10 = \frac{1.5(2.0)^{1.4}(9.81)^{0.45}(1.7)^{0.45}\left(10^{-6}\right)^{0.1}\left(1.49\times10^{-4}\right)^{0.2}(18)^{0.13}}{D^{1.4}D^{0.85}}$$

Solving for D:

$$D^{2.25} = 0.0884$$

$$\mathrm{D} = (0.0884)^{0.444} = 0.341 \text{ m } (1.12 \text{ ft})$$

To determine the power required for the impeller, the curves of Fig. 8 may be used. Computing the Reynolds number:

$$N_{\mathrm{Re}} = \frac{D^2 N}{\upsilon} = \frac{(0.341 \text{ m})^2(10 \text{ rev / sec})}{10^{-6} \text{ m}^2/\text{s}} = 1.16\times10^6$$

From Fig. 8, the power number becomes essentially constant at a value of 2.5 for a six-bladed turbine. Therefore

$$N_{\mathrm{P}} = \frac{g_c P}{\rho N^3 D^5} = 2.5$$

Solving for P:

$$P = \frac{2.5\left(1000 \text{ kg / m}^3\right)(10 \text{ rev / s})^3(0.34 / \text{m})^5}{9.81 \text{m/s}} = 1.175\times10^3 \text{ kg}\cdot\text{m/s}$$

$$= 11.5 \text{ kW (15.5 hp)}$$

Power consumption may be decreased by reducing the speed of rotation and using a larger diameter impeller. Using computations identical to those above for $N_c = 5$ rev/sec and 2.5 rev/sec, the following results are obtained:

Speed	Diameter	Power
600 rpm	0.341 m (1.12 ft)	11.5 kW (15.5 hp)
300 rpm	0.464 m (1.52 ft)	6.7 kW (8.7 hp)
150 rpm	0.631 m (2.07 ft)	3.9 kW (5.2 hp)

Selection of actual speed, diameter, and horsepower can be made from commercial units keeping in mind the specific requirements of the process. Because the clay may tend to agglomerate or may require dispersion of clusters of particles initially present, a relatively high impeller tip speed may be advantageous. Also, to obtain uniformity of suspension, higher power input is required than that predicted by the equation. A suitable unit for this application might be a 0.61 m (2.0 ft) impeller turning at 200 rpm. The power required would be:

$$P = \frac{2.5\left(1000 \text{ kg / m}^3\right)(3.33 \text{ rev / s})(0.61 \text{ m})^5}{9.81 \text{ m/s}^2} = 795 \text{ kg}\cdot\text{m / s} = 7.8 \text{ kW(10.5 hp)}$$

The impeller should be positioned approximately one-third of the slurry depth above the bottom of the tank.

NOMENCLATURE

C	Clearance of impeller from vessel bottom, m (ft)
C_D	Coefficient or drag
d	Pipe diameter, m (ft)
D	Impeller diameter, m (ft)
D_p	Particle diameter, m (ft)
D_v	Diffusion coefficient, m^2/s (ft^2/s)
F_D	Drag force, kg (lb)
G	Velocity gradient, m/s/m (ft/s/ft)
$\bar{G}$	Mean velocity gradient, m/s/m (ft/s/ft)
h_f	Head loss, m (ft)
I_s	Intensity of segregation
K_0	Reciprocal eddy size, 1/m (l/ft)
l	Impeller blade length, m (ft)
L	Length, m (ft)
L_s	Scale of segregation
M	Weight of slurry in vessel, kg (lb)
n	Number of impeller blades
N	Impeller speed, rev/s
N_c	Critical impeller speed for solids suspension, rev/s
N_{Fr}	Froude number
N_P	Power number
N_Q	Flow number
N_{Re}	Reynolds number
N_{Sc}	Schmidt number
p	Propeller pitch
P	Power, kg m/s (ft lb/s)
q_f	Inlet flow, m^3/s
q	Circulation flow within the vessel due to the impeller, m^3/s
Q	Impeller discharge, m^3/s
R	Weight ratio of solid to liquid
t	Time, s
t'	Time, min
T	Tank diameter, m (ft)
u	Velocity, m/s (ft/s)
u'	Root-mean-square velocity, m/sec (ft/s)
u	Average velocity, m/s (ft/s)
V	Volume of vessel, m (ft)
w	Blade width, m (ft)
W	Power dissipation per unit volume of fluid, kg-m/s/m^3(ft-lb/s/ft^3)
a	Liquid depth in vessel, m (ft)
α	Angle of resultant impeller discharge vector, degrees
β	Impeller blade angle, degrees
ϕ	Mixing time, min
μ	Absolute viscosity, kg/s-m (lb/s-ft)

μ' Dynamic viscosity, kg-s/m (lb-s/ft)
υ Kinematic viscosity, m/s (ft/s)
ρ Liquid density, $kg/m^3(lb/ft^3)$
ρ_p Particle density, $kg/m^3(lb/ft^3)$
τ Shear, kg/m (lb/ft)
ω Speed of revolution, rev/s

REFERENCES

1. P. V. Danckwerts, *Insights into Chemical Engineering (Selected Papers of P. V. Danckwerts)*, Pergamon Press, New York, 1981.
2. R. S. Brodkey, Fluid motion and mixing. In: *Mixing: Theory and Practice*. V. W. Uhl, and J. B. Gray, eds., Academic Press, New York 1996.
3. O. Levenspiel, *Chemical Reaction Engineering*, 2nd ed., John Wiley & Sons, New York, 1972.
4. G. R. Marr, Jr. and E. F. Johnson, The dynamical behavior of stirred tanks. *Chem. Eng. Prog. Symp.* **36**, 109–118 1961.
5. I. Zweitering, The degree of mixing in continuous flow systems. *Chem. Engrg. Sci.* **11**, 1–15 1959.
6. T. R. Camp, Flocculation and flocculation basins. *Transactions ASCE* **120**, 1–16 1955,
7. H. E. Hudson, and J. P. Wolfner, Design of mixing and flocculation basins, *JAWWA* **59**, 1257–1268 (1967).
8. J. Beak, Jr. and R. S. Miller, Turbulent transport in chemical reactors. *Chem. Engrg. Prog. Sym.* **25**, 23–28 (1959).
9. AIChE, *AIChE Equipment Testing Procedure, Mixing Equipment (Impeller Type), A Guide to Performance evaluation.* AIChE, New York, 1988.
10. P. L. Bates, P. L. Fondy, and J. G. Fenic, Impeller characteristics and power. In: *Mixing: Theory and Practice.* V. W. Uhl and J. B. Gray, eds., Academic Press, New York, 1996.
11. J. H. Rushton, E. W. Costich, and H. J. Everett, Power characteristics of mixing impellers. *Chem. Engrg. Prog.* **46**, 395–404 (1950).
12. G. B. Tatterson, *Scaleup and Design of Industrial Mixing Processes.* McGraw-Hill, New York, 1994.
13. J. J. Ulbrecht and G. K. Patterson, *Mixing of Liquids by Mechanical Agitation*, Gordon and Breach, New York, 1985.
14. F. P. O'Connell and D. E. Mack, Simple turbines in fully baffled tanks, power characteristics. *Chem. Engrg. Prog.* **46**, 358–362 (1950).
15. F. S. Hirsekorn and S. A. Miller, Agitation of viscous solid-liquid suspensions. *Chem. Engrg. Prog.* **49**, 459–467 (1953).
16. S. R. Qasim, E. M. Motley, and G. Zhu, *Water Works Engineering: Planning, Design and Operation*, Prentice-Hall, New Jersey, 2000.
17. R. D. Letterman (ed.), *Water Quality and Treatment, A Handbook of Community Water Supplies*, 5th Ed., McGraw-Hill, New York, 1999.
18. Metcalf and Eddy, Inc. (ed.), *Wastewater Engineering: Treatment Disposal and Reuse*, 4th ed., McGraw-Hill, New York, 2002.
19. J. B. Gray, Flow patterns, fluid velocities, and mixing in agitated vessel. In: *Mixing: Theory and Practice*, V. W. Uhl and J. B. Gray, eds., Academic Press, New York, 1966.
20. V. A. Mhaisalkar, R. Paramasivam, and A.G. Bhole, An innovative technique for determining velocity gradient in coagulation-flocculation process. *Wat. Res.* **20**, 1307–1314 (1986).
21. J. G. van de Vusse, Mixing by agitation of miscible liquids Part I. *Chem. Engrg. Sci.* **4**, 178–200 (1955).
22. K. W. Norwood, and A. B. Metzner, Flow patterns and mixing rates in agitated vessels. *AIChE* **6**, 432–442 (1960).

23. R. D. Biggs, Mixing rates in stirred tanks. *AIChE* **2**, 636–646 (1963).
24. E. A. Fox and V. E. Gex, Single-phase blending of liquids. *AIChE* 2, 539–544 (1956).
25. V. A. Mhaisalkar, R. Paramasivam, and A. G. Bhole, Optimizing physical parameters of rapid mix design for coagulation-flocculation of turbid waters. *Wat. Res.* **25**, 43–52 (1991).
26. S. A. Craik, D. W. Smith, M. Chandrakanth, and M. Belosevic, Effect of turbulent gas-liquid contact in a static mixer on *Cryptosporidium parvum* oocyst inactivation by ozone. *Water Res.* **37**, 3622–3631 (2003).
27. J. T. Lee, and J. Y. Choi, In-line mixer for feed forward control and adaptive feedback control of pH processes, *Chem. Engrg. Sci*, **55**, 1337–1345 (2000).
28. Th. N. Zwietering, Suspending of solid particles in liquid by agitators. *Chem. Engrg. Sci.* **8**, 244–253 (1958).
29. D. C. Hopkins and J. J. Ducoste, Characterizing flocculation under heterogeneous turbulence, *J. Colloid Interface Sci.* **264**, 184–194 (2003).
30. C. H. Kan, C. P. Huang, and J. R. Pan, Time requirement for rapid-mixing in coagulation, *Colloids Surfaces A: Physicochem. Engrg. Aspects* **203**, 1–9 (2002).
31. G. M. Fair, and J. C. Geyer, *Elements of Water Supply and Wastewater Disposal*, 5th ed., John Wiley & Sons, New York, 1958.
32. V. A. Mhaisalkar, R. Paramasivam, and A. G. Bhole, Optimizing physical parameters of rapid mix design for coagulation-flocculation of turbid waters. *Wat. Res.* **25**, 43–52 (1991).
33. J. M. Zalc, E. S. Szalai, M. M. Alvarez, and F. J. Muzzio, Using CFD to understand chaotic mixing in laminar stirred tanks. *AIChE* **48**, 2124–2134 (2002).
34. G. Baldi, R. Conti, and E. Alaria, Complete suspension of particles in mechanically agitated vessels. *Chem. Engrg. Sci.* **33**, 21–25 (1978).
35. R. K. Geisler, C. Buurman, and A. B. Mersmann, Scale-up of the necessary power input in stirred vessels with suspendsions. *Chem. Engrg. J.* **51**, 29–39 (1993).
36. K. P. Recknagle and A. Shekarriz, Laminar impeller mixing of Newtonian and non-Newtonian fluids: experimental and computational results. 1998 ASME Fluids Engineering Division Summer Meeting, Washington, DC, 1998.
37. S. J. Khang, and O. Levenspiel, New scale-up and design method for stirrer agitated batch mixing vessels. *Chem. Engrg. Sci.* **31**, 569–577 (1976).
38. A. W. Nienow, Suspension of solid particles in turbine agitated baffled vessels. *Chem. Engrg. Sci.* **23**, 1453–1459 (1968).
39. G. Rossi, The design of bioreactors. *Hydrometallurgy* **59**, 217–231 (2001).
40. R. N. Sharma, and A. A. Shaikh, Solids suspension in stirred tanks with pitched blade turbines. *Chem. Engrg. Sci.* **58**, 2123–2140 (2003).
41. J. M. T. Vasconcelos, S. S. Alves, and J. M. Barata, Mixing in gas-liquid contactors agitated by multiple turbines. *Chem. Engrg. Sci.* **50**, 2343–2354 (1995).
42. J. J. McKetta (ed.) *Unit Operations Handbook.* Marcel Dekker, New York, (1993).
43. K. J. Rogers, M. G. Milobowski, and B. L. Wooldridge, Perspectives on ammonia injection and gaseous static mixing in SCR retrofit application. *EPRI-DOE-EPA Combined Utility Air Pollutant Control Sympasium.* Atlanta, Georgia, 1999.
44. M. Fleischli, Make it short! *Sulzer Technical Review* **4**, 4101–4102 (2003).
45. N. Harnby, M. F. Edwards, and A. W. Nienow, eds., *Mixing in the Process Industries.* 2nd ed., Butterworth-Heinemann, Oxford, UK, 1992.
46. P. J. Fry, D. L. Pyle, and C. D. Rielly, eds., *Chemical Engineering for the Food Industry.* Blackie Academic & Professional, London, UK, 1997.
47. M. Zlokarnik, *Stirring, Theory and Practice.* Wiley-VCH, New York, 2001.
48. R. J. McDonough, *Mixing for the Process Industries*, Van Nostrand Reinhold, New York, 1992.

4
Coagulation and Flocculation

Nazih K. Shammas

CONTENTS

1. INTRODUCTION

Coagulation and flocculation constitute the backbone processes in most water and advanced wastewater treatment plants. Their objective is to enhance the separation of particulate species in downstream processes such as sedimentation and filtration. Colloidal particles and other finely divided matter are brought together and agglomerated to form larger size particles that can subsequently be removed in a more efficient fashion.

The traditional use of coagulation has been primarily for the removal of turbidity from potable water. However, more recently, coagulation has been shown to be an effective process for the removal of many other contaminants that can be adsorbed by colloids such as metals, toxic organic matter, viruses, and radionuclides (1,2). Enhanced coagulation is an effective method to prepare the water for the removal of certain contaminants in order to achieve compliance with the EPA (Environmental Protection Agency) newly proposed standards. These contaminants include arsenic (3,4), emerging

From: *Handbook of Environmental Engineering, Volume 3: Physicochemical Treatment Processes*
Edited by: L. K. Wang, Y.-T. Hung, and N. K. Shammas © The Humana Press Inc., Totowa, NJ

pathogens such as *Cryptosporidium* and *Giardia* (5), and humic materials (6–9). Humic substances are the precursors of THMs (trihalomethanes) and other DBPs (disinfection byproducts) formed by disinfection processes.

Amirtharaja and O'Melia (10) divided the coagulation process into three distinct and sequential steps:

1. Coagulant formation
2. Particle destabilization
3. Interparticle collisions

The first two steps are usually fast and take place in a rapid-mixing tank. The third step, interparticle collisions, is a slower process that is achieved by fluid flow and slow mixing. This is the process that causes the agglomeration of particles and it takes place in the flocculation tank.

Coagulation is usually achieved through the addition of inorganic coagulants such as aluminum- or iron-based salts, and/or synthetic organic polymers commonly known as polyelectrolytes. Coagulant aids are available to help in the destabilization and agglomeration of difficult and slow to settle particulate material.

2. APPLICATIONS OF COAGULATION

2.1. *Water Treatment*

1. Enhancing the effectiveness of subsequent treatment processes
2. Removal of turbidity
3. Control of taste and odor
4. Coagulation of materials causing color
5. Removal of bacteria and viruses
6. Removal of *Giardia* and *Cryptosporidium*
7. Coagulation of NOM (natural organic matter), humic materials which are the precursors of THMs and other DBPs
8. Removal of arsenic and radionuclides

2.2. *Municipal Wastewater Treatment*

1. Improving efficiency of primary treatment plants
2. Obtaining removals intermediate between primary and secondary treatments
3. Tertiary treatment of secondary effluents for water reuse
4. Handling of seasonal loads
5. Meeting seasonal requirements in receiving streams
6. Conditioning of biosolids before dewatering

2.3. *Industrial Waste Treatment*

1. Improving removals from secondary effluents
2. Removal of metals
3. Treatment of toxic wastes
4. Control of color
5. Handling seasonal wastes
6. Providing treatment to meet stream and disposal requirements at lower capital cost

2.4. *Combined Sewer Overflow*

1. Removal of particulate matter and BOD (biochemical oxygen demand)
2. Handling irregular occurrence of storm events

3. Preventing treatment upset by varying water quality
4. Meeting seasonal requirements in receiving streams

2.5. *Factors to Be Considered in Process Selection*

1. Flexibility of operation in response to variations in quality and quantity.
2. Low capital costs.
3. High operating costs.
4. Large volumes of sludge for disposal.
5. Skilled operation for optimum treatment.

3. PROPERTIES OF COLLOIDAL SYSTEMS

Colloids are very small particles that have extremely large surface area. Colloidal particles are larger than atoms and ions but are small enough that they are usually not visible to the naked eye. They range in size from 0.001 to 10 μm resulting in a very small ratio of mass to surface area. The consequence of this smallness in size and mass and largeness in surface area is that in colloidal suspensions (1):

1. Gravitational effects are negligible, and
2. Surface phenomena predominate.

Because of their tremendous surface, colloidal particles have the tendency to adsorb various ions from the surrounding medium that impart to the colloids an electrostatic charge relative to the bulk of surrounding water (11). The developed electrostatic repulsive forces prevent the colloids from coming together and, consequently, contribute to their dispersion and stability.

3.1. *Electrokinetic Properties*

The electrokinetic properties of colloids can be attributed to the following three processes (1,10):

1. Ionization of groups within the surface of particles.
2. Adsorption of ions from water surrounding the particles.
3. Ionic deficit or replacement within the structure of particles.

Organic substances and bacteria acquire their surface charges as a result of the ionization of the amino and carboxyl groups as shown below:

$$R{\pm}NH_3^+ \rightarrow R{\pm}NH_2 + H^+ \tag{1}$$

$$R{\pm}COOH \rightarrow R{\pm}COO^- + H^+ \tag{2}$$

The resulting charge on the surface of such particles is a function of the pH. At high pH values or low hydrogen ion concentrations, the above reactions shift to the right and the colloid is negatively charged. At a low pH, the reactions shift to the left, the carboxyl group is not ionized, and the particle is positively charged due to the ionized amino group. When the pH is at the isoelectric point, the particle is neutral, i.e., neither negatively nor positively charged. Proteinaceous material, containing various combinations of both amino and carboxyl groups, are usually negatively charged at pH values above 4 (10).

Oil droplets adsorb negative ions, preferably hydroxides (OH^-), from solution and, consequently, they develop a negative charge (11). Some other neutral particles adsorb selected ions from their surrounding medium such as calcium (Ca^{2+}) or phosphate (PO_4^{3-}) ions rendering them either positively or negatively charged, respectively.

Clays and other colloidal minerals may acquire a charge as a result of a deficit or imperfection in their internal structure. This is known as isomorphic replacement (10). Clays consist of a lattice formed of cross-linked layers of silica and alumina. In some clays there are fewer metallic atoms than nonmetallic ones within the mineral lattice producing a negative charge. In others, higher valency cations may be replaced by lower valency cations during the formation of the mineral lattice that renders the clay particles negatively charged. Examples of such imperfection include (a) the substitution of an aluminum ion (Al^{3+}) by either Mg^{2+} or Fe^{2+} and (b) the replacement of Si^{3+} cation by Al^{3+}. According to Amirtharaja and O'Melia (10), the type and strength of the charge resulting from this imperfection in the clay structure are independent of the surrounding water properties and pH. This is in contrast to the first two processes discussed above, in which both pH and ionic makeup of the surrounding solution play a big role in determining the sign and magnitude of the acquired charge on colloidal particles.

3.2. *Hydration*

Water molecules may also be sorbed on the surface of colloids, in addition to or in place of, other molecules or ions. The extent of this hydration depends on the affinity of particles for water. Colloidal particles that have water-soluble groups on their surface such as hydroxyl, carboxyl, amino, and sulfonic exhibit high affinity for hydration and cause a water film to surround the particles. Such colloids are classified as hydrophilic (water loving) particles. On the other hand, colloids that do not show affinity for water and do not have bound water films are classified as hydrophobic (water hating).

3.3. *Brownian Movement*

Colloids exhibit a continuous random movement caused by bombardment by the water molecules in the dispersion medium. This action, called Brownian movement, imparts kinetic energy to the particles that tends to cause an increase in the frequency of collisions, thus promoting coagulation. Elevated temperature increases molecular velocity resulting in more kinetic energy and more intense Brownian movement.

3.4. *Tyndall Effect*

Because colloidal particles have an index of refraction different from water, light passing through the dispersion medium and hitting the particles will be reflected. The turbid appearance due to this interference with the passage of light is termed the Tyndall effect. However, it should be noted that this might not always be the case. Water-loving, hydrophilic, colloids may produce just a diffuse Tyndall cone or none at all. The reason for this behavior can be attributed to the bound water layer surrounding colloids. These particles will have an index of refraction not very different from that of the surrounding water. Hence, the dispersed phase and the dispersion medium behave in a similar fashion toward the passage of light.

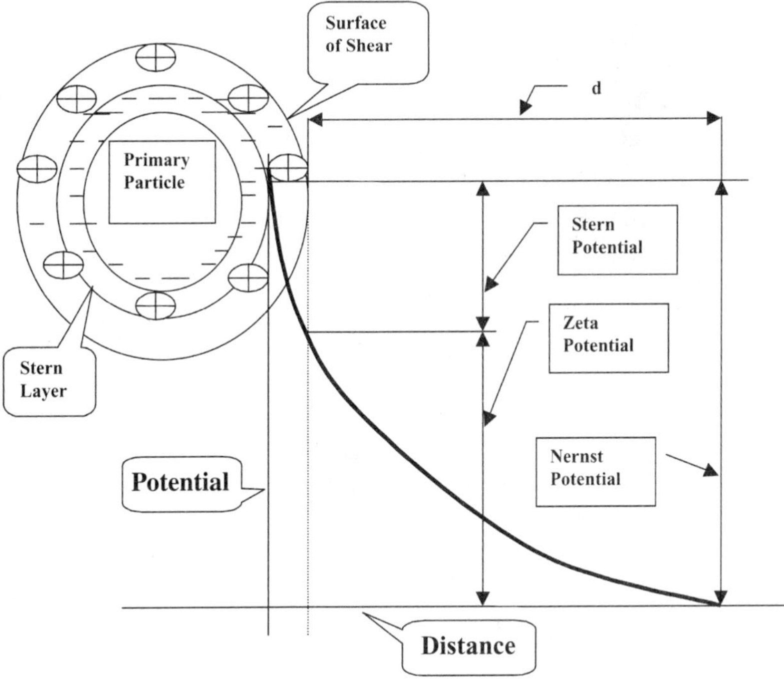

Fig. 1. The electrical potential of a negatively charged colloidal particle.

3.5. Filterability

Colloids are small enough to pass through ordinary filters, such as paper and sand, but are large relative to ions in size, diffuse very slowly, and will not pass through membranes. As a result, colloidal particles can be readily removed by ultrafiltration but require coagulation prior to their efficient removal by ordinary filtration.

4. COLLOIDAL STRUCTURE AND STABILITY

The stability of colloidal particulate matter is dependent on their electrokinetic property. Colloidal particles acquiring similar primary charges develop repulsive forces that keep them apart and prevent their agglomeration. The primary electrical charges could be either negative or positive. However, the majority of colloids that exist in aqueous systems are negatively charged. A colloidal system as a whole does not have a net charge. Negative primary charges on colloidal particles are balanced by positive counter-ions near the solid–liquid interface and in the adjoining dispersion medium. In a similar fashion, positively charged particles are counterbalanced by negative ions present in the surrounding water. This natural inclination toward achieving electrical neutrality and counterbalance of charges results in the formation of an electric double layer around colloidal particles.

The electric double layer, which comprises the charged particle and surrounding counter-ions, is illustrated in Fig. 1. The total potential at the surface of the primary charged particle is termed the Nernst potential. The dense layer of counter-ions fixed on the surface of the primary particle is called the Stern layer. The outer limit of this layer is defined by the surface of shear that separates the mobile portion of the colloid from the surrounding mixture of diffuse ions. In an electric field, the ions within the surface of shear will move with the particle as a unit. The concentrated counter-ions within the surface of shear reduce the net charge on the particle by an amount that is usually referred to as the Stern potential. Consequently, the potential is maximum at the surface of the primary particle, the Nernst potential, that decreases rapidly through the Stern layer resulting in a net overall charge on the particle at the surface of shear called the Zeta potential. This potential determines the extent of repulsion between similarly charged particles and is commonly considered to be the major cause of the stability of a colloidal system. Further away from the surface of shear both the concentration and potential gradients continue decreasing, but at a more gradual drop, until the potential approaches the point of electrical neutrality in the surrounding solution.

The counter-ions of the Stern layer are concentrated in the interfacial region owing to electrostatic attraction. However, these ions tend to be more loosely attached, as they are located at distances further away from the particle surface as a result of the potential gradient. Consequently, any thermal agitation may cause these less strongly held ions to diffuse away toward the bulk of the dispersion medium. These two opposite forces, electrostatic attraction and diffusion, give rise to the distribution of the potential over distance such that the highest concentration of counter-ions occurs at the particle interface and drops gradually with increasing distance. When the dispersion medium contains low concentration of ions (low ionic strength), the diffuse layer will be spread over a wide distance, *d*, as shown in Fig. 1. On the other hand, when the dispersion medium possesses a high ionic strength, the diffuse layer would get compressed, become thinner, and eventually extend far less distance into the bulk of the solution. Detailed analysis of the theory of the double layer and stability of colloids can be found in Verwey and Overbeek (12), Morel (13), O'Melia (14), and Elimelech and O'Melia (15).

When two similar primary charge particles drift toward each other, their diffuse layers start to interact leading to the production of a repulsive electrostatic force. The resulting repulsion between the approaching particles increases as the particles get closer. Such charged particles may not be able to collide at all if their charges are high enough. Ultimately, as illustrated in Fig. 2, the colloidal stability depends on the relative strength of the above electrostatic forces of repulsion and the forces of attraction.

The forces of attraction are due to van der Waals' forces. All colloidal particles, irrespective of their composition, sign or magnitude of charge, or the composition of the dispersion medium, possess such attractive forces. They arise from the following:

(a) Electronegativity of some atoms is higher than for others in the same molecule.
(b) Vibration of charges within one atom creates a rapidly fluctuating dipole.
(c) Approaching particles induce vibrations in phase with each other.

The above results in an attractive force between the two oppositely oriented dipoles. The magnitude of the force varies inversely with distance between particles, increasing rapidly with decreasing distance (see Fig. 2). If particles come close enough for these forces to take over, they will adhere.

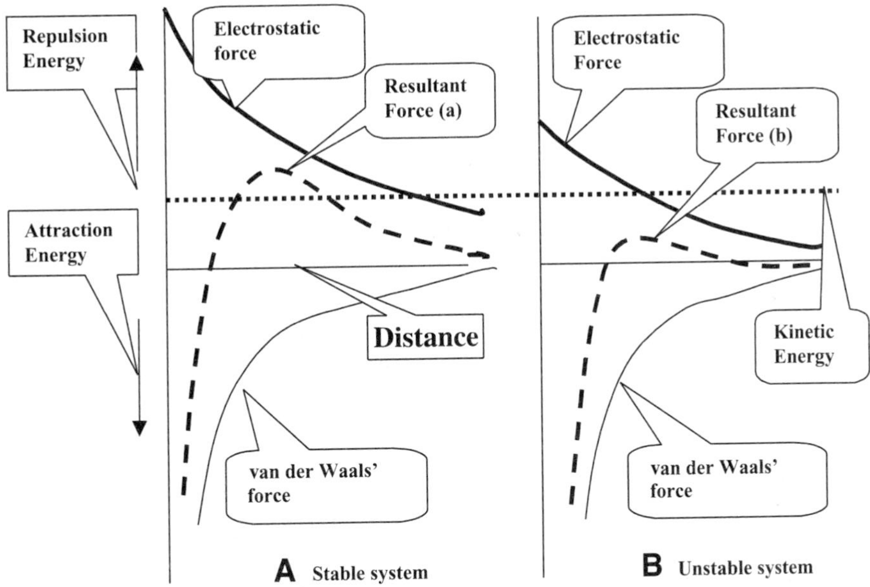

Fig. 2. Effect of interparticle forces on the stability of a colloidal system.

The other factor, in addition to van der Waals' attractive forces, tending to destabilize a colloidal system is Brownian movement. This is due to the random motion of colloids brought about by their bombardment by molecules of the dispersion medium. The outcome of the movement is to impart kinetic energy to the colloidal particles. Higher energy particles moving in a random fashion tend to collide eventually.

Figure 2 illustrates the relationship of forces that exist between colloidal particles as a function of the separation distance. The net resultant force is obtained by the summation of the respective electrostatic repulsive force and van de Waals' attractive force. When the resultant repulsion energy exceeds the kinetic energy (Fig. 2A), the particles will not coagulate and the dispersion is stable. When the kinetic energy is larger than the repulsion energy (Fig. 2B), the dispersion is unstable and the particles will coagulate. Consequently, if it is required to destabilize and coagulate a stable dispersion, then the electrostatic repulsion energy between the particles must be lowered and/or the kinetic energy of the particles must be raised.

5. DESTABILIZATION OF COLLOIDS

Destabilization of colloidal particles is accomplished by coagulation through the addition of hydrolyzing electrolytes such as metal salts and/or synthetic organic polymers. Upon being added to the water, the action of the metal salt is complex (10,16–18). It undergoes dissolution, the formation of complex highly charged hydrolyzed metal coagulants (hydroxyoxides of metals), interparticle bridging, and the enmeshment of particles into flocs. Polymers work either on the basis of particle destabilization or bridging between the particles.

The destabilization process is achieved by the following four mechanisms of coagulation:

1. Double-layer compression.
2. Adsorption and charge neutralization.
3. Entrapment of particles in precipitate.
4. Adsorption and bridging between particles.

5.1. Double-Layer Compression

When high concentrations of simple electrolytes are introduced into a stabilized colloidal dispersion, the added counter-ions penetrate into the diffuse double layer surrounding the particles rendering it denser and hence thinner and smaller in volume. The addition of counter-ions with higher charges, such as divalent and trivalent ions, will result in even steeper electrostatic potential gradients and more rapid decrease in charge with distance from the surface of the particles. The net repulsive energy (see Fig. 2) would become smaller or even would be completely eliminated, allowing the particles to approach each other and agglomerate.

A mathematical model that describes this coagulation mechanism is explained in detail in Verwey and Overbeek (12). The prediction of this model is in agreement with what is known as the Schultze–Hardly rule. This rule states that the coagulation of colloidal particles is achieved by ions of added electrolytes, which carry opposite charge to that of the colloids, and that the destabilization capability of the ions rises sharply with ion charge. Table 1 (1) illustrates the relative effectiveness of various electrolytes in the coagulation of negatively and positively charged colloids. For example, the relative power of Al^{3+}, Mg^{2+}, and Na^{+} for the coagulation of negative colloids is shown to vary in the ratio of 1000:30:1. A similar ratio is observed for the relative capability of PO_4^{3-}, SO_4^{2-}, and Cl^{-} for the coagulation of positively charged colloids.

5.2. Adsorption and Charge Neutralization

For all practical purposes, the ability of a chemical substance to destabilize and coagulate colloidal particles is the result of a combination of several mechanisms. Long-chained organic amines are often mentioned as being typical coagulants that function by adsorption and electrostatic neutralization (1,10,16). The positively charged organic amine molecules ($R–NH_3^{+}$) are easily and quickly attached to negatively charged colloidal particles. The charge on the particles gets neutralized and the electrostatic repulsion is decreased or eliminated resulting in the destabilization of the colloids and hence their agglomeration. The organic amines are hydrophobic because there is a lack of interaction between the CH_2 groups in their R–chain and the surrounding water. As a result, these positively charged ions are driven out of the water and get adsorbed on the particulate interface. An overdose of $R–NH_3^{+}$ counter-ions, however, can lead to charge reversal from negative to positive and the restabilization of the dispersion system.

When coagulants such as metal salts are added to water, they dissociate yielding metallic ions, which undergo hydrolysis and form positively charged metallic hydroxyoxide complexes (19). The commonly used coagulants, trivalent salts of aluminum and iron, produce numerous species because the hydrolysis products themselves tend to polymerize to give polynuclear metallic hydroxides (20,21). Examples of aluminum salt polymers are $Al_6(OH)_{15}^{3+}$ and $Al_7(OH)_{17}^{4+}$ and of iron salt polymers are $Fe_2(OH)_2^{4+}$ and $Fe_3(OH)_4^{5+}$. When such polyvalent complexes possessing high positive charges get adsorbed on to the surface of the negatively charged colloids, the result is again a

Table 1
Relative Coagulation Power of Electrolytes

	Relative power of coagulation	
Electrolyte	Positive colloid	Negative colloid
NaCl	1	1
Na_2SO_4	30	1
Na_3PO_4	1000	1
$BaCl_2$	1	30
$MgSO_4$	30	30
$AlCl_3$	1	1000
$Al_2(SO_4)_3$	30	>1000
$FeCl_3$	1	1000
$Fe_2(SO_4)$	30	>1000

neutralization of the charges, decrease in the repulsion energy, and destabilization of the colloids. In a similar fashion to what occurs with the organic amines, an overdose of metallic salts could reverse the colloidal charge and restabilize the particles.

5.3. *Entrapment of Particles in Precipitate*

When the coagulants alum [$Al_2(SO_4)_3$] or ferric chloride ($FeCl_3$) are added in high enough concentration, they will react with hydroxides (OH^-) to form metal hydroxide precipitates, $Al(OH)_3$ or $Fe(OH)_3$ respectively. The colloidal particles get entrapped in the precipitates either during the precipitate formation or just after. This type of coagulation by enmeshment of colloids in precipitates is commonly called sweep coagulation (10,22,23).

There are three elements that influence this coagulation mechanism (16):

1. Oversaturation: The rate of precipitation is a function of oversaturation with the metal hydroxide. To obtain fast precipitation and efficient sweep coagulation, high concentrations of $Al(OH)_3$ or $Fe(OH)_3$ are required.
2. Presence of anions: The rate of precipitation is improved by the presence of various anions in water. The most effective anions in this respect are the sulfate ions.
3. Concentration of colloids: The rate of precipitation is also improved with higher concentration of colloidal particles. The reason for this is that the colloids themselves could act as nuclei for the formation of precipitates. In this case, it can be concluded that lower rather than higher coagulant dosage will be required to coagulate water having higher colloidal particle concentration (22).

5.4. *Adsorption and Bridging between Particles*

Polymers destabilize colloidal particles through the formation of bridges that extend between them. The polymers have reactive groups that bind to specific sites on the surface of the colloidal particles. When a group on a polymer molecule attaches to a colloid, the remainder of the long-chain molecule extends away into the water. Once the extended portion of the polymer gets attached to another colloidal particle, the two particles become tied together or bridged by the polymer (24,25). If no other particle is available or if there is an overdose of polymer, the free extended portions of the polymer molecule would wrap around the same original particle, which could effectively bring about the restabilization of the colloid. Restabilization can also occur due to aggressive mixing or extended agitation, which may break the interparticle bridging and allow the folding back of the freed polymer portions around the same original particle (24).

6. INFLUENCING FACTORS

Many factors affect the coagulation process. In addition to mixing that will be explained in greater detail in separate sections, the following discussion covers the most important factors.

6.1. Colloid Concentration

Colloidal concentration has a large impact on both the required dosage and the efficiency of the coagulation process itself. The dosage of coagulants required for the destabilization of a colloidal dispersion is stoichiometrically related to the amount of colloidal particles present in solution (21). However, for dilute colloidal systems, the rate of coagulation is very slow because of the small number of colloids in suspension and, therefore, not enough contact between particles is available. Under such conditions increasing the concentration of particulate matter by the addition of a coagulant aid or recycling of settled sludge would improve the coagulation rate. Application of a large coagulant dosage to a dilute colloidal suspension would result in a greater chance of restabilizing the colloids.

6.2. Coagulant Dosage

The effect of aluminum and iron coagulant dosage on coagulation, as measured by the extent of removing particles causing turbidity in water, has been studied and evaluated in great detail by Stumm and O'Melia (21) and O'Melia (16). They divided the relationship into four zones starting with the first low-dosage zone and increasing the dosage progressively to the highest dosage that is applied in zone four:

Zone 1: Not enough coagulant is present for the destabilization of the colloids.
Zone 2: Sufficient coagulant has been added to allow destabilization to take place.
Zone 3: Excess concentration of coagulant can bring about charge reversal and restabilization of particles.
Zone 4:Oversaturation with metal hydroxide precipitate entraps the colloidal particles and produces very effective sweep coagulation.

The range of coagulant dosage that triggers the start, end, or elimination of any of the above zones is dependent on colloidal particle concentration and pH value.

6.3. Zeta Potential

The zeta potential represents the net charge of colloidal particles. Consequently, the higher the value of the zeta potential, the greater is the magnitude of the repulsive power between the particles and hence the more stable is the colloidal system. The magnitude of the zeta potential is determined from electrophoretic measurement of particle mobility in an electric field.

Table 2
Optimum pH Values for Metallic Coagulants

Coagulant	pH
Aluminum sulfate	4.0 to 7.0
Ferrous sulfate	8.5 and above
Ferric chloride	3.5 to 6.5 and above 8.5
Ferric sulfate	3.5 to 7.0 and above 9.0

6.4. Affinity of Colloids for Water

Hydrophilic (water-loving) colloids are very stable. Because of their hydration shell, chemicals cannot readily replace sorbed water molecules and, consequently, they are difficult to coagulate and remove from suspension. The stability of hydrophilic dispersions depends more on their "love" for water than on their electrostatic charge. It has been estimated that suspensions containing such particles require 10–20 times more coagulant than what is normally needed to destabilize hydrophobic particles (26). Typical examples are the color-producing material in surface water and organic colloids present in wastewater. On the other hand, examples of hydrophobic (water-hating) particles are metal oxides that can be easily coagulated and destabilized. However, the bulk of colloidal particles in turbid water usually exhibit a mixture of hydrophobic–hydrophilic properties resulting in suspensions that are intermediate in the degree of their difficulty to coagulate.

6.5. pH Value

pH is a measure of H^+ and OH^- ion concentration. The presence of these ions in the potential-determining layer may cause particle charge to be more positive or less negative at pH values below the isoelectric point. At high pH values above the isoelectric point the reverse effect takes place, whereby particle charge becomes more negative or less positive. The isoelectric point is the pH value at which charge is most nearly neutralized. The isoelectric point for aluminum hydroxide is around pH 8. It varies with the ionic strength in solution but is normally in the pH range of 7 to 9 (10).

The solubility of colloidal dispersions is affected radically by pH. $Al(OH)_3$ is amphoteric in nature and is soluble at low and high pH. The greatest adsorption occurs in the pH range where there is minimum solubility. Examples of optimum pH ranges for metallic salts are shown in Table 2 (27). Amirtharajah and Mills (28) reported that optimal coagulation with alum takes place at pH values near 5 and 7. At these points, the positively charged aluminum hydroxide neutralizes the negatively charged turbidity-producing colloidal particles, resulting in zero zeta potential. However, in the pH range from 5 to 7 the colloidal particles are restabilized due to charge reversal brought about by excess adsorption of the positively charged aluminum hydroxide species. pH also plays a part in affecting the amount of aluminum residual in the treated water (29).

The influence of pH on the polymer's behavior and effectiveness in coagulation is particularly important because of the interaction between pH and the charge on the electrolyte. The extent of charge change with pH is a function of the type of active group on the polymer (carboxyl, amino, etc.) and the chemistry of those groups.

6.6. Anions in Solution

As explained below, one of the constraints in using alum and iron as coagulants is the occurrence of charge reversal and restabilization of colloids. However, this behavior can be suppressed or eliminated in the presence of high concentrations of anions such as sulfate, silicate, and phosphate (30). It was found that background concentration of SO_4^{2-} in excess of 10 to 14 mg/L has the ability to prevent restabilization. Coagulation with alum is brought about by various species of positively charged aluminum hydroxyoxides. Aluminum hydroxide possesses its lowest charge and lowest solubility at its isoelectric point that lies in the pH range of 7 to 9 (10). As a result, when the alum dosage is increased within this pH range, sweep coagulation takes place due to the formation of the aluminum hydroxide precipitate. However, at lower pH values (5–7), higher dosages of alum will tend to increase the positively charged alum species that get adsorbed on particles' interface leading to charge reversal and the restabilization of the colloidal particles. Similar concepts and conclusions are applicable to iron coagulants.

6.7. Cations in Solution

The presence of divalent cations, such as Ca^{2+} and Mg^{2+}, in raw water is commonly considered not only to be helpful in the coagulation of negatively charged colloidal clay particles by anionic polymers but also to be necessary. Three reasons have been suggested to be behind this beneficial effect (31):

1. Compression of the colloidal double layer.
2. Reduction of the colloidal negative charge and minimization of repulsive potential.
3. Reduction in the range of repulsive barrier between adsorbed polymers.

6.8. Temperature

Coagulation by metallic salts is adversely affected by low temperature (29,32). However, the effect has been reported to be more pronounced in using alum, hence the recommendation to switch to iron salts when operating under low water temperatures (32,33). Another alternative option is to add bentonite as a coagulation aid. The addition of the negatively charged clay particles will enable the coagulation process to proceed as a result of charge neutralization rather than by sweep coagulation (10). The increase in rate and effectiveness of coagulation at higher temperatures can be attributed to the following:

1. Increase in velocity of molecules and hence in kinetic energy.
2. Increase in rate of chemical reactions.
3. Decrease in time of floc formation.
4. Decrease in viscosity of water.
5. Alteration in the structure of the flocs resulting in larger agglomeration.

7. COAGULANTS

Coagulants, i.e., chemicals that are added to the water to achieve coagulation, should have the following three properties (34):

1. Trivalent metallic cations or polymers whose effectiveness as coagulants has been determined.
2. Nontoxic and without adverse physiological effects on human health.

3. Insoluble or low solubility in the pH ranges common in water-treatment practice. This is necessary in order to have an efficient coagulation process and to be able to leave the lowest possible residual of the chemical in the treated water.

The most commonly used coagulants in water and wastewater treatment include aluminum sulfate (alum), ferric chloride, ferric sulfate, ferrous sulfate (copperas), sodium aluminate, polyaluminum chloride, and organic polymers.

7.1. *Aluminum Salts*

The chemistry of metallic salts is a complex one. It involves dissolution, hydrolysis, and polymerization reactions (16,27,35–37).

Dissolution. All metal cations in water are present in a hydrated form as aquocomplexes. The simple aluminum variety Al^{3+} does not exist as such in an aqueous solution. Rather, the aluminum species is present in the aquometal form as $Al(H_2O)_6{}^{3+}$:

$$Al_2(SO_4)_3 + 12H_2O \rightarrow 2Al(H_2O)_6{}^{3+} + 3SO_4{}^{2-} \quad (3)$$

Hydrolysis. The aquometal ions formed in the dissolution of alum in water are acidic or proton donors. This is demonstrated by the following hydrolytic reactions:

$$Al(H_2O)_6{}^{3+} + H_2O \rightarrow Al(H_2O)_5(OH)^{2+} + H_3O^+ \quad (4)$$

$$Al(H_2O)_5(OH)^{2+} + H_2O \rightarrow Al(H_2O)_4(OH)_2{}^{+} + H_3O^+ \quad (5)$$

$$Al(H_2O)_4(OH)_2{}^{1+} + H_2O \rightarrow Al(H_2O)_3(OH)_3 + H_3O^+ \quad (6)$$

$$Al(H_2O)_3(OH)_3 + H_2O \rightarrow Al(H_2O)_2(OH)_4{}^{1-} + H_3O^+ \quad (7)$$

Polymerization. The hydroxocomplexes formed as products of hydrolysis may combine to form a variety of hydroxometal polymers such as $Al_6(OH)_{15}{}^{3+}$, $Al_7(OH)_{17}{}^{4+}$, $Al_8(OH)_{20}{}^{4+}$, and $Al_{13}(OH)_{34}{}^{5+}$.

The net result of adding alum to an aqueous environment is the formation of large positively charged complexes that are insoluble and the generation of hydrogen ions. The actual exact variety of species present in water, following the addition of the coagulant, is determined by both pH and the extent of the applied dosage. As the dosage is increased such that it exceeds the solubility of alum in water, hydrolysis takes place. At further increase in dosage, a variety of hydroxocomplexes are formed, followed by the production of hydroxometal polymers, and finally the formation of the aluminum hydroxide precipitates.

Since the dissolution of alum in water increases the concentration of hydrogen ions, the net effect is a drop in pH or the consumption of present alkalinity:

$$Al_2(SO_4)_3 \cdot 14.3H_2O + 3Ca(HCO_3)_2 \rightarrow \underline{2Al(OH)_3} + 3CaSO_4 + 14.3H_2O + 6CO_2 \quad (8)$$

As shown in the above reaction, each additional mole of alum dosage consumes 6 moles of alkalinity (as $HCO_3{}^-$) and produces 6 moles of carbon dioxide. This means that each mg/L of alum will decrease water alkalinity by 0.50 mg/L (as $CaCO_3$) and will produce 0.44 mg/L of carbon dioxide. As long as adequate natural alkalinity (buffering capacity) is present in water, the tendency of alum to lower the pH does not create an operational problem.

When natural alkalinity is not sufficient to react with the alum dosage, lime or soda ash can be added to cover the deficit:

$$Al_2(SO_4)_3 \cdot 14.3H_2O + 3Ca(OH)_2 \rightarrow \underline{2Al(OH)_3} + 3CaSO_4 + 14.3H_2O \tag{9}$$

$$Al_2(SO_4)_3 \cdot 14.3H_2O + 3Na_2CO_3 + 3H_2O \rightarrow \underline{2Al(OH)_3} + 3Na_2SO_4 + 14.3H_2O + 3CO_2 \tag{10}$$

Lime is the most commonly used chemical because of its lower cost. However, soda ash has an advantage over lime in that it does not increase water hardness. The optimum pH for coagulation with alum is around 6 with an effective operational range between pH 5 and 8.

7.2. *Iron Salts*

Ferric salts (ferric chloride and ferric sulfate) when added to water behave in a similar fashion to alum. As illustrated in the following reactions for ferric sulfate, the dissolution, hydrolysis, and polymerization reactions are identical to that of alum:

$$Fe_2(SO_4)_3 + 12H_2O \rightarrow 2Fe(H_2O)_6^{\ 3+} + 3SO_4^{\ 2-} \qquad \text{Dissolution} \tag{11}$$

$$Fe(H_2O)_6^{\ 3+} + H_2O \rightarrow Fe(H_2O)_5(OH)^{2+} + H_3O^+ \qquad \text{Hydrolysis} \tag{12}$$

$$Fe_2(OH)_2^{\ 4+} \quad \text{(example of a ferric dimer)} \qquad \text{Polymerization}$$

Iron salts are also acidic in solution and consume alkalinity:

$$Fe_2(SO_4)_3 + 3Ca(HCO_3)_2 \rightarrow \underline{2Fe(OH)_3} + 3CaSO_4 + 6CO_2 \tag{13}$$

If natural alkalinity is not sufficient, lime or soda ash can be added:

$$Fe_2(SO_4)_3 + 3Ca(OH)_2 \rightarrow \underline{2Fe(OH)_3} + 3CaSO_4 \tag{14}$$

$$Fe_2(SO_4)_3 + 3Na_2CO_3 + 3H_2O \rightarrow \underline{2Fe(OH)_3} + 3Na_2SO_4 + 3CO_2 \tag{15}$$

Ferric coagulants may have some advantages when coagulating certain types of water (38). First, coagulation is effective over a wider pH range, usually from pH 4 to 9. However, best performance is between pH 3.5 and 6.5 and above 8.5. Second, a strong and heavy floc is produced, which can settle rapidly. Third, ferric salts are more effective for removing color, taste, and odor-producing matter.

7.3. *Sodium Aluminate*

The main difference between sodium aluminate and other common coagulants, alum, and iron salts, is its being alkaline rather than acidic in solution. It reacts with the natural carbon dioxide acidity and produces aluminum hydroxide floc:

$$2NaAlO_2 + CO_2 + 3H_2O \rightarrow \underline{2Al(OH)_3} + Na_2CO_3 \tag{16}$$

Sodium aluminate is most commonly used in combination with alum in the treatment of boiler water. This combination in water also produces aluminum hydroxide floc in a similar fashion to the previous reaction of sodium aluminate with carbon dioxide acidity:

$$6NaAlO_2 + Al_2(SO_4)_3 \cdot 14.3H_2O \rightarrow \underline{8Al(OH)_3} + 3Na_2SO_4 + 2.3H_2O \qquad (17)$$

Sodium aluminate can be produced by dissolving alumina in sodium hydroxide. The main deterrent to the wide scale use of this coagulant is its relatively high cost (38).

7.4. Polymeric Inorganic Salts

Polymeric ferric and aluminum salts are increasingly being used to coagulate turbid waters (33,37). Polyaluminum chloride (PACl) is used on a large scale in the treatment of potable water in Japan, France, and Germany. Although commercial preparations are available, they are more commonly prepared on site by the addition of a base to neutralize concentrated solutions of ferric and aluminum salts. The polymerization is affected by (10):

1. The concentration of the salt solution.
2. The type and concentration of the base solution.
3. Ionic strength.
4. Temperature.

7.5. Organic Polymers

Synthetic organic polymers are long-chain molecules composed of small subunits or monomeric units. Polymers that contain ionizable groups such as carboxyl, amino, or sulfonic groups are called polyelectrolytes. Polymers without ionizable groups are nonionic. On the other hand, polyelectrolytes may be cationic (contains positive groups), anionic (contains negative groups), or ampholytic (contains both positive and negative groups).

Polymers function as excellent coagulants due to their ability to destabilize particles by charge neutralization, interparticle bridging, or both. Anionic and nonionic polymers destabilize negatively charged colloidal particles through their bridging effect. Cationic polymers, on the other hand, are able to destabilize and coagulate such particles by both charge neutralization and interparticle bridging.

Factors that play a role in the effectiveness of polymers in accomplishing their function as coagulants include the following (10,16):

1. Polymer properties
   a. Functional groups on polymers. The type of groups is important for specific bonding to sites on the surface of colloidal particles.
   b. Charge density.
   c. Molecular weight and size. Large size is important for the effectiveness of anionic and nonionic polymers.
   d. Degree of branching.

2. Solution characteristics
   a. pH. It can affect the charge on both polymers and colloidal particles.
   b. Concentration of divalent cations (Ca^{2+}, Mg^{2+}). These are necessary to enable anionic polymers to effectively destabilize negatively charged colloids.

Accurate and precise control of dosage is very important for feeding of polymers in treatment plants. There is a narrow range for maximum performance. Concentrations lower than necessary will not produce effective coagulation, whereas over dosing of polymers will results in charge reversal and restabilization of the colloidal system. Also polymers are more expensive compared to metallic salts. However, this is usually more than compensated for by the lower polymer dosage as well as the reduced sludge production (38).

7.6. Coagulation Aids

Coagulation aids are sometimes used to achieve optimum conditions for coagulation and flocculation. The aim is to obtain faster floc formation, produce denser and stronger flocs, decrease the coagulant dosage, broaden the effective pH band, and improve the removal of turbidity and other impurities. Coagulant aids include four typical types:

1. **Alkalinity addition.** Alkalinity must be added to waters that may not have sufficient natural alkalinity to react with the acidic metallic coagulants to produce a good floc. Alkalinity is commonly supplemented in the form of the hydroxide ion by the addition of hydrated lime, $Ca(OH)_2$, or in the form of the carbonate ion by the addition of soda ash, Na_2CO_3.
2. **pH adjustment.** Acids and alkalis are used to adjust the pH of the water to fall within the optimal pH range for coagulation. pH reduction is usually accomplished by the addition of sulfuric and phosphoric acids. Increasing the pH is achieved by the addition of lime, sodium hydroxide, and soda ash.
3. **Particulate addition.** The addition of bentonite clays and activated silica (sodium silicate treated with sulfuric acid or alum) is very useful in coagulating low turbidity waters. These coagulant aids, when added in sufficient amounts, can increase the particulate concentrations to such an extent that more rapid coagulation will take place. In addition both activated silica and clays serve as weighing agents that produce denser and better settling floc.
4. **Polymers.** The use of organic polymers has recently replaced activated silica as a coagulation aid. Polymers can produce the same impact on coagulation, and they are applied at much lower concentrations and are easier to use. Anionic and nonionic polymers are used with ferric and aluminum salts to provide the interparticle bridging for effective coagulation. Polymers will tend to produce stronger and faster settling flocs and can reduce the metallic salt dosage that would have been required without polymers.

8. COAGULATION CONTROL

Theoretical analysis of coagulation is essential for understanding the process, for knowing how it works and what it can achieve as well as for discerning how to obtain the maximum performance out of it. There are four types of colloidal systems (16):

1. **Type I: High colloidal concentration, low alkalinity.** This is the least complicated system to treat. At low pH 4–6 levels metallic salts in water produce positively charged hydroxometal polymers. These in turn destabilize the negatively charged colloids by adsorption and charge neutralization. The high concentration of particulate material provides an ample opportunity for contact and building of good flocs. As a result, one has to determine only one variable—the optimum coagulant dosage.
2. **Type II: High colloidal concentration, high alkalinity.** Destabilization can also be accomplished, as in Type I, by adsorption and charge neutralization. However, in order to overcome the high alkalinity, there are two possible approaches. One alternative is to feed a high coagulant dosage that is sufficient to consume the excess alkalinity as well as to form

the positively charged hydroxometal polymers. The second alternative is to add an acid to lower the pH before feeding the coagulant. In this case one has to determine two variables—the optimum coagulant dosage and optimum pH.

3. **Type III: Low colloidal concentration, high alkalinity.** Because of the low chance of interparticle contacts due to the low colloidal concentration, the feasible approach in this case is to achieve sweep coagulation by feeding a high coagulant dosage that results in the entrapment of the colloidal particles in the metal hydroxide precipitate. A second alternative approach is to add a coagulant aid that will increase particle concentration and hence the rate of interparticle contact. A lower coagulant dosage will then be needed to achieve coagulation by charge neutralization.
4. **Type IV: Low colloidal concentration, low alkalinity.** This is the most difficult case to handle. The low colloidal concentration and depressed rate of interparticle contacts do not allow effective coagulation by adsorption and charge neutralization. On the other hand, the low alkalinity and low pH of the suspension do not enable rapid and effective destabilization by sweep coagulation. Coagulation in this system can be achieved by the addition of a coagulation aid (increase colloidal concentration), addition of lime or soda ash (increase alkalinity), or the addition of both but at lower concentrations.

However, because the process is so complex and the number of variables is so large, in most cases it is not feasible either to predict the best type of coagulant and optimum dosage or the best operating pH. The most practical approach is to simulate the process in a laboratory setting using the jar test. Other available alternatives and/or supplementary techniques include the zetameter (electrophoretic measurement) and the streaming current detector.

8.1. Jar Test

The jar test is the most valuable tool available for developing design criteria for new plants, for optimizing plant operations, and for the evaluation and control of the coagulation process. A jar test apparatus is a variable speed, multiple station or gang unit that varies in configuration depending on the manufacturer. The differences, such as the number of test stations (usually six), the size (commonly 1000 mL) and shape of test jars (round or square), method of mixing (paddles, magnetic bars, or plungers), stirrer controls, and integral illumination, do not have an appreciable impact on the performance of the unit.

The jar test can be run to select each of the following:

1. Type of coagulants.
2. Dosage of coagulants.
3. Coagulant aid and its dosage.
4. Optimum operating pH.
5. Sequence of chemical addition.
6. Optimum energy and mixing time for rapid mixing.
7. Optimum energy and mixing time for slow mixing.

The detailed procedure for the setting up, running, and interpreting a jar test is explained in various publications (39–42). Basically, for dosage optimization, samples of water/wastewater are poured into a series of jars, and various dosages of the coagulant are fed into the jars. The coagulants are rapidly mixed at a speed of 60–80 rpm for a period of 30–60 s then allowed to flocculate at a slow speed of 25–35 rpm for a period of 15–20 min. The suspension is finally left to settle for 20–45 min under quiescent conditions. The appearance and size of the floc, the time for floc formation, and the settling characteristics are noted. The supernatant is analyzed for turbidity, color, suspended solids, and pH. With this information in hand, the optimum chemical dosage is selected on the basis of best effluent quality and minimum coagulant cost.

8.2. Zetameter

The zeta potential measures the net charge of the colloidal particle, and it is dependent on the distance through which the charge is effective:

$$\zeta = 4\pi\delta q/\varepsilon \tag{18}$$

where ζ = zeta potential, q = charge at the shear surface, δ = thickness of diffuse layer, and ε = dielectric constant of the liquid.

Most naturally occurring colloidal particles are negatively charged. The more negative the charge, the higher the zeta potential, and the greater will be the repulsive force between the particles and hence the greater is the stability of the system. The reverse is also true. As the zeta potential approaches zero, the charges become so low that the repulsion becomes less effective and conditions become ideal for flocculation. The relationship between colloidal stability and zeta potential is shown in Table 3 (43).

The zeta potential of a given suspension can be determined by using the Helmholtz–Smoluchowski equation:

$$\zeta = K\mu v/\varepsilon E \tag{19}$$

where ζ = zeta potential (mV), K = constant, μ = viscosity, v = measured velocity of colloids (µm/s), ε = dielectric constant, and E = applied electric field gradient (V/cm).

Because the dielectric constant and viscosity are temperature dependent, the measurement should be made at the water operating temperature, otherwise a correction must be applied to compensate for the temperature difference. For example, at 25°C, the zeta potential can be found from the following relationship (1):

$$\zeta = 12.9\, v/E \tag{20}$$

The zeta meter, an instrument used for the determination of the zeta potential, is based on electrophoretic mobility measurement. An electric field is applied across an electrophoresis cell containing the colloidal suspension. The transfer of the negatively charged particles toward the anode is observed through a microscope. A prism situated between the eyepiece of the microscope and the cell is rotated until the colloidal particles appear to be stationary. At this point the prism rotation exactly cancels the transfer velocity of the particles. The unit is provided with an averaging computer equipped for digital read-out in millivolts of zeta potential.

Measurements of zeta potential can provide a good indication of a coagulant effectiveness in charge neutralization, and hence can help in the control and optimization of the coagulation process. However, the reader must be reminded that although this technique is helpful it cannot replace the jar test. The reason being that the zeta potential is unable to predict the enmeshment of particles that leads to sweep coagulation.

Table 3
Degree of Coagulation as a Function of Zeta Potential

Degree of coagulation	Zeta potential (mV)
Maximum	+3 to 0
Excellent	−1 to −4
Fair	−5 to −10
Poor	−11 to −20
Virtually none	−21 to −30

8.3. *Streaming Current Detector*

The streaming current detector (SCD) serves the same function as the zeta meter in that it measures the colloidal charge. This instrument consists of a piston that slides inside a cylinder. Two electrodes are attached to the ends of the cylinder to transmit the alternating (streaming) current that is generated from a colloidal suspension. When the piston is moved in a reciprocating motion up and down inside the cylinder, the water sample moves into and out of the annular space between the piston and cylinder. The alternating current generated at the electrodes is directly proportional to the charge on the colloidal particles.

Amirtharajah and O'Melia (10) reported the presence of a strong correlation between the charge measurements performed by the streaming current detector and the zeta meter. Either instrument can be used to evaluate the extent of charge neutralization. An advantage of the streaming current detector is its suitability for on line installation. This gives the treatment plant operator the feasibility to have automatic feedback control for adjusting the chemical dosage and hence, allows an effective and optimized coagulation process. However, it should be noted that one drawback of the SCD is that its measurements are a function of the water pH so that variations in pH require corresponding adjustments in the charge readings.

9. CHEMICAL FEEDING

Feeding of coagulants and coagulation aids includes chemical handling, storage, measurement, and transport of the required quantities to the mixing equipment. Chemicals can be purchased and fed either in a dry form or in solution. The physicochemical properties of the chemical will necessarily have an impact on the type of equipment needed for feeding it into the water treatment stream. Dry feeders have an edge over solution feeders because less equipment and labor are involved. The dry-feed system consists of a hopper, a measuring or proportioning system—volumetric or gravimetric—a dissolving basin, and chemical conveying lines to the proper point of application. For solution feeders, a solution of a predetermined concentration of the coagulant is prepared in dedicated storage tanks. A metering liquid feeder is used to deliver the required flow rate of the chemical solution to the point of application.

Chemical metering equipment must be able to maintain accurate feed rates that can easily be adjusted according to demand. Hence, most chemical feeders used in water/wastewater treatment plants are of the positive displacement type. There are several varieties of these on the market. Positive displacement pumps are used for solution feeders while screw, vibrating trough, rotary, and belt-type gravimetric feeders are used for dry-feed systems.

1. **Positive displacement pumps.** These include the plunger pump, the gear pump, and the diaphragm pump. All three varieties produce a constant chemical flow rate for a predetermined specific pump setting. However, the plunger pump is most widely used because of its accuracy and ease of adjusting the piston stroke. The rate of chemical output is easily calculated from the fixed volume of discharge per stroke and the number of strokes per minute.
2. **Screw feeder.** The screw feeder is located directly below the hopper. The unit maintains the desired chemical dosage by adjusting the speed and the duration of time the screw rotates as it delivers the chemical to the discharge point.
3. **Vibrating trough feeder.** The vibrating trough controls its rate of chemical delivery by the magnitude and the time interval of vibration.
4. **Rotary feeder.** The rotary feeder receives its chemical input from a hopper located above the rotating gear of the feeder. The gear has teeth that maintain a fixed amount of chemical between them. The chemical discharge rate is controlled by the speed of the rotor and its running duration.
5. **Belt-type gravimetric feeder.** A balance and a vibrating trough allow this feeder to maintain a constant weight of chemical on a moving belt. The rate of chemical feed is controlled by the amount of chemical on the belt and the speed and duration of belt travel.

When selecting a chemical feeder, the following factors should be taken into consideration:

a. Sufficient capacity of operating range for present and future expected feeding rates.
b. Accuracy of the unit in maintaining uniform feeding rates.
c. Repeatability of the unit when reverting to a previous setting.
d. Ease and difficulty of calibration, operation, and maintenance.
e. Resistance of the system to corrosion.
f. Provision for dust suppression.
g. Availability of reasonably priced spare parts.
h. Safety consideration in operation and maintenance.
i. Length of unit useful life.
j. The fixed initial cost and the yearly cost of operation and maintenance.

10. MIXING

Once the chemicals have accurately been measured and conveyed to the point of application, they should be thoroughly and rapidly dispersed in the water to be treated. Rapid mixing should then be followed by a slow mixing process in which the already destabilized particles and chemical precipitates are given a chance to come in contact and agglomerate into larger and heavier rapid-settling floc particles. The theory of mixing is quite complex and was extensively covered in Chapter 3. An excellent reference source on this subject can be found in the AWWARF book *Mixing in Coagulation and Flocculation* (43). A practical design approach to rapid-mix and flocculation is discussed below.

The degree of mixing is measured by the velocity gradient, G, which is a function of the power input into the water (40):

$$G = \sqrt{P/\mu V} \tag{21}$$

where G = velocity gradient (s^{-1}), P = power input [J/s or N·m/s (ft·lb/s)], V = volume of water, m^3 (ft^3), and μ = dynamic (absolute) viscosity of water [Pa·s, N·s/m^2, kg/m·s (lb.s/ft^2)].

The required power needed for mixing can be provided either by mechanical or through hydraulic means. The power dissipated by a paddle can be determined as follows:

$$P = F_D v_r \tag{22}$$

and

$$F_D = C_D A \rho v_r^2 / 2 \tag{23}$$

where F_D = drag force (N or kg·m/s^2), v_r = relative velocity of the paddle with respect to the water [m/s (ft/s)] = 0.50–0.75 of the velocity of the paddle, v, C_D = drag coefficient, A = area of the paddle in a plane perpendicular to the direction of motion [m^2 (ft^2)], and ρ = water density [kg/m^3 (lb·s^2/ft^4)].

Substitute the expression for F_D from Eq. (23) into Eq. (22):

$$P = C_D A \rho v_r^3 / 2 \tag{24}$$

Substitution of Eq. (24) into Eq. (21) yields Eq. (25), which defines the velocity gradient generated by paddle mixing:

$$G = \sqrt{C_D A \rho v_r^3 / 2\mu V} \tag{25}$$

The drag coefficient, C_D, is a function of the paddle dimensions and flow conditions (Reynolds number). The commonly used value for C_D is 1.8. For a Reynolds number, R_N, of 10^5, the drag coefficient can be calculated from the following expression (44):

$$C_D = 0.008\,R + 1.3 \tag{26}$$

where R_N = Reynolds number = $\rho v_r D/\mu = v_r D/\nu$, dimensionless number, $R = b/D$, b = length of blade [m (ft)], and D = width of blade [m (ft)].

In baffled tanks where interparticle contact is achieved by hydraulic mixing, the dissipated power is a function of the head loss in the tank:

$$P = Q\gamma h_f = Q\rho g h_f \tag{27}$$

where Q = water flow rate [m^3/s (ft^3/s)], h_f = head loss in the flocculator [m (ft)], γ = water specific weight [N/m^3 (lb/ft^3)], and g = acceleration of gravity [9.81 m/s^2 (32.2 ft/s^2)].

Substitution of Eq. (27) into Eq. (21) yields Eq. (28) that defines the velocity gradient generated by hydraulic mixing in a baffled tank:

$$G = \sqrt{Q\gamma h_f / \mu V} = \sqrt{Q\rho g h_f / \mu V} = \sqrt{\gamma h_f / \mu t} = \sqrt{g h_f / \nu t} \tag{28}$$

where $t = V/Q$ = detention time in the tank (s), and ν = Kinematic viscosity of water [m^2/s (ft^2/s)].

The hydraulic head loss in a baffled tank, h_f, is equal to the number of bends in the tank multiplied by the head loss at each bend. The head loss at each change in direction (bend) of either over-and-under or around-the-end flow pattern can be calculated as a function of the velocity head. Hence the total h_f in a baffled tank can be calculated from the following relationship:

$$h_f = nK\,v^2/2g \tag{29}$$

where n = number of bends in tank, K = constant = 2.0–4 depending on geometry of bend, and v = water velocity [m/s (ft/s)].

11. RAPID MIX

The purpose of rapid mixing is to achieve instantaneous, uniform dispersion of the chemicals through the water body. Instantaneous flash mixing is not only sufficient, but is desirable because (16):

1. The production of effective coagulant species greatly depends on being able to achieve a uniform dispersion of the added chemicals.
2. Rates for the formation of monohydroxocomplexes and other hydroxometal polymers are very rapid.
3. The adsorption rate for the various coagulant products is also very fast.

Extended mixing times are generally unwarranted, as these reactions will be completed in less than 1 s. However, when dealing with fragile colloidal particles such as in wastewater and biosolids treatment, it may be prudent to achieve the required dispersion through a less intense mixing over a longer time interval.

The efficiency of a rapid mix is based on the power imparted to the water, which is measured in terms of the velocity gradient G and the contact time t. Recommended detention times range from less than 1 s up to 1 min and G values from 700 to 4000 s^{-1}. Table 4 shows typical design values for G and t (10,11,45). In many instances the product of G and t, Gt, is used for the design of rapid-mix units. Recommended Gt values can range from 10,000 to 40,000 (43). Furthermore, some researchers (46) have recommended an empirical relationship that relates G to the coagulant concentration in addition to contact time:

$$GtC^{1.46} = 5.9 \times 10^6 \tag{30}$$

where C = coagulant concentration (mg/L).

Rapid chemical dispersion is generally obtained by some type of a stirring device, air injection or with hydraulic turbulence (10,11,43):

1. **Mechanical mixers.** Mechanical agitation is the most commonly used method in water treatment. It is flexible in accommodating operational variations, reliable, and very effective in achieving uniform dispersion of the added chemicals. A mechanical mixing device can be classified as a propeller, an impeller, or a turbine. Conventional rapid-mixing tanks are usually designed for a detention time of 10–60 s with velocity gradients of 700–1000 s^{-1}. Power requirement ranges between 0.9 and 1.2 hp/MGD.
2. **In-line blenders and pumps.** In-line mixers are becoming increasingly popular because of their low cost and compact installation as well as their flexibility when provided with variable speed drives. Typical velocity gradients generated in these devices are 3000–5000 s^{-1} with residence times of 0.5–1.0 s.
3. **Hydraulic jumps.** A hydraulic jump can easily be formed downstream of the Parshall flume (a device for flow rate measurement) at the inlet to the water treatment plant. This can be done inexpensively by a sudden drop in the bottom of the flume effluent channel.

Table 4
G and t Values for Rapid Mixing

Contact time t (s)	Velocity gradient, G (s^{-1})
0.5–1.0 (in-line blenders)	4,000
10–20	1,000
21–30	900
31–40	800
41–60	700

Although the operational flexibility is rather limited, it has the advantage of not using any mechanical equipment and being economical. Typical velocity gradient, G, and contact time values are 800 s^{-1} and 2 s, respectively.

4. **Pneumatic mixers.** Pneumatic systems employ various aeration devices such as diffusers, air jets, or other injection devices to achieve the required turbulence. The velocity gradient can be readily controlled by adjusting the airflow rate.
5. **Static mixers.** Static in-line mixers are installed in a pipe immediately after the chemical feeding point. These devices are compact with absence of moving parts. They produce the required turbulence and velocity gradient as a result of a hydraulic head loss developed in the fixed geometric design of vanes within the pipe. This significant head loss requires an extra power input of 0.5–1.0 hp/MGD.

12. FLOCCULATION

The agglomeration of particles is a function of their rate of collisions. The function of flocculation is to optimize the rate of contact between the destabilized particles, hence increasing their rate of collision and bringing about the attachment and aggregation of the particles into larger and denser floc. Thus, the flocculation process allows the colloidal particles to come together and build into larger flocs that are more amenable to separation by settling, or filtration.

Slow mixing can be achieved mechanically or hydraulically. Mechanical flocculation devices may be paddle wheel flocculators, flat blade turbines, and axial flow propellers. Shafts that carry the mixers can be placed in either a horizontal or vertical position. The most common paddle flocculator consists of a shaft with steel arms that support wooden, plastic, or steel blades. The paddle shafts can be positioned perpendicular or parallel to the direction of flow. The paddles rotate slowly at 1–4 rpm.

Mixing by hydraulic means is most commonly used in horizontal flow flocculators. Vertical baffles are used to change the flow direction, hence dissipating sufficient amount of energy (head loss) to create the required velocity gradients for mixing. There are two flow patterns to choose from: over-and-under and around-the-end configurations. This type of flocculator with channeled flow created by the baffles behaves very close to an ideal plug flow reactor with minimal short-circuiting. Design horizontal velocities for these tanks range between 0.8 to 1.7 ft/s (0.25 to 0.50 m/s). An advantage of using hydraulically mixed flocculators is the absence of mechanical equipment, which explains their popularity in developing countries.

Optimal mixing must be provided in order to bring particles into contact and keep them from settling in the flocculation tank. Below a minimum time and velocity gradient,

Table 5
Typical G and GT Values for Flocculation

Type	G (s^{-1})	Gt
Low turbidity, color removal coagulation	20–70	60,000–200,000
High turbidity, solids removal coagulation	50–150	90,000–180,000
Softening, 10% solids	130–200	200,000–250,000
Softening, 30% solids		300,000–400,000

no proper flocculation occurs, and increasing t and G beyond floc formation will shear the floc particles resulting in their breakup into smaller flocs. Recommended design range for G is 20–70 s^{-1} for a contact period of 20–30 min. Because proper floc formation is a function of G and t, the parameter usually used to define effective flocculation is the product Gt. Common Gt values are in the range of 2×10^4 to 2×10^5. Table 5 shows typical recommended values for G and Gt (45). Furthermore, Amirtharajah and O'Melia (10) have reported an empirical relationship that relates G to the coagulant concentration C in addition to contact time:

$$G^{2.8}\, Ct = 4.4 \times 10^6 \tag{31}$$

Flocculation basins are usually designed with multiple compartments to minimize short-circuiting and to facilitate the incorporation of zones of reduced energy input and tapered velocity gradients. The tapered feature may be provided by varying the rotational speed (variable-speed drives), the paddle size, the number of paddles, and the diameter of the paddle wheels. A typical example of such design is the provision of three flocculation compartments having G values of 60, 40, and 20 s^{-1} respectively. Typical design data for flocculation tanks (38,47) and properties of water (38) are given in Tables 6–8.

Table 6
Flocculation Tank Design Data

Parameter	Typical value
Velocity gradient, G	20–80 s^{-1}
Detention time, t	20–30 min
Gt value	20,000–150,000
Configuration	Rectangular
Length to width ratio	4:1
Maximum stage volume	12,500 ft^3 (304 m^3)
Depth	12 ft (3.6 m)
Horizontal mixing	V between 1,860 and 10^6 ft^3 (53 and 28,000 m^3)
Vertical mixing	V between 18,000 and 25,000 ft^3 (509 and 707 m^3)
Variable-speed motors	60% efficient
Freeboard and mixing apparatus	Require 20% of tank volume
Vertical mixers	Three-blade propeller impeller with R_N max of 10^4
Horizontal paddles mixers	Eight (four arms with two paddles)
Total paddle-blade area	Less than 20% of tank cross-sectional area
Paddle tip velocity	Less than 2 ft/s (0.61 m/s) for weak floc, and 4 ft/s (1.22 m/s) for strong floc

Table 7
Properties of Water in Metric Units

Temperature (°C)	Specific weight γ (kN/m^3)	Mass density ρ (kg/m^3)	Absolute viscosity μ ($\times 10^{-3}$ kg/m·s)	Kinematic viscosity ν ($\times 10^{-6}$ m^2/s)
0	9.805	999.8	1.781	1.785
5	9.807	1000.0	1.518	1.518
10	9.804	999.7	1.307	1.306
15	9.798	999.1	1.139	1.139
20	9.789	998.2	1.002	1.003
25	9.777	997.0	0.890	0.893
30	9.764	995.7	0.798	0.800
40	9.730	992.2	0.653	0.658
50	9.689	988.0	0.547	0.553
60	9.642	983.2	0.466	0.474
70	9.589	977.8	0.404	0.413

13. DESIGN EXAMPLES

13.1. *Example 1 (Metric Units)*

A rapid-mix tank is designed to treat 100,000 m^3/d of turbid water. If the detention time is 30 s, and the water operating temperature is 15°C, find

1. The required volume of tank.
2. The required mixing power.

Solution

1. The volume of tank is

$$V = Qt = 100{,}000(30)/(60 \times 60 \times 24) = \underline{34.7\ \text{m}^3}$$

Table 8
Properties of Water in English Units

Temperature (°F)	Specific weight γ (lb/ft^3)	Mass density ρ (lb·s^2/ft^4)	Absolute viscosity μ ($\times10^{-5}$ lb·s/ft^2)	Kinematic viscosity ν ($\times10^{-5}$ ft^2/s)
32	62.42	1.940	3.746	1.931
40	62.43	1.938	3.229	1.664
50	62.41	1.936	2.735	1.410
60	62.37	1.934	2.359	1.217
70	62.30	1.931	2.050	1.059
80	62.22	1.927	1.799	0.930
90	62.11	1.923	1.595	0.826
100	62.00	1.918	1.424	0.739
110	61.86	1.913	1.284	0.667
120	61.71	1.908	1.168	0.609
130	61.55	1.902	1.069	0.558
140	61.38	1.896	0.981	0.514
150	61.20	1.890	0.905	0.476

2. The power, P, is given by

$$G = \sqrt{P/\mu V} \quad \text{or} \quad P = \mu V G^2$$

From Table 7, the value of μ at 15°C is 1.139×10^{-3} kg/m·s. From Table 4, the value of G for $t = 30$ s is 900 s^{-1}

$$P = \mu V G^2 = (1.139 \times 10^{-3})(34.7)(900)^2 = 32{,}014\ N \cdot m/s = 32{,}014 \text{ W} = \underline{32 \text{ kW}}$$

Since 1 hp = 745.7 W:

$$P = 32{,}014/745.7 = \underline{43 \text{ hp}}$$

13.2. *Example 2 (English Units)*

A rapid-mix tank is designed to treat 26 MGD of turbid water. If the detention time is 30 s and the water operating temperature is 60°F, find

1. The required volume of tank.
2. The required mixing power.

Solution

1. The volume of tank is

$$V = Qt = 26 \times 10^6 (30)/(60 \times 60 \times 24) = 9028 \text{ gallons}$$

$$V = 9028/7.48 = \underline{1207 \text{ ft}^3}$$

2. The power, P, is given by

$$G = \sqrt{P/\mu V} \quad \text{or} \quad P = \mu V G^2$$

From Table 8, the value of μ at 60°F is 2.359×10^{-5} lb·s/ft². From Table 4, the value of G for $t = 30$ s is 900 s^{-1}:

$$P = \mu V G^2 = (2.359 \times 10^{-5})(1207)(900)^2 = 23{,}063 \text{ ft} \cdot \text{lb/s}$$

Since 1 hp = 550 ft·lb/s

$$P = 23063/550 = \underline{42 \text{ hp}}$$

13.3. *Example 3 (Metric Units)*

A flocculation basin is 24 m long, 4.5 m wide, and has a working water depth of 3.6 m. The net input power for slow mixing is 0.50 kW. Is this basin adequate for treating 25,000 m³/d of water for turbidity removal? Assume a water temperature of 10°C.

Solution

Check for the design parameters G, t, and Gt. From Table 6 for flocculation tank design, the recommended ranges for G, t, and Gt are

G	20–80 s^{-1}
t	20–30 min
Gt	20,000–150,000

The velocity gradient is given by

$$G = \sqrt{P/\mu V}$$

From Table 7, the value of μ at 10°C is 1.307×10^{-3} kg/m·s.

Volume of basin,

$$V = 24 \times 4.5 \times 3.6 = 389 \text{ m}^3$$

$$G = \sqrt{0.50 \times 1000/(1.307 \times 10^{-3})(389)} = \underline{31 \text{ s}^{-1} \text{ satisfactory}}$$

The retention time is given by

$$t = V/Q = 389 \times 24 \times 60/25{,}000 = \underline{23 \text{ min satisfactory}}$$

The Gt value is

$$Gt = 31 \times 23 \times 60 = \underline{42{,}8000 \text{ satisfactory}}$$

Since G, t and Gt values are within the recommended limits, the flocculation basin is considered to be adequate for treating a daily flow of 25,000 m^3 of water.

13.4. *Example 4 (English Units)*

A flocculation basin is 80 ft long, 15 ft wide, and has a working water depth of 12 ft. The net input power for slow mixing is 0.65 hp. Is this basin adequate for treating 6.5 MGD of water for turbidity removal? Assume a water temperature of 50°F.

Solution

Check for the design parameters G, t, and Gt. From Table 6 for flocculation tank design, the recommended ranges for G, t, and Gt are

G	20–80 s^{-1}
t	20–30 min
Gt	20,000–150,000

The velocity gradient is given by

$$G = \sqrt{P/\mu V}$$

From Table 8, the value of μ at 50°F is 2.735×10^{-5} lb·s/ft^2.

Volume of basin,

$$V = 80 \times 15 \times 12 = 14{,}400 \text{ ft}^3$$

$$G = \sqrt{0.65 \times 550/(2.735 \times 10^{-5})(14{,}400)} = \underline{30 \text{ s}^{-1} \text{ satisfactory}}$$

The retention time is given by

$$t = V/Q = 14{,}400 \times 7.48 \times 24 \times 60/6.5 \times 10^6 = \underline{24 \text{ min satisfactory}}$$

The Gt value is

$$Gt = 30 \times 24 \times 60 = \underline{43{,}200 \text{ satisfactory}}$$

Because *G*, *t*, and *Gt* values are within the recommended limits, the flocculation basin is considered to be adequate for treating a daily flow of 6.5 MG of water.

13.5. Example 5 (Metric Units)

A water treatment plant has a flow of 30,000 m^3/d. The detention times in rapid-mix and flocculation tanks are 40 s and 30 min, respectively. The water temperature is 20ºC. Assume that the optimal velocity gradient for rapid mixing is 800 s^{-1} and for slow mixing is 40 s^{-1}. Determine:

1. Size of rapid-mix tank.
2. Power requirement for rapid mixing.
3. Size of flocculation tank.
4. Power requirement for slow mixing.

Solution

1. Size of rapid-mix tank:

$$V = Qt = (30,000/24 \times 60 \times 60)(40) = \underline{13.9\ \text{m}^3}$$

2. Power requirement for rapid mixing: From Table 7, the value of μ at 20ºC is 1.002 × 10^{-3} kg/m·s. The power, *P*, is given by

$$G = \sqrt{P/\mu V}$$

or

$$P = \mu VG^2 = (1.002 \times 10^{-3})(13.9)(800)^2 = 8,914\ \text{W} = \underline{8.9\ \text{kW}} = 8.9/0.746 = \underline{12\ \text{hp}}$$

3. Size of flocculation tank:

$$V = Qt = (30,000/24 \times 60)(30) = \underline{625\ \text{m}^3}$$

4. Power requirement for slow mixing: The power, *P*, is given by

$$G = \sqrt{P/\mu V}$$

or

$$P = \mu VG^2 = (1.002 \times 10^{-3})(625)(40)^2 = 1,002\ \text{W} = \underline{1.0\ \text{kW}} = \underline{1.34\ \text{hp}}$$

13.6. Example 6 (English Units)

A water treatment plant has a flow of 8.0 MGD. The detention times in rapid-mix and flocculation tanks are 40 s and 30 min, respectively. The water temperature is 70ºF. Assume that the optimal velocity gradient for rapid mixing is 800 s^{-1} and for slow mixing is 40 s^{-1}. Determine:

1. Size of rapid-mix tank.
2. Power requirement for rapid mixing.
3. Size of flocculation tank.
4. Power requirement for slow mixing.

Solution

1. Size of rapid-mix tank:

$$V = Qt = (8.0 \times 10^6/24 \times 60 \times 60)(40) = 3700\ \text{gallons} = 3700/7.48 = \underline{495\ \text{ft}^3}$$

2. Power requirement for rapid mixing: From Table 8, the value of μ at 70°F is 2.050×10^{-5} lb·s/ft². The power, P, is given by

$$G = \sqrt{P/\mu V}$$

or

$$P = \mu V G^2 = (2.050 \times 10^{-5})(495)(800)^2 = 6494 \text{ ft} \cdot \text{lb/s} = 6494/550 = \underline{11.8 \text{ hp}}$$

3. Size of flocculation tank:

$$V = Qt = (8.0 \times 10^6/24 \times 60)(30) = 167{,}000 \text{ gallon} = 167{,}000/7.48 = \underline{22{,}300 \text{ ft}^3}$$

4. Power requirement for slow mixing: The power, P, is given by

$$G = \sqrt{P/\mu V}$$

or

$$P = \mu V G^2 = (2.050 \times 10^{-5})(22{,}300)(40)^2 = 731 \text{ ft} \cdot \text{lb/s} = 731/550 = \underline{1.33 \text{ hp}}$$

13.7. *Example 7 (Metric Units)*

A flocculation basin designed to treat 50,000 m³/d of water is 21 m long, 15 m wide, and 3.60 m deep. The paddle-wheel units consist of four horizontal shafts that rotate at 4 rpm. The shafts are located perpendicular to the direction of flow at mid-depth of the basin. Each shaft is equipped with four paddle wheels 3 m in diameter and each wheel has four blades 3.30 m long and 150 mm wide with two blades located on each side of the wheel. The blades are 300 mm apart. Assume the water velocity to be 30% of the velocity of the paddles and that the water temperature is 10°C. Determine:

1. The power input to the water.
2. The velocity gradient.
3. The retention time.
4. The Gt value.

Solution

1. The power, P, is given by Eq. (24):

$$P = C_D A \rho v_r^{\,3}/2$$

where from Eq. (26),

$$C_D = 0.008R + 1.3 = 0.008(3.30/0.15) + 1.3 = 1.5$$

From Table 7 at 10°C, ρ = 999.7 kg/m³:

$$\begin{aligned} A &= \text{area of outer blades} = \text{area of inner blades} \\ &= (0.15 \times 3.30) \times 2 \text{ blades} \times 4 \text{ shafts} \times 4 \text{ paddles} \\ &= 15.8 \text{ m}^2 \end{aligned}$$

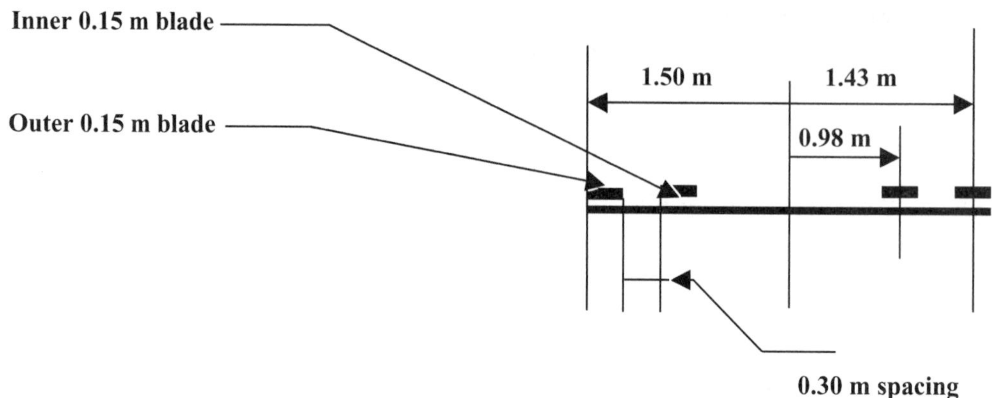

Fig. 3. Location of blades for Example 7.

$$v_r = 0.70\ (2\pi rn)$$
$$v_{r1} = 0.70(2 \times 3.14 \times 1.43 \times 4/60) = 0.42 \text{ m/s}$$
$$v_{r2} = 0.70(2 \times 3.14 \times 0.98 \times 4/60) = 0.29 \text{ m/s}$$

Hence

$$\begin{aligned} P &= C_D A \rho v_r^3/2 \\ &= (1.5)(15.8)(999.7)\left(v^3{}_{r1} + v^3{}_{r2}\right)/2 \\ &= 11{,}850\left(0.42^3 + 0.29^3\right) = 1200 \text{ W} = \underline{1.20 \text{ kW}} \\ &= 1.20/0.746 = \underline{1.60 \text{ hp}} \end{aligned}$$

2. The velocity gradient, G, is given by Eq. (21):

$$G = \sqrt{P/\mu V}$$

From Table 7, the value of μ at 10°C is 1.307×10^{-3} kg/m·s

The volume of basin,

$$V = 21 \times 15 \times 3.6 = 1{,}130 \text{ m}^3$$
$$\text{G} = \sqrt{1200/1.307 \times 10^{-3} \times 1{,}130} = \underline{28 \text{ s}^{-1}}$$

3. Retention time is given by

$$V/Q = 1{,}130 \times 24 \times 60/50{,}000 = \underline{33 \text{ min}}$$

4. The Gt value is given by

$$\begin{aligned} Gt &= (G)(\text{Retention time in seconds}) \\ &= (28)(33 \times 60) = \underline{55{,}000} \end{aligned}$$

13.8. *Example 8 (English Units)*

A flocculation basin designed to treat 13 MGD of water is 70 ft long, 50 ft wide, and 12 ft deep. The paddle-wheel units consist of four horizontal shafts that rotate at 4 rpm. The shafts are located perpendicular to the direction of flow at mid-depth of the basin. Each shaft is equipped with four paddle wheels 10 ft in diameter and each wheel has four blades 11 ft long and 6 inches wide with two blades located on each side of the wheel. The blades are 1 ft apart. Assume the water velocity to be 30% of the velocity of the paddles and that the water temperature is 50°F. Determine:

1. The power input to the water.
2. The velocity gradient.
3. The retention time.
4. The *Gt* value.

Solution

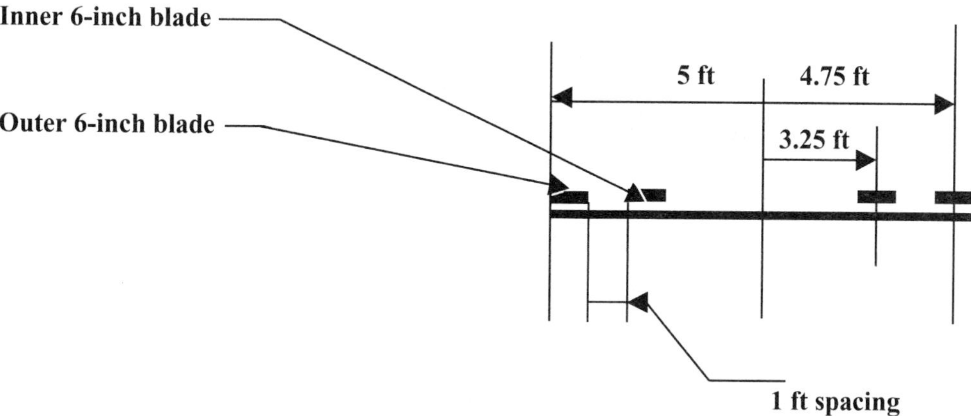

Fig. 4. Location of blades for Example 8.

1. The power, *P*, is given by Eq. (24):

$$P = C_D A \rho v_r^{\,3}/2$$

where from Eq. (26),

$$C_D = 0.008R + 1.3 = 0.008(11/0.5) + 1.3 = 1.5$$

From Table 8 at 50°F, $\rho = 1.936$ lb·s^2/ft^4:

$$\begin{aligned} A &= \text{area of outer blades} = \text{area of inner blades} \\ &= (0.5 \times 11) \times 2 \text{ blades} \times 4 \text{ shafts} \times 4 \text{ paddles} \\ &= 176 \text{ ft}^2 \end{aligned}$$

$$v_r = 0.70(2\pi rn)$$
$$v_{r1} = 0.70(2 \times 3.14 \times 4.75 \times 4/60) = 1.39 \text{ ft/s}$$
$$v_{r2} = 0.70(2 \times 3.14 \times 3.25 \times 4/60) = 0.95 \text{ ft/s}$$

Hence

$$\begin{aligned} P &= C_D A \rho v_r^{\,3}/2 \\ &= (1.5)(176)(1.936)\left(v^3_{r1} + v^3_{r2}\right)/2 \end{aligned}$$

$$= 255.6\left(1.39^3 + 0.95^3\right) = \underline{905\ \text{ft}\cdot\text{lb/s}}$$

$$= 905/550 = \underline{1.65\ \text{hp}}$$

2. The velocity gradient, G, is given by Eq. (21):

$$G = \sqrt{P/\mu V}$$

From Table 8, the value of μ at 50°F is 2.735×10^{-5} lb·s/ft^2.

The volume of basin,

$$V = 70 \times 50 \times 12 = 42{,}000\ \text{ft}^3$$

$$G = \sqrt{905/2.735 \times 10^{-5} \times 42{,}000} = \underline{28\ \text{s}^{-1}}$$

3. Retention time is given by

$$V/Q = 42{,}000 \times 7.48 \times 24 \times 60/13 \times 10^6 = \underline{35\ \text{min}}$$

4. The Gt value is given by

$$Gt = (G)(\text{Retention time in seconds})$$
$$= (28)(35 \times 60) = \underline{59{,}000}$$

13.9. Example 9 (Metric Units)

A hydraulically mixed flocculation basin is to be designed for a water-treatment plant that has a capacity of 100,000 m^3/d. The flocculator is to be of an around-the-bend baffled basin. Assume a water temperature of 10°C, a detention time of 30 min, and a velocity gradient of 40 s^{-1}. Determine

1. The required head loss in the channeled basin.
2. The required number of channels.
3. The basin dimensions.

Solution

1. The head loss, h_f, can be determined from Eq. (28):

$$G = \sqrt{\gamma h_f/\mu t}$$

$$G^2 = \gamma h_f/\mu t$$

From Table 7, the value of μ at 10°C is 1.307×10^{-3} kg/m·s. From Table 7, the value of γ at 10°C is 9.804 kN/m^3:

$$h_f = G^2\mu t/\gamma = (40)^2\left(1.307 \times 10^{-5}\right)(30 \times 60)/9.804 \times 1000 = \underline{0.38\ \text{m}}$$

2. The required number of channels. From Eq. (29), the head loss is given by

$$h_f = nK\,v^2/2g$$

Assuming a K value of 2 (usual values 2.0 to 4 depending on geometry of bend)

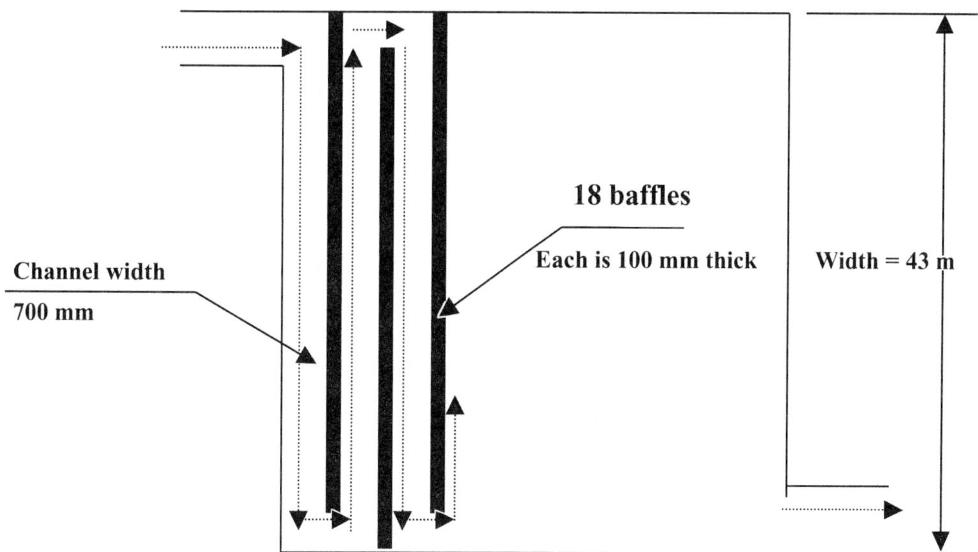

Fig. 5. Plan of flocculation basin for Example 9.

$$h_f = n(2)v^2/2g = nv^2/g$$
$$n = h_f g/v^2$$

Using a velocity of 0.45 m/s (common design values range between 0.25 and 0.50 m/s):

$$n = 0.38 \times 9.8/(0.45)^2 = 18$$

Hence

Number of channels = Number of bends + 1 = 18 + 1 = 19

3. Basin dimensions

$$\text{Required length of flow} = vt = 0.45 \times 30 \times 60 = 810 \text{ m}$$
$$\text{Length per channel} = 810/19 = 43 \text{ m} = \text{width of basin}$$
$$\text{Area of channel cross-section} = Q/v = 100{,}000/0.45 \times 24 \times 60 \times 60 = 2.57 \text{ m}^2$$

Using a 3.6 m deep basin,

$$\text{Width of channel} = 2.57/3.6 = 0.71 \text{ m} = 700 \text{ mm}$$
$$\text{Length of basin} = 19 \times 0.70 + 18 \times 0.100 = 15.1 \text{ m, use } 15 \text{ m}$$

13.10. *Example 10 (English Units)*

A hydraulically mixed flocculation basin is to be designed for a water-treatment plant that has a capacity of 26 MGD. The flocculator is to be of an around-the-bend baffled basin. Assume a water temperature of 50°F, a detention time of 30 min, and a velocity gradient of 40 s^{-1}. Determine:

1. The required head loss in the channeled basin.
2. The required number of channels.
3. The basin dimensions.

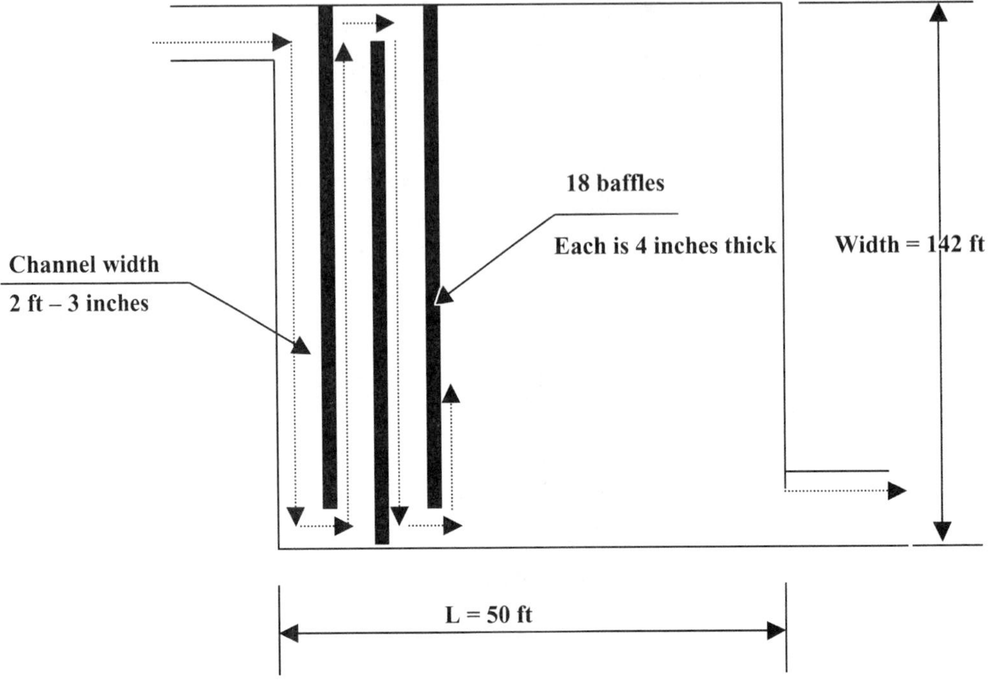

Fig. 6. Plan of Flocculation basis for Example 10.

Solution

1. The head loss, h_f, can be determined from Eq. (28):

$$G = \sqrt{\gamma h_f / \mu t}$$

$$G^2 = \gamma h_f / \mu t$$

From Table 8, the value of μ at 50°F is 2.735×10^{-5} lb·s/ft². From Table 8, the value of γ at 50°F is 62.41 lb/ft³:

$$h_f = G^2 \mu t / \gamma = (40)^2 \left(2.735 \times 10^{-5}\right)(30 \times 60)/62.41 = \underline{1.26 \text{ ft}}$$

2. The required number of channels. From Eq. (29), the head loss is given by

$$h_f = nK\, v^2 / 2g$$

Assuming a K value of 2 (usual values from 2.0 to 4 depending on geometry of bend)

$$h_f = n(2)\, v^2 / 2g = n v^2 / g$$

$$n = h_f g / v^2$$

Using a velocity of 1.5 ft/s (common design values range between 0.8 to 1.7 ft/s)

$$n = 1.26 \times 32.2/(1.5)^2 = 18$$

Hence

Number of channels = Number of bends + 1 = 18 + 1 = <u>19</u>

3. Basin dimensions

$$\begin{aligned}
&\text{Required length of flow} = vt = 1.5 \times 30 \times 60 = 2{,}700 \text{ ft} \\
&\text{Length per channel} = 2700/19 = \underline{142 \text{ ft} = \text{width of basin}} \\
&\text{Area of channel cross-section} = Q/v = 26 \times 10^6/1.5 \times 7.48 \times 24 \times 60 \times 60 = 26.8 \text{ ft}^2
\end{aligned}$$

Using a 12 ft deep basin,

$$\begin{aligned}
&\text{Width of channel} = 26.8/19 = 2.23 \text{ ft} = \underline{2'\text{-}3''} \\
&\text{Length of basin} = 19 \times 2.25 + 18 \times 4/12 = 48.75 \text{ ft, use } \underline{50 \text{ ft}}
\end{aligned}$$

NOMENCLATURE

A	Area, m^2 (ft^2)
b	Length of blade, m (ft)
C	Coagulant concentration, mg/L (ppm)
C_D	Drag coefficient
D	Width of blade, m (ft)
E	Applied electric field gradient, V/cm
F_D	Drag force, N or kg·m/s^2
g	Acceleration of gravity, 9.81 m/s^2 (32.2 ft/s^2)
G	Velocity gradient, s^{-1}
h_f	Head loss, m (ft)
K	Constant
n	Number of bends in flocculation tank
P	Power input, J/s or N·m/s (ft·lb/s)
q	Charge at the shear surface, C (Coulomb)
Q	Water flow rate, m^3/s (ft^3/s)
R	b/D = ratio of length to width of blade
R_N	Reynolds number = $\rho v_r D/\mu = v_r D/\nu$, dimensionless number
t	Detention time in tank, s or min
v	Water velocity, m/s (ft/s)
V	Volume, m^3 (ft^3)
v_r	Relative velocity of paddle with respect to water, m/s (ft/s)
δ	Thickness of diffuse layer, μ
ε	Dielectric constant
γ	Water specific weight, N/m^3 (lb/ft^3)
μ	Dynamic (absolute) viscosity, Pa·s, N·s/m^2, kg/m.s (lb·s/ft^2)
ν	Kinematic viscosity of water, m^2/s (ft^2/s)
ρ	Water density, kg/m^3 (lb·s^2/ft^4)
ζ	Zeta potential, mV

REFERENCES

1. C. N. Sawyer, P. L. McCarty, and G. E. Parkin, *Chemistry for Environmental Engineering,* 4th ed., McGraw-Hill, New York, NY, 1994.
2. V. C. Rao, et al., *J. Amer. Water Works Assoc.* **80** (2), 59 (1988).
3. R. C. Cheng, et al. *J. Amer. Water Works Assoc.* **86** (9), 79 (1994).
4. U.S. Environmental Protection Agency, *Arsenic Removal from Drinking Water by Coagulation/Filtration and Lime Softening Plants,* National Service Center for Environmental Publications, Cincinnati, OH, 2000.
5. G. S. Logsdon, et al. *J. Amer. Water Works Assoc.* **77** (2), 61 (1985).
6. W. R. Knocke, S. West, and R. C. Hoehn, *J. Amer. Water Works Assoc.* **78** (4), 189 (1986).
7. R. E. Hubel and J. K. Edzwald, *J. Amer. Water Works Assoc.* **79** (7), 98 (1987).
8. E. Lefebre and B. Legube, *Wat. Res.,* **25** (8), 939 (1991).
9. U.S. Environmental Protection Agency, *Enhanced Coagulation and Enhanced Precipitative Softening Guidance Manual,* EPA/815/R-99/012, National Service Center for Environmental Publications, Cincinnati, OH, 1999.
10. A. Amirtharajah and C. R. O'Melia, in *Water Quality and Treatment* AWWA, 5th ed., American Water Works Association, Denver, CO, 1999.
11. T. D. Reynolds, *Unit Operations and Processes in Environmental Engineering,* Brooks/Cole Engineering Division, Monterey, CA, 1982.
12. E. J. W. Verwey and J. Th. G. Overbeek, *Theory of the Stability of Lyophobic Colloids,* Elsevier, Amsterdam, 1948.
13. F. M. M. Morel, *Principles of Aquatic Chemistry,* Wiley-Interscience, New York, NY, 1983.
14. C. R. O'Melia, in *Aquatic Chemical Kinetics,* W. Stumm ed., Wiley-Interscience, New York, NY, 1990.
15. M. Elimelech and C. R. O'Melia, *Env. Sci. Tech.* **24**, 1528 (1990).
16. C. R. O'Melia, in *Physicochemical Processes for water Quality Control,* W. J. Webber, Jr., ed., Wiley-Interscience, New York, NY, 1972, pp. 61–109.
17. S. K. Dentel and J. M. Gossett, *J. Amer. Water Works Assoc.* **80**, 187 (1988).
18. I. Licsko, *Wat. Sci. Tech.* **36**, 103 (1997).
19. A. Cornelissen, et al. *Wat. Sci. Tech.* **36,** 41 (1997).
20. W. Stumm and J. J. Morgan, *J. Amer. Water Works Assoc.* **54**, 971 (1962).
21. W. Stumm and C. R. O'Melia, *J. Amer. Water Works Assoc.* **60**, 514 (1968).
22. R. F. Packham, *J. Coll. Interface Sci.* **20**, 81 (1965).
23. D. H. Bache, et al. *Wat. Sci. Tech.* **36**, 48 (1997).
24. V. K. LaMer and T. W. Healy, *Rev. Pure Applied Chem.* **13**, 112 (1963).
25. A. S. Michaels, *Industrial Engineering Chemistry,* **46** 1485 (1954).
26. M. J. Hammer, *Water and Wastewater Technology,* John Wiley and Sons, New York, NY, 1986.
27. T. J. McGhee, *Water Supply and Sewerage,* 6th Ed. McGraw-Hill, New York, NY, 1991.
28. A. Amirtharajah and K. M. Mills, *J. Amer. Water Works Assoc.* **74**, 210 (1982).
29. J. E. Van Benschoten, et al. *J. Amer. Water Works Assoc.* **120**, 543 (1994).
30. J. Boisvert, et al. *Wat. Res.* **31**, 1939 (1997).
31. A. P. Black, F. B. Birkner, and J. J. Morgan, *J. Amer. Water Works Assoc.* **57**, 1547 (1965).
32. J. K. Morris and W. R. Knocke, *J. Amer. Water Works Assoc.* **76**, 74 (1984).
33. A. Leprince, et al. *J. Amer. Water Works Assoc.* **76**, 93 (1984).
34. L. D. Mackenzie and D. A. Cornwell, *Introduction to Environmental Engineering*, PWS Publishers, Boston, MA, 1985.
35. W. Stumm, in *Principles and Applications of Water Chemistry,* S. D. Faust, and J. V. Hunter, eds., Wiley, New York, 1967.
36. W. Stumm , H. Huper, and R. L. Champlin, *Environ. Sci. Tech.* **1**, 221 (1967).

37. J. Y. Bottero, et al. *J. Phys. Chem.* **84**, 2933 (1980).
38. W. Viessman, Jr. and M. J. Hammer, *Water Supply and Pollution Control,* 5th ed., Harper Collins College Publishers, New York, NY, 1993.
39. A. P. Black, et al. *J. Amer. Water Works Assoc.* **39**, 1414 (1957).
40. T. R. Camp, *J. Amer. Water Works Assoc.* **60**, 656 (1968).
41. J. E. Singley, *Proc. Amer. Water Works Assoc. Annual Conf.,* St. Louis, MO, 1981.
42. H. E. Hudson, Jr. and E. G. Wagner, *J. Amer. Water Works Assoc.* **73**, 218 (1981).
43. American Water Works Association Research Foundation, *Mixing in Coagulation and Flocculation*, AWWARF, Denver, CO, 1991.
44. B. R. Munson, D. F. Young, and T. H. Okiishi, *Fundamentals of Fluid Mechanics*, Wiley, New York, NY, 1994.
45. M. L. Davis and D. A. Cornwell, *Introduction to Environmental Engineering,* PWS Publishers, Boston, MA, 1985.
46. R. D. Letterman, et al., *J. Amer. Water Works Assoc.* **65**, 716 (1973).
47. R. A. Corbitt, *Standard Handbook of Environmental Engineering,* McGraw-Hill, New York, NY, 1990.

5
Chemical Precipitation

Lawrence K. Wang, David A. Vaccari, Yan Li, and Nazih K. Shammas

CONTENTS

1. INTRODUCTION

Chemical precipitation in water and wastewater treatment is the change in form of materials dissolved in water into solid particles. Chemical precipitation is used to remove ionic constituents from water by the addition of counter-ions to reduce the solubility. It is used primarily for the removal of metallic cations, but also for removal of anions such as fluoride, cyanide, and phosphate, as well as organic molecules such as the precipitation of phenols and aromatic amines by enzymes (1) and detergents and oily emulsions by barium chloride (2).

Major precipitation processes include water softening and stabilization, heavy metal removal, and phosphate removal. Water softening involves the removal of divalent cationic species, primarily calcium and magnesium ions. Heavy metal removal is most widely practiced in the metal plating industry, where soluble salts of cadmium, chromium, copper, nickel, lead, zinc, and many others, need to be removed and possibly recovered. Phosphate removal form wastewater is used to protect receiving surface waters from eutrophication (plant growth stimulated by nutrient addition).

From: *Handbook of Environmental Engineering, Volume 3: Physicochemical Treatment Processes*
Edited by: L. K. Wang, Y.-T. Hung, and N. K. Shammas © The Humana Press Inc., Totowa, NJ

Competing processes for ion removal include ion exchange, electroprecipitation, and reverse osmosis. The disadvantages of these processes relative to chemical precipition are higher capital costs and, in the case of the latter two, higher energy costs for operation. Their advantage is that all these processes are better adapted to metal recovery and recycle than chemical precipitation is. Chemical precipitation has the advantage of low capital cost and simple operation. Its major disadvantages are its operating costs from the chemical expense and the cost of disposing of the precipitated sludge that is produced (3).

Most metals are precipitated as hydroxides, but other methods such as sulfide and carbonate precipitation are also used. In some cases, the chemical species to be removed must be oxidized or reduced to a valence that can then be precipitated directly. Phosphate can be removed by precipitation as iron or aluminum salts, and fluorine can be removed using calcium chloride (2).

Precipitation processes should be distinguished from coagulation and flocculation. Coagulation is the removal of finely divided non-settleable solid particles, especially colloids, by aggregation into larger particles through the destabilization of the electric double layer (4). Flocculation is the formation of yet larger particles by the formation of bridges between coagulated particles through the adsorption of large polymer molecules and by other forces. Both coagulation and flocculation, which often occur together, result in particles that can be removed by sedimentation or filtration (for details, the reader is referred to the Coagulation and Flocculation, Chapter 4, in this book). Coagulation and flocculation occur subsequent to and concomitant with the precipitation processes as it is usually applied in waste treatment.

2. PROCESS DESCRIPTION

Precipitation is a chemical unit process in which undesirable soluble metallic ions and certain anions are removed from water or wastewater by conversion to an insoluble form. It is a commonly used treatment technique for removal of heavy metals, phosphorus, and hardness. The procedure involves alteration of the ionic equilibrium to produce insoluble precipitates that can be easily removed by sedimentation. Chemical precipitation is always followed by a solids separation operation that may include coagulation and/or sedimentation, or filtration to remove the precipitates. The process can be preceded by chemical reduction in order to change the characteristics of the metal ions to a form that can be precipitated.

3. PROCESS TYPES

The chemical equilibrium relationship in precipitation that affects the solubility of the component(s) can be achieved by a variety of means. One or a combination of the following processes induces the precipitation reactions in a water environment.

3.1. *Hydroxide Precipitation*

Dissolved heavy metal ions can be chemically precipitated as hydroxide for removal by physical means such as sedimentation or filtration. The process uses an alkaline agent to raise the pH of the water that causes the solubility of metal ions to decrease and thus precipitate out of the solvent. The optimum pH at which metallic hydroxides are

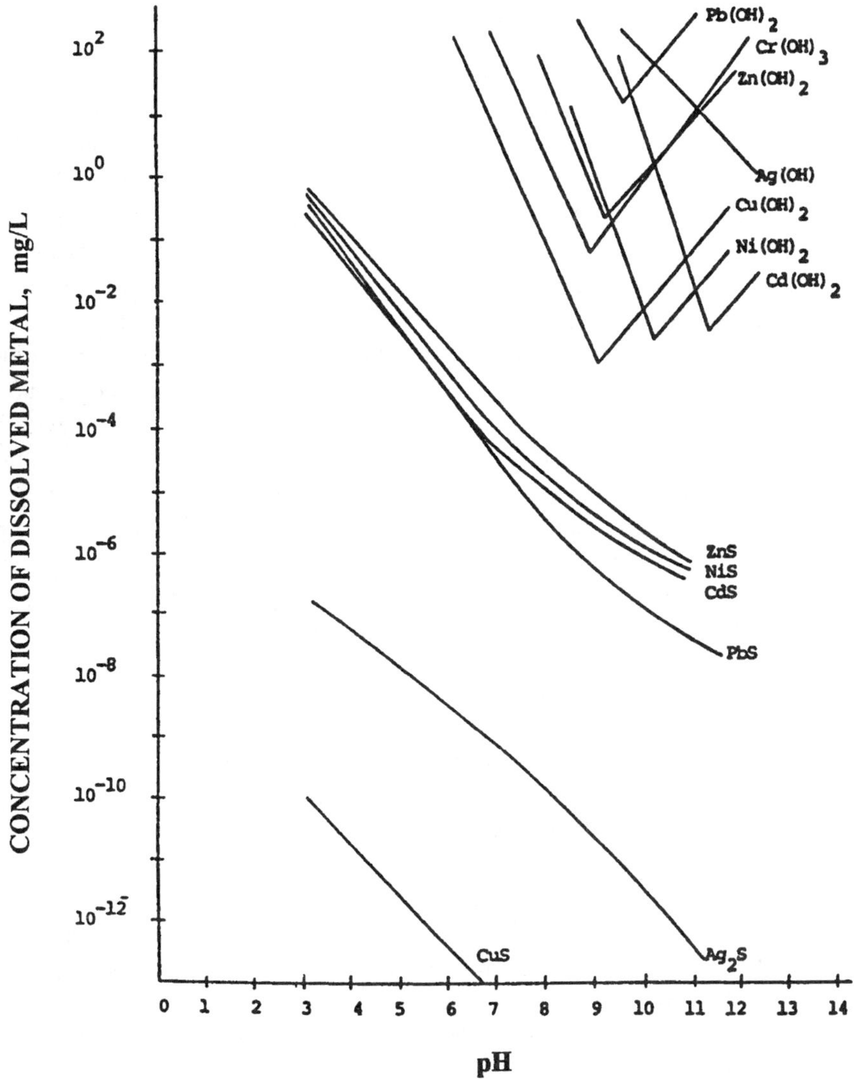

Fig. 1. Solubility of metal hydroxides and sulfides as a function of pH.

least soluble varies with the type of metal ion as shown in Fig 1. A simple form of the hydroxide precipitation reaction may be written as

$$M^{2+} + 2(OH)^{-} = M(OH)_2 \quad (1)$$

The product formed is an insoluble metal hydroxide. If the pH is below the optimum of precipitation, a soluble metal complex will form:

$$M^{2+} + OH^{-} = M(OH)^{+} \quad (2)$$

Hydroxide precipitation is also affected by the presence of organic radicals that can form chelates and mask the typical precipitation reaction:

$$M^{2+} + OH^- + nR = M(R)_n OH^+ \quad (3)$$

Reagents commonly used to effect the hydroxide precipitation include alkaline compounds such as lime or caustic soda (sodium hydroxide). Lime in the form of quicklime or un-slaked lime, CaO, and hydrated lime, $Ca(OH)_2$, can be used. Lime is generally made into wet suspensions or slurries before introduction into the treatment system. The precise steps involved in converting lime from the dry to the wet stage will vary according to the size of the operation and the type and form of lime used. In the smallest plants, bagged hydrated lime is often charged manually into a batch-mixing tank with the resulting "milk-of-lime" (or slurry) being fed by means of a solution feeder to the treatment process. Where bulk hydrate lime is used, some type of dry feeder charges the lime continuously to either a batch or continuous mixer. A solution feeder transfers lime to the point of application. With bulk quicklime, a dry feeder is also used to charge a slaking device, where the oxides are converted to hydroxides, producing a paste or slurry. The slurry is then further diluted to milk-of-lime before being fed by gravity or pumping into the process. Dry feeders can be of the volumetric or gravimetric type. Caustic soda, in the form of 6–20% aqueous solution, is fed directly to the treatment system and does not require any dispensing and mixing equipment. The treatment chemicals may be added to a flash mixer or rapid-mix tank, or directly to the sedimentation device. Because metal hydroxides tend to be colloidal in nature, coagulation agents may also be added to facilitate settling.

3.2. *Sulfide Precipitation*

Both "soluble" sulfides such as hydrogen sulfide or sodium sulfide and "insoluble" sulfides such as ferrous sulfide may be used to precipitate heavy metal ions as insoluble metal sulfides. Sodium sulfide and sodium bisulfide are the two chemicals commonly used, with the choice between these two precipitation agents being strictly an economic one. Metal sulfides have lower solubilities than hydroxides in the alkaline pH range and also tend to have low solubilities at or below the neutral pH value (Fig. 1).

The basic principle of sulfide treatment technology is similar to that of hydroxide precipitation. Sulfide is added to precipitate the metals as metal sulfides and the sludge formed is separated from solution by gravity settling or filtration. Several steps enter into the process of sulfide precipitation:

1. Preparation of sodium sulfide. Although there is often an abundant supply of this product from by-product sources, it can also be made by reduction of sodium sulfate. The process involves an energy loss in the partial oxidation of carbon (such as that contained in coal) as follows:

$$\begin{gathered} Na_2SO_4 + 4C = Na_2S + 4CO_2\ (\text{gas}) \\ \text{Sodium sulfate} + \text{carbon} = \text{metallic sulfide} + \text{carbon dioxide} \end{gathered} \quad (4)$$

2. Precipitation of the pollutant metal (M) in the waste stream by an excess of sodium sulfide:

$$\begin{gathered} Na_2S + M\,SO_4 = MS(\text{precipitate}) + Na_2SO_4 \\ \text{Sodium sulfide} + \text{metallic sulfate} = \text{metallic sulfide} + \text{sodium sulfate} \end{gathered} \quad (5)$$

3. Physical separation of the metal sulfide in thickeners or clarifiers, with reducing conditions maintained by excess sulfide ion.
4. Oxidation of excess sulfide by aeration:

$$Na_2S + 2O_2 = Na_2SO_4 \tag{6}$$

Sodium sulfide + oxygen = sodium sulfate

Because of the toxicity of both the sulfide ion and hydrogen sulfide gas, the use of sulfide precipitation may require both pre- and posttreatment and close control of reagent additions. Pretreatment involves raising the pH of water to between 7 and 8 to reduce the formation of obnoxious hydrogen sulfide gas. The pH adjustment may be accomplished at essentially the same point as the sulfide treatment, or by addition of a solution containing both sodium sulfide and a strong base (such as caustic soda). The posttreatment consists of oxidation by aeration or chemical oxidation to remove excess sulfide, a toxic substance.

A recently developed and patented process to eliminate the potential hazard of excess sulfide in the effluent and the formation of gaseous hydrogen sulfide uses ferrous sulfide as the sulfide source. The fresh ferrous sulfide is prepared by adding sodium sulfide to ferrous sulfate. The ferrous sulfide slurry formed is added to water to supply sufficient sulfide ions to precipitate metal sulfides, which have lower solubilities than ferrous sulfide. Typical reactions are

$$FeS + Cu^{2+} = CuS + Fe^{2+} \tag{7}$$

Ferrous sulfide + copper ion = insoluble copper sulfide + iron ion

$$FeS + Ni(OH)_2 = Fe(OH)_2 + NiS \tag{8}$$

Ferrous sulfide + nickel hydroxide = ferrous hydroxide + insoluble nickel sulfide

A detention time of 10–15 min is sufficient to allow the reaction to go to completion. Ferrous sulfide itself is also a relatively insoluble compound. Thus, the sulfide ion concentration is limited by the solubility of ferrous sulfide, which amounts to about 0.02 mg/L, and the inherent problems associated with conventional sulfide precipitation are minimized.

3.3. *Cyanide Precipitation*

Cyanide precipitation, although a method for treating cyanide in wastewater, does not destroy the cyanide molecule, which is retained in the sludge that is formed. Reports indicate that during exposure to sunlight, the cyanide complexes can break down and form free cyanide. For this reason the sludge from this treatment method must be disposed of carefully. Cyanide may be precipitated and settled out of wastewater by the addition of zinc sulfate or ferrous sulfate, which forms zinc ferrocyanide or ferro- and ferri-cyanide complexes. In the presence of iron, cyanide will form extremely stable cyanide complexes.

3.4. *Carbonate Precipitation*

Carbonate precipitation may be used to remove metals either by direct precipitation using a carbonate reagent such as calcium carbonate or by converting hydroxides into carbonates using carbon dioxide. The solubility of most metal carbonates is intermediate between hydroxide and sulfide solubilities; in addition, carbonates form easily filtered precipitates.

3.5. Coprecipitaion

In coprecipitaion, materials that cannot be removed from solution effectively by direct precipitation are removed by incorporating them into particles of another precipitate, which is separated by settling, filtration, or flotation.

3.6. Technology Status

Chemical precipitation of metal hydroxides is a classical water and wastewater treatment technology and is used by most industrial waste treatment systems. Chemical precipitation of metals in the carbonate form is used in water softening and in commercial applications to permit metals recovery and water reuse. Full-scale commercial sulfide precipitation units are operational at numerous industrial installations. Cyanide precipitation is used at several coil-coating plants.

4. CHEMICAL PRECIPITATION PRINCIPLES

Chemical precipitation processes perform by adjusting concentrations and other conditions so that the ionic constituents that are to be removed change from a dissolved ionic phase to a solid salt. Precipitation of salts is a fairly rapid process, and thus tends to be close to equilibrium. Accordingly, the remaining concentration of the ionic species in solution is controlled by the solubility of the solid phases present, and the theory of precipitation processes is described mostly by the principles of solubility equilibria.

4.1. Reaction Equilibria

Reactions in closed systems proceed to minimum in the Gibbs free energy, G. Thus, G is a reaction potential. At constant temperature and pressure the free energy of a dilute chemical species increases as a function of the natural logarithm of its concentration:

$$G_i = G_i\hat{\text{I}} + RT\ln\left(C_i/C\hat{\text{I}}\right) \tag{9}$$

where G_i = standard free energy in joules (J); $G_i\hat{\text{I}}$ = standard free energy in joules (J) at $C°$ for species i; R = universal gas constant = 8.314 J/K·mol = 1.99 cal/K·mol; T = absolute temperature in kelvins (K = °C + 273); $C°$ = reference concentration (mol/L); C_i = concentration of species i (mol/L).

The standard free energy is a characteristic of the species involved, and can be found in the chemical reference literature (5). Consider a reaction in which a moles of species a combines with b moles of species B to form d moles of species AB:

$$a[\text{A}] + b[\text{B}] = d[\text{AB}] \tag{10}$$

As this reaction proceeds, A and B disappear, which tends to reduce the total free energy of the system, and AB is created, which increases G. The reaction will only occur if the total free energy due to all the species present decreases. That is, the decrease in G due to the reduction in the concentration of A and B must be greater than the increase in G from the production of AB. Because G varies logarithmically with concentration, as the concentration of A and B decrease, the rate at which G decreases with concentration becomes greater. The converse occurs with AB—as its concentration increases, the rate of increase decreases. Eventually, these tendencies will approach a balance, and the reaction will stop. Mathematically, based on Eq. (9) this can be shown to occur when

$$\Delta G_T \hat{\mathrm{I}} = -RT \ln\left(C_{i,\mathrm{eq}}/C\hat{\mathrm{I}}\right)^{n_i} = 0 \tag{11}$$

where $C_{i,\mathrm{eq}}$ = concentration of species i at equilibrium (mol/L); n_i = stoichiometric coefficient of the species i (positive for products, negative for reactants), ΔGT^o = standard free energy change for the reaction [J (joule)].

Thus, when the concentrations are such that the standard free energy change for the reaction is equal to zero, the reaction is at equilibrium.

This can be simplified by defining the equilibrium constant K_c:

$$K_c = \left(C_{i,\mathrm{eq}}/C\hat{\mathrm{I}}\right)^{n_i} = \exp(-\Delta G/RT) \tag{12}$$

where K_c = standard concentration equilibrium constant.

For the reaction described in Eq. (10) above, the equilibrium constant is

$$K_c = a[\mathrm{A}]b[\mathrm{B}]/d[\mathrm{AB}] \tag{13}$$

This gives a condition on the concentrations that will exist at equilibrium. K_c can be computed from the standard free energies of the individual species, or it can be found in tabulations for particular reactions (6). When combined with other conditions, such as mass balance and charge balance considerations, the concentrations of the individual species can be computed. Furthermore, if other reactions occur and other species are present, as long as enough conditions are defined, the concentrations of all the species present can be calculated. Examples will be given below.

Some caution should be exercised in using solubility product constants from the literature. Often a range of values is found. Sources of error include insufficient approach to equilibrium, difficulty in measuring soluble metal species, ionic strength effects, and the presence of complexing substances (7,8). The latter two interferences are discussed below.

4.2. Solubility Equilibria

In precipitation we are concerned with reactions in which the reactants are dissolved ionic species and the products are solid salts. The prototypical reaction given above can be considered to represent a precipitation reaction in which the ions A and B combine to form precipitate AB, and the equilibrium constant given above would apply. However, AB is present in a different phase than A and B, and its activity cannot be expressed as a concentration.

This is handled in two equivalent ways. Because the activity of the solid phase is fairly constant (although it may vary slightly due to crystal structure and incorporation of impurities), it can either be defined as a reference value and has a unit activity, or its activity can simply be lumped into the equilibrium constant. In either case the equilibrium constant will now be called the "solubility product constant," K_{sp}. In the case of AB, its solubility product constant is

$$K_{\mathrm{sp,\,AB}} = [\mathrm{A}][\mathrm{B}] \tag{14}$$

In general, the solubility product constant is equal to the product of the concentrations or activities of each ionic species formed in the dissociation, raised to the power of its respective stoichiometric coefficient,

$$A_aB_bC_c = a[A] + b[B] + c[C]$$

$$K_{sp} = [A]^a[B]^b[C]^c \tag{15}$$

For example the solubility product for NaH_2,PO_4 dissociated into one Na^+ ion, two H^+, and one PO_4^{3-} is,

$$K_{sp} = [Na^+][H^+]^2[PO_4^{3-}].$$

4.2.1. *Solubility Equilibria Example*

Compute the solubility of CaF_2 in pure water given its $K_{sp} = 2.95 \times 10^{-11}$.

Solution

$$K_{sp} = [A]^a[B]^b[C]^c \tag{15}$$

$$K_{sp} = [Ca^{2+}][F^-]^2 = 2.95 \times 10^{-11} \tag{16}$$

Since two moles of F^- go into solution for every mole of Ca^{2+}, a concentration condition can be written:

$$2[Ca^{2+}] = [F^-] \tag{17}$$

There are two unknowns and two equations. Assuming X to represent the calcium concentration, then according to Eq. (17) the fluoride concentration should be 2X. Now substitute these values into Eq. (16) and solve for X:

$$X(2X)^2 = 2.95 \times 10^{-11}$$
$$4X^3 = 2.95 \times 10^{-11}$$
$$X = 1.95 \times 10^{-4}\ \text{mol/L} = [Ca^{2+}]$$
$$[F^-] = 2X = 3.89 \times 10^{-4}\ \text{mol/L}$$

The number of moles of CaF_2 that dissolves is 1.95×10^{-4} and its molecular weight is 78.1. The resulting concentration of calcium fluoride is

$$(1.95 \times 10^{-4}\ \text{mol/L}) \times (78.1\ \text{g/mol}) \times (1000\ \text{mg/g}) = 15.2\ \text{mg/L}$$

4.3. *Ionic Strength and Activity*

The above development assumes an "ideally dilute" solution, that is, one in which dissolved ions interact only with the water in the solution. A more precise analysis takes into account interactions with other ions, whether of the same or different species. This is related to the concentrations of ions present, which is measured by the "ionic strength" *I*. In most situations increasing ionic strength decreases the effective concentration of all species present. That is, an ionic species present behaves as though its concentration is lower than it actually is due to "interference" between dissolved constituents. As a result, the solubility of a salt usually increases a small amount with ionic strength.

The effect of ionic strength is taken into account in calculating solubilities by replacing concentrations in the solubility product expression with corresponding "activities," which are the effective concentrations. The activity of a solute depends on the ionic strength and on characteristics of the solutes, particularly its ionic size and charge. The activity is computed as the product of an activity coefficient, γ and concentration:

$$a_i = \gamma_i[A_i] \tag{18}$$

where a_i = activity (dimensionless number), $[A_i]$ = molar concentration mol/L, γ_i = activity coefficient, i = ith species.

To compute the activity, first calculate the ionic strength, I, then obtain the activity coefficient. The ionic strength is defined as

$$I = \tfrac{1}{2}\sum[A_i]Z_i^2 \tag{19}$$

where Z_i = the charge on species i.

If some or all dissolved species are not known, the ionic strength can be estimated from the total dissolved solids by using the following relationship (9):

$$I = (2.5\times10^{-5})(\text{TDS}) \tag{20}$$

where TDS = total dissolved solids (mg/L). The dependency of the activity coefficient upon the ionic strength can be approximated in several ways (6). The simplest which is good up to an ionic strength of 0.1, is

$$\log\gamma = -AZ^2\left[I^{1/2}/\left(1+I^{1/2}\right)\right] \tag{21}$$

where A = activity parameter = approx 0.5 in water at 25°C. Thus, at infinite dilution ($I = 0.0$) the activity coefficient is 1.0. For an ionic strength of 0.1, the activity coefficient of an ion with a charge of +1 would be 0.989, while for an ion with a charge of +2 it would be 0.955. More accurate expressions would take into account the temperature and the ionic size.

4.4. Ionic Strength Example

What would be the solubility of CaF_2 in a solution with an ionic strength of 0.1?

Solution

The solubility product expression with the activity coefficient included would become

$$K_{sp} = \gamma_{Ca}[Ca^{2+}]\gamma_F^2[F^-]^2 \tag{22}$$

This is rearranged to give

$$K_{sp}/\gamma_{Ca}\gamma_F = [Ca^{2+}][F^-]^2 = 2.95\times10^{-11}/(0.989)(0.955) \tag{23}$$

which can be solved as in the previous example using the same concentration condition:

$$3.12\times10^{-11} = 4X^3$$
$$X = [Ca^{2+}] = 1.98\times10^{-4}\ \text{mol/L}$$
$$[F^-] = 3.97\times10^{-4}\ \text{mol/L}$$

The solubility of CaF_2 becomes 15.49 mg/L. This is an increase of 1.9% over the dilute solution.

4.5. Common Ion Effect

In the previous example the concentration condition assumes that there are no other sources of calcium or fluoride. If another salt is added to the system, there are two ways in which the solubility of calcium fluoride can be affected. First, if the added salt does not contain any ions that participated in the CaF_2 precipitation reaction, namely, Ca^{2+} or F^-, then it will have a relatively small effect due to the change in ionic strength. This would occur if, for example, NaCl were added to the solution.

The second type of effect would occur if the added salt contained Ca^{2+} or F^- ions. For example, if solid $CaSO_4$ were present with the CaF_2, the calcium concentration would no longer be one-half the fluoride concentration. A new concentration condition must be developed to take into account the added calcium from $CaSO_4$. The added Ca^{2+}, which can be related to the SO_4^{2-} in solution, adds a third unknown to the system. The third equation comes from the solubility product expression for $CaSO_4$. The methodology of analysis for such a system is illustrated in the example shown in the following section.

4.6. Common Ion Effect Example

Compute the solubilities of CaF_2 and $CaCO_3$ when both solids are in equilibrium in the same solution. Assume an ideal dilute solution. K_{sp} for $CaCO_3$ is 4.82×10^{-9}.

Solution

Solubility of $CaCO_3$, calculated using the same method as for CaF_2, is

$$6.94 \times 10^{-5} \text{ mol/L} = 6.9 \text{ mg/L} \tag{23}$$

When both solids are present together, the concentration condition must reflect the contribution to the calcium ions in solution from the calcium fluoride and from the calcium carbonate. Thus, there must be one-half mole of calcium ions for every mole of fluoride, plus one mole for each mole of carbonate:

$$\left[Ca^{2+}\right] = \tfrac{1}{2}\left[F^-\right] + \left[CO_3^{2-}\right] \tag{24}$$

This is solved together with the solubility product expressions for each salt, giving three unknowns and three equations.

Solving the solubility product expressions for the anion concentrations and substituting them into the concentration condition can eliminate the anions:

$$\left[F^-\right] = \left(K_{sp,CaF_2} / \left[Ca^{2+}\right]\right)^{1/2} \tag{25}$$

$$\left[CO_3^{2}\right]^- = K_{sp,\,CaCO_3} / \left[Ca^{2+}\right] \tag{26}$$

Substituting these into Eq. (24) results in:

$$\left[Ca^{2+}\right] = \tfrac{1}{2}\left(K_{sp,CaF_2} / \left[Ca^{2+}\right]\right)^{1/2} + K_{sp,CaCO_3} / \left[Ca^{2+}\right] \tag{27}$$

This expression can be solved manually by a successive approximation method, or by any numerical root-finding technique such as the secant method or Newton's method (10), or by using a programmable pocket calculator, which has a built-in root finding capability. The solution in this case is

$$\left[Ca^{2+}\right] = 2.10 \times 10^{-4} \text{ mol/L}$$

This result can be substituted into the expressions for fluoride and carbonate concentrations, which also yield the solubilities of the respective salts. Substituting this concentration of $[Ca^{2+}]$ into Eqs. (25) and (26) yields

$$\left[F^{-}\right] = 3.75 \times 10^{-4} \text{ mol/L}$$
$$\left[CO_3^{\ 2-}\right] = 2.29 \times 10^{-5} \text{ mol/L}$$

Hence, the solubilities of their respective salts are,

$$14.6 \text{ mg/L for } CaF_2$$
$$2.3 \text{ mg/L for } CaCO_3$$

Note that the common ion effect has reduced the solubility of both compounds present in the same solution.

This example illustrates the method of computing the solubility when there are competing precipitation reactions. This is commonly the case in natural waters and in wastewaters. The general procedure is the same in all cases: expressions describing the equilibria for all reactions occurring are written down, including equilibrium with the atmosphere, if necessary; additional conditions describing the mass balance, and possible charge balance, are included. If the number of unknowns equals the number of independent equations, then there is enough information describing the system to solve the equations simultaneously. In computing the solubility of a particular metal, it may be necessary to consider several soluble species as well as several possible precipitates.

4.7. Soluble Complex Formation

Metals are acids in the Lewis sense. As such, they compete with protons and with each other for available bases, such as hydroxides. This competitive tendency increases with increasing valence of the metal and decreasing size of the metal atom. The "naked" metal ion, such as Fe^{3+} or Cd^{2+}, is rapidly hydrated in water, forming what is called the aquo complex. Under appropriate conditions the aquo complexes will have a tendency to combine with hydroxides to form hydroxo complexes. For example, cadmium species can have charges ranging from plus two, Cd^{2+}, to minus two, $Cd(OH)_4^{\ 2-}$.

Metals of higher valence (four or more) may also tend to form "oxo complexes." For example, hexavalent chromium forms the anionic oxo complex chromate, $CrO_4^{\ 2-}$, or dichromate, $Cr_2O_7^{\ 2-}$.

In general, as the oxidation state of the metal increases and the radius decreases, the pH at which hydroxo complexes dominate over aquo, or oxo dominate over hydroxo, decrease. It must be noted that the effect of valence is much stronger than that of size.

The hydroxo and oxo complexes are in equilibrium with any precipitates present and can be modeled using the solubility product expression just as the aquo ions. However, other complexing agents may be present, and in many cases the solubility should be experimentally determined. For example, in natural waters copper can form complexes with carbonate ions, with ammonium ions, and with dissolved organic matter. In plating solutions hexametaphosphate and cyanide are added specifically because their complexation behavior is desired for purposes of controlling the plating process.

Hydroxo complexes are also strongly removed by adsorption on solids, whether those solids are precipitates of the particular metal or not.

Different metal ions can form soluble complexes with each other. Copper, in particular, seems to be implicated often in this type of behavior. Solutions of zinc cyanide and copper cyanide resulted in significantly increased zinc solubility and a shift in the pH of minimum solubility (11). Dilute mixtures of copper and chromium plating solutions also resulted in increased copper solubility.

Complexes other than those described here may be important, particularly polynuclear complexes (6). Some of these may be intermediates in the formation of precipitates, and thus are not at equilibrium. Consequently, some metal ion solutions "age" over a period of days or weeks, changing their precipitation behavior.

There is a tendency for solids with a smaller solubility product to be more likely to form soluble complexes (6). This has been observed with metal sulfides to the point that those with the lowest solubility products exhibited the greatest solubility. Thus, solubility based on the solubility product expression must be used carefully, and the formation of the soluble complexes often should be taken into account.

4.8. *pH Effect*

The most important competing reaction is the dissociation of water into H^+ and OH^-, primarily because metal hydroxides are fairly insoluble. Another reason for this is the formation of soluble complexes, which has been discussed above. Thus, the solubility of the metal usually decreases as pH increases (hydroxide concentration increases), until the formation of soluble hydroxide complexes becomes significant, and then the total solubility begins to increase with pH. In general, there will be either a single pH of minimum metal solubility, or a pH range of minimum solubility. Table 1 indicates pH ranges of minimum solubility for several metal species.

4.9. *Solubility Diagrams*

The solubility of metal precipitates is controlled largely by the pH (see Fig. 1). A graphical solubility diagram can concisely express the relationship between pH and solubility. The development of a solubility diagram requires knowledge of the solubility product constants for the dissolution of each solid phase present into each soluble species. The total solubility is just the sum of the solubilities due to each dissolution reaction. The reader is referred to the literature (7) for details and illustrations on the construction of solubility diagrams.

5. CHEMICAL PRECIPITATION KINETICS

Once the possibility of precipitation is established by the equilibrium considerations described above, it is necessary to determine the factors that govern the rate of precipitation. Process design hinges upon the rate, as the slower the process occurs, the larger the size of reactors necessary to accomplish the degree of conversion desired.

Table 1
Metal Hydroxide Minimum Solubility Ranges

Metal	Solubility (mg/L)	pH range
Al^{3+}	0.00055	6.2
Cd^{2+}	0.60	10.5–13.0
Cr^{3+}	0.04	7.0
Cu^{2+}	0.03	7.5–11.5
Fe^{3+}	0.00006	7.0–10.0
Pb^{2+}	16.5	10.0
Zn^{2+}	0.13	9.5

The driving force for precipitation can be expressed as the degree of oversaturation (9):

$$DF = C/C_s \tag{28}$$

where DF = driving force, C = solution concentration (mol/L or mg/L) and Cs = saturation concentration (mol/L or mg/L). If DF is less than 1, the solution is undersaturated and no precipitation will occur. At DF = 1.0, the solution is saturated and will be in equilibrium with any salt present. If DF is greater than 1, the solution is supersaturated, and if it is much greater than 1, then salt should precipitate out of solution. This is termed the "labile" regime. However, if the solution is only slightly supersaturated, precipitation may occur very slowly, if at all. A solution in this regime is called "metastable." The boundary between the labile and metastable regimes is somewhat arbitrary, and will depend on the particular salt involved. Metastable solutions have been observed for DF as high as 10 (9). The presence of a seed crystal can cause rapid precipitation of metastable solutions to occur.

The formation of a precipitate consists of three steps (6): (a) nucleation; (b) crystal growth, and (c) aging.

5.1. Nucleation

Nucleation is the condensation of ions to very small particles. This process requires a DF significantly different from 1. It depends on a mutual attraction between dissolved neutral salt molecules. In natural waters, nucleation will be promoted by the presence of a foreign particle, termed a heteronucleus. It acts by adsorbing solute molecules, decreasing the DF needed for precipitation to occur at a significant rate.

5.2. Crystal Growth

Crystal growth is the depositing of material upon previously formed nuclei. Growth proceeds by a sequence of steps: transport to the crystal surface by convection and diffusion, adsorption onto the surface, and reaction, or formation of the crystal lattice bonds.

The rate of crystal growth is usually limited either by the diffusion step or by the reaction step, and mostly by the former. The rate depends on which step is limiting. For diffusion-limited crystal growth, the rate law is approximately first order:

$$dC/dt = -ks(C - Cs) \tag{29}$$

where k = rate coefficient, dependant upon amount of mixing and s = crystal surface area.

If the process is reaction-rate-limited, the rate expression may be other than first order, and will not depend on mixing. For example, sodium chloride has been found to be first order, silver chloride is second order, and silver chromate is third order.

5.3. Aging

"Ripening", or aging, refers to slow changes in the crystal structure that occur over time. Fresh precipitates are small and have a relatively disordered structure with more crystal defects and inclusions of impurities. A slow process of re-solution and precipitation effects a rearrangement into larger, pure crystals having a lower solubility. In fact, any finely divided precipitate is not, strictly specking, in equilibrium. True thermodynamic equilibrium will minimize the surface area of the crystals, ultimately resulting in the formation of a single, large crystal. Thus, solubility product constants measured using fresh precipitates may be larger than values obtained with ripened solids.

Also, as mentioned above, slow-forming polynuclear complexes may be intermediates in the formation of precipitates. Thus, even the solution may need to age in order to achieve equilibrium.

Complexing agents present in solution may affect the rate of precipitation even if they do not have an effect upon the solubility. Organic complex formers, in particular, may slow precipitation, as well as influence the crystal form that results (6).

5.4. Adsorption and Coprecipitation

As described above, precipitation from metastable solutions can be promoted by the presence of foreign solid particles owing to adsorption. Adsorption can be primarily of a physical nature, due to van der Waals forces or pi bonds, or chemical. Metal ions adsorb primarily by ion exchange, which is a form of chemisorption. Metals are also strongly removed by activated carbon adsorption.

Solids with oxide surfaces can act as weak acids and bases in solution (13) and are protonated and deprotonated in response to pH, ionizing the surface. The surface ions function as ion exchange sites. Increasing the pH increases the adsorption of cations and decreases adsorption of anions. The adsorption capacity will change from 0% to 100% of the adsorbent's total capacity over a narrow range of one or two pH units. The location of this "pH adsorption edge" also depends on the concentration of the adsorbent. The presence of competing adsorbents will either shift the pH adsorption edge or reduce the capacity of the adsorbent for a particular ion. Complexing agents can either increase or decrease adsorption. They may decrease adsorption by stabilizing the ion in solution. Alternatively, they may increase it by forming complexes that adsorb stronger than the ion alone. For example, cyanide can strongly increase adsorption of nickel ions at high pH values.

Coprecipitation refers to the simultaneous removal of an ion with the precipitation of another, with which it does not form a salt. The mechanism may be the inclusion of one ion as impurity in the crystal structure of the other or due to adsorption on the surfaces of the other's crystals. Thus, coprecipitaion can be used to remove metals at concentrations that are already below their minimum solubility. For example, precipitation of ferric nitrate at a pH above 7.0 was found to remove approximately 95% of 5.0×10^{-7} mol/L solution of cadmium (13). Another application is the removal of radium using lime-soda ash softening.

Coprecipitation has been used to explain the decreased solubility of zinc in the presence of trivalent chromium and divalent and/or tetravalent nickel from plating solutions (11). Mixed copper and nickel-plating solutions had substantially lower solubilities at pH below 10. Copper and chromium-plating solutions with 200–500 mg/L of each metal also resulted in reduced copper solubility, but at lower initial concentrations, higher solubility resulted, apparently due to complex formation.

6. DESIGN CONSIDERATIONS

6.1. General

The equipment required for chemical precipitation can be divided according to three main functions: chemical handling and feeding; mixing and contact between chemicals and wastewater for the formation of the precipitate; and separation of precipitate from the treated water. In some cases mixing and separation are both carried out in a single tank.

6.2. Chemical Handling

From a storage and handling point of view, the chemicals used in precipitation have a wide range of physical and chemical properties. They include liquids, solids, and gases, and acidic, neutral, and basic species. Many are available in alternate forms, or may require on-site preparation. Solids are usually mixed with water to form a slurry or solution before being added to the water stream. In liquid form many of the chemicals are corrosive, so storage containers, pipes, and pumps must be made of resistant materials.

Ferrous chloride is commonly used as a coagulant, and therefore can be used in coprecipitaion. It may be supplied either as a crystalline solid or as a viscous solution ready for immediate application. It is a Lewis acid and is therefore corrosive. Alum is another acidic coagulant with similar properties. Besides being used as a coagulant, both of these salts are used in precipitation of phosphate.

Hydroxide precipitation is preformed by raising the pH with either sodium hydroxide (caustic soda) or lime. Sodium hydroxide is a solid that must be dissolved in water, a process accompanied by the release of a considerable amount of heat. Lime is available as calcium oxide (quicklime), which is prepared by heating calcium carbonate. Before application it must be slaked, forming calcium hydroxide, or slaked lime. This can be dosed as slurry.

Carbonate precipitation is performed by the addition of either a soluble carbonate salt such as sodium carbonate (soda ash), or by directly contacting the solution with carbon dioxide gas. CO_2 is conveniently available from combustion processes. The higher dosages necessary for treating plating wastes may require the use of soda ash. In applications such as water softening, both soda ash and carbon dioxide gas may be used, the latter to save on the costs of the former. In this case gas–liquid contacting equipment is required. This may take the form of compressors combined with subsurface diffusers or spargers, or as one of a variety of turbines or surface-aeration devices (14).

Gas-contact equipment may also be used for some sulfide precipitation processes. For example, the calcium sulfide insoluble precipitation method involves the on-site production of the calcium sulfide by bubbling hydrogen sulfide through lime slurry. Alternatively, it can be formed by mixing $Ca(OH)_2$ with NaHS.

Chlorine is used in precipitation processes as a preliminary step for destruction of cyanide by the alkaline chlorination process. Chlorine may be supplied either as a gas or a compressed liquid requiring evaporation before application. Specialized equipment is available for metering chlorine gas and forming a solution in water for dosing the wastewater. Similarly, sulfur dioxide gas is used for preliminary reduction processes, such as for chromium reduction.

6.3. Mixing, Flocculation, and Contact Equipment

Precipitation equipment must be designed to ensure that the solid-formation process proceeds rapidly and produces solids that are easy to handle. That is, the solids should be easy to separate from water by sedimentation and/or filtration, and the resulting sludge should be amenable to further concentration processes such as thickening, centrifugation, or pressure dewatering.

These properties will depend on the shear and concentration history of the solids (15). The chemicals added should be rapidly mixed with the water so that the dosage concentration, which would result in optimal solids formation, is quickly achieved throughout the volume. The level of mixing is measured in terms of a parameter called the velocity gradient, G. The velocity gradient, in turn, is related to the design of the mixing equipment and the amount of power applied per unit volume of liquid. The unit for G is inverse time.

Design parameters for precipitation processes may be similar to those for coagulation processes (for details, the reader is referred to Chapter 4 on coagulation and flocculation). Initial mixing equipment, also called flash mixers, are designed with a detention time of from 10 to 30 s, and with a velocity gradient on the order of 300 s^{-1}. Where pumps are used to introduce the water stream to the process, the chemical can be introduced just ahead of the pump, and the mixing caused by the pump may be sufficient for this purpose.

Subsequent to flash mixing, the mixture should be gently agitated to encourage the flocculation reaction. Flocculation is a process where solid particles of precipitate agglomerate into larger particles, which are easier to remove. The agitation promotes flocculation by enhancing particle-to-particle contact. The level of agitation determines the maximum particle size distribution. There is a balance between the formation of flocs by particles contact and the breaking up of those flocs by shear forces. Also, sufficient time must be allowed for the equilibrium to be approached.

The design of the flocculation stage is based on G, the detention time t, and the solids volume concentration C. In precipitation processes, C is usually low unless the solid-contact process referred to below is used. The basic design parameters are the products Gt and GCt, with constraints on the individual values of G, t, and C. The velocity gradient may be up to 100 s^{-1}, Gt is usually between 0.3 and 1.5×105, and GCt ranges from 10 to 100. The detention time may be as low as 15 min or as high as 1 h or more.

Different precipitates may have very different flocculation properties and require very different design parameters. More granular flocs, such as those produced by sulfide precipitation, will benefit less from long flocculation times. In some cases the turbulence and detention time in the inlet of clarifiers is sufficient for flocculation to occur. Polymer flocculants may be added to improve floc formation.

In the solid-contact process the solids volume concentration is maintained at a very high level either by recycling solids into the system or by operating with a sludge blanket in an upflow clarifier so that the entire flow must percolate through the blanket. Such a unit is highly efficient, and is the process of choice for precipitation.

6.4. *Solids Separation*

After the formation of the precipitate, water must be separated from the mixture, and the concentrated slurry is segregated for ultimate disposal.

Initial separation is usually carried out by sedimentation. In batch operations this only involves turning off the mixing apparatus and allowing sufficient time for settling of the precipitate. In continuous operations, the key design parameter is the hydraulic loading rate, which is the flow divided by the surface area available for settling, Q/A. The unit for hydraulic loading is flow per unit area, which is equivalent to a velocity unit. Effectively, the hydraulic loading at which a process is operated is the minimum settling velocity a particle can have and still be removed. Thus, precipitates that form very fine particles will require a lower hydraulic loading rate to be removed. For example, alum sludge can be efficiently removed at a hydraulic loading of 500 to 600 gpd/ft^2, whereas iron sludge requires 700 to 800 gpd/ft^2 for equivalent removal, and lime needs 1400 to 1600 gpd/ft^2.

Treatment of wastes with low concentration of metals may result in fine, poorly flocculate solids mixtures, particularly if the solid-contact process is not used. Sedimentation can be improved by the use of lamellae separators. These are inclined parallel plates held in the clarification section of a settling tank, which effectively increase the surface area available for settling.

The flocculant nature of many precipitates precludes efficient capture by sedimentation. Once most of the solids have been removed in this way, it is usually necessary to polish the effluent by filtration. Rapid sand filtration is commonly employed for this purpose

6.5. *Design Criteria Summary*

Chemical precipitation treatment can either be a batch or a continuous operation; with batch treatment being favored when the treated water flows are small. In batch treatment, the equipment usually consists of two tanks, each with a capacity to treat the total water volume expected during the treatment period. These systems can be economically designed for flows up to 190,000 L/d (50,000 gpd).

Batch treatment tanks serve the multiple functions of equalizing the flow, acting as reactors, and acting as settlers. For a typical treatment operation, water is stirred, and a homogeneous sample is taken and analyzed to determine the chemical dosage requirements. The chemicals are then added, mixed, and stirred for about 10 min. After the reaction is complete, the solids are allowed to settle for a few hours. The clear liquid is then decanted and discharged. Settled sludge is retained to serve as a seed for crystal growth for the next batch, but must be drawn off periodically for disposal. For larger daily flows, a typical continuous flow treatment system consists of a chemical feed system, flash mixer, flocculator, settling unit, and, in some cases, a filtration system. A control system is used to regulate the chemical feed to the process. For high-speed mixing, residence times of 10–30 s have been reported as satisfactory and a mixing time of as much as 2 min has been recommended for two parallel units. For development of good floc characteristics, residence times of 15–30 min have been suggested.

The chemical dosage for precipitation can be determined on the basis of the treated water alkalinity and acidity, desired pH level to be maintained for the process, and stoichiometric requirements. An alternative is to estimate the chemical dosage on the basis of jar tests. These jar tests react the water with a series of chemical doses, with the optimum dose selected on the basis of observed and measured removal effectiveness. Dosage determined on the basis of stoichiometric requirements may have to be increased by up to four times the stoichiometric amount as a result of chemical interactions, solubility variances, mixing effects, and multivalent competition.

7. PROCESS APPLICATIONS

The principles described above find application in a number of ways. The particular reactions that are used in common precipitation technologies depend not only on the ion to be removed and the type of counter-ion used, but also on the presence of competing reactions and facilitating reactions (e.g., oxidation or reduction). The most important types of precipitation processes will be described here.

Chemical precipitation can be used to remove metal ions such as aluminum, antimony, arsenic, beryllium, cadmium, chromium, cobalt, copper, iron, lead, manganese, mercury, molybdenum, tin, and zinc. The process is also applicable to any substance that can be transformed into an insoluble form, for example, fluorides, phosphates, soaps, and sulfides.

7.1. Hydroxide Precipitation

This is the workhorse of precipitation processes. Many metal cations are removed easily as the hydroxide. The process consists of simply raising the pH to the range of minimum solubility with a strong base such as sodium hydroxide (NaOH), lime (CaO), or slaked lime [$Ca(OH)_2$].

The base dosage can be determined by titration or by calculation based on equilibrium considerations, if the water to be treated is characterized well enough. Of course, if the buffer capacity of the water is too great, the chemical dosage required may be large enough that some other precipitation process might be more economical, or even some other type of process entirely, such as ion exchange or reverse osmosis.

Waters containing mixtures of metals to be removed present special problems, because the pH corresponding to optimum removal for all species may not coincide. In this case either a tradeoff must be made between removals of the various metals, or the treatment must be applied in stages, each one optimized for removal of a particular metal or group of metals.

Hydroxide precipitation and particularly the use of lime to cause chemical precipitation has gained widespread use in industrial waste treatment because of its ease of handling, its economy, and its effectiveness in treatment of a great variety of dissolved material. Industries and utilities using hydroxide precipitation include:

(a) Inorganic chemicals manufacturing.
(b) Metal finishing.
(c) Coil coating.
(d) Copper forming.
(e) Aluminum forming.
(f) Foundries.
(g) Explosives manufacturing.
(h) Steam electric power plants.
(i) Photographic equipment and supplies.
(j) Pharmaceutical manufacturing.
(k) Rubber processing.
(l) Porcelain enameling.
(m) Battery manufacturing.
(n) Ion and steel manufacturing.
(o) Nonferrous metal manufacturing.
(p) Coal mining.
(q) Electrical and electronic components.
(r) Ore mining and dressing.
(s) Publicly owned treatment works.

The most common treatment configuration is pH adjustment and hydroxide precipitation using lime or caustic following by settling for solids removal. Most plants also add a coagulant or flocculent prior to solids removal.

7.2. Carbonate Precipitation

Carbonate precipitation has long been the method of choice for the removal of calcium hardness from water. More recently, carbonate has been proposed for the removal of heavy metals from wastewater. The reason for this is that hydroxide precipitation may yield large sludge volumes, which may be difficult to settle and filter, and is also due to the additional precipitation of gypsum if sulfate is present and lime is used. In some cases carbonate sludges settle and filter better than hydroxide sludges, and the treatment can be carried out at a lower pH.

Patterson (7) has shown that carbonate may in some cases decrease the solubility of metals, while increasing it in others. The effect depends on the particular metal and the pH at which the treatment is carried out. The effect may be calculated by applying the equilibrium approach discussed earlier, although the actual results may differ somewhat from the theoretical values.

Cadmium was predicted to have a minimum solubility of 0.011 mg/L at pH 9 and a total carbonic species concentration $C_T = 7.3$ mg/L. This is both a lower solubility and lower pH than obtained with hydroxide alone: 0.60 mg/L cadmium concentration at pH from 10.5 to 13.0. Furthermore, experimental values were somewhat lower, and showed a further reduction above pH 9. It was also found that cadmium solubility was much more sensitive to total carbonic species concentration than to pH. All of these factors favor carbonate for removal of cadmium over hydroxide precipitation.

The situation is quite different for copper. Both theory and experiment show an increase in solubility, particularly at pH less than about 9–10. Above pH 10, carbonate had little effect.

Lead shows even more complicated behavior with respect to carbonate. Experimental results indicate a strong sensitivity to carbonate concentration, such that even trace amount greatly reduced the solubility. At higher levels of carbonate, the effect was very different at different pH values. For example, at pH 6 increasing total carbonic species concentration decreased lead solubility, while at pH 9 it increased it. From theoretical predictions the minimum solubility occurs at a pH near 9 with total carbonic species concentration, C_T at about 10^{-4} mol/L.

Zinc solubility is reduced significantly by the presence of carbon dioxide, especially at pH less than 9. The experimentally measured effect is less than that theoretically predicted, but is still substantial (7). It is possible that zinc carbonate is formed quite slowly, and thus may not be practical under treatment conditions.

Overall, however, Patterson (7) makes the following conclusions concerning the use of carbonate precipitation: there is no advantage in using carbonate precipitation for zinc or nickel removal, and the sludges produced are not denser or easier to filter. Cadmium can be precipitated at a lower pH as its carbonate than as its hydroxide, and will result in sludge that can be flitted at about twice the rate. Lead can also be removed by carbonate precipitation at a lower pH than it could by hydroxide, and produces a denser sludge with better filtration characteristics. Treatment of wastes containing a mixture of metals was found to behave equivalently to those with single metals, with respect to both hydroxide and carbonate precipitation.

Carbonate precipitation is sometimes used to precipitate metals, especially where precipitated metals are to be recovered. Carbonate ions also appear to be particularly useful in precipitating lead and antimony. Coprecipitaion is used for radium control in the uranium industry (a subcategory of Ore Mining and Dressing). Radium sulfate ($RaSO_4$) is coprecipitated by addition of barium chloride, which in the presence of sulfate ion forms barium sulfate precipitates. Coprecipitation of molybdate anion, which is not removed effectively by hydroxide or sulfide precipitation, can be carried out by addition of ferric sulfate or ferric chloride, which forms ferric hydroxide precipitates at an acid pH. Vanadium is also subject to coprecipitaion with ferric hydroxide.

7.3. *Sulfide Precipitation*

Metal sulfide precipitation has the advantages of lower metal solubility, smaller sludge volume, and insensitivity to the presence of chelating agents. Preliminary studies have also indicated that sulfide sludges have fewer tendencies to leach metal ions than hydroxide sludges do. However, there are the disadvantages of odor and toxicity control, and contamination of the effluent with sulfide, which exerts an oxygen demand on receiving waters. The higher chemical cost of sulfide treatment is balanced by its ability to attain higher treatment efficiencies.

Sulfide is also capable of acting as a reducing agent to convert hexavalent chromium to the trivalent form. Under alkaline conditions the chromium can then be removed as the hydroxide precipitate. If ferrous sulfide is used, the products of the reduction and precipitation are chromic and ferric hydroxide sludges and elemental sulfur.

Two basic types of sulfide dosing exist: the soluble-sulfide method and insoluble-sulfide method (12). Odor problems and sulfide contamination of the effluent are caused by excessive sulfide dosage, and are mainly a problem with the soluble-sulfide delivery method. In this method Na_2S or NaHS is added to the wastewater. Dosage control is achieved by analysis of residual soluble metal concentration. Feedback control by specific ion electrodes for sulfide has been demonstrated in the laboratory. The precipitated particles are very small and require coagulants and flocculants for efficient sedimentation.

The insoluble-sulfide delivery method overcomes the control problem for sulfide precipitation. The source of sulfide ions is a sparingly soluble metal sulfide with a low solubility. This salt liberates its sulfide, as the metal to be removed consumes it by precipitation. Two forms of this method have been developed: the ferrous sulfide process by Permutit Company and a calcium sulfide process developed by Kim and Amodeo (12).

In the Permutit process, called Sulfex, FeS is produced onsite by mixing $FeSO_4$ with NaHS. H_2S gas emission from this part of the process may require control measures using NaHS. The dosage is determined by jar tests, and usually requires two to four times the stoichiometric amount of FeS. This may result in large chemical costs and in the generation of large amounts of sludge, almost three times as much as a hydroxide precipitation process (12).

The calcium sulfide process is another insoluble-sulfide method. The CaS is available in a stable, dry, solid form. When mixed with water, it forms an equimolar solution of calcium hydroxide and calcium bisulfate, $Ca(HS)_2$. Passing H_2S gas through lime slurries can also produce the solution on-site. The generation process can be controlled by pH control, as can the dosage to the precipitation process itself. This process has been demonstrated at the pilot and full scale for treatment of wastes from a wire manufacturing plant containing dissolved copper in an aqueous emulsion.

In chemical industry, sulfide precipitation use has mainly been to remove mercury, lead, and silver from wastewater, with less frequent use to remove other metal ions. Sulfide precipitation is also used to precipitate hexavalent chromium (Cr^{6+}) without prior reduction to the trivalent state (Cr^{3+}), as is required in the hydroxide process. Sulfide precipitation is being practiced in the following industries:

(a) Photographic equipment and supplies.
(b) Inorganic chemicals manufacturing.
(c) Coal mining.
(d) Textile mills.
(e) Nonferrous metals manufacturing.
(f) Ore mining and dressing.

Most of the chlor-alkali industry (subcategory of the Inorganic Chemicals Manufacturing) is applying this technology to remove lead or mercury from its waste streams. Most metal sulfides are less soluble than hydroxides and, in general, the precipitates are frequently more dependably removed from wastewater. Sulfide precipitation has potential for use as a polishing treatment after hydroxide precipitation and sedimentation to remove residual metals. This way one can obtain higher treatment efficiencies of sulfide removal at a lower chemical cost.

7.4. Cyanide Precipitation

Cyanide precipitation can be used when cyanide destruction is not feasible because of the presence of cyanide complexes that are difficult to destroy. This technology is being used in the Coil Coating industry.

7.5. Magnesium Oxide Precipitation

Magnesium oxide (MgO) treatment produces a MgO–hydroxide sludge that has a lower solubility than hydroxide sludge alone (16). The sludge is also relatively compact, and tends to cement together upon standing, which inhibits re-suspension of metal ions. The chemical costs are higher than for hydroxide precipitation, but as for sulfide removal, the process can be used after conventional treatment with lime.

7.6. Chemical Oxidation–Reduction Precipitation

Occasionally, metal ions present in a wastewater are in an oxidation state that is quite soluble. Changing the oxidation state in these cases may result in an ion with low solubility with respect to an appropriate treatment process. Examples of this are the reduction of hexavalent chromium (Cr^{6+}) to the trivalent state (Cr^{3+}), and the oxidation of ferrous ion (Fe^{2+}) and manganese (Mn^{2+}) to ferric (Fe^{3+}) and manganese (Mg^{3+}) forms, which form insoluble precipitates. In an unusual process, sodium borohydride has been used to reduce heavy metals to the elemental state, yielding a compact precipitate for lead, silver, cadmium, and mercury. Reduction is best done in the absence of organic compounds (16). For details, the readers are refered to the chapter dealing with chemical reduction and oxidation.

7.7. Lime / Soda-Ash Softening

This is one of the most common softening processes for removal of hardness from potable or industrial water. Because the process is discussed in Chapter 6 (Recarbonation and Softening) in detail, it will not be covered here.

7.8. Phosphorus Precipitation

Phosphorus must often be removed from wastewater to protect surface water from its fertilizing effect. Phosphorus is present in water in three main forms: phosphate, PO_4^{3-}, is the simplest, and is the form most easily removed by precipitation, phosphate can be condensed into the polymeric polyphosphate form, and organic forms may also be present. These complex forms are also removed during precipitation, but the mechanisms are complicated and include adsorption onto other flocs.

Phosphate precipitation is carried out primarily with aluminum, iron, and calcium cations (19). Theoretical stoichiometric dosages are not reliable indications of actuarial requirements, so dosages must be empirically measured.

Aluminum combines with phosphate to form aluminum phosphate, $AlPO_4$. The source of aluminum ions can be alum (aluminum sulfate) or sodium aluminate ($Na_2Al_2O_4$). Alum usage consumes alkalinity and therefore may decrease the pH. Low alkalinity wastewater may require addition of alkalinity. Sodium aluminate, on the other hand, contributes to alkalinity, tending to increase the pH.

The optimum pH for removal of phosphate by aluminum is between 5.5 and 6.5. The minimum solubility of aluminum phosphate is approx 0.01 mg/L PO_4^{3-}–phosphorous at pH 6.

The stoichiometric requirement is 1 mole of aluminum ions per mole of phosphate. However, higher dosage is usually required for adequate removal efficiency. For example, for municipal wastewater, a 38% excess was found to provide only 75% phosphorus reduction, while 72% excess was required to remove 85% of the phosphorus, and 130% excess was necessary to remove 95%. The removal of phosphorus with aluminum has also been found to help with the removal of copper, chromium, and lead (20).

Iron can be used for phosphate removal in either the ferrous (Fe^{2+}) or ferric (Fe^{3+}) forms. Both have been found experimentally to give similar results with similar molar ratios, in spite of their differing stoichiometry. However, the two forms have distinct pH optima, the optimum for precipitation as ferric phosphate is at pH from 4.5 to 5.0, although significant removal is obtained at higher pH. The optimum for precipitation with ferrous ion is at about pH 8, and remains good as low as pH 7.

Calcium can be added to water in the form of lime to precipitate phosphate as hydroxyapatite. The removal efficiency increases with pH, achieving 80% removal at pH below 9.5.

7.9. *Other Chemical Precipitation Processes*

A wide variety of less common applications have been developed. The removal of ammonium fluoride using limestone has been studied (21). Calcium chloride has also been used for fluoride removal (3). Mercury was removed using a process combining sodium sulfide and ferric chloride in two stages (22). Humic substances have been reported to be removed in lime softening (23). Heavy metals can be removed from municipal wastewater, without removing the organic suspended and settleable solids, by coprecipitaion on sand grains with calcium carbonate and calcium hydroxyapatite in an upflow expanded bed (24). This enables the organics to be removed subsequently in conventional processes and yield sludge with lower metal concentrations.

8. PROCESS EVALUATION

8.1. *Advantages and Limitations*

Chemical precipitation has proven to be an effective technique for removing many industrial wastewater pollutants. It operates at ambient conditions and is well suited to automatic control. The use of chemical precipitation may be limited because of interference of chelating agents and other chemical interference possible when mixing wastewater and treatment chemicals, or because of the potentially hazardous situation involved with the storage and handling of chemicals.

Hydroxide precipitation is most commonly used in industry and produces a high-quality effluent when applied to many waste streams (particularly when followed by flocculation and filtration). Often, coprecipitaion of a mixture of metal ions will result in residual metal solubilities lower than those that could be achieved by precipitating each metal at its optimum pH. Some common limitations of the hydroxide process are as follows:

1. The theoretical minimum solubility for different metals occurs at different pH values (Fig. 1). For a mixture of metal ions, it must be determined whether a single pH can produce sufficiently low solubilities for the metal ions present in the wastewater.
2. Hydroxide precipitates tend to resolubilize if the solution pH is increased or decreased from the minimum solubility point; thus, maximum removal efficiency will not be achieved unless the pH is controlled within a narrow range.
3. The presence of complexing ions, such as phosphates, tartrates, ethylenediaminetetraacetic acid (EDTA), and ammonia may have adverse effects on metal removal efficiencies when hydroxide precipitation is used.

4. Hydroxide precipitation usually makes recovery of the precipitated metals difficult because of the heterogeneous nature of most hydroxide sludges.

Lime for hydroxide precipitation has gained widespread use because of its ease of handling, economy, and treatment effectiveness for a great variety of dissolved materials. However, if there is sulfate ion present in the wastewater, gypsum (calcium sulfate) will be formed. This increases sludge production, may cause a scaling problem in pipelines, and may clog dual media filters. Using caustic soda is more expensive but it generally eliminates the scaling problem. Total dissolved solids will increase in wastewater treated with caustic soda as a result of the formation of sodium salt.

Sulfide precipitation has been demonstrated to be an effective alternative to hydroxide precipitation for removing various heavy metals from industrial wastewaters. The major advantage of the sulfide precipitation process is that because of the extremely low solubility of metal sulfides, very high metal removal efficiencies can be achieved. Additional advantages of sulfide precipitation are as follows:

1. The sulfide process has the ability to remove chromate and dichromate without preliminary reduction of chromium to its trivalent state.
2. The high reactivity of sulfides with heavy metal ions and the insolubility of metal sulfides over a broad pH range are attractive features compared with the hydroxide precipitation process.
3. Sulfide precipitation, unlike hydroxide precipitation, is relatively insensitive to the presence of most chelating agents and eliminates the need to treat these wastes separately.

The major limitations of the sulfide precipitation process are the evolution of toxic hydrogen sulfide fumes and the discharge of treated wastewater containing residual levels of sulfide. Other factors include:

1. Sulfide reagent will produce hydrogen sulfide fumes when it comes into contact with acidic wastes. Maintaining the pH of solution between 8 and 9.5 and providing ventilation of treatment tanks can control this problem.
2. As with hydroxide precipitation, excess sulfide ion must be present to drive the precipitation reaction to completion. Because the sulfide ion itself is toxic, sulfide addition must be carefully controlled to maximize heavy metals precipitation with a minimum of excess sulfide to avoid the necessity of posttreatment. Where excess sulfide is present, aeration of the effluent stream would be necessary to oxidize residual sulfide to the less harmful sodium sulfate (Na_2SO_4).
3. The cost of sulfide precipitation is high in comparison with hydroxide treatment, and disposal of metallic sulfide sludges may pose problems.

The use of ferrous sulfide (insoluble-sulfide process) as a source of sulfide reduces or virtually eliminates the problem of hydrogen sulfide evolution. The use of ferrous sulfide, however, requires reagent consumption considerably higher than stoichiometric and produces significantly larger amounts of sludge than either the hydroxide or soluble-sulfide treatment processes.

8.2. *Reliability*

Hydroxide and sulfide chemical precipitation are highly reliable, although proper monitoring and control are required. The major maintenance needs involve periodic upkeep of equipment for monitoring, automatic feeding, and mixing, and other hardware.

8.3. Chemicals Required

1. Hydroxide precipitation: Quicklime, CaO, hydrated lime, $Ca(OH)_2$, and liquid caustic soda, NaOH.
2. Sulfide precipitation: sodium sulfate, Na_2SO_4, sodium sulfide, Na_2S, and ferrous sulfate, $FeSO_4$.
3. Cyanide precipitation: zinc sulfate, $ZnSO_4$, and ferrous sulfate, $FeSO_4$.
4. Carbonate precipitation: calcium carbonate, $CaCO_3$, and carbon dioxide, CO_2.

8.4. Residuals Generated

Chemical precipitation generates solids that must be removed in a subsequent treatment step, such as sedimentation or filtration. Sulfide sludges are less subject to leaching than hydroxide sludges. However, the long-term impacts of weathering and of bacterial and air oxidation of sulfide sludges have not been evaluated.

8.5. Process Performance

The performance of chemical precipitation depends on several variables. The most important factors affecting precipitation effectiveness are:

1. Maintenance of an alkaline pH throughout the precipitation reaction and subsequent settling.
2. Addition of a sufficient excess of treatment ions to drive the precipitation reaction to completion.
3. Addition of an adequate supply of sacrificial ions (such as ion or aluminum) to ensure precipitation and removal of specific target ions.
4. Effective removal of precipitated solids.

Proper control of pH is absolutely essential for favorable performance of precipitation/sedimentation technologies. This is clearly illustrated by solubility curves for selected metal hydroxides and sulfides as shown in Fig. 1. Hydroxide precipitation is effective in removing arsenic, cadmium, trivalent chromium, copper, iron, manganese, nickel, lead, and zinc. Sulfide treatment is superior to hydroxide treatment for removal of several metals and is very effective for removal of mercury and silver. As shown by theoretical solubilities of hydroxides and sulfides of selected metals (Table 2), sulfide precipitation is highly effective in removal of cadmium, cobalt, copper, iron, mercury, manganese, nickel, silver, tin, and zinc. Estimated achievable maximum 30-d average concentrations of several heavy metals under different chemical precipitation and solid-removal technologies are shown in Table 3. The estimated achievable concentration is based on the performance data reported in literatures (27–39).

9. APPLICATION EXAMPLES

9.1. Example 1

It has been known that the chemical precipitation process is technically and economically feasible for treating the following industrial effluents (25–44):

(a) Foundries.
(b) Metal finishing.
(c) Iron and steel manufacturing.
(d) Textiles.

Table 2
Theoretical Solubilities of Hydroxides, Carbonates and Sulfides of Selected Metals in Pure Water

	Solubility of metal ion (mg/L)		
Metal	As hydroxide	As carbonate	As sulfide
Cadmium (Cd^{2+})	2.3×10^{-5}	1.0×10^{-4}	6.7×10^{-10}
Chromium (Cr^{3+})	8.4×10^{-4}		No precipitate
Cobalt (Co^{2+})	2.2×10^{-1}		1.0×10^{-8}
Copper (Cu^{2+})	2.2×10^{-2}		5.8×10^{-13}
Iron (Fe^{2+})	8.9×10^{-1}		3.4×10^{-5}
Lead (Pb^{2+})	2.1	7.0×10^{-3}	3.8×10^{-9}
Manganese (Mn^{2+})	1.2		2.1×10^{-3}
Mercury (Hg^{2+})	3.9×10^{-4}	3.9×10^{-2}	9.0×10^{-2}
Nickel (Ni^{2+})	6.9×10^{-3}	1.9×10^{-1}	6.9×10^{-8}
Silver (Ag^{+})	13.3	2.1×10^{-1}	7.4×10^{-12}
Tin (Sn^{2+})	1.1×10^{-4}		3.8×10^{-9}
Zinc (Zn^{2+})	1.1	7.0×10^{-4}	2.3×10^{-7}

(e) Steam electric power plants.
(f) Inorganic chemicals manufacturing.
(g) Ore mining and dressing.
(h) Porcelain enameling.
(i) Paint and ink formulation.
(j) Coil coating.
(k) Nonferrous metals manufacturing.
(l) Aluminum forming.
(m) Battery manufacturing.
(n) Electrical and electronic components.
(o) Copper coating.
(p) Organic and inorganic wastes.
(q) Auto and other laundries.

Conduct a literature search, and present the treatability data sheets for the chemical industries listed above.

Solution

The treatability data sheets shown in Appendices A–Q provide performance data from studies on the above industries and/or waste streams using chemical precipitation in combination with various solids separation processes including flocculation, sedimentation and filtration.

9.2. *Example 2*

The chemical precipitation process is commonly used in conjunction with either sedimentation (such as Appendices A, B1, B2, C, D, F, G1 G2, J1, J2, K1, K2, L–Q) or filtration (such as Appendices B3, E1, and E2) for removal of chemically produced precipitates (or flocs). List and explain other combinations of separation processes that can be used in conjunction with chemical precipitation. The processes and their combinations should be feasible from both technical and economical viewpoints.

Table 3
Estimated Achievable 30-d Averages of Final Concentrations for Various Applied Technologies (26)

	Final concentrations(mg/L)					
	Lime precipitation followed by filtration	Lime precipitation followed by filtration	Sulfide precipitation followed by filtration	Ferrite coprecipitation followed by filtration	Soda ash addition followed by sedimentation	Soda ash addition followed by filtration
Antimony, Sb	0.8–1.5	0.4–0.8				
Arsenic, As	0.5–1.0	0.5–1.0	0.05–0.1			
Beryllium, Be	0.1–0.5	0.01–0.1				
Cadmium, Cd	0.1–0.5	0.05–0.1	0.01–0.1	< 0.5		
Copper, Cu	0.05–1.0	0.4–0.7	0.05–0.5	< 0.5		
Chromium, Cr (III)	0.0–0.5	0.05–0.5		0.01		
Lead, Pb	0.3–1.6	0.05–0.6	0.05–0.4	0.20	0.4–0.8	0.1–0.6
Mercury, Hg (II)	0.01–0.05		< 0.01			
Nickel, Ni	0.2–1.5	0.05–0.2	0.1–0.5			
Silver, Ag	0.4–0.8	0.2–0.4	0.05–0.5			
Selenium, Se	0.2–1.0	0.1–0.5				
Thallium, Tl	0.2–1.0	0.1–0.5				
Zinc, Zn	0.5–1.5	0.4–1.2	0.02–1.2	0.02–0.5		

Solution

Other feasible process combinations include at least the following (45–90,92,93):

(a) Chemical precipitation + sedimentation + filtration.
(b) Chemical precipitation + flotation.
(c) Chemical precipitation + flotation + filtration.
(d) Chemical precipitation + ultrafiltration (UF).
(e) Chemical precipitation + ultrafiltration + reverse osmosis (RO).
(f) Chemical precipitation + fabric filtration (cartridge filter) + RO.
(g) Chemical precipitation + physical chemical sequencing batch reactor (PCSBR).
(h) Chemical precipitation + PCSBR + filtration.
(i) Chemical precipitation + PCSBR + membrane filtration (NF, UF, and/or RO).
(j) Chemical precipitation + filtration + granular activated carbon (GAR).
(k) Chemical precipitation + flotation + GAR.
(l) Chemical precipitation + biological treatment process.
(m) Biological treatment process + chemical precipitation.
(n) Chemical precipitation + sedimentation + filtration + ion exchange.
(o) Chemical precipitation + flotation + filtration + ion exchange.

Each process combination mentioned above may include chemical coagulation, if necessary, and can be a conventional continuous process, or an innovative physical chemical sequencing batch process (PCSBR) (87,96).

The most common filtration process is rapid sand filtration, although slow sand filtration is an option. Membrane filtration includes nanofiltration (NF), microfiltration (MF), ultrafiltration (UF), and reverse osmosis (RO), of which at least one is needed in the process system.

The flotation process can be one or more of the following: dissolved air flotation, dispersed air flotation, electroflotation, ion flotation, precipitate flotation, and foam separation.

The biological treatment process can be, but is not limited to, activated sludge, trickling filter, rotating biological contactors, sequencing batch reactor, and membrane bioreactor.

Conventional sequencing batch reactor (SBR) is a biological treatment process, while innovative physicochemical sequencing batch reactor (PCSBR) is a newly developed physicochemical treatment process (87,96).

9.3. *Example 3*

Explain the difference between the chemical precipitation process and the chemical coagulation process.

Solution

Chemical precipitation is a chemical unit process in which undesirable soluble metallic ions and certain anions are converted to insoluble form and then removed from water or wastewater. It is a commonly used treatment technique for removal of heavy metals, phosphorus, and hardness. The procedure involves alteration of the ionic equilibrium to produce insoluble precipitates that can be removed easily by a solid-separation operation that may include coagulation/flocculation and/or sedimentation, flotation, filtration, or a membrane process to remove the precipitates.

It is important to note that undesirable pollutants to be removed by chemical precipitation are soluble cations and/or anions. Chemical coagulation and flocculation, however, are terms often used interchangeably to describe the physiochemical process of suspended particles aggregation resulting from chemical additions to water or wastewater. Technically, coagulation involves the reduction of electrostatic surface charges and the formation of complex hydrous oxide. Coagulation is essentially instantaneous in that the only time required is that necessary for dispersing the chemicals in solution. Flocculation is the time-dependant physical process of the aggregation of solids into particles large enough to be separated by sedimentation, flotation, filtration, or any of the membrane processes. For particles in the colloidal and fine supercolloidal size ranges (less than 1–2 μm), natural stabilizing forces (electrostatic repulsion and physical repulsion by absorbed surface water layers) predominate over the natural aggregating forces (van der Waals forces) and the natural mechanism that tends to cause particles contact (Brownian motion). The purpose of coagulation is to overcome the above repulsive forces and, hence, to allow small particles to agglomerate into larger particles, so that gravitational and inertial forces will predominate and effect the settling of the particles. The process can be grouped into two sequential mechanisms: (a) Chemically induced destabilization of the repulsive surface related forces, thus allowing particles to stick together when contact between particles is made, and (b) chemical bridging and physical enmeshment between the non-repelling particles, thus allowing for the formation of large particles.

NOMENCLATURE

A	activity parameter = approximately 0.5 in water at 25°C
a_I	activity (dimensionless number)
$[A_i]$	molar concentration, mol/L or mg/L
C	solution concentration, mol/L or mg/L
C°	reference concentration, mol/L or mg/L
C_i	concentration of species i, mol/L or mg/L
$C_{i,\text{eq}}$	concentration of species i at equilibrium, mol/L or mg/L
Cs	saturation concentration, mol/L or mg/L
C_T	total carbonic species concentration, mol/L or mg/L
DF	driving force
G	velocity gradient
G_i	standard free energy, J
G_i°	standard free energy at C° for species i, J
ΔG_T°	standard free energy change for the reaction, J
γ_i	activity coefficient
i	ith species
k	rate coefficient, dependant upon amount of mixing
K_c	standard concentration equilibrium constant
K_{sp}	solubility product constant
n_i	stoichiometric coefficient of the species i (positive for products negative for reactants)
R	universal gas constant = 8.314 J/K-mol = 1.99 cal/K-mol
s	crystal surface area
t	detention time
T	absolute temperature in kelvins, K = °C + 273
TDS	total dissolved solids, mg/L
Z_i	the charge on species i.

REFERENCES

1. S. C. Atlow, *Biotechnol. Bioeng.* **26**, 599 (1984).
2. B. Gomulka and E. Gomolka, *Effluent Water Treat. J. (G.B.)* **24**, 119 (1985).
3. D. Biver and A. Degols, *Tech. de l' Eau (Fr.)*, 428/429, **31**, (1982); (abstr) WRC Info., 10, 83-0524 (1983).
4. V. K. La Ver, *J. Colloid Science* **19**, 291–293 (1964).
5. R. C. Weast, ed. *CRC Handbook of Chemistry and Physics*, CRC Press, Boca Raton, FL, 1980.
6. W. Stumm and J. J Morgan, *Aquatic Chemistry*, Wiley- Interscience, New York, 1970.
7. J. W. Patterson, Effect of carbonate ion on precipitation treatment of cadmium, copper, lead and zinc, *Proceeding of 13th Annual Purdue Industrial Waste Conference*, May, 8, 1975.
8. J. S. Patterson, H. E. Allen, and J. J. Scala, *JWPCF* **49**, 2397 (1977).
9. L. D. Benefield, J. F. Judkins, and B. L. Weand, *Process Chemistry for Water and Wastewater Treatment*, Prentice-Hall, NJ, 1980.
10. S. D. Conte and L. deBoor, *Elementary Numerical Analysis*, McGraw-Hill, New York, NY, 1988.
11. K. J. Yost and A. Scarfi, *JWPCF* **51**, 1878 (1979).
12. B. M. Kim and P. A. Amodeo, *Environ. Prog.* **2**, 175 (1983).
13. M. M. Benjamin, *J. Water Pollut. Control Fed.* **54**, 1472 (1982).
14. G. M. Fair, J. C. Geyer, and D. A. Okun, *Water and Wastewater Engineering*, Vol. 2. John Wiley & Sons, New York, NY, 1968.
15. US EPA, *Process Design Manual for Suspended Solids Removal*, EPA 625/1-75-003. US Environmental Protection Agency Technology, Washington, DC, 1975.
16. D. W. Grosse, *JAPCA*, **36**, May (1986).
17. D. Walker, *Water Serv.*, **86**, 165 (1982).
18. T. E. Larson and A. M. Buswell, *JAWWA* **34,** 1667 (1942).
19. US EPA, *Process Design Manual for Phosphorus Removal.* US Environmental Protection Agency Technology, Washington, DC, 1971.
20. D. B. Aulenbach, *Proc. of the 16th Mid-alt Ind. Waste Conf. Proc.* (June 24–26, 1984). Penn. State Univ., University Park, PA, vol. 16, P.318
21. P. Ekdunge and D. Simonsson, *J. Chem Technol. Biotechnol. (Sweden)*, **34A**, 1 (1984); (abstr.) *WRC Info.* **11**, 84–1502 (1984).
22. Osaka Soda Co, Ltd. *Japan Kokai Tokkyo koho (Jap.)*, Pat No. 83-49 490 (1983); Chem. Abstr., **99**, 127894u (1983).
23. S. J. Randtke, *JAWWA* **74**, 192 (1982).
24. J. Huang, P. M. McCole, and R. K. Breuer, *Proceedings of the 30th Annual Purdue Industrial Waste Conference,* May 8, 1975.
25. US EPA, *Development Document for Effluent Limitations Guidelines and Standards for the Battery Manufacturing Point Source Category.* EPA 440/1-80/067. US Environmental Protection Agency, Washington, DC, 1980.
26. US EPA, *Development Document for Effluent Limitations Guidelines and Standards for Inorganic Chemicals Manufacturing Point Source Category.*EPA-440/1-79/007. US Environmental Protection Agency, Washington, DC, 1980.
27. US EPA, *Development Document for Effluent Limitations Guidelines and Standards for the Auto and other Laundries Point Source Category.* US Environmental Protection Agency, Washington, DC, 1980.
28. US EPA, *Development Document for Existing Source Pretreatment Standards for the Electroplating Point Source Category.* EPA-440/1-79/003. US Environmental Protection Agency, Washington, DC, 1979.
29. US EPA, *Development Document for Effluent Limitations Guidelines and Standards for the Iron and Steel Manufacturing Point Source Category*; general. EPA-440/1-80/024-b. US Environmental Protection Agency, Washington, DC, 1980.

30. US EPA, *Development Document for Effluent Limitations Guidelines and Standards for the Iron and Steel Manufacturing Point Source Category; Coke Making Subcategory, Sintering Subcategory, Iron Making Subcategory*, Volume II. EPA-440/1-80/024-b. US Environmental Protection Agency, Washington, DC, 1980.
31. US EPA, *Development Document for Effluent Limitations Guidelines and Standards for the Iron and Steel Manufacturing Point Source Category; Hot Forming Subcategory*, Volume IV. EPA-440/1-80/024-b. US Environmental Protection Agency, Washington, DC, 1980.
32. US EPA, *Development Document for Effluent Limitations Guidelines and Standards for the Coil Coating Point Source Category*. EPA-440/1-81/071-b. US Environmental Protection Agency, Washington, DC, 1981.
33. US EPA, *Development Document for Effluent Limitations Guidelines and Standards for the Foundries (metal molding and casting) Point Source Category*. EPA-440/1-80/070-a. US Environmental Protection Agency, Washington, DC, 1980.
34. US EPA, *Development Document for Effluent Limitations Guidelines and Standards for the Metal Finishing Point Source Category*. EPA-440/1-80/091-a. US Environmental Protection Agency, Washington, DC, 1980.
35. US EPA, *Development Document for Effluent Limitations Guidelines and Standards for the Textile Mills Point Source Category*. EPA-440/1-79/022-b. US Environmental Protection Agency, Washington, DC, 1979.
36. US EPA, *Development Document for Effluent Limitations Guidelines and Standards for the Aluminum Forming Point Source Category*. EPA-440/1-80/073-a. US Environmental Protection Agency, Washington, DC, 1980.
37. US EPA, *Development Document for Effluent Limitations Guidelines and Standards for the Electrical and Electronic Components Point Source Category*. EPA-440/1-80/075-a. US Environmental Protection Agency, Washington, DC, 1980.
38. US EPA, *Process Design Manual for Upgrading Existing Wastewater Treatment Plants*. EPA-625/1-71-004a. US Environmental Protection Agency, Washington, DC, 1971.
39. US EPA, *Control and Treatment Technology for the Metal Finishing Industry, Sulfide Precipitation. Summary report*. EPA-625/8-80/003 US Environmental Protection Agency, Washington, DC, 1980.
40. N. Shammas and N. DeWitt, *Water Environment Federation, 65th Annual Conference, Proc. Liquid Treatment Process Symposium*, New Orleans, LA, September 20–24, 1992, pp. 223–232.
41. N. Shammas and M. Krofta, *A Compact Flotation - Filtration Tertiary Treatment Unit for Wastewater Reuse*, Water Reuse Symposium, AWWA, Dallas, TX, February 27, pp. 97–109, 1994.
42. M. Krofta, D. Miskovic, N. Shammas, and D. Burgess, *Pilot Scale Applications of a Primary-Secondary Flotation System on Three Municipal Wastewaters*. Specialist Conference on Flotation Processes in Water and Sludge Treatment, Orlando, FL, April 26–28, 1994.
43. M. Krofta, D. Miskovic, N. Shammas, D. Burgess, and L. Lampman, *IAWQ 17th Biennial International Conference*, Budapest, Hungary, July 24–30, 1994.
44. N. Shammas, *Proc. International Conference: Rehabilitation and Development of Civil Engineering Infrastructure Systems—Upgrading of Water and Wastewater Treatment Facilities*. Organized by The American University of Beirut and University of Michigan, Beirut, Lebanon, June 9–11, 1997.
45. L. K. Wang, *Water and Sewage Works*, **119**, p. 123–125 (1972).
46. L. K. Wang, *Journal American Water Works Association,* **65**, 355–358 (1973).
47. M. H. Wang, T. Wilson, and L. K. Wang, *Journal of American Water Resources Association*, **10**, 283–294 (1974).
48. L. K. Wang, *Resource Recovery and Conservation.* **1**, 67–84 (1975).
49. M. H. Wang, T. Wilson, and L. K. Wang, *Journal of the Environmental Engineering Division, Proceedings of the American Society of Civil Engineers*, **100**, 629–640 (1974).

50. L. K. Wang, *Treatment of a Wastewater from Military Explosive and Propellant Production Industry by Physicochemical Processes*. US Defense Technical Information Center, Alexandria, VA, AD-A027329, 1976.
51. L. K. Wang, *Separation and Purification Methods,* **6**(1), 153–187 (1977).
52. L. K. Wang, *The Canadian Journal of Chemical Engineering*, **60**, 116–122 (1982).
53. M. Krofta and L.K. Wang, *First Full Scale Flotation Plant in U.S.A. for Potable Water Treatment*, US Department. of Commerce, National Technical Information Service, Springfield, VA, PB82-220690, 1982, p. 67.
54. M. Krofta and L. K. Wang, (1984) *ASPE Journal of Engineering Plumbing*, 1–16, NTIS-PB83-107961, 1984.
55. M. Krofta and L. K. Wang, *Civil Engineering for Practicing and Design Engineers*, **3**, 253–272 (NTIS-PB83-171165) (1984).
56. L. K. Wang, *American Institute of Chemical Engineers National Conference*, Houston, TX, NTIS-PB83-232843 (1983).
57. M. Krofta and L. K. Wang, *American Institute of Chemical Engineers National Conference*, Houston, TX, NTIS-PB83-232850, 1983.
58. M. Krofta and L. K. Wang, *Recent Advances in Titanium Dioxide Recover, Filler Retention and White Water Treatment*, US Dept of Commence, National Technical Information Service, 1983, PB83-219543.
59. L. K. Wang, *Removal of Extremely High Color from Water Containing Trihalomethane Precursor by Flotation and Filtration*, US Dept. of Commerce, National Technical Information Service, 1982, PB83-240374.
60. M. Krofta and L. K. Wang, *Design of Innovative Flotation Waste Wastewater Treatment Systems for A Nickel-Chromium Plating Plate*, US Dept. of Commerce, National Technical Information Service, 1984, PB88-200522/AS.
61. M. Krofta and L. K. Wang, *Waste Treatment by Innovative Flotation-Filtration and Oxygenation-Ozonation Process*, US Department of Commerce, National Technical information Service, 1984, PB85-174738-AS.
62. M. Krofta and L. K. Wang, *Development of Innovative Electroflotation Water Purification System Service for Single Families and Small Communities,* US Department of Commerce, National Technical Information Service, Springfield, VA, 1984, PB85-207595/AS.
63. M. Krofta and L. K. Wang, *Proceedings of the American Water Works Association, Water Reuse Symposium III*, Vol. 2, 1984, pp. 881–898.
64. L. K. Wang, *Removal of Arsenic from Water and Wastewater*. US Department of Commerce, National Technical Information Service, Springfield, VA, 1984, PB86-169299.
65. L. K. Wang, *Development of New Treatment System Consisting of Adsorption Flotation and Filtration*, US Department of Commerce, National Technical Information Service, 1984, PB85-209401/AS.
66. W. Layer and L. K. Wang, *Water Purification and Wastewater Treatment with Sodium Aluminate*, US Department of Commerce, National Technical Information Service, Springfield, VA, 1984, PB85-214-492/AS.
67. L. K. Wang, *Removal of Arsenic and Other Contaminants from Storm Run-off Water by Flotation, Filtration, Adsorption and Ion Exchange*. US Department of Commerce, National Technical Information Service, 1984, PB88-200613/AS.
68. L. K. Wang, *OCEESA Journal* **1**, 15–18, 1984, NTIS-PB85-167229/AS.
69. M. Krofta and L. K. Wang, *Proceedings of the Powder and Bulk Solids Conference*, Chicago, 11, 1985.
70. M. Krofta and L. K. Wang, *Treatment of Scallop Processing Wastewater by Flotation, Adsorption and Ion Exchanging*, Technical Report #LIR/05-85/139, Lenox Institute of Water Technology, Lenox, MA, 1985.
71. M. Krofta, and L. K. Wang, *Proceedings of the 41st Industrial Waste Conference*, Lewis Publishers, Inc., Chelsea, MI, 1987, pp. 67–72,

72. L. K. Wang, *Tertiary Wastewater Treatment.* US Department of Commerce, National Technical information Service, Springfield, VA, 1987, PB88-168133/A.
73. M. Krofta and L. K. Wang, *Technical Association of Pulp and Paper Industry Journal, (TAPPI Journal)* **70**, 92–96, 1987.
74. M. Krofta and L. K. Wang, *Municipal Waste Treatment by Supracell Flotation, Chemical Oxidation and Star System*, US Department of Commerce, National Technical Information Service, 1986, PB88-200548/AS.
75. L. K. Wang, *Design Operation and Maintenance of the Nation's Largest Physicochemical Waste Treatment Plant*, Volume 1, Lenox Institute for Research, MA, Report #LIR/03-87-248, 1987.
76. L. K. Wang, *Design Operation and Maintenance of the Nation's Largest Physicochemical Waste Treatment Plant,* Volume 2, Lenox Institute for Research, MA, Report #LIR/03-87-249, 1987.
77. L. K. Wang, *Design Operation and Maintenance of the Nation's Largest Physicochemical Waste Treatment Plant*, Volume 3, Lenox Institute for Research, MA, Report #LIR/03-87-250, 1987.
78. M. Krofta and L. K. Wang *Proceedings of the Joint Conference of American Water Works Association and Water Pollution Control Federation*, Cheyenne, WY, 1987 (NTIS-PB88-200563/AS).
79. M. Krofta and L. K. Wang, *Proceedings of the 42nd Industrial Waste Conference*, pp. 185, 1987.
80. L. K. Wang, *Preliminary Design Report of a 10-MGD Deep Shaft-Flotation Plant for the City of Bangor*, US Department of Commerce, National Technical Information Service, 1987, PB88-200597/AS.
81. L. K. Wang, *Preliminary Design Report of a 10-MGD Deep Shaft-Flotation Plant for the City of Bangor, Appendix*, US Department of Commerce, National Technical Information Service, 1987, PB88-200605/AS.
82. M. Krofta and L. K. Wang, *Proceedings of the 1988 Food Processing Conference*, Georgia Institute of Technology, Atlanta, GA, 1988.
83. L. K. Wang, *Proceedings of the 44th Industrial Waste Conference*, 1990, pp. 655–666.
84. L. K. Wang, *Proceedings of the 44th Industrial Waste Conference*, 1990, pp. 667–673.
85. L. K. Wang, *Proceedings of New York–New Jersey Environmental Exposition*, Belmont, MA, Oct. 1990.
86. L. K. Wang, *Water Treatment*, **8**, 7–16 (1993).
87. L. K. Wang, *Water Treatment*, **10**, 121–134 (1995).
88. L. K. Wang, *Water Treatment*, **10**, 261–282 (1995).
89. L. K. Wang, *OCEESA Journal*, **13**, 12–16 (1996).
90. L. K. Wang and S. Kopko, *City of Cape Coral Reverse Osmosis Water Treatment Facility.* US Department of Commerce, National Technical Information Service, Springfield, VA, 1997, Publication # PB97-139547.
91. L. D. Benefield and J. M. Morgan, Chemical Precipitation, in *Water Quality and Treatment.* R. D. Letterman, ed. McGraw-Hill, Inc., NY, 1999, pp. 10.1–10.57.
92. L. K. Wang, C. Yapijakis, Y. Li, and Y.T. Hung, *OCEESA Journal* **20**, No. 2 (2003).
93. L. K. Wang, *The State-of-the-Art Technologies for Water Treatment and Management.* United Nations Industrial Development Organization (UNIDO), Vienna, Austria. UNIDO Training Manual No. 8-8-95, 1995.
94. US EPA. *Chemical Precipitation.* US Environmental Protection Agency, Washington, DC, Sept. 2000, EPA832-F-00-018.
95. Ondeo Nalco Co. *Chemical Precipitation.* Ondeo Nalco Co., Naperville, IL, 2004 (www.ondeonalco.com).
96. L. K. Wang, Y. T. Hung and N. K. Shammas (eds.). *Advanced Physicochemical Treatment Processes.* Humana Press, Totowa, NJ, 2005.

APPENDIXES

APPENDIX A

TREATMENT TECHNOLOGY: Chemical Precipitation With Sedimentation (Alum, NaOH, H_2SO_4)

Data source: Effluent Guidelines
Point source: Foundry Industry
Subcategory: Aluminum foundry-die casting
Plant: 574-C

Data source status:
Not specified
Bench scale ___
Pilot scale ___
Full scale x

Pretreatment/treatment: Emulsion Breaking/Chem. Ppt.

DESIGN OR OPERATING PARAMETERS

Wastewater flow rate: 4.3 L/s
Chemical dosages(s): Unspecified
Mix detention time: Unspecified
Flocculation detention time: Unspecified
Unit configuration: Continuous operation

Type of sedimentation: Basin

REMOVAL DATA

Sampling: Unspecified — Analysis: Data set 2 (V.7.3.12)

Pollutant/parameter	Concentration Influent	Concentration Effluent	Percent removal	Detection limit
Toxic pollutants, µg/L:				
Cyanide	BDL	BDL	NM	10
Lead	200	150	25	10
Zinc	1,300	40	97	10
Bis(2-ethylhexyl) phthalate	5,500	32	99	10
Butyl benzyl phthalate	690	BDL	99*	10
Di-n-butyl phthalate	74	BDL	93*	10
Diethyl phthalate	730	BDL	99*	10
2,4-Dimethylphenol	41	BDL	88*	10
Phenol	16	BDL	69*	10
p-Chloro-m-cresol	110	62	44	10
Anthracene/phenanthrene	BDL	BDL	NM	10
Benzo(a)pyrene	53	BDL	91*	10
Chrysene	780	10	99	10
Fluoranthene	370	BDL	98*	10
Fluorene	800	BDL	99*	10
Naphthalene	160	BDL	97*	10
Pyrene	80	BDL	94*	10
Chloroform	BDL	BDL	NM	10
Methylene chloride	BDL	39	NM	10
1,1,1-Trichloroethane	ND	51	NM	10
Acenaphthalene	20	BDL	75	10
Benzo(a)anthracene	ND	BDL	NM	10
Tetrachloroethylene	ND	30	NM	10
Trichloroethylene	ND	21	NM	10
Xylene	75	BDL	93*	10
Chromium	<100	<150	NM	10

BDL, below detection limit.
NM, not meaningful.
ND, not detected.
*Approximate value.

APPENDIX B1

TREATMENT TECHNOLOGY: Chemical Precipitation With Sedimentation

Data source: EGD Combined Data Base
Point source: Metal finishing
Subcategory: Common metals; precious metals; hexavalent chromium; cyanide, oils
Plant: 36040

Data source status:

Not specified	
Bench scale	___
Pilot scale	___
Full scale	x

Pretreatment/treatment: Chem. Ox.(CN), Chem. Red.(Cr)/Chem. Ppt., Sed.(clarifier)

DESIGN OR OPERATING PARAMETERS

Wastewater flow rate: 107,000 m^3/day
Chemical dosage(s): Unspecified
Mix detention time: Unspecified
Flocculation detention time: Unspecified
Unit configuration: Batch chem. ox. (CN); continuous chem. red. (Cr); clarifier - continuous operation

Type of sedimentation: Clarifier

REMOVAL DATA(a)

Sampling: 24-hr composite, flow proportion

Analysis: Data set 1(V.7.3.13)(a)

Pollutant/parameter	Concentration Influent	Concentration Effluent	Percent removal	Detection limit
Classical pollutants, mg/L:				
pH, minimum	6.8			
pH, maximum	7.1	9.1	NM	
Fluorides	4.5	5.7	NM	0.1
Phosphorus	2.2	0.07	97	0.003
TSS	100	11	89	5.0
TDS	960	1,500	NM	5.0
Iron	1.3	·0.07	95	0.005
Tin	0.08	0.06	25	
Oil and grease	20	BDL	88*	5.0
Gold	0.04	0.17	NM	
Toxic pollutants, µg/L:				
Cadmium	5.0	5.0	0	2.0
Chromium	26,000	530	98	3.0
Copper	5,900	69	99	1.0
Lead	53	BDL	72*	30
Nickel	120,000	1,400	99	6.0
Zinc	910	18	98	1.0
Cyanide, total	330	57	83	5.0
Hexavalent chromium	24,000	11	>99	5.0

Blanks indicate data not available.
BDL, below detection limit.
NM, not meaningful.
* Approximate value.
(a) Plant data are a three-day average.
(b) Original source of data: Electroplating Pretreatment 1976-1977 (HS).

APPENDIX B2

TREATMENT TECHNOLOGY: Chemical Precipitation With Sedimentation

Data source: EGD Combined Data Base
Point source: Metal finishing
Subcategory: Common metals; hexavalent chromium; cyanide; oil
Plant: 33024

Data source status:

Not specified	
Bench scale	
Pilot scale	
Full scale	x

Pretreatment/treatment: Chem. Ox.(CN), Chem. Red.(Cr)/Chem. Ppt., Sed.(clarifier)

DESIGN OR OPERATING PARAMETERS

Wastewater flow rate: 303,000 m³/day
Chemical dosage(s): Unspecified
Mix detention time: Unspecified
Flocculation detention time: Unspecified
Unit configuration: Clarifier - continuous operation

Type of sedimentation: Clarifier

REMOVAL DATA

Sampling: 8-hr composite, flow proportion

Analysis: Data set 1(V.7.3.13)(a)

Pollutant/parameter	Concentration Influent	Concentration Effluent	Percent removal	Detection limit
Classical pollutants, mg/L:				
pH, maximum	8.5	8.5		
Fluorides	23	18	22	0.1
Phosphorus	0.80	2.1	NM	0.003
TSS	250	42	83	5.0
Tin	0.12	0.15	NM	
Iron	2.5	0.18	93	0.005
Manganese	0.07	0.14	NM	0.005
Oil and grease	22	18	18	5.0
Aluminum	22	2.2	90	0.005
BOD	14	21	NM	
COD	82	90	NM	
Toxic pollutants, µg/L:				
Cadmium	95	5.0	95	2.0
Chromium	340	70	79	3.0
Copper	1,600	160	90	1.0
Lead	47	18	62	30
Nickel	96	19	80	6.0
Zinc	12,000	1,100	91	1.0
Cyanide, total	1,000	40	96	5.0
Hexavalent chromium	5.0	5.0	0	5.0
Mercury	1.0	1.0	0	0.1
Silver	2.0	4.0	NM	0.1/1.0

Blanks indicate data not available.
NM, not meaningful.
(a) Original source of data: M&MPM Composite Sampling 1975 (HS).

APPENDIX B3

TREATMENT TECHNOLOGY: Chemical Precipitation With Filtration

Data source: EGD Combined Data Base
Point source: Metal finishing
Subcategory: Common metals; hexavalent chromium; cyanide; oils
Plant: 36041

Data source status:
Not specified ___
Bench scale ___
Pilot scale ___
Full scale x

Pretreatment/treatment: Chem. Ox. (CN), Chem. Red. (Cr)/Chem. Ppt., Filter

DESIGN OR OPERATING PARAMETERS

Wastewater flow rate: 229,000 m^3/day
Chemical dosage(s): Unspecified
Mix detention time: Unspecified
Media (top to bottom): Unspecified
Unit configuration: Batch chem. ox.; batch chem. red.; continuous chem. ppt. and filter

Filtration rate (hydraulic loading):

REMOVAL DATA

Sampling: 24-hr composite, flow proportion (unspecified)   Analysis: Data set 1(V.7.3.13)(a)

Pollutant/parameter	Concentration Influent	Concentration Effluent	Percent removal	Detection limit
Classical pollutants, mg/L:				
pH, maximum	11	11		
Fluorides	2.5	3.9	NM	0.1
Phosphorus	1.2	0.05	96	0.003
TSS	520	10	98	5.0
TDS	1,400	1,600	NM	5.0
Iron	5.8	0.25	96	0.005
Tin	2.0	0.14	93	
Oil and grease	46	5.0	89	5.0
Toxic pollutants, µg/L:				
Cadmium	42	6.0	86	2.0
Chromium	12,000	610	95	3.0
Hexavalent chromium	5.0	5.0	0	5.0
Copper	7,500	440	94	1.0
Lead	140	32	77	30
Nickel	2,600	44	98	6.0
Zinc	13,000	140	99	1.0
Cyanide, total	2,000	400	80	5.0

Blanks indicate data not available.
NM, not meaningful.
(a)Original source of data: Electroplating Pretreatment 1976-1977(HS).

APPENDIX C

TREATMENT TECHNOLOGY: Chemical Precipitation With Sedimentation (Lime)

Data source: Effluent Guidelines
Point source: Iron and steel
Subcategory: Combination acid
Plant: I

Data source status:

Not specified	___
Bench scale	___
Pilot scale	___
Full scale	x

Pretreatment/treatment: Neutral./Chem. Ppt., Sed.

DESIGN OR OPERATING PARAMETERS

Wastewater flow rate: 69.4 L/s
Chemical dosages(s): Unspecified
Mix detention time: Unspecified
Flocculation detention time: Unspecified
Unit configuration: Continuous operation

Type of sedimentation: Settling lagoon

REMOVAL DATA

Sampling: Unspecified Analysis: Data set 2 (V.7.3.5)

Pollutant/parameter	Concentration Influent	Concentration Effluent	Percent removal	Detection limit
Classical pollutants, mg/L:				
TSS	560	130	77	
Oil and grease	0.7	1.5	NM	
Dissolved iron	62	24	61	
Fluoride	33	9.1	72	
Toxic pollutants, µg/L:				
Chromium	17,000	1,800	89	
Copper	150	ND	100	
Nickel	6,000	5,200	13	
Zinc	750	240	68	

Blanks indicate data not available.
ND, not detected.
NM, not meaningful.

APPENDIX D

TREATMENT TECHNOLOGY: Chemical Precipitation With Sedimentation (Lime)

Data source: Effluent Guidelines
Point source: Textile mills
Subcategory: Knit fabric finishing
Plant: Unspecified

Pretreatment/treatment: Unspecified/Chem. Ppt.

Data source status:

Not specified	
Bench scale	x
Pilot scale	
Full scale	

DESIGN OR OPERATING PARAMETERS

Wastewater flow rate: Unspecified
Chemical dosages(s): Unspecified
Mix detention time: Unspecified
Flocculation detention time: Unspecified
Unit configuration: Unspecified

Type of sedimentation: Unspecified

REMOVAL DATA

Sampling: Unspecified **Analysis: Data set 2 (V.7.3.32)**

Pollutant/parameter	Concentration Influent(a)	Concentration Effluent	Percent removal	Detection limit
Toxic pollutants, µg/L:				
Cadmium	10	ND	>99	
Chromium	930	80	91	
Copper	500	30	94	
Lead	100	ND	>99	
Nickel	50	ND	>99	
Silver	50	ND	>99	
Zinc	3,200	110	97	

Blanks indicate data not available.
ND, not detected.
(a)Sample taken from aeration basin at plant.

APPENDIX E1

TREATMENT TECHNOLOGY: Chemical Precipitation With Filtration (Lime)

Data source: Effluent Guidelines
Point source: Steam electric
Subcategory: Cooling tower blowdown
Plant: 5604

Data source status:
Not specified
Bench scale x
Pilot scale
Full scale

Pretreatment/treatment: Unspecified/Chem. Ppt., Filtration

DESIGN OR OPERATING PARAMETERS

Wastewater flow rate: Unspecified
Chemical dosages(s): Add to pH >11.0
Mix detention time: Unspecified
Unit configuration: Jar test

REMOVAL DATA

Sampling: Unspecified Analysis: Data set 2 (V.7.3.31)

Pollutant/parameter	Concentration		Percent removal	Detection limit
	Influent	Effluent		
Toxic pollutants, µg/L:				
Antimony	5	3	40	
Arsenic	7	<1	>86	
Chromium	2	<2	NM	
Copper	180	48	73	
Nickel	6	12	NM	
Silver	3	4	NM	
Zinc	780	140	82	
Beryllium	<0.5	<0.5	NM	
Cadmium	<0.5	<0.5	NM	
Lead	<3	<3	NM	
Mercury	<0.2	<0.2	NM	
Selenium	<2	<2	NM	
Thallium	<1	<1	NM	
Vanadium	24	77	NM	

Blanks indicate data not available.
NM, not meaningful.

APPENDIX E2

TREATMENT TECHNOLOGY: Chemical Precipitation With Filtration (Lime)

Data source: Effluent Guidelines
Point source: Steam electric
Subcategory: Ash transport water
Plant: 1226

Data source status:
- Not specified ___
- Bench scale x
- Pilot scale ___
- Full scale ___

Pretreatment/treatment: Sed. (ash pond)/Chem. Ppt., Filtration

DESIGN OR OPERATING PARAMETERS

Wastewater flow rate: Unspecified
Chemical dosages(s): Add to pH >11.0
Mix detention time: Unspecified
Unit configuration: Jar test

REMOVAL DATA

Sampling: Unspecified Analysis: Data set 2 (V.7.3.31)

Pollutant/parameter	Concentration Influent	Concentration Effluent	Percent removal	Detection limit
Classical pollutants, mg/L:				
TOC	<20	<20	NM	
Toxic pollutants, µg/L:				
Antimony	7	10	NM	
Arsenic	9	1	89	
Cadmium	2.0	2.0	0	
Chromium	6	11	NM	
Copper	14	10	29	
Lead	4	<3	>25	
Mercury	<0.2	0.3	NM	
Nickel	5.5	6.0	NM	
Selenium	8	8	0	
Silver	0.5	0.4	20	
Zinc	7	2	71	
Beryllium	<0.5	<0.5	NM	
Thallium	<1	<1	NM	
Vanadium	78	78	0	

Blanks indicate data not available.
NM, not meaningful.

APPENDIX F

TREATMENT TECHNOLOGY: Chemical Precipitation With Sedimentation (Lime)

Data source: Effluent Guidelines
Point source: Inorganic chemicals
Subcategory: Hydrofluoric acid
Plant: 705

Data source status:

Not specified	
Bench scale	
Pilot scale	
Full scale	x

Pretreatment/treatment: Unspecified/Sed.

DESIGN OR OPERATING PARAMETERS

Wastewater flow rate: Unspecified
Chemical dosages(s): Unspecified
Mix detention time: Unspecified
Flocculation Detention time: Unspecified
Unit configuration: 30 to 35% of effluent recycled, remaining effluent neutralized and discharged

Type of sedimentation: Unspecified

REMOVAL DATA

Sampling: 72-hr composite and grab

Analysis: Data set 1 (V.7.3.15)

Pollutant/parameter	Concentration, (a) Influent	Effluent	Percent removal	Detection limit
Toxic pollutants, µg/L:				
Antimony	10	1.9	81	
Arsenic	40	<9.7	>76	
Cadmium	9.7	1.6	84	
Chromium	390	47	88	
Copper	290	19	93	
Lead	50	23	54	
Mercury	5.8	0.48	92	
Nickel	560	<9.7	>98	
Thallium	2.6	1.1	58	
Zinc	240	53	78	

Blanks indicate data not available.
(a)Values are for combined wastes from HF and AlF_3, concentrations are calculated from pollutant flow in m^3/Mg and pollutant loading in kg/Mg.

APPENDIX G1

TREATMENT TECHNOLOGY: Chemical Precipitation With Sedimentation (Lime)

Data source: Effluent Guidelines
Point source: Ore mining and dressing
Subcategory: Base-metal mine
Plant: Plant 3 of Canadian pilot plant study

Pretreatment/treatment: Unspecified/Chem. Ppt.

Data source status:
- Not specified ___
- Bench scale ___
- Pilot scale x
- Full scale ___

DESIGN OR OPERATING PARAMETERS

Wastewater flow rate: Unspecified
Chemical dosages(s): Unspecified
Mix detention time: Unspecified
Flocculation detention time: Unspecified
Unit configuration: Two-stage lime addition

Type of sedimentation: Unspecified

REMOVAL DATA

Sampling: One year

Analysis: Data set 4 (V.7.3.23)

Pollutant/parameter	Concentration Influent(a)	Concentration Effluent(b)	Percent removal	Detection limit
Toxic pollutants, µg/L				
Copper	19,000	60	99	
Lead	1,300	150	88	
Zinc	110,000	350	99	

Blanks indicate data not available.
(a)Average value for raw minewater influent to pilot plant.
(b)Effluent qualities during periods of optimized steady operation.

APPENDIX G2

TREATMENT TECHNOLOGY: Chemical Precipitation With Sedimentation (Lime)

Data source: Effluent Guidelines
Point source: Ore mining and dressing
Subcategory: Lead/zinc mine
Plant: 3113

Data source status:
Not specified ___
Bench scale ___
Pilot scale x
Full scale ___

Pretreatment/treatment: None/Chem. Ppt.

DESIGN OR OPERATING PARAMETERS

Wastewater flow rate: Unspecified
Chemical dosages(s): Unspecified
Mix detention time: Unspecified
Flocculation detention time: Unspecified
Unit configuration: Unspecified
pH in clarifier: 9.1-9.7

Type of sedimentation: Unspecified

REMOVAL DATA

Sampling: 4 days Analysis: Data set 4 (V.7.3.23)

Pollutant/parameter	Concentration Influent(a)	Effluent	Percent removal	Detection limit
Classical pollutants, mg/L:				
TSS	112	33	71	
Toxic pollutants, µg/L:				
Cadmium	230	25	89	
Copper	1,500	100	93	
Lead	88	100	NM	
Zinc	71,000	<20	>99	

Blanks indicate data not available.
NM, not meaningful.
(a)Average of seven observations.

APPENDIX H

TREATMENT TECHNOLOGY: Chemical Precipitation With Sedimentation (Lime, Polymer)

Data source: EGD Combined Data Base
Point source: Porcelain
Subcategory: Alum
Plant: 33077

Data source status:
- Not specified ___
- Bench scale ___
- Pilot scale ___
- Full scale x

Pretreatment/treatment: Equal./Chem. Ppt., Sed. (tube/plate settler)

DESIGN OR OPERATING PARAMETERS

Wastewater flow rate: 965 m^3/day
Chemical dosages(s): Lime: 47,200 kg/yr; polymer: 320 kg/yr
Mix detention time: Unspecified
Flocculation detention time: Unspecified
Unit configuration: Continuous operation (16 hr/day)

Type of sedimentation: Tube/plate settler

REMOVAL DATA

Sampling: 16-hr composite, flow proportion (one hr)
Analysis: Data set 2 (V.7.3.16)

Pollutant/parameter	Concentration Influent	Concentration Effluent	Percent removal	Detection limit
Classical pollutants, mg/L:				
pH, minimum	8.9	9.4		
pH, maximum	10.5	10.0		
Fluorides	1.8	2.0	NM	0.1
Phosphorus	12	0.89	92	0.003
TSS	53	ND	>99	5.0
Iron	2.0	0.038	98	0.005
Titanium	1.2	ND	>99	
Manganese	0.017	ND	>99	0.005
Phenols, total	0.006	ND	>99	0.005
Aluminum	1.2	ND	>99	0.04
Barium	.23	0.20	13	
Toxic pollutants, µg/L:				
Cadmium	2,900	57	98	2.0
Chromium, total	11	ND	>99	3.0
Copper	4.0	ND	>99	1.0
Lead	1,200	ND	>99	30
Zinc	220	540	NM	1.0
Cyanide, total	160	ND	>99	5.0
Selenium	300	ND	>99	

Blanks indicate data not available.
ND, not detected.
NM, not meaningful.

APPENDIX I

TREATMENT TECHNOLOGY: Chemical Precipitation With Sedimentation (Alum, Aluminum Sulfate, Polymer)

Data source: Effluent Guidelines
Point source: Paint manufacturing
Subcategory: Unspecified
Plant: 24

Data source status:
- Not specified ___
- Bench scale ___
- Pilot scale ___
- Full scale x

Pretreatment/treatment: Neutral., Oil Sep./Chem. Ppt., Sed.

DESIGN OR OPERATING PARAMETERS

Wastewater flow rate: 0.26-0.52 L/s
Chemical dosages(s): Unspecified
Mix detention time: Unspecified
Flocculation detention time: Unspecified
Unit configuration: Batch operation
Type of sedimentation: Unspecified

REMOVAL DATA

Sampling: Grab and composite — Analysis: Data set I (V.7.3.25)

Pollutant/parameter	Concentration(a) Influent	Concentration(a) Effluent	Percent removal	Detection limit
Classical pollutants, mg/L:				
BOD(5)	16,000	1,100	25	
COD	36,000	11,000	69	
Total phenol	0.20	0.15	25	
Total solids	41	3	93	
Toxic pollutants, μg/L:				
Ethylbenzene	1,900	460	75	
Toluene	2,900	2,900	0	
Chloroform	48	26	40	
Methylene chloride	130,000	13,000	90	
1,1,2-Trichloroethane	<7	<11	NM	
1,1,1-Trichloroethane	380	<170	<55	
Phenol	ND	<10	NM	
Bis(2-ethylhexyl) phthalate	<10	ND	>99	
Tetrachloroethylene	740	ND	>99	
Trichloroethylene	<10	ND	>99	

Blanks indicate data not available.
NM, not meaningful.
(a)Average of three samples, except total phenol: two samples.

APPENDIX J1

TREATMENT TECHNOLOGY: Chemical Precipitation With Sedimentation (Sodium Hydroxide; Lime)

Data source: EGD Combined Data Base
Point source: Coil coating
Subcategory: Alum
Plant: 13029
References: 3-113
Pretreatment/treatment: None/Chem. Red. (Cr), Chem. Ppt., Sed. (tube/plate)

Data source status:
Not specified ___
Bench scale ___
Pilot scale ___
Full scale x

DESIGN OR OPERATING PARAMETERS

Wastewater flow rate: 3,930 L/day
Chemical dosages(s): NaOH: 8,700 kg/yr; $Ca(OH)_2$: 4,300 kg/yr
Mix detention time: Unspecified
Flocculation detention time: Unspecified
Unit configuration: Tube/plate settler-continuous operation

Type of sedimentation: Tube/plate settler

REMOVAL DATA

Sampling: 24-hr composite, flow proportion (one hr) Analyses: Data set 2 (V.7.3.9)

Pollutant/parameter	Concentration: Influent stream 200	201	Average	Effluent	Percent removal	Detection limit
Classical pollutants, mg/L:						
pH, minimum	11	3.1	7.0	8.3		
pH, maximum	11	5.4	8.2	8.7		
Fluorides	0.43	340	170	44	74	0.1
Phosphorus	91		46	1.3	97	0.003
TSS	970	99	530	37	93	5.0
Iron	0.61	14	7.3	0.1	99	0.005
Oil and grease	2,800	8.0	1,400	20	98	5.0
Phenols, total	0.14	ND	0.07	0.2	71	0.005
Aluminum	970	99	530	5.1	99	0.04
Manganese	1.5	0.76	1.1	0.011.	99	0.005
Toxic pollutants, µg/L:						
Cadmium	3.0	8.0	5.5	ND	>99	2.0
Chromium	180	660,000	330,000	2,500	99	3.0
Copper	210	230	220	10	95	1.0
Lead	60	170	115	ND	>99	30
Nickel	ND	190	95	ND	>99	6.0
Zinc	280	38,000	19,000	69	>99	1.0
Anthracene	BDL	BDL	BDL	ND	NM	10
Bis(2-ethyhexyl) phthalate	220	62	140	BDL	96*	10
Fluorene	BDL	BDL	BDL	ND	NM	10
Di-n-butyl phthalate	12	BDL	BDL	ND	NM	10
Diethyl phthalate	410	68	240	3.0	99	10
Hexavalent chromium	ND	290,000	140,000	ND	>99	5.0
Naphthalene	ND	BDL	BDL	BDL	NM	10

Blanks indicate data not available.
BDL, below detection limit.
ND, not detected.
NM, not meaningful.
*Approximate value.

APPENDIX J2

TREATMENT TECHNOLOGY: Chemical Precipitation With Sedimentation (Lime)

Data source: EGD Combined Data Base
Point source: Coil coating
Subcategory: Steel
Plant: 46050

Data source status:
- Not specified ___
- Bench scale ___
- Pilot scale ___
- Full scale _x_

Pretreatment/treatment: Ion Exch./Chem. Red. (Cr), Equal., Chem. Ppt., Coag. Floc. (polymer), Sed. (tank)

DESIGN OR OPERATING PARAMETERS

Wastewater flow rate: 156 m^3/day
Chemical dosages(s): Unspecified
Mix detention time: Unspecified
Flocculation detention time: Unspecified
Unit configuration: Batch (8 hr/day) Chem. Red. (Cr); continuous (24 hr/day) Chem. Ppt.

Type of sedimentation: Tank

Hydraulic detention time: 16.0 hr

REMOVAL DATA

Sampling: Influent: (201,202) continuous 24-hr composite, time proportion (one hr); effluent: batch (unspecified) composite, time proportion (three hr); (205) continuous (unspecified) composite, flow proportion (one day); (253) batch-unspecified composite, flow proportion (one day)

Analysis: Data set 2 (V.7.3.9)

Pollutant/parameter	Concentration: Influent stream(a) 201	202	253	205	Avg.	Effluent	Percent removal	Detection limit
Classical pollutants, mg/L:								
pH, minimum	7.0	4.3	7.5	2.0*	5.2*	7.0		
pH, maximum	7.4	5.8	7.5	6.9	6.9	7.0		
Fluorides	1.0	78	2.6	0.78	21	10	52	0.1
Phosphorus	22	11		0.6	17	1.6	90	0.003
TSS	160	70	870	110	150	8.0	95	5.0
Iron	0.85	1.4	7.2	0.60	1.1	0.17	84	0.005
Oil and grease	10	1.4	ND	ND	6.4	11	NM	5.0
Phenols, total		ND	ND	0.005*	BDL	0.020	NM	0.005
Manganese	0.73	1.05	3.6	2.4	1.1	0.16	85	
Toxic pollutants, µg/L:								
Chromium	ND	130	620,000	ND	18,000	24	>99	3.0
Copper	14	ND	43	11	11	3.0	73	1.0
Lead	180	ND	56	ND	110	ND	>99	30
Nickel	150	32,000	20,300	ND	9,100	1,400	85	6.0
Zinc	5,300	65,000	370,000	230	31,000	440	98	1.0
Cyanide, total	43	ND	ND	ND	BDL	ND	NM	5.0
1,1,1-Trichloroethane	1.2	ND	ND	ND	BDL	ND	NM	0.1
Bis(2-ethyhexyl) phthalate	23	200	15	BDL	68	40	41	10
Di-n-butyl phthalate	BDL	ND	ND	ND	BDL	ND	NM	10
Diethyl phthalate	BDL	330	15	ND	91	40	56	10
Trichloroethylene	0.5	ND	ND	0.6	36	ND	>99	0.1
Phenanthrene	ND	ND	ND	ND	ND	BDL	NM	10
Hexavalent chromium	ND	60	330,000	ND	9,500	ND	>99	5.0
Acenaphthylene	ND	BDL	ND	ND	BDL	BDL	NM	10
Anthracene	ND	ND	ND	ND	ND	BDL	NM	10

Blanks indicate data not available.
BDL, below detection limit.
ND, not detected.
NM, not meaningful.
(a)Influent streams 202, 201 and 205 are coded as continuous raw waste streams, stream 253 is coded as batch.

APPENDIX K1

TREATMENT TECHNOLOGY: Chemical Precipitation With Sedimentation (Lime)

Data source: Effluent Guidelines
Point source: Nonferrous metals
Subcategory: Columbium/tantalum
Plant: Unspecified

Data source status:
Not specified ___
Bench scale ___
Pilot scale ___
Full scale x

Pretreatment/treatment: None/Chem. Ppt.

DESIGN OR OPERATING PARAMETERS

Wastewater flow rate: Unspecified
Chemical dosages(s): Unspecified
Mix detention time: Unspecified
Flocculation detention time: Unspecified
Unit configuration: Unspecified

Type of sedimentation: Unspecified

REMOVAL DATA

Sampling: 24-hour and 72-hour composite and grab

Analysis: Data set 2 (V.7.3.22)

Pollutant/parameter	Concentration Influent	Concentration Effluent	Percent removal	Detection limit
Classical pollutants, mg/L:				
COD	16	8	50	
TSS	900	10	99	
Fluoride	4.5	2.5	44	
Aluminum	9.0	0.2	98	
Calcium	550	230	58	
Iron	120	0.3	>99	
Manganese	17	0.2	99	
Toxic pollutants, µg/L:				
Cadmium	25	2	92	
Copper	110,000	700	99	
Nickel	60,000	500	99	
Zinc	27,000	200	99	

Blanks indicate data not available.

APPENDIX K2

TREATMENT TECHNOLOGY: Chemical Precipitation With Sedimentation (Lime)

Data source: Effluent Guidelines
Point source: Nonferrous metals
Subcategory: Tungsten
Plant: Unspecified

Data source status:
Not specified ___
Bench scale ___
Pilot scale ___
Full scale x

Pretreatment/treatment: None/Chem. Ppt.

DESIGN OR OPERATING PARAMETERS

Wastewater flow rate: Unspecified
Chemical dosages(s): Unspecified
Mix detention time: Unspecified
Flocculation detention time: Unspecified
Unit configuration: Unspecified

Type of sedimentation: Unspecified

REMOVAL DATA

Sampling: 24-hour and 72-hour composite and grab

Analysis: Data set 2 (V. 7. 3. 22)

Pollutant/parameter	Concentration Influent	Concentration Effluent	Percent removal	Detection limit
Classical pollutants, mg/L:				
COD	300	53	82	
TSS	300	150	50	
Chloride	25,000	19,000	24	
Aluminum	3	0.5	83	
Iron	50	2	96	
Toxic pollutants, µg/L:				
Arsenic	7,000	80	99	
Cadmium	200	80	60	
Chromium	2,000	50	98	
Copper	5,000	70	99	
Lead	20,000	200	99	
Nickel	1,000	100	90	
Zinc	2,000	600	70	

Blanks indicate data not available.

APPENDIX L

TREATMENT TECHNOLOGY: Chemical Precipitation With Sedimentation

Data source: Effluent Guidelines
Point source: Aluminum forming
Subcategory: Unspecified
Plant: J

Data source status:
Not specified ___
Bench scale ___
Pilot scale ___
Full scale x

Pretreatment/treatment: None/Equal., Chem Ppt., Sed.

DESIGN OR OPERATING PARAMETERS

Wastewater flow rate: Unspecified
Chemical dosages(s): Unspecified
Mix detention time: Unspecified
Flocculation detention time: Unspecified
Unit configuration: Unspecified

Type of sedimentation: Clarifier

REMOVAL DATA

Sampling: Three 24-hour or one 72-hour composite   Analysis: Data set 2 (V.7.3.7)

Pollutant/parameter	Concentration Influent	Concentration Effluent	Percent removal	Detection limit
Classical pollutants, mg/L:				
Oil and grease	86	15	99	
Suspended solids	450	710	NM	
COD	260	280	NM	
TOC	75	74	1	
Phenol	0.003	0.002	33	
pH, pH units	2.8	3.7	NM	
Toxic pollutants, µg/L:				
Chromium	900,000	790,000	12	5
Copper	2,200,000	2,200,000	0	9
Cyanide	BDL	BDL	NM	100
Lead	3,200	1,000	69	20
Mercury	<1	<1	NM	0.1
Nickel	2,600	2,400	8	5
Zinc	2,000,000	1,800,000	10	50
Fluoranthene	10	ND	>99	10
Methylene chloride	260	15	93	10
2,4-Dinitrophenol	37	ND	>99	10
N-nitrosodiphenylamine	67	ND	>99	10
Chrysene	10	ND	>99	10
Anthracene/phenanthrene	<26	BDL	NM	10
Pyrene	16	ND	>99	10

Blanks indicate data not available.
BDL, below detection limit.
ND, not detected.
NM, not meaningful.

APPENDIX M

TREATMENT TECHNOLOGY: Chemical Precipitation With Sedimentation

Data source: EGD Combined Data Base
Point source: Battery
Subcategory: Lead
Plant: 20993

Data source status:
Not specified ___
Bench scale ___
Pilot scale ___
Full scale x

Pretreatment/treatment: Equal., Screen/Chem. Ppt., Sed. (clarifier), Polishing Lagoon

DESIGN OR OPERATING PARAMETERS

Wastewater flow rate: Influent: 561,000 m^3/day; effluent: 552,000 m^3/day
Chemical dosages(s): Sodium hydroxide: 227,000 kg/yr
Mix detention time: Unspecified
Flocculation detention time: Unspecified
Unit configuration: Continuous operation (24 hr/day)
Hydraulic detention time: 10.2 L/hr/m^2

Type of sedimentation: Clarifier
Hydraulic loading rate: 693 L/hr/m^2
Hydraulic detention time: 7.0 hr
Weir loading rate: Unspecified
Type of sedimentation: Polishing lagoon
Hydraulic loading rate: 120 hr

REMOVAL DATA

Sampling: 24-hr composite, flow proportion (one hr) Analysis: Data set 2 (V.7.3.8)

Pollutant/parameter	Concentration Influent	Concentration Effluent	Percent removal	Detection limit
Classical pollutants, mg/L:				
pH, minimum	2.0	8.7		
pH, maximum	2.4	9.1		
TSS	14	11	21	5.0
TDS	880	2,000	NM	5.0
Iron	16	0.92	94	0.005
Oil and grease	BDL	BDL	NM	5.0
Manganese	120	44	63	
Strontium	33	27	18	
Toxic pollutants, µg/L:				
Chromium	57	5.0	91	3.0
Copper	78	14	82	1.0
Lead	1,400	130	91	30
Nickel	36	9.0	75	6.0
Zinc	120	ND	>99	1.0
1,1,1-Trichloroethane	0.1*	0.1*	NM	0.1
Bis(2-ethylhexyl)phthalate	10	BDL	50	10
Butyl benzyl phthalate	ND	BDL	NM	10
Methylene chloride	BDL	BDL	NM	1.0

Blanks indicate data not available.
BDL, below detection limit.
ND, not detected.
NM, not meaningful.
*Approximate value.

APPENDIX N

TREATMENT TECHNOLOGY: Chemical Precipitation With Sedimentation (Sodium Carbonate)

Data source: Effluent Guidelines	Data source status:
Point source: Electrical and electronic components	Not specified ___
Subcategory: Unspecified	Bench scale ___
Plant: 30172	Pilot scale ___
	Full scale x

Pretreatment/treatment: None/Chem. Ppt.

DESIGN OR OPERATING PARAMETERS

Wastewater flow rate: 3.77 m^3/day
Chemical dosages(s): Sodium carbonate
Mix detention time: Unspecified
Flocculation Detention time: Unspecified
Unit configuration: 6,610 liter tank

Type of sedimentation: Unspecified

REMOVAL DATA

Sampling: Three 24-hour composites   Analysis: Data set 2 (V.7.3.11)

Pollutant/parameter	Concentration(a)		Percent removal	Detection limit
	Influent	Effluent		
Classical pollutants, mg/L:				
Oil and grease	11	14	NM	
TOC	<1	160	NM	
BOD	<1	<1	NM	
TSS	190	17	91	
Phenol	0.01	0.08	NM	
Fluoride	160	76	52	
pH, pH units	<2	7.3	NM	
Calcium	88	29	67	
Magnesium	31	18	42	
Sodium	640	13,000,000	NM	
Aluminum	12	0.68	94	
Manganese	5.9	0.55	91	
Vanadium	0.16	0.024	85	
Boron	350	400,000	NM	
Barium	200	12,000	NM	
Molybdenum	1.6	0.17	89	
Tin	3.0	0.39	87	
Yttrium	17	<0.008	>99	
Cobalt	2.6	<0.120	>95	
Iron	1,900	0.38	99	
Titanium	0.31	0.043	86	
Palladium	0.32	<0.003	>99	
Tellurium	0.29	0.013	96	
Platinum	0.09	0.02	78	
Gold				
Toxic pollutants, μg/L:				
Antimony	92	<15	>84	
Arsenic	250	10	96	
Beryllium	4	<1	>75	
Cadmium	1,100	<5	>99	
Chromium	4,700	27	99	
Copper	<50	48	NM	
Lead	890,000	1,900	99	
Mercury	1	<1	NM	
Nickel	18,000	640	96	
Selenium	<20	<4	NM	
Silver	60	<2	>97	
Thallium	2		NM	
Zinc	1,500,000	11,000	99	
Cyanide	<5	<5	NM	

Blanks indicate data not available.
NM, not meaningful.
(a)Values presented as "less than" the reported concentration are below detectable limits. They are not reported as BDL because the detection limits are variable in this industry.

APPENDIX O

TREATMENT TECHNOLOGY: Chemical Precipitation With Sedimentation (Lime)

Data source: EGD Combined Data Base
Point source: Copper
Subcategory: Pickle
Plant: 6070

Data source status:
- Not specified ___
- Bench scale ___
- Pilot scale ___
- Full scale x

Pretreatment/treatment: None/Chem. Ppt., Sed. (clarifier)

DESIGN OR OPERATING PARAMETERS

Wastewater flow rate: 3,000 m^3/day
Chemical dosages(s): Unspecified
Mix detention time: Unspecified
Flocculation detention time: Unspecified
Unit configuration: Continuous operation (24 hr/day)

Type of sedimentation: Clarifier

REMOVAL DATA

Sampling: 24-hr composite, flow proportion (one hr) Analysis: Data set 1 (V.7.3.13)

Pollutant/parameter	Concentration Influent	Concentration Effluent	Percent removal	Detection limit
Classical pollutants, mg/L:				
pH, minimum	1.0	5.0		
pH, maximum	3.2	7.0		
Fluorides	0.80	10	NM	0.1
Phosphorus	5.0	0.86	83	0.003
TSS	18	18	0	5.0
Iron	13	0.27	98	0.005
Oil and grease	4.0	1.0	75	5.0
Phenols, total	0.01	0.01	0	0.005
TOC	12	10	17	
Manganese	0.77	0.32	58	0.005
Toxic pollutants, µg/L:				
Chromium	200	23	88	3.0
Copper	9,400	220	98	1.0
Lead	430	ND	>99	30
Nickel	320	300	6	6.0
Zinc	74,000	1,400	98	1.0
1,1,1-Trichloroethane	0.1*	ND	NM	0.1
Chloroform	BDL	BDL	NM	1.0
Bis(2-ethylhexyl)phthalate	BDL	ND	NM	10
Trichloroethylene	0.2	ND	>99	1.0
Toluene	1.0	ND	>99	1.0
Phenanthrene	BDL	BDL	NM	10
Anthracene	BDL	BDL	NM	10
Benzene	1.0	ND	>99	1.0
Naphthalene	BDL	BDL	NM	10

Blanks indicate data not available.
BDL, below detection limit.
ND, not detected.
NM, not meaningful.
*Approximate value.

APPENDIX P

TREATMENT TECHNOLOGY: Chemical Precipitation With Sedimentation (Alum)

Data source: Government report
Point source: Organic and inorganic wastes
Subcategory: Unspecified
Plant: Reichhold Chemical, Inc.

Data source status:
Not specified ___
Bench scale ___
Pilot scale _x_
Full scale ___

Pretreatment/treatment: Equal./Chem. Ppt.

DESIGN OR OPERATING PARAMETERS

Wastewater flow rate: Unspecified
Chemical dosages(s): 650 mg/L (alum)
Mix detention time: Unspecified
Flocculation Detention time: Unspecified
Unit configuration: Unspecified

Type of sedimentation: Unspecified

REMOVAL DATA

Sampling: 24-hour composite   Analysis: Data set 2 (V.7.3.35)

Pollutant/parameter	Concentration Influent	Concentration Effluent	Percent removal	Detection limit
Classical pollutants, mg/L:				
BOD_5	2,400	2,200	8	
COD	3,600	3,500	3	
Total phenol	320	220	31	
Total phosphorus	49	43	12	
SS	140	28	80	
TS	4,600	4,300	6	
DS	4,400	4,300	2	
Sulfate	750	830	NM	
Sulfite	40	10	75	
Iron	40	ND	>99	
Nitrate	320	310	3	

Blanks indicate data not available.
ND, not detected.
NM, not meaningful.

APPENDIX Q

TREATMENT TECHNOLOGY: Chemical Precipitation With Sedimentation (Alum, Polymer)

Data source: Effluent Guidelines
Point source: Auto and other laundries
Subcategory: Power laundries
Plant: N

Data source status:
- Not specified ___
- Bench scale ___
- Pilot scale ___
- Full scale x

Pretreatment/treatment: Screen, Equal./Chem. Ppt.

DESIGN OR OPERATING PARAMETERS

Wastewater flow rate: 15.1 m^3/d
Chemical dosages(s): Alum-2,800 mg/L, polymer-200 mg/L
Mix detention time: Unspecified
Flocculation detention time: Unspecified
Unit configuration: Circular clarifier, 4.92 m^3 with mix tank

Type of sedimentation: Clarifier
Hydraulic detention time: 0.33 day

REMOVAL DATA

Sampling: Composite and grab   Analysis: Data set 1 (V.7.3.1)

Pollutant/parameter	Concentration Influent	Concentration Effluent	Percent removal	Detection limit
Classical pollutants, mg/L:				
BOD(5)	160	57	64	
COD	240	130	46	
TOC	63	40	37	
TSS	40	46	NM	
Oil and grease	15	4	73	
Total phenol	0.038	0.028	26	
Total phosphorus	7.0	1.6	77	
Toxic pollutants, µg/L:				
Cadmium	51	12	76	2
Chromium	39	34	13	4
Copper	140	31	78	4
Lead	71	66	7	22
Nickel	55	50	9	36
Silver	14	11	21	5
Zinc	610	240	61	1
Phenol	ND	2	NM	0.07
Toluene	5	3	40	0.1
Tetrachloroethylene	2	100	NM	
Trichloroethylene	0.5	12	NM	0.5
Cyanide	<2	<2	NM	
Chloroform	ND	70	NM	5
Methyl chloride	ND	38	NM	0.4
Chlorodibromomethane	BDL	ND	NM	0.9
Bis (2-ethylhexyl)phthalate	ND	67	NM	0.04
Butyl benzyl phthalate	ND	36	NM	0.03
Di-n-butyl phthalate	ND	7	NM	0.02
Di-n-octyl phthalate	ND	5	NM	0.89

Blanks indicate data not available.
BDL, below detection limit.
ND, not detected.
NM, not meaningful.

6
Recarbonation and Softening

Lawrence K. Wang, Jy S. Wu, Nazih K. Shammas, and David A. Vaccari

CONTENTS

1. INTRODUCTION

According to Wang (1), recarbonation is defined as "(a) the process of introducing carbon dioxide, CO_2, as a final stage in the lime-soda ash softening process in order to convert carbonates to bicarbonates and thereby stabilize the solution against precipitation of carbonates, (b) the diffusion of carbon dioxide gas through liquid to replace the carbon dioxide gas removed by the addition of lime, or (c) the diffusion of carbon dioxide through a liquid to render the liquid stable with respect to precipitation or dissolution of alkaline constituents." The process is accomplished by bubbling gases containing carbon dioxide (CO_2) through water. This chapter introduces the recarbonation process, its closely related lime/soda-ash softening process, and various applications of recarbonation.

2. PROCESS DESCRIPTION

Water, as it is found in nature, usually contains some CO_2, which comes most likely from the decomposition of organic matter or from the atmosphere. The CO_2 hydrolyzes according to Eq. (1):

From: *Handbook of Environmental Engineering, Volume 3: Physicochemical Treatment Processes*
Edited by: L. K. Wang, Y.-T. Hung, and N. K. Shammas © The Humana Press Inc., Totowa, NJ

$$CO_2 + H_2O \leftrightarrow H_2CO_3 \quad (1)$$

to form carbonic acid (H_2CO_3). In turn the H_2CO_3 dissociates

$$H_2CO_3 \leftrightarrow H^+ + HCO_3^- \quad (2)$$

to form bicarbonate ion (HCO_3^-) which in turn dissociates further

$$HCO_3 \leftrightarrow H^+ + CO_3^{2-} \quad (3)$$

to form the carbonate ion (CO_3^{2-}).

The degree of the two dissociations is dependent primarily on the pH and somewhat on temperature and the ionic content of the water. Fair et al. (2) and Rich (3) have described these distributions in detail. For practical purposes, it is reasonable to assume that below a pH of about 4.5 all of the carbon is in the form of CO_2 or H_2CO_3. Above this pH, the CO_2 content decreases logarithmically to approach a value of zero at a pH of about 8.3. The shift of the carbon distribution above a pH of 4.5 is from CO_2 to HCO_3^-. At a pH of about 7.5, the CO_3^{2-} begins to become apparent and the CO_3^{2-} increases as the HCO_3^- decreases, the latter approaching a value of zero at a pH of about 11.5. The HCO_3^- and CO_3^{2-} are equal in molar concentration at a pH of about 10.

Water containing CO_2 or H_2CO_3 moving through limestone formations dissolves the quite insoluble calcium carbonate, $CaCO_3$,

$$H_2CO_3 + CaCO_3 \leftrightarrow Ca(HCO_3)_2 \quad (4)$$

to form the quite soluble calcium bicarbonate, $Ca(HCO_3)_2$, which, in reality, exists as the ions Ca^{2+} and $2HCO_3^-$. Similarly, magnesium carbonate ($MgCO_3$) is dissolved to form the Mg^{2+} and $2HCO_3$. Other salts of calcium and magnesium such as sulfates (SO_4^{2-}), chlorides (Cl^-), nitrates (NO_3^-), etc., are also dissolved in water. Similarly, other cations are dissolved such as sodium and potassium. The ionic content of the water is dependent on the CO_2 in the water, the soil or rocks over or through which the water passes, and the time of contact of the water.

Water containing too much (depending on intended use of the water) Ca^{2+} and Mg^{2+} may react with soap to form curds or may form insoluble precipitates (scale)—this water is termed "hard." If the so-called lime/soda ash process is used to soften the water, recarbonation may well be needed. The softening and recarbonation reactions and processes are described in succeeding sections of this chapter.

Natural water, particularly surface water, may contain suspended matter that does not settle out readily. This suspended matter usually has the characteristics of colloids, that is, they carry a surface electrostatic charge. The terms applied are "turbidity" in the case of the mineral and possibly some organic colloids and "color" in the case of most of the organic colloids such as leachates from decaying matter.

To aid in the removal of such colloids, which are usually negatively charged, cations that form hydrous gel-like positively charged colloids may be introduced into the water (for details refer to Chapter 4 Coagulation and Flocculation). Salts of trivalent aluminum and ferric iron are usually used and do very well at the pH values resulting from the addition of aluminum or iron salt. That is, no pH adjustment is usually necessary.

However, the sludges formed with the gel-like floc and the turbidity or color colloids do not dewater easily, and disposal may become a nuisance. Until recently, such sludges were commonly returned to surface waters. Now, however, such sludge disposal is prohibited in most developed countries.

Another gel-like positively charged colloid that may be used is the one formed from magnesium and is probably in the form of magnesium hydroxide [$Mg(OH)_2$]. The solubility product of $Mg(OH)_2$ is such that the pH must be quite high before a good insoluble precipitate is formed. Lime is usually used to raise the pH. Both the lime and the $Mg(OH)_2$ can be recovered from the sludge and recycled, thus removing the sludge disposal problem. The most common method of separation of the magnesium from the calcium in the sludge is recarbonation. The processes are described in succeeding sections.

Phosphorous is often the limiting nutrient in algal growth in water. The addition of phosphorus may stimulate such growth that is usually undesirable. Domestic sewage contains considerable phosphorous and in some instances it may be desirable to remove the phosphorous. A common method for such removal is precipitation as calcium phosphate, $Ca_3\,(PO_4)_2$. Lime (CaO) is the most convenient and economical source of calcium and its use results in high pH values in the treated wastewater. Reduction of that high pH may be accomplished by recarbonation. This is also described in succeeding sections.

3. SOFTENING AND RECARBONATION PROCESS CHEMISTRY

If water is softened by precipitation using the lime or lime-soda process, recarbonation is often necessary to prevent after precipitation on the filters or in the distribution piping.

Fair et al. (2), Rich (3), Clark et al. (4), Sawyer et al. (5), AWWA (30), Weber (7) and Benefield and Morgans (17) all describe the commonly accepted reactions of water softening using the lime-soda process.

The insoluble precipitates sought are calcium carbonate, $CaCO_3$, and magnesium hydroxide, $Mg(OH)_2$. The values of their solubility products are

$$\left(\mathrm{Ca}^{2+}\right)\left(\mathrm{CO_3}^{2-}\right) = K_{sp} = 5\times10^{-9}$$

$$\left(\mathrm{Mg}^{2+}\right)\left(\mathrm{OH}^{-}\right)^2 = K_{sp} = 9\times10^{-12}$$

It is assumed that the lime will react with CO_2, Ca^{2+}, and Mg^{2+} to the extent of the bicarbonates present in the following order:

$$\mathrm{CO_2 + CaO \leftrightarrow CaCO_3(s)} \tag{5}$$

$$\mathrm{Ca^{2+} + 2HCO_3^- + CaO + H_2O \leftrightarrow 2CaCO_3(s) + 2H_2O} \tag{6}$$

$$\mathrm{Mg^{2+} + 2HCO_3^- + CaO + H_2O \leftrightarrow MgCO_3 + CaCO_3(s) + 2H_2O} \tag{7a}$$

$$\mathrm{MgCO_3 + CaO + H_2O \leftrightarrow Mg(OH)_2(s) + CaCO_3(s)} \tag{7b}$$

If the cations Ca^{2+} and Mg^{2+} are in excess of the bicarbonate ion, soda ash (Na_2CO_3) is used for the removal of the Ca^{2+} and soda ash and lime for the removal of Mg^{2+}. Assuming the anion to be SO_4^{2-}, for example,

$$Ca^{2+} + SO_4{}^{2-} + Na_2CO_3 \leftrightarrow CaCO_3(s) + 2Na^+ + SO_4{}^{2-} \quad (8)$$

$$Mg^{2+} + SO_4{}^{2-} + CaO + H_2O \leftrightarrow CaSO_4 + Mg(OH)_2(s) \quad (9)$$

Because the retention time in settling basins is usually inadequate to permit complete formation of the insoluble compounds $CaCO_3$ and $Mg(OH)_2$ resulting in after precipitation that is not desirable and because you may run out of carbonate in solution if part of the hardness is due to permanent hardness, recarbonation is often employed:

$$CO_2 + Ca^{2+} + +2OH^- \leftrightarrow CaCO_3 + H_2O \quad (10)$$

$$CO_2 + CaCO_3 + H_2O \leftrightarrow Ca^{2+} + 2HCO_3{}^- \quad (11)$$

and

$$CO_2 + Mg(OH)_2 \leftrightarrow Mg^{2+} + CO_3{}^{2-} + H_2O \quad (12a)$$

or more completely

$$CO_2 + Mg^{2+} + CO_3{}^{2-} + H_2O \leftrightarrow Mg^{2+} + 2HCO_3{}^- \quad (12b)$$

Two phenomena are occurring simultaneously, the reduction in pH, shifting the $CO_3{}^{2-}$ to $HCO_3{}^-$ and OH^- to H_2O and also the addition of CO_2, provides the extra necessary $HCO_3{}^-$ to shift the $CO_3{}^{2-}$ to the $2HCO_3{}^-$.

Coincidently with the softening processes will be the removal of turbidity and color. If the magnesium hardness can be considered negligible, only enough lime and/or soda ash may be added to neutralize the CO_2 and precipitate the Ca^{2+} as $CaCO_3$ as indicated by Eq. (5), (6), and possibly (8). Following the chemical additions, flocculation and settling are carried out. The flocculation (crystal growth) period may be about 30 min and settling period several hours. Because, as stated above, the precipitation is incomplete, recarbonation may be used as indicated by Eq. (11). Recarbonation is a fast reaction and may be completed in a tank with a detention period of about 15 min.

According to Culp and Culp (8), the single-stage softening process results in a significant residual of both calcium and magnesium hardness—the amount may be about 80–100 mg/L expressed as $CaCO_3$.

If the magnesium hardness is not considered negligible, the softening and recarbonation processes may become more complex. Sawyer et al. (5) has described the following arrangement as first-stage recarbonation followed by second-stage treatment and finally second-stage recarbonation. The first-stage treatment consists of adding lime in accordance with Eq. (5), (6), (7a), (7b), and (9) plus an excess of about 35 mg/L of CaO to ensure precipitation of the Mg^{2+} as $Mg(OH)_2$. The final pH should be about 10.8 and the precipitate formed will consist of $CaCO_3$ and $Mg(OH)_2$. The first-stage recarbonation will follow Eq. (10) and (12a). Sawyer (5) suggests that in this recarbonation the pH should not be reduced below a value of about 9.5 to prevent conversion of $CO_3{}^{2-}$ to $HCO_3{}^-$. The water now contains the remaining magnesium hardness, that portion of the $Mg(OH)_2$ which did not precipitate and was converted by recarbonation to $MgCO_3$, the unprecipitated or unsettled $CaCO_3$ and the Ca^{2+} residual from Eq. (9) plus possibly original $CaSO_4$. The second-stage treatment consists of adding Na_2CO_3 and the reaction follows that of Eq. (8). This second-stage sludge is practically pure $CaCO_3$. This softened water will now be supersaturated with $CaCO_3$. Recarbonation to a pH of about 8.6 will convert the CO_3^{2-} to HCO_3^- in accordance with Eq. (11).

Waters softened by the lime/soda ash process will have a residual hardness (Ca^{2+} plus Mg^{2+}) of about 40 mg/L expressed as $CaCO_3$.

There are, of course, modifications to the two types of processes described above. The two-stage process may provide water with a residual hardness of about 40 mg/L, although a hardness of about 80–100 mg/L may be acceptable for a domestic water supply. Thus, a split treatment may be used so that a portion of the water may be softened and a portion bypassed. The mixing of the two waters will result in a reduction in the pH of the softened water reducing considerably or possibly eliminating completely the need for recarbonation.

Because both $CaCO_3$ and $Mg(OH)_2$ are less soluble in hot water than in cold, softening water that must be heated anyway, a process or boiler feed water, after heating will result in a reduction in chemical needs and also residual hardness.

4. LIME/SODA ASH SOFTENING PROCESS

In the previous section the process chemistry of recarbonation and softening was discussed. This section introduces various hardness-causing substances and the practical application of the lime/soda ash softening process for hardness removal.

One of the most common problems resulting from the presence of minerals in water is caused by their precipitation. This occurs either as a hard deposit called scale on water-conveying or in treatment equipment, particularly pipes and filters, or in combination with soap to reduce their effectiveness and make them difficult to rinse. This can be caused by any multivalent metallic cations, but is usually associated with divalent metal cations, and most particularly with calcium and magnesium.

This problem is termed "hardness." The hardness is usually measured as the total calcium and magnesium concentration in mg/L as calcium carbonate. This is called the "total hardness."

In the neutral pH range and at ambient temperatures, a concentration greater than 100 mg/L as calcium carbonate is considered hard. Lime/soda-ash precipitation of water hardness is theoretically capable of reducing hardness to about 25 mg/L, but only to 50–80 mg/L in practical terms. This is usually adequate for domestic and many industrial uses.

In lime/soda-ash softening, it is necessary to add slaked lime, $Ca(OH)_2$, or CaO to raise the pH, converting the free CO_2, carbonic acid and bicarbonate forms of dissolved carbon dioxide to carbonate ions. At pH greater than 10.8, the calcium will then precipitate as calcium carbonate, and magnesium as its hydroxide. Assuming slaked lime (calcium hydroxide) is used in the process, the reactions are as follows (H_2CO_3 represents the free carbon dioxide in solution, plus the carbonic acid):

$$H_2CO_3 + Ca(OH)_2 = CaCO_3\ (s) + 2H_2O \tag{13}$$

$$2HCO_3^- + Ca(OH)_2 = CaCO_3\ (s) + 2H_2O + CO_3^{2-} \tag{14}$$

$$Mg^{2+} + 2HCO_3^- + 2Ca(OH)_2 = 2CaCO_3\ (s) + Mg(OH)_2 + 2H_2O \quad (15)$$

$$Mg^{2+} + 2HCO_3^- + 2Ca(OH)_2 = 2CaCO_3\ (s) + Mg(OH)_2 + 2H_2O \quad (16)$$

$$Mg^{2+} + Ca(OH)_2 = Mg(OH)_2 + Ca^{2+} \quad (17)$$

However, there might not be enough carbonate present in the water naturally to precipitate all the calcium in the wastewater as well as that contributed by the lime. The portion of the hardness with an equivalent carbonate concentration is termed "temporary hardness" or carbonate hardness, because the complementary ion for the precipitation of the calcium is already present. Operationally, this is measured as the total hardness minus the alkalinity, because in the neutral pH range the alkalinity mostly consists of bicarbonate ions expressed as equivalents of calcium carbonate. The remaining hardness is called "permanent hardness" or noncarbonate hardness (NCH).

These reactions remove much of the carbonate hardness. In order to complete the treatment, some form of carbonate must be added. Soda ash, $Na_2(CO_3)$, provides this. The reaction involving soda ash is

$$Ca^{2+} + Na_2CO_3 = CaCO_3 + 2Na^+ \quad (18)$$

This reaction removes the calcium noncarbonate hardness (CaNCH), which in equivalents is equal to the equivalents of calcium minus the equivalents of carbonate species. Magnesium noncarbonate hardness (MgNCH) can also be defined as equal to the equivalents of magnesium present if there are fewer equivalents of carbonate species than of calcium, or, the MgNCH is equal to the total NCH if the carbonates exceed calcium equivalents.

If there is magnesium noncarbonate hardness, soda ash must be added to precipitate the calcium from the lime used in Eq. (17). The resulting reaction is a combination of Eqs. (17) and (18):

$$Mg^{2+} + Ca(OH)_2 + Na_2CO_3 = Mg(OH)_2 + CaCO_3 + 2Na^+ \quad (19)$$

Soda ash is more expensive than lime, and its use adds to the sodium concentration in the finished water. Bubbling carbon dioxide gas through the water in a recarbonation step after lime treatment and before adding soda ash can reduce the soda ash requirement. The use of caustic soda (sodium hydroxide) instead of lime also reduces the soda ash requirement because it does not add calcium, which needs to be removed.

There are several variations to the basic softening process. In the simplest scheme, the precipitation is carried out in a single stage in which the lime and soda ash are added together. This may be followed by pH adjustment either by the addition of a strong acid such as sulfuric acid, or by recarbonation in one or two stages. In another variation the lime is added in a first stage, followed by recarbonation, then the soda ash is dosed in a second stage, also followed by recarbonation for pH adjustment.

The dosages of lime and soda ash can be computed based on either stoichiometric or equilibrium considerations. The equilibrium calculation is performed either by direct use of the reactions involved, which can be quite complex, or with the aid of Caldwell-Lawrence diagrams.

Table 1
Lime and Soda Ash Equivalent Dosages

		Equivalents of	
Constituent to be removed	Equation	Lime	Soda Ash
Carbonic acid	(13)	1	0
Calcium carbonate hardness	(15)	1	0
Magnesium carbonate hardness	(16)	2	0
CaNCH	(18)	0	1
MgNCH	(19)	1	1

The stoichiometric approach assumes the reactions go to completion and thus may overestimate the required dosage. This is not a real problem, because in practice it is necessary to increase the dosage by 5–10% above the stoichiometric amount in order to ensure rapid formation of precipitate. Algae have been observed to reduce lime requirement by consuming carbon dioxide.

Table 1 summarizes the number of equivalents of lime and soda ash needed to treat water in one stage without recarbonation. The equivalents are the stoichiometric values based on the above equations (bicarbonate to be removed is included in the carbonate hardness).

5. WATER STABILIZATION

Hard water is undesirable because of its tendency to deposit hard deposits called scale in pipes and elsewhere in water supply systems. Softening removes this scale-forming tendency. However, there is an opposite extreme to this behavior, and that is the tendency of water to corrode metals exposed to it. The presence of scale protects metal from corrosion. The desirable situation in most cases is to allow a thin scale to be deposited, but keep the water close to or at the saturation point for scale (calcium carbonate) formation so that the rate of growth of the scale is practically nil.

The corrosiveness or scaling tendency of water may be expressed in more than one way. The qualitative expression simply indicates whether the water is undersaturated, in equilibrium, or oversaturated with respect to calcium carbonate. In the former case there could be a corrosion problem, in the latter scale may deposit, and if equilibrium exists, it depends on whether a layer of scale has previously been deposited. This qualitative expression is called the Langelier saturation index. There are several quantitative measures that indicate not only whether scale will form, but also how much scales a particular water supply is capable of producing. An example of a quantitative method is the determination by using the Caldwell–Lawrence diagrams (17,18). Only the Langelier index will be discussed here. For a description of the use of Caldwell–Lawrence diagrams, the reader is referred to the reference by Benefield and Morgan (17).

The Langelier index (LI) takes into account the pH, the calcium concentration, the alkalinity, and the temperature and ionic strength of the water to determine its saturation state. LI is defined as:

$$LI = pH - pHs \tag{20}$$

Table 2
Constants for Langelier Index

		$pK_2' - pK_s'$			
Ionic strength	TDS (mg/L)	0°C	10°C	20°C	25°C
0.000	0	2.45	2.23	2.02	1.94
0.002	80	2.62	2.40	2.19	2.11
0.005	200	2.71	2.49	2.28	2.20
0.010	400	2.81	2.59	2.38	2.30
0.015	600	2.88	2.66	2.45	2.37
0.020	800	2.93	2.71	2.50	2.42

where pH_s is pH of saturation, that is the pH water, with the same calcium concentration, alkalinity, temperature, and ionic strength, would have if it was in equilibrium with calcium carbonate. When LI is less than zero, protective scale would not be deposited, any existing scale would be removed, and the risk of corrosion exists. When LI is greater than zero, a protective scale would form. However, it should be noted that the actual numerical value of LI is not by itself an indication of the magnitude of these effects.
The pH of saturation, pH_s, can be computed as follows. Assuming that it falls between 6.5 and 9.5, and that the total dissolved solids concentration is less than 2000 mg/L:

$$pH_s = pCa^{2+} + (pK_2' - pK_s') + p(Alk) \tag{21}$$

where Ca^{2+} = molarity of calcium, K_2' = the dissociation equilibrium constant for bicarbonate, adjusted for temperature and ionic strength, K_s' = solubility product constant for calcium carbonate, at the given temperature and ionic strength, (Alk) = alkalinity, in equivalents per liter and the *p* function, as usual, indicates the negative logarithm of a value. Values for $pK_2' - pK_s'$ are given for various temperatures and total dissolved solids concentrations (TDS) in Table 2.

A procedure should be used if the calculated value for pH_s is outside the range 6.5–9.5, or if the ionic strength is above that for which data are given in the table. An example of the calculation of the Langelier index is presented in Section 8.

6. OTHER RELATED PROCESS APPLICATIONS

6.1. *Chemical Coagulation Using Magnesium Carbonate as a Coagulant*

Black et al. (9), Thompson et al. (10,11), Culp (12), Kinman (13), and Wang (19) report on magnesium carbonate coagulation including recovery of the magnesium carbonate. Coincidently, this coagulation process results in water softening. As stated above, the softening process results in removal of turbidity. Thus, actually, the processes are essentially identical but control is directed toward the major objective, softening or coagulation.

Hard turbid waters although they can occur, are not very common because the hydrogeologic systems that produce hardness are not likely to produce turbidity and vice versa. Many of the hard waters are from ground sources and are not turbid. Most of the turbid waters are surface waters and have not been in contact with soil or rocks long enough to dissolve out hardness constituents.

$Mg(OH)_2$ is formed by the addition of $MgCO_3$ and lime to water. The reaction is

$$MgCO_3 + CaO + H_2O \leftrightarrow Mg(OH)_2 + CaCO_3 \quad (22)$$

$Mg(OH)_2$ and to some extent $CaCO_3$ coagulate the turbidity and color colloids and the resulting sludge contains the $CaCO_3$, $Mg(OH)_2$, and turbidity and color. The floc formed, settles rapidly and the effectiveness is equal to or greater than the alum coagulation (10).

If the water is soft, there is, of course, little if any softening; however, the addition of the Ca^{2+} and CO_3^{2-}does stabilize the water and renders it less corrosive. If the water is hard, the addition of lime results in the reduction of carbonate hardness.

The major advantage of the magnesium carbonate coagulation process over the alum coagulation process is the possibility of recovering the $MgCO_3$ for reuse. If the original Mg^{2+} content of the water is substantial, and as described above, the process coincidentally results in softening, more magnesium carbonate will be produced in the recovery process than was added. This increased amount may be marketed. As described below lime may also be recovered and in that recovery process CO_2 for recarbonation may be produced.

6.2. *Recovery of Magnesium as Magnesium Carbonate*

The sludge produced from the magnesium carbonate coagulation process and similarly from the softening process will contain $Mg(OH)_2$, $CaCO_3$, and turbidity and color colloids. By using carbon dioxide (CO_2) and, if necessary froth flotation, separation of these three major components in the sludge is possible. Subsequent dewatering, and, in the case of $CaCO_3$ drying and calcining, can be employed. The processes used are described below.

According to Black et al. (9) the magnesium is selectively dissolved from either the softening or the $Mg(OH)_2$ coagulation precipitates by recarbonation in accordance with Eqs. (12a) and (12b). The solution of $Mg(HCO_3)_2$ is removed from the still undissolved $CaCO_3$ and if turbidity is present, by settling or filtration. The clarified $Mg(HCO_3)_2$ solution can be flushed to the sewer or can be recycled where it reacts again with lime to produce $Mg(OH)_2$ or it can be heated as shown in Eq. (23).

$$Mg(HCO_3)_2 + 2H_2O \xrightarrow{35\pm45^\circ\,C\ (air)} MgCO_3 \cdot 3H_2O \quad (23)$$

The precipitation of $MgCO_3{\cdot}3H_2O$ is complete in about 90 min (13). The excess $MgCO_3{\cdot}3H_2O$ produced can be dried for use elsewhere if there is a market, or put into landfill.

6.3. *Recovery of the Calcium Carbonate as Lime*

As shown in Eqs. (6), (7a), and (7b), $CaCO_3$ is produced due to the addition of CaO. Recovery of this $CaCO_3$ for production of CaO by calcining can be accomplished as shown in Eq. (24):

$$CaCO_3 \longrightarrow CaO + CO_2 \quad (24)$$

Calcining consists of tumbling the sludge down an inclined rotating kiln. If there is substantial magnesium present in the sludge, the end product will be a mixture of CaO and magnesium salts. Normally the magnesium is not desired. Therefore, the above-described process of prior selective magnesium removal may well be necessary. The

$CaCO_3$ sludge from the second-stage softening process would normally not require magnesium or turbidity removal.

Also, if there is a significant amount of turbidity in the sludge, and if the CaO is recycled, there will be a build up of turbidity rendering the CaO less useful. Turbidity (due to the presence of insoluble material) may be separated from the $CaCO_3$ by froth flotation (13).

The amount of lime that can be recovered may be about twice what is needed; therefore, lime may be marketed. The CO_2 produced in the calcining process is a valuable source for the recarbonation processes.

6.4. Recarbonation of Chemically Treated Wastewaters

Adjustment of pH of wastewaters could be purposeful, incidental, or inadvertent. Examples are given below to illustrate the three cases in consecutive order. The pH is usually adjusted to values above 10.8 in order to shift the ammonia-ammonium ion equilibrium to the ammonia side to permit ammonia stripping by aeration. Aluminum and iron salts precipitate in hydrous oxide forms reducing the pH, thus often necessitating the addition of lime to restore the pH necessary for good coagulation and/or precipitation. To remove phosphorous from wastewaters massive doses of lime may be added to first precipitate the CO_3^{2-}, then the phosphate and in so doing raise the pH above 11 (8). Adjustment of the pH in the treatment of industrial wastewaters are for many reasons too numerous to attempt to illustrate herein.

Recarbonation would be necessary when the pH is sufficiently high to interfere with secondary treatment processes, to result in after precipitation of calcium or magnesium salts in pipe, channels, tanks, filters, etc., or to cause stripping of adsorbed compounds from carbon adsorption columns.

Recarbonation may be utilized in one- or two-step processes. If the pH of the water treated for phosphorous removal is excessively above 10 as a result of lime treatment, reduction of the pH to about 9.3 using primary recarbonation will result in precipitation of the calcium as calcium carbonate, $CaCO_3$. Secondary recarbonation may be used to reduce the pH to the neutral range. If a single-step recarbonation were used to reduce the pH from above 10 to about 7, most of the calcium would remain in solution.

If a secondary treatment process is to follow the chemical precipitation step, it may be possible to avoid the secondary recarbonation step and treat the effluent in a completely mixed activated sludge reactor (14). Section 8.7 shows how recarbonation is applied in a well-known "two-stage tertiary lime treatment" (16).

7. PROCESS DESIGN

Recarbonation is used to prevent after precipitation of $CaCO_3$ and $Mg(OH)_2$ in softened waters. Also it is used for selective magnesium separation, and for pH adjustment (reduction) prior to secondary treated wastewater.

7.1. Sources of Carbon Dioxide

According to Culp and Culp (8) the usual sources of carbon dioxide are (a) stack gases from fuel combustion containing about 10% CO_2, (b) stack gases from the calcining furnace containing about 16% CO_2, and (c) commercial liquid carbon dioxide. The burning of 1000 ft^3 (28.3 m^3) of natural gas produces about 115 lb (52.3 kg) and

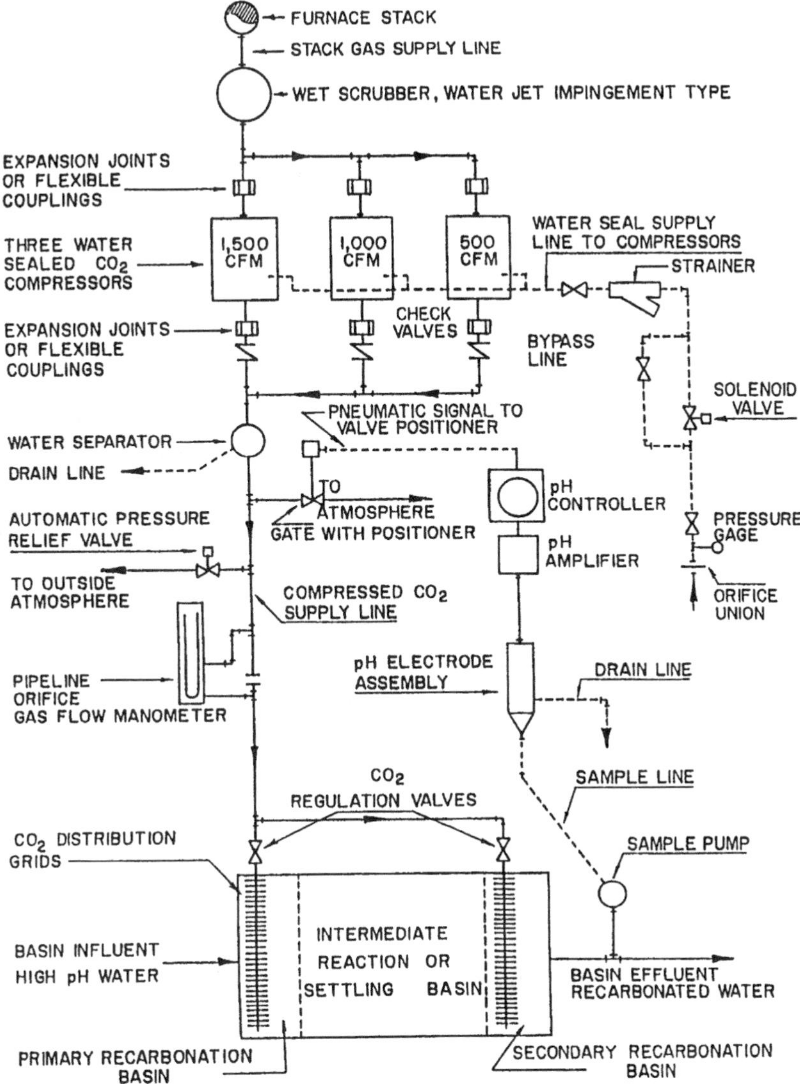

Fig. 1. Typical recarbonation system using stack gas (32).

artificial gas about 80 lb (36.4 kg) of CO_2. Kerosene and No. 2 fuel oil will yield about 20 lb (9 kg) of CO_2, coke will yield about 3 lb of CO_2 for each pound burned (3 kg for each kilogram). Commercial liquid CO_2 contains about 99.5% CO_2.

Stack gases usually contain significant particulate matter, and before their use as a CO_2 source, the stack gases should be wet scrubbed. The scrubber also cools the gases to about 110°F (43.3°C). The gases are then compressed and sent to the distribution system to be added to water for recarbonation (see Fig. 1). The normal combustion gases are about 6–18% CO_2 depending on the fuel and amount of excess air used in the combustion furnace. As stated above, the off gases from the limekiln will be richer in CO_2. The compressors are subject to clogging and are exposed to warm, moist, and corrosive gases.

When fuel is burned for CO_2 production only, it may be burned in pressurized combustion chambers. Usually, air and fuel are pressurized prior to combustion, eliminating the need for compressing the moist warm gases. If natural gas is used as the fuel, submerged combustion may be utilized. Gas and the combustion air are compressed and the burning actually takes place under water. This is a simple process with minimum maintenance problems.

Haney and Hamman (15) discussed in detail the use of carbon dioxide in its various forms—gas, liquid, or solid—in recarbonation. The CO_2 is fed as a gas; however, it may be purchased and stored in any one of the three states—gas, liquid, or solid.

At normal atmospheric temperature and pressure, CO_2 is a gas with a density about 1.5 times that of air. When compressed and cooled to the correct temperature, it liquefies; the liquid in turn can become a solid upon further compressing and cooling. Table 3 gives the physical properties of CO_2 and for comparison properties of chlorine gas are included. Figure 2 is the phase diagram of CO_2 and Fig. 3 shows the vapor pressure–temperature relationships of both CO_2 and chlorine.

Carbon dioxide gas may be purchased in 20 and 50 lb (9.2 and 22.7 kg) containers. Cylinders should not be stored at temperatures above 125°F (52°C). At normal temperatures the gas may be withdrawn from either containers at about 4 lb (2 kg) per hour.

Carbon dioxide may be stored in low temperature–low pressure containers with capacities of 0.5 to 100 tons (454 to 9100 kg). These containers are designed to maintain the CO_2 at a temperature near 0°F (–18°C) and a pressure of 300 pounds per square inch gage, psig (21,000 kg/m^2). The bulk storage tanks are normally equipped with safety devices to automatically control the temperature, and thus the pressure, with refrigeration systems and pressure relief valves.

The liquid CO_2 withdrawn from the tanks passes through steam or electrical vaporizers, pressure relief valves, a differential pressure transmitter, control valves and then to the recarbonation basin. The feeding of the CO_2 in the recarbonation basin is the same as the feeding of flue gases.

The main advantages of the carbon dioxide feed rather than the use of flue gases is the flexibility of feed rate, accuracy of control, and overall gas transfer efficiency. Cost may or may not be an advantage. The design engineer will need to make that determination in each instance.

7.2. Distribution Systems

Except in the cases of submerged combustion, the gases are delivered from the compressor or pressurized furnace to submerged distributors in the recarbonation chamber (see Fig. 1). The submerged distributor is at least 8 ft (2.4 m) below the water surface. The distributor is usually a perforated pipe (32).

Even though about 90% of the CO_2 in the bubbles will be absorbed, the recarbonation reactions are not instantaneous; the reactions may need about 15 min for completion. Therefore, a basin with at least a 15 min retention time should be provided. If the recarbonation step results in the formation of significant precipitation (e.g., first- and second-stage softening reactions or primary recarbonation of water or wastewaters), the reaction time may be included in the precipitation period.

Table 3
Physical Properties of Carbon Dioxide and Chlorine (32)

Property	Carbon dioxide	Chlorine
Molecular symbol	CO_2	Cl_2
Molecular weight	44.01	70.91
Specific gravity		
@ 32°F (0°C) and 1 atm pressure	1.53	2.48
Critical temperature		
°F	87.8	291
°C	31	143.9
Critical pressure		
psia	1072	1118
kg/m^2	753,723	786,066
Freezing point @ 1 atm		
°F	–109.3	–149.8
°C	–78.4	–101
Density lb/ft^3		
Solid @ 109.3°F	97.6	—
Liquid @ (0°F), 291 psig	63.7	94.8
Liquid @ (70°F), 838 psig	47.4	86.8
Density kg/m^3		
Solid @ (–78.4°C)	1563.5	—
Liquid @ (–17.7°C), kg/m^3 gage	1020.5	1518.7
Liquid @ (21.1°C), kg/m^3 gage	759.4	1390.6
Vapor pressure psig		
Saturated liquid		
@ 0°F	291	13.9
@70°F	838	85.4
Vapor pressure kg/m^3 gage		
Saturated liquid		
@ (–17.7°C)	4661.9	222.7
@ (21.1°C)	13424.8	1368.1
Gas (STP), liquid–volume ratio	520:1	457:1
Heat of vaporization, BTU/lb		
Solid @ –109.3°F	247	—
Liquid @ 0°F	120	115
Liquid @ 70°F	64.1	99.9
Heat of vaporization kcal/g		
Solid @ (–78.4°C)	62.2	—
Liquid @ (–17.70°C)	30.2	29.0
Liquid @ (21.1°C)	16.2	25.2
Viscosity centipoise		
Gas @ 70°F (68.4°C), 1 atm	0.015	0.013
Liquid @ 0°F (–18°C)	0.14	0.43
Solubility in H_2O g/100 g		
@ 68°F (20°C), 1 atm	0.169	0.729

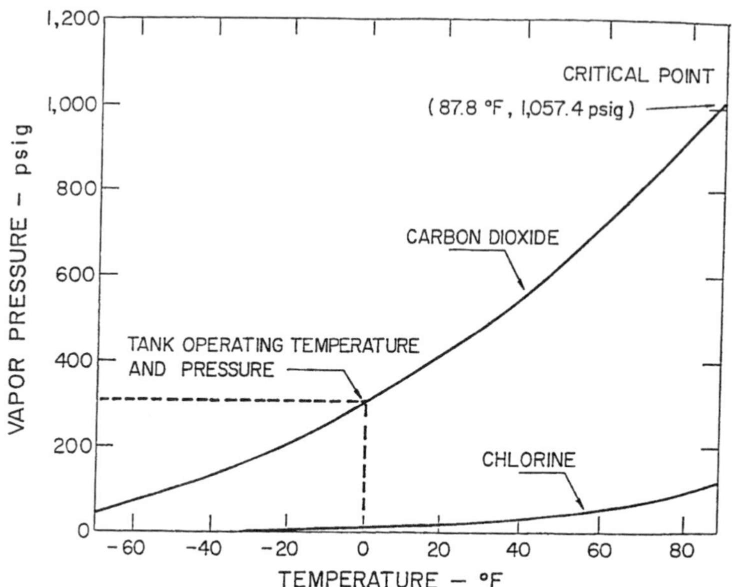

Fig. 2. Carbon dioxide and chlorine vapor pressure versus temperature (32).

7.3. Carbon Dioxide Quantities

According to the reactions indicated by Eqs. (10) and (11), one molecule of CO_2 forms calcium carbonate from calcium hydroxide and another molecule of CO_2 forms calcium bicarbonate from calcium carbonate. All alkalinity, OH^-, CO_3^{2-} and HCO_3^- are expressed as $CaCO_3$ which has a molecule weight of 100. Carbon dioxide has a molecular weight of 44. Therefore, the weight of CO_2 in mg/L per mg/L of OH^- or CO_3^{2-} is 44/100 = 0.44.

In terms of pounds of CO_2 per MG, the CO_2 requirement is $8.33 \times 0.44 = 3.66$ lb (or use 3.7 lb) for each of the two steps, or 7.32 lb total (or use 7.4 lb total). In terms of kg of CO_2 per million liters, the CO_2 requirement is 0.44 kg for each of the two steps or 0.88 kg total.

7.4. Step-by-Step Design Approach

7.4.1. Selection of Detention Time and Desired pH

Recarbonation is a unit process that has long been used in lime-softening water-treatment plants and lime wastewater-treatment plants. In water treatment, recarbonation is usually practiced ahead of the filters to prevent calcium carbonate deposition on the grains, which will result in shortening of the filter runs. Recarbonation is also used to lower the pH of the lime-treated water to the point of calcium carbonate stability to avoid deposition of calcium carbonate in pipelines.

More recently, with the increased use of lime treatment of wastewaters, recarbonation has been more widely used in wastewater treatment. Recarbonation, in wastewater treatment, is mainly used to adjust the pH following lime treatment for such applications as phosphorus removal, ammonia stripping, or chemical clarification.

Recarbonation may be practiced as either a two-stage or a single-stage system. Two-stage recarbonation consists of two separate treatment steps. In the first stage, sufficient carbon dioxide is added in the primary recarbonation stage to lower the pH of the

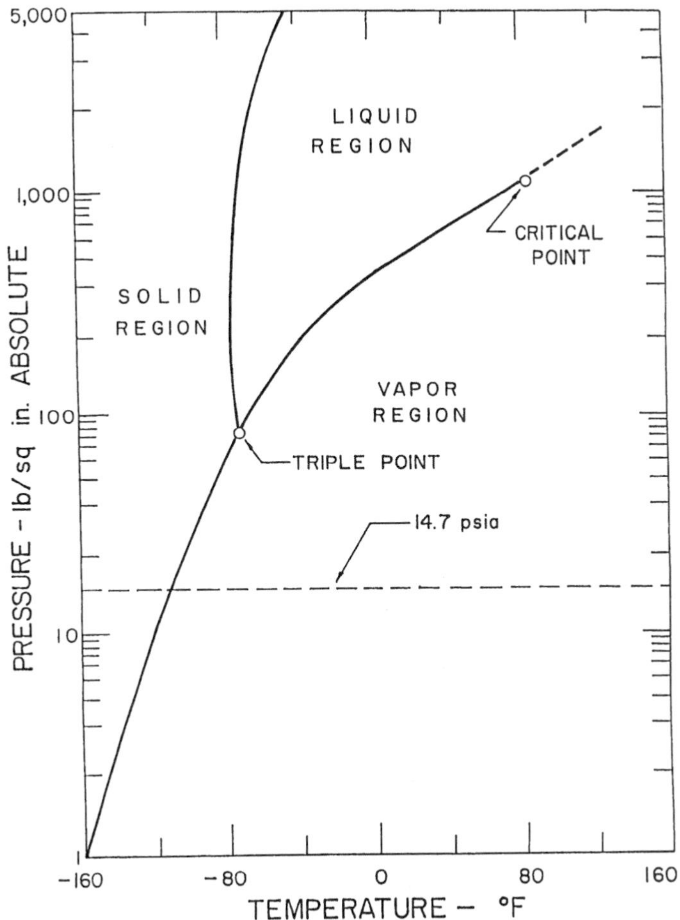

Fig. 3. Phase diagram for carbon dioxide (32).

wastewater to pH 9.3, which is near the minimum solubility of calcium carbonate. The sludge produced, which is mainly calcium carbonate, is then removed through settling and recalcined if recovery of the lime is desired. The time required to complete the reaction is normally 15–30 min. In the second stage, carbon dioxide is added to lower the pH to a value of pH 7. It is possible, however, to add sufficient carbon dioxide to lower the pH from 11 to 7 in a single stage. Single-stage recarbonation eliminates the need for an intermediate settling basin, which is needed in the two-stage system. However, single-stage recarbonation normally results in an increase in the calcium hardness of the water.

7.4.2. Selection of Recarbonation Process Reactions

The reactions involved in the recarbonation process may be simplified as follows:

$$Ca(OH)_2 + CO_2 \rightarrow CaCO_3 + H_2O \qquad (25)$$

$$CaCO_3 + CO_2 + H_2O \rightarrow Ca(HCO_3)_2 \qquad (26)$$

Dosage of CO_2 required for converting calcium hydroxide to calcium carbonate:

$$CO_2(\text{lb/MG}) = 3.7(\text{OH}^{\pm}\text{ alkalinity in mg/L as CaCO}_3) \tag{27}$$

Dosage of CO_2 required for converting calcium carbonate to calcium bicarbonate:

$$CO_2(\text{lb/MG}) = 3.7(\text{CO}_3^{2-}\text{ alkalinity in mg/L as CaCO}_3) \tag{28}$$

7.4.3. Selection of Process Design Input Data

Influent flow data should include (a) average flow and (b) peak flow. Influent characteristics data should include (a) total alkalinity, mg/L as $CaCO_3$, (b) hydroxide alkalinity, mg/L as $CaCO_3$, (c) carbonate alkalinity, mg/L as $CaCO_3$, and (d) pH value.

7.4.4. Process Design Parameters

Process design parameters that need to be decided are (a) contact time, 15–30 min, (b) carbon dioxide dosage, lb/MG, and (c) desired effluent pH.

7.4.5. Design Procedure for a Two-Stage Process System

1. Adjust the primary stage pH to 9.3.
2. Calculate the primary stage tank volume:

$$V = \frac{Q(t)10^6}{60(24)} \tag{29}$$

where V = tank volume (gal), Q = influent flow (MGD), and t = contact time (min).

3. Calculate the primary stage CO_2 requirement in ft³/min/MGD:

$$CO_2 = \frac{(3.7)(\text{OH}^-)(Q)}{0.116(1440)} \tag{30}$$

where CO_2 = carbon dioxide requirement (ft³/min/MGD), OH^- = hydroxide alkalinity (in mg/L as $CaCO_3$), Q = influent flow (MGD), 0.116 = density of CO_2 (lb/ft³), and 1440 = min/d.

4. Adjust the secondary stage pH to 7.
5. Calculate the secondary stage tank volume:

$$V = \frac{Q(t)10^6}{60(24)} \tag{29}$$

where V = tank volume (gal), Q = influent flow (MGD), and t = contact time (min).

6. Calculate the second stage CO_2 requirement in ft³/min/MGD:

$$CO_2 = \frac{(3.7)\left(\text{CO}_3^{2-} + \text{OH}^-\right)(Q)}{0.116(1440)} \tag{31}$$

where CO_3^{2-} = carbonate alkalinity (mg/L as $CaCO_3$), OH^- = hydroxide alkalinity (mg/L as $CaCO_3$), Q = influent flow (MGD), 0.116 = density of CO_2 (lb/ft³), and 1440 = min/d.

7.4.6. Design Procedure for a Single-Stage Process System

1. Adjust the single-stage recarbonation pH to 7.
2. Calculate the single-stage tank volume.

$$V = \frac{Q(t)10^6}{60(24)} \tag{29}$$

where V = tank volume (gal), Q = influent flow (MGD), and t = contact time (min).

3. Calculate the single-stage CO_2 requirement in ft³/min/MGD:

$$CO_2\left(ft^3/min/MGD\right)=\frac{\left[7.4\left(OH^-\right)+3.7\left(CO_3^{2-}\right)\right](Q)}{0.116(1440)} \tag{32}$$

where OH^- = hydroxide alkalinity (mg/L as $CaCO_3$), CO_3^{2-} = carbonate alkalinity (mg/L as $CaCO_3$), Q = influent flow (MGD), 0.116 = density of CO_2 (lb/ft³), and 1440 = min/d.

$$CO_2(lb/MG) = 7.4(OH^-) + 3.7(CO_3^{2-}) \tag{32a}$$

7.4.7. *Process Design Output Data*

The output data should include (a) volume of tank (MG), (b) carbon dioxide requirement (ft³/min/MGD), (c) final pH, and (d) contact time (min).

8. DESIGN AND APPLICATION EXAMPLES

8.1. *Example 1*

A water supply has been treated with lime to a pH of 11.5 and with alkalinities (as mg/L $CaCO_3$) of OH^- = 320, CO_3^{2-} = 100, HCO_3^- = 0. Determine the total CO_2 requirement in lb/MG (1 lb/MG = 0.12 kg/million L) to reduce the pH to about 8.3 by a single-stage recarbonation process.

Solution

The total weight of the stack gas is

$$(7.4\times 320)+(3.7\times 100)=2700 \text{ lb/MG} \quad (325 \text{ kg/million L})$$

If a stack gas of 12% of CO_2 is used, the weight of the required stack gas is

$$2700/0.12=22{,}500 \text{ lb/MG} \quad (2700 \text{ kg/million L})$$

If the stack gas at 60°F (15.5°C) and one atmosphere pressure has a density of 0.116 lb/ft³ (1.85 kg/m³), the volume of gas required is

$$22{,}500/0.116=194{,}000 \text{ ft}^3 \quad \left(5500 \text{ m}^3\right)$$

If the gas used is at 110°F (43.3°C), the volume correction is

$$(110+460)/(60+460)=1.1$$

and the required volume is

$$195{,}000\times 1.1=214{,}000 \text{ ft}^3 \quad \left(6060 \text{ m}^3\right)$$

Assuming 80 % gas transfer efficiency, the design requirement is

$$214{,}000/0.8=\left(268{,}000 \text{ ft}^3\right) \quad \left(7580 \text{ m}^3\right)$$

8.2. *Example 2*

If the CO_2 requirement for recarbonation is 2710 lb/MG (325 kg/million L) and if the stack gas used is 12% CO_2 by weight, determine the volume of gas required at 60°F (15.5°C) and at an elevation of 6300 ft (1920 m) above sea level (11.6 psia or 56.6 kg/m²).

Solution

The total weight of stack gas is

$$2710/0.12 = 22{,}600 \text{ lb/MG} \quad (2712 \text{ kg/million L})$$

If at 60°F (15.5°C) and at sea level the stack gas has a density of 0.116 lb/ft^3 (1.85 kg/m^3), the density of the gas at the same temperature and at 6300 ft (1920 m) would be

$$(11.6/14.7)0.116 = 0.0915 \text{ lb/ft}^3 \quad \left(1.49 \text{ kg/m}^3\right)$$

The volume of the gas required per day would be

$$22{,}600/0.0915 = 247{,}000 \text{ ft}^3 \quad \left(6990 \text{ m}^3\right)$$

8.3. Example 3

Determine the required piping and diffusion system for a 1 MGD plant using the volume of gas determined in Example 1 above at 60°F (15.5°C), and at an elevation of 6300 ft (1920 m) above sea level, assuming 80% gas transfer efficiency.

Solution

From Example 1, the gas volume is 247,000 ft^3 (6990 m^3) per day, which is equivalent to a flow rate of

$$Q_g = 171 \text{ ft}^3/\text{min} \quad \left(4.86 \text{ m}^3/\text{min}\right)$$

Assume that the gas pipeline is 2 in. (50 mm) in diameter, 400 ft (122 m) long, three elbows, six T's and one globe valve. Also let us assume that the gas is diffused at 8 ft (2.44 m) below the water surface through 150 3/16 in. (5 mm) orifices. The air pressure in the piping is 10 psi gage.

The head loss in an air pipe can be estimated by the Darcy-Weisbach equation modified for airflow (8):

$$\Delta p = \frac{F}{38{,}000} \frac{LTQ_g^{\,2}}{pD^5} \tag{33}$$

$$F = \frac{0.048\, D^{0.027}}{Q_g^{\,0.148}} \tag{34}$$

where Δp = pressure drop in psi, L = pipe length in ft, T = absolute temperature of the gas (= F° + 460), Q_g = gas flow (ft^3/min), p = absolute pressure of the gas in psi (line pressure + 14.7), and D = pipe diameter (in.).

The head loss in elbows and T's can be estimated by:

$$L = \frac{7.6D}{1 + \dfrac{3.6}{D}} \tag{35}$$

where L = equivalent length of pipe (ft) and D = pipe diameter (in.). The head loss in globe valves can be determined by:

$$L = \frac{11.4D}{1 + \dfrac{3.6}{D}} \tag{36}$$

where L = equivalent length of pipe (ft) and D = pipe diameter (in.).

Using Eq. (35), the head losses in the 3 elbows and 6 T's are equivalent to

$$L = \frac{9 \times 7.6 \times 2}{1 + \frac{3.6}{2}} = 49 \text{ ft of pipe}$$

Using Eq. (36), the head loss in one globe valve is equivalent to

$$L = \frac{11.4 \times 2}{1 + \frac{3.6}{2}} = 8 \text{ ft of pipe}$$

The total head loss in fittings is equivalent to 49 + 8 = 57 ft of pipe. Total equivalent pipe length is equal to (400 ft + 57 ft) or 457 ft.

The head loss in the orifices is computed by the standard orifice equation; with an orifice coefficient of 0.8, the head loss is 5.4 in. of water or 0.20 psi.

The head loss in the piping is estimated as follows:

$$F = \frac{0.048(2)^{0.027}}{(171)^{0.148}} = \frac{0.048 \times 1.019}{2.140} = 0.023 \tag{34}$$

$$L = 400 + 57 = 457 \text{ ft}$$
$$T = 60 + 460 = 520$$
$$p = 10 + 14.7 = 24.7$$

$$\Delta p = \frac{0.023 \times 457 \times 520 \times (171)^2}{38{,}000 \times (10 + 14.7)(2)^5} = 5.32 \text{ psi} \tag{33}$$

Eight feet of water is equivalent to 8 x 0.434 = 3.47 psi. Total head or pressure loss in the system is

$$0.20 + 5.32 + 3.47 = 9.0 \text{ psi}$$

The compressor theoretical horsepower requirement is,

$$P = 0.22 Q_g \left[(p/14.7)^{0.283} - 1 \right] \tag{37}$$

where P = theoretical horsepower (HP), Q_g = gas flow (ft^3/min), p = absolute gas pressure (psia = psig + 14.7):

$$P = 0.22 \times 171 \left[(24.7/14.7)^{0.283} - 1 \right]$$
$$= 6 \text{ HP}$$

At 80% efficiency, the required compressor horsepower is

$$6/0.8 = 7.5 \text{ HP}$$

8.4. *Example 4*

Calculate the lime and soda ash dosages for one-stage treatment of hard water. Analysis of the water shows the following:

Hardness = 280 mg/L as $CaCO_3$
Magnesium = 21 mg/L

Alkalinity = 170 mg/L as $CaCO_3$
Carbon dioxide = 6 mg/L
pH (assumed near neutral)

Solution

First, it is necessary to convert these to equivalent concentrations in milliequivalents per liter by dividing milligrams per liter by the gram-equivalent weight:

Total hardness = 280 mg/L / 50 mg/meq = 5.60 meq/L
Magnesium = 21 mg/L / 12.2 mg/meq = 1.72 meq/L
Alkalinity = 170 mg/L/ 50 mg/meq = 3.40 meq/L
Carbon dioxide = 6 mg/L / 22 mg/meq = 0.27 meq/L

Assuming all the hardness is present as calcium and magnesium, the equivalents of calcium is equal to the total hardness minus the magnesium, or 5.60–1.72 = 3.88 meq/L. In neutral water virtually all the alkalinity is present as bicarbonate. In this example the carbon dioxide is given, although in practice it is usually computed from the alkalinity and a measurement of the pH.

The calcium carbonate hardness is equal to the bicarbonate present, as long as that is less than the total calcium. Thus, the CaNCH is the balance 3.88–3.40 = 0.44 meq/L. Because there is no carbonate left over, all the magnesium present is noncarbonate hardness.

The number of equivalents of chemicals needed to soften the water can be determined by combining the above information with the stoichiometric requirements shown in Table 1:

	Required dosage (meq/L)	
Constituent to be removed	Lime	Soda ash
Carbonic acid	0.27	0.00
Calcium carbonate hardness	3.40	0.00
Magnesium carbonate hardness	0.00	0.00
CaNCH	0.44	0.44
MgNCH	1.72	1.72
Total dosage required	5.83	2.16

Finally, the dosage in pounds for a flow of 1 MGD can be computed by multiplying by the equivalent weight (37.5 for slaked lime and 53 for soda ash) by 8.34:

Lime dosage: 5.83 meq/L × 37.5 mg/meq × 8.34 × 1 = 1,820 lb/MG
Soda ash: 2.16 meq/L × 53.0 mg/meq × 8.34 × 1 = 955 lb/MG

Five to 10% excess should be applied to ensure rapid and complete precipitation. Furthermore, the dosage should be divided by the fractional purity of the supplied chemical (e.g., if the lime is 80% pure, the actual lime dosage should be 1820/0.80 = 2,280 lb).

8.5. Example 5

What is the Langelier index for neutral water with calcium hardness of 100 mg/L, alkalinity equal to 150 mg/L as calcium carbonate, and TDS equal 800 mg/L, at a temperature of 10°C?

Solution

The number of moles of calcium per liter is

$$100 \text{ mg/L } CaCO_3 \times 1 \text{ mol } CaCO_3/100 \text{ g} \times 1 \text{ g}/1000 \text{ mg} = 0.001\ M$$

$pCa^{2+} = 3.0$

The alkalinity must be expressed in equivalents per liter. The equivalent weight of calcium carbonate is 50, thus,

$$(\text{Alk}) = 150 \text{ mg/L} / (50 \text{ eq/g}) \times (1 \text{ g}/1000 \text{ mg}) = 0.003 \text{ eq/L}$$

$$p(\text{Alk}) = 2.52$$

The value of $pK_2' - pK_s'$ is found from Table 2 (using TDS = 800 mg/L and temperature = 10°C) to be 2.71.

Entering the above values into Eq. (21) for the pH of saturation:

$$pH_s = pCa^{2+} + \left(pK_2' - pK_s'\right) + p(\text{Alk})$$

$$pH_s = 3.0 + 2.71 + 2.52 = 8.23$$

and from Eq. (20), LI = pH – pH_s = 7.0 – 8.23 = –1.23

The negative index indicates that the water is "corrosive," or unstable. Note also that this water would actually be considered moderately hard. At 90 (C the value for $pK_2' - pKs'$ is 1.43, hence pH_s = 6.95 and Langelier Index will become positive (LI = 0.05), thus scaling would occur in boilers using this water supply.

8.6. Example 6

Conventional lime/soda-ash softening process uses a sedimentation tank for clarification. What are the other options for clarification (or solid–water separation) in a softening process?

Solution

Innovative processes, such as dissolved air flotation (21,25), physicochemical sequencing beach reactors (20,21), and membrane processes (22,23) can replace the conventional sedimentation clarifier for solid–water separation in a softening process system. The readers are referred to other chapters for details.

8.7. Example 7

Recarbonation is frequently applied to a "two-stage tertiary lime treatment process" for treating wastewaters. Explain what the "two-stage tertiary lime treatment process" is, and how recarbonation functions in the process.

Solution

The flow diagram of a "two-stage tertiary lime treatment process" is shown in Fig. 4 (16). Lime treatment of secondary effluent for the removal of phosphorus and suspended solids is essential for environmental protection. Calcium carbonate and magnesium hydroxide precipitate at high pH along with phosphorus hydroxyapatite and other suspended solids. In the two-stage system, the first stage precipitation generally is controlled around a pH of 11, which is approximately one pH unit higher than that used in the single-stage process. After precipitation and clarification in the first stage, the wastewater is recarbonated with carbon dioxide, forming a calcium carbonate precipitate, which is removed in the second clarification stage.

Lime is generally added to a separate rapid-mixing tank or to the mixing zone of a solid-contact or sludge-blanket clarifier. After mixing, the wastewater is flocculated to allow for the particles to increase in size to aid in clarification. The clarified wastewater is

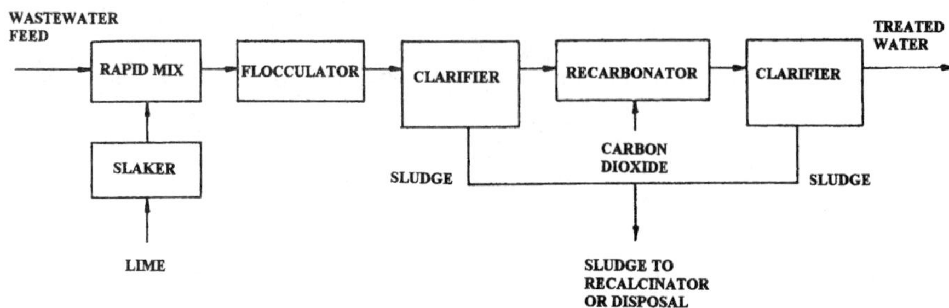

Fig. 4. Application of recarbonation in a two-stage tertiary lime treatment process.

recarbonated in a separate tank following the first clarifier, after which it is re-clarified in a second clarifier. Final pH adjustment may be required to meet allowable discharge limits.

Treatment systems can consist of separate units for flash mixing, flocculation, and clarification, or they can consist of specially designed solid-contact or sludge-blanket units, which contain flash mix, flocculation, and clarification zones in one unit. The calcium carbonate sludge formed in the second stage can be recalcined. Final effluent can be neutralized with sulfuric acid, as well as other acids.

8.8. *Example 8*

Recarbonation is frequently applied to a "two-stage lime-soda ash softening process system," and to a "single-stage lime-soda ash softening process system" for hardness removal. Please answer the following:

1. Discuss the hardness classification scale.
2. Present the flow diagram of a two-stage lime-soda ash softening process system.
3. Present the flow diagram of a single-stage lime-soda ash softening process system.

Solution

1. The hardness may be classified as follows:

Hardness description	Hardness range (mg/L as $CaCO_3$)
Soft	0–75
Moderately Hard	76–150
Hard	151–300
Very Hard	>300

2. The flow diagram of a two-stage lime-soda ash softening process system is shown in Fig. 5. Note that recarbonation is applied twice. In the first stage, sedimentation follows recarbonation for solid–water separation. In the second stage, filtration follows recarbonation for solid–water separation. (17)

Alternatively, dissolved air flotation may replace sedimentation in the first stage, and membrane filtration (such as ultrafiltration) may replace conventional sand filtration for final solid–water separation.

3. The flow diagram of a single-stage lime/soda-ash softening process system is shown in Fig. 6. Recarbonation plays an important role for precipitation of hardness. To protect the filter and enhance filtration efficiency, usually there is an intermediate solid-water separation process (either sedimentation or dissolved air flotation)

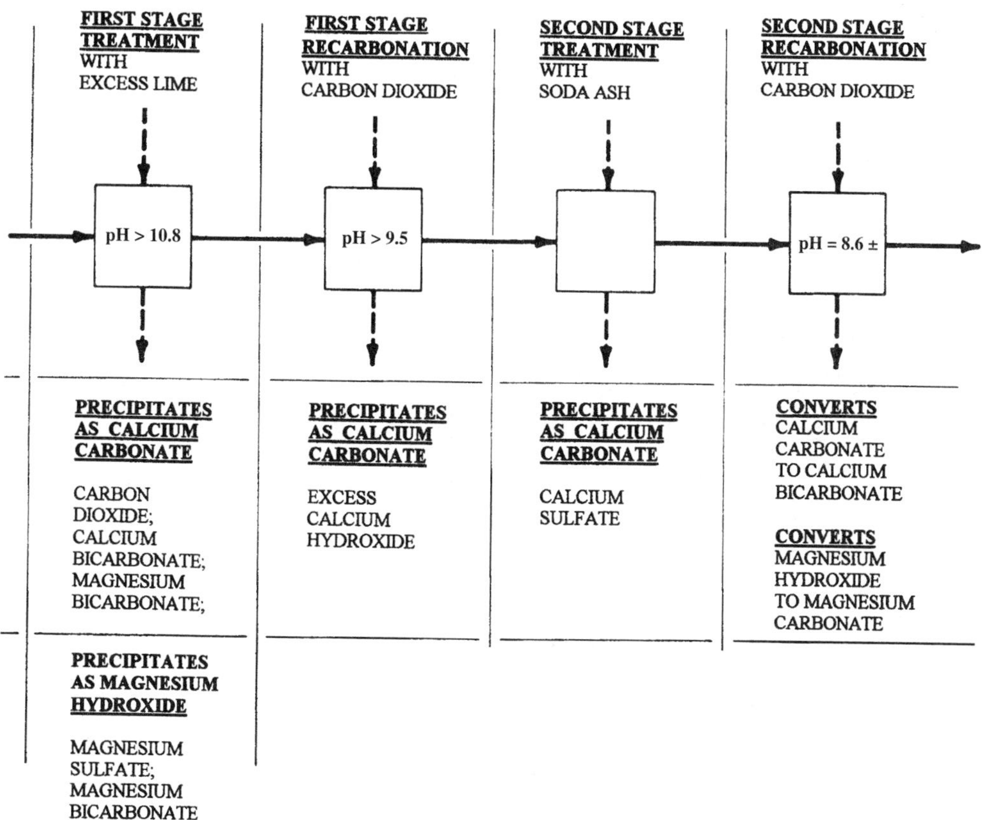

Fig. 5. Two-stage lime/soda ash softening process

8.9. Example 9

How do you determine the required quantities of lime and soda ash in a lime-soda ash softening process system?

Solution

The amount of lime required for softening water is determined by the amounts of free carbon dioxide and of magnesium that are present. The CO_2 amounts to 44% of the bicarbonate alkalinity. On the basis of the molecular weights of the substances, the amount of pure CaO required to react with 1 mg/L of free and half-bound CO_2 is 10.75 lb per MG of water, and the amount of pure CaO required to precipitate 1 mg/L of magnesium is 19.23 lb per MG of water. For instance, the ratio of the molecular weights of CO_2 and CaO is 44.01: 56.08, or 1:1.27, and for each mg/L of CO_2 present, 1.27 mg/L of CaO will be required. Since 1 gal of water weighs 8.34 lb, the weight of lime required for 1 MG of water is $1.27 \times 8.34 = 10.6$ lb.

$$\begin{aligned} \text{CaO Dosage} &= (CO_2\ \text{mg/L})(56/44) \\ &= 1.27\ \text{mg/L CaO} \\ &= 1.27\ (8.34\ \text{lb/MG}) \\ &= 10.6\ \text{lb/MG} \end{aligned}$$

If hydrated lime, $Ca(OH)_2$, or impure lime is used, a greater weight will be required. Because the molecular weight of CaO is $40.0 + 16.0 = 56.0$ and the molecular weight of $Ca(OH)_2$

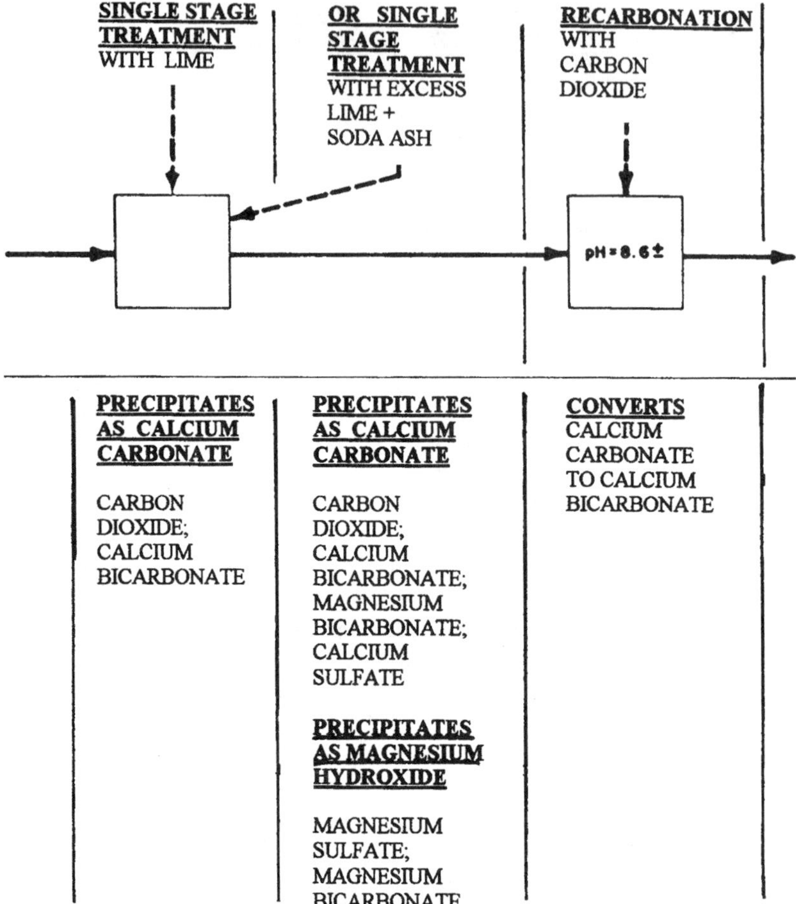

Fig. 6. Single-stage lime-soda-Ash softening process.

is 40.0 + 32.0 + 2.0 = 74.0, then 74.0 lb of hydrated lime is equivalent to 56.0 lb of CaO. If 100 lb CaO is needed, the required weight of $Ca(OH)_2$ is 74.0/56.0 × 100 = 132 lb.

As determined by the molecular weights, the amount of soda ash required to remove 1 mg/L of noncarbonate hardness from 1 MG of water is 8.84 lb. Because it is customary to remove only a part of the noncarbonate hardness, only the weight of the portion that is to be removed is multiplied by 8.34 to determine the required amount of soda ash in pounds.

Soda ash dosage, mg/L = (noncarbonate hardness, mg/L) (106/100)
=1.06 mg/L Na_2CO_3 = 1.06 × 8.34 lb/MG = 8.84 lb/MG

8.10. Example 10

Determine the amounts of chemicals required for treating the following water supply by the lime/soda ash softening process: free CO_2, 2 mg/L, alkalinity, 60 mg/L, noncarbonate hardness, 90 mg/L, and total magnesium, 12 mg/L. Assume that it is possible to removal all but 35 mg/L of carbonate hardness with lime, and the finished water is expected to have a total hardness of 85 mg/L.

Solution

The amount of noncarbonate hardness that may be left in the water is 85 – 35 = 50 mg/L and the noncarbonate hardness to be removed is 90–50 = 40 mg/L.

The amount of free and half-bound CO_2 in the water is

$$2 + (60 \times 0.44) = 28.4 \text{ mg/L}$$

The amount of lime needed to neutralize the CO_2 in 1 MG of water is

$$28.4 \times (56/44) \times 8.34 = 301 \text{ lb}$$

Also, the amount of lime needed to react with the magnesium in 1 MG of water is

$$12 \times (56/24.3) \times 8.34 = 231 \text{ lb}$$

The total amount of pure lime needed per MG is

$$301 + 231 = 532 \text{ lb}$$

The amount of soda ash needed to remove the noncarbonate hardness from 1 MG of water is

$$40 \times (106/100) \times 8.34 = 354 \text{ lb}$$

8.11. Example 11

Discuss the essential equipment for recarbonation in Example 7 (two-stage tertiary lime treatment process) and Example 8 (two-stage lime/soda ash-softening process and single-stage lime-soda ash softening process).

Solution

A typical installation for producing and feeding carbon dioxide can be found in a book entitled, *Water Supply and Waste Disposal*, by Hardenbergh and Rodie (24). Essential equipment for recarbonation includes a gas producer (coke burner, lime stone reactor), a scrubber, a moisture trap, a drier, a compressor, and a carbonation chamber with diffuser pipes. The compressed gas is applied under a low pressure to the water by means of a grid of pipes perforated with holes 3/32 or 1/8 in. in diameter and about 6 in. apart, or by some other type of diffuser. A carbonation chamber having a depth of about 10 ft and a displacement period (detention time) of 15–30 min will be sufficient (24).

The gas from a well-designed coke burner contains from 10% to as high as 19% of CO_2. Also, 1 ft^3 of CO_2 at 70°F and atmospheric pressure weighs 0.1145 lb. When the amount of water to be treated and its phenolphthalein and methyl-orange alkalinities are known, the compressor size required for a recarbonation unit can be determined.

The City of Kent, OH, offers a tour of its recarbonation equipment and basins (31). The City of Kent's recarbonation process equipment has a capacity of 6 MGD, and a detention time of 1.35 h.

8.12. Example 12

Determine the compressor capacity of a recarbonation plant shown in Fig. 7, assuming the carbonate alkalinity is 100 mg/L (phenolphthalein alkalinity = 50 mg/L and methyl-orange alkalinity = 110 mg/L), the softening plant capacity is 1.2 MGD, and the coke burner produces a gas that contains 12.5% of CO_2 by weight.

Solution

The reaction by which calcium carbonate and carbon dioxide form calcium bicarbonate may be written as follows:

$$CaCO_3 + CO_2 + H_2O \rightarrow Ca(HCO_3)_2$$

A definite amount of CO_2 is required to react with a given quantity of $CaCO_3$. The ratio of the weight of CO_2 to the weight of $CaCO_3$ is 44:100. Therefore, 1 mg/L $CaCO_3$ requires 0.44 mg/L of CO_2. Because 1 gal of water weighs 8.34 lb, the weight of CO_2 required to react with 1 mg/L of $CaCO_3$ is 0.44 × 8.34 = 3.7 lb per MG of water. An additional 25% is added to allow for losses in the scrubber or drier. Thus, the weight of CO_2 generally provided for 1 mg/L of $CaCO_3$ is 3.7 × 1.25 = 4.6 lb per MG.

Because the phenolphthalein alkalinity is less than one-half of the methyl-orange alkalinity, the carbonate alkalinity is twice the phenolphthalein alkalinity = 2 × 50 = 100 mg/L.

The required weight of CO_2 per MG of water is

$$100 \times 4.6 = 460 \text{ lb}$$

and the daily requirement for 1.2 MG is

$$460 \times 1.2 = 552 \text{ lb}$$

The required weight of CO_2 per minute is

$$552/1440 = 0.383 \text{ lb/min}$$

Because each cubic foot of gas evolved contains 12.5% CO_2, the weight of CO_2 in a cubic foot of gas is

$$0.1145 \times 0.125 = 0.0143 \text{ lb}$$

Then, the required capacity of the compressor is

$$0.383/0.0143 = 26.8 \text{ ft}^3\text{/min}$$

8.13. Example 13

Discuss (a) the various sources of carbon dioxide gases available for recarbonation, and (b) some additional applications of recarbonation.

Solution

(a) Carbon dioxide can be produced by burning coke purchased in cylinders, or collected from industrial stacks (26) for reuse in wastewater recarbonation system (see Example 7, Fig. 4). For potable water applications only clean carbon dioxide from a coke burning plant or cylinders can be used. Details of a recarbonation plant using liquid CO_2 cylinders can be found in the literature (30,33).

(b) Additional applications of recarbonation include recovery of milk protein from dairy factory wastewater and recovery of protein from tannery wastewater (26).

8.14. Example 14

Briefly discuss an innovative recarbonation chamber.

Solution

Conventional recarbonation chamber is shown in Fig. 7. When dissolved air flotation (DAF) (19, 25) is used to replace sedimentation for solid–water separation, carbon dioxide gas can be introduced through the gas-dissolving pressure tank (27). Consequently, the entire carbonation chamber (together with the diffusers) can be eliminated, resulting in cost saving.

Table 4
Comparison of Common Stability Indices

	Stability index		
Stability characteristics	Langelier index (LI)	Aggressive index (AI)	Ryznar index (RI)
Highly aggressive	<–2.0	<10.0	>10.0
Moderately aggressive	–2.0 to <0.0	10.0 to <12.0	6.0 to <10.0
Nonaggressive	>0.0	>12.0	<6.0

8.15. *Example 15*

Explain (a) how the Langelier index (LI) can be applied to water stabilization control, (b) the purpose of water stabilization control, and (c) other stability indices.

Solution

(a) The day-to-day operation of the water stabilization process consists mainly of operating and maintaining chemical feed facilities. This is essential in order to be able to control the water quality properly. The following water quality characteristics should be examined in water distribution systems:

pH
alkalinity
hardness
temperature
carbon dioxide
color
chloride
sulfate
fluoride
cadmium
iron
zinc
copper
lead
dissolved oxygen
conductivity
total dissolved solids

The gathered water quality data can then be used for calculating the Langelier Index ($LI = pH - pH_s$) according to Section 5 and Example 5.

The pH (actual) is the measured pH of the water, and the pH_s is the theoretical pH at which the water will be saturated with calcium carbonate. pH_s is calculated using a formula that takes into account calcium ion concentration, alkalinity, pH, temperature, and the total dissolved solids. If the LI equals zero, the water is considered stable. If the LI is a positive value, calcium carbonate will precipitate and the water has scale-forming tendencies. If the LI is a negative value, calcium carbonate will be dissolved and the water has corrosive tendencies.

(b) The purpose of water stabilization control is to prevent corrosion and scaling in the water distribution system. The physical condition of the distribution system usually gives the most accurate indication of the need for water stabilization. Neither corrosion nor scaling will be desirable.

(c) Other stability indices include: Aggressive Index (AI) and Ryznar Index (RI).

$$AI = pH + \log_{10}(Alk) + \log_{10}(Ca) \qquad (38)$$

$$RI = 2pH_s - pH = pH - 2(LI) \qquad (39)$$

where pH = actual pH, pH_s = saturation pH, Alk = total alkalinity in mg/L $CaCO_3$, and Ca = calcium hardness as mg/L $CaCO_3$. A comparison of LI, AI and RI is given in Table 4

8.16. Example 16

A 1-MGD two-stage lime treatment process is to be designed for a water supply system. Determine the pH, tank volumes, and CO_2 requirements of both primary and secondary stages. Analysis of the water shows the following:

Hydroxide alkalinity = 50 mg/L as $CaCO_3$
Carbonate Alkalinity = 150 mg/L as $CaCO_3$

Assume a contact time of 15 min (Note: between 15 to 30 min.) for each stage.

Solution

1. Adjust primary stage pH to 9.3.
2. Calculate the primary stage tank volume:

$$V = \frac{Q(t)10^6}{60(24)} \tag{29}$$

$$V = \frac{(1.0)(15)10^6}{60(24)} = 10{,}416 \text{ gal}$$

3. Calculate the primary stage CO_2 requirement.

$$CO_2 = \frac{3.7(\text{OH}^-)Q}{0.116(1440)} \tag{30}$$

$$CO_2 = \frac{3.7(50)(1.0)}{0.116(1440)} = 1.1 \text{ ft}^3/\text{min/MGD}$$

4. Adjust secondary stage pH to 7, and select contact time of 15 min.
5. Calculate the secondary stage tank volume.

$$V = \frac{Q(t)10^6}{60(24)} \tag{29}$$

$$V = \frac{(1.0)(15)10^6}{60(24)} = 10{,}416 \text{ gal}$$

6. Calculate the secondary stage CO_2 requirement.

$$CO_2 = \frac{3.7\left(\text{CO}_3^{2-} + \text{OH}^-\right)Q}{0.116(1440)} \tag{31}$$

$$CO_2 = \frac{3.7(150 + 50)(1.0)}{0.116(1440)} = 4.4 \text{ ft}^3/\text{min/MGD}$$

NOMENCLATURE

AI	Aggressive Index
D	pipe diameter, in. (mm)
Δp	pressure drop in psi (kg/m^2)
K_2'	the dissociation equilibrium constant for bicarbonate
K_s'	solubility product constant for calcium carbonate

K_{sp}	solubility product
L	pipe length or equivalent length, ft (m)
LI	Langelier index
NCH	noncarbonate hardness, mg/L as $CaCO_3$
p	absolute pressure, pounds per square inch, psia (kg/m^2) = psig + 14.7
pH	actual pH
pH_s	saturation pH
P	horsepower, HP
Q	water flow, MGD (million L/d)
Q_g	gas flow, ft^3/min (m^3/min or L/s)
RI	Ryznar Index
t	contact time, min
T	absolute temperature = F° + 460 (°C + 276)
TDS	total dissolved solids, mg/L
V	tank volume, gal or ft^3 (L or m^3)

ACKNOWLEDGMENTS

This chapter has been written in honor of Prof. Marvin L. Granstrom who was the former Chairman of the Department of Civil and Environmental Engineering, Rutgers University, New Brunswick, New Jersey. Prof. Granstrom received his PhD degree from Harvard University and has educated over 100 PhD students from around the world in his life long teaching career at Rutgers. Dr. Lawrence K. Wang and Dr. Jy S. Wu are among those graduates from Rutgers, who have received supervision and guidance from Dr. Granstrom. The authors of this chapter salute our great professor for the dedication of his entire academic career to promote environmental engineering education.

REFERENCES

1. L. K. Wang, *Environmental Engineering Glossary*, Calspan Corporation, Buffalo, NY, 1974.
2. G. M. Fair, J. C. Geyer and D. A. Okun *Water and Wastewater Engineering*, Volume 2, John Wiley and Sons, New York, 1968.
3. L. G. Rich *Unit Processes of Sanitary- Engineering*, John Wiley and Sons, New York, 1963.
4. J. W. Clark, W. Viessman and M. J. Hammer, *Water Supply and Pollution Control*, International Textbook Company, Scrontes, PA, 1971.
5. C. N. Sawyer, P. L. McCarty and G. F. Parkin, *Chemistry for Environmental Engineering*, McGraw-Hill, New York, 1994.
6. Am. Water Works Assoc. (1990), *Water Quality and Treatment* 4th ed., McGraw-Hill, NY.
7. W. J. Weber Jr. *Physicochemical Processes for Water Quality Control*, Wiley-Interscience, New York, 1972.
8. R. L. Culp and G. L. Culp, *Advanced Wastewater Treatment*, Van Nostrand Reinhold, New York, 1972.
9. A. P. Black, B. S. Shuey, and P. J. Fleming *J. Am. Water Works Assoc.* **63**, 616 (1971).
10. C. G. Thompson, J. E. Singely and A. P. Black *J. Am. Water Works Assoc.* **64**, 11 (1973) and 64, 93 1972 (Parts I and II).
11. C. G. Thompson, *Proceedings of Minimizing an Recycling Water Plant Sludge*, Am. Water Works Association, 1973.
12. G. L. Culp and R. L. Culp, *New Concepts in Water Purification*, Van Nostrand Reinhold Co, New York, 1974.

13. R. N. Kinman, *Proceedings of Minimizing and Recycling Water Plant Sludge*, Am. Water Works Association, 1973.
14. Metcalf and Eddy *Wastewater Engineering*, 3rd edition McGraw Hill, New York, 1991.
15. P. D. Haney and C. L. Hamann, *Jour. Am. Water Works Assoc.* **61**, 512-521 (1969).
16. US EPA, *Innovative and Alternative Technology Manual*, US Environmental Protection Agency, Washington, DC (1980). EPA 403/9-78-009.
17. L. D. Benefield and J. M. Morgans (1999). Chemical precipitation,In *Water Quality and Treatment.* R. D. Letterman, ed. American Water Works Association, McGraw-Hill Book Co., NY, 1999, pp.10.1–10.60.
18. M. J. Humenick Jr *Water and Wastewater Treatment.* Marcel Dekker, New York, 1997, p. 31.
19. L. K. Wang *Treatment of Groundwater by Dissolved Air Flotation Systems Using Sodium Aluminate and Lime as Flotation Aids.* US Department of Commerce, National Technical Information Service, Springfield, VA, 1985, PB85-167229/AS.
20. L. K. Wang, L. Kurylko, and M. H. S. Wang *Sequencing Batch Liquid Treatment.* US Patent 5354458. US Patent & Trademarks Office, Washington, DC, 1984.
21. L. K. Wang, P. Wang, and N. Clesceri *Water Treatment*, **10**, 121–134 (1995).
22. L.K. Wang, M.X. Krougzek and U. Kounitson *Case Studies of Cleaner Production and Site Remediation.* United Nations Industrial Development Organization (UNIDO), Vienna, Austria. UNIDO-Registry No. DTT-5-4-95, 1995.
23. L. K. Wang and S. Kopko, *City of Cape Coral Reverse Osmosis Water Treatment Facility.* US Department of Commerce, National Technical Information Service, Springfield, VA, 1997, PB97-139547.
24. W. A. Hardenbergh and E. R. Rodie *Water Supply and Waste Disposal.* International Textbook Co., Scranton, PA, 1993.
25. R. Gregory, T. F. Zabel, and J. K. Edzwald Sedimentation and flotation. In *Water Quality and Treatment.* R.D. Letterman ed. American Water Works Association, McGraw-Hill Book Co., NY, 1999, 7.1–7.82.
26. L. K. Wang. and S. L. Lee *Utilization and Reduction of Carbon Dioxide Emissions: An Industrial Ecology Approach.* Annual Conference of CAAPS, St. Johns University, New York, 2001.
27. L. K. Wang, L. Kurylko, and M. H. S. Wang. *Gas Dissolving and Releasing Liquid Treatment System.* US Patent 5167806. US Patent & Trademarks Office, Washington, DC, 1992.
28. Editor, 2002–2003 Annual Buyer's Guide; *Water Quality Products.* **7**, 72–73 (2002).
29. Editor, *Environmental Protection.* **14**, 115–116 (2002).
30. AWWA. *Water Treatment.* Volume 2, American Water Works Association, Denver, CO, 1984.
31. City of Kent, Ohio. Virtual Tour of Recarbonation Equipment and Basins. City Hall, Kent, OH, 2003. www.kentwater.org/recarbonation.htm.
32. M. L. Granstrom. Recarbonation. Lenox Institute of Water Technology (formerly Lenox Institute for Research), Lenox, MA. LIR/01-87/215. 29p, 1987.
33. L. K. Wang, Y. T. Hung and N. K. Shammas (eds.) *Advanced Physicochemical Treatment Processes.* Humana Press, Totowa, NJ, 2005.

7
Chemical Oxidation

Nazih K. Shammas, John Y. Yang, Pao-Chiang Yuan, and Yung-Tse Hung

CONTENTS

1. INTRODUCTION

Chemical oxidation is a process involving the transfer of electrons from an oxidizing reagent to the chemical species being oxidized. In water and wastewater engineering, chemical oxidation serves the purpose of converting putrescible pollutant substances to innocuous or stabilized products. Chemical oxidation processes take place in natural waters and serve as an important mechanism in the natural self-purification of surface waters. Oxidative removal of dissolved iron and sulfide pollutants in aerated waters is a prominent example. The degradation of organic waste materials represents an even more important phenomenon associated with natural water self-purification. It is well known that the efficacy of natural water organic oxidations is due to the presence of microorganisms, which serve to catalyze a highly effective utilization of dissolved oxygen as an oxidant. In fact, such microorganism-catalyzed processes have been optimized and developed into the various forms of so-called "biological processes" in high concentration organic waste treatment applications. The subject of biochemical oxidation processes is thus covered in a different book that deals with biological treatment processes.

Under the usual environmental engineering terminology, the subject of chemical oxidation is not normally considered to include biological oxidation processes. Without the inclusion of biological processes, applications of chemical oxidation to water and

From: *Handbook of Environmental Engineering, Volume 3: Physicochemical Treatment Processes*
Edited by: L. K. Wang, Y.-T. Hung, and N. K. Shammas © The Humana Press Inc., Totowa, NJ

wastewater treatment are largely utilized in specialized industrial uses. This is due to the fact that in view of the low cost public water requirements, the unit treatment cost by chemical oxidation is rather high, either because the oxidizing reagent is costly or because the oxidation efficiency is low. The emphasis on practical applications of chemical oxidation techniques is therefore placed under the following conditions:

1. Final polishing of comparatively high-quality influent to achieve compliance with potable water standards.
2. Treatment of low volume industrial wastewaters containing highly toxic contaminants.
3. Treatment of residue wastes isolated within a concentrated phase, such as the waste brines and wastewater sludges.
4. Removal of objectionable waste constituent characteristics without requiring complete oxidation of pollutants, such as in the case of disinfection and taste and odor controls.

A variety of reagents and oxidation systems are found to be cost-effective within the above-listed operational constraints.

Ozone and chlorine may be singled out as two important chemical oxidants finding wide applications especially in water and wastewater disinfection. The topics of ozonation and halogenation are therefore covered in other chapters. In this chapter, we shall define the concept of oxidation and treat the theory and principles of chemical oxidation. Selected oxidation systems of practical importance, other than the above-mentioned ozonation and halogenation processes, are discussed.

1.1. Dissolved Oxygen and Concept of Oxidation

Putrescible substances are known to comprise the most frequently occurring classes of pollutants in natural water systems. These substances have a most objectionable effect on water quality in that their decomposition often causes a depletion of dissolved oxygen in water. Dissolved oxygen is, in turn, essential to the existence of upper trophic aquatic organisms and is widely accepted as a most important indicator of the quality of a water system or its state of pollution. An analysis of oxygen balance in the aquatic environment shows that oxygen transfer from the atmosphere normally constitutes the most important oxygen source, whereas pollutional material consumption constitutes the major sinks. Chemical reactions giving rise to such consumption of oxygen are known as oxidation processes.

To account for proper materials balance, every chemical change of a specific nature must be accompanied by a process of opposite effect. Thus, the actual consumption of molecular oxygen is more properly termed as reduction, while the accompanying degradation of putrescible pollutants is defined as oxidation. In other words, oxidation and reduction must occur as coupled processes. A most functionally acceptable definition of oxidation–reduction is given in terms of electron transfer between reacting species. Each overall oxidation–reduction reaction may be considered to comprise two half-reactions, neither of which can occur independently. One of the half-reactions involves a loss of electrons, and it is defined as oxidation. The other half-reaction, involving the gain of electrons, is defined as reduction. Chemical species serving as potential electron acceptors are regarded as oxidants. Those functioning as potential electron donors are known as reductants. An illustration is given by the reaction between calcium and chlorine to yield calcium chloride as follows:

$$\mathrm{Ca} \leftrightarrow \mathrm{Ca}^{2+} + 2e \qquad \text{(oxidation)} \tag{1}$$

$$\mathrm{Cl}_2 + 2e \leftrightarrow 2\mathrm{Cl}^- \qquad \text{(reduction)} \tag{2}$$

$$\mathrm{Cl}_2 + \mathrm{Ca} \leftrightarrow \mathrm{CaCl}_2 \text{ or } \left(\mathrm{Ca}^{2+}, 2\mathrm{Cl}^-\right) \tag{3}$$

For the overall Reaction (3), two electrons are transferred from a calcium atom to each of two chlorine atoms. In terms of the component half-reactions, calcium is oxidized as illustrated by Reaction (1), and chlorine is reduced as shown by Reaction (2).

Inasmuch as the control of putrescible pollutants in natural waters depends on the supply of dissolved oxygen, chemical oxidation with molecular oxygen is generally recognized as a most important process in natural water quality control. Half-reactions ascribable to the consumption of molecular oxygen as oxidant in aqueous systems may be given as follows:

$$\mathrm{O}_2 + 4\mathrm{H}^+ + 4e \leftrightarrow 2\mathrm{H}_2\mathrm{O} \tag{4}$$

$$\mathrm{O}_2 + 2\mathrm{H}_2\mathrm{O} + 4e \leftrightarrow 4\mathrm{OH}^- \tag{5}$$

As a matter of common usage, Reaction (4) is normally designated when the aqueous medium is acidic (pH <7), and Reaction (5) is specified when the solution is alkaline (pH >7). In fact, however, Reactions (4) and (5) are completely equivalent in that the H^+ and OH^- ions in an aqueous system are related by equilibrium ionization process as follows:

$$\mathrm{H}_2\mathrm{O} \leftrightarrow \mathrm{H}^+ + \mathrm{OH}^- \tag{6}$$

Consumption of H^+ ions would promote H_2O dissociation giving rise to an increase in OH^- ions. Conversely, the generation of excess OH^- ions would lead to partial neutralization giving rise to a decrease in H^+ ions. In any event, the net result as expressed by either Reactions (4) or (5) is that the molecular oxygen is converted to a form combined with hydrogen atoms giving either H_2O or OH^- as the final reaction product.

1.2. The Definition of Oxidation State

Although oxidation–reduction must always occur as coupled reaction processes, it is not always a simple matter to sort out the specific half-reactions. For the illustrative Reaction (3) involving the combination of calcium with chlorine, the respective oxidation and reduction half-reactions are unambiguously defined because the product calcium chloride is known to be composed of Ca^{2+} and Cl^- ions in a crystalline state. Thus, as a result of the overall reaction, each calcium atom has suffered a loss of two electrons, while each chlorine atom has gained an electron. For the conversion of molecular oxygen to H_2O or OH^- as shown in Reactions (4) or (5), it is not obvious that a gain of electrons can be ascribed to the reacting oxygen atoms. In the case of reactions involving covalent bonded and multielement compounds, uncertainties in the assignment of electron gains or losses can be especially difficult to unravel. In order to simplify the task required for oxidation–reduction assessment, the concept of oxidation state incorporating a set of sometimes-arbitrary rules has been devised.

Each atom of a specific molecule is assigned a value of oxidation state, signifying the number of electrons either in excess or in deficit of that atom in its normal atomic state.

By a generally accepted convention, Roman numerals are used to designate the assigned oxidation state values. On the basis of the octet rule that for each atom (other than hydrogen and helium) there are eight valence electrons available for covalent bonding, the value of oxidation state may range from–VII to +VII. The general rules for assignment of oxidation state values are as follows:

1. The oxidation state value is zero for any atom in its uncombined atomic or elemental state.
2. For a covalently bonded species, the electrons being shared between a pair of atoms is assigned completely to the one with higher electronegativity.
3. For a covalent bond between two like atoms or ones of the same electronegativity, the bonding electrons are assigned with equal division to the two sharing atoms.
4. The sum of oxidation values of all atoms in a chemical species must be equal to the net ionic charge associated with that species

Rigorous practice of the above-stated rules for assignment of oxidation state values would require considerable understanding of the nature of chemical bonding including the concept of electronegativity. Fortunately for environmental engineering applications, a set of simplified guidelines may be used to derive satisfactory assignment of oxidation state values. For elements present in chemical species commonly found in water and wastewater, the order in terms of decreasing electronegativity may be given as $O > Cl > Br > I > N > S > C > H >$ metals. In. general, the oxidation state of oxygen atoms in any species other than a homomolecular compound may be assigned a value of –II and that of hydrogen atoms may be assigned a value of +I. In bonding with less electronegative elements, a nitrogen atom may be assigned an oxidation state value of –III, and similarly a sulfur atom may be assigned a value of –II. The oxidation state value of atoms of any remaining elements is then derived by using the previously stated rule 4 governing the sum of oxidation state values. To aid in the understanding of the guidelines for oxidation state assignment, a number of examples are illustrated in Table 1.

A notable exception to the application of the simplified guidelines for the oxidation state assignment is the case of hydrogen peroxide (H_2O_2). Obviously the assignment of –II for oxygen atoms and +I for hydrogen atoms in H_2O_2 would give a net oxidation state sum of –II in violation of the previously stated rule 4. In this case, it is necessary to understand that the hydrogen peroxide molecule may be represented structurally as H–O–O–H. A rigorous application of rules for oxidation state assignment will then give the values of –I for oxygen atoms and +I for hydrogen atoms.

Another complication arises in the case of complex molecules where a number of different atoms of a single element may take part in different bonding structures and therefore may be assigned different oxidation state values. Take acetic acid (CH_3COOH) for example. One of the carbon atoms is bonded to three hydrogen atoms and one carbon atom, and it should be properly assigned an oxidation state of –III. The other oxygen atom is bonded to a carbon atom, a singly bonded oxygen atom, and a doubly bonded oxygen atom; and thus it should have a properly assigned oxidation state of +III. By application of the simplified guideline of assigning –II for oxygen atoms and +I for hydrogen atoms, the oxidation state value for carbon atoms in acetic acid is, however, computed as zero. This latter value represents an average of the values for the two carbon atoms derived by the rigorous assignment rules. For the purpose of materials balance in oxidation–reduction assessment, the use of an averaged oxidation state value for atoms of a given element is quite

Table 1
Oxidation State Assignment for Selected Chemical Compounds

Chemical species	Oxidation state and value		
Nitrogen compounds:			
NH_3	N = –III	H = + I	
N_2	N = 0		
N_2O	N = + I	O = –II	
NO_2^-	N = +III	O = –II	
NO_2	N = +IV	O = –II	
NO_3^-	N = +V	O = –II	
Carbon compounds:			
CH_4 (methane)	C = –IV	H = +I	
CH_3OH (methanol)	C = –II	H = +I	O = –II
C_6H_6 (benzene)	C = –I	H = +I	
H_2CO (formaldehyde)	C = 0	H = +I	O = –II
HCOOH (formic acid)	C = +II	H = +I	O = –II
HCO_3^-, CO_3^{2-}, CO_2	C = +IV	H = +I	O = –II
Sulfur compounds:			
H_2S	S = –II	H = +I	
$S_2O_3^{2-}$	S = +II	O = –II	
SO_2, SO_3^{2-}	S = +IV	O = –II	
SO_4^{2-}	S = +VI	O = –II	
S-C-N compounds:			
CN^-	N = –III	C = +II	
CNO^-, NCO^-	N = –III	C = +IV	O = –II
SCN^-	N = –III	C = +VI	S = –II
Chlorine compounds:			
HCl	Cl = –I	H = +I	
Cl_2	Cl = 0		
HOCl	Cl = +I	H = +I	O = –II
ClO_2	Cl = +IV	O = –II	
ClO_3^-	Cl = +V	O = –II	
Metal–oxygen compounds:			
MnO_2	Mn = +IV	O = –II	
MnO_4^-	Mn = +VII	O = –II	
$Cr(OH)_3$	Cr = +III	O = –II	H = +I
CrO_4^{2-}, $Cr_2O_7^{2-}$	Cr = +VI	O = –II	

satisfactory. Therefore, the application of simplified guidelines for oxidation state assignment is generally recommended. It should be noted, however, that because of the implicit average of values for several different atoms, fractional values of oxidation states might often be encountered. A prominent example is that of phenol (C_6H_5OH) for which the oxidation state of carbon atoms as derived by the simplified guidelines is given as –2/3.

2. THEORY AND PRINCIPLES

Chemical oxidation may be regarded as a unit process of environmental engineering, applicable to the removal or inactivation of putrescible contaminants. Inasmuch as chemical oxidation often contributes to a relatively high fraction of the total water or wastewater treatment cost, it is applied only under especially justifiable conditions. In addition, optimization of chemical oxidation systems may be regarded as a most important factor in the cost- effectiveness of the overall treatment design. The intelligent application of chemical oxidation processes must therefore be considered as an important design objective. Consequently, it is necessary to understand the fundamental principles involved in the chemical reactions governing the effective removal of putrescible pollutants.

One of the basic treatment design optimization requirements lies in the determination of proper reagent dosage. In the case of chemical oxidation, dosage computations must be based on the stoichiometry of pertinent chemical reactions. The extent of reaction and the conditions favorable to effective pollutant removal must be derived by chemical thermodynamic evaluations. In the proper design of a reactor, the residence time and the corresponding retention volume are of utmost importance. These design parameters are to be derived by understanding the kinetics of the pertinent reactions. The stoichiometry, thermodynamics, and kinetics of chemical oxidation constitute the fundamental principles of importance to the understanding of the desired process optimization.

2.1. Stoichiometry of Oxidation–Reduction Processes

It has been noted that free electrons do not exist to a measurable extent in aqueous solutions, and so oxidation and reduction must take place in a coupled and balanced manner. For a stoichiometric balance of oxidation–reduction processes, it is therefore necessary to satisfy the following two types of materials conservation criteria:

1. Elemental materials conservation. There cannot be either a net gain or a net loss of the total number of atoms of any given chemical element in the overall reaction.
2. Electron balance. The number of electrons gained by the oxidant must be equal to that lost by the oxidized elements.

The simplest method for balancing oxidation–reduction equations is to employ the half-reaction approach. Materials balance in terms of individual chemical elements is derived in each of the oxidation and reduction half- reactions. Electron balance is then easily achieved by coupling the pertinent half-reactions in appropriate ratios to ensure complete balance of the gain and loss of electrons. This method is easily applied to reactions of ionic species, but not straightforwardly to oxidative degradations of covalently bonded organic molecules. In the latter case, it would be more convenient to determine electron balance in the overall reaction by application of the concept of oxidation states. A series of examples are given below to demonstrate the available methods for oxidation–reduction assessment and stoichiometric reaction balance.

(a) Removal of dissolved iron in water by aeration:

Oxidation: $$\mathrm{Fe^{2+} + 3H_2O \leftrightarrow Fe(OH)_3 + 3H^+} + e \qquad \text{(Multiply by 4)}$$

Reduction: $$\mathrm{O_2 + 4H^+} + 4e \leftrightarrow \mathrm{2H_2O} \qquad \text{(Multiply by 1)}$$

Overall Reaction: $$\mathrm{4Fe^{2+} + O_2 + 10H_2O \leftrightarrow 4Fe(OH)_3 + 8H^+}$$

In the above example, the electron gain and loss balance is achieved by using a 4:1 ratio of the respective oxidation and reduction half-reactions. The latter are individually balanced to satisfy the elemental materials balance requirements. The overall reaction is then derived by addition of the individual reactant and product species in the two half-reactions. Any species appearing on both sides of the equation are appropriately adjusted to give only the excess number either as reactant or as product.

(b) Removal of dissolved manganese by permanganate oxidation:

Oxidation: $Mn^{2+} + 2H_2O \leftrightarrow MnO_2 + 4H^+ + 2e$ (Multiply by 3)

Reduction: $MnO_4^- + 4H^+ + 3e \leftrightarrow MnO_2 + 2H_2O$ (Multiply by 2)

Overall Reaction: $3Mn^{2+} + 2MnO_4^- + 2H_2O \leftrightarrow 5MnO_2 + 4H^+$

In the above example, a 3:2 ratio of the designated oxidation and reduction half-reactions is found necessary to achieve the desired stoichiometric balance. Both the dissolved manganese (+II) ions and the soluble permanganate (+VII) reagent are removed from water by mutual reaction with the formation of an insoluble manganese dioxide (+IV). Manganese ions at different oxidation states are involved as oxidants and reductants.

(c) Partial conversion of dissolved ammonia to dichloramine:

Oxidation: $NH_3 + 4Cl^- \leftrightarrow NHCl_2 + 2HCl + 4e$ (Multiply by 1)

Reduction: $Cl_2 + 2e \leftrightarrow 2Cl^-$ (Multiply by 2)

Overall Reaction: $NH_3 + 2Cl_2 \leftrightarrow NHCl_2 + 2HCl$

A 1:2 ratio of the designated oxidation and reduction half-reactions is found to be sufficient to achieve the desired stoichiometric balance. However, the half-reactions as shown above illustrate a grossly erroneous representation of the chemical reaction actually taking place. Chloride ions do not take part as reaction intermediates in the actual overall reaction. Thus, it would be more proper to balance the overall reaction directly by application of the oxidation state concept as follows:

Reactant conversion: $NH_3 + Cl_2$ to give $NHCl_2 + 2HCl$

Oxidation state: −III 0 +I −I −I

Electron gain or loss: −4 +2 (× 2)

Balance reaction: $NH_3 + 2Cl_2 \leftrightarrow NHCl_2 + 2HCl$

In the above method, the ratio of oxidant and reductant molecules is determined from the electron balance criterion. Elemental materials balance is then applied to achieve the complete reaction balance.

(d) Oxidative degradation of phenol by ozonation:

$$\text{Oxidation:} \quad C_6H_5OH + 11H_2O \leftrightarrow 6CO_2 + 28H^+ + 28e \quad \text{(Multiply by 1)}$$

$$\text{Reduction:} \quad O_3 + 2H^+ + 2e \leftrightarrow H_2O + O_2 \quad \text{(Multiply by 14)}$$

$$\text{Overall Reaction:} \quad C_6H_5OH + 14O_3 \leftrightarrow 6CO_2 + 3H_2O + 14O_2$$

The half-reaction approach is again used to illustrate the utility of the method for balancing oxidation–reduction reactions. The direct reaction balance approach is demonstrated below:

Reactant conversion:	$C_6H_5OH + O_3$ to give $CO_2 + H_2O + O_2$
Oxidation state:	$-2/3 \quad 0 \quad +IV \quad -II \quad 0$
Electron gain or loss:	$-14/3\ (\times 6) \quad +2\ (\times 14)$
Balance reaction:	$C_6H_5OH + 14O_3 \leftrightarrow 6CO_2 + 3H_2O + 14O_2$

To balance the above reaction, it is assumed that a molecule of oxygen is produced for each ozone molecule consumed. The remaining stoichiometric reaction balance is then achieved by elemental materials balance.

2.2. *Thermodynamics of Chemical Oxidation*

Chemical stoichiometry as expressed in a balanced overall equation does not signify the extent to which a net chemical change will take place. Frequently, it is necessary to evaluate the detailed reaction routes (reaction mechanism) before the extent of reaction can be predicted. Take, for example, the oxidative degradation of phenol as expressed in the following equation:

$$C_6H_5OH + 14O_3 \leftrightarrow 6CO_2 + 3H_2O + 14O_2 \quad (7)$$

This reaction is known to take place through many intermediate steps. More likely, the net reaction will not proceed according to the stoichiometry expressed in Reaction (7). The actual net reaction will depend on the energetics associated with the change as well as the rate at which the individual intermediate reaction step will proceed under specified conditions. Interpretations of the rate of the reaction (kinetics) will require complete understanding of the pertinent reaction mechanisms and will be difficult to cover in a generalized fashion. Energetics of the reaction, however, can be derived from knowledge of the initial and final states of a chemical system. The science of chemical energetics interpretations is commonly known as thermodynamics. Insofar as it is relevant to the subject matter of environmental science and engineering, a brief discussion of the thermodynamics of chemical oxidation is presented here.

For a stoichiometrically balanced chemical equation, the reaction may be assumed to be driven in the forward direction. There will be a net energy change associated with a specific extent of the reaction. When the extent of reaction is given in terms of the number of moles of reactants in the stoichiometric ratios as represented in the equation, the corresponding energy change is known as the standard free energy of reaction; denoted as ΔF°. Take, for example, Reaction (7) representing the oxidative degradation of phenol by ozonation; the associated standard free energy of reaction may be computed from known free energies of formation of the respective reactant and product molecules.

A value of $\Delta F^\circ = -1150$ kcal is thus computed. By convention, the negative sign denotes energy being liberated for the specified reaction in the forward direction. Thus, the ozonation of phenol is highly exothermic and may be expected to proceed to completion on energetic considerations alone.

Because oxidation–reduction processes are defined to involve an electron transfer from the oxidized species to the oxidant, these processes may be considered as analogous to electrochemical reactions. In fact, many oxidation–reduction reactions of ionic reactants in aqueous solutions may be simulated by coupled electrochemical half-cells with oxidation and reduction taking place at separated electrodes. The value of potential difference between the anode (oxidation) and cathode (reduction) serves to indicate the magnitude of the driving force for the specified oxidation–reduction couple. It is desirable to correlate the cell potential with the alternate indicator of the driving force in terms of the free energy of reaction. This is achieved by defining a standard cell potential so that

$$\Delta F^\circ = -nFE^\circ$$

where ΔF° = the standard free energy of reaction (kcal), n = the number of electrons transferred for the specified chemical reaction, F = the Faraday's constant found to have a value of 96,500 coulombs, and E° = the standard cell potential for reaction (V).

Using this interconversion relationship, the standard cell potential for Reaction (7) is then computed to be $E^\circ = 1.78$ V. Because the ozonation of phenol cannot be simulated in an electrochemical reaction cell, the computed E° strictly serves only as a hypothetical indicator. In general, it is more convenient to use cell potential values to indicate the energetic feasibility of oxidation–reduction processes involving ionic reactants. For non-ionic reactants, the concept of free energy remains more logically suited to denoting the energetic feasibility of specific reaction processes.

One advantage of the standard cell potential approach for thermodynamic consideration of oxidation–reduction processes lies in that separate half-cell or electrode potentials can be derived for individual oxidation or reduction half-reactions. Standard oxidation potentials at 25°C for a selected list of electrode reactions commonly encountered in aqueous chemical processes are compiled in Table 2.

Take, for example, the reaction involving the removal of dissolved manganese by permanganate oxidation,

$$3Mn^{2+} + 2MnO_4^- + 2H_2O \leftrightarrow 5MnO_2 + 4H^+ \tag{8}$$

The separation of the overall reaction into two half-cell reactions has been illustrated earlier. Using the convention of expressing each of the half-cell reactions as an oxidation process, the pertinent electrode reactions may be given as follows:

$$Mn^{2+} + 2H_2O \leftrightarrow MnO_2 + 4H^+ + 2e \qquad E^\circ = -1.22 \text{ V}$$

$$MnO_2 + 2H_2O \leftrightarrow MnO_4^- + 4H^+ + 3e \qquad E^\circ = -1.69 \text{ V}$$

The standard cell potential as represented by Reaction (8) is then derived by combining the two electrode potentials, giving a value of $E^\circ = 0.47$ V.

The standard cell potential for a specified oxidation–reduction process, such as Reaction (8), represents the cell potential at a condition when the activity (or concentration) of each

Table 2
Standard Oxidation Potentials of Selected Electrode Reactions

Electrode reaction	E° (V)
$H_2 \leftrightarrow 2H^+ + 2e$ [Base for the standard electrode potential]	0.00
$Ag \leftrightarrow Ag^+ + e$	–0.799
$2Br^- \leftrightarrow Br_2$ (aq) + 2e	–1.09
$2HCN \leftrightarrow (CN)_{2+} 2H^+ + 2e$	–0.37
$(CN)_2 + 2H_2O \leftrightarrow 2\ HCNO + 2H^+ + 2e$	–0.33
$HCOOH \leftrightarrow CO_2 + 2\ H^+ + 2e$	0.20
$H_2C_2O_4 \leftrightarrow 2CO_2 + 2\ H^+ + 2e$	0.49
$Ce^{3+} \leftrightarrow Ce^{4+} + e$	–1.70
$2Cl^- \leftrightarrow Cl_2$ (aq) + 2e	–1.359
$Cl_2 + 2H_2O \leftrightarrow 2HClO + 2\ H^+ + 2e$	–1.63
$HClO + H_2O \leftrightarrow HClO_2 + 2H^+ + 2e$	–1.64
$HClO_2 \leftrightarrow ClO_2 + H^+ + e$	–1.27
$2Cr^{3+} + 7H_2O \leftrightarrow Cr_2O_7^{2-} + 14H^+ + 6e$	–1.33
$Cr(OH)_3 + 5OH^- \leftrightarrow CrO_4^{2-} + 4H_2O + 3e$	0.13
$Cu \leftrightarrow Cu^{2+} + 2e$	–0.337
$Fe^{2+} \leftrightarrow Fe^{3+} + e$	–0.771
$Fe(OH)_2 + OH^- \leftrightarrow Fe(OH)_3 + e$	0.56
$2I^- \leftrightarrow I_2 + 2\ e$	–0.536
$Mg \leftrightarrow Mg^{2+} + 2e$	2.37
$Mn^{2+} + 2H_2O \leftrightarrow MnO_2 + 4H^+ + 2e$	–1.22
$Mn^{2+} + 4H_2O \leftrightarrow MnO_4^- + 8H^+ + 5e$	–1.51
$MnO_2 + 2H_2O \leftrightarrow MnO_4^- + 4H^+ + 3e$	–1.69
$MnO_4^{2-} \leftrightarrow MnO_4^- + e$	–0.6
$2NH_4^+ \leftrightarrow N_2H_5^+ + 3H^+ + 2e$	–1.27
$N_2H_5^+ + 2H_2O \leftrightarrow 2NH_3OH^+ + H^+ + 2e$	–1.42
$2NH_3OH^+ \leftrightarrow N_{2+} 2H_2O + 4\ H^+ + 2e$	1.87
$HNO_2 + H_2O \leftrightarrow NO_3^- + 3H^+ + 2e$	–0.94
$2H_2O \leftrightarrow H_2O_2 + 2H^+ + 2e$	–1.77
$2H_2O \leftrightarrow O_2 + 4H^+ + 4e$	–1.229
$4OH^- \leftrightarrow O_2 + 2H_2O + 4e$	–0.401
$O_2 + H_2O \leftrightarrow O_3 + 2H^+ + 2e$	–2.07
$H_2S \leftrightarrow S + 2H^+ + 2\ e$	–0.14
$2S + 3H_2O \leftrightarrow S_2O_3^{2-} + 6H^+ + 4e$	–0.50
$S_2O_3^{2-} + 3H_2O \leftrightarrow 2H_2SO_3 + 2\ H^+ + 4e$	–0.40
$H_2SO_3 + H_2O \leftrightarrow SO_4^{2-} + 4H^+ + 2e$	–0.17
$SO_3^{2-} + 2(OH)^- \leftrightarrow SO_4^{2-} + H_2O + 2e$	0.93
$2S_2O_3^{2-} \leftrightarrow S_4O_6^{2-} + 2e$	–0.09

of the reactant or product species is at unit molarity. A positive value for the cell potential indicates that it is energetically favorable for the reaction to proceed in the forward direction. As the reaction proceeds or when the activity (or concentration) of the reacting species deviates from unit molarity, the actual cell potential will be shifted accordingly. It may be recalled that the cell potential for a specified reaction is related to the free energy of reaction by the relationship

$$\Delta F^{\circ} = -nFE^{\circ} \quad \text{or} \quad \Delta F = -nFE$$

From thermodynamic considerations of chemical reactions in aqueous solution, it has been further established that

$$\Delta F = \Delta F^\circ + RT \ln Q \tag{9}$$

where R = the gas constant, T = the absolute temperature, and Q = the reactant quotient consisting of a ratio of product concentration terms to reactant concentration terms with each raised to the power of appropriate stoichiometric ratios.

Taking Reaction (8), for example, Eq. (9) may be expressed as

$$\Delta F = \Delta F^\circ + RT \ln \left[\mathrm{MnO_2}\right]^5 \left[\mathrm{H^+}\right]^4 / \left[\mathrm{Mn^{2+}}\right]^3 \left[\mathrm{MnO_4^-}\right]^2 \left[\mathrm{H_2O}\right]^2$$

For aqueous solutions, the concentration of water is little changed and the activity of water is always regarded as unity. The product manganese dioxide is insoluble in water with a concentration always limited by its solubility product and so its activity is again taken by convention as unity. Thus, the above relationship may be simplified as follows:

$$\Delta F = \Delta F^\circ + RT \ln \left[\mathrm{H^+}\right]^4 / \left[\mathrm{Mn^{2+}}\right]^3 \left[\mathrm{MnO_4^-}\right]^2$$

It may be recalled that, according to the empirically stated law of mass action, the rates of reaction in the forward and reverse directions of a reversible process, such as Reaction (8), will depend on the concentrations of the pertinent reactant species. An equilibrium condition where the forward and reverse reaction rates are equal is reached when the reactants and products concentration relationship are such that

$$\left[\mathrm{H^+}\right]^4 / \left[\mathrm{Mn^{2+}}\right]^3 \left[\mathrm{MnO_4^-}\right]^2 = K$$

where K = the equilibrium constant, is satisfied. From the thermodynamic point of view, the condition where there is no net driving force for reaction in either the forward or the reverse direction is represented by $\Delta F = 0$. Substituting these conditions in Eq. (9), we obtain

$$\Delta F = 0 = \Delta F^\circ + RT \ln K \quad \text{and} \quad \Delta F^\circ = -RT \ln K$$

Taking into account the relationship between the cell potential and the free energy of reaction in oxidation–reduction processes, the following set of equations are obtained:

$$E = E^\circ - (RT/nF) \ln Q \tag{10}$$

$$E^\circ = (RT/nF) \ln K \tag{11}$$

For an oxidation electrode process, the reactant quotient Q may be replaced by [oxidized]/[reduced], representing the ratio of reactant concentrations in the oxidized and reduced states, respectively. Substituting known values for the gas constant and Faraday's constant and assuming a reaction temperature of about 300° K, Eq. (10) is transformed to the following:

$$E = E^\circ - (0.06/n) \log \{[\text{oxidized}]/[\text{reduced}]\} \tag{12}$$

The above equation is popularly known as the Nernst equation, and it is widely applied in the assessment of electrochemical and oxidation–reduction processes.

Barring kinetic limitations, the standard cell potential computed for a specified oxidation–reduction reaction is useful in predicting the limit to which the reaction will precede under fixed initial conditions. Taking Reaction (8), for example, $E^{\circ} = 0.47$ V. According to Equation (12),

$$E = 0.47 - (0.06/6)\log\left\{[\mathrm{H^+}]^4 / [\mathrm{Mn^{2+}}]^3[\mathrm{MnO_4^-}]^2\right\}$$

The reaction reaches equilibrium ($E = 0$) when

$$[\mathrm{H^+}]^4 / \left([\mathrm{Mn^{2+}}]^3[\mathrm{MnO_4^-}]^2\right) = 10^{47}$$

$$[\mathrm{Mn^{2+}}]^3[\mathrm{MnO_4^-}]^2 = 10^{-47}[\mathrm{H^+}]^4$$

If we assume that an equivalent amount of the permanganate reagent had been added so that the residue [Mn^{2+}] and [MnO_4^-] concentration values are about the same, then one obtains

$$[\mathrm{Mn^{2+}}] = 10^{-9.4}[\mathrm{H^+}]^{0.8}$$

This residue concentration value is given in terms of molar concentration. To convert this residue value into the more commonly used water quality unit of mg/L manganese, a multiplication factor of 55×10^3 (atomic weight of manganese in mg) must be used to give

$$[\mathrm{Mn^{2+}}] = 2.2 \times 10^{-5}[\mathrm{H^+}]^{0.8}\ \mathrm{mg/L}$$

Thus, even when the pH of the waste solution is as low as 1, the manganese residue should be less than 2.2×10^{-5} mg/L provided the manganese dioxide solubility is below that value. On the basis of chemical thermodynamics alone, permanganate oxidation presents indeed a very effective method for manganese removal. Insofar as the standard cell potential values are available, analogous computations may be conducted to predict the thermodynamic feasibility of any given oxidation–reduction couple in application to water pollution control.

2.3. *Kinetic Aspects of Chemical Oxidation*

On the basis of thermodynamic considerations alone, most of the commonly occurring elemental constituents in water would exist at highly oxidized valence states as long as dissolved oxygen is transported from the atmosphere. Nevertheless, reduced-form pollutional species are often found in natural waters, because thermodynamic equilibrium is seldom reached or maintained within the dynamic aquatic systems. The failure to reach equilibrium states is, in turn, attributed to kinetic limitations. Such kinetic limitations are also responsible for the fact that aeration or oxygen gas applications have not been established as generally effective methods for oxidative removal of water pollutants. Other types of active oxygen reagents, such as ozone, hydrogen peroxide, and peroxy acids, are subjected to kinetic limitations to somewhat lesser degrees. Thus, kinetic considerations are to be recognized as the most critical process design factors in the application of chemical oxidation principles to water and wastewater engineering.

Chemical kinetics is a branch of the chemical science dealing with studies of chemical reaction rates as they are affected by individual physical parameters within the reaction environment. Whereas thermodynamic principles are applicable as long as the initial and the final chemical states are known, kinetic understanding require a knowledge of detailed process mechanisms whereby a chemical system is transformed from one state to another. Time scales associated with each of the individual step changes are important. Even for apparently simple reactions, such as the oxidation of hydrogen to yield water, the intervening transition processes are often very complex. Solvent-solute interactions and impurity catalytic effects, in particular, often further complicate aqueous phase reactions. Thus, a set of easily applicable guidelines to aid in an apriori evaluation of the kinetics of a given reaction cannot be made available. Empirical approaches based on experimental measurements of reaction rates and correlations of the effects of varying environmental parameters must be relied on for specific process optimization.

In the application of empirical kinetic evaluations, it remains necessary to have an understanding of basic physical parameters, which may affect the rate of a chemical reaction. In the case of homogeneous reactions, important parameters are reactant and catalyst concentrations and reaction temperature. As a simply stated theory of chemical kinetics, the rate of transition in a discrete reaction step of an overall reaction may be described in terms of the frequency of encounter of reacting molecules and the energetic states of the encountering species. The reaction temperature effect is ascribed to the fact that the average energy of reacting molecules is greater at higher temperatures. The frequency of encounter effect may be simply stated as a form of the widely known law of mass action.

It is to be emphasized that kinetic equations are unrelated to stoichiometric relationships and any resemblances found for specific chemical reactions are purely incidental. This is illustrated by the oxidation of ferrous ion in water with dissolved oxygen. The stoichiometric relationship may be represented by,

$$2Fe^{2+} + {}^{1}/_{2}O_2 + 5H_2O \leftrightarrow 2Fe(OH)_3 + 4H^+ \tag{13}$$

The rate equation is found as (1),

$$-d[Fe^{2+}]/dt = k[Fe^{2+}][OH^-]^2[O_2] \tag{14}$$

The oxidation rate is then said to be first order with respect to ferrous ion and dissolved oxygen concentrations and second order with respect to the hydroxyl ion concentration. The law of mass action in application to kinetic evaluation must be modified to state that the reaction rate is proportional to the "active masses" of participating reactants. "Active masses" are, in turn, understood to represent the activated species taking part in a rate-determining transition step within a complex sequence of reactions. The concentration relationship between the "active mass" and the conventional form reactant may be affected by a variety of environmental factors including the temperature, pH of solution, ionic strength of medium, and the presence of catalytic or inhibiting constituents. The empirically determined rate equations may thus involve dependence on reactant concentrations to the fractional order. Once the rate relationship under a fixed set of conditions is determined, treatment design parameters including reagent dosage and retention time can be computed accordingly.

In practical waste treatment systems, homogeneous reaction conditions are seldom encountered but may be closely approximated by the stirred batch reactor. Ineffective mixing or dispersion of one or more reactants may easily lead to further rate limitations. In the application of gaseous reagents, such as oxygen and ozone with limited solubilities, reactant dosages must be continuously replenished by external additions. Interfacial transfers of gaseous reagents are often slow and may present rate restrictions on the desired oxidation. Interfacial gas transfer is reasonably well explained by a liquid film transport mechanism. An equilibrium governed by the reagent solubility is established at the gas-liquid interface within a thin film surface layer of the liquid. Under quiescent conditions, the dissolved gas is transported from this surface film layer to the bulk solution by molecular diffusion. At increased turbulence levels, the surface film is frequently displaced, and the transport of the dissolved gas to the bulk solution is promoted. The gas transfer rate may be expressed as

$$dC/dt = k_1 A\left(C_s - C\right) \tag{15}$$

where k_1 = the liquid film renewal rate constant, A = the interfacial surface area [m^2 (ft^2)], C_s = the saturation solution concentration (mg/L), C = the bulk solution concentration of the gas (mg/L), and t = time (h).

Enhanced gas dissolution is therefore favored by increased contact surface areas as well as increased turbulence levels. Design parameters based on these considerations may be regarded as the most important factors in treatment process optimization when oxygen or ozone is employed as active oxidant.

In a majority of waste treatment applications, reactor systems are designed with some or all of the reactants under constant flow. These may be represented as continuous stirred tank reactors or plug flow reactors. The former is, in essence, a series of batch reactors with stepped concentration gradients. The latter is characterized by a continuous concentration gradient in the direction of flow.

Oxygen balance in natural waters is perhaps a most prominent example of complex interrelationships of diverse kinetic factors governing the time rate concentration changes of an aqueous constituent in a flow reactor system. A generalized relationship describing the rate of dissolved oxygen concentration changes at a fixed point is given as follows:

$$\partial C/\partial t = \varepsilon\left(\partial^2 C/\partial x^2\right) - U\left(\partial C/\partial x\right) \pm \Sigma S \tag{16}$$

where ε = the turbulent diffusion coefficient, U = the linear flow velocity [m/s (ft/s)], x = the distance along the direction of flow [m (ft)], and ΣS = the net oxygen concentration change due to a combination of oxygen sources and sinks (mg/L).

In the absence of a significantly large concentration gradient along the x direction, the last term in Eq. (16) would be mainly responsible for the dissolved oxygen balance. A detailed analysis of natural water oxygen balance is beyond the scope of this chapter. However, the kinetic factors so involved are relevant to process design consideration in the application of chemical oxidation techniques to water and wastewater treatment. Pertinent discussions are provided in subsequent considerations of specific techniques involving active oxygen reagents.

3. OXYGENATED REAGENT SYSTEMS

Chemical oxidation in water and wastewater treatment serves the purpose of converting putrescible pollutants to innocuous or stabilized products. The significant pollutant chemical element is converted to a higher oxidation state as a result of oxidation. In aqueous systems, it is often observed that the pollutant element undergoing oxidation will result in a chemical state bound to an increased number of oxygen atoms. Thus, molecular oxygen and active oxygenated reagents may be considered as favorable oxidants for water and wastewater treatment. As it is defined, the process of chemical oxidation involves, in general, a transfer of electrons from the chemical species being oxidized to the oxidants. Favorable oxidants are not limited to molecular oxygen and active oxygenated reagents. Transition metal elements in their higher oxidation states are very effective as oxidants, and these are discussed in another section of this chapter. Halogens, in general, possess favorable oxidation potentials. In view of the special importance of chlorine (a member of the halogen family) in water and wastewater engineering, the subject of halogenation is also covered in a separate chapter. Among the active oxygenated reagents, ozone has made great gains as an effective reagent in recent water and wastewater treatment technology developments, and it is expected to be of even greater importance in the near future. Only chemical oxidation reactions associated with water aeration, peroxygen reagent treatment, and high-temperature wet oxidation are addressed in this section.

3.1. *Aeration in Water Purification and Waste Treatment*

Aeration is widely recognized as one of the most important unit processes in environmental pollution control engineering. Aeration serves the function of increasing the level of dissolved oxygen in water, which is essential to support the life of aquatic organisms and to limit the proliferation of noxious pollutants. Normal aquatic organisms life cycles under aerobic conditions serve an extremely important function in the natural processes of water quality purification. The assimilative and respiratory activities associated with the production and survival of aquatic organisms serve, in essence, a catalytic role to bring about an oxidative removal of reduced-form pollutants. These natural biological oxidation principles are adopted and optimized in the design of biological waste treatment systems. The importance of biological processes in wastewater treatment has merited the treatment of this subject in a separate book. Only non-biologically related chemical oxidation due to aeration will be discussed here.

As discussed previously, the application of aeration to oxidative removal of water pollutants is frequently limited by a lack of sufficiently fast reaction rates. The use of suitable rate-accelerating catalysts to achieve more practical treatment system designs is sometimes feasible. Transition metal ions, which are among the most widely known groups of catalysts, may present subsequent removal difficulties when these are used in homogeneous catalytic process designs. Even when the metal ions are incorporated onto heterogeneous catalyst support, care must be taken to ascertain that such ions are not leached into the aqueous phase. Copper sulfate is therefore one of the few metal ion reagents finding application as catalyst in water treatment aeration processes, principally for the oxidative removal of iron and manganese. Among the other known practical systems, aeration has also found at least limited applications in the oxidative treatment of sulfide, cyanide, and petroleum wastes.

In the dissolved oxygen oxidation of aqueous ionic pollutants, it is generally known that the reaction rates are much greater in alkaline solutions. Although an understanding of detailed reaction mechanisms of such oxidation processes is not available, a reasonable explanation may be given in term of OH^- ions serving a catalytic role as electron transfer bridging species. The kinetics of ferrous ion oxygenation (1) is perhaps the most extensively studied among the aqueous ionic species oxidation processes. As noted earlier, the oxidation rate may be expressed as

$$-d[Fe^{2+}]/dt = k[Fe^{2+}][OH^-]^2[O_2] \tag{14}$$

This kinetic relationship would be consistent with the hypothetical postulation of a rate-determining process based on the following bridging structure:

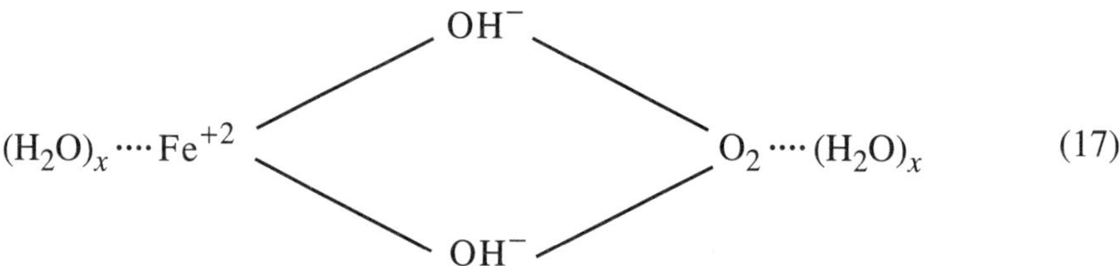

(17)

To be consistent with the known rate relationship, two bridging OH^- groups are invoked for each ferrous ion oxidized. One of the OH^- groups may serve the role of electron transport to the reacting oxygen molecule. The other may be retained as a complexing ligand with the resulting ferrous ion to serve as an added driving force for the reaction. Stoichiometric balance requires that four ferrous ions be oxidized for each oxygen molecule consumed. The intermediate oxygen species following electron attachment in Reaction (17) must take part in the oxidation of three additional ferrous ions. As long as the subsequent reactions take place at rates much greater than that in Reaction (17), the overall rate would remain in accord with Eq. (14). The suggested mechanism as illustrated by Reaction (17) is also consistent with the reported retarding effects in the presence of organic ligands such as humic and tannic acids (2). Analogous bridging mechanisms may be postulated in the molecular oxygen reaction with other types of ionic pollutants, such as Mn^{2+}, sulfide, and cyanide.

Stoichiometric relationships for molecular oxygen reactions with Fe^{2+}, Mn^{2+}, HS^-, and CN^- may be expressed respectively as follows:

$$2Fe^{2+} + {}^1/_2O_2 + 5H_2O \leftrightarrow 2Fe(OH)_3 + 4H^+ \tag{18}$$

$$Mn^{2+} + {}^1/_2O_2 + H_2O \leftrightarrow Mn(OH)_2 + 2H^+ \tag{19}$$

$$HS^- + 2O_2 \leftrightarrow H^+ + SO_4{}^{2-} \tag{20}$$

$$2CN^- + 2O_2 + H_2O \leftrightarrow 2HCO_3{}^- + N_2 \tag{21}$$

In each of the above reactions, the product tends to be more acidic in comparison with the oxidized reactant. Thus, a decrease in pH may be expected as the reaction progresses. As the oxidation rate is highly dependent on the OH^- ion concentration, a drop in the treatment efficiency may be expected to accompany the lowering of the pH of solution. This points out the need for attention to another important treatment system design parameter, namely that the reaction system should be adequately buffered. Natural waters are usually already well buffered. In industrial waste treatment applications, it is often necessary to add buffering reagents. The carbonate–bicarbonate buffer couple will serve the purpose without added concern on the treated effluent water quality.

Heterogeneous catalytic oxidation systems are not yet proven practical in demonstrated treatment systems. Catalytic oxidation coupled with adsorption on granular activated carbon, however, does show good promise at least in terms of kinetic efficiencies. In application to cyanide detoxification, for example, the cyanide is converted to cyanate and ammonia. In the presence of cupric ion catalyst, conversion to ammonia with low cyanate yield is favored. Copper catalyst is consumed and precipitated as basic copper carbonate and new reagent or recovered reagent must be continuously added to the reactor.

The application of aeration to oxidative degradation of petroleum wastes is effective only under autocatalytic conditions. Free radical chain reaction mechanisms as shown below are necessary:

$$RH + O_2 \rightarrow R^{\bullet} + HO_2^{\bullet} \tag{22}$$

$$R^{\bullet} + O_2 \rightarrow RO_2^{\bullet} \tag{23}$$

$$RO_2^{\bullet} + RH \rightarrow RO_2H + R^{\bullet} \tag{24}$$

Reaction (22), known as the initiation process, may be promoted by the presence of transition metal catalysts. Reactions (23) and (24) are responsible for chain propagation and sustenance of the overall reaction. Chain termination takes place usually by radical recombination or disproportionation. The intermediate hydroperoxide (RO_2H) is more rapidly oxidized in comparison with the original hydrocarbon (RH) pollutant. Autocatalysis arises from the fact that some of the intermediate products are more readily converted to free radicals by processes analogous to Reaction (22). Carefully controlled kinetic studies (3) indicate that aqueous petroleum waste catalytic oxidation can be achieved within reasonable time scales only at high temperatures above 150°C. In the catalytic oxidation of benzene (4), for example, phenol is observed as an intermediate product with the following kinetic relationship:

$$dx/dt = k(1-x)x^2 \tag{25}$$

where $x = [C_6H_5OH]/[C_6H_5OH]_{max}$ and max with denotes the limiting concentration of phenol that is reached when the rate of phenol oxidation exceeds that of benzene conversion to phenol.

In terms of benzene degradation, the rate would increase exponentially, but the incubation period is highly temperature dependent. Effective treatment at ambient temperatures would seem improbable.

An autooxidation system for refinery waste treatment, however, has been reported in the literature (5). A spray tower with air blowing design is employed. The effluent is treated in a once through system with a 3 to 6 min residence period. Up to 50% COD removal was reported. It is not known to what extent the COD removal is accomplished due to air stripping or by surface retention on the fill sections of the spray tower. On the other hand, the effectiveness of the reported system may be because the residual petroleum waste being treated consisted mainly of easily oxidizable carbonyl fractions. More comprehensive studies are needed to determine the efficacy of aeration in the oxidative treatment of petroleum wastes.

3.2. *Hydrogen Peroxide and Peroxygen Reagents*

Hydrogen peroxide is technically very effective as an oxidizing agent for the treatment of many common types of water pollutants. However, its relatively high cost has limited its practical usage in large-scale waste treatment systems. Peroxygen reagents, including hydrogen peroxide, are, however, very useful as oxidants in the treatment of sometimes difficult-to-handle wastes.

Pure H_2O_2 is a pale blue, syrupy liquid with a boiling point of 152.1°C and a freezing point of –0.89°C. Because of its strong oxidizing power and its ready decomposition, it is normally sold commercially as aqueous solutions. It is produced commercially by electrolysis of sulfuric acid or ammonium sulfate–sulfuric acid solutions. The peroxydisulfate produced at the anode is recovered and subsequently hydrolyzed by a two-step reaction to yield hydrogen peroxide. The H_2O_2 is rapidly removed together with excess water by distillation at high temperature and reduced pressure. The dilute distillate is concentrated by further distillation to give commercial grade products containing 28–35% H_2O_2 by weight. Alternatively, H_2O_2 may be produced by organic oxidation processes. Hydrogen peroxide is generated as a co-product when an anthraquinol is oxidized with molecular oxygen to yield anthraquinone. It is recovered by extraction with water and concentrated by distillation.

Hydrogen peroxide is somewhat more easily ionized than water, dissociating as shown below:

$$H_2O_2 \leftrightarrow H^+ + HO_2^- \qquad K_{20\hat{\imath}} = 1.5 \times 10^{-12} \tag{26}$$

It is very effective in the oxidation of ionic inorganic pollutants in terms of both oxidation potential and reaction kinetics. Its effectiveness toward oxidative degradation of organic pollutants appears, however, to depend largely on an intermediate dissociation to yield OH^- and HO_2^- radicals. Thus, hydrogen peroxide is most effective as an oxidizing agent in the presence of ferrous ions which tend to induce H_2O_2 decomposition to give OH^- radicals. The combination of H_2O_2 with ferrous sulfate is popularly known as Fenton's reagent.

Principal reactions involved in the Fenton's reagent system are as follows (6):

$$Fe^{2+} + H_2O_2 \rightarrow Fe^{3+} + OH^\bullet + OH^- \tag{27}$$

$$OH^\bullet + H_2O_2 \rightarrow H_2O + HO_2^\bullet \tag{28}$$

$$HO_2^\bullet + H_2O_2 \rightarrow H_2O + O_2 + OH^\bullet \tag{29}$$

Reactions (28) and (29) constitute a chain process, so that only a small amount of ferrous ion would be needed to catalyze the decomposition of large quantities of hydrogen peroxide. Because organic species are readily susceptible to autooxidation, the initiation of chain reactions by OH^- radicals will usually lead to effective degradation of organic pollutants. Decomposition of H_2O_2 in the presence of ferric ions is also possible above 65°C via the following reaction:

$$Fe^{3+} + H_2O_2 \rightarrow Fe^{2+} + HO_2^{\bullet} + H^+ \quad (30)$$

The ferrous ion produced by Reaction (30) will, in turn, react rapidly by Reaction (27). It is found that the OH^- radical production reactions take place effectively only within the pH range of 3–5. Thus, the effective pH range for the application of Fenton's reagent to organic pollutant oxidation is similarly limited.

As noted earlier, hydrogen peroxide is a relatively expensive chemical reagent. Studies of its utility in water pollutant oxidation applications have generally centered on the treatability of refractory organics (7). The treatment of phenolic and ABS-detergent wastes, for example, has been studied extensively. More effective hydrogen peroxide reagent utilization is achieved when a significant level of dissolved oxygen is present in the waste being treated. A high efficiency greater than 90% of pollutant removal is routinely achieved. Approximately an equivalent weight of the oxidant is required for the corresponding waste constituent removal. Apparently, an autooxidation-type chain process is not involved. Ferrous sulfate in the amount of about 1.5 times the weight of H_2O_2 was found necessary to achieve rapid oxidation and efficient oxidant utilization. This tends to support the postulation that OH^- radicals are the active species involved in the oxidative reaction processes.

Hydrogen peroxide is also known to be effective in the treatment of oxidizable inorganic wastes including sulfides, sulfites, and cyanides (8). Under neutral conditions, hydrogen sulfide is oxidized to elemental sulfur:

$$H_2S + H_2O_2 \leftrightarrow S + 2H_2O \quad (31)$$

In alkaline solutions, the reaction will proceed further to give sulfate as the final product:

$$SH^- + 4H_2O_2 \leftrightarrow SO_4^{2-} + 4H_2O + H^+ \quad (32)$$

Highly specific oxidant utilization efficiencies with respect to sulfide removal are readily achieved. The effectiveness of H_2O_2 in the conversion of sulfite or sulfur dioxide to sulfate may be attested to by the prevalent usage of H_2O_2 in SO_2 analysis applications. Cyanide removal by oxidation with H_2O_2 is known to proceed by the following reactions:

$$CN^- + H_2O_2 \leftrightarrow CNO^- + H_2O \quad (33)$$

$$CNO^- + 2H_2O \leftrightarrow CO_2 + NH_3 + OH^- \quad (34)$$

A pH range of 8 to 10 is most suitable for the above cyanide oxidation reaction. The rate of treatment can be greatly accelerated by the presence of trace catalytic metal ions such as copper. Reaction (33) proceeds much more rapidly than Reaction (34), and so cyanate is usually generated as the major product.

A plating waste cyanide treatment process based on reactions with hydrogen peroxide and formaldehyde has been marketed commercially (9). This reagent system is reported to result in effective metal ion as well as cyanide removal. Persulfuric acid (H_2SO_5) is another peroxygen reagent found to be effective in the conversion of cyanide to cyanate, as shown by Reaction (35):

Table 3
Specific Gravity of Aqueous H_2O_2 Solutions in g/cm³ at 25°C (77°F)

Weight % H_2O_2	0	1	2	3	4	5	6	7	8	9
0	0.9970	1.0004	1.0039	1.0074	1.0109	1.0144	1.0179	1.0214	1.0250	1.0286
10	1.0322	1.0358	1.0394	1.0431	1.0468	1.0505	1.0542	1.0579	1.0616	1.0654
20	1.0692	1.0730	1.0768	1.0806	1.0845	1.0884	1.0923	1.0962	1.1001	1.1040
30	1.1080	1.1120	1.1160	1.1200	1.1240	1.1281	1.1322	1.1363	1.1404	1.1445
40	1.1487	1.1529	1.1571	1.1613	1.1655	1.1698	1.1741	1.1784	1.1827	1.1871
50	1.1915	1.1959	1.2003	1.2047	1.2092	1.2137	1.2182	1.2227	1.2273	1.2319
60	1.2365	1.2411	1.2458	1.2505	1.2552	1.2599	1.2617	1.2695	1.2743	1.2791
70	1.2839	1.2888	1.2937	1.2986	1.3035	1.3085	1.3135	1.3185	1.3236	1.3287
80	1.3338	1.3390	1.3442	1.3494	1.3546	1.3598	1.3651	1.3704	1.3758	1.3812
90	1.3866	1.3921	1.3976	1.4031	1.4086	1.4141	1.4197	1.4253	1.4309	1.4365
100	1.4422									

$$CN^- + H_2SO_5 \leftrightarrow CNO^- + H_2SO_4 \tag{35}$$

In view of the relatively high cost of hydrogen peroxide, it is natural to find that the important areas of applications are centered around waste treatment systems requiring only low levels of reagent consumptions Thus, hydrogen peroxide is often used for:

(a) Odor control.
(b) Reduction of activated sludge bulking.
(c) Improvement of suspended solids settling
(d) Supplemental addition of oxygen.

In addition to its effectiveness in the oxidative degradation of a wide variety of deleterious pollutants, hydrogen peroxide reagents present an advantage in their ease of handling and application. Hydrogen peroxide is easily dosed in the form of aqueous reagents of varying concentrations. The specific gravity of aqueous hydrogen peroxide solutions at 25°C as a function of the reagent concentration is tabulated in Table 3. The required reagent dosage is therefore easily applied by volume measurements. The use of hydrogen peroxide as a supplemental reagent in both municipal and industrial waste treatment plants may be expected to grow significantly as the effluent discharge standards are stringently applied.

3.3. *High-Temperature Wet Oxidation*

In a preceding discussion of autooxidation processes, it has been noted that in the aqueous phase oxidation of organic species, such as benzene and phenol, oxidation reactions with molecular oxygen will proceed at measurable rates only at temperatures exceeding about 150°C. The effective rate of oxidation is enhanced with increasing reaction temperature. The upper temperature is, however, limited by the critical temperature of water at 374°C. The critical temperature is the limiting temperature above which the liquid phase cannot exist regardless of the externally applied pressure.

For reaction at temperatures higher than 100°C, higher than atmospheric pressures are required to maintain the reaction medium in a liquid state. At 374°C, for example, the critical pressure required for liquid water is known to be 217.7 atmosphere (219 bar or 3,200 lb/in.^{2}). Another limiting factor in the reactor temperature design lies in the fact that many organic pollutants have high vapor pressures and also low solubilities in water. At relatively high temperatures, some of these pollutants will be preferably transferred into the vapor phase by means of the steam distillation phenomenon. In the gas phase, these species will undergo oxidation at much slower rates. On the other hand, suspended solids or liquid emulsions may be oxidized effectively as a part of the aqueous phase reactions. Thus, the highest possible design temperature as determined by practical pressure limitations may not represent the optimum reaction temperature for a given waste system.

High-temperature wet oxidation is widely applicable to the treatment of industrial wastes containing dissolved organics and putrescible biosolids, which are not easily dewatered (10). The optimum temperature and pressure required for effective oxidation will depend on the characteristics of the waste constituents, as well as the degree of treatment desired. For complete oxidative degradation of pollutants to innocuous products, temperatures ranging from 200°C to 300°C and pressures from 500 psi to 4000 psi are often employed. For the primary purpose of biosolids (sludge) treatment, on the other hand, a temperature of 200°C and pressure at about 500 psi will usually be sufficient.

In the wet oxidation system, which is illustrated in Fig. 1, the waste liquor, slurry, or sludge is fed with a positive displacement pump cocurrent with compressed air through a reactor, which is heated to a predetermined initiation temperature. Under continuous operation conditions, the heat generated due to the exothermic oxidation process is sufficient to maintain the reactor at the desired temperature. In fact, excess energy is often available for either process steam or electric power generation. The electric power is, in turn, mainly used to drive the compressor for air injection into the oxidation reactor. A continuous flow reactor with recycle ratio based on a required effective residence time is used to achieve the desired degree of treatment. With respect to pollutant degradation, the degree of treatment is conveniently measured in terms of residual COD (Chemical Oxygen Demand) of the effluent. For the purpose of sludge stabilization, the treatment is continued until a desired value of specific resistance to filtration is reached. The waste liquor being treated is delivered to a storage tank and preheated to about 60–80°C. Sludge masses are comminuted to particles less than ¼ in. sizes before admission to the storage tank. The metered mixture of waste liquor and compressed air is passed through a series of heat exchangers heated with the treated effluent. During start-up, the incoming mixture to the reactor is heated to the reaction temperature necessary to sustain autooxidation at a sufficiently fast rate. In the continuous operation cycle, the effluent will exit at a higher temperature, and it is thus used to heat the influent liquor. The effluent stream following thermal energy reduction through a heat exchanger is then passed through a gas–liquid separator where the product gases and excess steam are removed from the liquid stream. For the purpose of energy recovery, the product gases including nitrogen and carbon dioxide are expanded to near atmospheric pressures in a mixed gas turbine. The excess steam is recovered in a series of reboilers. Process steam pressurized to about 20–120 psi, together with high-pressure steam at about 650 psi, are recoverable. The high-pressure steam is depressurized by expansion to about 120 psi in a steam turbine. Both the gas turbine and the steam turbine are mounted concentrically to drive the air

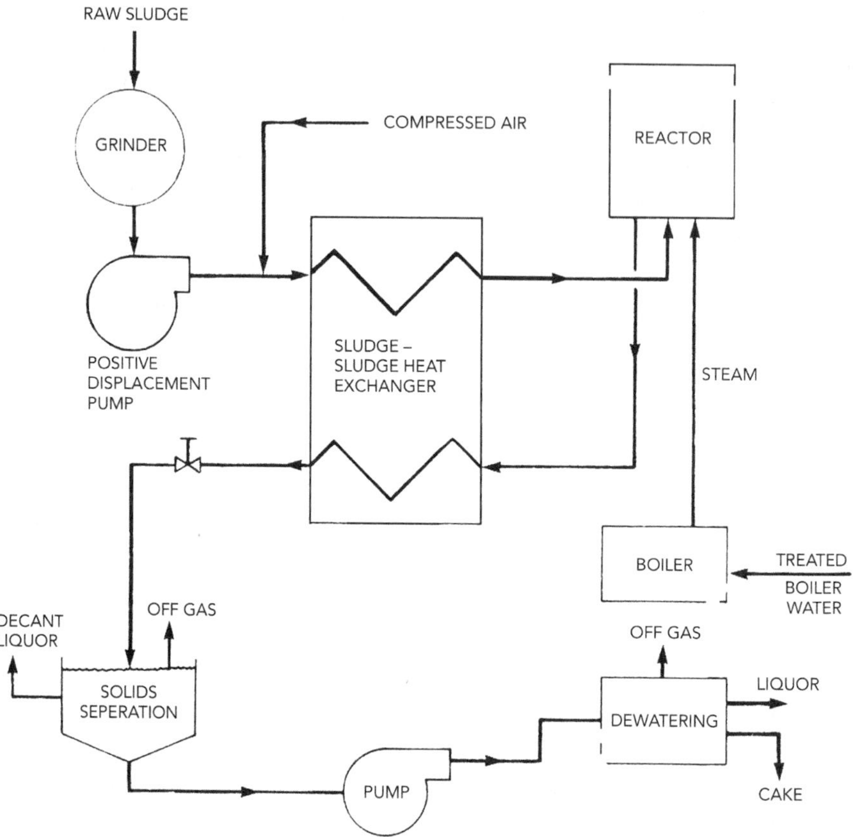

Fig. 1. Flow scheme for biosolids oxidative conditioning.

compressor to provide compressed air for the oxidation system. When the fuel value of the waste stream is sufficiently large, this auxiliary-generated power supply would be self-sufficient to drive the air compression and wet oxidation system.

The thermal oxidation and conditioning of biosolids has the following advantages:

a. It produces biosolids with excellent dewatering characteristics.
b. The processed sludge does not need chemical conditioning.
c. The process disinfect the sludge, rendering it free from pathogenic microorganisms.
d. It is suitable for sludges that cannot be stabilized biologically because of the presence of toxic matter.
e. It is insensitive to changes in biosolids composition.

However, the oxidation process also has some disadvantages including:

a. The process has high capital cost.
b. It requires skilled operators.
c. It produces odorous gas stream that must be treated
d. Possible scale formation that requires acid washing.

In normal combustion systems, the supporting air is usually supplied free by natural draft designs. The derivable energy is therefore computed by means of the fuel calorific

Table 4
Energy Equivalence of Air Consumption in Wet Oxidation

Fuel material	Specific calorific value (kwh/kg)	Stoichiometric air-fuel equivalence (kwh/kg)	Calorific equivalence of air (kwh/kg)
Specific chemicals			
Hydrogen	39.4	34.3	1.15
Ethylene	13.8	14.8	0.94
Carbon	9.10	11.5	0.79
Acetic acid	4.10	4.60	0.89
Oxallic acid	0.78	0.77	1.01
Pyridine	9.60	10.9	0.88
Lactose	4.60	4.90	0.94
Casein	6.80	7.60	0.90
Petroleum fuel oil	12.5	14.0	0.90
Composite waste			
Sulfite pulping	5.10	5.70	0.90
Semichemical pulping	3.80	4.10	0.93
Primary wastewater sludge	5.00	5.80	0.86
Secondary wastewater sludge	4.20	5.10	0.82

values. In the wet oxidation system, the fuel supplied generally consists of waste streams with negative values. The supply of air must be provided at high pressures entailing considerable costs. Thus, the available energy from a wet oxidation system may be more logically measured in terms of the calorific equivalence of air being consumed. Air consumption calorific equivalence values for a number of pure chemical compounds and some typical waste materials are computed and illustrated in Table 4. For a wide variety of waste systems, the calorific value derived from the consumption of 1 kg of air tends to fall within the range of 0.8–0.95 kwh. The energy from a wet oxidation system can thus be estimated even in application to unknown waste streams.

As noted above, a major cost of the wet oxidation process is associated with the air or oxygen supply system. The extent of pollutant oxidation is controlled by the equilibrium values determined as a function of the reaction temperature. The injection of oxygen in excess of that required for maintaining the equilibrium oxidation rate will not aid in increased treatment efficiency. An undersupply of oxygen, on the other hand, will definitely impede the rate of oxidation. Thus, a cost-effective design of the wet oxidation system rests heavily on the proper selection of the air supply rate. In consideration of the stoichiometric oxygen requirement, it is to be realized that in the wet oxidation process the nitrogen constituent of the organic proteinaceous waste is usually released in the ammonia form. The carbon and hydrogen constituents are oxidized respectively to carbon dioxide and water. Furthermore, in most practical applications, only a partial oxidation of the organic waste is either necessary or achieved. The oxygen requirement may therefore be estimated according to the following stoichiometric relationship (11):

$$\begin{aligned} C_aH_bO_cN_d + 0.5(ny + 2s + r - c)O_2 \rightarrow\ & nC_wH_xO_yN_z + sCO_2 \\ & + rH_2O + (d - nz)NH_3 \end{aligned} \tag{36}$$

where $r = 0.5[b - nx - 3(d - nz)]$, $s = a - nw$, $C_aH_bO_cN_d$ = the empirical formula of the organic constituents in the influent, and $C_wH_xO_yN_z$ = the empirical formula of the organic constituents in the effluent.

Equation (36) is, however, very cumbersome and difficult to apply. In practice, it is therefore, recommended that the influent and effluent COD values be determined. The difference between these values is then used as a measure of the oxygen consumption in the wet oxidation treatment.

The wet air oxidation system is especially suitable to the treatment of difficult-to-dewater organic sludge wastes. In commercial application, it is known as the Zimmerman process. In industrial waste treatment applications, the efficacy of the Zimmerman process has been demonstrated for many different types of pulp and paper mill wastes as follows:

a. Magnesium-base spent sulfite liquor.
b. Calcium-base spent sulfite liquor.
c. Ammonium-base spent sulfite liquor.
d. Sodium-base spent sulfite liquor.
e. Kraft sulfate waste liquor.
f. Semichemical sulfite waste liquor.

In each of these cases, the organic constituents are converted to innocuous carbon dioxide and water. The inorganic constituents are oxidized to sulfate chemicals and can be recovered and recycled to the pulping process. High-temperature and high-pressure systems are normally employed.

Wet air oxidation applications to industrial waste treatment and process clinical recovery may be illustrated by depicting the process flow designed for the desired organic constituent removal and caustic soda recovery from a pulp mill waste stream (12). The wet oxidation reactor is operated somewhat in excess of 300°C and at a pressure of 2500–3000 psi in order to achieve 95–98% oxidation of the putrescible constituents. The effluent liquor is practically colorless. Dregs containing the majority of inorganic impurities present in the fiber, including silica, magnesia, alumina, and calcium carbonate, are easily removed by sedimentation. The remaining organic fraction consists mainly of short-chain aliphatic acid salts such as sodium acetate. The recovered liquor is therefore essentially a concentrated solution of sodium carbonate and some sodium bicarbonate. It is recausticized by addition of lime and recycled for use in the Kraft pulping process.

For applications to wastewater biosolids (sludge) oxidation, the important objectives are to achieve microorganism destruction, odor removal, and solid filterability (13). Low degrees of oxidation in the order of 30% COD reduction or less are often sufficient to produce the stated objectives. With the destruction of the fibrous and cellular structure of the primary or secondary raw biosolids, a silt-like residue with good settling and drainage qualities will result. Such a residue is easily dewatered and subsequently disposed by landfill or incineration. Relatively low temperatures at about 200°C and pressures below 600 psi are therefore suitable for biosolids treatment. Considerable savings in both equipment and operating costs of the oxidation system are thus realized.

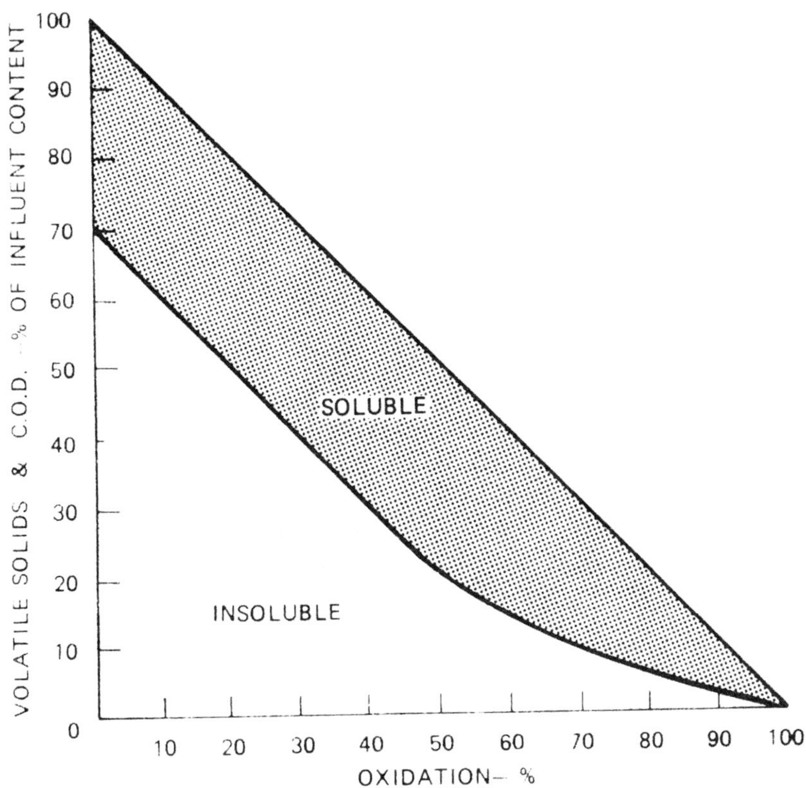

Fig. 2. Volatile solids and COD content of oxidized biosolids.

The US EPA (Environmental Protection Agency) compiled sufficient data to be able to represent the results of wet oxidation, as shown in Fig. 2, for typical wastewater biosolids, showing volatile solids content or COD content in the solid phase and the total sludge as a function of oxidation in both phases (14). The vertical distance between the two curves is the content in the liquid phase. Up to about 50% total oxidation, reduction in the volatile solids or COD in the liquid phase are minimal; above 50%, the volatile solids and COD of both phases are reduced to low values. At 80% total oxidation, about 5% of the original total volatile solids in the sludge are in the solid phase and 15% are in the liquid phase.

An example of wet air oxidation application to municipal wastewater biosolids treatment is given by the plant in operation at Levittown, PA (14), which is illustrated in Fig. 3. The oxidation reactor is operated at a temperature of only 150–175°C and a pressure of about 300 psi. The oxidized sludge is brown in color with uniform consistency, and it is easily filterable. The filter cake has an earthy odor, and it is sterile and nontoxic to plant growth. It is thus useful as a mulch material with characteristics somewhat superior to peat moss. The liquid fraction of the oxidation effluent has a deep reddish brown color, but it is virtually odorless when cooled. The filtrate is returned to the wastewater treatment plant influent, and it is not found to present any detrimental effects.

In the example of wet oxidation application to pulp mill waste treatment, energy recovery via steam and power generation is practiced. For the case of municipal

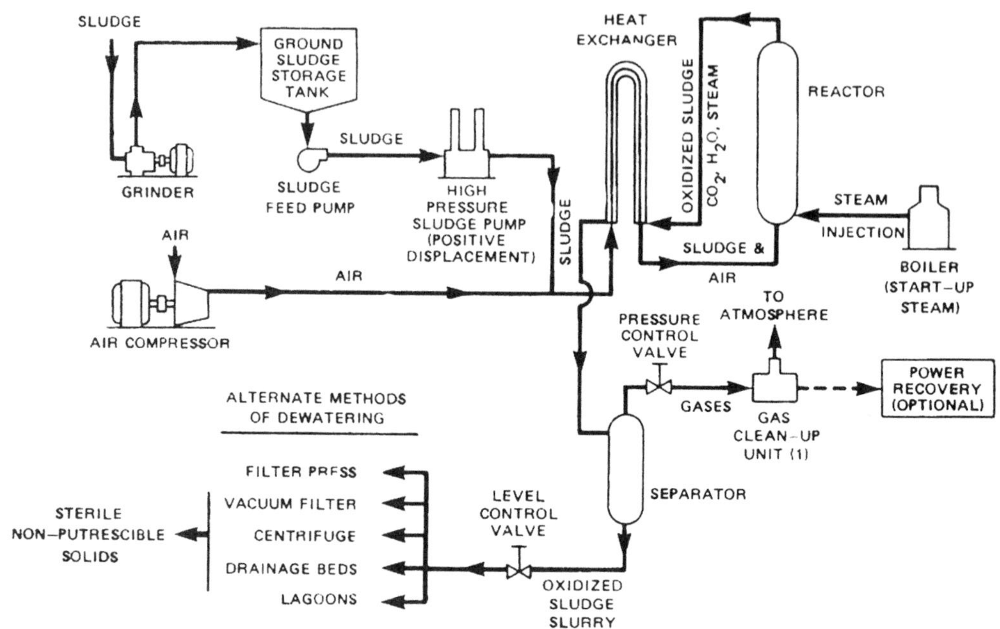

Fig. 3. Flow scheme for biosolids wet air oxidation.

wastewater biosolids treatment as illustrated by the process flow diagram given in Fig. 3, energy recovery other than pre-heating the influent waste is not included. This is dictated by the economics of the respective systems. As may be recalled, the capitalization and operating cost of a wet oxidation system is largely related to the required system pressure and the quantity of air supplied. The available energy of a given system is, in turn, measurable as calorific equivalence of air being consumed. As a general rule (15), steam generation with the reactor effluent can usually be justified only in the case of oxidation systems consuming more than 500 compressor horsepower. Above 1500 compressor horsepower, direct drive of the compressor by steam and gas turbine power generation becomes attractive. For plants justifying the installation of greater than 2500 compressor horsepower, the steam and power generated will generally exceed the plant operation requirement. The electrical energy requirements (14) are shown in Fig. 4.

In a modified application of the Zimmerman process, wet oxidation is coupled with activated carbon adsorption to provide tertiary organic waste removal (16). Powder-activated carbon is known to be effective for the removal of low concentrations of organic pollutants in advanced water treatment applications. The only drawback is that the reagent cost is relatively high, and a recovery and regeneration of the spent carbon is difficult to achieve. Selective carbon regeneration by wet oxidation at about 200°C with the reuse of spent carbon exceeding 23 cycles has been demonstrated. The regenerated carbon had increased ash content but did not display a measurable decrease in adsorptive efficiency. A 5% carbon make-up was found necessary in each regeneration cycle. The treatment system is shown capable of producing high-quality effluents.

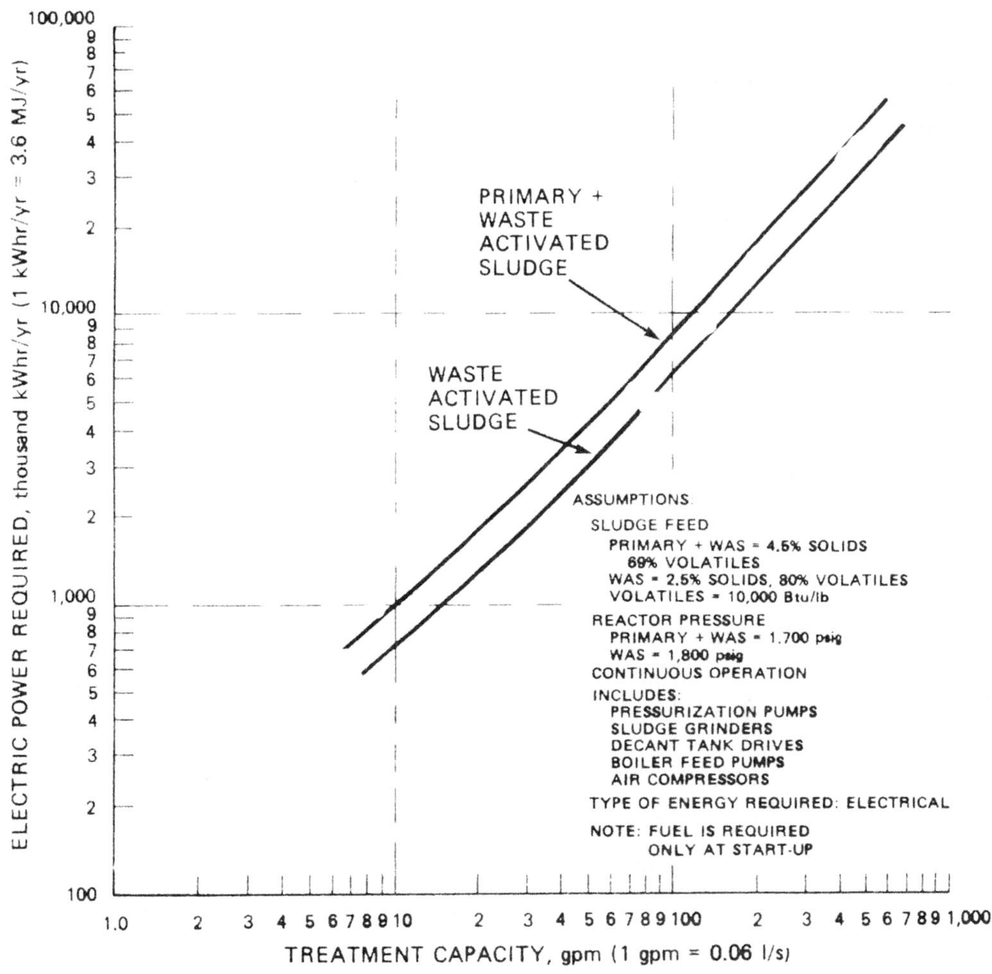

Fig. 4. Electrical energy requirements for wet air oxidation.

Application of the system is limited largely by cost considerations in that inexpensive water supplies are still abundantly available.

Thus far, the conditioning of municipal wastewater biosolids to improve the sludge filtration and disposal has been the single largest application of wet oxidation. Even in this area of application, wet oxidation has not gained widespread acceptance. Its drawback lies mainly in cost considerations in that it is an inherently expensive process in terms of initial capital investment. As a method of ultimate disposal, it can rarely compete in cost with deep well injection. In terms of more practical comparisons, wet oxidation may be regarded as serving waste treatment functions in competition with biological stabilization and sludge incineration. An economic comparison with biological treatment will favor wet oxidation only when the wastewater BOD (Biochemical Oxygen Demand) is extremely high (>10,000 mg/L) or if the waste constituent is toxic or only marginally biodegradable. Wet oxidation will compete favorably with incineration only when the pollution control and fuel consumption costs become important factors. The inherent nature of wet oxidation minimizes residual waste and air pollution problems. As pollution discharge limitations are more stringently implemented, increased reliance on wet oxidation may be expected (17).

4. TRANSITION-METAL ION OXIDATION SYSTEMS

Transition-metal ion species are known to be effective catalysts for the oxidation of both organic and inorganic constituents of aqueous systems. As noted in preceding discussions, oxidation–reduction reactions are, in essence, electron transfer processes. Transition-metal atoms are those with expandable valence shell electrons and can make available a varying number of electrons. These atoms may therefore exist at varying stages of oxidized states and are easily accessible to electron exchange reactions. At a higher oxidized state, the metal ion will serve as an electron acceptor or oxidant. At a lower oxidized state, the metal ion will be a ready electron donor or reductant. The complimentary metal ion pairs; e.g., Fe(II)–Fe(III) or Co(II)–Co(III), can therefore serve effectively as electron transfer agents or oxidation–reduction catalyst systems. Some of the transition metal elements, such as chromium and manganese will exist at very high oxidation states as oxygenated anions. Their anionic species, e.g., chromate and permanganate, are especially potent oxidants.

Cationic transition-metal ion systems are mainly effective as oxidation catalysts. The application of such catalysts to wastewater treatment is, however, limited by the toxicity of transition metals to aquatic organisms. To satisfy the commonly accepted water quality standards, transition-metal contents must generally be controlled within stringent limits. As oxidation reagents, anionic transition-metal species are much more effective. Only dichromate and permanganate are widely used in waste oxidation applications.

4.1. Chromic Acid Oxidation

Chromic acid is an extremely important reagent in both organic and inorganic oxidation reactions (18). In terms of oxidation potential, chromic acid is most effective as an oxidant only under acidic conditions. In aqueous systems, the chromate ion species will exist in two chemical forms in accordance with the following equilibrium relationship:

$$Cr_2O_7^{2-} + H_2O \leftrightarrow 2HCrO_4^- \tag{37}$$

The dissociation of dichromate to monomeric bichromate is favored at dilute concentrations in accordance with the mass action law. The dichromate ion has a characteristic orange color, whereas the bichromate ion appears red. The chemical reagent available in crystalline form is usually isolated from concentrated solutions, and it is thus obtained commercially as the orange colored dichromate. When dissolved into dilute aqueous solution, the reagent will spontaneously dissociate into the red colored bichromate. It has further been established that the monomeric bichromate is, in fact, the effective oxidant in the oxidation of organic constituents. The bichromate oxidation half-reaction may be presented as:

$$HCrO_4^- + 7H^+ + 3e \leftrightarrow Cr^{3+} + 4H_2O \tag{38}$$

The product chromic ion obtained in Reaction (38) may be removed from solution by pH adjustment to alkaline conditions and precipitation as $Cr(OH)_3$.

Chromic acid reagent is well known in wastewater engineering in that it is used as an oxidant in the standard method for chemical oxygen demand (COD) measurement. The oxidation of ionic inorganic species, such as Fe(II), Mn(II), sulfide, and cyanide, will usually proceed at effective rates at ambient temperatures. Organics oxidation rates may vary widely according to the nature of the pollutant species. In general, the oxidation of organics will proceed stepwise from hydrocarbons to the various oxygenated stages of alcohols, carbonyls, acids, and finally carbon dioxide. The rate of reaction with hydrocarbons is very slow. With oxygenated organics, the oxidation rate is much faster due to the fact that a chromate ester intermediate can be formed to facilitate electron transfer from chromate to the organic species. In order to achieve effective oxidation of even the more resistant organic constituents, the COD test is thus carried out at an elevated temperature by refluxing the reaction solution for a prolonged period. The oxidation rate is further promoted by the addition of silver ion in the form of silver sulfate as a catalyst.

In the COD test, the oxidation of organic pollutants may be represented by the generalized reaction,

$$C_nH_m + lHCrO_4^- + kH^+ \leftrightarrow nCO_2 + jH_2O + lCr^{3+} \tag{39}$$

To achieve an electron exchange balance, it is required that

$$l = 4n/3 + m/3$$

For stoichiometric balance in oxygen atoms,

$$j = 10n/3 + 4m/3$$

For hydrogen atom stoichiometric balance,

$$k = 2j - l = 16n/3 + 7m/3$$

We may keep in mind a previously stated guideline of always assigning an oxidation state value of –II for oxygen atoms in chemically combined form with other elements. Reaction (39) can thus be applied also to oxygenated organics by adding equal increments of H_2O to C_nH_m on the left side and to the water on the right side of the equation. A simple illustrative example is given as follows:

$$C_2H_4 + 4HCrO_4^- + 20H^+ \leftrightarrow 2CO_2 + 12H_2O + 4Cr^{3+} \tag{40}$$

If the organic species to be oxidized has a molecular formula of C_2H_6O (*e*.g., ethanol or methyl ether), the reaction may be balanced by simply adding an additional H2O to Reaction (40):

$$C_2H_6O + 4HCrO_4^- + 20H^+ \leftrightarrow 2CO_2 + 13H_2O + 4Cr^{3+}$$

The above stoichiometric reaction balance permits computation of COD corresponding to specific quantities of any known organic constituents. More important, it illustrates the chemical and water quality principles associated with the standard COD measurement technique.

By accepted convention, COD is expressed as equivalent oxygen consumption in the unit of milligrams oxygen per liter of wastewater. An excess of chromic acid reagent is used to achieve complete degradation of oxidizable pollutants. The remaining chromic acid reagent is measured by titration with ferrous ammonium sulfate as shown by,

$$3Fe^{2+} + HCrO_4^- + 4H^+ \leftrightarrow 3Fe^{3+} + Cr^{3+} + 4H_2O \quad (41)$$

The number of equivalents of chromic acid consumed is then converted to oxygen consumption by the factor of 8 mg COD for each milliequivalent oxidant consumed. COD is generally designated as a measure of the oxygen consumption potential of the waste when discharged into surface waters. The presence of any species stable to oxygen in natural waters but oxidizable by chromic acid would give rise to erroneous COD measurements. Chloride ion in water is an important example as it can be oxidized by chromic acid in accordance with Reaction (42),

$$6Cl^- + 2HCrO_4^- + 14H^+ \leftrightarrow 3Cl_2 + 2Cr^{3+} + 8H_2O \quad (42)$$

This chloride interference can be eliminated by the addition of mercuric sulfate to the chromic acid reaction system whereby the chloride ion would be bound into non-dissociated mercuric chloride,

$$Hg^{2+} + 2Cl^- \leftrightarrow HgCl_2 \quad (43)$$

Thus, the COD data as measured by chromic acid oxidation represent indications of putrescible organics as well as air-oxidizable inorganic species in the water sample.

Chromium is generally regarded as a toxic metal, and its discharge in effluent waters must be stringently limited. Thus, chromic acid is not commonly employed as a water or wastewater treatment reagent. In fact, the use of chromic acid as a water-conditioning chemical in corrosion prevention applications is gradually being curtailed. Chromate discharges in industrial process waste stream, such as spent plating solutions and metal finishing rinse waters, are considered among the difficult industrial waste problems. The reduction of chromate to chromic(III) ion followed by a precipitation removal of the latter as $Cr(OH)_3$ at suitable pH conditions is the prevalent method for chromium discharge control. Chromate reduction with SO_2 is illustrated as follows:

$$2HCrO_4^- + 3SO_2 + 2H^+ \leftrightarrow 2Cr^{3+} + 3SO_4^{2-} + 2H_2O \quad (44)$$

In the above chromate waste treatment scheme, sulfur dioxide is oxidized to sulfate. Other analogous oxidation–reduction systems are employed in chromate waste control applications.

4.2. Permanganate Oxidation

Permanganate is effective as an oxidizing agent in either acid or alkaline conditions (19). Under acid conditions, the oxidation half-reaction will proceed as,

$$MnO_4^- + 8H^+ + 5e \leftrightarrow Mn^{2+} + 4H_2O \quad (45)$$

with manganese (II) ion as the end product. Under neutral or alkaline conditions, the half-reaction will proceed by,

$$MnO_4^- + 2H_2O + 3e \leftrightarrow MnO_2 + 4OH^- \quad (46)$$

Table 5
Water Pollutants Removal by Permanganate Oxidation

Pollutant species	$KMnO_4$ dose, weight ratio to pollutant
Pesticides	
Vapona	1.50
Phosdrin	1.10
Endothal	0.55
2,4-D	0.70
Odorous industrial chemicals	
Benzonitrile	0.70
Methyl acrylate	2.00
Ethyl acrylate	2.00
Methyl methacrylate	1.00
Ethyl methaclylate	1.00
m-Chlorophenol	12.0
o-Chlorophenol	8.00
p-Chlorophenol	7.00
Phenol	6.00
m-Cresol	7.20
o- Cresol	6.00
p- Cresol	6.00

with manganese dioxide as the end product. Manganese(II) salts are soluble in water, while manganese dioxide is insoluble. In neutral or alkaline systems, therefore, the application of permanganate oxidation treatment will not introduce any deleterious constituents in the dissolved state. Alkaline or neutral permanganate oxidation is thus a popular process in industrial waste treatment applications.

Permanganate reagent is most commonly available as crystalline potassium permanganate. It is deeply purple in color, and this color is useful as a built-in indicator of dosage requirement. Alkaline or neutral conditions are preferable for permanganate oxidation applications in that the byproduct would be in the form of manganese dioxide. The hydrated manganese dioxide is insoluble in water and can subsequently be removed by filtration. Similar to chromic acid reactions, permanganate is generally capable of rapid oxidation of ionic inorganic constituents. In the case of organic pollutants, the oxidation rate may vary widely according to the structure of the individual chemical species. Many of the organics in water may be virtually non-reactive toward permanganate under mild treatment conditions. Permanganate oxidation of organics is known to involve oxygen transfer through ester-type bridging transition structure. The more highly oxygenated or unsaturated organic species are thus easily accessible to facilitate permanganate reactions. These types of organic constituents, such as the phenols and humic and tannic acids, are mainly responsible for organic sources of color, taste, and odor in industrial wastewaters. Permanganate oxidation is therefore generally effective in color, taste, and odor applications (20). A listing of some specific pesticides and odorous pollutants treatable by permanganate oxidation is illustrated in Table 5.

Another important application of permanganate treatment is the control of dissolved iron and manganese in potable water supplies derived from groundwater sources. The corresponding chemical reactions are as follows:

$$3Fe^{2+} + MnO_4^- + 7H_2O \leftrightarrow 3Fe(OH)_3 + MnO_2 + 5H^+ \quad (47)$$

$$3Mn^{2+} + 2MnO_4^- + 2H_2O \leftrightarrow 5MnO_2 + 4H^+ \quad (48)$$

The dissolved Fe(II) and Mn(II) are converted to the insoluble hydrated oxides of these metals at higher oxidation states. The precipitates are subsequently removed by coagulation and filtration. Permanganate oxidation of iron and manganese is highly effective and easily performed. Intermediate manganese ions are effective as autooxidation catalysts causing secondary oxidations by dissolved oxygen in the waste system. Permanganate dosage required for complete treatment is therefore often found to be considerably less than the theoretical amount computed by Reactions (47) or (48).

Potassium permanganate is sometimes used for the control of toxic inorganic anionic species, such as sulfide and cyanide. Depending on the treatment reaction conditions, sulfide is converted to a mixture of sulfate and elemental sulfur. Sulfate is normally the predominant product. Elemental sulfur would be adsorbed on the hydrated manganese dioxide floc and can subsequently be removed by filtration. Cyanide oxidation is expected to proceed in two stages:

$$3CN^- + 2MnO_4^- + H_2O \leftrightarrow 3CNO^- + 2MnO_2 + 2OH^- \quad (49)$$

$$2CNO^- + 2MnO_4^- + 2H_2O \leftrightarrow N_2 + 2CO_2 + 2MnO_2 + 4OH^- \quad (50)$$

Reaction (49) is much more rapid in comparison to Reaction (50). For the most part, the cyanide will be converted to the considerably less toxic cyanate. Only under more drastic treatment conditions will the conversion to molecular nitrogen take place significantly. Permanganate is generally effective for the removal of inorganic COD-relating pollutants in industrial wastewaters. Only a relatively high reagent cost has prevented wide applications of permanganate oxidation in wastewater engineering.

Permanganate oxidation has also been applied successfully to the control of odorous air pollutants. The most effective treatment design is based on wet scrubbing of the exhaust air with an aqueous potassium permanganate solution. Such a system is marketed commercially under the tradename of the Cairox Method (21). The potassium permanganate reagent solution is usually kept within the pH range of 8–10. Most organic oxidations by permanganate are favored under alkaline conditions where the OH^- often serves effectively as a catalyst agent. For air pollutant conversion alone, complete degradation to innocuous products is not necessary. To minimize reagent dosage requirements, it is often desirable to restrict the degree of oxidation merely to the stage where the product can be removed from the exhaust air stream by dissolution in the water phase. Thus, the pH of scrubber solution may be maintained below about 10 to avoid unnecessary secondary consumption of the permanganate reagent.

In the aqueous permanganate scrubbing application to odorous air emission control, pollutant oxidation reactions take place effectively only in the liquid phase (22). The well established practice for odor reduction is based on 1% potassium permanganate solution scrubbing with about 1 s contact time at a pH of 8.5. More effective odor control can be achieved by process designs permitting scrubber contact time period in excess of 1 s.

5. RECENT DEVELOPMENTS IN CHEMICAL OXIDATION

Throughout the last two decades, the chemical oxidation process has been put into use in various water and wastewater treatment processes as well as in some other applications. These are:

a. Ozone processes.
b. Ultraviolet processes.
c. Wet oxidation.
d. Supercritical water oxidation.
e. Biological oxidation.

5.1. Ozone (O_3) Processes

As early as 1886, the ability of O_3 to destroy impurities and make microorganisms inactive was well known (23). Since 1893, it has been used as a disinfection agent for drinking water in Europe. By 1986, over 1000 potable water facilities using ozone had been reported. Similarly, the United States began to use the ozone process in order to control taste and odor in effluent water at Philadelphia, PA in 1940. Based on the US Environmental Protection Agency's (USEPA) research in the 1970s, ozone was found to be a feasible disinfection technology. Since then, much research has been done on surface and groundwater, as well as industrial wastewaters. Kanzelmeyer and Admas (24) studied the recovery of copper from solutions containing direct blue 80, a common metal-complex dye. This process was examined using ion exchange preceded by oxidative pretreatment employing ozone alone and in combination with hydrogen peroxide. It was discovered that ozonation followed by ion exchange could be a viable method for removing copper from textile wastewater containing copper complex dyes. Decomposition of monocrotophos in aqueous solution by the hydrogen peroxide–ozone process was studied under various pHs and H_2O_2 to O_3 molar ratios by Ku and Wang (25). Simplified steps and consecutive reactions based on elemental mass balances was found to accurately describe the temporal behaviors of reacting species during the decomposition of monocrotophos in aqueous solution by the H_2O_2 process. Brain E. Reed and other scholars (26) summarized the different physicochemical processes that included ozone oxidation. Also, Gracia and other researchers (27) used Mn^{2+} as a catalyst during ozonation of raw river water, which facilitated reductions in the content of organic matter. The authors characterized raw and ozonated water and reported sizable removal of organic material at two different ozone to carbon weight ratios.

The influence of water quality parameters (dissolved organic matter and alkalinity) on the efficiency of the ozone/hydrogen peroxide (O_3/H_2O_2) advanced oxidation processes, relative to corresponding conventional ozonation process, was investigated by Aero and Van Gunten (28). In natural waters with a high natural organic matter content ($\geq$ 3 mg/L), O_3 decomposition is controlled by radical-type chain reactions. The overall OH^- radical oxidation capacity is nearly unaffected if conventional ozonation is replaced by O_3/H_2O_2. In contrast, when the organic matter content in the natural water is low ($\leq$ 1 mg/L), the addition of H_2O_2 considerably enhances the oxidation capacity by OH^- radicals. Because a constant ratio between the OH^- radical and O_3 concentration was found during both treatments, it is possible to predict the oxidation of micropollutants with O_3 and OH^- radicals if the rate constants of both oxidants are known. The oxidation of atrazine during ozonation and combined process O_3/H_2O_2 was accurately predicted in natural waters. Liang and others (29) did the same combination of ozone and hydrogen peroxide study as a treatment alternative for removing methyl tertiary butyl ether (MTBE), a common fuel oxygenate. The investigators studied the effects of oxidation of ozone and peroxide on MTBE in Santa Monica, CA, groundwater. An experiment conducted in a large-scale semi-batch reactor demonstrated that peroxide at concentration of 1.0 mg/L with applied ozone doses of $\leq$ 10 mg/L, was consistently more effective in oxidizing MTBE than was the ozone alone.

Wojtenko, et al. (30) studied the performance of ozone as a disinfectant for combined sewer overflow (CSO). The purpose of the study was to minimize the number of disease-causing microorganisms released into receiving waters. This work presented the results of an innovative review of effectiveness of ozone for CSO disinfection along with its advantages and disadvantages. The ozone was found to have a relatively high disinfection power, released fewer by-products, and was non-reactive with ammonia. Hsu et al. (31) did a study on ozone decolorization of solutions, which were prepared by mixing two and three various dye samples, in a newly developed gas-induced reactor. Decolorization kinetics of the mixed-dye solutions, the ozone utilization rate, and chemical oxygen demand (COD) removal were determined. They reported greater than 90% decolorization at 90% ozone utilization rate. The removal of COD was approx 33% at 96% decoloration. Beltran, et al. (32) carried out a study on the ozonation of the herbicide alachlor in distilled and in surface water. The influence of pH, water type, and free-radial-inhibiting substances on the removal rate of alachlor was investigated. Alachlor was primarily removed from water through hydroxyl radical oxidation because low pH and the presence of natural substances such as carbonates significantly reduced its removal rate. Hydrogen peroxide was formed in low concentrations during ozonation. The rate constants of the reactions between alachlor and ozone and alachlor and the hydroxyl radicals were 2.8 and 3.2×10^{10} L/mole, respectively. Because combined UV radiation and ozonation constitutes one of the possible ozonation processes to treat alachlor, the quantum yield of this herbicide was determined and found to be 0.177 mole/Einstein, which is a significant value compared to other different herbicides.

5.2. *Ultraviolet (UV) Processes*

Mary A. Parmelee (33), section editor of the *Journal of AWWA*, stated that the expectations of new regulations being applied to the disinfection of drinking water would lead to an increase in the adoption of UV as a water treatment solution. UV treatment is currently applied to approx 1% of the 40 billion gal (150 Mm^3) of drinking water provided daily by public water utilities in the United States.

Many researchers studied the performance of ultraviolet combined with hydrogen peroxide, which is not the case for a combination of UV and ozonation. However, Viraraghavan and Sapach (34) investigated the advanced oxidation of pentachlorophenol (PCP) in water using hydrogen peroxide and ultraviolet light. The authors reported that the initial concentration of PCP and H_2O_2, contact time and UV light all affected the degradation of PCP in the H_2O_2–UV system. Andreozzi et al. (35) studied the oxidation of metol (*N*-methyl-*p*-aminophenol) in aqueous solution by means of a combination of UV–H_2O_2 in the pH range 3.0–9.0. The results of the investigation indicated that pH, H_2O_2, and substrate concentration as well as the presence of oxygen significantly influenced the systems' behavior. Toxicity tests showed that the H_2O_2 photolytic process was capable of reducing the toxicity of metol aqueous solutions. Gen-Shuh Wang and other investigators (36) used a batch reactor to evaluate the advanced oxidation process of the UV–H_2O_2 system for control of natural organic matter in drinking water. The light sources used included a 450 W high-pressure mercury vapor lamp and sunlight. Both quartz and Pyrex filters were used to control the wavelength and energy of UV light applied to the aqueous system. The results showed that non-organic material oxidation and H_2O_2 decomposition followed first-order and zero-order reaction kinetics, respectively. The optimum H_2O_2 dose was found to be 0.01% for the oxidation of humic acids. Bose and Maddox (37,38) reported on the use of five types of oxidation processes (UV, UV–H_2O_2, ozone, ozone–H_2O_2, and UV-ozone) to degrade 1,3,5-trinitrotriazacyclohexane (RDX), a widely used explosive that contaminates groundwater and other environmental media. Degradation rates of RDX by the various advanced oxidation processes were determined and the effects of process parameters on the degradation rates were examined.

5.3. *Wet Oxidation*

Lei et al. (39,40) and Thomesen and Kilen (41,42) are the two main groups who have studied the application of the wet oxidation process to textile wastewater and quinoline, respectively. The first group reported on a series of wet oxidation experiments using two new oxidation methods that were used for treatment of dying wastewater concentrates from the membrane-separation processes. The first method partially or totally replaced oxygen with a strong oxidant, while the second introduced a catalyst. Both methods were found to achieve higher initial reaction temperatures and pressures when compared with conventional wet oxidation techniques. In 2000, an extensive series of experiments were performed to identify suitable catalysis to increase the reaction rate of wet-air oxidation of textile wastewater at relatively mild temperatures and pressures. The wastewater types that were treated included natural-fiber desizing wastewater, synthetic-fiber desizing wastewater, and printing and dying wastewater. Experimental results indicated that all catalysts tested in this investigation significantly increased the chemical oxygen demand (COD) and total organic carbon (TOC) removal rates as well as total COD and TOC removals. Of all catalysts tested, copper salts were the most effective. Anions in the salt solutions played a role in the catalytic process. Nitrate ions were more effective than sulfate ions. Similarly, copper nitrates were more effective than copper sulfates. A mixture of salts containing different metals performed better than any single salt.

Thomesen (41) studied the high-temperature and high-pressure wet oxidation reaction of quinoline as a function of initial concentration, pH, and temperature. At neutral to acidic pH, high pressure wet oxidation was effective in the oxidation of quinoline for temperatures in excess of 240°C. However, under alkaline conditions, the reaction was much slower. The author also studied the wet oxidation of deuterium-laced quinoline as a method of verifying and quantifying the reaction products. In further studies by

Thomesen and Kelin (42), 15 reaction products were identified and quantitatively determined, accounting for 70% of the carbon present after treatment. Pressure, reaction time, and temperature were major factors of this study. Wet oxidation made quinoline more toxic to *Nitrosomonas* and *Nitrobacter* under conditions of low oxygen, low pressure and long reaction times. The reaction products were reported to be highly digestible in activated sludge treatment. Following combined wet oxidation and biological treatment, the effluent showed low toxicity toward nitrosomonas and no toxicity toward nitrobacter.

5.4. *Supercritical Water Oxidation*

Supercritical water describes a state of water that has been heated above its critical temperature (374°C) and compressed greater than its critical pressure (221 bar). The high temperature favors the gas phase and the high pressure favors the liquid phase. As a result, supercritical water is a non-ploar fluid in which the gas and liquid phase are indistinguishable from each other. These are characteristics that make supercritical water an excellent solvent for both organic compounds and oxygen. The supercritical water oxidation process destroys the hazardous organic portion of waste via oxidation. Bench-scale testing has been conducted at Los Alamos National laboratory since 1995. A supercritical water oxidation system may be installed on the Department of Energy (DOE) site to dispose of the hazardous materials (43)

Lin et al. (26,44) reported on an investigation of the oxidation kinetics of 2,4-dichlorophenol (DCP) in supercritical water. Condensation byproducts such as tricholophenols and dibenzo-*p*-dioxine from supercritical water oxidation of DCP at 673 K were determined. The authors also reported experimental results in which the presence of sodium or iron cations in the oxidation of DCP under supercritical water conditions suppressed the formation rate of condensation byproducts and enhanced the conversion reactions.

5.5. *Biological Oxidation*

An investigation on Cr^{6+} reduction was conducted by Philip et al. (26,45) using bacteria isolated from soil samples receiving electroplating wastewater. The chromium reduction capacity of this isolates was compared with that of *Pseudomonas aeruginosa* and *Bacillus circulans* (a laboratory isolate from garden soil). *Bacillus* coagulants, isolated and identified from chromium-polluted soil, gave the maximum reduction potential among all of the organisms studied. The oxidation of ferrous ions by *Leptospirillum* bacteria was reported by Van Scherpenzeel et al. (26,46). Lampron et al. (26,47) examined effects of combining zero-valent iron with anaerobic microorganisms for the reductive dehalogenation of TCE. The authors reported that TCE in a reactor containing both iron and anaerobic bacteria was degraded faster than in reactors containing iron or anaerobic microorganisms alone.

6. EXAMPLES

6.1. *Example 1*

In Table 2, there are listed standard oxidation potentials for two electrode reactions representing the oxidation of iron from the +II state to the +III states as follows:

$$\mathrm{Fe^{2+} \leftrightarrow Fe^{3+} + e} \qquad E^\circ = -0.771$$
$$\mathrm{Fe(OH)_2 + OH^- \leftrightarrow Fe(OH)_3 + e} \qquad E^\circ = 0.56$$

Are these two relationships, both representing the same oxidation process, mutually consistent?

Solution

The standard electrode potential is defined as the value corresponding to the reversible electrode reaction only when the concentrations of all reactants are present at unit molar activity. The actual electrode potential under a specific set of conditions is given by the Nernst equation:

$$E = E^\circ - (0.06/n)\log\{[\text{oxidized}]/[\text{reduced}]\} \tag{12}$$

For the first reaction, therefore,

$$E = -0.771 - (0.06/1)\log\{[\mathrm{Fe^{3+}}]/[\mathrm{Fe^{2+}}]\}$$

The standard state in the second reaction requires that $[OH^-] = 1\ M$ and $Fe(OH)_2$ and $Fe(OH)_3$ form precipitates in equilibrium with the solution. The solubility products for $Fe(OH)_2$ and $Fe(OH)_3$ are, respectively, as follows:

$$[\mathrm{Fe^{3+}}][\mathrm{OH^-}]^3 = 1\times10^{-36}$$

$$[\mathrm{Fe^{2+}}][\mathrm{OH^-}]^2 = 1.64\times10^{-14}$$

Thus, under the standard state condition for the second reaction,

$$[\mathrm{Fe^{3+}}] = 1\times10^{-36}\,M$$

$$[\mathrm{Fe^{2+}}] = 1.64\times10^{-14}\,M$$

Specifying the above conditions with respect to the first reaction and substituting the values into the Nernst equation, one obtains,

$$\begin{aligned} E &= -0.771 - 0.06\ \log 10^{-36}/1.64\times10^{-14} \\ &= -0.771 - 0.06(-36 + 13.8) \\ &= -0.771 + 1.322 \\ &= 0.561\ \mathrm{V} \end{aligned}$$

It is seen that both reactions do correspond to the same oxidation process, but the standard electrode potentials are different because the conditions are defined differently.

6.2. Example 2

Aeration is known to be effective for the oxidation of Fe^{2+} to Fe^{3+} in natural waters and eventual removal as the hydroxide precipitate. The corresponding electrode reactions may be represented as follows:

$$\mathrm{4Fe^{2+} \leftrightarrow 4Fe^{3+} + 4}e \qquad E^\circ = -0.771$$
$$\mathrm{O_2 + 2H_2O + 4}e \mathrm{\leftrightarrow 4OH^-} \qquad E^\circ = 0.401$$

$$\mathrm{4Fe^{2+} + O_2 + 2H_2O \leftrightarrow 4Fe^{3+} + 4OH^-} \qquad E^\circ = -0.370$$

Does the negative value of standard potential for the overall reaction indicate that the iron oxidation reaction is thermodynamically unfavorable?

Solution

The thermodynamic feasibility of the reaction depends on the potential of the reaction under the prevailing conditions of reactant concentrations not usually at unit activity. Thus, the actual reaction potential should be computed according to the Nernst equation,

$$\begin{aligned} E &= E^\circ - (0.06/n)\log\{[\text{oxidized}]/[\text{reduced}]\} \\ &= E^\circ - (0.06/4)\log\{[\text{Fe}^{3+}]^4 + [\text{OH}^-]^4/[\text{Fe}^{2+}]^4\} \\ &= -0.37 - (0.06/4)\times 4\log\{[\text{Fe}^{3+}] + [\text{OH}^-]/[\text{Fe}^{2+}]\} \\ &= -0.37 - 0.06\log\{[\text{Fe}^{3+}] + [\text{OH}^-]/[\text{Fe}^{2+}]\} \end{aligned} \quad (12)$$

Since the solubility product for $Fe(OH)_3$ is 1×10^{-36},

$$[\text{Fe}^{3+}][\text{OH}^-]^3 = 1\times 10^{-36}$$

so $[Fe^{3+}]$ and $[OH^-]$ cannot coexist at unity activity. If one assumes a neutral pH, where

$$[\text{OH}^-] = [\text{H}^+] = 10^{-7} \text{ and } [\text{Fe}^{3+}] = [\text{Fe}^{2+}]$$

the reaction potential is then given as

$$\begin{aligned} E^\circ &= -0.37 - 0.06\,\log 10^{-7} \\ &= 0.05 \end{aligned}$$

Under the above assumed conditions, oxidation of Fe^{2+} to Fe^{3+} is then thermodynamically favorable. As the OH^- concentration is increased, Fe^{3+} concentration will be decreased in accordance with the $Fe(OH)_3$ solubility limitations. Even at pH 7, Fe^{3+} can only equal to 10^{-15} *M* or less than 10^{-10} mg/L.

An alternate approach to compute the reaction potential under another set of possible conditions is to allow both the Fe^{3+} and Fe^{2+} ions to be in equilibrium with the respective hydroxide precipitates. Thus,

$$[\text{Fe}^{3+}][\text{OH}^-]^3 = 1\times 10^{-36}$$

$$[\text{Fe}^{2+}][\text{OH}^-]^2 = 1.64\times 10^{-14}$$

Taking the ratio of the two solubility products

$$[\text{Fe}^{3+}][\text{OH}^-]/[\text{Fe}^{2+}] = 1\times 10^{-36}/1.64\times 10^{-14}$$

Substitution of the above relationship in Nernst equation

$$E^\circ = -0.37 - 0.06\,\log\{[\text{Fe}^{3+}]\times[\text{OH}^-]/[\text{Fe}^{2+}]\}\times\{[\text{OH}^-]^2/[\text{OH}^-]^2\}$$

will give

$$\begin{aligned} E^\circ &= -0.37 - 0.06\,\log\{1\times 10^{-36}/1.64\times 10^{-14}\} \\ &= -0.37 + 1.32 \\ &= 0.95 \end{aligned}$$

The oxidation reaction is therefore highly favorable. The above-specified condition is reasonable, however, only under strongly alkaline conditions so that the Fe^{2+} concentration needs not to be unreasonably high.

6.3. Example 3

In the case of wet air oxidation application to waste biosolids treatment, a major source of energy consumption lies in the supply of compressed air to support the desired oxidation processes. On the other hand, an inherent chemical energy associated with the oxidation reactions is released during the treatment process. Based on a purely thermodynamic analysis, will there be a net energy generation or consumption associated with wet oxidation?

Solution

As summarized in Table 4, the energy derived from oxidation of organic waste materials may be expressed in terms of calorific equivalence for the air consumed. For the majority of types of waste normally treated in high-temperature wet oxidation systems, an average of 0.9 kwh energy equivalence is released for each kilogram of air consumed. The work required to deliver 1 kg of air at the reactor pressure of the designed wet-oxidation system may be calculated by assuming adiabatic compression of air from atmospheric pressure to the desired process pressure. The applicable equation of state relating the pressure-volume relationship of dry air may be taken as,

$$PV^r = K \qquad (\text{where } r = 1.4)$$

from the work relationship,

$$W = -\int_{V_1}^{V_2} PdV$$

Substituting the value of P from the equation of state and integrating, one obtains the work required for adiabatic compression as,

$$W = \frac{P_2V_2 - P_1V_1}{r-1}$$

One kilogram of air under ambient conditions may be assumed to have a volume of 0.8 m^3 with an atmospheric pressure of 101,200 N/m^2. If we assume a wet-oxidation reactor pressure of 2,000 psi (1.38×10^7 N/m^2), 1 kg of air would be expected to compress adiabatically to a volume of 0.024 m^3. The energy consumed for the compression of 1 kg air is then calculated to be

$$W = \left(1.38\times10^7\right)(0.024) - (101{,}200)(0.8)/1.4-1$$

$$W = 331{,}200 - 80{,}960/0.4$$

$$W = 625{,}600\,\text{Nm or J}$$

$$W = 625{,}600/3.6\times10^6\,\text{kwh}$$

$$W = 0.17\,\text{kwh}$$

This 0.17 kwh of required energy is only 19% (0.17/0.9 × 100) of the chemical energy available from the oxidation process. It is to be realized, however, that the above computation is based strictly on the theoretical energy and work relationships associated with the pertinent chemical and physical processes. Energy recovery, transfer and conversion losses are inevitable and will often make the energy recovery from wet oxidation only marginally profitable.

6.4. *Example 4*

Complex and refractory organic waste materials are effectively treated by chemical oxidation. However, it is often not possible nor is it necessary to convert the pollutant species to the ultimate oxidation products of carbon dioxide and water. Take, for example, phenol as the pollutant and potassium permanganate as the oxidation agent. Illustrate the likely course of oxidation and calculate the required dosage of permanganate reagent required for an industrial waste source of 10,000 gpd with a phenol content of 100 mg/L.

Solution

The oxidation of phenol is known to take place by a stepwise mechanism. Intermediate products, such as catecol, *o*-qinone, muconic acid, maleic acid, and fumaric acid, are often detected. Under the most commonly employed oxidation conditions, however, oxalic acid may be expected as a major final product. The stoichiometry of permanganate oxidation may thus be represented as follows:

$$8KMnO_4 + C_6H_5OH \leftrightarrow 8MnO_2 + 2K_2C_2O_4 + 2K_2CO_3 + 3H_2O$$

It is assumed that for each mole of phenol oxidized, two moles each of oxalic acid and carbon dioxide are produced. Under neutral or alkaline conditions, the above products will remain in solution as the potassium salts of oxalate and carbonate. The equation is balanced by considering that, for the six-carbon atoms of the phenol molecule, the oxidation states are −2/3 and for the carbon atoms of oxalic acid and carbonate the oxidation states are, respectively, +III and +IV. Thus, for each molecule of phenol oxidized a loss of 24 electrons is involved. In the conversion of permanganate to manganese dioxide, there is a gain of three electrons for each of the manganese atoms.

On the basis of the reaction stoichiometry, 8 moles of potassium permanganate would be required for the removal of 1 mole of phenol. The molecular weight of potassium permanganate is 158 and that of phenol is 94. Therefore, 8 × 158/94 = 13.5 g of the permanganate reagent will be required for the oxidation of 1 g phenol.

The required dosage will therefore be

$$10{,}000\,\text{gpd} \times 8.34\,\text{lb/gal} \times 100 \times 13.5 \times 10^{-6} = 113\,\text{lb/d}$$

It should be recognized that the computed dose requirement is highly dependent on the stoichiometry assumed. In practice, batch tests are advisable to determine the optimum reagent dosage.

NOMENCLATURE

A	surface area, m^2 (ft^2)
C	solute concentration, mg/L
Cs	the saturated solution concentration, mg/L
E	cell potential, volt
E^o	the standard cell potential for reaction, volt
ε	the turbulent diffusion coefficient, cm^2/s
F	the Faraday's constant = 96,500 coulombs
ΔF	free energy, cal/mole
ΔF^o	the standard free energy, cal/mole
K	the equilibrium constant
k	specific rate constant

k_1 the oxygen transfer coefficient, $1/cm^2 \cdot s$

n the number of electrons transferred in a chemical reaction

Q the reactant quotient consisting of a ratio of product concentration terms to reactant concentration terms with each raised to the power of appropriate stoichiometric ratios.

R the gas constant = 1.98 cal/mole °K

ΣS the net oxygen concentration change due to a combination of oxygen sources and sinks, mg/L

T the absolute temperature, °K

t time, s or hr

U the linear flow velocity, m/s (ft/s)

x the distance along the direction of flow, m·(ft)

REFERENCES

1. W. Stumm and G. F. Lee, *Ind. Eng. Chem.* **53**, 143 (1961).
2. W. K. Oldham and E. F. Gloyna, *Journal AWWA* **61**, 610 (1969).
3. A. Sadana and J. R. Katzer, *Ind. Eng. Chem.* **13**, 127 (1974).
4. E. T. Enisor and D. I. Metelitsa, *Russ. Chem. Revs.* **37**, 656 (1968).
5. V. Prather, *Journal WPCF* **42**, 596 (1970).
6. T. A. Turney, *Oxidation Mechanisms*, Butterworth, Washington, DC, 1965.
7. H. R. Eisenhauer, *Proc. Int. Conf. Water for Peace*, p. 163.
8. W. H. Kibbble, C. W. Raleigh, and J. A. Shepherd, *Industrial Waste*, Nov./Dec., 41 (1972).
9. DuPont, E. I. de Nemououors & Co., *Kastone Peroxygen Compound.*
10. G. H. Teletzke, *Chem. Eng. Progress*, **60**, 33 (1964).
11. L. G. Rich, *Units Processes of Sanitary Engineering*, John Wiley & Sons, New York, 1963.
12. J. E. Morgan, *TAPPI Non-wood Plant Fiber Conference*, 1973.
13. E. Hurwitz, G. H. Teletzke, and W. B. Gitchel, *Water and Sewage Works*, 298 (1965).
14. US Environmental Protection Agency, *Process Design Manual for Sludge Treatment and Disposal.* EPA 625/1-79-011, Municipal Environmental Research Laboratory, Cincinnati, OH, 1979.
15. R. B. Ely, *Poll. Eng.* **5**, 37 (1973).
16. W. B. Gitchel, J. A. Meidl, and W. Burant, Jr., *Chem. Eng. Progress.* **71**, 90 (1975).
17. Anon., *Environ. Sci. Tech.* **9**, 300 (1975).
18. R. Stewart, *Oxidation Mechanisms*, W. A. Benjamin, Inc., New York, 1964.
19. J. W. Ladbury and C. F. Gullies, *Chem. Revs.* **58**, 403 (1958).
20. S. B. Humphrey and M. A. Eikleberry, *Water and Sewage Works* (1962).
21. Carus Chemical Co., Inc., *The Cairox Method.*
22. H. S. Posselt and A. H. Reidies, *I & EC Prod. Res. Dev.* **4**, 48 (1965).
23. US Environmental Protection Agency, *Micro-straining and Disinfections of Combined Sewer Overflows—Phase 1* Report. EPA-11023 EVO 06/70. US EPA, Washington, DC, 1970.
24. T. J. Kanzelmeyer and C. D. Adams, *Water Environ. Res.* **68**, 222–228 (1996).
25. Y. Ku and W. Wang, *Water Environ. Res.* **71**, 18–22 (1999).
26. B. E. Reed, M. R. Matsumoto, R. Viadreo, Jr., R. L. Sega, Jr., R. Vaughan, and D. Mascioia, *Water Environ. Res.* **71**, 584–618 (1999).
27. R. Gracia, J. Aragues, and J. Ovelleiro, *Water Res. (G. B.)* **32**, 57 (1998).
28. J. L. Acero and U. V. Gunten, *Journal of AWWA* October, 90–100 (2001).
29. S. Liang, R. S. Yates, D. V. Davis, S. J. Pastor, L. S. Palencia, and J. M. Jeanne-Marie Bruno, *Journal of AWWA*, June, 110–120 (2001).
30. I. Wojtenko, M. K. Stinson, and R. Field, *Crit. Rev. Env. Sci. Tec.* **31**, 295–309 (2001).

31. Y. C. Hsu, J. T. Chenm, H. C. Yang, J. H. Chen, and C. F. Fang, *Water Environ. Res.* **73**, 494–503 (2001).
32. F. J. Beltran, M. Gonzalez, F. J. Rivas, and B. Acedo, *Water Environ. Res.* **72**, 659–697 (2000).
33. M. A. Parmelee *Journal AWWA*, September, 56 (2001).
34. T. Viaraghaven and R. Sapach, *Proc. 30th Mid-Atl. Ind Waste Conference.*, Technonic Publishing, Lancaster, PA, 1998, p. 775.
35. R. Andreozzi, V. Caprio, A. Insola, and R. Marotta, *Water Res. (G. B.)* **34**, 463–472 (2000).
36. H.S. Wang, S.T. Hsieh, and C.S. Hong, *Water Res. (G. B.)* **34**, 3882–3887 (2000).
37. W. Glazz and D. Maddox, *Water Res. (G. B.)* **32**, 997 (1998).
38. P. Bose, W. Glazz and D. Maddox, *Water Res. (G. B.)* **32**, 1005 (1998).
39. L. Lei, X. Hu and P. Yue, *Water Res. (G. B.)* **32**, 2753 (1998).
40. L. Lecheng, G. Chen, X. Hu, and P.L. Yue, *Water Environ. Res.* **72**, 147–151 (2000).
41. A. Thomesen, *Water Res. (G. B.)* **32**, 36 (1998).
42. A. Thomesen and H. Kilen, *Water Res. (G. B.)* **32**, 3353 (1998).
43. W. Julie, *New and Innovative Technologies for Mixed Waste Treatment*, University of Michigan, School of Natural Resources and Environment for EPA office of Solid Waste Permits and State Programs Division, 1997, U-915074-01-0.
44. K. Lin, P. H. Wang, and M. Li, *Chemisophere (G. B.)* **36**, 2075 (1998).
45. L. Philip L. Iyengar and C. Venkobachar, *J. Environ. Eng.-ASCE* **124**, 1165 (1998).
46. D. Van Scherpenzeel, M. Boon, C. Ras, G. Hansford, and J. Heijnen, *Biotechnology Progress* **14**, 425 (1998).
47. K. Lampron, X. Chad, and P. Chiu, *Proceeding, 30th Mid-Atlantic, Industrial waste Conference.* Technomic Publishing, Lancaster, PA. 1998, p. 448.

8
Halogenation and Disinfection

Lawrence K. Wang, Pao-Chiang Yuan, and Yung-Tse Hung

Contents

1. INTRODUCTION

Fluorine, chlorine, bromine, and iodine are the elements of the halogen family. Fluorine, in its oxidizing form, has no practical value in water or wastewater treatment systems. Of the other three, chlorine is by far the commonly used and thus will receive most of attention in this chapter.

A good history of the discovery, development, and methods of manufacture of chlorine are given by White (1). Chlorine is collected as a gas during production, then purified, liquefied, and transported as a liquid under pressure. It is usually converted back into the gaseous state prior to introduction into the water system. Because of its usage as both a gas and a liquid, the physical and chemical properties of both states are included herein. The properties of chlorine gas is given in Table 1. The properties of both liquid chlorine and liquid bromine are given in Table 2. At ordinary temperatures bromine vaporizes to yield a toxic and irritating gas; however, its usage in water or wastewater is as a liquid; therefore, its properties as a gas are not of particular interest herein.

From: *Handbook of Environmental Engineering, Volume 3: Physicochemical Treatment Processes*
Edited by: L. K. Wang, Y.-T. Hung, and N. K. Shammas © The Humana Press Inc., Totowa, NJ

Table 1
Properties of Chlorine Gas

Symbol $C1_2$, Atomic weight 35.457, Atomic number 17

Isotopes 33, 34, 35, 36, 37, 38, 39

Density at 0°C (32°F) and one atmosphere
3.214 kg/m^3 (0.2006 lb/ft^3)

Liquifaction temperature at one atmosphere
–34.5°C (–30.1°F)

Specific heat at constant pressure of one atmosphere and 15°C (59°F)
0.115 kg-cal/kg/°C
(0.115 Btu/lb/°F)

Thermal conductivity at 0°C (32°F)
0.06 kg-cal/h/m^2/°C/m
(0.0042 Btu/h/ft^2/°F/ft)

Solubility in water at 20°C (68°F)
7. 291 kg/m^3
(60.84 lb/1000 gal)

Iodine, in its elemental form, is a solid at ordinary temperatures. It sublimates readily; however, its use in water or wastewater is limited to the solid crystalline form (or as an iodide oxidized to iodine by chlorine), therefore; its properties as a solid only are of interest. These properties are given in Table 3.

Chlorine dioxide is a gas at ordinary temperatures and pressures. This gas is greenish-yellow and more irritating and toxic than chlorine. Its density is 2.4 that of air. The gas can be compressed to a liquid with a boiling point of 11°C and a melting point of –59°C. The gas is about five times as soluble in water as chlorine. The gas is very explosive

Table 2
Properties of Liquid Chlorine and Liquid Bromine

	Chlorine	Bromine
Symbol	$C1_2$	Br_2
Atomic weight	35.457	79. 916
Atomic number	17	35
Specific gravity	1.41 (20°C) (68°F)	3.12
Boiling point	–34.5°C (–30.1°F)	58.78°C
Freezing point	–100.98°C (–149.76°F)	–7.3°C (18.9°F)
Latent heat of vaporization	68.77 kg-cal/kg (123.8 Btu/lb) at –34.1°C (–2 9.3°F)	
Heat of fusion	22.89 kg-cal/kg (41.2 Btu/lb) at –101.5°C (–150.7°F)	
Critical temperature	144°C (291.2°F)	
Critical pressure	786.319 kg/m^2 (1118.36 psia)	
Critical density	573 g/L (35.77 lb/ft^3)	

Table 3
Properties of Solid Iodine

Symbol I_2,
Atomic weight 126.92,
Atomic number 53
Specific gravity 4.93,
Melting point 114°C (237.2°F)
Boiling point 184°C (363°F)

and is thus not transported as a gas but transported as a solution. In most instances, it is generated at the site of usage.

Halogenation (most particularly chlorination) is one of the most common and certainly one of the most valuable processes used in environmental pollution control. The most important process is that of disinfection of both potable water supplies and wastewater effluents. Because of the necessary reuse of water—the river or lake receives wastewater effluent and downstream or across the lake, the same water (diluted usually) becomes the source of a potable water supply—special care must be taken to stop the transmission of disease from upstream residents to those downstream. Both wastewater treatment and the potable water treatment processes reduce the number of pathogenic organisms; however, the final barrier to disease transmission is disinfection by chemical means and usually this is by halogenations and particularly chlorination.

The ability of the halogens to kill organisms is utilized for algae control in reservoirs, swimming pools, sedimentation tanks, circulating water-cooling systems, etc. Also halogens are used in household-type cleansing of such things as milk and other food handling equipment, dishwashing, toilet and locker scrubbing, clothes washing, deodorizing space, etc.

Because of their high oxidizing power and because of reactivity as a substitution product, particularly on nitrogenous matter, the halogens are commonly used for many processes other than disinfection. The oxidation characteristic is utilized for conversion of the relatively soluble reduced (ferrous or manganous) forms to the very insoluble oxidized (ferric and magnetic) forms of iron and manganese. The two elements are commonly found in water and unless removed may cause problems of staining and precipitate formation at the customers' sink. Also, halogens may be used to oxidize hydrogen sulfide, H_2S (the gas in water giving the rotten-egg odor), to a more highly oxidized sulfur form to remove the odor problem. Similarly, halogens are used to convert under special conditions, the exceedingly toxic cyanide ion (CN^-) to the relatively non-toxic cyanate form (CNO^-). Cyanide compounds are commonly used in the plating industry.

Chlorine reacts readily with nitrogenous compounds, particularly ammonia, to form substitution products known as chloramines. These chloramines may in turn be oxidized by additional chlorine to yield as end products nitrogen gas or nitrogen oxides. This removes the nitrogen from any further reactions thus, in effect, removing nitrogen from the water. This may be desirable if the nitrogen in the form of ammonia causes a toxicity problem in receiving water, or, if in the form of nitrates, causes excess algae growth.

The halogens, versatile or useful as they may be, are not without their shortcomings and even pitfalls. All of the compounds may be exceedingly dangerous to handle and special care must be taken to protect handlers and the communities in the vicinity of the manufacturers and along the routes of transportation. Techniques for proper handling have been developed and accident rates are very low.

Chlorine compounds are, even in low concentration, toxic to many forms of aquatic life. Fish living in water with concentrations of chlorine commonly recommended as potable water supply residuals soon die. Other aquatic organisms, both plant and animal, may have similar responses. The toxic effects vary, of course, with concentration, temperature, period of exposure, and type of organism. The problem is serious in water receiving chlorinated treatment plant effluent. Limits of permissible chlorine concentration in the water and thus in the effluents are being established. This is in conflict with the stated need of disinfection of sewage treatment plant effluent. Resolution of the conflict is not yet accomplished.

Chlorinated hydrocarbons have been found in water sources in sufficient concentration to cause some concern and alarm. It is not certain if these compounds got into the water as the compound or were formed as the result of chlorination of wastewater or of potable water. The evidence seems to show that the compounds were not formed as a result of disinfection practices. At this time the need for disinfection seems to outweigh the possible concern of toxic compounds formed by the disinfection processes.

Also, of course, the halogens do cause an odor in the water. At very low concentrations, the odor may be imperceptible and, also, the consumer seems to learn to ignore the odor. However, under certain conditions such as with phenolic-type compounds in the water, the addition of chlorine may result in a water so odorous that it is not palatable. In such cases special techniques of disinfection must be employed. On the other hand, halogens in water may reduce or even eliminate taste and odor problems. The operator may adjust the location of feeding the halogen and the dosage and also may use such other chemical additives as ammonia or sulfur dioxide to remove odor problems. No rules of procedure for eliminating odors in water using halogens have been established—success, if possible, is dependent on the ingenuity of the plant operator and his staff.

2. CHEMISTRY OF HALOGENATION

The halogens as a group are powerful oxidizing agents with decreasing oxidation power with increasing atomic weights. It must be emphasized that neither oxidation potential nor free energy reflect the disinfecting properties of these compounds.

2.1. Chlorine Hydrolysis

Chlorine when dissolved reacts rapidly with water to form hypochlorous acid and hydrochloric acid in accordance with Eq. (1):

$$Cl_2 + H_2O \Leftrightarrow HOCl + HCl \tag{1}$$

The equilibrium constant of this hydrolysis reaction is expressed as,

$$K_H = (H^+)(Cl^-)(HOCl)/Cl_2 \tag{2}$$

The value for K_H is 3×10^4 at 15°C and ranges from 1.5×10^4 at 0°C to about 4.4×10^4 at 25°C. It may be shown that at pH values above about 3 and total chlorine concentrations less than 1000 mg/L, there is a very little molecular chlorine.

2.2. *Chlorine Dissociation*

The important product of the reaction, Eq. (1), is hypochlorous acid—a weak acid that dissociates according to Eq. (3):

$$\mathrm{HOCl} \Leftrightarrow \mathrm{H} + \mathrm{OCl}^- \tag{3}$$

to form the hypochlorite ion. The dissociation or acidity constant, K_a, is

$$K_a = (\mathrm{H}^+)(\mathrm{OCl}^-)/(\mathrm{HOCl}) = 2.5 \times 10^{-8} \text{ at } 20°\mathrm{C} \tag{4}$$

The value of K_a ranges from 1.5×10^{-8} at 0°C to 2.7×10^{-8} at 25°C.The HOCl is a better disinfectant than the OCl^- ion; therefore, the distribution between the two is of major interest. The percentage of HOC1 at any pH can be calculated from Eq. (4) and rearranged to yield Eq. (5):

$$\%(\mathrm{HOCl}) = 100/\left[K_a(\mathrm{H}^+) + 1\right] \tag{5}$$

Hypochlorite salts such as NaOCl dissociate completely in water:

$$\mathrm{NaOCl} \Leftrightarrow \mathrm{Na}^+ + \mathrm{OCl}^- \tag{6}$$

The OCl^- must then come into equilibrium with the HOCl molecule in accordance with Eq. (4).

It is common practice to designate the HOCl molecule and the OCl^- ion as free available chlorine (FAC), and chlorine atoms in the oxidizing state and combined with nitrogenous matter are referred to as combined available chlorine (CAC).

2.3. *Chlorine Reactions with Nitrogenous Matter*

Nitrogen in water may be classified as inorganic or organic. The former consists of ammonia, nitrite, and nitrate. The organic nitrogen may be classified as amino acids or proteins. The reactions of chlorine with the inorganic nitrogen species is fairly well defined; however, with the organic nitrogen such definitions are not yet available.

2.3.1. *Inorganic Nitrogen*

Ammonia may exist in water as the molecule or as the ammonium ion. The hydrolysis reaction for the ion is

$$\mathrm{NH}_4^+ + \mathrm{H}_2\mathrm{O} \Leftrightarrow \mathrm{NH}_3 + \mathrm{H}^+ \tag{7}$$

The equilibrium constant for the reaction is

$$K_a = (\mathrm{H}^+)(\mathrm{NH}_3)/\mathrm{NH}_4^+ = 5.0 \times 10^{-10} \text{ at } 20°\mathrm{C} \tag{8}$$

The ammonium ion predominates at the pH values ordinarily found in water.

Chlorine reacts with the ammonium ion or ammonia in a stepwise manner; the reactions between molecules are illustrated by Eqs. (9)–(11):

$$\underset{\text{monochloramine}}{\mathrm{NH}_3 + \mathrm{HOCl} \leftrightarrow \mathrm{NH}_2\mathrm{Cl} + \mathrm{H}_2\mathrm{O}} \tag{9}$$

$$NH_2Cl + HOCl \leftrightarrow \underset{\text{dichloramine}}{NHCl_2} + H_2O \tag{10}$$

$$NHCl_2 + HOCl \leftrightarrow \underset{\substack{\text{trichloramine}\\ \text{or nitrogen}\\ \text{trichloride}}}{NCl_3} + H_2O \tag{11}$$

There may also be reactions between ions,

$$NH_4^+ + OCl^- \leftrightarrow NH_2Cl + H_2O \tag{12}$$

or reactions between molecules and ions,

$$NH_3 + OCl^- \leftrightarrow NH_2Cl + OH^- \tag{13}$$

$$NH_4^+ + HOCl \leftrightarrow NH_2Cl + H^+ \tag{14}$$

The nitrogen trichloride formed is unstable at normal water pH levels; however, when present in recently chlorinated water such as in swimming pool systems, the nitrogen trichloride may be an eye irritant.

The distribution between monochloromine and dichloramine is hypothesized to occur as a result of an equilibrium resulting from Eq. (15), according to Moore (2):

$$2NH_2Cl + H^+ \leftrightarrow NH_4^+ + NHCl_2 \tag{15}$$

for which

$$K = (NH_4)^+(NHCl_2)/[(H^+)(HH_2Cl)] = 6.7 \times 10^5 \text{ at } 25°C \tag{16}$$

For a weight ratio of chlorine to ammonia nitrogen of 5 to 1, the distribution of monochloromine to dichloromine is presented in Table 4, [Weber (3)]. Results of other investigators are given by White (1). The other results are similar to those given in Table 4.

Other research has shown that the equilibrium described by Eq. (16) may be insufficient to describe the distribution. Morris (2) and Fair et al. (4) concluded that the distribution of chloromine is largely dependent on the rates of their formation. At a pH greater than 6 and a molar ratio of chlorine to ammonia of less than unity, the formation of monochloramine follows the second-order reaction

$$d(NH_2Cl)/dt = K(NH_3)(HOCl) \tag{17}$$

A similar kinetic equation for the formation of dichloramine is not available. It is known, of course, that dichloramine is formed when the molar ratios of chlorine to ammonia exceeds unity. There is a further consideration in the chloramine distribution, that is, stability. Granstrom (5) showed that after formation monochloromine disproportionates to form dichloromine. This takes place by two parallel reactions. The first is apparently first-order resulting from the hydrolysis of monochloromine to form hypochlorous acid:

$$NH_2Cl + H_2O \leftrightarrow HOCl + NH_3 \tag{18}$$

Table 4
Chloramine Species Distribution as a Function of pH

pH	NH_2Cl (%)	$NHCl_2$ (%)
5	16	84
6	38	62
7	65	35
8	85	15
9	94	6

This hypochlorous acid in turn reacts with another monochloromine to form dichloromine, Eq. (10). There is also a second-order reaction, which appears to be acid-catalyzed, and is, therefore, pH- and buffer-dependent:

$$NH_2Cl + Acid \leftrightarrow NH_2Cl \cdot Acid \tag{19}$$

$$2NH_2Cl \cdot Acid \leftrightarrow NHCl_2 + NH_3 + 2Acid \tag{20}$$

This disproportionation is of importance because it permits the hypothesis that the break-point reactions (described below) may take place through the decomposition of dichloramine, Morris et al. (6).

The reaction of free chlorine with the nitrite ion,

$$NO_2^- + HOCl \leftrightarrow NO_3^- + H^+ + Cl^- \tag{21}$$

is a rapid reaction. Nitrites will not react with chloramines in the pH range of 6–9 [Hulbert (7)]. Figure 1 shows the proportions of mono- and dichloramines in water with equimolar concentrations of chlorine and ammonia (81).

2.3.2. *Organic Nitrogen*

Standard Methods (8) gives two methods for determination of organic nitrogen. The one is for albuminoid nitrogen, which is roughly equivalent to the nitrogen in the amino acids. The other is for total organic nitrogen, which includes the nitrogen in the proteinaceous matter as well as the amino acids. The reactions with amino acids may be illustrated by Eq. (22):

$$R\text{-}NH_2\text{-}COOH + HOCl \leftrightarrow RNHCl\text{-}COOH + H_2O \tag{22}$$

resulting in the formation of chlorinated amino acids. Borchardt (9) has shown that only glycine of the amino acids is further oxidized by HOCl to exhibit a similarity to ammonia in the break-point reaction. The other chlorinated amino acids remain as formed. As discussed below, these compounds may be considered in the general category of "toxic."

The reactions of chlorine with organic nitrogen compounds more complex than the amino acids—in decreasing order of molecular size and complexity: proteins, proteoses, peptones, polypoptides, dipeptides, and alpha amino acids—are not well known. The chlorine molecule (HOCl) will react with the amino groups on the organic nitrogen; however, many of the reactions are apparently slow. Taras (10) found that the total organic nitrogen consumed after chlorination follows a reasonably well-defined pattern. The ammonia—chlorine reaction is rapid and with molecular excess of chlorine the

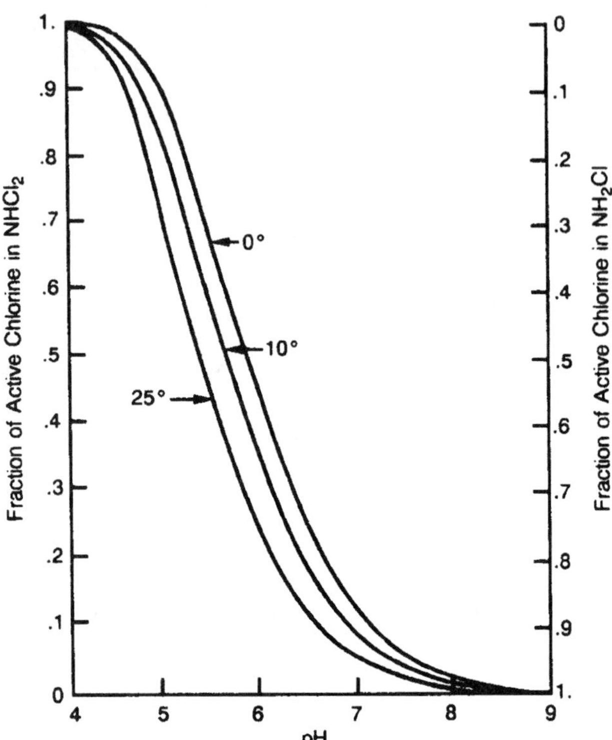

Fig. 1. Proportions of monoamines and dichloramines in water with equimolar concentrations of chlorine and ammonia (US EPA & NAS).

chloromines thus are found to disappear within 1 h. The amino acid–chlorine reaction is also rapid; however, the rate of disappearance with an excess of chlorine is usually slow. This was later explained by Borchardt (9). Chlorine–nitrogenous compounds formed from chlorine- and nitrogen-containing organic compounds more complex than amino acids did not exhibit break-point-type behavior:

$$NHCl_2 \leftrightarrow H^+ + NCl_2^- \tag{23}$$

In the presence of the hydroxyl ion (OH^-)

$$NCl_2^- \leftrightarrow NCl + Cl^- \tag{24}$$

$$NCl + OH^- \leftrightarrow NOH + Cl^- \tag{25}$$

or

$$(NCl_2)^- + (OH)^- \leftrightarrow NCl(OH)^- + Cl^- \tag{26}$$

$$NCl(OH)^- \leftrightarrow NOH + Cl^- \tag{27}$$

The nitroxyl radical NOH may decompose by one of three possible paths. The first is the dimerization,

$$2NOH \leftrightarrow H_2N_2O_2 \tag{28}$$

and the hyponitrous acids decomposes to

$$H_2N_2O_2 \leftrightarrow N_2O + H_2O \tag{29}$$

or if it is assumed that the end product of the breakpoint reaction is nitrogen gas.

$$NHCl_2 + H_2O \leftrightarrow NOH + 2H^+ + 2Cl^- \tag{30}$$

and

$$NOH + NH_2Cl \leftrightarrow N_2 + H_2O + H^+ + Cl^- \tag{31}$$

or

$$NOH + NHCl_2 \leftrightarrow N_2 + HOCl + H^+ + Cl^- \tag{32}$$

and

$$NOH + H_2O \leftrightarrow NO_3^- + 4H^+ + Cl^- \tag{33}$$

A third possible break-point reaction might be

$$H_2N_2O_2 + HOCl \leftrightarrow 2NO + H_2O + H^+ + Cl^- \tag{34}$$

It thus appears as though the ammonia–chlorine break-point reaction is complex and not clearly understood. Still less understood is the glycine–chlorine reaction system.

Even though in water or wastewater it is usual for most of the nitrogenous matter. To be in the form of ammonia, a considerable amount may be in more complex forms. Thus, a more typical reappoint curve for natural conditions would not show the pattern illustrated in Fig. 1, but would appear more like Fig. 2. There may be a significant demand for chlorine by the organic carbon in these or other compounds. This demand results in the reduction of chloritie to the chloride form. In fact, the break point might not be apparent. The shape of Fig. 2 might change significantly with time. Thus, the plateau shown on Fig. 2 denotes a minimum residual of both free and combined chlorine and beyond the plateau the proportion of free chlorine would, of course, increase. As shown below, it may be very desirable, if not necessary, for good oxidation and disinfection to have free available chlorine in a public water supply.

2.4. Chlorine Reactions with Other Inorganics

Chlorine reacts with inorganic carbon similar to its reaction with organic carbon (98,99),

$$C + 2Cl_2 + 2H_2O \leftrightarrow 4H^+ + 4Cl^- + CO_2 \tag{35}$$

The H^+ released consumes approximately 2.1 mg/L of alkalinity to each mg/L of chlorine consumed. At pH levels above approx 8.5, chlorine converts cyanide to cyanate,

$$Cl_2 + 2OH^- + CN^- \leftrightarrow CNO^- + 2Cl^- + H_2O \tag{36}$$

Complete destruction of cyanide by chlorine is usually carried out at pH values of 8.5 to 9.5 to form nitrogen gas:

$$5Cl_2 + 10OH^- + 2CN^- \leftrightarrow 2HCO_3^- + 10Cl^- + N_2 + 4H_2O \tag{37}$$

Hydrogen sulfide is fairly common in groundwater and may be evolved in sewers. Depending on the pH, the reactions of chlorine may be to form sulfate or elemental sulfur. The reaction forming the sulfate is

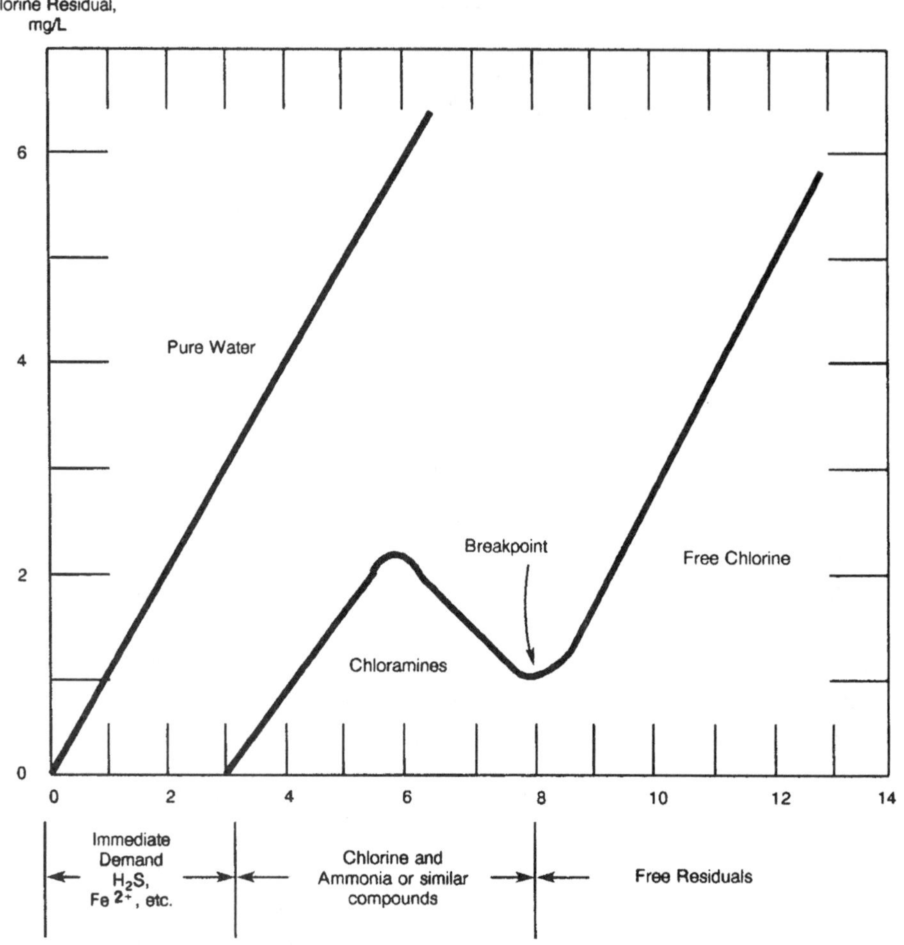

Fig. 2. Graphical representation of the break-point chlorination reaction. The straight line at the left shows that chlorine residual is proportional to chlorine dosage in pure water. When impurities are present, they exert a chlorine demand (US EPA).

$$H_2S + 4Cl_2 + 4H_2O \leftrightarrow 10H^+ + SO_4^{2-} + 8Cl^- \tag{38}$$

The reaction forming the sulfur is

$$H_2S + Cl_2 \leftrightarrow S + 2H^+ + 2Cl^- \tag{39}$$

with a substantial excess of chlorine at pH levels below 6.4 all sulfide is converted to sulfate, at pH of 7 about 70% is converted to sulfate and 30% to sulfur and at pH from 9 to 10 the conversion is about 50–50.

To remove ferrous ions from water and/or to convert ferrous to ferric ion for use as a coagulant, the reaction of chlorine with iron is illustrated:

$$2Fe^{2+} + Cl_2 + 6HCO_3^- \leftrightarrow 2Fe(OH)_3 + 2Cl^- + 6CO_2 \tag{40}$$

This reaction is rapid and takes place over a wide pH range.

Manganese in the manganous form may be difficult to remove from a water supply. Chlorination with free available chlorine (combined available chlorine is ineffective) can be accomplished in the pH range of 7–10 with the higher pH levels more effective. The reaction is

$$Mn^{2+} + Cl_2 + 4OH^- \leftrightarrow MnO_2 + 2Cl^- + 2H_2O \quad (41)$$

The reaction may take 2–4 h for completion. If the manganese is in an organic complex, the reaction may be slower or even unpredictable.

Methane gas is oxidized by chlorine to form carbon tetrachloride:

$$CH_4 + 4Cl_2 \Leftrightarrow CCl_4 + 4H^+ + 4Cl^- \quad (42)$$

This is one of the compounds, that may be considered as toxic in a water supply.

2.5. *Chlorine Dioxide (ClO_2) Applications*

Because chlorine dioxide does not react with phenols, and in fact destroys preformed chlorophenols, it has been found useful as a water disinfectant in circumstances where phenols in water are a problem. It does not react with ammonia or amino acids in water to form chloramines and it is not affected by increases in pH. Its use in water supply is limited because of the high costs compared to chlorine.

2.6. *Chlorine Dioxide Generation*

There are three historical techniques for generating chlorine dioxide. However, some of these procedures can result in excess free chlorine being present. Free chlorine can oxidize chlorine dioxide to form chlorate ions, which are difficult to remove from solution. Consequently, the current recommended approach to chlorine dioxide generation is to maximize its yield while minimizing the presence of free chlorine (thus minimizing the formation of chlorate ion). For water disinfection, chlorine dioxide can be generated using several reaction schemes, such as the reaction of aqueous hypochlorous acid with dissolved chlorite ion:

$$2NaClO_2 + HOCl \rightarrow NaCl + NaOH + 2ClO_2 \quad (43)$$

Chlorine dioxide also can be generated by the reaction of solid sodium chlorite in solution with mineral acid, with chlorine, or with hypochlorous acid. The reaction for chlorine and/or hypochlorous acid with chlorite ion is:

$$2ClO_2{}^- + Cl_2\,(g) \rightarrow 2ClO_2\,(g) + 2Cl^- \quad (44a)$$

$$2ClO_2^- + HOCl \rightarrow 2ClO_2\,(g) + Cl^- + (OH)^- \quad (44b)$$

These reactions involve the formation of the unsymmetrical intermediate Cl_2O_2:

$$Cl_2 + ClO_2^- \rightarrow Cl_2O_2 + Cl^- \quad (45)$$

At high concentrations of both reactants, the intermediate is formed very rapidly. Elemental chlorine formed by Eq. (46a) is recycled by means of Eq. (45) . Thus, primarily chlorine dioxide is produced as a result:

$$2Cl_2O_2 \rightarrow 2ClO_2 + Cl_2 \quad (46a)$$

or

$$Cl_2O_2 + ClO_2 \rightarrow 2ClO_2 + Cl^- \tag{46b}$$

On the other hand, at low initial reactant concentrations, or in the presence of excess hypochlorous acid, primarily chlorate ion is formed in the following reactions:

$$2Cl_2O_2 + H_2O \rightarrow ClO_3^- + Cl^- + 2H^+ \tag{47}$$

and

$$Cl_2O_2 + HOCl \rightarrow ClO_3^- + Cl^- + H^+ \tag{48}$$

Therefore, high concentrations of excess chlorite ion favor the second-order reactions, Eqs. (46a) and (46b), and chlorine dioxide is formed. At low concentrations, the second order disproportionation process becomes unimportant, and Eqs. (47) and (48) produce chlorate ion rather than chlorine dioxide. The reasons for the production of chlorate ion are related to the presence of high concentrations of free chlorine and the rapid formation of the Cl_2O_2 intermediate, which, in turn, reacts with the excess hypochlorous acid to form the unwanted chlorate ion.
The stoichiometry of the undesirable reactions which forms chlorate ion is

$$ClO_2^- + HOCl \rightarrow ClO_3^- + Cl^- + H^+ \tag{49}$$

$$ClO_2^- + Cl_2 + H_2O \rightarrow ClO_3^- + 2Cl^- + 2H^+ \tag{50}$$

Accordingly, the most effective way to minimize chlorate ion formation is to avoid conditions that result in low reaction rates (e.g., high pH values and/or low initial reactant concentrations, and the presence of free hypochlorous acid). Clearly, the reaction forming chlorate ion (Eq. 6) will be more troublesome in dilute solutions. On the other hand, whenever treatment by chlorine dioxide (which forms chlorite ion in the process) is followed by the addition of free chlorine (HOCl with a pH of 5–8), the unwanted chlorate ion will also be formed.

About 70% of the chlorine dioxide added to drinking water is converted to chlorite ion. Therefore, 1.2–1.4 mg/L chlorine dioxide is the maximum practical dosage to meet the currently recommended maximum total oxidant residual of 1 mg/L. Nearly all of the chlorine dioxide ion added as a primary oxidant/disinfectant is converted to chlorite ion. Because of differences in the nature of water constituents that exert demands for chlorine dioxide, this ratio should be individually determined for each water supply.

A stabilized aqueous solution of chlorine dioxide is available. It is manufactured by International Dioxide Inc. of New York, NY, and marketed as a 5% solution with the name of Anthium Dioxide. It is sold in 52-gal drums and has a shelf life claimed to be 1 yr. Chlorination process equipment is commercially available (77).

2.7. *Chlorine Dioxide Reaction with Nitrogenous Matter*

Chlorine dioxide does not react with ammonia. However, it does react with other compounds. The complete oxidative potential of chlorine dioxide is 2.5 times that of chlorine:

$$ClO_2 + 5e^- \leftrightarrow Cl^- + 2O^{2\pm} \tag{51}$$

whereas

$$HOCl + 2e^- \leftrightarrow Cl^- + OH^- \quad (52)$$

However, the most common reaction product is chloride and the electron change is just one.

$$ClO_2 + e^- \leftrightarrow ClO_2^- \quad (53)$$

2.8. Chlorine Dioxide Reactions with Phenolic Compounds and Other Substances

In 1944, chlorine dioxide was used at the Niagara, NY water-treatment plant for the first recorded use in a water works (13). Other plants in the area soon adopted the use of chlorine dioxide: Woodlawn, Tonawanda, North Tonawanda, Lockport, and Port Colbourne all of New York. The problem at these plants was the development of strong chlorophenolic taste and odors when chlorine was added to the water which had been contaminated by industrial discharges of phenols. The results were encouraging and many other plants have adopted the use of chlorine dioxide (11).

The practice has evolved of prechlorinating the water and then adding the chlorine dioxide as a postdisinfectant. Thus, the chlorine demand is satisfied by the relatively cheap chlorine and any chlorophenols formed are destroyed by the relatively expensive chlorine dioxide. Chorine dioxide has been found useful in control of tastes and odors resulting from algae and from decaying vegetable matter (14–18).

Manganese in water supplies either as the manganous ion or as adsorbed or completed manganese on organic compounds may be difficult to remove by either oxidation to the manganic state as the destruction of the complex. Free available chlorine may accomplish the oxidation or destruction slowly even at elevated pH levels, i.e., greater than 10. The after precipitation of manganese in the distribution system may lead to black deposits, encrustation of pipes, and debris in the water.

Chlorine dioxide reacts more rapidly a chlorine with the manganese and may prove to be the chemical of choice. The postulated reaction is

$$2ClO_2 + Mn^{2+} + 4OH^- \leftrightarrow MnO_2 + 2ClO_2^- + 2H_2O \quad (54)$$

2.9. Bromine Hydrolysis

The use of bromine as a disinfectant has been limited to swimming pools. Its use then has some advantage in that bromine or its compounds are not as irritating to the eyes as is chlorine and that, unlike monochloromine, monobromamine is a good bactericide. Bromine is soluble in water to about 3.5%. It hydrolyzes to form hypobromous acid, The equilibrium constant is 5.8×10^{-9}:

$$Br_2 + H_2O \leftrightarrow HOBr + H^+ + Br^- \quad (55)$$

2.10. Bromine Dissociation

Similar to hypochlorous acid, the hypobromous acid dissociates, to form the hypobromite ion:

$$HOBr \leftrightarrow H^+ + OBr^- \quad (56)$$

The dissociation constant is 2×10^{-9} at 25°C.

2.11. *Bromine Reactions with Nitrogenous Matter*

Bromine reacts in water to form monobromamine and dibromamine. A stable tribromamine does not exist:

$$OBr^- + NH_3 \leftrightarrow NH_2Br + OH^- \tag{57}$$

$$OBr^- + NH_2Br \leftrightarrow NHBr_2 + OH^- \tag{58}$$

Monobromamine is almost as strong a bactericide as free bromine (19). This can be explained by the presence of monobromammonium ion formed in the reaction:

$$NH_2Br + H^+ \leftrightarrow NH_3Br^+ \tag{59}$$

for which the equilibrium constant is $K = 3.2 \times 10^{-7}$ at 25°C. The monobromammonium ion releases a positively charged bromine which has strong oxidizing properties:

$$NH_3Br^+ \leftrightarrow NH_3 + Br^+ \tag{60}$$

Because monobromamine is a strong bactericide, there is no need to attempt break-point bromination.

2.12. *Iodine Hydrolysis*

There are several iodine species, which must be considered simultaneously: elemental iodine I_2, hypoiodus acid HIO, per iodide or tri-iodide I_3^-, and iodated ion, IO_3^-. The elemental iodine hydrolysis

$$I_2 + H_2O \leftrightarrow HIO + H^+ + I^- \tag{61}$$

for which the hydrolysis constant is 3×10^{-13} at 25°C. Both elemental iodine and hypoiodus acid are effective germicides.

2.13. *Iodine Dissociation*

Hypoiodus acid dissociates to form the hypoiodite ion:

$$HIO \leftrightarrow H^+ + IO^- \tag{62}$$

for which the dissociation constant is $K = 4.5 \times 10^{-13}$

The degree of dissociation is negligible, which is fortunate because the hypoiodite ion is an ineffective germicide. The distribution of the three forms, elemental iodine, hypoiodus acid, and hypoiodite ion, are a function of pH as shown in Table 5, according to Chang (20).

There is the possibility of the formation of the bactericidally ineffective triiodite ion, I_3^-,

$$I_2 + I^- \leftrightarrow I_3^- \tag{63}$$

Chang (20) reports that the amount of I_3^- formed can be ignored when iodine crystals, iodine tablets, or tincture of iodine is used. Furthermore, there is the possibility of the formation of the iodate ion $(IO_3)^-$:

$$3HIO + 3(OH)^- \leftrightarrow IO_3^- + 2I^- + 3H_2O \tag{64}$$

This is not likely to be a problem unless the pH rises above about 8.0. Also, this is fortunate because the iodate ion is apparently an ineffective germicide.

Table 5
Effect of pH on Iodine Hydrolysis[a]

pH	I_2 (%)	HIO (%)	IO^- (%)
5	99	1	0
6	90	10	0
7	52	48	0
8	12	88	0.005

[a]Note: Total iodine residual 0.5 mg/L.

2.14. *Iodine Reactions with Nitrogenous Matter*

Iodine does not react with ammonia to form iodamines; it oxidizes the ammonia. Thus, less iodine is required to provide a residual. Iodine is less likely than chlorine to form organic substitution products. Thus, for the same dosages it provides a greater oxidizing residual then does chlorine—this means that the disinfecting capability is longer lasting.

3. DISINFECTION WITH HALOGENS

A recent review of the problems of disinfection was presented by Morris (21). Increased reuse of water introduces potentially more resistant pathogens (viruses), which require free available chlorine for disinfection, and also introduces more ammonia and organic compounds, which make the attainment of a free chlorine residual more difficult. Furthermore, since Morris' paper, there has been established, because of toxicity to aquatic life, limitations on permissible chlorine residuals in streams and thus in sewage plant effluents. Thus, the processing of community- or industry-used water to provide a hygienic water supply is becoming more difficult.

Table 6 was prepared by Morris (21). From it, it is seen that if viruses are to be taken as potential pathogens, which they are, only free residual chlorination or ozone of those listed are acceptable disinfectants. Bromine iodine and chlorine dioxide could also have been included. Furthermore, Morris suggests that for virus inactivation, specific free chlorination of 0.5–2.0 mg/L could be required, depending on pH, temperature, and time of contact. This range was further reduced to a suggested value of about 1.0 mg/L free chlorine for 10–30 min.

The maintenance of about 1.0 mg/L of free chlorine for 10–30 min might provide a residual of free chlorine in the distribution system. A free-chlorine residual in the distribution system would provide limited protection against a massive contamination; however, such a residual would tend to prevent slime growths in the system. Also, with a monitoring program within the distribution system, the sudden disappearance of chlorine residual would indicate contamination.

3.1. *Modes and Rate of Killing in Disinfection Process*

It is postulated that chemical disinfection is accomplished by destruction of cell protein, particularly of the cell enzyme system. To gain access to the interior of the cell, the chemical must pass through the cell wall. It appears that a neutral molecule can pass more readily through the wall than a charged ion and is, consequently, a better

Table 6
Relative Germicidal Activities of Disinfecting Materials[a]

	Type of organism			
Germicide	Bacteria	Enteric cysts	Amoebic viruses	Bacterial spores
O_3	0.0001	1.0	0.1	0.20
HOCl as Cl_2	0.02	10	0.4	10
OCl^- as Cl_2	2	10^3	>20	$>10^3$
NH_2Cl as Cl_2	5	20	10^2	4×10^2
Free Cl, pH 7.5	0.04	20	0.8	20
Free Cl, pH 8	0.1	50	2	50

[a]Note: Numbers represent the concentration in mg/L required to kill or inactivate 99% of the organisms within 10 min at 5°C. The tests were conducted with clean water (11,56).

disinfectant. Hypochlorous acid (HOCl) is a much better disinfectant than hypochlorite ion $(OCl)^-$.

The disinfection processes are not instantaneous, and the die-away can be represented quantitatively by a rate equation. The one most commonly used is referred to as Chick's law, which is written

$$-dN/dt = KN \tag{65}$$

where $-dN/dt$ is the die-off rate; K is the rate constant and N is the number of organisms surviving to time t. Integration equation between limits of N_0 at $t = 0$ and N at $t = t$, yields

$$\text{In } N/N_0 = -Kt \tag{66}$$

$$N = N_0 e^{-K't} \tag{67}$$

Conversion of $\ln_e$ to $\log_{10}$

$$N = N_0 10^{-K't} \tag{68}$$

where K′ = 0.434K.

A semi-log plot of N vs t should yield a straight line with a slope K. It is quite common to observe that the disinfection process does not follow Chick's law. Either an autocatalytic-type phenomenon increases the rate with time or a retardant-type phenomenon decreases the rate. Several methods for evaluating the changes in rates are given by Weber (3) .

3.2. *Disinfection Conditions*

3.2.1. *Temperature*

The rates of disinfection are affected by conditions over which the engineer or operator have little control—these are temperature, pH, and organic matter. The effects of temperature are usually included in the evaluation of the reaction rate constant K. The Arrhenius equation is commonly used to describe the change in K with temperature in the temperature region of 20°C

$$K_1 = K_2 \beta^{T_1 - T_2} \tag{69}$$

where K_1 and K_2 are reaction rate constants at temperatures T_1, and T_2, β is an empirical constant.

3.2.2. pH

Most microorganisms are killed by either an acid condition, pH< 3, or an alkaline condition, pH > 11. In most situations the pH of the solution being disinfected lies in an intermediate range of pH, i.e., between 4 and 10. The range of pH reflects a million-fold range in hydrogen ion concentration. Changes in pH change the state of ionization of the amino acids making up the protein of the cell structure and thus alter the response of the cell to the disinfectant. Also, as noted elsewhere in this section, the pH affects the molecular–ionic relationship of the disinfectants themselves. The overwhelming effect of pH on disinfection, within normal ranges of pH, appears to be the effect on the disinfectant. With pH, as with most studies of disinfection, quantification of the effect of environmental conditions on a rational or analytical basis is not possible. Empirical studies are necessary.

3.2.3. Organic Matter

Water or wastewater being disinfected usually contains some organic matter, which usually reduce the effectiveness of the disinfection process. Organisms may adhere to a piece of organic matter, perhaps even to the extent of "clumping"; this may protect the inner organisms from the disinfectant. This may be particularly serious in the case of viruses. Also, the organic matter may react very rapidly with the disinfectant, effectively removing it from the solution.

3.3. *Disinfection Control with Biological Tests*

Disinfection control is either with biological tests or by measurement of disinfectant residuals. Ozone, which is an excellent disinfectant, normally shows no residual a few minutes after the application. Ozone is discussed in chapter 9.

The numbers of identifiable pathogenic organisms in a water system is normally small. The concentration of these small numbers to a reasonable density for detection and also the methodologies for laboratory detection and enumeration are normally beyond the capabilities of most water or wastewater treatment facility or stream survey programs. Thus, an indicator organism procedure was adopted long ago and has not been changed in the 1975 the US Environmental Protection Agency (US EPA), Interim Primary Drinking Water Standards (22). The indicator organisms are those, in the category of coliform. These include not only organisms from the intestinal tract of warm blooded animals including humans (primarily *E. coil*), but also soil organisms (primarily *A. aerogenes*). The coliform group includes all of the aerobic facultative anaerobic, non-spore forming, Gram negative, rod-shaped bacteria which ferment lactose with the production of gas at 35°C within 48 h. There is other organisms which ferment lactose with gas production, and so, if desired, a confirmatory test can be used to differentiate between coliforms and non-coliforms after the production of gas in the lactose broth, according to Standard Methods (8).

Further differentiation between those of fecal origin and those of soil or plant origin can be made (8). The Interim Primary Drinking Water Standards (22) do not require such differentiation. However, the US EPA Proposed Criteria for Water Quality (23) does indicate for public raw water supplies different standards for the total coliform and the fecal coliform content. Those standards are 10,000/100 mL for total coliform and 2,000/100 mL for fecal coliform. A high level of either total or fecal coliforms is good, although not infallible, evidence of recent sewage contamination. The coliform group is also a good indicator because their absence is good evidence of the absence of pathogenic etneric bacteria. The E. coli has about the same resistance to the actions of various disinfection procedures as the pathogenic bacteria.

The presence of coliforms indicates recent contamination by sewage and in sewage there are enteric viruses (24), many of these have been shown to be actually or potentially infectious. McDermott (24) postulates that waterborne enteric virus infections are grossly underestimated. The actual number of cases may be at least several hundred thousand per year in the United States. However, the absence of coliforms does not indicate the absence of enteric viruses. If water or sewage is disinfected minimally, that is, to eliminate the coliforms only, there would be little effect on the enteric viruses. Table 6 shows that the minimum required dosage ratio to eliminate viruses compared to bacteria is 20 to 1. White (25) suggests this ratio should be 100 or more to 1.

This leads to the tentative conclusion that enteric viruses, potentially pathogenic, may exist in public water supplies without detection. This is true; in one example 9% of all distribution systems sampled in Massachusetts indicated positive virus contamination. Similar findings were reported from other localities (22).

All this indicates the need for a rapid detection system for viruses in water. None is currently available that might be suitable for use in a water or wastewater or stream survey system. Because of this, the US EPA does not propose a limit of acceptability in water supply sources (23) or in the finished water (22). As a consequence of this lack of ability to detect and enumerate viruses in water under ordinary operating circumstances, the single most reliable indicator of disinfection is the presence of free available chlorine residual.

3.4. *Disinfectant Concentration*

As stated above, environmental conditions are important in disinfection rates and effectiveness. However, for a given set of environmental conditions, the concentration of the disinfectant is important if a specified degree of disinfection is to be accomplished in a given time period. The form of equation that relates concentration and time to a given level of effectiveness of kill is

$$C^n t = \text{constant} \tag{70}$$

where C = concentration of disinfectant, t = time, and n = constant for a particular disinfectant. If C is expressed in mg/L of hypochlorous acid and t is time in minutes for a 99% kill of several organisms, the data are presented in Table 7, according to Weber (3).

4. CHLORINE AND CHLORINATION

Chlorine is the only halogen at this time used extensively in environmental control processes. Because of its extensive use, appropriate technology, and practice have been

Table 7
Concentration-Time Factors for HOCl[a]

Organisms	$C^n t$
Adenovirus 3	0.098
E. coli	0.24
Poliomyelitus virus 1	1.20
(*Coxsackie virus* A2)	6.30

[a]The equation plots as a straight line on log-log paper. $n = 0.89$, a constant for a particular disinfectant, t = time for 99% kill of organisms, and C = concentration of disinfectant.

developed including disinfection. Chlorination is a halogenation process involving the addition of chlorine for disinfection and is also one of chemical oxidation processes.

Chemical oxidation is a process of electron transfer in which the material being oxidized loses an electrons(s) and at the same time there must be a reduction—the gaining of an electron(s). Disinfection is a chemical process for destruction of pathogenic organisms. Detailed terminologies concerning halogenation, chlorination, chemical oxidation, and disinfection can be found from Section 12.3, Glossary of Halogenation, Chlorination, Oxidation, and Disinfection. A few important technical terms are presented below (91):

- *Available chlorine.* All chlorine having capacity for oxidation.
- *Free available chlorine.* Chlorine in the form of the molecule (Cl_2), hypochlorous acid (HOCl), and hypochlorite ion (OCl^-).
- *Combined available chlorine.* All available chlorine not in the free available chlorine state and also combined with other elements or compounds such as but not limited nitrogen or nitrogenous matter
- *Free residual chlorination.* Chlorination to provide free available chlorine.
- *Combined residual chlorination.* Chlorination to provide combined available chlorine. This may include the addition of ammonia.
- *Chlorine demand.* The amount of chlorine needed to the level needed to show an incipient residual.
- *Plain chlorination.* The application of chlorine to an otherwise untreated water supply as it enters the distribution system or pipeline leading thereto.
- *Prechlorination.* The application of chloride to a water prior to some other form of treatment.
- *Postchlorination.* The application of chlorine to a water subsequent to any other treatment.
- *Rechlorination.* The application of chlorine to water at one or more points in the distribution system following previous chlorination.
- *Dechlorination.* The practice of removing all or part of the total available chlorine.

Values of available chlorine of several common chlorine compounds are shown in Table 8.

4.1. *Chlorine Compounds and Elemental Chlorine*

Chlorine is distributed for use in its elemental form or in the form of compounds. For certain small operations such as swimming pools, small water supplies and package sewage treatment plants, the compounds are found to be of use because of the ease of handing. The most common compounds used are solutions of sodium hypochloride (NaOCl) or calcium hypochlorite $Ca(OCl)_2$ as a powder. The solution of NaOCl is readily feed by a solution feeder; however, its cost includes the cost of transportation of large

Table 8
Percentage of Available Chlorine

Compound	Mol. Wt.	Mol. of equivalent chlorine	% Cl_2 (wt) actually present	% Cl_2 (wt) available
Cl_2	71	1	100	100
HOCl	52.5	1	67.7	135.4
NaOCl	74.5	1	47.7	95.4
$Ca(OCl)_2$	143	2	49.6	99.2
NH_2Cl	51.5	1	69.0	138.0
$NHCl_2$	86	2	82.5	165.0
NCl_3	120.5	3	88.5	265.5

amounts of water and is consequently high. Industrial bleaches contain 15% or less of available chlorine. The common household bleach is usually much weaker.

Calcium hypochloride in the solid form may be fed by a dry feeder but usually it is used to form a solution (1–2%) which is then fed by a solution feeder. A typical package has 70% calcium hypochlorite, a maximum of 2% moisture, a maximum of 0.5% of oxides of heavy metals. The remainder of about 30% consists of mixtures of calcium, chloride, chlorate, hydroxide, and carbonate plus sodium chloride. The 70% calcium hypochlorite corresponds to about 70% available chlorine; see Table 8.

Chlorine is a gas at normal ambient conditions therefore, elemental chlorine. Under pressure, it readily shifts its state to that of a liquid. It is as a liquid that the element is shipped, stored, and fed. Because of the pressure requirement, the container must be structurally strong and resistant to damage. The container and its use must meet ICC and Coast Guard regulations. Containers commonly used are 100- and 150-lb cylinders; 1-ton containers; 16-, 30-, 55-, 85-, and 90-ton tank cars; 15- to 16-ton tank trucks, and barges varying from 55- to 1100-ton capacity. Each type of container is equipped with pressure relief safety valves and the container is designed to withstand pressures in excess of that of the safety valves. The cylinders and ton containers are used directly connected to consuming process. The larger containers are usually emptied, under pressure, in the liquid form to storage containers on the plant site.

4.2. *Chlorine Feeders*

There are two standard ways of feeding chlorine gas to a water supply—the direct feed and the solution feed. The direct feed systems consist of metering the dry chlorine gas and conducting it under pressure directly to the water supply; this method is used only for emergency use or for small installations not having a convenient water supply necessary for use in the solution feeder. The solution feeder mixes the dry chlorine gas with a minor flow of water making a concentrated chlorine solution, which is the fed to the water supply.

There are two basic types of chlorine solution feeders. The one is a vacuum system and the other a mechanical diaphram pressure system. An injector water supply at a pressure of 25–300 psi must be available. Solution feed systems are available with capacities as low as 0.1 lb per 24 h and others with capacities to 8000 lb per 24 h. Within the capacity range of each individual feeder, the operator can select the feed rate desired.

Because evaporation of chlorine from the liquid state requires considerable heat energy, there is a maximum rate of evaporation, under a given ambient temperature, before the evaporated gas re-condenses and even forms chlorine "ice". Therefore, when a high rate of chlorine withdrawal for a given container is required (usually ton containers or tank cars), the chloride is withdrawn from the bottom of the container in the liquid state and then evaporated by a heating system. The most common heating system includes a sealed chlorine pressure chamber immersed in hot water. Other models are steam heated or heated by water piped to the chlorine chamber.

4.3. Chlorine Handling Equipment

The dry chlorine gas or pure liquid chlorine is relatively noncorrosive and may be piped or otherwise handled in extra heavy wrought-iron or steel pipe. Cast or malleable iron pipe and fittings should be avoided. Soft seamless copper tubing and type K copper water tubes are also suitable.

Chlorine solutions are extremely corrosive and special materials are required. Stoneware, glass, concrete, or porcelain and special alloys are satisfactory. For low-pressure service lines, chemical hard rubber hose, unplasticized polyvinyl chloride, Saran, Teflon, Haveg, and similar materials may be used. For high-pressure service liners of silver or platinum or as alloy "C" or rubber linings in common metal pipes may be used. Other materials such as chemical rubber hose, plastic-lined pipe, polyethylene plastic pipes are also satisfactory. All service lines should be as short as possible.

Valves are necessary at such locations as the chlorine source, at manifold systems, and at other points that permit isolation of any of the units in the system. Check valves and pressure-reducing values may be necessary. The number of valves should be kept to a minimum. Valves up to 1.5 in. should be globe or angle pattern, 600 lb ASA, outside screw and yoke, with forged body and bonnet and renewable or hard-faced seats; valves 2 in. and over may be 300-lb ASA with cast steel bodies. Equipment manufacturers or the Chlorine Institute will provide detailed information as required.

Chlorine containers must be housed in a separate room from the feeders. This room should be well ventilated and equipped with an emergency exhaust fan located near the floor and during cold weather periods heated to about 65–70°F. The intake vents should be near the ceiling of the room. During hot weather, the containers should not be permitted to be in the direct sunlight. If the container is out of doors, shade protection is required.

Scales should be provided to each container in use to permit the operator to check on the rate of feed and to determine when the container is exhausted. Emergency equipment consisting of gas masks, emergency showers, eye bubblers, and alarms should be provided. Emergency repair equipment should be conveniently available and tanks of alkaline absorbents of chlorine gas should be provided. The alkaline material may be caustic soda, soda ash, or hydrated lime slurries

4.4. Measurement of Chlorine Residuals

Standard Methods (8) provides the details of measurement procedures for chlorine residuals. White (1) has given detailed evaluation of several methods. Because it is desirable, if not necessary, to distinguish between free available and combined available chlorine, the procedures may include several steps. Measurements may be made by calorimetric and oxidation–reduction methods.

On-line chlorine residual measurement procedures usually involve the amperometric procedures. The magnitude of the residual measured may be used as a signal to increase or decrease the feed rate of the chlorination devices. This automatic control is commonly used in larger plants.

4.5. Chlorine Dosages

For potable water supplies the commonly accepted minimum bacteria residuals of chlorine, AWWA (26), are: 0.2 mg/L as free residual at pH values of about 9.2 or less increasing to about 0.3 mg/L at pH 9.5 and 0.5 at pH 10. The minimum combined residuals are 2 mg/L at pH levels below about 7 and rising logarithmically to 2.5 at pH 8, 2.8 at pH 9, and 3.0 at pH 10. It should be noted that these minimums are merely a guide to operation and should not replace culture or microscopic analyses.

The US EPA (22) does not specify chlorine residuals except when such residuals are a substitute in part or in total for the coliform test. When the substitution is in part, the minimum free chlorine residual is 0.2 mg/L in representative locations in the distribution system. When the substitution is total (permissible only for water systems serving less than 4900 people), the corresponding minimum is 0.3 mg/L.

From the above discussion, it appears that the present position is that coliform standards do provide a reasonably sure protection against enteric pathogenic bacterial infections. However, if the potential hazard of enteric waterborne viral infections are to be considered, it will be necessary to provide free residual chlorination in the water treatment process and probably to continue through the distribution system. As discussed above, it may not be easy to attain a free residual in water contaminated with nitrogenous matter.

The disinfection of wastewater is even more difficult because the presence of nitrogenous matter, particularly ammonia, results in the formation of chloramines when chlorine is added. The chloramines are effective in the destruction of the enteric bacteria if the concentration and exposure time are adequate. However, under even very good procedures of wastewater disinfection, the destruction by disinfection of enteric viruses is inadequate. A detailed review of research in wastewater disinfection prior to about 1972 is presented by White (1), who concluded at the time of the publication of his book in 1972 that wastewater disinfection practices in the United States were generally inadequate against all types of pathogenic organisms.

Referring to Figs. 1 and 2 it is seen that up to a molar concentration ratio of chlorine to ammonia of one, chloramines, primarily monochloramine, are formed and these reactions are very rapid, < l s. As the ratio increases, there is a decrease in the chloramines; however, the time for the reactions resulting in the decrease, may be rather slow—minutes to hours. And during this time, free available chlorine is present. Beyond the break point both free and combined available chlorine are present in a more or less pronounced ratio. The molecular weight of ammonia, expressed as nitrogen, is 14 and the molecular weight of chlorine is 7%. Therefore, the actual weight of the chlorine reacting is five times the weight of ammonia.

There is ordinarily little if any decrease in the ammonia content of a raw sewage as it passes through a primary or even a secondary sewage treatment plant. Therefore, the ammonia content may be, on the average about 15 mg/L. To reach the break point in the

Table 9
Probable Amounts of Chlorine Required to Produce a Chlorine Residual of 0.5 mg/L after 15 Min in Sewage and Sewage Effluents

Type of sewage or effluent	Probable amounts of chlorine (mg/L)
Raw Sewage, depend on age or staleness	6–24
Settled sewage, depending on age or staleness	3–18
Chemically precipitated sewage, depending on strength	3–12
Trickling filter effluent depending on performance	3–9
Activated sludges effluent, depending on performance	3–9
Intermittent sand-filter effluent depending on performance	1–6

Source: Fair et al. (4).

chlorine dosage would be about 150 mg/L or 1250 lb per million gallons. Furthermore, the residual chlorine, both free and combined, would be substantial and would have a strong odor and be toxic to organisms in the receiving water course. Consequently, break-point chlorination of sewage effluents is not ordinarily practiced.

Chlorination of sewage effluents is, of course, commonly practiced. The range of dosages ordinarily applied are shown in Table 9 which was taken from Fair et al. (4). During the course of sewage treatment there is a very substantial drop in numbers of organisms. Treatment through an ordinary secondary process will reduce both bacterial and virus counts by 90–95%. If this is followed by chlorination at the levels shown in Table 9, the enteric pathogen bacteria reduction would be substantially increased, even toward 100% if proper procedures were employed. However, the enteric virus content might still be substantial and break point chlorination to a free residual is required. If the ammonia content of sewage is reduced to 1–2 mg/L, break-point chlorination is practical, and in fact essential if the waste water is to be reused, according to Culp (27) and White (1).

4.6. Chlorination By-Products

The contractors of the US Environmental Protection Agency (US EPA) have been conducting experiments to develop methods of identifying by-products associated with the use of chlorine as a disinfectant. Various by-products have been isolated and identified in field studies. To date, six compounds (or in some cases groups of compounds) have been identified:

1. Trihalomethanes (four compounds); greater than 100 ppb.
2. Dihaloacetonitriles (three compounds); less than 10 ppb.
3. Chloroacetic acids (three compounds); from less than 10 ppb to greater than 100 ppb.
4. Chloral hydrate; greater than 100 ppb.
5. Chloropicrin; less than 10 ppb.
6. 1,1,1-Trichloropropanone; less than 10 ppb.

These by-products comprise 30–60% of the total organic halogen (TOH) in drinking water. The levels of common chlorination by-products found in drinking water are also given above. Through research, it was found that moving chlorination to a point just before filtration minimized the production of total trihalomethanes (TTHMs) while maintaining microbiological quality. The assimilable organic carbon of the finished water, which is a measure of the biological quality, was also found to be reduced. Therefore, the overall quality of the water improved under this scenario.

Although the individual concentrations of the disinfection by-products (DBPs) vary with pH, temperature, and chlorine concentration, research has shown that the removal of TTHM precursors seems to remove the formation potential for the other individual DBPs. This finding confirms earlier studies and opens up the possibility of precursor removal as an effective means of controlling chlorinated DBP (59–61).

An alternative to chlorine is the use of chloramines or chlorine dioxide for minimized DBP. Studies have shown that these approaches can be effective in maintaining microbiological integrity and minimizing by-product formation. This issue and the associated technologies will become extremely important if the THM standard is ultimately lowered. The US EPA SDWA amendments accelerated the schedule for setting MCLs for contaminants in drinking water. Contaminant limits are set according to the removal potential of existing technologies that can be obtained at a reasonable cost.

5. CHLORINE DIOXIDE DISINFECTION

As discussed above chlorine dioxide (ClO_2) does not react with nitrogenous matter and therefore would remain available in its initial form for wastewater disinfection. Furthermore, chlorine dioxide is not affected by dissociation as is hypochloric acid. Chlorine dioxide has been found to be a good disinfectant and its extended use is anticipated.

White (1) has reviewed the research and plant experiences between 1944 and 1968. Two specific studies will be reported here. The one by Benarde et al. (28) showed that at a pH of 8.5, sterile swage with a BOD of 165, innoculated with E. coli when treated with 5 mg/L of chlorine gave a 90% kill in 5 min, whereas a 2.0 mg/L of dosage of chlorine dioxide gave a 100% kill in 30 s. The second study also by Benarde et al. (29) was conducted to show the relative efficiencies of free chlorine and chlorine dioxide at several pH values, i.e., 4.0, 6.45, and 8.42. At pH 8.5, the chlorine dosage or the detention time required for 99% kill of E. coli is much greater than the chlorine dioxide dosage. The conclusions reached were that (a) in the neutral pH range the bactericidal efficiency of chlorine dioxide is comparable to that of free chlorine, (b) at a pH value of 8.5, chlorine dioxide is superior, and (c) chlorine dioxide may be superior to chlorine in the disinfection of sewage.

6. BROMINE AND BROMINATION

Bromine is the least effective germicide of the halogens. As discussed above, it is a liquid at ordinary ambient temperatures and as such it is very hard to handle. It is much more expensive than chlorine. Koski (30) reported that liquid bromine when compared to chlorine required from about 1.5 to 2.5 times as much depending upon the test organism. McKee (31) reported that in settled sewage the bromine requirement was 45 mg/L compared to 8 for chlorine for equivalent disinfection.

Bromine is being used for swimming pools. It is less initiating to the eyes than chlorine and there is no chance of the formation of nitrogen trichloride. Bromine may be purchased as a stick also containing chlorine known as bromo-chloro-dimethyl hydantoin (Dihalo). It is useful for small pools, but is not economical for larger installations. The bromine residual should be about 1.0–2.0 mg/L

7. IODINE AND IODINATION

Iodine is not competitive with chlorine for the disinfection of water and wastewater under ordinary conditions. It is relatively expensive and may have some undesirable physiological effects.

Of the several species of testable iodine, the elemental form, I_2, and hypoiodous acid, HIO, are the two important disinfectants. The relative amounts of these two species are discussed above. The efficiencies of these two species on the reduction of *E. coli*, cysts and a virus are shown a diagram prepared by Chang (32).

For disinfecting individual or small water supplies a compound was developed by Chang, Morris, and coworkers (33,34) for the US Army. The product is named globaline and contains 20 mg tetraglycine hydroperiodide of which 40% is I_2 and 20% is iodide along with 85 mg acid pyrophosphate. One tablet yields 8 mg/L of I_2 per liter of water. With a contact time of 10 min, it eliminates not only bacteria and viruses, but also the cysts of *Endamoeba histolytica*. This is one of the few if not the only chemical disinfection procedure intended to kill the cysts. Attempts to use iodine as a swimming pool disinfectant have been quite unsuccessful. The normal procedure for adding iodine to swimming pool waters is by chlorination of iodide to form I_2:

$$2HOCl + 2KI \leftrightarrow I_2 + 2KCl + 2OH^- \tag{71}$$

and

$$I_2 + H_2O \leftrightarrow HIO + H^+ + I^- \tag{72}$$

Iodine does not react with nitrogenous matter, which is an advantage in its disinfecting properties when compared to chlorine. However, if the nitrogenous matter—ammonia and urea principally—are available for algae growth, this growth does occur.

According to Koski (30), for disinfection equivalent to that commonly expected from the use of chlorine in swimming pools, the iodine residual should be about 1.0 mg/L. Unfortunately iodine colors swimming pool waters. For iodine residuals less than 0.3 mg/L, the water color is a clear light green; at 0.3–0.5 mg/L, the color is an emerald green; and above 0.6 mg/L it has a yellow to brown tint.

With the disadvantages cited, it is not likely that iodine will be found very useful for swimming pool water disinfection.

8. OZONE AND OZONATION

Ozone has been found to be a practical disinfectant for wastewater. A good review is presented by McCarthy (35). The processes of ozonation are discussed in Section 10.5, in chapter 9, and in the literature (69–75,80). Ozonation process equipment is commercially available (75).

9. COST DATA

The use of chloride or some substitute in public water supplies is mandated in most states. Also, in many states chlorination of sewage treatment plant effluents is mandated. The disinfection processes cost money. Such costs are initial end replacement costs of tanks, buildings, scales, plumbing system, pumps, etc.; costs for materials, power and supplies, primary chlorine; and costs for labor, both operation and maintenance.

The procedures for computing costs are beyond the scope of this volume. However, estimates of certain costs are presented by Black and Veatch (36). Cost data change rapidly with time and vary from one section of the United States to another. Bleckor and Cedman (37), Bleck and Nichols (38), and Wang et al. (86,87) presented detailed procedures for computing environmental process equipment costs and for adjusting changes with time and with location. The cost data and procedures referenced above were developed primarily for wastewater treatment facilities. However, they are equally applicable to potable water treatment facilities as well. These data and procedures are to be used as guides in estimating costs only. Detailed information on costs at a particular location or at a particular time should be sought for each project under consideration. It should be noted that all cost data in the literature are outdated. Appropriate cost indices, such are Chemical Engineering Equipment Indices (83,85,87), Engineering News Record Cost Indices (84,86,87), and US Army Corps Engineers Indices (88) can be used to update the past cost data to the current cost.

10. RECENT DEVELOPMENTS IN HALOGENATION TECHNOLOGY

10.1. Recent Environmental Concerns and Regulations

Protection of public water supplies relies heavily on the use of disinfectants. Disinfectants are used to maintain a residual in the distribution system to prevent any health problems and to maintain the water quality standards. Since the new regulation requirements, the water industry has been looking for alternative chemicals or techniques to replace chlorine. In this section, instead of studying halogenation technology, we present techniques to reduce halogenation by-products. Different techniques include (a) chlorine dioxide, (b) chloramines, (c) coagulant, (d) ozonation, (e) organic disinfectants, and (f) ultraviolet light (40–45,65–71,97–98). To comply with the upcoming stringent law, the techniques were tested by different plants. In the past, we have used chlorine to disinfect the finished drinking water, but then it may produce trihalomethanes (THMs) and other products. These can be potential carcinogens. This includes most of the halogens, especially the chlorine (61).

Chlorine is a major halogen used in water treatment for controlling microbial quality. Marhaba (39) described the US Environmental Protection Agency (US EPA) initiated and negotiated the rule-making process for the Disinfectant/Disinfections By Products (D/DBPs) Rule in 1992. Owing to the complexity of the problems, US EPA had to draw on the expertise of others to prepare the rule. The regulation was proposed in two steps. Stage 1 of the D/DBPs Rule was proposed in 1994 and became effective in December 1998. It lowered the total THM (TTHM) maximum contaminant level (MCL) from 0.100 to 0.0800 mg/L and three other classes of DBPs. The rule also set maximum residual disinfectant levels (MRDL) for three disinfectants. To provide necessary data for stage 2 of the D/DBP regulations, the Information Collection Rule (ICR) (begun July 1, 1997, ended December 1998) was proposed in 1994 with stage 1 of the D/DBP Rule. Stage 2 was re-proposed in 2000 and required even lower MCLs for DBPs than those proposed in stage 1. The 1996 Amendments to the Safe Drinking Water Act (SDWA) require US EPA to promulgate the stage 2 Rule by May 2002. Stage 1, proposed

Table 10
Proposed Disinfectant Level on Disinfectant Residuals and DBPs

Parameter	Effective Stage 1 (mg/L)	Anticipated Stage 2 (mg/L)
MRDL for chlorine	4.0	4.0
MRDL for chloramines	4.0	4.0
MRDL for chlorine dioxide	0.8	0.8
MCL for TTHM	0.08	0.04
MCL for five haloacetic acids (HAAs)	0.06	0.02
MCL for bromate ion	0.01	
MCL for chlorite ion	1.0	

in 1994 and promulgated in 1999, provided maximum contaminant levels (MCLs) for the sum of five haloacetic acids (HAAs) at 0.6 mg/L, BrO_3^- at 0.010 mg/L, and brominates trihalomethane (THMs) at 0.08 mg/L. Stage 2 MCLs of 0.040 mg/L for TTHMs and 0.020 mg/L for HAAs were proposed. Table 10 gives a summary of the proposals according to the affected parameters.

10.2. Chlorine Dioxide

Chlorine dioxide is widely used as an alternative to chlorine for treating drinking water (56–64,78). Numerous chlorine dioxide generation technologies have recently been developed to improve the conversion efficiency and purity of chlorine dioxide (50). Water utilities use chlorine dioxide for peroxidation, control of taste and odor problems, and inactivation of common pathogens. Because chlorine dioxide is an oxidizing agent that does not chlorinate, it is often used for lower THM concentrations in finished water to meet levels established by the US EPA. Rittmann and Tenney (46) evaluated the effectiveness of generating chlorine dioxide gas using Eka Chemical's method of sodium chlorate/hydrogen peroxide/sulfuric acid technology as it compares to Rio Linda chlorine gas/25% sodium chlorite system process at the El Paso Water plant. The objective of the trial was to compare the effectiveness of the sodium chlorate technology to the sodium chlorite technology. They found both techniques worked equally well, effluents were able to comply with total oxidant level regulations. The Eka Chemical Generator produced chlorite free and chlorine free chlorine dioxide that may be beneficial in further reduction of THM levels. Sung et al. (51) established semimechanistic models that described the formation of disinfection by-products including THMs and haloacetic acids. They were developed for the Massachusetts Water Resources Authority for unfiltered surface water using chlorine. They found the DBP concentration in a transmission system changes in response to downstream treatment processes. Elshorbagy et al. (53) used a new approach to characterize and model the kinetics of THM species using nonlinear optimization. The approach combined site-specific water-quality trends with stoichiometric expressions based on an average representative bromine content factor of the source. The model is capable of modeling chlorine, total THM, and the four THM species in water distribution systems subjected to different varying loading conditions.

10.3. Chloramines

Owing to the D/DBP rule, many water utilities may be switching from chlorine to alternative disinfectants. Chloramines have become the disinfectant of choice to replace free chlorine in distribution systems because they produce fewer DBPs while controlling the re-growth of bacteria. Controlling nitrification is essential if chloramines are to be a viable alternative disinfectant scheme for distribution systems in all types of environments. El-Shafy and Grunwald (49) studied the formation of THMs and its species from the reaction of chlorine with humic acid substances. This has caused much attention because of their carcinogenic and dangerous health effects. They found residual chlorine in water entering the distribution pipelines was on average 0.75 mg/L and decreased with distance until it reaches zero. The low velocity and large volume of reservoirs increased the residence time and correspondingly provided conditions for more chlorine decay and accordingly an increase in THM formation. The residence time and decay of chlorine were used as good predictors for the formation of THM and Chloroform in this study. They found there is a linear correlation between the cumulative chlorine decay and the cumulative THMs formed in the pipeline. McGuire et al. (48) study on problems associated with nitrification and presents laboratory and field evidence for using the chlorite ion (ClO_2^-) to control nitrification in distribution systems. Laboratory experiments showed that even low dosages of ClO_2^- (0.05 mg/L) can inactive 3–4 logs of ammonia-oxidizing bacteria (AOB) over several hours.

Higher concentrations of ClO_2^- inactive all of the AOB in as little as 30 minutes. Laboratory and field studies of five Texas distribution systems suggest that chlorite ion has significant potential for controlling nitrification in chloraminated water. Diehl et al. (44) performed batch experiments on three diverse organic halogen (DOX), THMs, haloacetic acids (HAAs), and cyanogens halides (CNX) during chloramination. The authors used and performed chloroamines to examine the effect of pH, mass ratio of chlorine to ammonia-nitrogen (Cl_2 to N), and bromide concentration on disinfections by-product (DBP) formation.

Formation of specific DBPs as well as the group parameter DOX was greatest at low pH and high Cl_2-to N ratios. Results followed the general tend of decreasing with increasing pH and decreasing Cl_2-to-N ratio. Bromide addition increased the concentration of bromine substitutes DBPs and DOX. These experiments demonstrated that because of dihaloacetic acid formation, HAA formation is more problematic during chloramination than THM formation. Because the specific DBPs measured in this research accounted for more than 35% of the DOX concentration, utilities may want to consider both specific DBPs and DOX in selecting appropriate chloramination conditions.

10.4. Coagulant

The evaluation of 16 sites, with optimized coagulation provide an assessment of the technique and illustrate its capabilities to meet the requirements of Disinfectants/Disinfections by-product rule (D/DBP), were done by Bell-Ajy et al. (47). Jar tests were used to determine the effectiveness of optimized coagulation for the removal of organic carbon, DBP precursors, particles, and turbidity when supernatant results were compared with conventional treatment. Jar-test results indicated that optimized coagulation could enhance the removal of organic carbon and DBP precursors.

The proper pH and coagulant $FeCl_3$ and alum were compared for optimized coagulation. Ferric (expressed as Fe^{3+}) resulted in better TOC removal than alum.

10.5. Ozone

Ozonation is one of the alternative techniques to replace traditional chlorine (69–73, 80,97,98). Although the use of ozone will not produce chlorinated THM, haloacetic acids or other chlorinated by products, it will react with nature organic material. Ozone and its primary reactive product, the hydroxyl free radial (OH^-), are strong oxidizers. The oxidation by-products typically include aldehydes, aldo and keto acids, carboxylic acids, and peroxide. Grosvener (54) presented a paper providing a detailed summary of ozonation and by-product formation chemistry, effective approaches toward the control of by-product formation, and DBP precursor removal technologies. Natural organic materials (NOM), a major component of total organic materials, is a complex matrix of total organic chemicals that can be derived from partial bacterial degradation of soil, living organisms, and plant detritus. The author's (54) conclusion stated that it is arduous to simply pave a procedural pathway toward ozonation by-product-free-drinking water. In other words, each treatment plant's influent may react differently to the similar process of optimization techniques, owing to variations in key parameters. Truly, exchanges will happen when O_3 is used. Even though lowering the usage of O_3 may decrease the forming of BrO_3^-, it may increase the forming of DBPs. Yet, higher O_3 doses leads to much BrO_3^- forming, especially toward high Br^- levels and in circulation pH. Furthermore, a proper use of the tradeoffs must associate with the control of the standards, as well as weighing the viruses and the protozoan cysts. Clearly, water sources responded very differently to O_3 treatment, and it is necessary for these studies to be performed thoroughly and efficiently. Wang (74) has developed a rapid method for ozone determination.

Through research, both GAC filtration and membrane filtration have proved to be very effective antagonists of DBPs. In fact, GAC filtration is effective in removing DBP precursors when biofilm is nurtured. It was previously believed that the only way to balance microbiological safety, minimize DBPs, and maintain high-quality drinking water was through improvement. Some changes when the BOM concentrations are low, conventional rapid filters, and GAC absorbers can be altered into hybrid biofilters, which eliminates chlorine in the feed water.

As well, membrane filtration technologies has become very capable in the removing of DBPs and precursors (76,96). Nonetheless, removing organic from a treatment plant is necessary to concentrate on membrane fouling. In the process, it would be beneficial to concentrate on high organic rejection and low inorganic rejection especially when the concentration of organic is high. The proper usage of quenching agents should also not be overlooked. Although ammonium sulfate can be an effective quenching agent for $HOBr/OBr^-$ in ozonated samples, agents such as sodium thiosulfate and sulfite have been carefully observed to destroy other brominated organic by-products.

10.6. Organic Disinfectants

Wang (65–71) has studied the use of various organic disinfectants for water purification, swimming pool water disinfection, and sludge disinfection. The major advantage of using organic disinfectants is that organic disinfectant will not be consumed easily by the target influent water, wastewater, or sludge containing organics.

10.7. Ultraviolet (UV)

Hartz (55) described the pilot study at Midway Sewer District, located south of Seattle, WA. Owing to new regulation requirements, the district commissioned an investigation of alternative methods of disinfection, a pilot study to determine the effectiveness of ultraviolet irradiation. The UV process involved subjecting the wastewater to light energy in which lamps are tuned to emit certain light frequencies (79). In the case of UV used for microorganism inactivation, the lamps are tuned to a specific emission wavelength, for low-pressure lamps, the frequency most effective for inactivation around 250 nm. A number of variables regarding the effectiveness of the UV systems included: (a) light intensity, (b) residence time, and (c) effluent requirements. The results were that the percentage of light transmission for this pilot trial was slightly lower than normal. It was indicated that trickling filter tended to produce a wastewater that has a lower percentage light transmission. Then, owing to the solid content contact unit following the trickling filter system, the residual turbidity is lowered and the light transmittance is slightly increased. The UV light transmission was about 62% for an unfiltered sample of the wastewater. Filtration of the wastewater sample improved the light transmission by 5%. The district has found this technique as a possible alternative. A full scale study will be performed in the future (98).

11. DISINFECTION SYSTEM DESIGN

11.1. Design Considerations Summary

Disinfection is the selective destruction of pathogenic organisms; sterilization is the complete destruction of all microorganisms. Disinfection may be considered as one of the most important processes in water and wastewater treatment. This practice used in water and wastewater treatment has resulted in the virtual disappearance of waterborne diseases.

Disinfection may be accomplished through the use of chemical agents, physical agents, mechanical means, and radiation. In wastewater treatment, the most commonly used disinfectant is chlorine; however, other halogens, ozone, and ultraviolet radiation, and organic disinfectants have been used.

The rate of disinfection by chlorine depends on several factors, including chlorine dosage, contact time, presence of organic matter, pH, and temperature. For potable water treatment, the readers are referred to the Federal Final Surface Water Treatment Rule (SWTR) for determination of the chlorine dosages for water disinfection (81).

The most common forms of chlorine used in water and wastewater treatment plants are calcium and sodium hypochlorites and chlorine gas. Hypochlorites are recommended for small treatment plants where simplicity and safety are more important than cost. Chlorine gas may be applied as a gas, or mixed with water to form a solution, a method used almost exclusively in municipal water and wastewater treatment. Figure 3 shows the distribution of hypochlorous acid and hypochlorite ions in water at different pH values and temperatures of 0°C and 20°C.

The design of the chlorine contact tank plays an important role in the degree of effectiveness produced from chlorination. Factors which must be considered in the design

$NaOCl$ + H_2O -----> $HOCl$ + $NaOH$
sodium hypochlorite + water -----> hydrochlorous acid + sodium hydroxide

$Ca(OCl)_2$ + H_2O -----> $2HOCl$ + $CaOH$
calcium hypochlorite + water -----> hypochlorous acid + calcium hydroxide

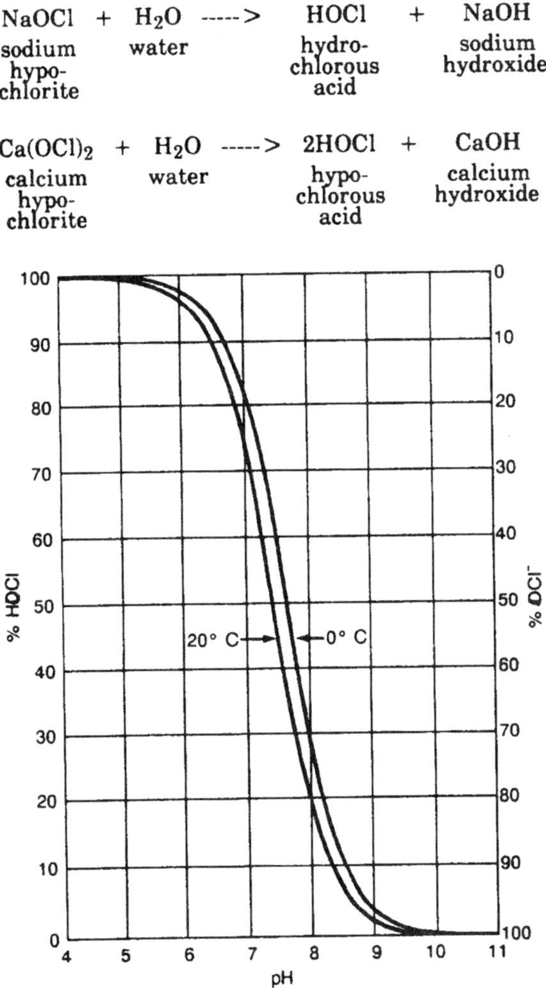

Fig. 3. Distribution of hypochlorous acid and hypochlorite ions in water at different pH values and temperatures of 0°C and 20°C (US EPA).

include method of chlorine addition, degree of mixing, minimization of short circuiting, and elimination of solids settling. To minimize short-circuiting, the basin outlet may be designed as a sharp-crested weir that spans the entire width of the basin outlet. The longitudinal baffling of a serpentine flow basin was superior to cross-baffling; a length-to-width ratio of 40 to 1 was necessary to reach maximum plug flow performance regardless of the type of baffling.

11.2. Wastewater Disinfection

For wastewater treatment, the recommended chlorine dosage for disinfection purposes is that which produces a chlorine residual of 0.5–1 mg/L after a specified contact time. Effective contact time of not less than 15 min at peak flow is recommended. Practical chlorine dosages recommended for wastewater disinfection and odor control are presented in below:

(a) Untreated wastewater (prechlorination) = 6–25 mg/L.
(b) Primary sedimentation = 5–20 mg/L.
(c) Chemical precipitation plant = 2–6 mg/L.
(d) Trickling filter plant = 3–15 mg/L.
(e) Activated sludge plant = 2–8 mg/L.
(f) Multimedia filter following activated sludge plant = 1–5 mg/L.

The required input data include (a) chlorine contact tank influent flow, MGD; (b) peak flow, MGD; and (c) average flow, MGD. The design parameters include (a) contact time at maximum flow, min; (b) length-to-width ratio; (c) number of tanks; and (d) chlorine dosage, mg/L.

The following is recommended design procedure. The first step is to select contact time at peak flow and calculate the volume of the contact tank:

$$\mathrm{VCT} = \left[Q_p (CT)(10^6)\right] / (24 \times 60) \tag{73}$$

where VCT = volume of contact tank (gal), Q_p = peak flow (MGD), and CT = contact time at maximum flow (min).

The second step in design is to select a side water depth and calculate surface area:

$$\mathrm{SA} = \mathrm{VCT}/[7.48 \times \mathrm{SWD}] \tag{74}$$

where SA = surface area (ft^2), VCT = volume of contact tank (gal), and SWD = side water depth = 8 ft. The third step in design is to select a length-to-width ratio and calculate dimensions by the following equations:

$$\mathrm{CTW} = [\mathrm{SA}/\mathrm{RLW}]^{0.5} \tag{75}$$

$$\mathrm{CTL} = \mathrm{SA}/\mathrm{CTW} \tag{76}$$

where CTW = contact tank width (ft), SA = surface area (ft^2), RLW = length-to-width ratio, and CTL = contact tank length (ft). The fourth step in design is to select chlorine dosage according to the recommended chlorine dosages in this section, and then calculate chlorine requirements:

$$\mathrm{CR} = (Q_a)(\mathrm{CD})(8.34) \tag{77}$$

where CR = chlorine requirement (lb/d), Q_a = average flow (MGD), and CD = chlorine dosage (mg/L). The fifth step in design is to calculate peak chlorine requirements by the following equation:

$$\mathrm{PCR} = (\mathrm{CR})(Q_p)/Q_a \tag{78}$$

where PCR = peak chlorine requirements (lb/d), CR = chlorine requirements (lb/d), Q_p = peak flow (MGD), and Q_a = average flow (MGD). The output data of wastewater disinfection design will be:

(a) Maximum flow (MGD).
(b) Average flow (MGD).
(c) Contact time (min).
(d) Volume of contact tank (gal).
(e) Average chlorine requirement (lb/d).

(f) Peak chlorine requirement (lb/d).
(g) Tank dimensions.

11.3. Potable Water Disinfection

Water treatment plants employ both primary and secondary disinfection: (a) Primary disinfection achieves the desired level of microorganism kill or inactivation; and (b) secondary disinfection ensures a stable residual concentration of disinfectant in the finished water to prevent microbial growth in the distribution system (63,64,81,82,89,90). Major primary disinfectants are chlorine, ozone, chlorine dioxide, and ultraviolet (UV) radiation; major secondary disinfectants are chlorine and monochloramine. Some disinfectants can be used for both processes.

The 1986 Amendments to the Safe Drinking Water Act (SDWA) require all public water suppliers to disinfect drinking water. In addition, inorganic and organic chemicals will be regulated by means of Maximum Contaminant Levels (MCLs). Since some disinfectants can produce chemical by-products, the dual objective of disinfection is to provide the required level of organism destruction while remaining within the MCLs for by-products set by the US EPA.

Chlorine has been the most widely used disinfectant in the United States; however, it produces trihalomethanes (THMs) and other halogenated organic compounds in drinking water. Because of this, water suppliers are beginning to utilize other disinfectants, such as ozone, chlorine dioxide, and monochioramine, or combinations of disinfectants, such as ozone followed by chlorine.

According to the Amendments to the SDWA, all public water suppliers, including those that rely on groundwater, will have to disinfect drinking water before distribution. To ensure compliance with all applicable regulations (both current and anticipated), the specific objectives of disinfection are to (a) ensure 99.9% (3 log) and 99.99% (4 log) inactivation of *Giardia lamblia* cysts and enteric viruses, respectively; (b) ensure control of other harmful microorganisms; (c) not impart toxicity to the disinfected water; (d) minimize the formation of undesirable disinfection by-products; and (e) meet the MCLs for the disinfectants used and by-products that may form (81).

Disinfection alone, or a combination of disinfection and filtration, can achieve the minimum mandatory removals and/or inactivations of 99.9% *Giardia* cysts and 99.99 percent enteric viruses. Primary disinfection systems that use ozone, chlorine, or chlorine dioxide can achieve greater than the above-stated inactivation of enteric viruses when 99.9% inactivation of *Giardia* cysts is attained. Therefore, achieving sufficient *Giardia* cyst inactivation can ensure adequate inactivation of both types of organisms. This is not the case, however, when using chloramination because it is such a poor virucide.

Conventional treatment, which includes coagulation, flocculation, sedimentation, and filtration, along with disinfection, can achieve 99.9% inactivation of *Giardia* cysts and 99.99% inactivation of enteric viruses when properly designed and operated. Direct filtration, slow sand filtration, and diatomaceous earth filtration systems, each combined with disinfection, have also achieved these reductions.

Groundwater systems that apply disinfection to comply with regulations may have to add filtration if they contain iron and manganese. Insoluble oxides form when chlorine, chlorine dioxide, or ozone are added to these systems; thus, filters would be needed for their removal. In addition, both ozonation and chlorination may cause flocculation of dissolved organics, thus increasing turbidity and necessitating filtration. The presence of such insolubles will require the use of secondary disinfection after filtration as well (81,89).

"*CT* values" indicate the effectiveness of disinfectants in achieving primary disinfection. They describe the attainable degree of disinfection as the product of the disinfectant residual concentration (in mg/L) and the contact time (in minutes). For chlorine, chlorine dioxide, or monochloramine, the contact time can be the time required for the water to move from the point at which the disinfectant is applied to the point it reaches the first customer (at peak flow). This is the total time the water is exposed to the chlorinous residual before being used. Ozone, however, has a short half-life in water; therefore, the contact time is considered the time water is exposed to a continuous ozone residual during the water treatment process only.

The US EPA Final Surface Water Treatment Rule (SWTR) states the following: Systems may measure "*C*" (in mg/L) at different points along the treatment train, and may use this value, with the corresponding "*T*" (in minutes), to calculate the total percent inactivation. In determining the total percentage inactivation, the system may calculate the *CT* at each point where "*C*" was measured and compare this with the $CT_{99.9}$ value (the *CT* value necessary to achieve 99.9% inactivation of *Giardia* cysts) in the rule for specified conditions (pH, temperature, and residual disinfectant concentration). Each calculated *CT* value (CT_{calc}) must be divided by the $CT_{99.9}$ value found in the SWTR tables to determine the inactivation ratio. If the sum of the inactivation ratios, or

$$\text{Summation of } CT_{\text{calc}}/\left(CT_{99.9}\right) \tag{79}$$

at each point prior to the first customer where *CT* was calculated is equal to or greater than 1.0, i.e., there was a total of at least 99.9% inactivation of *Giardia lamblia*, the system is in compliance with the performance requirement of the SWTR.

For groundwater not under direct influence of surface water, *CT* is determined in the same manner using enteric viruses or an acceptable viral surrogate as the determinant microorganism, since *Giardia* cysts will not be present.

Table 11 presents the *CT* values required to attain 1-log reductions of *Giardia* cysts, for four disinfectants. As shown, lower temperatures require higher *CT* values; with chlorine, an increase in pH also increases necessary *CT* values. If more than one disinfectant is used, the percentage inactivation achieved by each is additive and can be included in calculating the total *CT* value.

When direct filtration is included in the water treatment process, disinfection credit can be taken by the filtration step for a 2-log inactivation of *Giardia* cysts and a 1-log inactivation of viruses. This means that the primary disinfectant must provide an additional 1-log inactivation of *Giardia* cysts and 3-log inactivation of viruses. In the specific instance of a conventional treatment process that includes coagulation, flocculation, sedimentation, and filtration, an inactivation credit of 2.5 logs for *Giardia* cysts and 2 logs for viruses may be taken. This means that the primary disinfectant must provide an additional 0.5 log inactivation of *Giardia* cysts but a 2-log inactivation of viruses.

Table 11
CT Values for Achieving 90% Inactivation of *Giardia lamblia* for Potable Water Disinfection

		Temperature					
Disinfectant	pH	≤1°C	5°C	10°C	15°C	20°C	25°C
Free chlorine[a] (2 mg/L)	6	55	39	29	19	15	10
	7	79	55	41	28	21	14
	8	115	81	61	41	30	20
	9	167	118	88	59	44	29
Ozone	6–9	0.97	0.63	0.48	0.32	0.24	0.16
Clorine dioxide	6–9	21	8.7	7.7	6.3	5	3.7
Chloramines[b] (preformed)	6–9	1270	735	615	500	370	250

Source: US EPA (81,89).

[a]*CT* values will vary depending on concentration of free chlorine. Values indicated are for 2.0 mg/L of free chlorine. *CT* calues for different free chlorine concentrations are specified in tables in the US EPA and UNIDO manuals.

[b]To obtain 99.99% inactivation of enteric viruses with performed chloramines requires *CT* values > 5000 at temperatures of 0.5,5,10, and 15°C.

If a water supply system does not use filtration, the 99.9% inactivation of *Giardia* and 99.99% inactivation of enteric viruses must be achieved by the primary disinfecting agents alone. Table 12 presents *CT* values for the four disinfectants for achieving 99.9% reductions of *Giardia* cysts. Table 13 presents the *CT* values for virus inactivation. Although groundwater disinfection regulations have not been finalized, these values will probably apply to systems treating groundwater determined by the state not to be under direct influence of surface water (81,89).

In the final SWTR, the *CT* values for ozone have been lowered to levels such that the *CT* values required to provide 0.5-log inactivation of *Giardia* cysts at the higher water temperatures are below those required to provide 2 or 3 logs of inactivation of enteric viruses. Consequently, the 2- or 3-log virus inactivation *CT* requirement becomes the pacing parameter for the amount of additional primary disinfection to be provided by ozone during conventional treatment, rather than the 0.5-log inactivation of *Giardia* cysts.

12. DESIGN AND APPLICATION EXAMPLES

12.1. *Example 1 (Wastewater Disinfection)*

Design a chlorine contact tank for wastewater disinfection based on the following data and equations.

Solution

The first step is to select contact time at peak flow and calculate volume of contact tank (82,90):

$$\text{VCT} = \left[Q_p(CT)(10^6)\right] / (24 \times 60) \tag{73}$$

Table 12
CT Values for Achieving 99.9% Inactivation of *Giardia lamblia*[a] for Potable Water Disinfection

Disinfectant	pH	Temperature ≤1°C	5°C	10°C	15°C	20°C	25°C
Free chlorine[b] (2 mg/L)	6	165	116	87	58	44	29
	7	236	165	124	83	62	41
	8	346	243	182	122	91	61
	9	500	353	265	177	132	88
Ozone	6–9	2.9	1.9	1.4	0.95	0.72	0.48
Clorine dioxide	6–9	63	26	23	19	15	11
Chloramines[c] (preformed)	6–9	3800	2200	1850	1500	1100	750

Source: US EPA (81,89).

[a]These *CT* values for free chlorine dioxide, and ozone will guarantee greater than 99.99% inactivation of enteric viruses.

[b]*CT* values will vary depending on concentration of free chlorine. Values indicated are for 2.0 mg/L of free chlorine. *CT* values for different free chlorine concentrations are specified in tables in the US EPA UNIDO manuals.

[c]To obtain 99.99% inactivation of enteric viruses with performed chloramines requires *CT* values > 5000 at temperatures of 0.5,5,10, and 15°C.

where VCT = volume of contact tank (gal), Q_p = peak flow (2 MGD), and CT = contact time at maximum flow (15 min); then

$$\text{VCT} = \left[Q_p(CT)(10^6)\right]/(24 \times 60) = 2(15)(106)/(24 \times 60) = 20{,}833 \text{ gal}$$

The second step in design is to select a side water depth and calculate surface area:

$$\text{SA} = \text{VCT}/[7.48 \times \text{SWD}] \tag{74}$$

where SA = surface area (ft^2), VCT = volume of contact tank (20,833 gal), and SWD = side water depth = 8 ft; then

$$\text{SA} = \text{VCT}/[7.48 \times \text{SWD}] = 20{,}833/[7.48 \times 8] = 348 \text{ ft}^2$$

The third step in design is to select a length-to-width ratio and calculate dimensions by the following equations:

$$\text{CTW} = [\text{SA}/\text{RLW}]^{0.5} \tag{75}$$

$$\text{CTL} = \text{SA}/\text{CTW} \tag{76}$$

where CTW = contact tank width (ft), SA = surface area (348 ft^2), RLW = length-to-width ratio = select 40, CTL = contact tank length (ft); then

$$\text{CTW} = [\text{SA}/\text{RLW}]^{0.5} = [348/40]^{0.5} = 2.95 \text{ ft}$$

$$\text{CTL} = \text{SA}/\text{CTW} = 348/2.95 = 118 \text{ ft}$$

Table 13
CT Values for Achieving Inactivation of Viruses at pH 6–9 for Potable Water Disinfection

		Temperature					
	Log inactivation	0.5°C	5°C	10°C	15°C	20°C	25°C
Free chlorine[a]	2	6	4	3	2	1	1
	3	9	6	4	3	2	1
	4	12	8	6	4	3	2
Ozone[b]	2	0.9	0.6	0.5	0.3	0.25	0.15
	3	1.4	0.9	0.8	0.5	0.4	0.25
	4	1.8	1.2	1.0	0.6	0.5	0.3
Clorine dioxide[c]	2	8.4	5.6	4.2	2.8	2.1	—
	3	25.6	17.1	12.8	8.6	6.4	—
	4	50.1	33.4	25.1	16.7	12.5	—
Chloramines[d]	2	1423	857	643	428	321	214
	3	2063	1423	1067	712	534	356
	4	2883	1988	1491	994	746	497

Source: US EPA.

[a]For inactivation of Hepatitis A Virus (HAV) at pH = 6, 7, 8, 9, and 10 at 5°C. *CT* values include a safety factor of 3.

[b]For inactivation of poliovirus at pH 7.2 and 5°C. *CT* values include a safety factor of 3.

[c]*CT* values for chlorine dioxide are based on laboratory studies at pH 6.0 and 5°C. *CT* values include a safety factor of 3.

[d]For inactivation of HAV for pH = 8.0, 5°C, and assumed to apply to pH in the range of 6.0–10.0. These *CT* values apply only for systems using combined chlorine where chlorine is added prior to ammonia in the treatment sequence. *CT* values given here should not be used for estimating the adequacy of disinfection in systems applying performed chloramines, or applying ammonia ahead of chlorine.

The fourth step in design is to select chlorine dosage according to the recommended chlorine dosages in this section, and then calculate chlorine requirements.

$$\mathrm{CR} = (Q_a)(\mathrm{CD})(8.34) \tag{77}$$

where CR = chlorine requirement (lb/d), Q_a = average flow (1 MGD), and CD = chlorine dosage, 8 mg/L; then

$$\mathrm{CR} = (Q_a)(\mathrm{CD})(8.34) = 1 \times 8 \times 8.34 = 66.7\ \mathrm{lb/d}$$

The fifth step in design is to calculate peak chlorine requirements by the following equation:

$$\mathrm{PCR} = (\mathrm{CR})(Q_p)/Q_a \tag{78}$$

where PCR = peak chlorine requirements (lb/d), CR = chlorine requirements (66.7 lb/d), Q_p = peak flow (2 MGD), and Q_a = average flow (1 MGD); then

$$\mathrm{PCR} = (\mathrm{CR})(Q_p)/Q_a = (66.7 \times 2)/(1) = 133.4\ \ \mathrm{lb/d}$$

Finally the output data of wastewater disinfection design will be:

(a) Maximum flow, 2 MGD
(b) Average flow, 1 MGD
(c) Contact time, 15 min
(d) Volume of contact tank, 20,833 gal
(e) Average chlorine requirement, 66.7 lb/d.
(f) Peak chlorine requirement, 133.4 lb/d.
(g) Tank dimensions: surface area = 348 ft^2; side water depth = 8 ft; length-width ratio = 40; contact tank length = 118 ft.

12.2. Example 2 (Potable Water Disinfection)

Chlorine, ozone, UV, chlorine dioxide and chloramines are the five most common disinfectants used in potable water disinfection. Discuss (a) the advantages and disadvantages of the five disinfectants; and (b) the desired points of disinfectant application at a water transmission-treatment-distribution system.

Solution

Table 14 summarizes the advantages and disadvantages of the five disinfectants. Table 15 summarizes the desired points of disinfectant application at a water supply system (98).

12.3. Example 3 (Glossary of Halogenation, Chlorination, Oxidation, and Disinfection)

The following glossary terms are collected from the leterature (91–94, 97–98):

Available Chlorine: All chlorine having capacity for oxidation.

Bacteria: Microorganisms often composed of single cells shaped like rods, spheres, or spiral structures.

Bromination: The unit process of adding a form of bromine to water or wastewater. Bromination is a halogenation process involving the addition of bromine for disinfection or oxidation.

Chemical oxidation: A process of electron transfer in which the material being oxidized loses an electrons(s) and coincidentally there must be a reduction—the gaining of an electron(s).

Chlorination: The unit process of adding a form of chlorine to water or wastewater. Chlorination is a halogenation process involving the addition of chlorine for disinfection and is also one of the chemical oxidation processes.

Chlorine Demand: The amount of chlorine needed to the level needed to show an incipient residual.

Chlorine Residual: The measurement of chlorine in water after treatment.

Clarification: Removal of bulk water from a dilute suspension of suspended solids by gravity sedimentation, dissolved air flotation, aided by chemical flocculating and/or precipitating agents.

Combined Available Chlorine: All available chlorine not in the free available chlorine state and also combined with other elements or compounds such as but not limited nitrogen or nitrogenous matter.

Combined Residual Chlorination: Chlorination to provide combined available chlorine. This may include the addition of ammonia.

Contact Time: The period of treatment (such as disinfection) in water or wastewater treatment.

Dechlorination: The practice of removing all or part of the total available chlorine.

Disinfection: Destruction of harmful microorganisms, usually by the use of bactericidal chemical compounds or UV light.

Disinfection By-products: Compounds created by the reaction of a disinfectant with organic compounds in water or wastewater.

Distribution System: A network of water pipes leading from a water treatment plant to customers' plumbing systems.

Filtration: The operation of separating suspended solids from a liquid (or gas) by forcing the mixture through porous media.

Table 14
Advantages and Disadvantages of Five Major Disinfectants

Disinfectant	Advantages	Disadvantages
Chlorine	Effective. Widely used. Variety of possible application points. Inexpensive. Appropriate as both primary and secondary disinfectant.	Harmful halogenated by-products. Potential conflict with corrosion control pH levels, when used as a secondary disinfectant.
Ozone	Very effective. Minimal harmful by-products identified to date. Enhances slow sand and GAC filters. Provides oxidation and disinfection in the same step.	Requires secondary disinfectant. Relatively high cost. More complex operations because it must be generated on-site.
Ultraviolet radiation	Very effective for viruses and bacteria. Readily available. No harmful residuals. Simple operation and maintenance.	Inappropriate for water with *Giarcia* cysts, high suspended solids, high colour, high turbidity, or soluble organics. Requires a secondary disinfectant.
Chlorine dioxide	Effective. Relatively low costs. Generally does not produce THMs.	Some harmful by-products. Low dosage currently recommended by US EPA may make it ineffective. Must be generated on-site.
Chloramines	Mildly effective for bacteria. Long-lasting residual. Generally does not produce THMs.	Some harmful by-products. Toxic effects for kidney dialysis patients. Only recommended as a secondary disinfectant. Ineffective against viruses and cysts.

Table 15
Desired Points of Disinfectant Application[a]

Chlorine	Toward the end of the water treatment process to minimize THM formation and provide secondary disinfection
Ozone	Prior to the rapid mixing step in all treatment proceses, except GAC and convention for GAC; post-sedimentation for conventional treatment. In addition, sufficient time for biodegration of the oxidation products of the ozonation of organic compounds is recommended prior to secondary disinfectant
Ultraviolet radiation	Toward the end of the water treatment process to minimize presence of other contaminants that interfere with this disinfection
Chlorine dioxide	Prior to filteration; to ensure low levels of ClO_2, ClO_2^-, and ClO_3^-, treat with GAC after disinfection
Monochloramines	Best applied toward the end of the process as a secondary disinfectant

[a]In general, disinfectant dosages will be lessened by placing the point of application toward the end of the water treatment process because of the lower levels of contaminants that would interfere with efficient disinfection. However, water plants with short detention time in clear wells and with nearby first customers may be required to move their point of disinfection upstream to attain the appropriate *CT* value under the Surface Water Treatment Rule.

Free Available Chlorine: Chlorine in the form of the molecule (Cl_2), hypochlorous acid (HOCl), and hypochlorite ion (OCl^-).
Free Chlorine: The sum of hypochlorous acid and hypochlorite ions expressed in terms of mg/L or ppm.
Free Residual Chlorination: Chlorination to provide free available chlorine.
Groundwater: The water that systems pump and treat from aquifers (natural reservoirs below the earth's surface).
Haloacetic Acids: A group of disinfection by-products that includes dichloroacetic acid, trichloroacetic acid, monochloroacetic acid, bromoacetic acid, and dibromoacetic acid.
Halogenation: A process of reaction wherein any members of the halogen group are introduced into aqueous organic compounds, either by simple addition or substitution. The unit process of adding any members of the halogen group to water or wastewater mainly for disinfection or oxidation.
Iodination: The unit process of adding a form of iodine to water or wastewater. Iodination is a halogenation process involving the addition of iodine mainly for disinfection.
Maximum Contaminant Level (MCL): The highest level of a contaminant that US EPA allows in drinking water. MCLs are set as close to Maximum Contaminant Level Goals (MCLGs) as feasible using the best available technology (BAT) and taking costs into consideration. MCLs are enforceable standards.
Maximum Contaminant Level Goal (MCLG): The level of a contaminant, determined by US EPA, at which there would be no risk to human health. This goal is not always economically or technologically feasible, and the goal is not legally enforceable.
Microbial Contamination: Contamination of water supplies with microorganisms such as bacteria, viruses, and parasitic protozoa.
Microorganisms: Tiny living organisms (or microbes) that can be seen only with the aid of a microscope. Some microorganisms can cause acute health problems when consumed in drinking water.
Organic Matter: Matter derived from organisms, such as plants and animals.
Ozonation: The unit process of adding ozone gas to water or wastewater. Ozonation is a chemical process for disinfection and/or oxidation.
Parasitic Protozoa: Single-celled microorganisms that feed on bacteria and are found in multicellular organisms, such as animals and people.
Pathogen: A disease-causing organism.
pH: A measure of the acidity or alkalinity of an aqueous solution.
Plain Chlorination: The application of chlorine to an otherwise untreated water supply as it enters the distribution system or pipeline leading thereto.
Postchlorination: The application of chlorine to a water subsequent to any other treatment.
Prechlorination: The application of chloride to a water prior to some other form of treatment.
Raw Water: Water in its natural state, prior to any treatment for drinking.
Rechlorination: The application of chlorine to water at one or more points in the distribution system following previous chlorination.
Risk Assessment: The process evaluating the likelihood of an adverse health effect, with some statistical confidence, for various levels of exposure.
Surface Water: The water sources open to the atmosphere, such as rivers, lakes, and reservoirs.
Trihalomethanes: A group of disinfection byproducts that includes chloroform, bromodichloromethane, bromoform, and dibromochloromethane.
Turbidity: The cloudy appearance of water caused by the presence of tiny particles. High levels of turbidity may interfere with proper water treatment and monitoring.
Ultraviolet Radiation: Radiation in the region of the electromagnetic spectrum including wavelengths from 100 to 3900 angstroms.
Viruses: Microscopic infectious agents, shaped like rods, spheres or filaments that can reproduce only within living host cells.
Waterborne Disease: Disease caused by contaminants, such as microscopic pathogens like bacteria, viruses, and parasitic protozoa, in water.
Watershed: The land area from which water drains into a stream, river, or reservoir.

NOMENCLATURE

β	empirical constant
C	concentration of disinfectant, mg/L
CD	chlorine dosage, mg/L
CR	chlorine requirement, lb/d
CT	constant time at maximum flow, min
CT_{values}	concentration time values
CT_{calc}	calculated CT value
$CT_{99.9}$	CT value for 99.9% kill of organisms
CTL	contact tank length, ft
CTW	contact tank width, ft
K	the equilibrium constant or decay constant
K_1	reaction rate constant at temperature T_1
K_2	reaction rate constant at temperature T_2
K_a	dissociation or acidity constant
K_H	hydrolysis reaction constant
n	constant for particular disinfectant
N	number of organisms surviving to time t
N_0	number of organisms at time 0
N_t	number of organisms at time t
PCR	peak chlorine requirement, lb/d
Q_a	average flow, MGD
Q_p	peak flow, MGD
RLW	length-to-width ratio
SA	surface area, ft^2
SWD	side water depth, ft
T_1	temperature 1
T_2	temperature 2
t	time
VCT	volume of the contact tank, ft^3

REFERENCES

1. G. C. White, *Handbook of Chlorination*, Van Nostrand Reinhold Company, NY, 1972.
2. E. W. Moore, *Water Sewage Works*, **93**, 130 (1951).
3. W. J. Weber, Jr., *Physicochemical Processes for Water Quality Control*, Wiley, New York, 1972.
4. G. M. Fair, J. C. Geyer, and D. S. Okun, *Water and Wastewater Engineering*, Vol. 2, Wiley, New York, 1968.
5. M. L. Granstrom, *Disproportionation of Monochloramine*, Ph.D. thesis, Harvard University, Boston, MA, 1952.
6. J. C. Morris, I. Weil, and R. N. Culver, *Kinetic Studies on the Break point with Ammonia and Glycerine*, Technical Report, Harvard University, Boston, MA, 1952.
7. R. Hulbert, *JAWWA*, **26**, 1638 (1934).
8. AWWA, WEF, and APHA, *Standard Methods for the Examination of Water and Wastewater*. Am. Public Health Assoc., Washington, DC, 2003.
9. J. A. Borchardt, *Symposium—Oxidation and Adsorption*, Dept. of Civil Engineering, Univ. of Michigan, Ann Arbor, MI, 1965.

10. M. Taras, *JAWWA* **45**, 47 (1953).
11. M. L. Granstrom and G. F. Lee, *JAWWA* **50**, 1453 (1958).
12. H. Dodgen and H. Taube, *J Am. Chem. Soc.* **71**, 2501 (1949).
13. R. N. Anston *JAWWA* **39**, 687 (1947).
14. R. J. Mounsey and C. Hagar *JAWWA* **38**, 1051 (1946).
15. R. Coote *Water Sewage Works* **97**, 13 (1950).
16. R. Harlock and R. Dowlin, R *Water Sewage Works*. **100**, 74 (1953).
17. W. C. Ringer and S. J. Cambell, *JAWWA* **47**, 740 (1955).
18. J. F. Maples, *Eff. Water Treatment J.* 370, (1965).
19. J. K. Johannesson, *Am. J. Public Health*, **50**, 1731 (1960).
20. S. L. Chang, *Am. J. Pharm* **48**, 17 (1958).
21. J. C. Morris, *Proc. of Joint Symposium, The Society for Water Treatment and Examination and the Water Research Association*, Univ. of Reading, England, 1970.
22. US EPA, *Interim Primary Drinking Water Standards*, U. S. Federal Register 40, 51, Ph llb90-ll998, Mar. 14, 1975. US Environmental Protection Agency, Washington, DC, 1975.
23. US EPA, *Proposed Criteria for Water Quality*, U. S. Environmental Protection Agency, Washington, DC, Vol. I, 1973.
24. J. H. McDermott, *JAWWA*. **66**, 693 (1974).
25. G. C. White, *JAWWA* **66**, 689 (1974).
26. AWWA *Water Treatment Plant Design*, American Water Works Association, Washington, DC, 1969.
27. R. L. Culp, *JAWWA*. **66**, 699 (1974).
28. M. A. Benarde, B. M. Israel, V. P. Oliveri, and M. L. Granstrom, *Applied Microbiology*, **13**, 776 (1965).
29. M. A. Benarde, W. B. Snow, V. P. Oliveri, and B. Davidson, *Applied Microbiology*, **15**, 257 (1967).
30. T. A. Koski, L. S. Stuart, and L. F. Ortenzio, *Applied Microbiology*, **14,** 68 (1966).
31. J. E. McKee, C. J. Brokaw, and R. T. McLaughlin, *JWPCF*, **32**, 795 (1960).
32. S. L. Chang, *J. Am. Pharm. Edu.*, **47**, 417 (1958).
33. J. C. Morris, S. L. Chang, G. M. Fair, and G. H. Conent, Jr., *Industrial Engineering Chemistry*. **45**, 1013 (1953).
34. S. L. Chang and J. C. Morris, *Industrial Engineering Chemistry* **45**, 1009 (1953).
35. J. J. McCarthy and C. H. Smitli, *JAWWA* **66**, 718 (1974).
36. Black and Veatch, Consulting Engineers, *Estimating Cost and Manpower Requirement for Conventional Wastewater Treatment Facilities*, US Environmental Protection Agency, Washington, DC, 1971.
37 H. G. Bleckor and T. W. Cedman, *Capital and Operating Cost of Pollution Control Modules, Vol. I*, US Environmental Protection Agency Washington, DC, 1973.
38. N. G. Bleckor and T. M. Nichols, *Capital and Operating Cost of Pollution Control Modules, Vol. II*, US Environmental Protection Agency Washington, DC, 1973.
39. T. F. Marhaba, *Water Engineering Management*, January, 30–34 (2000).
40. T. Governor, *Water Engineering Management*, February, 30–33 (1999).
41. A. Lafarge, *Water Engineering Management*, May, 24–26 (1998).
42. P. Westerhoff, J. Debroux, G. L. Amy, D. Gatel, V. Mary, and J. Cavard, *JAWWA,* **92**, 89–102 (2000).
43. F. W. Pontius, *JAWWA,* **92**, 14–22 (2000).
44. A. C. Diehl, G. E. Speitel Jr., J. M. Symons, S. W. Krasner, C. J. Hwang, and S. E. Barrett, *JAWWA*, **92**, 76–89 (2000).
45. W. Sung, B. Reilley-Matthews, D. K. O'Day, and K. Horrigan, *JAWWA,* **92**, 53–63 (2000).
46. D. Rittmann and J. Tenney, *Water Engineering Management*, 22–30 (1998).
47. K. Bell-Ajy, E. Mortezn, D. V. Ibrahim, and M. Lechevallier, *JAWWA*, **92**, 44–53 (2000).

48. M. J. McGuire, N. I. Lieu, and M. S. Peatube, *JAWWA* **91**, 52–61 (1999).
49. M. A. El-Shafy and A. Grunwald, *Water Research* **34**, 3453–3459 (2000).
50. G. Gorden, *JAWWA* **91**, 163–174 (1999).
51. W. Sung, B. Reilley-Matthews, D. K. O'Day, and K. Horrigan, *JAWWA*, **92**, 53–63 (2000).
52. D. C. Alicicia, G. E. Speitel, Jr., J. M. Symons, S. W. Krasner, C. J. Hwang, and S. E. Barrett, *JAWWA*. 92, 76–90 (2000).
53. E. E. Walid, H. Abu-Qdais, and M. K. Elsheamy, *Water Research*, **34**, 3431–3439 (2000).
54. T. Grosvenor, *Water Engineering Management*, 30–39 (1999).
55. K. Hartz, *Water Engineering Management*, August, 21–23 (1999).
56. M. L. Granstrom, *Halogenation*. Technical Report LIR-01-87-224. Lenox Institute of Water Technology (formerly Lenox Intitute for Research), Lenox, MA, 1987.
57. M. Krofta and L. K. Wang, *Water Treatment by Disinfection, Flotation and Ion Exchange Process Systems*. US Department of Commerce, National Technical Information Service, Springfield, VA, PB82-213349, 1982.
58. L. K. Wang, *JAWWA*, **74**, 304–310, (1982).
59. M. Krofta and L. K. Wang, *Over One-Year Operation of Lenox Water Treatment Plant—Part 1*, US Department of Commerce, National Technical Information Service, Springfield, VA, PB83-247270, July, 1983, pp. 1–264.
60. M. Krofta and L. K. Wang, *Over One-Year Operation of Lenox Water Treatment Plant—Part 2*, US Dept of Commerce, National Technical Information Service, Springfield, VA. PB83-247288, July, 1983, pp. 265–425.
61. M. Krofta and L. K. Wang, *Removal of Trihalomethane Precursors and Coliform Bacteria by Lenox Flotation-Filtration Plant*, Water Quality and Public Health Conference, US Department of Commerce, National Technical Information Service, Springfield, VA, Technical Report PB83-244053, 1983, pp. 17–29.
62. M. Krofta and L. K. Wang, *Development of a New Water Treatment Process for Decreasing the Potential for THM Formation*, US Department of Commerce, National Technical Information Service, Springfield, VA, PB81-202541, 1987.
63. L. K. Wang, *Standards and Guides of Water Treatment and Water Distribution Systems*, US Department of Commerce, National Technical Information Service, Springfield, VA, PB88-177902/AS, 1987.
64. L. K. Wang, *Drinking Water Standards and Regulations*, US Department of Commerce, National Technical Information Service, Springfield, VA, PB88-178058/AS, 1987.
65. L. K. Wang, *J. New England Water Works Association*, **89**, 250–270 (1975).
66. L. K. Wang, *Water Resources Bulletin, Journal of American Water Resources Association*, **11**, 919–933, (1975).
67. L. K. Wang, *Proceedings of the Third National Conference on Complete Water Reuse*, June 1976, pp. 252–258.
68. L. K. Wang, *Industrial and Engineering Chemistry, Product Research and Development*, **14**, 308–312 (1975).
69. L. K. Wang, *Water and Sewage Works*, **125**, 30–32, (1978).
70. L. K. Wang, *Water and Sewage Works*, **125**, 58–62, (1978).
71. L. K. Wang, *Water and Sewage Works*, **125**, 99–104, (1978).
72. L. K. Wang, *Pretreatment and Ozonation of Cooling Tower Water, Part I*, US Department of Commerce, National Technical Information Service, Springfield, VA. PB84-192053, 1983.
73. L. K. Wang, *Pretreatment and Ozonation of Cooling Tower Water, Part II*, US Department of Commerce, National Technical Information Service, Springfield, VA, PB84-192046, 1983.
74. L. K. Wang, *A Modified Standard Method for the Determination of Ozone Residual Concentration by Spectrophotometer*, US Department of Commerce, National Technical Information Service, Springfield, VA, PB84-204684, 1983.
75. Editor, *Environmental Protection* **13**, 149 (2002).

76. P. S. Carturight, Water Conditioning and Purification, 45, 68–72, (2003).
77. Editor *Water Quality Products*, **8** (2003).
78. J. L. Cleasby and G. S. Logsdon, *Water Quality and Treatment*, R. D. Letterman, ed., McGraw Hill, New York, 1989, pp. 8.1–8.92.
79. B. Shipe, *Water Conditioning and Purification*, **45**, 34–36 (2003).
80. R. Nathanson, *Water Quality Products* **8**, (2003).
81. US EPA *Technologies for Upgrading Existing or Designing New Drinking Water Treatment Facilities*. EPA/625-4-89/023. US Environmental Protection Agency, Washington, DC, 1989.
82. US Army, *Engineering and Design—Design of Wastewater Treatment Facilities Major Systems*. Engineering Manual No. 1110-2-501. US Army, Washington, DC, 1978.
83. Editor, *Chem. Eng. Equipment Indices*, McGraw-Hill Publishing, New York, 2003.
84. Editor, *Engineering News Record. ENR Cost Indices*, McGraw-Hill Publishing, New York, 2003.
85. Editor, *Chem. Eng.* **107**, 410 (2000).
86. J. C. Wang, D. B. Aulenbach, and L. K. Wang, Energy and Cost Models. In *Clean Production* (Misra, K. B., ed.), Springer-Verlag, Berlin, Germany, 1996, pp. 685–720.
87. L. K. Wang, N. C. Pereira, and Y. T. Hung, (eds.) *Advanced Air and Noise Pollution Control*. Humana Press, Totowa, NJ, 2004.
88. US ACE, *Civil Works Construction Cost Index System Manual*, No. 1110-2-1304, US Army Corps of Engineers, Washington, DC, PP 44 (PDF file is available on the Internet; 2000-Tables Revised 31 March 2003 at http://www.nww.usace.army.mil/cost), 2003.
89. L. K. Wang, *The State-of-the-Art Technologies for Water Treatment and Management*. United Nations Industrial Development Organization (UNIDO), Vienna, Austria. UNIDO Training Manual No. 8-8-95, 1995.
90. HES, *Recommended Standards for Water Works: Policies for the Review and Approval of Plans and Specifications for Public Water Supplies*. Committee of the Great Lakes/Upper Mississippi River Board of Sanitary Engineers. Health Education Service, Albany, NY, 2000.
91. L. K. Wang, *Environmental Engineering Glossary*, Calspan Corp. Buffalo, NY, 420 p, 1974.
92. WEF, *Glossary of Water Environment Terms*. Water Environment Federation, Washington, DC, 2003.
93. M. J. Pidwirny, *Fundamentals of Physical Geography*. Department of Geography, Okanagan University, Canada (http://www.geog.ouc.bc.ca/glossary), 2003
94. D. Roy, E. Englebrecht, and E.S.K. Chian, *JAWWA*, **74**, 660–664 (1982).
95. Sobsey, M. D. (1988). *Detection and Chlorine Disinfection of Hepatitus A in Water*. CR-813-024. US Environmental Protection Agency, Washington, DC.
96. L. K. Wang and S. Kopko, *City of Cape Coral Reverse Osmosis Water Treatment Facility*, US Department of Commerce, National Technical Information Service, Springfield, VA. PB97-139547. 1997 [Proceedings of Advances in Filtration and Separation Technology, American Filtration and Separation Society, Vol. 11, pp. 499–506 1997].
97. L. K. Wang, Y. T. Hung and N. K. Shammas (eds.), *Physicochemical Treatment Processes*. Humana Press, Totawa, NJ. 2004.
98. L. K. Wang, N. K. Shammas and Y. T. Hung (eds.), *Advanced Physicochemical Treatment Processes*. Humana Press, Totawa, NJ. 2005
99. M. H. S. Wang (formerly M. H. Sung). Adsorption of Nitrogen and Phosphorus Using a Continuous Mixing Tank with Activated Carbon. Master thesis. University of Rhode Island, Kingston, RI. 1968.
100. L. K. Wang, Y. T. hung, H. H. Lo and C. Yapijakis (eds.) *Handbook of Industrial and Hazardous Wastes Treatment*. Marcel Dekker Inc., NYC, NY. 2004.

9
Ozonation

Nazih K. Shammas and Lawrence K. Wang

CONTENTS

1. INTRODUCTION

1.1. General

The history of ozone has been documented by several authors (1,2). Experiments conducted in 1886 showed that ozonized air can sterilize polluted water. The first drinking water plant to use ozone was built in 1893 at Oudshoorn, Holland. The French studied the Oudshoorn plant, and, after pilot testing, constructed an ozone water plant at Nice, France in 1906. Because ozone has been used at Nice since that time, Nice is often referred to as "the birthplace of ozonation for drinking water treatment" (3).

In Europe there is a strong commitment to attain a water of the highest chemical quality. There are over 1000 European drinking water plants that use ozone at one or more points in the treatment process. In contrast to the widespread use of ozone for water treatment in Europe, very few European wastewater ozone disinfection systems exist. Currently, there are more ozone disinfection systems in use at US wastewater plants than at US water plants or European wastewater plants.

The first US wastewater plant to use ozone for disinfection was Indiantown, FL, which began operation in 1975 (4). By 1980 about 10 wastewater-treatment plants using ozone for disinfection had been constructed. Unfortunately, some of these earlier ozone disinfection facilities have chosen to abandon ozone disinfection for one or more of the following reasons (5):

From: *Handbook of Environmental Engineering, Volume 3: Physicochemical Treatment Processes*
Edited by: L. K. Wang, Y.-T. Hung, and N. K. Shammas © The Humana Press Inc., Totowa, NJ

(a) Excessive high cost of operation.
(b) Equipment problems.
(c) Excessive maintenance cost.
(d) Inability to attain performance objectives without major modifications.

Despite these early setbacks, by 1985 many facilities, over 40 wastewater-treatment plants, have used or were using ozone disinfection.

1.2. *Alternative Disinfectants*

New water-treatment goals for disinfection by-products (DBP) and for microbial inactivation have increased the need for new disinfection technologies. Water and wastewater systems will need to use disinfection methods that are effective for killing pathogens without forming excessive DBP (6). Ozone is an attractive alternative. This technology has evolved and improved in recent years, thereby increasing its potential for successful application. In August 1997, the US Environmental Protection Agency (US EPA) listed ozone as a "compliance" in the requirements of the Surface Water Treatment Rule for all three sizes of drinking water systems (7). Many of the existing facilities using chlorination may be required to upgrade their present disinfection systems to install alternative disinfection processes.

2. PROPERTIES AND CHEMISTRY OF OZONE

2.1. *General*

Ozone (O_3) is a molecule that can co-exist with air or high-purity oxygen, or can dissolve in water. It is a very strong oxidizing agent and a very effective disinfectant. Ozone is a colorless gas that has an odor most often described as the smell of air after a spring electrical thunderstorm. Some people also refer to the odor as similar to the smell of watermelons. Actually, ozone owes its name to its odor. The word ozone is derived from the Greek word ozein (3). Ozone is an extremely unstable gas. Consequently, it must be manufactured and used onsite. It is the strongest oxidant of the common oxidizing agents. Ozone is manufactured by passing air or oxygen through two electrodes with high, alternating potential difference.

2.2. *Physical Properties*

O_3 is an unstable gas that is produced when oxygen molecules are dissociated into atomic oxygen and subsequently collide with another oxygen molecule (8). The energy source for dissociating the oxygen molecule can be produced commercially or can occur naturally. Some natural sources for ozone production are ultraviolet light from the sun and lightning during a thunderstorm. Ozone may be produced by electrolysis, photochemical reaction, radiochemical reaction, or by "electric discharge" in a gas that contains oxygen (9). The electric discharge principle has been used in most commercial applications and in all known water- and wastewater-treatment applications.

At ordinary temperatures ozone is a blue gas, but at typical concentrations its color is not noticeable unless it is viewed through considerable depth (9). The stability of ozone is greater in air than in water but is not excessively long in either case. The half-life of residual ozone in water has been reported by Grunwell et al. (10) to range from

8 min to 14 h depending on the phosphate and carbonate concentration of the water. In the absence of phosphates and carbonates, water having a pH of 7 was found to have an ozone half-life of 8 min. The residual ozone concentration in water is decreased rapidly by both aeration and agitation of the liquid (9).

Temperature significantly affects the stability of ozone in air or oxygen. In a clean vessel at room temperature the half-life of ozone may range from 20 to 100 h (9). At 120°C (248°F) the half-life is only 11–112 min and at 250°C (482°F) only 0.04–0.4 s. This characteristic of ozone is important for design because cooling of the ozone generators is necessary. Also, good room ventilation is necessary in case an ozone leak occurs, and ozone contained in the off-gas must be destroyed.

Gaseous ozone is explosive at an ozone concentration of 240 g/m^3 (20% by weight in air) (9). Fortunately, the maximum gaseous ozone concentration typically found in water or wastewater ozone disinfection systems does not exceed 50 g/m^3 (4.1% by weight in air). If, however, a medium that can adsorb and concentrate ozone is inappropriately located in the system, then explosive ozone concentrations could develop.

Ozone solubility in water is important because ozone disinfection is dependent on the amount of ozone transferred to the water. Henry's law relative to ozone systems states that the mass of ozone that will dissolve in a given volume of water, at constant temperature, is directly proportional to the partial pressure of the ozone gas above the water (10). Mathematically, Henry's law is expressed as follows:

$$H = P/C \quad \text{or} \quad \mathrm{P} = \mathrm{HC} \tag{1}$$

where P = partial pressure of the gas above the liquid (atm), C = molar fraction of the gas in water at equilibrium with the gas above the water, and H = Henry's law constant (varies with temperature) (atm/mole fraction).

2.3. *Chemical Properties*

Ozone is a very strong oxidizing agent, having an oxidation potential of 2.07 V (1). Ozone will react with many organic and inorganic compounds in water or wastewater. These reactions are typically called "ozone demand" reactions. They are important in ozone disinfection system design because the reacted ozone is no longer available for disinfection. Waters or wastewaters that have high concentrations of organics or inorganics may require high ozone dosages to achieve disinfection. It is very important to conduct pilot plant studies on these wastewaters during ozone disinfection system design in order to determine the ozone reaction kinetics for the level of treatment prior to ozone disinfection.

In most instances, the oxidation reactions produce an end product that is less toxic than the original compound (1,2). Numerous studies have been completed describing the reactions with ozone and various inorganic and organic compounds (11,12). A brief summary of these reactions is given below.

2.3.1. *Reactions with Inorganic Compounds*

The inorganic compounds that most commonly react with ozone in a wastewater-treatment plant are sulfide, nitrite, ferrous, manganous, and ammonium ions. Other reactions may also occur if the wastewater characteristics are affected by an industrial contribution or by in-plant recycle loads. Ozone reactions with various inorganic compounds are summarized below (2):

(a) *Sulfide*: The degree of oxidation of sulfide depends on the amount of ozone used and the contact time. Organic sulfides will oxidize to sulfones, sulfoxides, and sulfonic acids at slower rates than the sulfide ion itself. The sulfide ion will oxidize to sulfur, to sulfite, and to sulfate.
(b) *Nitrogen compounds*: Organic nitrites, nitroso compounds, and hydroxylamines will be oxidized depending on the amount of ozone used and the contact conditions. The oxidation reaction of ammonia is first order with respect to the concentration of ammonia and is catalyzed by OH^- over the pH range 7–9 (13). At an initial ammonium concentration of 28 mg/L as N, a pH of 7.0, and a contact time of 30 min, an 8% reduction of ammonium was reported. At a pH of 7.6, a 26%, at 8.4 a 42%, and at 9.0 a 70% reduction was observed. Narkis et al. (14) report that total oxidation of organic nitrogen and ammonia was never achieved, even at a pH of 12, and at a pH of 6 nitrates were not produced. Nitrite ion is oxidized very rapidly to nitrate ion. This reaction can have a significant effect on ozone disinfection capability when incomplete nitrification occurs. Venosa (15) reported that as much as 2 mg/L of ozone was required to oxidize 1 mg/L of nitrite–nitrogen.
(c) *Iron and manganese*: The reaction with ozone and the ferrous and manganous ions will form an insoluble precipitate. The ferrous ion will be oxidized to ferric, which will react with OH^- to form an insoluble precipitate. Similarly, manganous ions will form manganic ions, which will react with OH^- to form a precipitate.
(d) *Cyanide*: Toxic cyanide ions are readily oxidized by ozone to the much less toxic cyanate ion. At low pH, cyanate ion hydrolyzes to produce carbon dioxide and nitrogen.

2.3.2. Reactions with Organic Compounds

Several investigators developed an in-depth analysis of the reactions for ozone with various organic compounds. These reactions were described by Miller et al. (1), and are summarized below:

(a) *Aromatic compounds*: Phenol reacts readily with ozone in aqueous solution. Oxalic and acetic acids are relatively stable to ozonation in the absence of a catalyst such as ultraviolet light or hydrogen peroxide. Cresols and xylenols undergo oxidation with ozone at faster rates than does phenol. Pyrene, phenanthrene, and naphthalene oxidize by ring rupture. Chlorobenzene reacts with ozone slower than does phenol.
(b) *Aliphatic compounds*: There is no evidence that ozone reacts with saturated aliphatic hydrocarbons under water- or wastewater-treatment conditions. There is no evidence that ozone oxidizes trihalomethanes. Ozone combined with ultraviolet radiation does oxidize chloroform to produce chloride ion, but no identified organic oxidation product. Unsaturated aliphatic or alicyclic compounds react with ozone.
(c) *Pesticides*: Ozonation of parathion and malathion produces paraoxon and malaoxon, respectively, as intermediates, which are more toxic than are the starting materials. Continued ozonation degrades the oxons, but requires more ozone than the initial reaction. Ozonation of heptachlor produces a stable product not yet identified. Aldrin and 2,4,5-T are readily oxidized by ozone, but dieldrin, chlordane, lindane, DDT, and endosulfan are only slightly affected by ozone.
(d) *Humic acids*: Humic materials are resistant to ozonation, requiring long ozonation times to produce small amounts of acetic, oxalic, formic, and terephthalic acids, carbon dioxide, and phenolic compounds. Ozonation of humic materials followed by immediate chlorination (within 8 min) has been shown to reduce trihalomethane formation in some cases. Ozonized organic materials generally are more biodegradable than the starting, unoxidized compounds.

2.4. *Advantages and Disadvantages*

It is important to note that ozone, like other technologies, has its own set of advantages and disadvantages that show up in differing degrees from one location to the next (16). Using ozone has the following advantages:

(a) Possesses strong oxidizing power and requires short reaction time, which enables the pathogens to be killed within a few seconds.
(b) Produces no taste or odor.
(c) Provides oxygen to the water after disinfecting.
(d) Requires no chemicals.
(e) Oxidizes iron and manganese.
(f) Destroys and removes algae.
(g) Reacts with and removes all organic matter.
(h) Decays rapidly in water, avoiding any undesirable residual effects.
(i) Removes color, taste, and odor producing compounds.
(j) Aids coagulation by destabilization of certain types of turbidity.

Among the disadvantages of using ozone are the following:

(a) Toxic (toxicity is proportional to concentration and exposure time).
(b) Cost of ozonation is high compared with chlorination.
(c) Installation can be complicated.
(d) Ozone-destroying device is needed at the exhaust of the ozone reactor to prevent toxicity.
(e) May produce undesirable aldehydes and ketones by reacting with certain organics.
(f) No residual effect is present in the distribution system, thus postchlorination may be required.
(g) Much less soluble in water than chlorine; thus, special mixing devices are necessary.
(h) It will not oxidize some refractory organics or will oxidize too slowly to be of practical significance.

3. APPLICATIONS OF OZONE

Ozone acts both as a very strong oxidizing agent and as a very effective disinfectant. Consequently, it has multiple uses in potable water treatment, wastewater renovation, cooling water towers, groundwater remediation, and industrial waste treatment. The following is a snapshot description of each of its applications (16–19). The oxidation function is discussed fully in a different chapter, while the emphasis in this chapter is on using ozone as a disinfectant.

3.1. *Disinfection Against Pathogens*

Transfer of ozone into the water is the first step in meeting the disinfection objective, because ozone must be transferred and residual oxidants produced before effective disinfection will occur (20). Once transferred, the residual oxidants, such as ozone, hydroxide, or peroxide must make contact with the organisms in order for the disinfection action to proceed. Design of an ozone system as primary treatment should be based on simple criteria, including:

(a) Ozone contact concentrations,
(b) Competing ozone demands
(c) Minimum contact concentration-time (*CT*) to meet the required inactivation requirements, in combination with US EPA recommendations. The *CT* requirement will be discussed in another section.

Systems that need to provide *CT* to comply with the Ground Water Disinfection Rule, but are also having problems with DBP or maintaining distribution system residuals, should consider using ozone as the primary disinfectant and then chloramines for distribution system protection.

Ozone has been observed to be capable of inactivating *Cryptosporidium* and there is significant interest in this aspect of its application (21). Similar findings have been reported for the control of cyanobacterial toxins (microcystins) under various bloom conditions (22). Available data indicate that a significant increase in ozone dose (at least 1.5 mg/L) and CT may be required as compared with past practices. Therefore, these needs in addition to continuous monitoring and ensuring a low total organic carbon (TOC) in the flow should be considered in planning.

3.2. Zebra Mussel Abatement

Zebra mussel (*Dreissena polymorpha*), has arrived in the United States by attaching itself to ships in the infested waters of its natural habitat. Subsequently, it was imported over by the ships to cause infestation in US waters. The growth of zebra mussels on raw water intake pipes decreases the capacity of water transmission to potable-water-treatment plants, cooling towers of power plants, and hatcheries. To restore the full water flow capacity, plant operators have resorted to mechanical means for removing the mussel infestation on intake pipes.

Ozonation seems to be the most promising among the chemical alternatives tested for controlling zebra mussel growth. An 11 MGD side stream ozonation process was designed for a fish culture plant in New England (23). It involved the pumping of high concentration ozone in water solution (15–25 mg/L) into the intake of the raw water pipe and blending it into the intake water to attain a final ozone concentration of 0.1–0.3 mg/L, which is sufficient to control the zebra mussel infestation.

3.3. Iron and Manganese Removal

The standard oxidation–reduction potential and reaction rate of ozone is such that it can readily oxidize iron and manganese in groundwater and in water with low organic content. Groundwater systems that have iron levels above 0.1 mg/L may have iron complaints if ozonation or chlorination is added. Excessive doses of ozone will lead to the formation of permanganate, which gives water a pinkish color. This soluble form of manganese (Mn) corresponds to a theoretical stoichiometry of 2.20 mg O_3/mg Mn.

3.4. Color Removal

Because humic substances are the primary cause of color in natural waters, it is useful to review the reactions of ozone with humic and fulvic acids. According to different authors, ozone doses of 1–3 mg O_3/mg C lead to almost complete color removal. The ozone dosages to be applied in order to reach treatment goals for color can be very high. It is interesting to note that when the ozone dosage is sufficient, the organic structure is modified such that the final chlorine demand can decrease.

Konsowa (24) found that ozone is efficient in the removal of color from textile dyeing wastewater. The rate of dye oxidation was determined to be a function of dye concentration, ozone concentration, ozone-air flow rate and pH. The products produced by the break down of the dye are nontoxic and can be removed by biological treatment.

3.5. *Control of Taste and Odor*

The National Secondary Drinking Water Regulations recommend that the threshold odors number (TON) be 3 or less in finished water. It has been shown that ozone can be effective in treating water for taste and odor problems, especially when the water is relatively free from radical scavengers.

It has also been observed that ozone, in combination with other downstream treatment processes, especially granular activated carbon (GAC) filtration, can greatly increase taste and odor treatment efficiency and reliability. Again, the causes of taste and odor compounds, as well as the source water to be treated, need to be carefully considered prior to designing a treatment system. Analysis and possibly pilot-scale experimentation may be required to determine the optimum choice of ozone and downstream treatment.

3.6. *Elimination of Organic Chemicals*

Ozone or advanced ozonation processes can remove many synthetic organic chemicals (SOC). This removal leads to the chemical transformation of these molecules into toxic or nontoxic by-products. Such transformation can theoretically lead to complete oxidation into carbon dioxide (CO_2); however, this is rarely the case in water treatment. Any observable reduction in total organic carbon (TOC) is due to either a small degree of CO_2 formation (for example, decarboxylation of amino acids) or the formation and loss of volatile compounds through stripping.

3.7. *Control of Algae*

Ozone, like any other oxidant, such as chlorine or chlorine dioxide, has a lethal effect on some algae or limits its growth. Ozone is also capable of inactivating certain zooplankton, e.g., mobile organisms, *Notholca caudata*. Such organisms must first be inactivated before they are removed by flocculation and filtration.

3.8. *Aid in Coagulation and Destabilization of Turbidity*

It is important to understand that the coagulating effects of ozone go beyond any direct oxidative effects on organic macro-pollutants. For this reason, one must be wary of studies claiming improved removal of organic matter when the data are based solely on color removal or ultraviolet (UV) absorption. Also, when studying the removal of DBP such as trihalomethanes, one must be careful to incorporate controls permitting the separate evaluation of ozone's direct effects. Finally, the coagulating effects of ozone may not be observed with all waters. Whenever considering the use of ozone as a coagulant aid, the preozonation effects should be critically evaluated in pilot studies incorporating the proper controls.

4. PROCESS AND DESIGN CONSIDERATIONS

4.1. *Oxygen and Ozone*

Ozone is a powerful oxidizing agent, second only to elemental fluorine among readily available chemical supplies. Because it is such a strong oxidant, ozone is also a powerful disinfectant. Unlike chlorine, ozone does not react with water to produce a disinfecting species. Instead, when exposed to a neutral or alkaline environment (pH above 6), UV light, or hydrogen peroxide, it decomposes in water to more reactive hydroxyl free radicals as shown in the equation below:

$$\underset{\text{ozone}}{O_3} + \underset{\text{water}}{H_2O} \rightarrow \underset{\text{oxygen}}{O_2} + \underset{\text{hydroxyl}}{2OH^{\bullet}} \tag{2}$$

This reaction is accelerated at pH values above 8.

Hoigne and Bader (25) found that ozone will react directly with solutes in the water, and that hydroxide ions (OH^-) and hydroxyl radicals ($OH^{\bullet}$) will provide a catalyst for the decomposition of ozone into intermediate compounds that are also reactive, such as peroxide ions (O_2^-) and the radical $HO_2^{\bullet}$. These results suggest that ozone disinfection is influenced by raw water chemistry characteristics, in addition to the better-known influences of wastewater pollutants. These influences are important to keep in mind when a comparison is made of ozone disinfection performance at different treatment plants.

Because ozone is unstable at ambient temperatures and pressures, it must be generated onsite and used quickly. Ozone is generated by applying energy to oxygen (pure oxygen or dried air). A high-energy electrical field causes oxygen to dissociate according to the equation below:

$$O_2 + e^- \rightarrow 2[O] + e^- \tag{3}$$

These oxygen "fragments" are highly reactive and combine rapidly with molecular oxygen to form the triatomic molecule, ozone:

$$2[O] + 2O_2 \rightarrow 2O_3 \tag{4}$$

The overall reaction that produces ozone is the sum of the above reactions:

$$3O_2 + e^- \leftrightarrow 2O_3 + e^- \tag{5}$$

This reaction is reversible; once formed, ozone decomposes to oxygen. This reverse reaction increases with temperature and occurs quite rapidly above 35°C. Because of this, ozone generators have cooling components to minimize ozone losses by thermal decomposition.

Ozone, as mentioned above, has a characteristic odor that is detectable even at low concentrations (0.01–0.02 ppm by volume). Higher levels may cause olfactory and other reaction fatigue, and much higher levels are acutely toxic. The longer the exposure to ozone, the less noticeable is the odor.

Ozone is only slightly soluble in water depending on the temperature and its concentration as it enters the ozone contactor. The higher the concentration of ozone generated, the more soluble it is in water. Increasing pressure in the ozone contactor system also increases its solubility.

4.2. Disinfection of Water by Ozone

The effectiveness of disinfectants in achieving disinfection is indicated by the *CT* values. They describe the attainable degree of disinfection as the product of the disinfectant residual concentration (*C*) in mg/L and the contact time (*T*) in minutes. For chlorine, chlorine dioxide, or monochloramine, the contact time can be the time required for the water to move from the point at which the disinfectant is applied to the point it reaches the first customer (at peak flow). This is the total time the water is exposed to the chlorinous residual before being used. Ozone, however, has a short half-life in water; therefore, the contact time is considered the time water is exposed to a continuous ozone residual during the water treatment process (26–28).

The Final Surface Water Treatment Rule (SWTR) (US EPA, 1989) states:

> Systems may measure "*C*" (in mg/L) at different points along the treatment train, and may use this value, with the corresponding "*T*" (in minutes), to calculate the total percent inactivation. In determining the total percent inactivation, the system may calculate the *CT* at each point where "*C*" was measured and compare this with the $CT_{99.9}$ value (the *CT* value necessary to achieve 99.9 percent inactivation of *Giardia* cysts) in the rule for specified conditions (pH, temperature, and residual disinfectant concentration). Each calculated *CT* value (CT_{calc}) must be divided by the $CT_{99.9}$ value found in the SWTR tables to determine the inactivation ratio. If the sum of the inactivation ratios, or
>
> $$\sum CT_{\text{calc}}/CT_{99.9} \tag{6}$$
>
> at each point prior to the first customer where *CT* was calculated is equal to or greater than 1.0, i.e., there was a total of at least 99.9 percent inactivation of *Giardia lamblia*, the system is in compliance with the performance requirement of the SWTR.

The final Guidance Manual for the SWTR recommend that systems determine contact time based on the time it takes water with 10% of an approximate tracer concentration (T_{10}) to appear at the sampling site at peak hourly flow. For groundwater not under direct influence of surface water, *CT* is determined in the same manner using enteric viruses or an acceptable viral surrogate as the determinant microorganism, since *Giardia* cysts will not be present.

Table 1 presents the *CT* values required to attain 1-log reductions of *Giardia* cysts for four disinfectants. As shown, lower temperatures require higher *CT* values; with chlorine, an increase in pH also increases the necessary *CT* values. If more than one disinfectant is used, the percentage inactivation achieved by each is additive and can be included in calculating the total *CT* value.

When direct filtration is included in the water treatment process, disinfection credit can be taken by the filtration step for a 2-log inactivation of *Giardia* cysts and a 1-log inactivation of viruses. This means that the primary disinfectant must provide an additional 1-log inactivation of *Giardia* cysts and 3-log inactivation of viruses. In the specific instance of a conventional treatment process that includes coagulation, flocculation, sedimentation, and filtration, an inactivation credit of 2.5 logs for *Giardia* cysts and 2 logs for viruses may be taken. This means that the primary disinfectant must provide an additional 0.5 log inactivation of *Giardia* cysts, but a 2-log inactivation of viruses.

If a water supply system does not use filtration, the 99.9% inactivation of *Giardia* and 99.99% inactivation of enteric viruses must be achieved by the primary disinfecting agents alone. Table 2 presents CT values for the four disinfectants for achieving 99.9% reductions of *Giardia* cysts. Table 3 presents the *CT* values for virus inactivation. These values also apply to systems treating groundwater determined by the State not to be under direct influence of surface water.

Table 1
CT Values for Achieving 90% Inactivation of *Giardia lamblia*

		Temperature					
Disinfectant	pH	≤ 1°C	5°C	10°C	15°C	20°C	25°C
Free chlorine (2 mg/L)	6	55	39	29	19	15	10
	7	79	55	41	28	21	14
	8	115	81	61	41	30	20
	9	167	118	88	59	44	29
Ozone	6–9	0.97	0.63	0.48	0.32	0.24	0.16
Chlorine dioxide	6–9	21	8.7	7.7	6.3	5	3.7
Chloramines (preformed)	6–9	1270	735	615	500	370	250

Source: US EPA 1989.

In the final SWTR (US EPA, 1989), the *CT* values for ozone have been lowered to levels such that the CT values required to provide 0.5-log inactivation of *Giardia* cysts at the higher water temperatures are below those required to provide 2 or 3 logs of inactivation of enteric viruses. Consequently, the 2- or 3-log virus inactivation *CT* requirement becomes the pacing parameter for the amount of additional primary disinfection to be provided by ozone during conventional treatment, rather than the 0.5-log inactivation of *Giardia* cysts.

4.3. *Disinfection of Wastewater by Ozone*

As mentioned above, transfer of ozone into the wastewater is the first step in meeting the disinfection objective. Once transferred, the residual oxidants must make contact with the microorganisms in order for the disinfection action to proceed. Therefore,

Table 2
CT Values for Achieving 99.9% Inactivation of *Giardia lamblia*

		Temperature					
Disinfectant	pH	≤ 1°C	5°C	10°C	15°C	20°C	25°C
Free chlorine (2 mg/L)	6	165	116	87	58	44	29
	7	236	165	124	83	62	41
	8	346	243	182	122	91	61
	9	500	353	265	177	132	88
Ozone	6–9	2.9	1.9	1.4	0.95	0.72	0.48
Chlorine dioxide	6–9	63	26	23	19	15	11
Chloramines (preformed)	6–9	3800	2200	1850	1500	1100	750

Source: US EPA 1989.

Table 3
***CT* Values for Achieving 90% Inactivation of Viruses at pH 6–9**

	Log inactivation	Temperatute					
		0.5°C	5°C	10°C	15°C	20°C	25°C
Free chlorine	2	6	4	3	2	1	1
	3	9	6	4	3	2	1
	4	12	8	6	4	3	2
Ozone	2	0.9	0.6	0.5	0.3	0.25	0.15
	3	1.4	0.9	0.8	0.5	0.4	0.25
	4	1.8	1.2	1.0	0.6	0.5	0.3
Chlorine dioxide	2	8.4	5.6	4.2	2.8	2.1	—
	3	25.6	17.1	12.8	8.6	6.4	—
	4	50.1	33.4	25.1	16.7	12.5	—
Chloramines	2	1,243	857	643	428	321	214
	3	2,063	1,423	1,067	712	534	356
	4	2,883	1,988	1,491	994	746	497

Source: US EPA, 1989.

similar requirements and kinetic relationships used for chlorine disinfectants can also be used for ozone disinfection.

4.3.1. Wastewater Treatment Prior to Ozonation

The US EPA Water Engineering Research Laboratory and other researchers have evaluated several treatment plant effluents to determine the relationship between ozone dosage and total coliform reduction (29–33). The most significant factor influencing the ozone dosage requirement to achieve a desired effluent total coliform concentration was the TCOD (total chemical oxygen demand) concentration of the effluent. For example, at five plants where the TCOD of the secondary effluent was less than 40 mg/L, a total coliform concentration of 1000 per 100 mL could be achieved with ozone dosages between 4 and 7 mg/L (33). However, when Meckes et al. (33) evaluated a plant, which treated a significant amount of industrial waste (TCOD = 74 mg/L), a dosage greater than 12 mg/L was projected in order for the process to meet the 1000 total coliforms per 100 mL limit.

Based on the results of the above researchers, it appears that there is no technical basis for excluding the use of ozone following any treatment scheme (primary, secondary, tertiary, or advanced). However, depending on the type of wastewater treated and/or the effluent disinfection requirement, the wastewater-treatment scheme may be an important economical consideration. A summary of the issues to consider when selecting the wastewater-treatment scheme prior to ozone disinfection is presented below:

(a) Required Effluent Target
  - First: To meet the former US EPA standard of 200 fecal coliforms per 100 mL, tertiary treatment may not be necessary.
  - Second: To meet more stringent standards, such as 14 fecal coliforms per 100 mL, tertiary treatment should be considered.
  - Third: To meet a standard of 2.2 total coliforms per 100 mL, advanced treatment processes prior to the ozone disinfection process may be required.

(b) Influent Coliform Concentration

First: Coliform removal is a function of transferred ozone dosage; thus, the influent coliform concentration will affect the amount of ozone dosage required to meet specific effluent criteria.

Second: Treatment processes that reduce the influent coliform concentration (such as filtration) will decrease the ozone dosage required to achieve a specific effluent standard.

(c) Wastewater Quality Characteristics

First: The ozone demand of the wastewater significantly increases the ozone dosage requirements. A plant with a large industrial contribution may have a large ozone dosage requirement. Pilot testing to establish ozone dosage requirements in these plants is highly recommended.

Second: Incomplete nitrification and a high concentration of nitrite-nitrogen will significantly increase the ozone demand and thus the ozone dosage requirement. The nitrite-nitrogen concentration preferably should be less than 0.15 mg/L to optimize disinfection performance.

4.3.2. *Ozone Dosage Design Considerations*

Proper sizing of the ozone generation equipment is important for meeting desired effluent criteria without excessive capacity that results in high capital costs. To properly establish the ozone production capacity applied ozone dosage (*D*) must be properly selected.

Both applied (*D*) and transferred (*T*) ozone dosages are important in ozone process design. Transferred ozone dosage is typically used for establishing the relationship between ozone dosage and disinfection performance. Once *T* and transfer efficiency (*TE*) are defined, D can be established.

4.3.2.1. Determination of Transferred Ozone Dosage

The transferred ozone dosage (*T*) required to achieve disinfection is dependent on the quality of the wastewater (i.e., potential for chemical reaction with ozone), the plant discharge criteria, and the disinfection performance capability of the ozone contact basin. Because of the variables involved, selection of transferred ozone dosage is probably the most difficult process design consideration. The preferred approach to establishing a design-transferred ozone dosage is to conduct a pilot-plant evaluation on the treated wastewater to be disinfected. The type of pilot-scale ozone generator used is not critical to overall results; however, the type of pilot-scale ozone contact basin must duplicate the proposed full-scale basin for the results to be applicable to full-scale design.

In practice, pilot testing has not been routinely accomplished. Dosage requirements have often been based on published pilot-plant or existing full-scale-plant operating data. However, these data are site specific and may not be directly applicable to other installations. In this section both reported data from existing plants and a rational approach for determination of transferred ozone dosage are discussed.

The transferred ozone dosage (*T*) requirement to achieve various levels of disinfection performance was evaluated by several investigators. Typically, transferred ozone dosages between 4 and 10 mg/L met the former US EPA fecal coliform standard of 200 per 100 mL (33) when the total COD concentration of the treated wastewater was less than 40 mg/L. Transferred ozone dosages greater than 10 mg/L were projected when the wastewater had a large industrial contribution and a COD concentration greater than

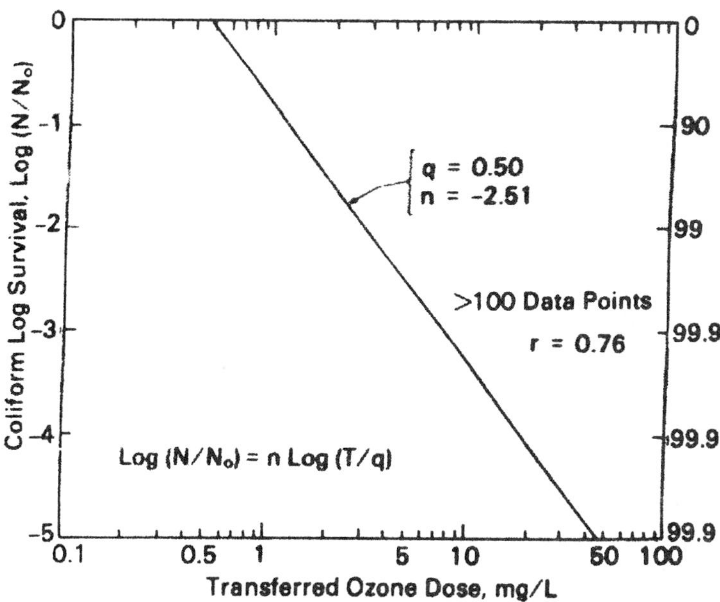

Fig. 1. Dose response curve for ozonation (US EPA).

70 mg/L. To meet a stringent standard of 2.2 total coliforms per 100 mL, a transferred ozone dosage between 36 and 42 mg/L was required when secondary treatment plant effluent was disinfected (29). A transferred ozone dosage between 15 and 20 mg/L was required when nitrified wastewater was disinfected.

The transferred ozone dosage that is required to achieve a desired concentration of coliform organisms in the effluent is dependent on the disinfection performance capability of the ozone contact basin, the demand for ozone in reactions not associated with disinfection, the influent coliform concentration, and the discharge coliform requirement. A change in any of these parameters can cause a significant change in the discharge coliform concentration. The approach to design presented in the remainder of this section allows for an independent evaluation of the effect of each parameter on transferred ozone dosage requirement.

The rational approach to design uses the relationship between coliform removal and transferred ozone dosage as reported by several investigators (30,33-35). A linear-log relationship was indicated for the data points over the range of transferred ozone dosage, although the slope and intercept of the individual lines are quite variable. An example relationship is presented in Fig. 1. It shows that total coliform reduction increased as the transferred ozone dosage increased. The equation of the regression line is:

$$\log(N/N_0) = n\log(T/q) \tag{7}$$

where T = transferred ozone dosage (mg/L), N = effluent coliform concentration (#/100 mL), N_0 = influent coliform concentration (#/100 mL), n = slope of dose/response curve, and q = X-axis intercept of dose/response curve (mg/L) = the amount of ozone transferred before measurable kill is observed.

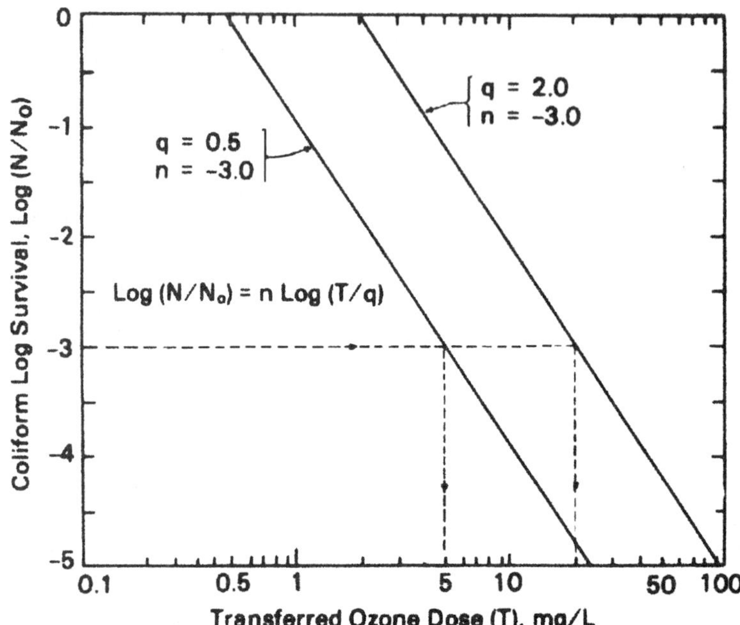

Fig. 2. Effect of *X*-axis intercept (*q*) on dosage response curve for ozonation (US EPA).

By mathematical rearrangement, the slope (n) of the dose/response curve for Stover's results (29) was calculated as –2.51, and the *X*-axis intercept (*q*) was 0.50 mg/L (Fig. 1). A wide range of slope and intercept data are indicated by the other researchers, but individual results appear to accurately describe each operating condition as indicated by the good correlation coefficient obtained in each study. Therefore, specific differences are assumed to contribute to the variation in results for the *X*-axis intercept (*q*) and slope (*n*).

It should be noted that the *X*-axis intercept (i.e., transferred ozone dosage at 100% coliform survival) of the dose/response curve is calculated from the data; it is not a measured value. It is improbable that a straight-line relationship occurs near the *X*-axis intercept, because some degree of coliform reduction would be expected to occur immediately as ozone is transferred to the wastewater. However, from a practical standpoint this reduction is insignificant and the *X*-axis intercept is defined as the transferred ozone dosage where effective disinfection begins to occur.

Note that the term, log-coliform survival, can also be described as percentage reduction of coliform organisms. In Fig. 1 both percentage reduction and log-coliform survival are shown.

In the presentation of this design approach, the *X*-axis intercept is called the initial ozone demand of the wastewater. Generally, initial ozone demand will increase as the quality of the wastewater deteriorates. Factors affecting initial ozone demand are organic and inorganic materials in the wastewater that are readily oxidized by ozone, such as iron, nitrite-nitrogen, and manganese; materials that affect the COD concentration; and other materials. A combination of these materials typically affects the initial ozone demand. Limited data are available to be able to quantify the ozone demand of a

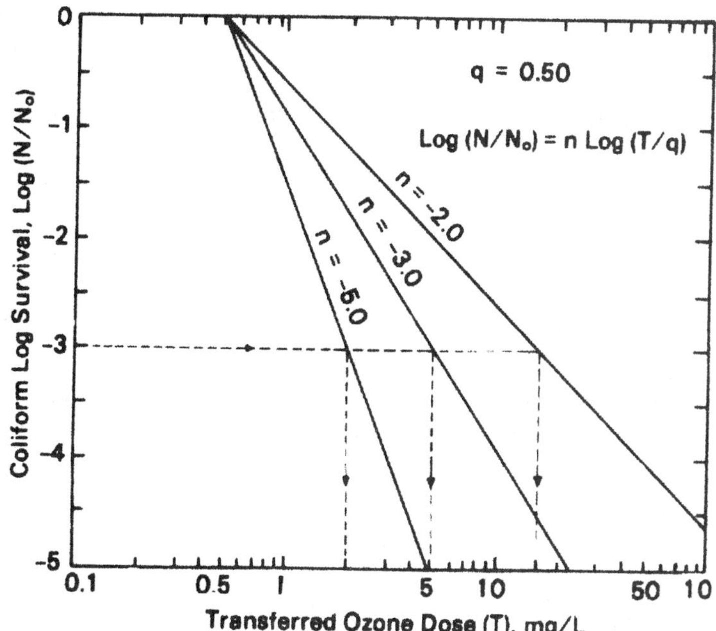

Fig. 3. Effect of slope (n) on dosage response curve for ozonation (US EPA).

particular wastewater; however, from Tables 6–12 some general trends are indicated based on wastewater COD concentration. The X-axis intercept for the studies of the wastewaters with a low COD (20–30 mg/L) were between 0.5 and 1 .0 mg/L; with a moderate COD (30–40 mg/L) between 1.0 and 2.0 mg/L; and with a high COD (74 mg/L) about 5 mg/L. These data may be used to estimate an X-axis intercept for ozone process design, but conservative estimates may be appropriate considering the limited database that is available. It is recommended that pilot or bench-scale testing be completed to better define the X-axis intercept (i.e., initial ozone demand).

The selection of the X-axis intercept will affect the transferred ozone dosage requirement, as shown in Fig. 2. For a high-quality wastewater with an initial ozone demand of 0.5 mg/L a projected transferred ozone dosage of 5 mg/L would be required to achieve a 3-log reduction in coliform organisms when the slope of the dose/response curve is –3.0. For a wastewater with an initial ozone demand four times greater (2.0 mg/L), the projected transferred ozone dosage is four times greater (20 mg/L) to meet the same level of disinfection. Ozone disinfection effectiveness is highly dependent on the initial demand for ozone. Wastewaters with a potential high initial ozone demand may not be good candidates for ozone disinfection systems.

The slope of the dose/response curve represents the change in coliform survival per mg/L transferred ozone dosage. The effect of slope on transferred ozone dosage required to achieve a 3-log reduction in coliform organisms is shown in Fig. 3, assuming the X-axis intercept is 0.5 mg/L. At a relatively steep slope of –5.0, only 2 mg/L transferred ozone dosage is required. At a flatter slope of –3.0, a projected transferred dosage of 5 mg/L is required and at a slope of –2.0 a dosage of 16 mg/L would be necessary.

The slope of the dose/response curve will become flatter when the disinfection performance capability of the ozone contact basin is poorer, or when ongoing chemical reactions with ozone reduce the effectiveness of the disinfectant. For wastewaters that have similar water quality characteristics and a similar initial ozone demand, it is anticipated that the long-term ozone reactions would be similar. Therefore, for these conditions and in the rational design approach, the different slopes are considered to be primarily a function of the disinfection performance capability of the contactor.

An overall review of the dose/response data available indicates general design criteria that may be used for a rational approach to the determination of the design transferred ozone dosage. For a good-quality secondary treatment plant effluent (COD less than 40 mg/L), an initial ozone demand of 1.0 mg/L appears reasonable. If a poorer quality wastewater is anticipated, a higher initial ozone demand should be selected. Conversely, a lower initial ozone demand can be selected if a high-quality wastewater is to be disinfected. Pilot plant results may be used to obtain a reasonably good estimation of the initial ozone demand.

The slope of the dose/response curve is more difficult to establish. Pilot-plant results were generally better than full-scale performance capability; however, the pilot plants had three stages, while the full-scale plants had only one and two. If the field-scale ozone contact basins are designed to match the performance capability of the pilot-scale units (i.e., multiple stages), then the steeper slopes, –4.0 to –5.0, may be used in design. Otherwise, a flatter slope of –3.0 appears justified.

A summary of applicable guidelines for determining the transferred ozone dosage requirement is presented below:

(a) The approach for determination of transferred ozone dosage may be used for ozone process design

First: The initial ozone demand can be estimated based on the quality of the wastewater treated. For a good-quality secondary treatment plant effluent (COD less than 40 mg/L and negligible nitrite nitrogen), an initial ozone demand of 1.0 mg/L appears reasonable.

Second: The slope of the dose/response curve can be based on design features that enhance contact basin disinfection capability. For a contact basin with good design features that emulate reported pilot-scale performance, a slope of –4.0 to –5.0 may be used. Otherwise, a flatter slope should be used.

Third: Influent coliform concentration should be determined based on existing data, if available, or on reported concentrations for similar plants.

Fourth: Effluent coliform concentration should be based on the most stringent design limitations.

(b) To properly establish the transferred ozone dosage requirement, pilot testing should be conducted for all wastewaters and especially for unique ozone disinfection applications such as:

First: Disinfection of "strong" or highly industrial wastewaters.

Second: Disinfection to achieve permit standards more stringent than the former EPA standard of 200 fecal coliforms per 100 mL.

Third: Disinfection using a type of ozone contact basin that does not have a proven record of performance.

(c) Literature-reported ozone dosages may be used for conventional applications of ozone disinfection

First: A transferred ozone dosage between 4 and 10 mg/L appears satisfactory to meet the former EPA standard of 200 fecal coliforms per 100 mL, when disinfecting a good-quality secondary or tertiary treatment plant effluent in a properly designed ozone contact basin.

Second: A transferred ozone dosage between 15 and 20 mg/L reportedly meets the stringent standard of 2.2 total coliforms per 100 mL, when disinfecting good quality tertiary plant effluent in a properly designed ozone contact basin.

Third: A transferred ozone dosage between 36 and 42 mg/L reportedly meets the stringent standard of 2.2 total coliforms per 100 mL, when disinfecting highly polished secondary treatment plant effluent in a properly designed ozone contact basin.

4.3.2.2. Determination of Applied Ozone Dosage

The applied ozone dosage is the mass of ozone from the generator that is directed to a unit volume of the wastewater to be disinfected. The following equation can be used to determine applied ozone dosage:

$$D = T \times 100/TE \tag{8}$$

where D = applied ozone dosage (mg/L), T = transferred ozone dosage (mg/L), and TE = Ozone transfer efficiency (%).

Several researchers (36-38) have evaluated the transfer efficiency *TE* of ozone into wastewater. All conclude that ozone transfer into wastewater can be described by the two-film theory. In this theory the mass transfer of ozone per unit time is a function of the two-film exchange area, the exchange potential, and a transfer coefficient. The exchange area for the bubble diffuser contactor is the surface area of the bubbles. The exchange potential is called the "driving force" and is dependent on the difference between saturation and residual ozone concentrations. However, in practice, contactor basins have not been designed utilizing this theory. The theoretical basis has been avoided because the design coefficients have not been well documented, not because the theory is unsound. When the design coefficients are well documented, contactor design indeed may be established using the two-film transfer model.

Ozone *TE* is primarily influenced by the physical characteristics of the contactor and the quality of the wastewater. At a given applied ozone dosage, a wastewater of poor quality will have a high ozone demand and the contactor will exhibit a high *TE*. The high *TE* is due to the disappearance of ozone in oxidation reactions (i.e., ozone demand reactions). The effect of water quality on transfer efficiency is illustrated in Fig. 4. The *TE* of the same contactor was higher when treating secondary quality wastewater than when treating tertiary quality effluent. The differences in *TE* were more pronounced as applied ozone dosage increased.

The chemical quality of the wastewater also affects ozone *TE*, especially pH and alkalinity. A high pH and/or a low alkalinity will cause a lower ozone residual (i.e., other factors being constant) because the hydroxyl radicals will be maximized. The lower residual will increase the exchange potential, or driving force, and will increase *TE*.

Wastewater quality will affect ozone *TE*, and is important to keep in mind when evaluating the performance of existing contactors, pilot-scale contactors, and newly installed contactors. Wastewater quality is typically not used, as a basis to modify the physical characteristics of the contactor. A summary of the important water quality considerations on ozone *TE* design is listed below:

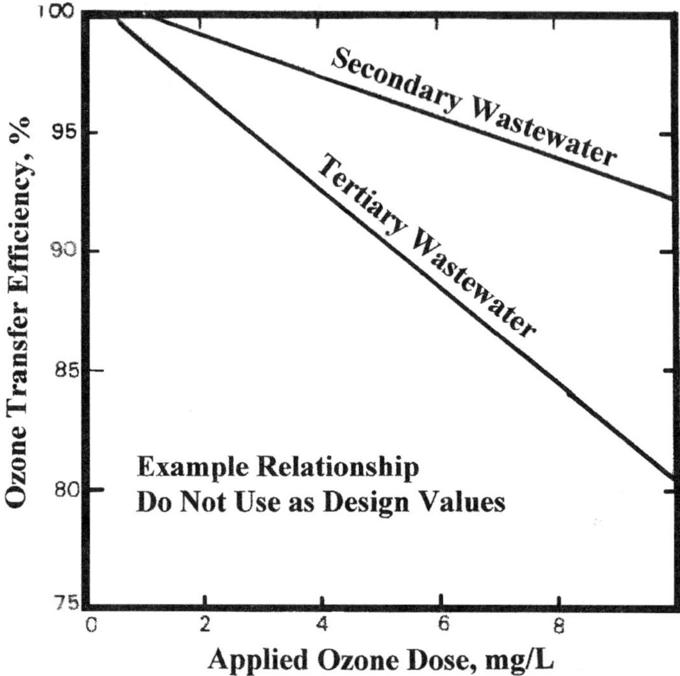

Fig. 4. Ozone transfer efficiency (*TE*) versus applied dosage (US EPA).

(a) Ozone *TE* will decrease as applied ozone dosage increases. A specified minimum design *TE* should be coupled with a specified applied ozone dosage.
(b) Ozone *TE* will increase as wastewater quality deteriorates (i.e. ozone demand increases). A specified minimum design *TE* should be coupled with a specified description of the wastewater quality.
(c) Ozone *TE* will increase as wastewater chemical quality favors the presence of hydroxyl radicals such as a high pH or low alkalinity. A comparison of *TE* of existing full-scale and pilot-scale results should consider differences in wastewater chemical quality.

The physical characteristics of the ozone contactor are the most important considerations for the design engineer because the engineer controls this element of the process. The most important physical characteristics for optimizing ozone *TE* are depth of the contactor and type and location of the diffusers. The contactors physical characteristics are discussed in another section of this chapter.

Results of disinfection by ozonation have been reported by various sources as follows:

(a) Secondary effluent: Dosage = 5–6 mg/L, contact time = < 1 min.
Final effluent: Less than 2 fecal coliforms/l00 mL.
(b) Secondary effluent: Dosage = 10 mg/L, contact time = 3 min.
Final effluent: 99% inactivation of fecal coliform.
(c) Secondary effluent: Dosage = 2–4 mg/L, contact time = 14 min.
Final effluent: Less than 200 fecal coliform/100 mL.
(d) Drinking water: dosage = 4 mg/L, contact time = 8 min.
Final effluent: sterilization of virus.

Table 4
Disinfectants and Disinfectants By-Products Listed in the DWPL (Drinking Water Priority List)

Disinfectant
Chlorine
Hypochlorite ion
Chlorine dioxide
Chlorite ion
Chlorate ion
Chloramine
Ammonia
Haloacetonitriles
Bromochloroacetonitrile
Dichloroacetonitrile
Dibromoacetonitrile
Trichloroacetonitrile
Trihalomethanes
Chloroform
Bromoform
Bromodichloromethane
Dichlorobromomethane
Halogenated acids, alcohols, aldehydes, ketones, and other nitriles
Others
Chloropicrin (trichloronitromethane)
Cyanogen chlorides
Ozone by-products

Source: US EPA.

4.4. Disinfection By-Products

Because chlorine produces undesirable by-products regulated by EPA, its use for pretreatment or primary disinfection of potable water and for disinfection of wastewater must be carefully scrutinized. In addition, regulation of some of the other halogenated by-products of chlorination listed in Table 4 may place even greater restrictions on chlorine use.

The alternative use of ozonation has generated much interest because of its ability to avoid the formation of halogenated organics inherent in the practice of chlorine treatment. However, raw water quality significantly affects ozonation results and could lead to the formation of other undesirable byproducts (39–41).

Disinfection by-products are formed by two basic mechanisms:

1. Reduction, oxidation, or disproportionation of the disinfecting agent.
2. Reaction of oxidation by the disinfectant with materials already in the water.

Reduction, oxidation, or disproportionation can occur when the disinfecting agent is added to water. Three examples of this reaction are the formation of chlorite and chlorate ions associated with chlorine dioxide, the formation of dissolved oxygen associated with ozone, and the formation of chloride ions associated with chlorine.

Oxidation of humic acids (in the water from organic materials) produces aldehydes, ketones, alcohols, and carboxylic acids upon the addition of ozone, chlorine, chlorine dioxide, or potassium permanganate. Halogenation of organic materials can occur in the presence of free chlorine to produce trihalomethanes and other halogenated organics. Chlorine can also form organic chloramines by reacting with nitrogen-containing organic compounds (amino acids and proteins). In addition, monochloramine can produce organic chloramines in the presence of organonitrogen compounds.

If bromide ion is present in the untreated water, it may be oxidized by ozone or chlorine (but apparently not by chlorine dioxide or chloramine) to form hypobromous acid, which in turn can brominate organic materials. Bromine-containing trihalomethanes, for example, are known to form in this manner (42).

By-products are also produced when oxidants, like ozone or chlorine, are used for oxidation purposes other than disinfection. For instance, breakpoint chlorination is sometimes used early in the water-treatment process to remove ammonia. In the presence of organic compounds considered precursors, the same by-products that are formed during chlorine disinfection are also formed in this oxidation step.

As another example, ozone is used as an oxidant to improve turbidity, color, taste, odor, or microflocculation; or to oxidize organic compounds, iron, or manganese. The addition of ozone early in the treatment process as an oxidant may produce the same by-products as when added later in an ozone disinfection process. Potassium permanganate, also used as an early oxidant, can produce oxidation by-products as well. The maximum concentration of by-products is usually produced when oxidants are used at the point in the treatment process where the concentration of organics capable of being oxidized is greatest and/or when large amounts of oxidizing agents are used for long contact times.

Even when oxidants are used in the treatment process for purposes other than disinfection, some degree of disinfection occurs. In some cases, especially in treatment processes involving ozone, chlorine dioxide, and chlorine under lower pH conditions, the primary disinfection requirement may be satisfied during the preoxidation procedure (prior to filtration).

Because oxidation is so important in determining disinfection by-products, a brief description of the chemistry of oxidation is provided in the following Section.

4.5. *Oxidation by Ozone*

The measure of an agent's ability to oxidize organic material is its oxidation potential (measured in volts of electrical energy). Oxidation potential indicates the degree of chemical transformation to be expected when using various oxidants. It gauges the ease with which a substance loses electrons and is converted to a higher state of oxidation. For example, if substance A has a higher oxidation potential than substance B, substance B theoretically can be oxidized by substance A. Conversely, a particular substance cannot oxidize another with a higher oxidation potential. For example, ozone and chlorine can oxidize bromide ions to hypobromous acid, but chlorine dioxide cannot. The oxidation potentials of common oxidants and disinfectants associated with water treatment are listed in Table 5.

An agent's effectiveness as a disinfectant is not always related to its effectiveness as an oxidant. For example, whereas ozone is a powerful oxidant and disinfectant, hydrogen peroxide and potassium permanganate are powerful oxidants but poor disinfectants. Chlorine dioxide and iodine are weak oxidants but strong disinfectants.

Table 5
Oxidation Potentials of Water Treatment Oxidants

Species		Oxidation Potential (V)
Hydroxyl free radical	$(OH)^{\bullet}$	2.80
Ozone[a]	O_3	2.07
Hydrogen peroxide	H_2O_2	1.76
Permanganate ion	MnO_4^-	1.68
Hypochlorous acid[a]	HOCl	1.49
Chlorine[a]	Cl_2	1.36
Hypobromous acid[a]	HOBr	1.33
Bromine[a]	Br_2	1.07
Hypoidous acid	HOI	0.99
Chlorine dioxide[a]	ClO_2(aq)	0.95
Iodine[a]	I_2	0.54
Oxygen	O_2	0.40

Source: US EPA.
[a]Excellent disinfecting agents.

Oxidation potential does not indicate the relative speed of oxidation nor how complete the oxidation reactions will be. Complete oxidation converts a specific organic compound to carbon dioxide and water. Oxidation reactions that take place during water treatment are rarely complete; therefore, partially oxidized organic compounds, such as aldehydes, ketones, acids, and alcohols, normally are produced during the relatively short reaction periods.

The behavior of a disinfectant as an oxidant will also depend on the particular organic compounds in the water supply. The level of total organic carbon (TOC) and the total organic halogen formation potential (TOXFP), when chlorine is used, indicate the likelihood that undesirable halogenated by-products will be formed. Simply monitoring the reduction in concentration of a particular organic compound, however, is insufficient to indicate how completely oxidation reactions are taking place. Unless a compound is totally oxidized to carbon dioxide and water, the TOC level may not change; therefore, the concentrations of oxidation products must also be measured. The TOXFP and the nonvolatile TOXFP, referred to as the nonpurgeable TOXFP (NPTOXFP), indicate the potential for halogenated by-products to be formed from a specific raw water source.

Ozone is the strongest disinfectant and oxidizing agent available for water treatment; however, it is an unstable gas and must be generated on site. In addition, it is only partially soluble in water, so efficient contact with the water must be established and excess ozone from the contactor must be handled properly. Other sections discuss the specifics of ozone generation and contacting methods. Ozone cannot be used as a secondary disinfectant because it is unable to maintain an adequate residual in water for more than a short period of time.

Although the capital costs of ozonation systems are high, their operating costs are moderate. Because of its high oxidation potential, ozone requires short contact times and low dosages for disinfection and oxidative purposes. As a microflocculation aid, ozone is added during or before the rapid mix step and its usage is followed by coagulation and direct or conventional filtration. Higher dosages are used to oxidize undesirable inorganic materials, such as iron, manganese, sulfide, nitrite, and arsenic; or to treat organic materials responsible for tastes, odors, color, and trihalomethane (THM) precursors.

Design criteria can be summarized as follows:

(a) Contact time: 1–90 min
(b) Dosage rate: 10–300 mg/L
(c) pH range: 5–11 (6–8 optimum)
(d) Ozone production:
From oxygen: 4.5 kwh/lb
From air: 7.5 kwh/lb

Ozone does not directly produce any halogenated organic materials, but if bromide ion is present in the raw water, it may do so indirectly. Ozone converts bromide ion to hypobromous acid, which can then form brominated organic materials (42). The primary by-products of ozonation are oxygen-containing derivatives of the original organic materials, mostly aldehydes, ketones, alcohols, and carboxylic acids. Ozone, however, produces toxic oxidation products from a few organic compounds. For example, the pesticide heptachlor forms high yields of heptachlor epoxide upon ozonation. Therefore, when selecting ozone for oxidation and/or disinfection purposes, one must know the specific compounds present in the raw water so as to provide the appropriate downstream treatment to cope with by-products. Researchers are continuing to study ozonation by-products and their potential health effects.

Even when ozone is used to oxidize rather than disinfect, primary disinfection is attained simultaneously provided contact times and dissolved ozone concentrations are appropriate. Consequently, both oxidation and primary disinfection objectives can be satisfied with ozone prior to filtration, after which only secondary disinfection is needed. There are two cases in which this one-step oxidation/disinfection with ozone is not feasible:

1. When high concentrations of iron or manganese are in the raw water,
2. When ozone is used for turbidity control.

In both of the above cases, measurement of the degree of disinfection (dissolved ozone concentrations) is impractical. When iron or manganese is in the water, ozonation precipitates dark insoluble oxides that interfere with the measurement of dissolved ozone. When ozone is used for turbidity control, such low dosages of ozone are used that a measurable concentration of dissolved ozone may never be attained. In these two cases, ozone oxidation and disinfection must occur separately.

After being partially oxidized by ozone, organic materials become more biodegradable and, therefore, more easily mineralized during biological filtration. Preozonation of water fed to slow sand filters increases the ease of biodegradation of organic materials and enhances biological removal of organic materials during GAC (granular activated carbon) filtration. The adsorptive efficiency of the GAC is extended because it only has to adsorb the organics unchanged by ozone, while the partially oxidized organics are biologically converted to carbon dioxide and water.

Primary disinfection (or oxidation) with ozone produces a significant amount of assimilable organic carbon (AOC) comprised of readily biodegradable aldehydes, acids,

Table 6
Treatment Technologies Evaluated by DWRF (Drinking Water Research Foundation)

Conyaminant (or contaminant classess)	Proven effective in field tests	Proven effective in pilot tests	Being evaluated as promising technologies
Volatile organic compounds	Carbon adsorption, packed tower, and diffused aeration	—	Ozone oxidation, reverses osmosis, ultraviolet treatment
Synthetic organic compounds	Carbon adsorption	—	Conventional treatment with powdered activated carbon, ozone oxidation, reverse osmosis, ultraviolet treatment
Nitrates	Ion exchange	Reverse osmosis	
Radium	Reverse osmosis, ion exchange	—	—
Radon	Carbon adsorption	—	Aeration
Uranium	Ion exchange	Reverse osmosis	—

Source: US EPA.

ketones, and alcohols. Many of these are also precursors of THM and TOX compounds. Consequently, if ozone disinfection is immediately followed by chlorination, higher levels of THM and TOX compounds may be produced than without ozonation.

The treatment technologies that have been evaluated by the Drinking Water Research Foundation (DWRD) are listed in Table 6. A matrix of five technologies applicable to the treatment of VOCs (volatile organic contaminants) and their removal efficiencies for 33 compounds is shown in Table 7. Ozone oxidation has been found to be an excellent application for 40% of the listed compounds. Ozone was shown to be effective in removing aromatic compounds, alkenes and certain pesticides. It is not effective, however, in removing alkanes.

4.6. *Advanced Oxidation Processes*

Ozone used in combination with ultraviolet (UV) radiation or hydrogen peroxide can adequately disinfect and, at the same time, oxidize many refractory organic compounds such as halogenated organics present in raw water. Although contact times for ozone disinfection are relatively short, they are quite long for oxidizing organic compounds. This combination process accelerates the oxidation reactions.

Advanced oxidation processes involve combining ozonation with UV radiation (UV_{254} bulbs submerged in the ozone contactor) and hydrogen peroxide (added prior to ozonation) or simply by conducting the ozonation process at elevated pH levels (between 8 and 10). Under any of these conditions, ozone decomposes to produce the hydroxyl free radical, which has an oxidation potential of 2.80 V compared with 2.07 V for molecular ozone. However, hydroxyl free radicals have very short half-lives, on

Table7
Performance Summary for Five Organic Removal Technologies

Organic Compounds	Removal Efficiency				
	Granular activated carbon adsorption[b]	Packed column aeration	Reverse osmosis thin film composite	Ozone oxidation (2–6 mg/L)	Conventional treatment
Volatile Organic Contaminants					
Alkanes					
Carbon tetrachloride	++	++	++	0	0
1,2-Dichlroethane	++	++	+	0	0
1,1,1- Trichlroethane	++	++	++	0	
1,2-Dichlropropane	++	++	++	0	0
Ethylene dibromide	++	++	++	0	0
Dibromochloropropane	++	+	NA	0	0
Alkenes					
Vinyl chloride	++	++	NA	++	0
Styrene	NA	NA	NA	++	0
1,1-Dichlroethylene	++	++	NA	++	0
cis- 1,2- Dichlroethylene	++	++	0	++	0
trans- 1,2- Dichlroethylene	++	++	NA	++	0
Trichloroethylene	++	++	++	+	0
Aromatics					
Benzene	++	++	0	++	0
Toluene	++	++	NA	++	0
Xylene	++	++	NA	++	0
Ethylbenzene	++	++	0	++	0
Chlorobenzene	++	++	++	+	0
o- Dichlorobenzene	++	++	+	+	0
p- Dichlorobenzene	++	++	NA	+	0
Pesticides					
Pentachlorophenol	++	0	NA	++	NA
2,4-D	++		NA	+	0

Alachlor	++	++	++	++	0
Aldicarb	NA	0	NA	NA	NA
Carbofuran	++	0	++	++	0
Lindane	++	0	NA	0	0
Toxaphene	++	++	NA	NA	0
Heptachlor	++	++	NA	++[c]	NA
Chlorodane	++	0	NA	NA	NA
2,4,5-TP	++	NA	NA	+	NA
Methoxychlor	++	NA	NA	NA	NA
Other					
Acrylamide	NA	0	NA	NA	NA
Epichlorohydrin	NA	0	NA	0	NA
PCBs	++	++	NA	NA	NA

Source: US EPA

[a]++ = Excellent 70–100%

+ = Average 30–69%

0 = Poor 0–29%

NA = Data not availabe or compound has not been tested by US EPA Drinking Water Research Division.

[b]Excellent removal category for carbon indicates compound has been demonstrated to be adsorbable onto GAC, in full- or pilot-scale applications, or in the labroatory with charachteristics suggesting GAC can be a cost-effective technology.

[c]Ozone oxidation of heptachlor produces a high yeild of heptachlor epoxide, which is not suitable for futher oxidation.

the order of microseconds, compared with much longer half-lives for the ozone molecule.

Many organic compounds that normally are stable under direct reaction with the ozone molecule can be oxidized rapidly by the hydroxyl free radical. Alkanes and chlorinated solvents such as trichloroethylene (TCE) and tetrachloroethylene can be destroyed rapidly and cost effectively by hydroxyl free radicals

4.6.1. *Ozone with High pH Levels*

Ozone, at low pH levels (less than 7), reacts primarily as the O_3 molecule by selective and sometimes relatively slow reactions. Ozone at elevated pH (above 8) rapidly decomposes into hydroxyl free radicals, which react very quickly. Many organic compounds that are slow to oxidize with ozone oxidize rapidly with hydroxyl free radicals.

The alkalinity of the water is a key parameter in advanced oxidation processes. This is because bicarbonate and carbonate ions are excellent scavengers for free radicals. Consequently, advanced oxidation processes are incompatible with highly alkaline water. In addition, carbonate ions are 20-30 times more effective in scavenging for hydroxyl free radicals than bicarbonate ions. Therefore, ozonation at high pH should be conducted below 10.3 at which level all bicarbonate ions convert to carbonate ions.

4.6.2. *Ozone with Hydrogen Peroxide*

The combination of ozone with hydrogen peroxide much more effectively reduces levels of trichloroethylene (TCE) and tetrachloroethylene than ozone alone. A significant advantage of the peroxide process over GAC (granular activated carbon) and PTA (packed tower aeration) is the absence of vapor controls.

4.6.3. *Ozone/Ultraviolet*

In ozone/ultraviolet (UV) treatment, ozone catalyzed by UV oxidizes organic substances. This process breaks down the saturated bonds of the contaminant molecules. Typical contact time is 0.25 h. A major advantage of this system is that it does not produce any THMs. These systems also do not require waste disposal because the contaminants are destroyed.

There is some concern about the completeness of the ozone/UV oxidation process and the intermediate breakdown products. If oxidation is incomplete, some of the compounds produced in the intermediate reactions may still be available to form THMs. The influent contaminant profile also affects the performance of these systems. However, if oxidation is followed by a biological filtration step, particularly GAC on sand or GAC adsorber, these oxidation products are mineralized into carbon dioxide and water. Consequently, THM formation potential and TOX formation potential are lowered.

5. OZONATION SYSTEM

The five major elements of an ozonation system are:

1. Air preparation or oxygen feed.
2. Electrical power supply.
3. Ozone generation.
4. Ozone contacting.
5. Ozone contactor exhaust gas destruction.

5.1. Air Preparation

Ambient air should be dried to a maximum dew point of –65°C before use in the ozonation system. Using air with a higher dew point will produce less ozone, cause slow fouling of the ozone production (dielectric) tubes, and cause increased corrosion in the ozone generator unit and downstream equipment. These last two factors result in increased maintenance and downtime of the equipment (2,43). Post-desiccant filters are installed to remove particulates less than 0.3–0.4 μm in diameter. Two-stage filtration is recommended. The first-stage filter removes particulates greater than 1 μm and the second stage removes particulates less than 0.3–0.4 μm in diameter (34,44,45).

Air feed systems can dry ambient air or use pure oxygen. Using pure oxygen has certain advantages that have to be weighed against its added expense. Most suppliers of large-scale ozone equipment consider it cost effective to use ambient air for ozone systems having less than 1,590 kg/d (3,500 lb/d) generating capacity. Above this production rate, pure oxygen appears to be more cost effective. Systems that dry ambient air consist of desiccant dryers, commonly used in conjunction with compression and refrigerant dryers for generating large and moderate quantities of ozone. Very small systems (up to 0.044 m^3/s) can use air-drying systems with just two desiccant dryers (no compression or refrigerant drying). These systems use silica gel, activated alumina, or molecular sieves to dry air to the necessary dew point (–65°C).

Ambient air-feed systems used for ozone generation are classified by low, medium, or high operating pressure. The most common type is a system that operates at medium pressures ranging from 0.7 to 1.05 kg/cm^2 (10 to 15 psig). High-pressure systems operate at pressures ranging from 4.9 to 7.03 kg/cm^2 (70 to 100 psig) and reduce the pressure prior to the ozone generator. Low- and high-pressure systems are typically used in small- to medium-sized applications. Medium- and high-pressure systems may be used in conjunction with most ozone generators and with most contacting systems. Low-pressure systems operate at subatmospheric pressures, usually created by a submerged turbine or other contactors producing a partial vacuum throughout the air preparation and ozone generation system. Creation of this vacuum results in ambient air being drawn into the ozonation system.

The decision to use a high-, medium-, or low-pressure air preparation system often is based on a qualitative evaluation of potential maintenance requirements, as well as an evaluation of capital and operating costs. High-pressure air pretreatment equipment generally has higher air compressor maintenance requirements, lower desiccant dryer maintenance requirements, and lower capital costs. At small- to medium-sized installations, the lower capital costs may offset the additional maintenance required for the air compressors and associated equipment, such as filters for the oil-type compressors. Schematic diagrams of low- and high-pressure feed gas pretreatment systems are shown in Figs. 5 and 6. It must be pointed out that Fig. 5 is also representative of a medium-pressure system, but may require a pressure-reducing valve (PRV) upstream from the ozone generator as shown in Fig. 6. Each diagram illustrates a dual component process, and depicts the desired flexibility for the provided equipment.

For many applications, pure oxygen is more attractive as feed gas than air for the following reasons:

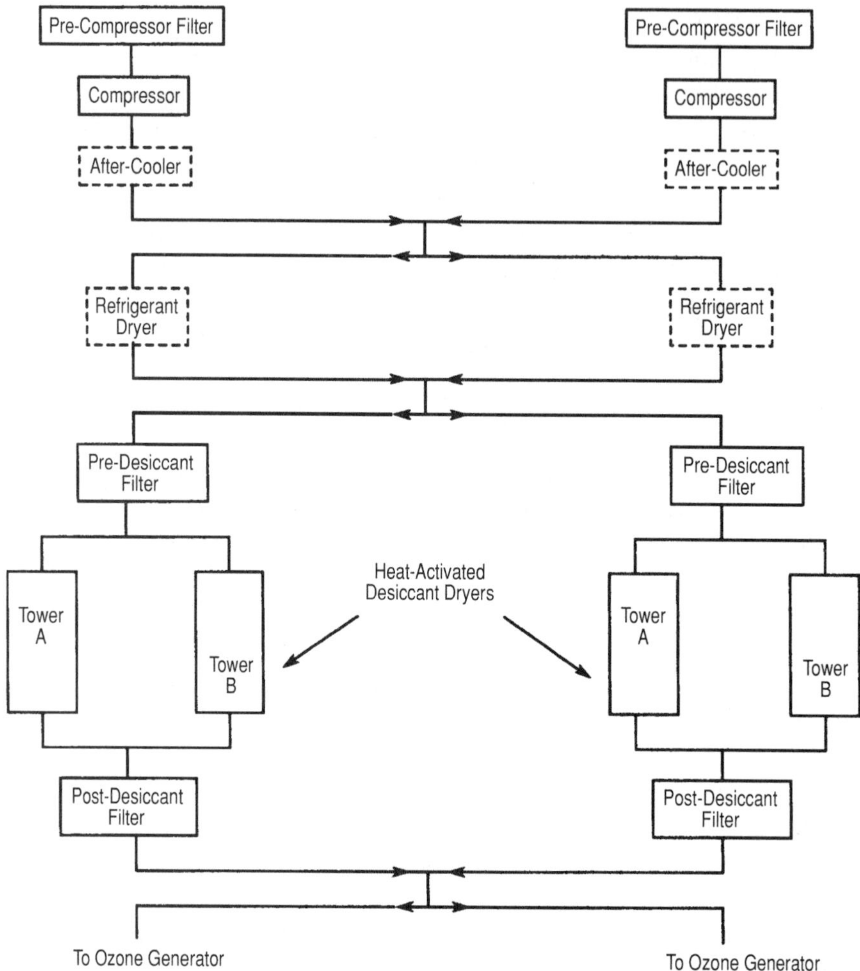

Fig. 5. Schematic diagram for low pressure air feed gas treatment system (US EPA).

(a) It has a higher production density (more ozone produced per unit area of dielectric).
(b) It requires lower energy consumption (energy supplied per unit area of dielectric).
(c) Essentially double the amount of ozone can be generated per unit time from oxygen than from air (for the same power expenditure); this means that ozone generation and contacting equipment can be halved in size when using oxygen, to generate and contact the same amount of ozone.
(d) Smaller gas volumes are handled using oxygen, rather than air, for the same ozone output; thus, costs for ancillary equipment are lower with oxygen feed gas than with air.
(e) If used in a once-through system, gas recovery and pretreatment equipment are eliminated.
(f) Ozone transfer efficiencies are higher due to the higher concentration of ozone generated.

However, the economic implications of these advantages must be weighed against the capital expenditure required for onsite oxygen production or operating costs associated with purchase of liquid oxygen produced off site. Oxygen can be purchased as a gas (pure or mixed with nitrogen) or as a liquid (at −183°C or below). Normally the purity of the

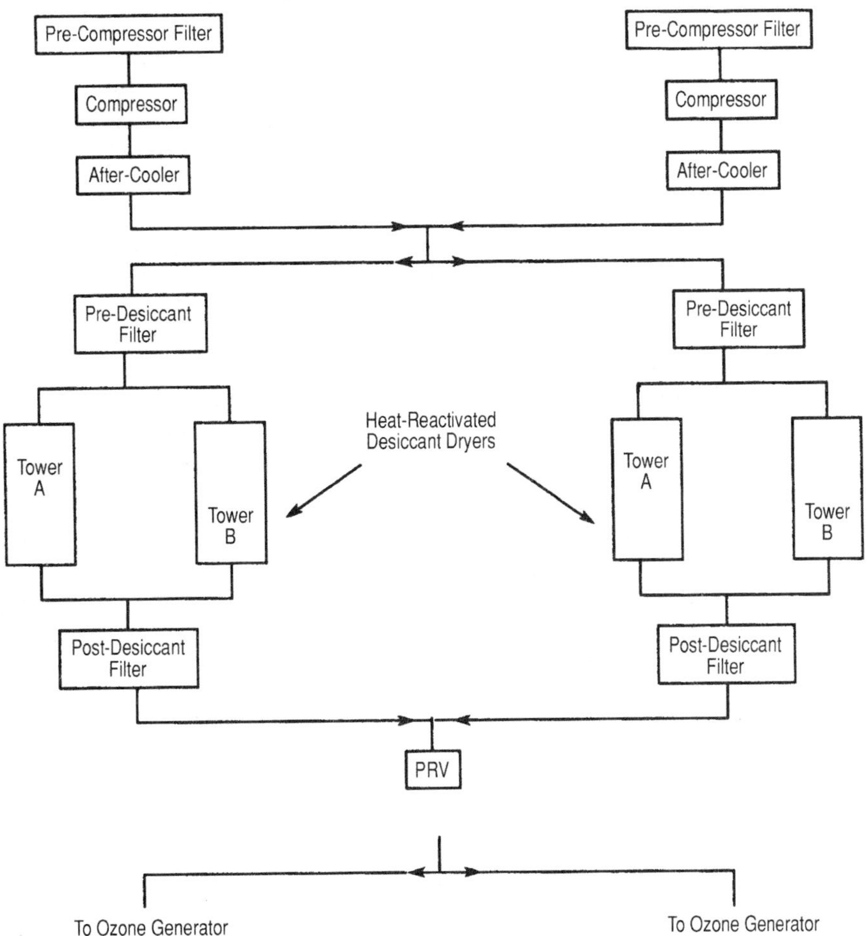

Fig. 6. Schematic diagram for high-pressure air feed gas treatment system (US EPA).

available oxygen gas is quite adequate, and no particular pretreatment is required. Its purity always should be better than 95%, and its dew point consistently lower than –60°C. When liquid oxygen is the oxygen source, it is converted to the gas phase in an evaporator, and then sent directly to the ozone generator. Purchasing oxygen as a gas or liquid is only practical for small- to medium-sized systems. Liquid oxygen can be added to dried air to produce oxygen-enriched air, as at the Tailfer plant serving Brussels, Belgium.

There are currently two methods for producing oxygen on site for ozone generation; pressure swing adsorption of oxygen from air and cryogenic production (liquefaction of air followed by fractional distillative separation of oxygen from nitrogen). Systems for the production of oxygen on site contain many of the same elements as air preparation systems discussed above, since the gas stream must be clean and dry in order to successfully generate ozone. Gaseous oxygen produced on site by pressure swing adsorption typically is 93–95% pure, while liquid oxygen produced cryogenically generally is 99.5% pure. In most plants utilizing on site production of ozone, a backup liquid oxygen storage system is included.

At smaller plants, oxygen can be separated from ambient air by pressure swing adsorption using molecular sieves. During the high-pressure phase, the sieves adsorb nitrogen and oxygen exits the system as product gas. When the pressure is reduced, the nitrogen desorbs and is removed from the vessel by purge gas. Precautions should be taken to avoid contamination of the oxygen prepared by this procedure with hydrocarbons, which are present as oils associated with the pressurized equipment. Pressure swing adsorption systems for producing oxygen are manufactured to produce from 90 to 27,000 kg/d (200 to 60,000 lb/d) of oxygen. This production range would supply ozone for water-treatment systems at 6% concentration in oxygen, for an applied ozone dosage of 4 mg/L, and production of water at the rates of about 0.02 to about 5 m^3/s (0.4 to about 120 MGD).

For cryogenic oxygen production, low-temperature refrigeration is used to liquefy the air, followed by column distillation to separate oxygen from nitrogen. Cryogenic systems are operationally sophisticated, and operating and maintenance expertise is required. Production of oxygen by the cryogenic technique is more capital intensive than by pressure swing adsorption, but generally operation and maintenance costs are lower. For oxygen requirements of about 18,000 to 18,000,000 kg/d (20 to 20,000 ton/d), cryogenic separation systems are quite practical. This would exclude their use for small water-treatment plants. The Los Angeles Aqueduct Filtration Plant that is treating about 26 m^3/s (600 MGD) is using a 3,600-kg/d (8,000-lb/d)-ozone system with cryogenically produced oxygen as a feed gas. Reuse or recycling of the oxygen-rich contactor off-gases is possible, and requires removal of moisture and possibly ammonia, carbon dioxide, and nitrogen before returning the gas to the generator.

5.2. *Electrical Power Supply*

The voltage or frequency supplied to the ozone generator is varied to control the amount and rate of ozone produced. Varying the power requires specialized supply equipment that should be designed for and purchased from the ozone generator manufacturer.

Ozone generators use high voltages (>10,000 V) or high-frequency electrical current (up to 2,000 Hz), necessitating special electrical design considerations (1,9,46). Electrical wires have to be properly insulated; high-voltage transformers must be kept in a cool environment; and transformers should be protected from ozone contamination, which can occur from small ozone leaks. The frequency and voltage transformers must be high-quality units designed specifically for ozone service. The ozone generator supplier should be responsible for designing and supplying the electrical subsystems.

5.3. *Ozone Generation*

Ozone used for water treatment is usually generated using a corona discharge cell. This technique produces concentrations of ozone sufficiently high (above 1% by weight) to solubilize enough ozone and to attain the requisite *CT* values necessary to guarantee disinfection of *Giardia* cysts. Ozone also can be generated by UV radiation techniques, but only at maximum concentrations of 0.25% by weight. At such low gas concentrations, it is not possible to solubilize sufficient ozone to guarantee disinfection of *Giardia* cysts.

The discharge cell as depicted in Fig. 7 consists of two electrodes separated by a discharge gap and a dielectric material, across which high voltage potentials are maintained. Oxygen-enriched air, pure oxygen, or air that has been dried and cooled flows

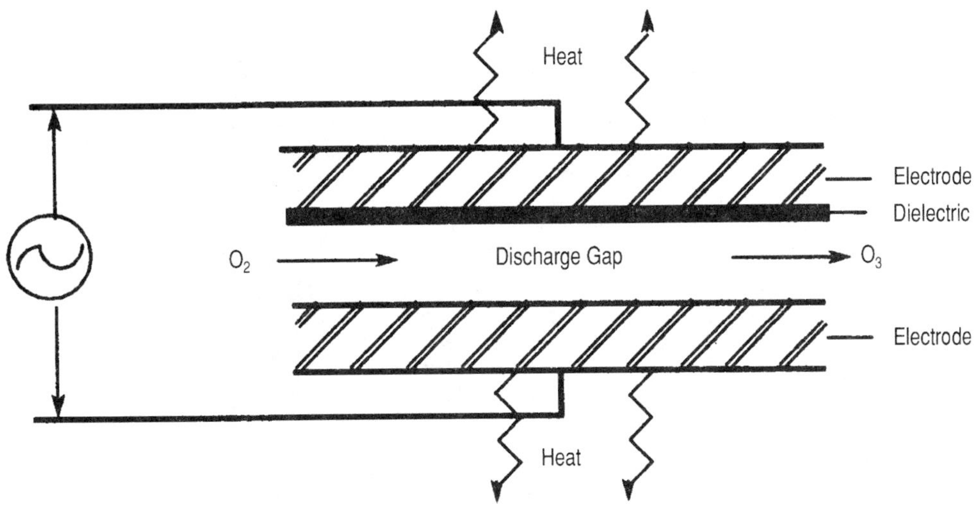

Fig. 7. Cross-section view of principal elements of a corona discharge ozone generator (US EPA).

between the electrodes and produces ozone. More recent designs use medium and high frequencies, rather than high voltages and low frequencies, to generate ozone. Figure 7 depicts the essential components of a corona discharge ozone generator. Ambient air, oxygen-enriched air, or pure oxygen is fed to the generator. If ambient air is used, the generator produces dry, cool air containing 1–3.5% ozone (by weight), which can be mixed with water. When pure oxygen is used, the concentration of ozone produced is approximately double that produced with ambient air (up to 8–9% by weight).

The most common commercially available ozone generators are (1):

(a) Horizontal tube, one electrode water cooled.
(b) Vertical tube, one electrode water cooled.
(c) Vertical tube, both electrodes cooled (water and oil cooled).
(d) Plate, water or air cooled.

The operating conditions of these generators can be subdivided into low -frequency (60 Hz), high voltage (>20,000 V); medium frequency (600 Hz), medium voltage (<20,000 V); and high frequency (>1,000 Hz), low voltage (<10,000 V). Currently, low-frequency, high-voltage units are most common, but recent improvements in electronic circuitry make higher-frequency, lower-voltage units more desirable.

Operating an ozone generator at 60–70% of its maximum production rate is most cost effective. Therefore, if the treatment plant normally requires 45 kg/d (100 lb/d) of ozone and 68 kg/d (150 lb/d) during peak periods, it is wise to purchase three generators, each designed for 27 kg/d (60 lb/d) and operate all three at about 65% capacity for normal production. This arrangement will satisfy peak demands and one generator will be on hand during off-peak periods for standby or maintenance.

5.4. *Ozone Contacting*

Because ozone is only partially soluble in water, once it has been generated it must contact water to be treated. Many types of ozone contactors have been developed for

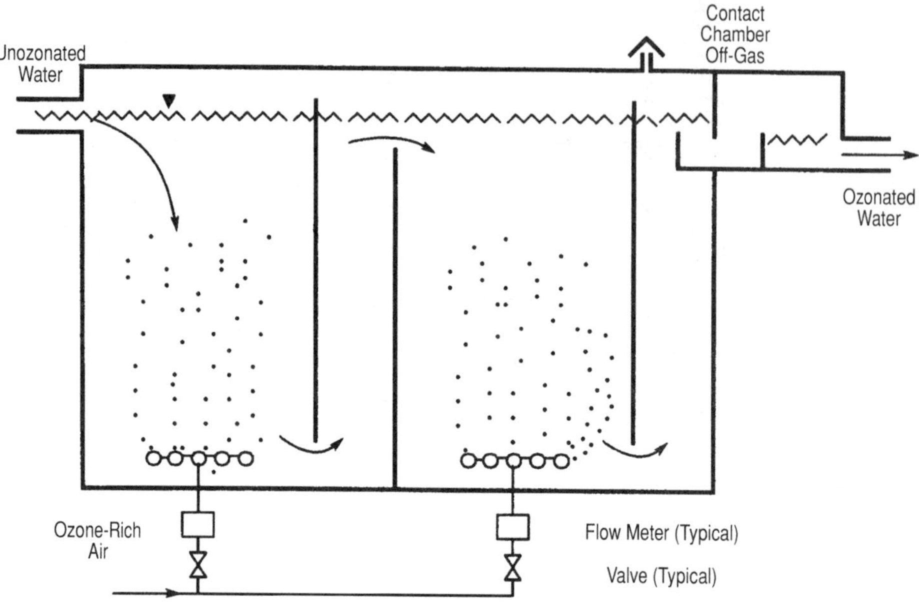

Fig. 8. Cross-section view of a two-stage bubble diffuser ozone contact basin (US EPA)

this purpose. Ozone contacting for disinfection has typically been accomplished in deep, multistage contactors that employ fine bubble diffusers. Newer alternatives have emerged that provide an option for small systems through the use of side-stream injection technologies that eliminate the need for fine bubble injection. Package units are available that include a gas separator that fuses and eliminates excess gas that results from ozone addition and a Venturi jet that is used to inject and blend the ozone with a solution feed stream. These systems allow the alternative of injecting ozone into an enclosed vessel or a pipe.

Ozone can be generated under positive or negative air pressure. If generated under positive pressure, the contactor most commonly used is a two-chamber porous plate diffuser, with a 4.8-m (16-ft) water column (Fig. 8). With this method, the ozone-containing air exits the ozone generator at approx 1.05 kg/cm^2 (15 psig) and passes through porous diffusers at the base of the column. Fine bubbles containing ozone and air (or oxygen) rise slowly through the column, ozone is transferred (dissolves), and oxidation and/or disinfection take place. The 4.8-m (16-ft) height maximizes the amount of ozone transferred from the bubbles as they rise in this type of porous diffuser contactor.

Other types of positive-pressure ozone contactors include packed columns, static mixers, and high-speed agitators. An atomizer that sprays water through small orifices into an ozone-containing atmosphere also can be used.

When ozone is generated under negative pressure, vacuum action draws the ozone mixture from the generators, providing contact as the gas mixes with the flowing water as with a submerged turbine (29,35). Other common methods of creating negative pressure use injectors or Venturi-type nozzles. These systems pump water past a small orifice (injector) or through a Venturi nozzle, thus creating negative-pressure.

The diffuser and packed tower contactors require no energy above that required to generate ozone at 1 kg/cm^2 (15 psig). The high-speed agitators, static mixers, and all the negative-pressure contactors require additional energy.

Ozone reactions are very fast to destroy or inactivate microorganisms; to oxidize iron, manganese, sulfide and nitrite ions, and some organics; and to lower turbidity levels. However, ozone oxidizes organic compounds such as humic and fulvic materials, as well as many pesticides and volatile organic compounds, quite slowly compared to these other solutes.

For disinfection, the initial dose of ozone is used to satisfy the ozone demand of the water. Once this demand is satisfied, a specific ozone concentration must be maintained for a specific period of time for disinfection. These two stages of ozonation are usually conducted in two different contacting chambers (Fig. 8). Approximately two-thirds of the total ozone required is added to the first chamber where the ozone demand of the water is met and the dissolved ozone reaches a residual level (typically 0.4 mg/L). The remaining ozone is applied in the second chamber, where it maintains the residual ozone concentration for the necessary contact time with water to attain the required CT value (47–49).

When ozone is added to water, its dissolved residual is not stable. Not only will ozone react with many contaminants in water, but owing to its short half-life, it will also decompose within minutes to oxygen. At the higher pH ranges (above 8), decomposition of molecular ozone into reactive intermediates (including the hydroxyl free radical) is accelerated. Consequently, it is not possible to monitor the residual ozone concentration at any single point in the treatment train and expect a single concentration level to hold steady from the point of gas/liquid mixing throughout the ozone treatment subsystem.

Therefore, it is important when using ozone for primary disinfection to monitor for dissolved ozone at a minimum of two points. In the event that two ozone contact chambers are utilized, the dissolved residual ozone can be monitored at the outlets of the two chambers. The average of these two numbers can be used as the C for calculation of CT values.

Absolute measurement of ozone contact time (T) is not simple, because the objective of adding ozone to water involves maximizing the mixing of a partially soluble gas with the liquid. The more complete the contacting, the shorter the actual residence time for the water in the contacting chamber. As a result, the more completely the gas and liquid media are mixed, the less the hydraulic residence time can be used as the value for T. Only when water flowing through an ozone contacting system approaches plug flow does the actual ozone contact time approach the hydraulic residence time.

The greater the number of ozone contact chambers that can be connected in series, the closer the water flow will approach plug flow. In such cases, the T_{10} (time for 10% of an added tracer to pass through the ozone contacting system) will approach 50% of the hydraulic detention time for water passed through the ozone contacting system. Recently published studies of the 26 m^3/s (600 MGD) Los Angeles Aqueduct Filtration Plant and of the design of the 5 m^3/s (120 MGD) Tucson water-treatment plant have confirmed that the actual hydraulic residence time (T_{10}) is approx 50% of the theoretical hydraulic residence time.

Design considerations that maximize ozone transfer and disinfection performance are as follows:

(a) The contact basin should be as deep as practical, preferably greater than 5 m (16 ft) at sea level and deeper at a higher elevation, such as 6 m (20 ft) at 2,440 m (8,000 ft). The maximum depth may be limited by the maximum pressure in the ozone generator, which is usually 103 kPa (15 psig).
(b) The bubbles formed by the porous stone diffusers should range between 2 and 3 mm in diameter.
(c) The contactor should have at least two independent trains with isolated off-gas compartments to allow for continuous operation during inspection and cleaning.
(d) The contactor should have features that simulate plug flow and reduce short-circuiting:
  First: A minimum of two or three, and preferably more, separate stages should be provided.
  Second: Each stage should be positively separated from the other stages. No chance for short-circuiting should exist (for example, through drain holes at the bottom of the walls separating the stages).
  Third: Each stage should be provided with a separate drain pit to aid in cleaning on a routine basis (e.g., once or twice per year).

(e) The contactor should have from 1.2 to 1.8 m (4 to 6 ft) of headspace to allow for foaming.
(f) Each set of diffusers should have a flow control valve on its piping and separate flow measurement. More diffusers should be located in the first stage to meet the higher demand for ozone in that stage, and thus provide capability to maintain a uniform residual oxidants concentration throughout all stages of the contact basin. The rest of the diffusers can be equally spaced in the remaining stages.
(g) The wastewater flow should be countercurrent to the ozonized airflow to maximize ozone transfer efficiency.
(h) The contact basins should be made of typical construction grade concrete, with ozone resistant (e.g., Hypalon) water stops.
(i) The contact basins should be covered and sealed as much as possible. Sealing with Sika-flex 1-A compound covered with coal tar epoxy or Teflon sheeting has been used in some cases (34,50). However, basin sealing is difficult to maintain, and periodic leaks through the ceiling of the contact basin may occur (50). It is suggested that the ozone contact basin be placed in a location where the entire roof of the basin is the open atmosphere. Also, the basin should have the capability to operate under negative pressure.
(j) Stainless-steel piping for ozonized gas flow must be provided for positive pressure ozone systems:
  First: Tungsten inert gas (TIG) welding is recommended.
  Second: Schedule 10 or better and type 304L or 316L stainless steel is recommended.
  Third: Flange-to-flange fittings, rather than threaded fittings, should be used in applications where welded connections are not made.

(k) Ozonized feed-gas and contact basin off-gas sample lines should be stainless steel tubing. Teflon tubing may be considered for short runs.

5.5. *Destruction of Ozone Contactor Exhaust Gas*

The ozone in exhaust gases from the contacting unit must be destroyed or removed by recycling prior to venting. The current Occupational Safety and Health Administration standard for exposure of workers during an 8-h workday is a maximum ozone concentration of 0.0002 g/m^3 (0.1 ppm by volume time-weighted average). Typical concentrations in contactor exhaust gases are greater than 1 g/m^3 (500 ppm by volume).

The four primary methods of destroying excess ozone are (51–56):

1. Thermal destruction (heating the gases to 300–350°C for 3 s).
2. Thermal/catalytic destruction.
3. Catalytic destruction (with metal catalysts or metal oxides).
4. Moist granular activated carbon (used extensively at European small plants treating less 2 MGD and for swimming pools).

When ozone is generated from air, destroying ozone in exhaust gases is more cost effective than recirculating the gases through the air preparation system and ozone generator. When ozone is generated from pure oxygen, destroying the ozone and discharging the excess oxygen can be more cost effective than destroying the excess ozone, and drying and recycling the excess oxygen. A number of "once-through" oxygen feed systems have been installed throughout the world since 1980 to generate ozone. The largest of these is at the 26 m^3/s (600 MGD) Los Angeles plant, which has been operating in this manner since 1987.

5.6. Monitors and Controllers

Proper monitors and controllers should be supplied with the ozone system, including (10,15,34,57):

(a) Gas-pressure and temperature monitors at key points in the air-preparation system. Simple pressure gauges and mercury thermometers are adequate.
(b) Continuous monitors/controllers for the dew point to determine the moisture content of the dried feed gas to the ozone generator. The monitors should sound an alarm and shut down the generator when high dew points are indicated. Equipment to calibrate the dew point monitor should be provided as well.
(c) Inlet/discharge temperature monitors for the ozone generator coolant media (water and/or oil, or air), and a means of determining whether coolant is actually flowing through the generator. An automatic system shutdown should be provided if coolant flow is interrupted or if its discharge pressure exceeds specified limits.
(d) Flow rate, temperature, and pressure monitors, and an ozone concentration monitor for the gas discharged from the ozone generator to determine the ozone production rate.
(e) Power input monitor for the ozone generator.

6. COSTS OF OZONATION SYSTEMS

The discussion of ozone system costs is divided into four sections covering equipment, installation, housing, and operating and maintenance costs.

6.1. Equipment Costs

Ozonation equipment to be purchased include air-preparation equipment (drying and cooling), an ozone generator, an ozone contactor, an ozone destruction unit, and instrumentation and controls. Because of the many differences in air pretreatment methods, ozone contacting, contactor exhaust gas handling, monitoring, and other operational parameters, equipment costs presented in this section should be considered as general guidelines only.

For generating large quantities of ozone, 45 kg/d (100 lb/d) and higher, air preparation, ozone generation, and contacting equipment costs run approx $1,300/lb ($1,950 in

Table 8
Equipment Cost of Ozonation Systems for Small Water Plants—Supplier A

	Size of treatment plant[c]					
	500,000 GPD		350,000 GPD		180,000 GPD	
Maximum dosage of ozone (mg/L) at peak flow	5	3	5	3	5	3
Daily ozone requirement (lb)	21	14	14	7	7	5
Contact chamber diameter [a](ft)	6	6	5	5	4	4
			Equipment costs			
Air preparation and ozone generation unit	$31,500	$25,000	$25,000	$22,000	$22,000	$19,500
Contact chamber with diffusers	11,500	11,500	10,200	10,200	9,900	9,900
Monitoring instrumentation[b]	15,000	15,000	15,000	15,000	15,000	15,000
Ozone destruction unit	6,700 (10 cfm)	5,000 (7 cfm)	5,000 (7 cfm)	4,200 (3 cfm)	4,200 (3 cfm)	4,200 (3 cfm)
Total Equipment Costs	$64,700	$56,500	$55,200	$51,400	$51,100	$48,600
Power Requirement (kWh)	13.3	10.1	10.1	5.0	5.0	3.65

[a]14 ft.high, four compartments, four diffusers, Derakane fiberglass reinforced plastic.
[b]Includes monitors for ozone in generator product, ozone in ambient plant air, ozone dissolved in water, and dew point monitor in air-preparation unit.
[c]1 lb = 0.4536 kg; 1 ft = 0.3048 m; 1 GPD = 0.003785 m^3/d.
Source: US EPA (1983).

2003 Dollars) of ozone generation capacity per day. This figure does not include ozone destruction, instrumentation, control, building, and installation costs. For smaller quantities of ozone, costs are higher per kilogram, but vary significantly from site to site. For plants serving less than 10,000 persons per day, 1.4–9.5 kg/d (3–21 lb/d), all items can be assembled into a single skid-mounted unit.

Small ozonation systems can use diffuser contactors, which are generally constructed of polyvinyl chloride (PVC) pipe standing on end or fiberglass reinforced plastic (FRP) tanks. Tables 8 and 9 list equipment cost estimates obtained from two ozonation system suppliers in 1983 for small water supply systems. All costs shown in the two tables should be multiplied by a factor of 1.5 in order to get the cost in year 2003 US dollars (see Table 10). Equipment costs are higher at higher dosages for a given flow rate.

Ozone Supplier A provides four monitors with the system: dew point in the air preparation unit, ozone output of the generator, ozone in the plant ambient air (in case of leaks), and dissolved ozone residual in the water. All are optional (but recommended) for optimum performance and minimal labor and downtime.

Table 9
Equipment Cost of Ozonation Systems for Small Water Plants - Supplier B

	Size of Treatment Plant[e]									
	0.1 MGD		0.2 MGD		0.3MGD		0.4 MGD		0.5MGD	
Maximum ozone dosage (mg/L), at peak flow	3		3		3		3		3	
Daily ozone requirement (lb/day)	3		6		7		12		14	
Equipment Costs	Pressure									
	Low[b]	High[c]	Low[b]	High[c]	Low[b]	High[c]	Low[b]	High[c]	Low[b]	High[c]
Air preparation and ozone generation[a]		\$17,500	\$33,200	\$30,200	\$38,500	\$35,000	\$43,000	\$40,000	\$49,800	\$46,800
Power requirements (kWh/lb of O_3 generated)	10.5	20	10.5	13.5	10.5	13.5	10.5	13.5	10.5	13.5
Ozone contactor with diffusers	\$8,500		\$12,000		\$16,000		\$21,000		\$29,000	
Ozone generation monitor[d]	4,000		4,000		4,000		4,000		4,000	
Chamber exhaust monitor	2,200		2,200		2,200		2,200		2,200	
Dew point monitor	3,500		3,500		3,500		3,500		3,500	
Total equipment costs		\$35,700	\$52,900	\$49,000	\$62,200	\$59,200	\$71,700	\$68,700	\$86,500	\$83,500

[a]Includes air preparation, ozone generation, ozone destruction, and system controls.

[b]Air-preparation unit includes air filters or separators, compressor delivering air at 8–12 psig to a refrigerative cooler and a dual tower desiccant dryer

[c]Same as low-pressure air, preparation system, except compressor delivers air at 80–120 psig. High-pressure system takes less space and requires less maintenance, but requires more energy.

[d]\$4,000 instrument in an automatic, continuous reading in-line monitor. A substitute is a \$2,000 instrument that is not automatic and utilizes wet chemistry.

[e]I lb = 0.4536 kg; 1 MGD = 0.044 m^3/s.

Source: US EPA (1983).

Table 10
US Yearly Average Cost Index for Utilities[a]

Year	Index	Year	Index
1967	100	1986	347.33
1968	104.83	1987	353.35
1969	112.17	1988	369.45
1970	119.75	1989	383.14
1971	131.73	1990	386.75
1972	141.94	1991	392.35
1973	149.36	1992	399.07
1974	170.45	1993	410.63
1975	190.49	1994	424.91
1976	202.61	1995	439.72
1977	215.84	1996	445.58
1978	235.78	1997	454.99
1979	257.20	1998	459.40
1980	277.60	1999	460.16
1981	302.25	2000	468.05
1982	320.13	2001	472.18
1983	330.82	2002	484.41
1984	341.06	2003	495.72
1985	346.12		

[a]Extracted from US ACE 2000 *Civil Works Construction Cost Index System Manual*, # 1110-2-1304, U.S. Army Corps of Engineers, Washington, DC, USA, PP 44 (PDF file is available on the Internet at http://www.nww.usace.army.mil/cost), (Tables Revised 31 March 2003).

Ozone Supplier B provided estimates for two types of air preparation equipment. One type operates at high pressures (5.6–8.4 kg/cm^2) (80–120 psig), the other at low pressures (0.56–0.84 kg/cm^2) (8–12 psig). The high-pressure units have lower capital costs, but require more energy for their operation. Supplier B offers two types of device to monitor ozone output from the generator. The automatic, inline continuous reading monitor costs \$4,000 (\$6,000 in 2003 dollars); the nonautomatic monitor, which requires wet chemistry calculations to determine ozone output costs \$2,000 (\$3,000 in 2003 dollars).

6.2. *Installation Costs*

Costs to install ozonation equipment include labor and material for piping and electrical wiring. Piping can be extensive–transporting water to and from the ozone generators (if they are water-cooled) and the contactor, transporting ozone-containing air to the contactor chamber, and transporting contactor off-gases to and from the ozone destruction unit.

The ozonation equipment suppliers estimate that for units producing up to 13.6 kg/d (30 lb/d) of ozone, installation costs average from \$9,705–\$16,175 (\$14,560–\$24,260 in 2003 dollars) for Supplier A and \$12,750 to \$21,250 (\$19,130 to \$31,880 in 2003 dollars) for Supplier B.

6.3. *Housing Costs*

Installation of the power supply, air preparation, ozone generation, and turbine contacting operations require an area of approx 3 × 5.1 m (10 × 17 ft). Diffuser contacting units are quite large and high (5.4 m), and are typically installed outside existing buildings or in the basement of buildings constructed for the other ozonation equipment. A building with the above dimensions can be constructed for about $6,000 ($9,000 in 2003 dollars).

6.4. *Operating and Maintenance Costs*

Operating costs for ozonation systems vary depending on:

(a) Oxygen use or air-preparation method—high or low pressure, or subatmospheric pressure desiccant systems with or without an air chiller.
(b) Generator cooling method—air or water-cooled. In northern climates, water at the plant is generally cold enough to be used as a coolant all year. Southern climates must refrigerate cooling water most of the year. Medium frequency generators require increased cooling.
(c) Contacting method—diffuser contactors do not require electrical energy as do the more compact turbine diffusers.
(d) Ozone dosage required.
(e) Contactor exhaust gas handling—thermal, catalytic, or GAC destruction systems.

Maintenance costs include periodic cleaning, repair, and replacement of equipment parts. For example, air preparation systems contain air filters that must be replaced frequently, and tube-type ozone generators normally are shut down for annual tube cleaning and other general maintenance. Tube cleaning can require several days of labor, depending on the number and size of ozone generators in the system. Tubes, which can be broken during cleaning or deteriorate after years of operation at high voltages (or more rapidly if the air is improperly treated), must be replaced periodically. Labor requirements, other than for periodic generator cleaning, include annual maintenance of the contacting basins and day-to-day operation of the generating equipment average 0.5 h/d.

7. SAFETY

Ozone is a toxic gas, and like chlorine can cause severe illness and death if inhaled in sufficient quantity. However, ozone systems have safety advantages not available with the chlorine disinfection process. Ozone is generated onsite, thus eliminating transportation hazards. Also, the generation system can be shut down if an ozone leak develops. Another safety advantage is the physical characteristic of ozone that allows it to be detected (smelled) at concentrations much lower than harmful levels.

The reported biological effects of exposure to ozone range from dryness of mouth and throat, coughing, headache, and chest restrictions at concentrations near the recommended limit, to more acute problems at higher concentrations. The Occupational Safety and Health Administration (OSHA), the American National Standards Institute/American Society for Testing and Materials (ANSI/ASTM), the American Conference of Government Industrial (ACGI), and the American Industrial Hygiene Association (AIHA), have reported on issues of ozone safety and the recommended ambient ozone exposure levels (59–64).

As with any other chemical, the Occupational Health and Safety Administration (OSHA) has established maximum contaminant inhalation guidelines for ozone in the work place as follows (60):

(a) No workers should be exposed to ozone concentrations in excess of 0.1 ppmv (0.2 mg/m^3) during an 8-h working day
(b) No workers should be exposed to a ceiling concentration of ozone in excess of 0.3 ppmv (0.6 mg/rn^3) for more than 10 min.

These recommended limits for ozone concentration are much higher than the concentrations at which ozone can typically be smelled. Generally, an individual can detect ozone at concentrations ranging from 0.01 to 0.05 ppmv (0.02 to 0.1 mg/m^3) (1). The more often a person is exposed to ozone, the higher the required concentration for detection.

The subject of safety in the design and operation of an ozone system should receive high priority (59). All ozone systems should be provided with an ambient ozone monitor or monitors, which are set up to measure the ozone concentration at potential ozone-contaminated locations within the plant. A single monitor may be installed, and the air from different locations pumped to the monitor for detection of ozone concentration. The monitors should be set up to sound an alarm when the ozone concentration reaches 0.1 ppmv, and should be set up to shut down the ozone system when the concentration exceeds 0.3 ppmv.

As with any toxic chemical, the operators should be trained concerning the potential hazards involved and the emergency operating procedures required if a problem occurs. Equipment that should be provided to assist the operator are as listed below:

(a) A self-contained breathing apparatus should be provided, and should be located at a place where access is not restricted by ozone in case an ozone leak occurs.
(b) An eye-washing sink should be provided to enable the operator to rinse ozone from the eyes, if needed.
(c) Safety manuals on performing artificial respiration should be provided.
(d) Separate ladders should be provided to enable the operator to enter the ozone contact chamber. Fixed steps in the contact basin should not be relied on.

The readers are referred to another reference (58–67) for additional information on the ozonation process.

NOMENCLATURE

C molar fraction of the gas in water at equilibrium with the gas above the water
C concentration (mg/L)
CT contact concentration-time [(mg/L)-min]
D applied ozone dosage (mg/L)
H Henry's law constant (varies with temperature) (atm/mole fraction)
N effluent coliform concentration (# /100 mL)
N_0 influent coliform concentration (# /100 mL)
n slope of dose/response curve
P partial pressure of the gas above the liquid (atm)
q X-axis intercept of dose/response curve (mg/L)
the amount of ozone transferred before measurable kill is observed

T transferred ozone dosage (mg/L)
T time (min)
T_{10} (time for 10% of an added tracer to pass through the ozone contacting system)
TE ozone transfer efficiency (%)

REFERENCES

1. G. W. Miller, et al., *An Assessment of Ozone and Chlorine Dioxide Technologies for Treatment of Municipal Water Supplies*, EPA-600/2-78-147, US Environmental Protection Agency, Cincinnati, OH, 1978.
2. R. G. Rice, et al., *Proceedings of the Seminar on The Design and Operation of Drinking Water Facilities Using Ozone or Chlorine Dioxide*, New England Water Works Association, Vol. 1, June, 1979.
3. US EPA, *Design Manual—Municipal Wastewater Disinfection*, US Environmental Protection Agency, EPA/625 1-86-021, Water Engineering Research Laboratory, Cincinnati, OH, 1986.
4. F. Novak, Presentation at the *EPA Seminar on the Current Status of Wastewater Treatment and Disinfection with Ozone*, September, 1977.
5. R. F. Weston Inc., *Factors Affecting the Operation and Maintenance of Selected Ozone and Ultraviolet Disinfection Systems*. Report for MERL, US EPA Contract No. 68-83-3019, February, 1983.
6. C. Gottschalk, J. A. Libra, and A. Saupe, *Ozonation of Water and Waste Water: A Practical Guide to Understanding Ozone and its Application*, Wiley-VCH, New York, NY, 2000.
7. US EPA, *Small System Treatment Technologies for Surface Water and Total Coliform Rules*, US EPA Office of Ground Water and Drinking Water, Washington, DC, 1998.
8. M. J. Klein, et al., *First International Symposium on Ozone for Water and Wastewater Treatment*, International Ozone Institute, 1975.
9. T. C. Manley and S. J. Niegowski, *Encyclopedia of Chemical Technology, 2nd ed.,* Wiley, New York, 1967, **14**, pp. 410–432.
10. J. Grunwell, et al., *IOA J. Ozone Science & Engineering,* **5**, 4 (1983).
11. R. L. Jolley, et al., *Proceedings of the National Symposium—Progress in Wastewater Disinfection Technology,* EPA-600/9-79-O1 **8**, US Environmental Protection Agency, Cincinnati, OH, 1979.
12. E. G. Fochtman and J. E. Huff, *Ozone-Ultraviolet Light Treatment of TNT Wastewate,* Second International Symposium on Ozone Technology, International Ozone Institute, 1976.
13. P. C. Singer and W. B. Zilli, *Ozonation of Ammonia in Municipal Wastewater,* First International Symposium on Ozone for Water and Wastewater Treatment, International Ozone Institute, 1975.
14. N. Narkis, et al., *J. Environmental Engineering Div.,* ASCE, 103 (EE5) 877–891, (1977).
15. A. D. Venosa, *J. Water Pollution Control Fed.,* **55**, 457–466, (1983).
16. NDWC Tech brief: Ozone, *Newsletter On Tap*, National Drinking Water Clearinghouse (NDWC), No. 12, December, (1999).
17. X. Paraskeva, Y. Panagiota, Z. Graham, J. D. Nigel, *Water Environment Regulation Watch*, November/December, (2002).
18. M. R. Collins, *Small Systems Water Treatment Technologies: State-of-the-Art Workshop*, American Water Works Association, Denver, CO, 1998.
19. Tri-State (2003), Ozonation Systems, *Energy Technologies*, Generation and Transmission Association, http://tristate.apogee.net/et/ewtwozs.asp, November, (2003).
20. S. Farooq, et al., *Criteria of Design of Ozone Disinfection Plants*, Forum on Ozone Disinfection, International Ozone Institute, 1976.

21. SNWA (2003), *Water Quality and Water Treatment: Ozonation*, Southern Nevada Water Authority, http://www.snwa.com/html/wq_treatment_ozonation.html, November.
22. S. J. Hoeger, D. R. Dietrich, and B. C. Hitzfeld Environmental Health Perspectives, 110, November (2002).
23. Bollyky Associates, Inc. *Zebra Mussel Control by Ozone Treatment*, http://www.bai-ozone.com/bai 11.htm, November, (2003).
24. A. H. Konsowa, *Desalination*, **158**, 233–240, (2003).
25. J. Hoigne and H. Bader, *Identification and Kinetic Properties of the Oxidizing Decomposition Products of Ozone in Water and its Impact on Water Purification*, Second International Symposium on Ozone Technology, International Ozone Institute, 1976.
26. B. Langlais, D. A. Reckhow, and D. R. Brink *Ozone in Water Treatment: Application and Engineering*. AWWA Research Foundation and Lewis Publishers, Denver, Co, 1991.
27. R. G. Rice, P. K. Overbeck, and K. Larson, In *Providing Safe Drinking Water in Small Systems: Technology, Operations, and Economics*, J. A. Cotruvo, G.F. Craun and N. Hearne eds. CRC Press LLC, Boca Raton, FL, 1999.
28. S. Vigneswaran and C. Visvanathan, *Water Treatment Processes: Simple Option*, CRC Press Inc., Boca Raton, FL, 1995.
29. E. L. Stover, et al., *High Level Ozone Disinfection of Municipal Wastewater Effluents*, US EPA Grant No. R8O4946, 1980.
30. P. W. Given and D. W. Smith, In: *Municipal Wastewater Disinfection*, Proceedings of Second National Symposium, Orlando, Florida, EPA-600/9-83-009, US Environmental Protection Agency, Cincinnati, OH, 1983.
31. H. B. Gan, et al., *Forum on Ozone Disinfection*, International Ozone Institute, 1976.
32. A. D. Venosa, et al., *Environment International*, **4**, 299–311, (1980).
33. M. C. Meckes, et al., *J. Water Pollution Control Fed.*, **55**, 1158–1162 (1983).
34. K. L. Rakness, et al., *J. Water Pollution Control Fed.*, **56**, 1152–1159 (1984).
35. A. D. Venosa, et al., In: *Progress in Wastewater Disinfection Technology*, Proceedings of the National Symposium, Cincinnati, Ohio, EPA-600/9-79-018, US Environmental Protection Agency, Cincinnati, OH, 1979.
36. A. G. Hill and H. T. Spencer, *First International Symposium on Ozone for Water and Wastewater Treatment*, International Ozone Institute, 1975.
37. E. J. Opatken, In: *Progress in Wastewater Disinfection Technology*, Proceedings of the National Symposium, Cincinnati, Ohio, EPA-600/9-79-018, US Environmental Protection Agency, Cincinnati, OH, 1979.
38. M. Rouston et al., *Ozone: Science and Engineering*, **2**, 337–344, (1981).
39. AWWA, *Controlling Disinfection By-Products*, American Water Works Association, Denver, CO, 1993.
40. G. C. Budd, G. S. Logdson, and B. W. Long, In: *Providing Safe Drinking Water in Small Systems: Technology, Operations, and Economics*, J. A. Cotruvo, G. F. Craun, and N. Hearne eds., CRC Press LLC, Boca Raton, FL, 1999..
41. O_3 Water Systems, *Successful Treatment of Potable Drinking Water with Ozone*, http://www.O_3water.com/Articles/guideto.htm, March, 2003.
42. T. Hatatatsu and Y. Suzuki, *WEFTEC 2000 Conference Proceedings*, October, 2000.
43. B. A. Hegg, et al., *Evaluation of Pollution Control Processes: Upper Thompson Sanitation District*, EPA-600/2-80-0 16, US Environmental Protection Agency, Cincinnati, OH, 1980.
44. A. J. Varas, *New York City's Ozone Demonstration Plant Design*, Presented at International Ozone Association Conference, Montreal, September 11, 1984.
45. R. Gerval, *Ozone Manual for Water and Wastewater Treatment*, Wiley, New York, 1982.
46. W. J. Masschelein, *Ozone Manual for Water and Wastewater Treatment*, Wiley, New York, 1982.
47. A. D. Venosa, *J. Water Pollution Control Fed.*, **56**, 137–142, (1984).

48. L. J. Bollyky and B. Siegel, *Water & Sewage Works*, **124**, 90–92, (1977).
49. J. Perrich et al., *AIChE Symposium Series*, **73**, No. 166, (1976).
50. W. L. LePage, *Case Histories of Mishaps Involving the Use of Ozone*, Presented at International Ozone Association Conference, Montreal, September, (1984).
51. C. Coste, In: *Ozone Manual for Water and Wastewater Treatment*, Wiley, New York, 1982.
52. M. Horst, In: *Ozone Manual for Water and Wastewater Treatment*, Wiley, New York, 1982.
53. K. Orgler, In: *Ozone Manual for Water and Wastewater Treatment* Wiley, New York, 1982.
54. P. Chapsal and X. Trailigaz, *Ozone Manual for Water and Waste-water Treatment*, Wiley, New York, 1982.
55. W. J. Masschelein, In: *Ozone Manual for Water and Wastewater Treatment*, Wiley, New York, 1982.
56. X. Eloret, *Ozone Technology for pollution Remediation*, http://www.eloret.com/area15.html, Eloret, Sunnyvale, CA, 2003.
57. G. Gordon and J. Grunwell, In: *Municipal Wastewater Disinfection*, Proceedings of Second National Symposium, Orlando, Florida, EPA-600/9-83-009, US Environmental Protection Agency, Cincinnati, OH, 1983.
58. A. D. Venosa, et al., *J. Water Pollution Control Fed.*, **57**, 929–934, (1985).
59. F. Damez, In: *Ozone Manual for Water and Wastewater Treatment*, Wiley, New York, 1982.
60. US OSHA, Title 29, Chapter XVII, 1910.1000, *US Occupational Safety and Health Administration*, Washington, DC.
61. US DOL, *Occupational Health Guideline for Ozone*, US Department of Labor, Washington, DC, 1978.
62. L. K. Wang, *Principles and Kinetics of Oxygenation-Ozonation Waste Treatment Systems.* U.S. Dept. of Commerce, National Technical Information Service, Springfield, VA, PB83-127704, Sept. 1982.
63. M. Krofta, L. K. Wang, L. Kurylko, and A. E. Thayer, *Pretreatment of Ozonation of Cooling Tower Water.* U.S. Dept. of Commerce, National Technical Information Service, Springfield, VA, PB84-192053; PB84-192046, 1983.
64. ASTM, *Standard Practices for Safety and Health Requirements Relating to Occupational Exposure to Ozone,* American Society for Testing and Materials, 1977.
65. L. K. Wang, Y. T. Hung, and N. K. Shammas (eds.), *Physicochemical Treatment Processes,* Humana Press, Totowa, NJ, 2004.
66. L. K. Wang, N. K. Shammas and Y. T. Hung (eds.), *Advanced Physicochemical Treatment Processes.* Humana Press, Totowa, NJ, 2005.
67. L. K. Wang, Y, T. Hung, H. H. Lo, and C. Yapijakis (eds.), *Handbook of Industrial and Hazardous Wastes Treatment,* Marcel Dekker, Inc., NYC, NY, 2004.

10
Electrolysis

J. Paul Chen, Shoou-Yuh Chang, and Yung-Tse Hung

CONTENTS

1. INTRODUCTION

The word "lysis" means to dissolve or break apart, so the word "electrolysis" literally means to break substances apart by using electricity. Michael Faraday first formulated the principle of electrolysis in 1820. The process occurs in an electrolyte, a watery solution or a salt melting that gives the ions a possibility to transfer between two electrodes. The electrolyte is the connection between the two electrodes, which are also connected to a direct current. If you apply an electrical current, the positive ions migrate to the cathode while the negative ions will migrate to the anode. At the electrodes, the cations will be reduced and the anions will be oxidized.

Electrolysis is the chemical decomposition and/or dissociation of organic and inorganic substances by an electrical current. The electrolytic cell contains an anode and a cathode, where separate oxidation and reduction reactions occur.

In the anode, there are the following oxidation reactions:

$$\mathrm{Me_1(insoluble)} = \mathrm{Me_1^{m+}(soluble)} + me^- \tag{1a}$$

$$4\mathrm{OH}^- = 2\mathrm{H_2O} + \mathrm{O_2} + 4e^- \tag{1b}$$

From: *Handbook of Environmental Engineering, Volume 3: Physicochemical Treatment Processes*
Edited by: L. K. Wang, Y.-T. Hung, and N. K. Shammas © The Humana Press Inc., Totowa, NJ

$$2Cl^- = Cl_2 + 2e^- \tag{1c}$$

where Me_1 is metal in the anode. The electrode metal enters into the reaction, losing a flow of electrons to the electrode.

In the cathode, the following reduction reactions occur:

$$Me_2^{m+}(\text{soluble}) + me^- = Me_2(\text{insoluble}) \tag{2a}$$

$$2H^+ + 2e^- = H_2(g) \tag{2b}$$

where Me_2^{m+} is the soluble ion in the solution.

Hydrogen gas produced at the cathode creates turbulence in the system, which can enhance the mixing. The gas also serves to transport the insoluble coagulated particles to the surface of the solution. Thus, a floating layer is formed at the liquid surface consisting of both hydrogen bubbles and entrapped suspended matter.

For an individual electrode, the net current flowing through the wire to the electrode is the difference between the anodic and cathodal currents developed by the electrode reactions at the electrode interface. The utilization of electrolytic precipitation for isolating a number of elements from aqueous solution has been practiced by analytical chemists since the 19th century (1). The principles can be grouped into two basic categories: (a) electrolysis at constant applied cell potential for the deposition of certain elements and (b) electrolysis at variable potential to maintain convenient current through the cell for the deposition of a large group of elements. It is obvious that such techniques can be applied to the area of wastewater treatment for the removal of undesirable elements. In addition to the dissolution of anodic metals or generation of oxygen, other strong oxidizing agents, e.g., chlorine, can be generated in the anode if the appropriate types of electrolyte are present in solution. For the generation of chlorine, high levels of chlorine are needed for the process to be economically feasible.

Electrochemical methods for treatment of wastewaters are not a new concept; as early as 1887, Eugene Hermite (2) described a method of treating sewage by electrolyzing after mixing with seawater. He found that in addition to deodorizing and disinfecting effects, $Mg(OH)_2$ was produced as a flocculant in the process, which helped to lower sewage solids effectively.

A series of techniques was then developed. Most of these required the use of sacrificial electrodes such as iron or aluminum to aid in sludge flocculation and the addition of chloride for the generation of chlorine for disinfection. Several studies utilizing such methods have been described in the literature and in patents, most of which are similar in principle with minor modifications in the process or design of the electrode (3,4). There has been a strong emphasis on development of electrochemical processes for wastewater treatment. Although the economic advantage of these processes over conventional secondary-tertiary processes has not been fully demonstrated, some studies show that electrolysis is a promising method (5–7).

Earlier electrolytic wastewater treatment had been directed toward the generation of $C1_2$ or O_3 for the deodorization and disinfection of wastewater, or toward controlling part of another treatment process (8–10). Most electrolytic odor control systems destroy malodorous emissions by oxidation. Ozone, although not strictly an electrolytic process in the usual sense, nevertheless must be produced at the treatment site by electrical

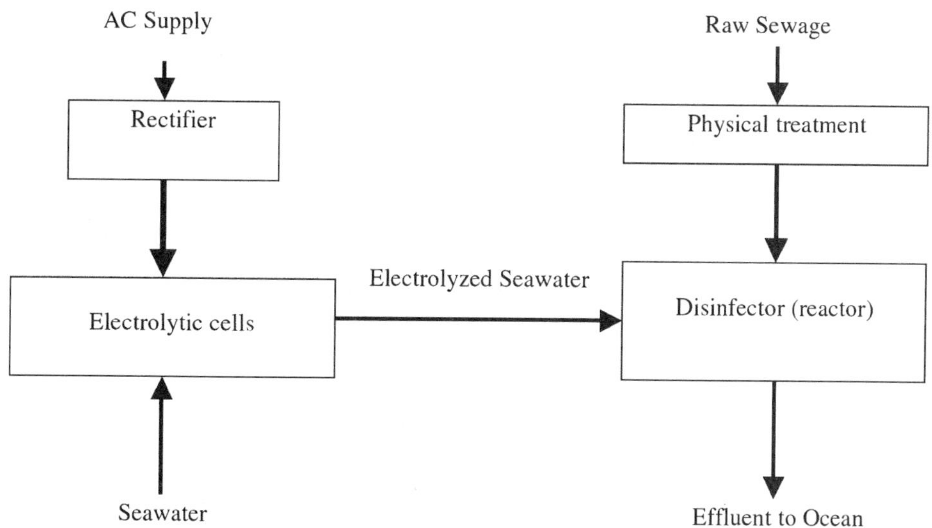

Fig. 1. Simplified flowsheet of processes used in CJB Guernsey plant.

means. The process, using a high-voltage alternating current (AC) silent discharge in air or oxygen gas, can eliminate shipping of dangerous materials. The large-scale use of electrolysis to treat sewage was demonstrated in a coastal area (Guernsey, England) as shown in Fig. 1 (3). The plant, which was constructed by CJB Process Ltd. in 1965, disinfected raw sewage with a dose of electrolyzed seawater that contained disinfectant (Cl_2) due to the reduction in the anode (Table 1) by the direct current (DC). CJB claimed

Table 1
List of Important Chemical Reactions in Electrocoagulation

Location	Reaction[a]
Anode	$4OH^- = 2H_2O + O_2 + 4e^-$
	$2Cl^- = Cl_2 + 2e^-$
	$Al = Al^{3+} + 3e^-$ (when Al is applied)
	$Fe = Fe^{2+} + 2e^-$ (when Fe is applied)
Cathode	$2H^+ + 2e^- = H_2(g)$
	$O_2 + 2H_2O + 4e^- = 4OH^-$
Solution phase	$nAl^{3+} + mOH^- = Al_n(OH)_m^{3n-m}$ (when Al is applied)
	$iFe^{2+} + jOH^- + kH_2O = Fe_i(H_2O)_k(OH)_j^{2i-j}$ (when Fe is applied)
	$4Fe^{2+} + 10H_2O + O_2 = 4Fe(OH)_3(g) + 8H^+$ (when Fe is applied)
	$Fe^{2+} + 4H^+ + O_2 = Fe^{3+} + 2H_2O$ (when Fe is applied)
	$iiFe^{3+} + jjOH^- + kkH_2O = Fe_{ii}(H_2O)_{kk}(OH)_{jj}^{3ii-jj}$ (when Fe is applied)

[a] i, j, k, m, ii, jj, RR, and $n = 0, 1, 2, \ldots$

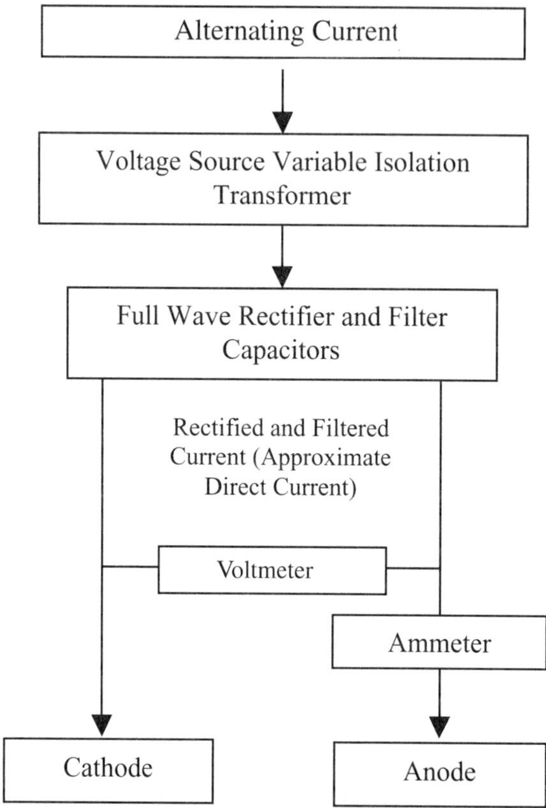

Fig. 2. Schematic of typical electrolytic cell.

that further improvement at Guernsey had halved the total processing time from 60 to 30 min, and reduced production costs to 2–3 cents per 1000 gal. The method, which had been used in operations on a much smaller scale at Sorrento, Italy, two years before the Guernsey plant, was shown to yield a final effluent that was aesthetically, hygienically; and economically acceptable.

In the United States, an electrocoagulation cell utilizing a metallic sacrificial anode as the generating source of coagulant was used by the General Electric Company as part of a shipboard waste-treatment system. Electrolytic waste treatments have been demonstrated over a considerable range of sizes and with various types of wastewaters in 1972. A similar approach was recently intensively studied by Stephenson and Tennant and Chen et al. (11,12). The electrocoagulation approach is used for treatment of food wastewater. These systems with increasing frequency may prove to be the optimum processing method where electric power is available and particularly where space and safety are prime considerations. Where electrolytes are neither present in nor available for addition to wastewaters, electrolysis may be severely handicapped by the high electrical resistance of water.

2. MECHANISMS OF ELECTROLYSIS

Electrochemical processes take place in an interfacial region between an electrode and an electrolyte where the mode of electric conduction changes from electronic to ionic. A schematic diagram of a typical electrolysis cell is shown in Fig. 2. The kinds of reactions that predominate depend on the direction of current flow. The majority of the chemical reactions that have been useful for wastewater treatment occur at the anode, where chemical oxidation takes place. In most instances, the main cathode reactions produce hydrogen in water-disinfection applications more often a hazard than a desideratum. If heavy metal ions are present in wastewater, they may also be deposited on the cathode by the electrolysis current. A process utilizing this principle has been reported for recovery of metals from waste solutions (13). The more toxic forms of chromium, i.e., Cr(VI), can be reduced to Cr(III) at a cathode, which is subsequently precipitated as $Cr(OH)_3$ and removed from the waste stream. Copper ion in wastewater can be reduced to its elemental form in the cathode, which can be reused.

Under normal circumstances the cathodic removal efficiency for very low concentrations of trace metals would be extremely low. The most important use of cathodic reactions in water purification is the raising of pH by the hydrogen ion removal. High pH precipitates $Mg(OH)_2$ in waters containing magnesium and facilitates the removal of insoluble phosphates and carbonates, as well as the removal of soluble phosphate and ammonia as $Mg(NH_4)PO_4$. It is useful to consider the anodic reactions for wastewater treatment as following.

Sacrificial anodes: Metal ions are generated in the aqueous phase directly from the electrode material. The most useful appear to be Fe^{2+}, Fe^{3+}, and Al^{3+}. The majority of the current is generated at the electrode surface by the loss of electrons in oxidizing and dissolving the anode metal. The fate of the metal ions after dissolution depends on concentrations of the anion species. Both Fe^{3+} and Al^{3+} are precipitated as metal hydroxides in neutral solutions to produce flocs for flocculating suspended solids (12). Direct reaction may also remove nutrients (e.g., phosphate) as insoluble precipitates (14).

Inert electrodes: The participants in the anode reactions all come from the solution as the anode material remains essentially unchanged. The reactions might include the oxidation of reduced metals, such as Fe(II) to Fe(III), or the discharge of negative ions, such as Cl^- or OH^-. The overall result of the discharge of Cl^- ions is usually the hydrolysis of $C1_2$ in the water. The discharge of OH^- usually results in the release of oxygen at the anode, a reaction that is, of course, favored by high pH. The production of oxygen (or ozone) at the anode will tend to lower the pH in the anode region just as, conversely, the production of H_2 (g) at a cathode raises the pH in that region.

3. ORGANIC AND SUSPENDED SOLIDS REMOVAL

The effective oxidation of organic removal in various waste streams by the electrolysis has been reported (12,14–24). It can be used for treatment of industrial wastewaters, landfill leachate, and domestic sewage. Successful examples include color removal and food wastewater treatment. Specific organic compounds such as phenols, tannic acid, lignin, EDTA, and aniline can be destructed by the approach.

3.1. *Organic and Suspended Solids Removal by Regular Electrolysis*

As far as the electrode choice is concerned, the important criteria are a high hydrogen potential for the cathodes and, to avoid electrode combustion, high oxygen overpotential for the anodes (19). The cathodes are based on carbon/PTFE, copper, steel, and iron; graphite has frequently been used as an anode due to its low cost as well as satisfactory treatment efficiencies. Recently, titanium or platinum electrodes covered with thin layers of electrodeposited metal such as IrO_2 and RuO_2 have been used. The anodes are normally based on platinum, sacrificial iron, and lead oxide. Platinum, ruthenium, and rhodium have been applied as electrocatalysts to enhance the electrolysis process. The treatment efficiency depends on many factors, including operating time, stability and concentration of target compounds, ionic strength, temperature, pH, size of anode, current and voltage applied, dissolved oxygen, and presence of competing organic compounds.

The treatment is due to two main factors: free radical production and anodic oxidation (e.g., lead dioxide anode). Aqueous free radicals are very reactive, powerful oxidants, and short lived; thus, they easily recombine to form water. One of the most reactive aqueous radical species is the hydroxyl radical ($HO^{\cdot}$). This radical has an electron affinity value of 136 kcal and is able to oxidize all organic compounds (18). The $HO^{\cdot}$ is the primary oxidant involved in the organic oxidation reaction because (a) oxidation extent is limited only by quantity of electric energy, and (b) the oxidation yields derived are greater than the one gram equivalent weight per Faraday relationship.

Additionally, the removal is due to the formation of metal hydroxide formed by the oxidation. The hydroxides [e.g., $Mg(OH)_2$)] floc carrying the insoluble organics to the liquid surface are responsible for the initial removal of organic suspended solids. The subsequent organic removal (primarily in soluble form) is due to the adsorption or the oxidation in the presence of chlorine and oxygen; however, the kinetics is much slower. Removal of suspended solids is found to be parallel to the BOD removal, which results from the effect of the metal hydroxide floc. However, the floc must be carefully controlled. Otherwise, it can cause a high suspended solid concentration in the effluent.

The electrolytic approach was applied for the reduction of both organic load and color in textile dye wastewater by using Tr/Pt as anode and stainless steel 304 as cathode (16). With 18 min reaction time, COD, BOD_5, and color removal percentages of more than 80% were achieved in the treatment of dyeing wastewater with TOC, COD, BOD_5, and color of 740 mg/L, 1250 mg/L, 450 mg/L, and 3450 ADMI color units, respectively. The electrochemical method was used to treat a wastewater from the cigaret industry (20). COD and BOD removal efficiencies of 56 and 84% were reported for the wastewater with a COD of 1180 and a BOD of 530 mg/L, respectively. With an increase in the surface area of anode, the removal was increased and reaction time decreased.

3.2. *Organic and Suspended Solids Removal by Electrocoagulation*

Recently, electrocoagulation for wastewater treatment has been widely studied and the results have appeared in the literature (16–28). It has been an "emerging technology" since 1906 when the first US patent was awarded. At that time, there were insufficient financial or regulatory incentives for industry to adopt the process. It has been available for almost a century; however, the design of an industrial electrocoagulation unit is still based on empirical knowledge.

This process is an extension of the electrolysis and combined with the concept of chemical coagulation. In the treatment, aluminum or iron plates are normally used as electrodes. When a direct current (DC) voltage is applied, the anodes sacrifice themselves to produce Al^{3+} or Fe^{2+} ions, which results in the formation of metal hydroxides [$Al(OH)_3$ and $Fe(OH)_2$] as good coagulants. A list of chemical reactions together with the products is showed in Table 1. Al^{3+} and OH^- ions generated by the electrode reactions react to form various monomeric species such as $Al(OH)^{2+}$, $Al(OH)_2^+$, $Al_2(OH)_2^{4+}$, and $Al(OH)_4^-$, as well as polymeric species such as $Al_6(OH)_{15}^{3+}$, $Al_7(OH)_{17}^{4+}$, $Al_8(OH)_{20}^{4+}$, $Al_{13}O_4(OH)_{24}^{7+}$, and $Al_{13}(OH)_{34}^{5+}$, which transform finally into $Al(OH)_3$. The electrogenerated ferric ions can form monomeric ions, ferric hydroxo complexes with hydroxide ions and polymeric species, depending on the pH range. These species include $FeOH^{2+}$, $Fe(OH)_2^+$, $Fe(H_2O)_2^+$, $Fe(OH)_4^-$, $Fe(H_2O)_4^-$, $Fe(H_2O)_5OH^{2+}$, $Fe(H_2O)_4(OH)_2^+$, $Fe(H_2O)_8(OH)_2^{4+}$, and $Fe_2(H_2O)_6(OH)_4^{2+}$, which transform finally into $Fe(OH)_3$, and other species shown in Table 1.

The metallic ions and the hydroxides are able to destabilize the finely dispersed organic and inorganic particles present in the water and wastewater (29). The destabilized particles then aggregate to form focs. At the same time, the tiny hydrogen bubbles produced at the cathode can float the flocs formed, achieving effective separation of particles from the waters. This process can remove soluble organic compounds, oily droplets, as well as organic/inorganic suspended solids. The electrolysis voltage and the reaction time must be carefully controlled; otherwise, too much floc can be generated, which can cause a high suspended solid concentration in the effluent as well as higher operational cost.

The electrolysis voltage is one of the most important variables. It is strongly dependent on the current density, the interelectrode distance, the conductivity of the water, and the surface state of electrodes. (26) Development of electrocoagulation has involved resolving the key issues of electrochemical cell design, electrode fouling, power supply, operating conditions, and providing the most suitable unit operations to support the process.

Compared with conventional coagulation, electrocoagulation has several advantages. It is more effective in destabilizing small colloidal particles (22). It can fulfill simultaneous chemical oxidation, coagulation, and flotation, with less sludge produced. The treatment system is very compact and thus suitable for installation where the available space is rather limited such as restaurants and hotels. In addition, the convenience of dosing control only by adjustment of current makes the automation of treatment easier.

Electrocoagulation is a good fit to treat wastewaters contaminated with emulsified oils, PAHs, poorly settling solids, poorly soluble organics, contaminants in general that add turbidity to water, as well as negatively charged metal species such as arsenic, molybdenum, and phosphate that form coprecipitates with iron or aluminum. Heavy metals and soluble organic compounds are removed from wastewater by this approach in association with the removal of emulsified and particulate solids. The following are the examples in the application of electrocoagulation.

Electrocoagulation was applied for the treatment of olive oil mill wastewater with the COD of 48,500 mg/L and the suspended solid of 1780 mg/L (27). The treatment system is illustrated in Fig. 3. Aluminum and iron were used; reactor voltage was 12 V, current density ranged between 10 and 40 mA cm^{-2}, and pH was 4, 6, 7, and 9. At a retention time of 30 min, COD removal percentages of 52 % and 42 % were achieved by the aluminum anode and the iron anode, respectively; the color removal was 90–97 %.

Chen and coworkers reported that oil and grease in the restaurant wastewater were treated successfully by electrocoagulation (25). The raw wastewater had COD of 1010–1700 mg/L and oil and grease of 505–1140 mg/L, respectively. The optimum

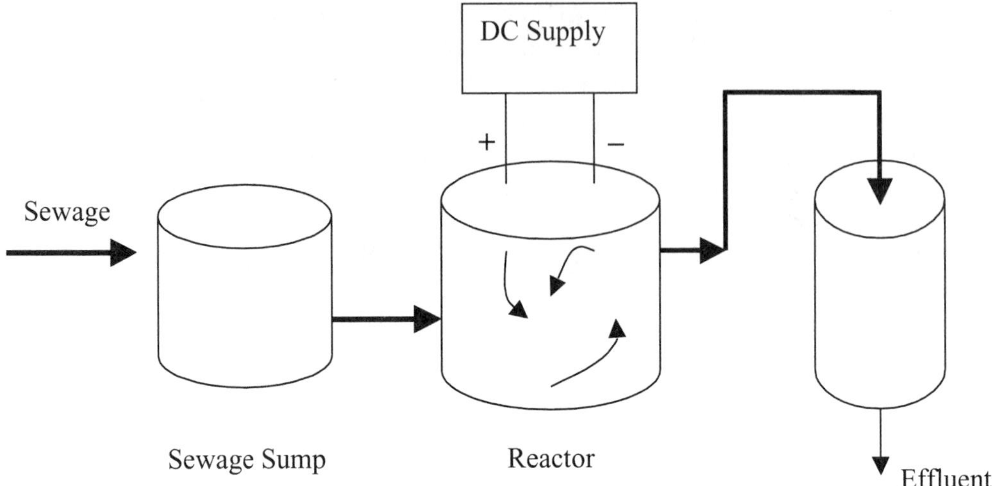

Fig. 3. Illustration of electrocoagulation system.

charge loading and current density were 1.67–9.95 F/m wastewater and 30–80 A/m^2. The removal efficiencies of COD and oil and grease exceeded 72% and 94%, respectively. The aluminum electrode consumption ranged from 17.7 to 106.4 g/m^3, while the power requirement was less than 1.5 kW h/m^3.

Treatment of textile wastewaters by electrocoagulation using iron and aluminum electrode materials was studied (28). The COD, the suspended solids, and the turbidity in the raw wastewater were 3422 mg/L, 1112 mg/L, and 5700 NTU, respectively. At initial pH of 6–8, the COD removal percentages of 50% and 70% were achieved by the aluminum anode and the iron anode, respectively; the turbidity treatment efficiency of 85–95 % was observed. It was found that iron was superior to aluminum as sacrificial electrode material, from COD removal efficiency and energy consumption points.

The process, while extraordinarily effective to remove a wide range of contaminants, is no wastewater treatment panacea. For example, the soluble BOD resulting from contaminants such as antifreeze or solvents cannot be removed directly. The factors that drive the costs of the treatment system include labor, electrode replacement, electrical power, amortization of capital, disposal of solids residuals, and chemicals. The requirements for anode replacement depend on the nature and the concentration of contaminants to be removed. The requirements for electrode (e.g., aluminum) are determined through treatability testing, preferably under continuous operations. Electrical power requirements depend on the concentration of aluminum or iron needed to treat a particular wastewater, on the configuration of the electrochemical cell, and on the conductivity of the wastewater. When needed, salts can be added to increase conductivity.

4. DISINFECTION

It is normally required that the effluents after secondary biological treatment be disinfected before being discharged into nearby waters. Chlorination is the major method of disinfection (14). Other methods such as ultraviolet radiation and ozonation are still more expensive or less convenient than chlorination for the disinfection of the effluents. It has been reported that the electrolysis can kill a large spectrum of microorganisms including bacteria and algae (30–32).

During the process, the effluents flow through the contactors equipped with electrodes on which DC is charged. The effluent quality, nature of the electrodes, energy input, and other operational conditions influence the effectiveness of the process. It has been reported that the process can generate a high disinfection efficiency within a short contact time for the secondary wastewater effluents and surface waters. Higher killing is normally observed in saline waters owing to the high content of chloride ions. The microbial killing is attributed to various functions, including electrochlorination, destruction caused by the electric field, and inactivation by strongly oxidative but short-lived intermediate radicals. The disinfection of microorganisms is due to oxidative stress and cell death caused by electrochemically generated oxidants, irreversible permeabilization of cell membranes by the applied electric field, and electrochemical oxidation of vital cellular constituents during exposure to electric current or induced electric fields.

Electrochlorination refers to the production of common disinfectant chemicals such as Cl_2 and NaClO in saline waters during the electrolysis process. Chlorine gas produced at the anode causes the disinfection of wastewaters, oxidation of organic matter, and bleaching of smaller degradation products. The following equations typify the reactions for this process:

$$2Cl^- = Cl_2 + 2e^- \tag{1c}$$

$$Cl_2 + H_2O = HOCl + H^+ + Cl^- \tag{3a}$$

$$HOCl = H^+ + OCl^- \tag{3b}$$

In dilute solutions and at pH levels above 4, the reaction shown in Eq. (3a) shifts to the right. The reaction shown in Eq. (3b) is also a function of pH and shifts to the right at pH > 7.5. The disinfection efficiency is generally close to 100 % for total coliform groups after a detention time of 30 min or longer with a sewage-to-seawater ratio of 9:1. However, the ratio varies depending on sewage strength. In order to determine how much electrolyzed seawater should be used, it is necessary to install an active-chlorine analyzer that effectively measures the strength of the sewage.

Electrochlorination for disinfection has long been recognized. Other mechanisms also contribute to the killing. An electric field can cause the inactivation and destruction of microorganisms by a series of electrochemical reactions inside the cells and by electromechanical compression of the cell membrane. A series of short-lived and energy-rich free radicals generated during electrolysis such as $OH^{\cdot}$ and $ClO_2^{-\cdot}$ plays a critical role in the killing. They are very unstable and difficult to detect; however, they are extremely reactive bactericidal agents and can provide nearly instant killing action. Chlorine ions in the saline solution can function as a catalyst to facilitate the generation of the short-lived free radicals and extend significantly their functional life by a factor as high as 10 (33). As a result, the efficiency of microbial destruction is dramatically improved. Considering the short contact time and relatively low chlorine produced, it is quite unlikely that the main killing is chlorination. Therefore, Cl^- may act as a catalyst in the formation of free radicals, rather than as a precursor to Cl_2 production.

A laboratory-scale electrolysis disinfector was used for the disinfection of saline primary and secondary effluent (with a salinity content of around 8% and a chloride content of 4,800–5,000 mg/L), and freshwater secondary effluent (with a chloride content of 110–130 mg/L) (30). A killing efficiency of 99.9% on total coliform bacteria was achieved when the contact time of less than 10 s and 20 s and the power consumption of 0.006 and 0.08 kWh/m^3, for the saline secondary effluent and the saline primary effluent, respectively. However, a similar degree of effectiveness for freshwater sewage effluent was not observed, even with a longer contact time and higher power input.

Few such studies have been conducted with viruses. It is expected that viruses are more resistant to electrochemical inactivation than bacteria. For example, bacteriophage, one of important viruses in water treatment, is not enveloped with a membrane and would be immune to inactivation processes involving irreversible membrane permeabilization. Even enveloped viruses are more resistant than bacteria due to their smaller size. In addition, viruses tend to be more resistant to commercially used chemical disinfectants (e.g., chlorine) than vegetative bacteria. Drees and coworkers reported that bacteriophages survived short exposures to various current magnitudes in an electrochemical cell better than bacteria at both low and high population density (32). The inactivation rate of bacteria exposed to a low current magnitude (5 mA) for an extended time ranged from 2.1 to 4.3 times greater than that of bacteriophages, indicating that bacteria are more sensitive to electrolysis than bacteriophages. It was found that electrochemically generated oxidants were a major cause of inactivation within the electrochemical cell.

Since viruses are more resistant than bacteria to the electrochemical inactivation, use of this technology in fields that affect human health (such as drinking water disinfection) must ensure the destruction of viruses, not just bacteria, in order to consider the treated medium safe. Another disadvantage of its application for water disinfection is the formation of undesirable disinfection by-products. These compounds are not only formed in the disinfection, but also in its application for organic removal if the chloride ions are present in the solution. The by-products include chloroform, chloromethane, bromodichloromethane, and 1,1,1–trichloroethane; they are suspected human carcinogens. Careful control and management are therefore important.

5. PHOSPHATE REMOVAL

Electrolysis can be applied for removal of phosphate from wastewaters. Aluminum and iron are used as sacrificial anodes. Two possible mechanisms for the removal are postulated: (1) adsorption of phosphate onto the metallic hydroxide floc [e.g., $Al(OH)_3$, $Fe(OH)_3$ in Table 1] generated and (2) precipitation such as $AlPO_4$ and $FePO_4$.

Sadek reported that phosphate was successfully removed by using aluminum anodes (5). The initial phosphate concentration was 30.5 mg/L, the current was 0.14 A, the voltage was 5 V and the contact time was 0.5 h. The removal efficiency of phosphate was 99%. However, the removal was less than 10% when a carbon anode was used. Phosphate removal from water by electrolysis was examined by using aluminum and iron electrodes (34). It was found that the aluminum electrodes were better than iron electrodes.

Campbell and coworkers used aluminum electrode to treat Kraft mill wastewaters; $CaC1_2$ was added as an electrolyte to minimize power consumption (35). They proposed the adsorption mechanism for the phosphate removal. Hemphill and Rogers used an electrolysis reactor with a series of lead cathodes and lead dioxide anodes to treat a wastewater containing phosphate (15). They found that the phosphate removal was due to the chemical coprecipitation.

6. AMMONIUM REMOVAL

As shown in Table 1, chlorine gas can be produced during the electrolysis of wastewaters that contain a sufficient amount of chloride ions. Cl_2 can oxidize ammonium/ammonia, organic compounds, nitrite ions, hydrogen sulfide, and other oxidizable substances. Break-point chlorination is therefore suggested as the nitrogen-removal mechanism (14,36).The following reactions take account of the ammonium removal:

$$2Cl^- = Cl_2 + 2e^- \tag{1c}$$

$$NH_4^+ + HOCl = NH_2Cl + H_2O + H^+ \tag{4a}$$

$$NH_2Cl + 2HOCl = NH_2Cl \uparrow + 2H_2O \tag{4b}$$

$$NHCl_2 + HOCl = NCl_3 \uparrow + H_2O \tag{4c}$$

$$NH_3 + HOCl = NH_2Cl + H_2O \tag{4d}$$

$$NH_2Cl + NHCl_2 + HOCl = N_2O + 4HCl \tag{4e}$$

$$2NH_2Cl + HOCl = N_2 + H_2O + 3HCl \tag{4f}$$

A leachate from a new municipal domestic waste landfill site was treated by using an electrolysis system, which has mean COD and BOD values of 53,300 and 30,300 mg/L, respectively (37). Ti/Pt electrodes were used. With a reaction time of 1 h and at pH 9, the COD, the ammonium, and the total phosphorus were reduced by 84%, 100% and 100%, respectively.

Chiang and coworkers studied the treatment of a low BOD/COD ratio landfill leachate by an electrolysis process (38). A ternary Sn-Pd-Ru oxide-coated titanium (SPR) anode was used; the current density was 15 A/m^2; the chloride concentration was 7500 mg/L; the contact time was 240 min; the initial COD, BOD, and ammonium concentrations were 4100–5000 mg/L, less than 1000 mg/L, and 2100–3000 mg/L, respectively. The removal percentages of COD and ammonium of 92%, and 100% in the leachate was achieved. Among the four anode materials (graphite, PbO_2/Ti, DSA®, and SPR anodes), the SPR anode having a high electrocatalytic activity gave the best chlorine/hypochlorite production efficiency and landfill leachate treatment efficiency.

7. CYANIDE DESTRUCTION

The alkaline chlorination process for the destruction of cyanide waste from the metal-processing industry is a widely accepted practice (39). At pH > 8, cyanide ion (CN^-) is oxidized stepwise to cyanogen (CNCl), cyanate ion (CNO^-), and finally to N_2 gas. The application of electrolytic practice for cyanide destruction requires longer periods of time to reduce cyanide to a level of 1 ppm, which is substantially longer than the time required for traditional chlorination practices.

The anode used in this process can be copper, stainless steel, or carbon steel. However, a current density from 30–80 A/ft^2 must be maintained. The cathode employed can be of carbon steel. Optimum operating temperature for this CN^- destruction process is approximately 200°F.

Cyanide is usually used in gold mining, electroplating, and metal finishing industries (40,41). It is normally complexed with metals such as copper. Conventional treatment of these wastewaters consists of oxidation of the cyanide by chlorine gas or hypochlorite and the subsequent removal of metal ions by precipitation. The pH and the chemical dose must be carefully controlled. The resulting sludge is difficult to handle because it contains metal hydroxides, which can be subject to easy dissolution. In the recent years, there is a series of studies on the simultaneous electrooxidation of cyanides and recovery of heavy metals as a metallic deposition on the cathode. Electrochemical oxidation, particularly if conducted in a reactor with plate electrodes, enabling the cathode to be reused, can be a technically and economically feasible alternative.

Szpyrkowicz and coworkers reported that both a direct electrooxidation process and an indirect electrooxidation (that has a chloride-rich medium) proved feasible in the simultaneous treatment of both cyanides and copper (41). The direct electrooxidation was preferable due to the lower-energy consumption. The direct electrooxidation under alkaline conditions results in formation of an electrocatalytic film on the anode. Simultaneous copper electrodeposition on the cathode is feasible and economically convenient at pH > 13; the process can be described by the pseudo-first-order kinetics. Energy consumed for copper electrodeposition proved to be inversely proportional to the initial Cu concentration. For wastewater containing 1100 mg/L Cu, 5.46 kWh is needed to eliminate 1 kg of metal. Under the optimum conditions, the total cyanide concentration was lowered from 250 to 7.9 mg/L.

8. METAL REMOVAL

Rapid industrialization and urbanization have poised a series of environmental problems in the last several decades. Among them, heavy metal contamination becomes more serious due to dramatic growth in the electronic industry. This triggers numerous studies on the metal removal, which leads to development of various techniques, such as precipitation, adsorption/biosorption, and ion exchange (42–44). These technologies have proved to be effective for metal removal; however, they cannot recover the valuable heavy metals that are subsequently reused.

Conversion of precious metal waste to useful materials has not widely studied in the past. Chemical reduction can provide a useful tool for the metal recovery. For example, hydrazine (N_2H_4) was found to be powerful in recovery of silver and copper (45). The use of the electrochemical approach to recover metal ions in the wastewater in their metallic state can be considered, as it is a relatively simple and clean method.

Electrochemistry deals with the charge transfer at the interface between an electrically conductive material and an ionic conductor as well as with the reactions within the electrolytes and the resulting equilibrium. Cathodic removal of heavy metals shows several benefits in terms of costs, safety, and versatility. The equipment needed is only an electroplating bath, an insoluble anode, and a suitable cathode. At the cathode, the metal ions are being reduced according to Eq. (2a). There are various competing reactions at the cathode and the most common reaction is when the H^+ ions are being reduced to hydrogen gas based on Eq. (2b). Although the metal to be recovered will be deposited at the cathode, careful selection of the anode is needed to ensure that it will be inert in the electrolyte. If it dissolves into the solution during the electrochemical deposition process, any metal recovery efforts will then be futile. A common reaction at the anode is shown in Eq. (1b).

Other than recovering metals in their metallic form, the electrochemical treatment of metal ion pollutants has several advantages: recovery of precious metals; no extra chemical reagents required; no sludge production; high selectivity; low operating cost; and possible disinfection of the wastewater. However, there are disadvantages: the deposition rate and the composition of the solution in some cases can cause the production of dendrites and loose or spongy deposits; and interference from the hydrogen evolution reaction or from dioxygen reduction has to be minimized.

There are a few studies available in the literature on metal removal by the electrochemical approach. Kongsricharoern and Polprasert used an electrochemical precipitation process to treat an electroplating wastewater containing Cr concentrations of 570–2100 mg/L (46). It was found that the Cr removal efficiencies were higher than 99% and the Cr concentrations in the treated effluent were less than 0.5 mg/L. Kusakabe et al. studied the simultaneous electrochemical removal of copper and organic wastes by using a packed-bed electrode cell (47). The final attainable concentration of copper in the effluent was as low as 3 ppm and the removal of COD ranged from 25% to 41%.

Chen and Lim studied conversion of soluble precious metals into a solid form for further reuse by using an electrochemical deposition approach (13). It was found that the metal recovery followed a first-order reaction kinetics. Distance between electrodes had no much impact on the recovery, while higher mixing led to faster kinetics. The presence of humic acid with lower concentration (< 20 ppm) did not have impact on the recovery. When its concentration was increased to 50 ppm, it decreased the metal reduction. Presence of EDTA and ionic strength slightly reduced the copper recovery rate. About 50% removal of humic acid and EDTA was achieved. In the competing environment, metal recovery was in the following descending order: silver > lead > copper. X-ray photoelectron spectroscopy (XPS) and scanning electron microscopy (SEM) analysis of the reduced metals demonstrated that the depositions were composed of mainly elemental metals together with their oxides, which were due to the oxidation.

General Environmental Corporation, Inc. (GEC), of Denver, CO has developed the CURE electrocoagulation technology for removal of low levels of the radionuclides uranium, plutonium, and americium as well as other contaminants in wastewater (48). The technology includes the coagulation and precipitation of contaminants by a direct-current electrolytic process followed by settling with or without the addition of coagulation-inducing chemicals as shown in Fig. 4. Treated water is discharged from the system for reuse, disposal, or reinjection. Concentrated contaminants in the form of sludge are placed in drums for disposal or reclamation. The CURE technology was demonstrated under the SITE Program at the U.S. Department of Energy's Rocky Flats Environmental Technology Site (formerly the Rocky Flats Plant) near Golden, CO. Approximately 4500 gal of wastewater containing low levels of the radionuclides uranium, plutonium, and americium were treated in August and September 1995. Water from the solar evaporation ponds was used in the demonstration. Six preruns, five optimization runs, and four demonstration runs were conducted over a 54-d period. The

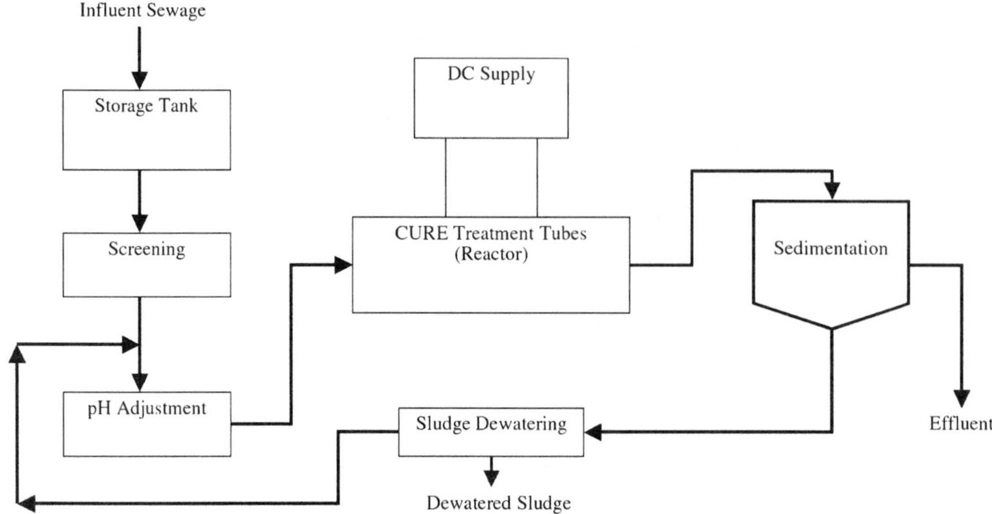

Fig. 4. Simplified process diagram of CURE electrocoagulation technology.

demonstration runs lasted 5.5–6 h each, operating the CURE system at approx 3 gal/min. Filling the clarifier took approx 2.5 h of this time. Once the clarifier was filled, untreated influent and effluent from the clarifier were collected every 20 min for 3 h. Because of the short run times, there is uncertainty whether the data represent long-term operating conditions. Results indicated that removal efficiencies for the four runs ranged from 32% to 52% for uranium, 63% to 99% for plutonium, and 69% to 99% for americium. Colorado Water Quality Control Commission (CWQCC) standards were met for plutonium and americium in some cases. CWQCC standards for uranium were not met. Arsenic and calcium concentrations were decreased by an average of 74% and 50 %, respectively, for the two runs for which metals were measured. Evaluation of the CURE electrocoagulation technology against the nine criteria used by the EPA in evaluating potential remediation alternatives indicates that the CURE system provides both long- and short-term protection of the environment, reduces contaminant mobility and volume, and presents few risks to the community or the environment. Potential sites for applying this technology include Superfund, US Department of Energy, US Department of Defense, and other hazardous waste sites where water is contaminated with radionuclides or metals. Economic analysis indicates that remediation cost for an 100 gal/min CURE system could range from about $0.003 to $0.009 per gallon, depending on the duration of the remedial action.

9. REMEDIATION OF NITROAROMATIC EXPLOSIVES-CONTAMINATED GROUNDWATER

Groundwater contamination by nitro compounds is associated principally with the military industry and practices as well as the mining industry (49). Modern explosives are nitrogen-containing organic compounds with the potential for self-oxidation to small gaseous molecules such as N_2 and CO_2. The important compounds are polynitroaromatic compounds, including 2,4,6-trinitrotoluene (TNT), 1,3,5-trinitrobenzenene

(TNB), dinitrotoluene (2,4-DNT, 2,6-DNT), dinitrobenzene (DNB), methyl-*N*,2,4,6-tetranitroaniline (Tetryl), hexahydro-1,3,5-trinitro-1,3,5-triazine (RDX) and 2,4,6-trinitrophenol (49–55). The United States ceased TNT production in the mid-1980s; however, subsurface contamination still exists due to the previous activities and demilitarization. TNT Contamination requires large amounts of water for its purification. The aqueous wastes, known as red water, contained up to 30 nitroaromatics in addition to TNT. Pink water, which is generated during loading, packing or assembling munitions, normally contains high concentrations of other nitroaromatic explosives. The US Department of Defense has identified more than 1000 sites with explosives contamination, of which >95% were contaminated with TNT and 87% exceeded permissible groundwater contaminant levels. Many sites became contaminated through open detonation and burning of explosives at army depots, evaluation facilities, artillery ranges, and ordnance disposal sites.

Nitroaromatic explosives are toxic, and their environmental transformation products, including arylamines, arylhydroxylamines, and condensed products such as azoxy and azo compounds, are even more toxic than the parent nitroaromatic. TNT, which is on the list of US EPA priority pollutants has been a known mutagen and can cause pancytopenia as a result of bone marrow failure (14). Aromatic amines and hydroxylamines are implicated as carcinogenic intermediates as a result of nitrenium ions formed by enzymatic oxidation (49).

Aromatic nitro compounds are environmentally persistent, and the remediation of contaminated groundwater is difficult. In addition, TNT has very low mobility due to its high adsorption onto soils. Remediation strategies must be considered on a site-by-site basis. The toxicity of nitroaromatics limits the applicability of bioremediation when concentrations are high, or the treatment process may produce recalcitrant reaction by-products. Therefore, chemical approaches can play an important in the remediation. The aromatic nitro compounds can be catalytically hydrogenated to amines in acidic solutions. However, hydrogenation is not a complete remediation method as the anilines are toxic and must be treated further. Noble metal catalysts (e.g., palladium) used in the laboratory are too expensive for waste treatment.

Electrolysis can be used to treat the aromatic nitro compounds due to the relatively simple equipment and operating cost. During the treatment, a series of intermediates and end-products is formed, which is very much dependent on the operational conditions such as current, voltage, and dissolved oxygen. Further treatment of toxic end-products such as anilines must be addressed and carefully handled. They may be treated by chemical and biological processes.

An electrochemical pilot-scale reactor was used to treat simulated munitions wastewater containing 100 mg/L of 2,4-dinitrotoluene (DNT). Experiments were conducted by using a glassy carbon (zero porosity) coated graphite cylinder as the cathode and a platinum wire as the anode (52,53). The dissolved oxygen concentration was less than 1.5 mg/L. It was shown that the rates of reduction of DNT increased with an increase in current or concentration of electrolyte. At an ionic strength of 0.027 *M* and with a current of 200 mA and a contact time of 14 d, a reduction of DNT at the 80% level was observed.

The electrochemical reduction of DNT and a mixture of TNT and RDX by an electrolysis approach was studied by Doppalapudi and coworkers (54). Glassy carbon rods as the cathode and platinum wire as the anode were used. The experimental results showed that the reduction of nitroaromatics followed pseudo-first-order rate kinetics, whose rate constants increased with an increase in current or stirring rate. The reduction rate was much higher under anoxic conditions than under oxic conditions.

Electrochemical degradation of DNT in a bench-scale reactor was studied by Jolas and coworkers (55). Rates of the degradation were quantified and range from 0.003 to 0.008 min^{-1} for graphite and 0.003 to 0.06 min^{-1} for titanium. The products of the degradation were identified to be DAT and azoxy dimers and in some cases they were quantified as well. An azoxy dimer is found only in experiments with dissolved oxygen and regioselectivity favoring the first reduction at the para position is evident based on product studies. The degradation studies utilizing a graphite rod cathode were shown to be mass transfer limited. On the other hand, the degradation studies by titanium mesh wire were not. The material of the electrode has a strong influence on the rates of degradation.

10. ELECTROLYSIS-STIMULATED BIOLOGICAL TREATMENT

In the recent years, electrolytically stimulated biological reduction of environmentally relevant compounds such as nitrogen has been studied (56–66). The electrolysis of water results in the production of hydrogen and oxygen gases at the cathode and the anode according to Eqs. (1b) and (2b), respectively. The focus of most of these studies has been on the use of hydrogen as an electron donor for the reduction of nitrate or chlorinated compounds. In addition, the oxygen produced can be used for the enhancement of bioremediation of both surface and subsurface waters. Microorganisms can be either cultured as a biofilm on the cathode surface or in suspension in a compartment separated from the anode in a divided electrochemical cell.

10.1. Nitrogen Removal

A maximum contamination level (MCL) of 10 mg NO_3-N/L is set by the US Environmental Protection Agency due to the health effects associated with the ingestion of nitrate. Nitrate can cause methemoglobinemia termed as blue-baby syndrome when it is ingested by infants. It may cause heart and behavioral problems. In addition, nitrosoamines are carcinogenic compounds that may be formed due to the presence of nitrate. In the United States and elsewhere, nitrate content above the MCL have been reported in many water supplies. The problem is becoming more serious nowadays. The removal of nitrate can be achieved by the traditional biological nitrification as well as chemical processes. Electrolysis approach as one of emerging technologies is reportedly applied for the enhancement of biological nitrification process.

In the autotrophic denitrifcation of nitrate to nitrogen gas by hydrogen gas produced by cathode as an electron donor, the following reactions occur (56):

$$2NO_3^- + 2H_2 = 2NO_2^- + 2H_2O \tag{4a}$$

$$2NO_2^- + 3H_2 + 2H^+ = N_2 + 4H_2O \tag{4b}$$

The complete reaction is therefore:

$$2NO_3^- + 5H_2 + 2H^+ = N_2 + 6H_2O \tag{5}$$

It has reported that higher denitrification efficiency is realized when hydrogen gas produced by electrolysis is used as an electron donor (58).

Islam and Suidan studied the 1-yr performance of denitrification in a biofilm-electrode reactor (57). A nitrate removal efficiency above 98% was achieved at a current of 20 mA. Current intensities between 20 and 25 mA resulted in the lowest effluent concentrations of nitrate. For higher current intensities, hydrogen inhibition and charge-induced repulsion caused decreased reduction efficiencies. Oxidation–reduction potential (ORP) was governed by both the concentrations of hydrogen and effluent nitrate. A total nitrate removal efficiency of 85% was achieved at a current level of 25 mA in the absence of any nutrient.

Cast and Flora studied the denitrification efficiency of two types of cathode material (stainless-steel rods wrapped with stainless-steel mesh, and graphite rods wrapped with polypropylene mesh) (56). They reported that the difference in performance of the two types of biologically active cathodes was insignificant. Exposure of the biologically active cathodes to copper inhibited the attached microorganisms and had a statistically significant reduction in denitrification efficiency.

Electrolytically stimulated bioreactor can be combined with chemical and physical processes, which yield high-quality water. Prosnansky and coworkers developed a complex system that is composed of a multicathode biofilm-electrode reactor and microfiltration for treatment of nitrate-contaminated water (60). The multicathode electrodes were composed of multiple-granular activated carbons. The carbons attached to each cathode to enlarge the surface area of the electrodes and to attach bacteria quickly and firmly. In the biofilm-electrode reactor, H_2 gas is produced through the electrolysis and serves as an electron donor in biological reduction of nitrate to nitrogen gas. The bioreactor with high denitrification rates and short hydraulic retention time (<20 min) can be operated. The denitrification rate was significantly enhanced. MF membrane with plate modules and a pore size of 0.2 μm successfully rejected the bacteria escaping from the bioreactor, so that the suspended solid in the effluent was below 1 mg/L.

10.2. Electrolytic Oxygen Generation

Cost of aerobic biological wastewater treatment is very much dependent on the oxygen supply. It becomes more serious when contaminated soils and groundwater are treated *in situ*. The long-term oxygen supply poses an engineering problem due to the limited solubility of oxygen in waters and the low mass transfer resistance of oxygen from the gaseous phase to the liquid phase. Air sparging, membrane oxygen diffusion, and hydrogen peroxide injection can be used.

A possible method for increasing dissolved oxygen (DO) levels is through electrolysis, where oxygen is produced using an electrolytic cell according to Eq. (1b). By applying a small potential difference across the cathode and anode of the cell, oxygen can be produced from the anode. This oxygen could serve as an electron acceptor for the biologically mediated oxidation of chemical contaminants (61,62).

Franz et al. studied the oxygen-generating characteristics and side reactions of an electrolytic cell assembly for bioremediation of contaminants in groundwater (61). The oxygen-generating capabilities of new electrolytic cells and cells with light and heavy calcium carbonate precipitates on the cathode were evaluated under current densities of 0.5–5.0 mA/cm^2. Higher current densities resulted in higher mass transfer coefficients ($K_L a$) and greater saturation oxygen concentrations (C_{sat}). As the cathodic deposits increased, the K_La tended to decrease and the C_{sat} tended to increase. It was found that hydrogen peroxide was generated at low concentrations (<1 mg/L) and at higher levels in the absence of chloride in the feed solution.

The electrolysis method was applied to a submerged biofilter process to improve the organic and nitrogen removal (62). Activated carbon was used as the electrodes and support material for the microbial growth in the bioreactor. It was reported that nitrification and denitrification rates were enhanced by supplying oxygen and hydrogen, respectively, due to the electrolysis of water. Higher electric current would cause an increase in the nitrification and denitrification rates.

REFERENCES

1. J. J. Lingane, *Electroanalytical Chemistry*, 2nd ed., Wiley-Interscience, New York, 1996.
2. I. I. W. Marson, *The Engineer*, pp. 591–592, April, 1965.
3. R. Eales, *Chemical Engineering,* 172–174, June (1968).
4. S. A. Michalek and F. B. Leitz, *J.W.P.C.F.*, **4**, 1697–1712, (1972).
5. S. E. Sadek, *An Electrochemical Method for Removal of Phosphates from Wastewaters*, Water Pollution Control Research Series No. 17010–01/70, US Department of the Interior, Federal Water Quality Administration, 1970.
6. C. P. C. Poon, Electrochemical Process of Sewage Treatment, Purdue Univ. Engineering Bulletin, *Proc. 28th Industrial Waste Conference*, Part I, Engineering Extension Series, pp. 281–292, May, 1973.
7. C. P. C. Poon and T. G. Brueckner, *J.W.P.C.F.*, **47**, 66–78, (1975).
8. R. McNabey and J. Wynne *Water and Wastes Engineering*, 46–48, (1971).
9. C. Nebel, P. C. Unangst, and R. D. Gottschling, Ozone Disinfection of Combined Industrial and Municipal Secondary Effluents, Part I—Laboratory Studies, Purdue Univ. Engineering Bulletin, *Proc. 27th Industrial Waste Conference*, Part II, Engineering Extension Series No. 141, pp. 1039–1055, May, 1972.
10. C. Nebel, R. D. Gottschling, and H. J. O'Neill, Ozone treatment of sewage plant odors. *First International Symposium on Ozone for Water and Wastewater Treatment.* R. G. Rice and M. E. Browning (Eds). Vol. 1–2: 445–449.
11. R. Stephenson and B. Tennant, *Environmental Science & Engineering, www.esemag.com,* January, 2003.
12. G. H. Chen, X. M. Chen, and P. L. Yue, *J. Environ. Eng. ASCE* **126**, 858 (2000).
13. L. L. Lim, Removal and Recovery of Heavy Metal Ions. B. Eng Thesis, National University of Singapore (2002).
14. Metcalf & Eddy Inc., *Wastewater Engineering: Treatment Disposal Reuse*, 4 edition, McGraw-Hill, New York, 2002.
15. L. Hemphill and R. Rogers, Electrochemical Degradation of Domestic Wastewater, Purdue Univ. Engineering Bulletin, *Proc. 28th Industrial Waste Conference*, Part I, Engineering Extension Series No. 137, pp. 214–223, May, 1973.
16. A. G. Vlyssides, D. Papaioannou, M. Loizidoy, P. K. Karlis, and A. A. Zorpas, *Waste Management*, **20**, 569–574 (2000).
17. C. Borras, T. Laredo, and B. R. Scharifker, *Electrochimica Acta*, **48**, 2775–2780 (2003).
18. Y. J. Li, F. Wang, G. D. Zhon, and Y. M. Ni, *Chemosphere*, **53**, 1229–1234 (2003).
19. M. De Francesco and P. Costamagna, *J Cleaner Production,* **12**, 159–163 (2004).

20. R. S. Bejankiwar, *Water Research*, **36**, 4386–4390 (2002).
21. R. Bellagamba, P. A. Michaud, C. Comninellis, and N. Vatistas, *Electrochem Communi*, **4**, 171–176 (2002).
22. E. Brillas, J. C. Calpe, and J. Casado, *Water Rese*, **34**, 2253–2262 (2000).
23. S. H. Lin and C. S. Lin, *Desalination* **120**, 185–195 (1998).
24. O. Simond, V. Schaller, and C. Comninellis, *Electrochimica Acta*, **42**, 2009–2012 (1997).
25. X.M. Chen, G. H. Chen and P. L. Yue *Separation and Purification Technology*, **19**, 65–76 (2000).
26. X. M. Chen, G.H. Chen, and P.L. Yue, *Chemical Engineering Science,* 57, 2449–2455 (2002).
27. H. Inan, A. Dimoglo, H. Simsek, and M. Karpuzcu, *Separation and Purification Technology* **36**, 23–31 (2004).
28. M. Kobya, O. T. Can, and M. Bayramoglu, *J Hazard Mat*, **100**, 163–178 (2003).
29. R. D. Letterman and A. Amirthafajah, In *Water quality and treatment, A handbook of community water supplies*, 5 edition, R.D. Letterman (ed) McGraw-Hill, New York, 1999.
30. X. Y. Li, F. Ding, P. S. Y. Lo, and S. H. P. Sin, *J Environ Eng-ASCE*, **128**, 697–704, (2002).
31. G. Patermarakis and E. Fountoukidis, *Water Research*, **24**, 1491–1496, (1990).
32. K. P. Drees, M. Abbaszadegan, and P. M. Maier, *Water Research*, **37**, 2291–2300, (2003).
33. M. Saran, I. Beck-Speier, B. Fellerhoff, and G. Bauer, *Free Radical Biology and Medicine* **26**, 482–490, (1999).
34. O. Grøterud and L. Smoczynski, *Water Research*, **20**, 667–669, (1986).
35. H. J. Campbell, Jr., F. E. Woodward, and D. O. Herer, Purdue Univ. Engineering Bulletin, Proc. *25th Industrial Waste Conference*, Part I, Engineering Extension Series No. 137, pp. 203–213, May, 1970.
36. W. J. Viessman and M. J. Hammer, *Water Supply and Pollution Control* 6th edition, Addison-Wesley, Sydney, 1998.
37. A. Vlyssides, P. Karlis, M. Loizidou, A. Zorpas, and D. Arapoglou, *Environmental Technology,* **22**, 1467–1476, (2001).
38. L. C. Chiang, J. E. Chang, and T. C. Wen, *Water Research,* **29**, 671–678, (1995).
39. W. W. Eckenfelder, *Industrial Water Pollution Control*, McGraw-Hill, New York, 1966, pp. 130–133.
40. S. C. Cheng, M. Gattrell, T. Guena and B. MacDougall, *Electrochimica Acta*, **47**, 3245–3256, (2002) .
41. L. Szpyrkowicz, Z. G. Francesco, S. N. Kaul, and A. M. Polcaro, *Industrial and Engineering Chemistry Research*, **39**, 2132–2139, (2000).
42. J. P. Chen and H. Yu, Lead removal from synthetic wastewater by crystallization in a fluidized-bed reactor. *Journal of Environmental Science and Health, Part A-Toxic/Hazardous Substances & Environmental Engineering*, **A35**, 817–835 (2000).
43. J. P. Chen and M. S. Lin, *Separation Science and Technology*, **35**, 2063–2081 (2000).
44. J. P. Chen, L. Hong, S. N. Wu, and L. Wang, *Langmuir*, **18**, 9413–9421 (2002).
45. J. P. Chen and L. L. Lim, *Chemosphere*, **49**, 363–370 (2002).
46. N. Kongsricharoern and C. Polprasert, *Water Science and Technology*, **34**, 109–116 (1996).
47. K. Kusakabe, H. Nishida, S. Morooka, and Y. Kato, *J Appl Electrochem*, **16**, 121–126 (1986).
48. US Environmental Protection Agency. CURE Electrocoagulation Technology, *Innovative Technology Evaluation Report*, Cincinnati, Ohio 45268EPA/540/R–96/502, September, 1998.
49. J. D. Rodgersa and N. J. Bunce, *Water Research,* **35**, 2101–2111, (2001).
50. J. L. Hintze and P. J. Wagner, *TNT Wastewater Feasibility Study: Phase 1 Laboratory Study.* Gencorp Aerojet Propulsion Division, US Army Missile Command/Production Base Modernization Activity, Contract Number DAAH01-91-C-0738, 1972, pp. 1–730.

51. J. D. Rodgers, W. Jedral, and N. J. Bunce, *AWMA 92nd Annual Meeting*, Proceedings (CD-ROM), St. Louis, MO, Session, Abs. No. 875, 1999.
52. R. Doppalapudi, D. Palaniswamy, G. Sorial, and S.Maloney, *Water Science and Technology* **47**, 173–178 (2003).
53. R. B. Doppalapudi, G. A. Sorial, and S. W. Maloney, *J Environ Engin-ASCE*, **129**, 192–201, (2003).
54. R. B. Doppalapudi, G. A. Sorial, S. W. Maloney, *Environmental Engineering Science*, **19**, 115–130, (2002).
55. J. L. Jolas, S. O. Pehkonen, and S. W. Maloney, *Water Environmental Research*, **72**, 179–188 (2000).
56. L. Cast Kerri and J. V. R. Flora, *Water Research*, **32**, 63–70, (1998).
57. S. Islam and M. T. Suidan, *Water Research*, **32**, 528–536, (1998).
58. V. Beschkov, S. Velizarov, S. N. Agathos, and V. Lukova, *Biochemical Engineering Journal*, **17**, 141–145 (2004).
59. Y. Sakakibara and T. Nakayama, *Water Research*, **35**, 768–778, (2001).
60. M. Prosnansky, Y. Sakakibara and M. Kuroda, *Water Research*, **36**, 4801–4810, (2002).
61. J. A. Franz, R. J. Williams, J. V. R. Flora, M. E. Meadows, and W. G. Irwin, *Water Research,* **36**, 2243–2254, (2002).
62. T. Tanaka and M. Kuroda, *ASCE J Environ Eng* **126**, 541–548 (2000).

11
Sedimentation

Nazih K. Shammas, Inder Jit Kumar, Shoou-Yuh Chang, and Yung-Tse Hung

CONTENTS

1. INTRODUCTION

1.1. Historical

Humans knew the technique of sedimentation since time immemorial. They had learned from Mother Nature that when river waters come to rest in ponds and lakes they become much clearer. Whenever they had to drink water from rivers carrying silt and other suspended particles, they let it rest a while before drinking. The technique of sedimentation was used in the construction of prehistoric water works. According to Babbit and Doland (1) the history of water works structures, which mainly consisted of a reservoir to store water and to remove suspended particles, and a system of carrying water from the reservoir to populated areas, are found in the excavations of prehistoric ruins. The remains of Lake Moeris, which was one of the largest reservoirs of the Nile Valley, indicate its construction dates back to about 2000 BC. The Romans are known to have had an elaborate system of aqueducts, reservoirs, public baths, and public fountains for

From: *Handbook of Environmental Engineering, Volume 3: Physicochemical Treatment Processes*
Edited by: L. K. Wang, Y.-T. Hung, and N. K. Shammas © The Humana Press Inc., Totowa, NJ

their water supply. Public water supplies in the United States date back to 1652 in Boston and to about 1732 at Schaefferstown, PA (2).

The modern usage of sedimentation tanks, which were constructed specifically for the purpose of removing suspended solids from water prior to filtration, started early in the 19th century. Later in the same century, the use of settling tanks for partial clarification of wastewater was also utilized. The sedimentation process experienced no changes until recently when the tube settlers were developed. The advent of tube settlers has led to a considerable reduction in the size of sedimentation tanks.

1.2. Definition and Objective of Sedimentation

Sedimentation is defined as a unit operation in which suspended particles are separated from a suspension by gravitational settling. The terms clarification and thickening of sludge apply to the same unit operation. Coagulation, which is discussed in Chapter 4, involves the addition of chemicals to induce faster aggregation and settling of initially finely divided suspended and colloidal particles. The objective of sedimentation is to remove settleable particles from suspensions either with or without the addition of chemicals. When no chemicals are added to the process, it is called plain sedimentation. Plain sedimentation is usually employed in wastewater treatment, whereas in water-treatment plants, sedimentation, in most cases, is preceded by chemical coagulation. Sedimentation is also employed, to a limited scale, in separating particulates from air streams.

1.3. Significance of Sedimentation in Water and Wastewater Treatment

Sedimentation is a major unit operation that is employed at almost every water- and wastewater-treatment plant. In water purification plants the turbidity of water must be reduced to levels as low as possible. Usually a turbidity of 1.5 units is considered desirable for water introduced into filters. This is achieved either by plain sedimentation or by coagulation followed by sedimentation. Plain sedimentation usually requires a long time to be effective in water-treatment operations, so chemicals are added to induce floc formation and the flocs then settle rapidly. At a wastewater-treatment plant, sedimentation in the primary settling tank reduces the load on subsequent treatment units. Sedimentation in secondary settling tanks is most important as it removes a large percentage of the suspended solids from the treated wastewater, which would otherwise be carried into the effluent and then into the streams to which the effluent is usually discharged.

Sedimentation in sludge thickeners is also very important in reducing the amount of water retained by sludge. As sludge is thickened, its volume is reduced considerably, with the result that a smaller sludge digester or less sludge dewatering equipment is required. For example, a 1% sludge thickened to 3% would occupy approximately one-third of the original volume, thereby reducing the sludge to be handled by over 66%.

2. TYPES OF CLARIFICATION

Clarification occurs in various sedimentation regimes, the particular regime of sedimentation would depend mainly on the nature of the settling particles and the concentration of particles in suspension. In general, there are four sedimentation regimes as shown in Fig. 1, plotted after Fitch (3).

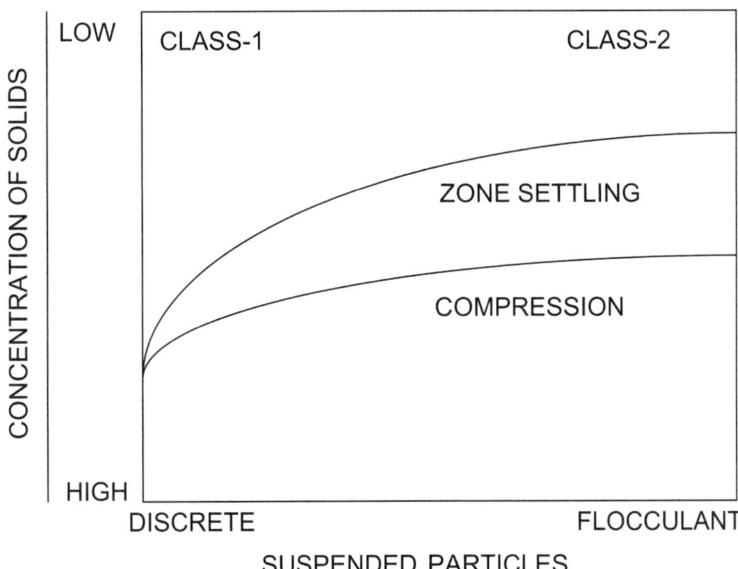

Fig. 1. Types of clarification.

Class 1 clarification is a type of clarification in which particles with little tendency to coalesce settle out of a dilute suspension. Such non-flocculating, discrete particles do not alter their size, shape, or weight during settling. Sedimentation of such particles in dilute suspensions is unhindered by the presence of other settling particles. If the particles flocculate, the sedimentation regime in a dilute suspension is called Class 2 clarification.

When the flocculant suspension is of an intermediate concentration instead of a dilute one, the particles are closer together and the interparticle forces hold the particles in a fixed position relative to each other. This results in subsidence of the mass of particles as a whole. Such a sedimentation regime is called zone settling. At still higher concentrations, the particles come in actual contact with one another and the weight of the particles is partly supported by the flocculated mass. This type of settling is called compaction or compression.

3. THEORY OF SEDIMENTATION

All particles suspended in a fluid of lower density tend to settle under the influence of gravity. Table 1 lists the settling velocities of suspended particles in still water. The rate of settling depends on the shape, size, and specific gravity of the particles, as well as the viscosity, temperature, and quiescence of the liquid.

Coarse and heavy particles settle rapidly and can be removed by storage in large tanks. Fine particles, 10 μm or less in diameter and with a density a little above that of water, cannot be economically removed by sedimentation alone. These particles require coagulation by the addition of chemicals to increase the effective particle size, which results in an increased rate of settling.

In the field of environmental engineering, the principles of sedimentation are applied in both water and wastewater treatment. In the treatment of surface waters, sedimentation is employed after coagulation and before filtration. In the treatment of wastewater,

Table 1
Settling Velocities of Suspended Particles in Water[a]

Diameter of particle (μm)	Approximate time required to settle 1 m	Typical material
10,000	1.2 s	Gravel
1,000	9 s	Coarse sand
100	2 min	Fine sand
10	2 h	Silt
1	6 d	Bacteria
0.1	800 d	Clay particles
0.01	250 yr	Color bodies

[a]Temperature 10°C, spherical particles, specific gravity = 2.65, same chemical characteristics, quiescent conditions.

sedimentation principles are employed in the design of grit chambers, primary settling tanks, secondary or final settling tanks, and gravity sludge thickeners. The theory of sedimentation for different types of clarification is explained below.

3.1. Class 1 Clarification

The sedimentation of a discrete particle in Class 1 clarification regime is a function of the properties of both the fluid and the particle. In falling freely through a quiescent fluid, a discrete particle would accelerate until the resistance of the fluid equals the impelling force acting on the particle. The velocity of the particles at this point would be constant and is called the terminal velocity of the particle. The impelling force acting on the particles is equal to the external force minus the buoyant force.

The external force

$$F_e = \frac{ma_e}{g_c} \tag{1}$$

where m = mass of the particle in [g (lb mass)], a_e = acceleration of particle from the external force [m/s^2 (ft/s^2)], and g_c = Newton's Law conversion factor [981 m-g/m-dyne (32.2 ft·lb mass/ft.·lb force)]

The buoyant force

$$F_b = \frac{\rho m a_e}{\rho_s g_c} \tag{2}$$

where ρ = fluid density [g/m^3 (lb·mass/ft^3)] and ρ_s = particle density [g/m^3 (lb·mass/ft^3)]

Therefore, the impelling force on the particle equals

$$F_e - F_b = \frac{ma_e}{g_c}\frac{\rho_s - \rho}{\rho_s} \tag{3}$$

The drag force F_d of the fluid is a function of the dynamic viscosity and mass density of the fluid, and of the roughness, size, shape, and velocity of the particle. Experimentally, it has been found that

$$F_d = \frac{c_d A_p \rho u^2}{2g_c} \tag{4}$$

where c_d = coefficient of drag, A_p = projected area of particle [m^2 (ft^2)], and u = linear velocity of particle [m/s (ft/s)]. At terminal velocity the impelling force would be equal to the drag force or

$$F_e - F_b = F_d \quad \text{and} \quad u = u_t \tag{5}$$

or

$$\frac{ma_e}{g_c}\frac{\rho_s - \rho}{\rho_s} = \frac{c_d A_p \rho u_t}{2g_c}$$

where u_t = the terminal velocity [m/s (ft/s)]. For spherical particles under the external force of gravity, the following relationships are obtained:

$$A_p = \frac{\pi d_p^{\ 2}}{4}, \quad m = \frac{\rho_s \pi d_p^{\ 3}}{6}, \quad a_e = g$$

where d_p = diameter of particle [m (ft)] and g = acceleration due to gravity [m/s^2 (ft/s^2)]. Substituting values of A_p, m, and a_e in eq. (5):

$$u_t = \left[\frac{4}{3}\frac{\rho_s - \rho}{\rho_s} \times \frac{g}{c_d} \times d_p\right]^{\frac{1}{2}} \tag{6}$$

The coefficient of drag c_d is a function of Reynolds number R_e. The relationship between c_d and R_e, as observed by Camp (4), is shown in Fig. 2. For spherical particles, the observational relationship between c_d and R_e can be approximated (5) by the following equation when R_e is less than 10,000:

$$c_d = \frac{24}{R_e} + \frac{3}{R_e^{\ 1/2}} + 0.34 \tag{7}$$

where R_e = Reynolds number. For viscous resistance at low Reynolds numbers ($R_e < 0.5$):

$$c_d = \frac{24}{R_e} \tag{8}$$

but

$$R_e = \frac{d_p \mu \rho}{\mu},$$

where μ = kinematic viscosity [m^2/s (ft^2/s)], therefore

$$u_t = \left[\frac{4}{3} \times \frac{g}{24} \times \frac{d_p u \rho}{\mu}\frac{\rho_s - \rho}{\rho_s} d_p\right]$$
$$= \left[\frac{1}{18} g \frac{d_p^2}{\mu}(\rho_s - \rho)\right]^{1/2} \tag{9}$$

which is Stoke's Law.

True spherical particles are seldom found in water or wastewater. The irregular shaped particles possess a greater surface area per unit volume than do spheres and

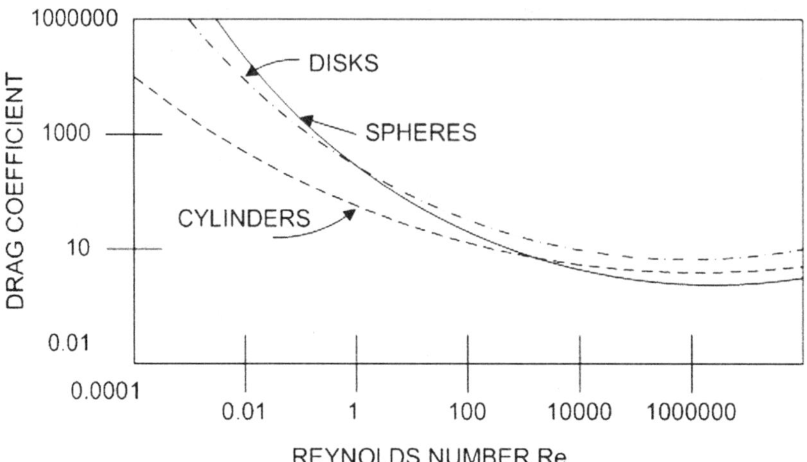

Fig. 2. Relationship of drag coefficient to Reynolds number.

therefore their rates of settling are less than that of the spheres of equal volume. The influence of irregularities in the shape of the particles is greatest at high values of R_e. According to McNown and Malaika (6), the settling velocities of rod-like and disc-like spheroidal particles are, respectively, 78% and 73% of the velocity of an equal volume sphere. In general, for particles of irregular shape,

$$\frac{A}{V} = \frac{6}{\psi d} \tag{10}$$

where A = area of the particle [m^2 (ft^2)], V = volume of the particle [m^3 (ft^3)], d = characteristic diameter of the particle [m (ft)], and ψ = the sphericity of the particle ($6/\psi$ is the shape factor).

The rate of clarification for Class 1 type of clarification can be obtained from settling column analysis. Consider a settling column of cross-sectional area A, filled with a dilute suspension of discrete particles of uniform size, shape, and specific gravity. Under quiescent conditions, the suspended particles will all settle with a terminal settling velocity u_t. If in time "t" a depth "H" in the column is completely clarified, the terminal settling velocity of the suspended particles would be:

$$u_t = \frac{H}{t} \tag{11}$$

where t = time [min] H = depth in column [m (ft)] and the rate of clarification would be:

$$q = \frac{H \times A}{t} = u_t \times A \tag{12}$$

in other words, if liquid is withdrawn from the column at a rate of q, it would contain no suspended solids.

Now, if instead of particles with the same settling velocity, we fill the column with particles having two different velocities, u_{t1} and u_{t2}, such that $u_{t1} > u_{t2}$. In time "t"

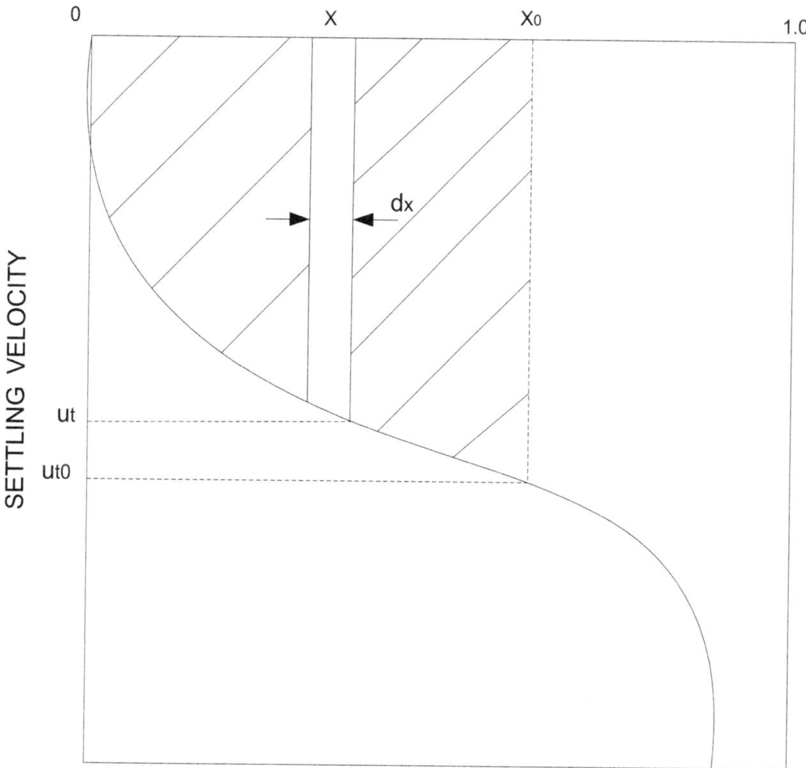

Fig. 3. Settling velocity analysis curve for discrete particles.

when the particles with u_{t2} velocity travel through a distance H_2, the particles with velocity u_{t1} would travel a distance of only H_1. If liquid at a rate of $q_2 = u_{t2} \times A$ is withdrawn from the column, it would have been removed. It is evident that the weight fraction of particles with u_{t1} velocity removed is equal to:

$$x_1 = \frac{H_1}{H_2} = \frac{u_{t1}}{u_{t2}} \tag{13}$$

where x = fraction of particles removed.

According to Camp (7) a settling velocity curve such as that shown in Fig. 3 can be drawn for a particular suspension, depicting the settling velocity and the corresponding fraction of particles with less than that stated velocity. Suppose it is required to find overall removals of suspended particles at a given clarification rate q_0 corresponding to a settling velocity of u_{t0} ($q_0 = U_{t0} \times A$) and it is also known from experimental settling column data that fraction of particles with less than settling velocity u_{t0} is x_0.

The fraction of particles with velocity greater than u_{t0} is $(1-x_0)$. This fraction will be completely removed. A portion of the fraction x_0 of the particles will also be removed. According to Fig. 3 and Eq. (14), this portion is equal to

$$\frac{1}{u_{t0}}\int_0^{x_0} u_t dx \tag{14}$$

the value of $\int_0^{x_0} u_t dx$ is obtained graphically.

The overall removals are, therefore, equal to

$$x_T = (1 - x_0) + \frac{1}{u_{t0}}\int_0^{x_0} u_t dx$$

3.2. *Class 2 Clarification*

Class 2 clarification refers to the settling of particles that do not act as discrete particles but coalesce or flocculate during settling from dilute suspensions. A large particle that is sinking tends to overtake and coalesce with slower moving particles. This leads to the formation of a new larger particle, which settles faster than the constituent particles. The flocculation of particles depends on the opportunity for contact, which varies with the depth of the basin, the overflow rate, velocity gradients in the system, the concentration of particles, and the range of particle size. Flocs formed in water and wastewater are relatively fragile. As they grow in size, velocity gradients across them are increased, which may break up the flocs at some limiting size. In general, this limiting size is not reached in flocculent suspensions entering settling tanks in water- and wastewater-treatment works. Flocculation, natural or chemically induced, occurring in settling tanks improves sedimentation.

To determine the effect of flocculation on sedimentation it is necessary to perform settling column analysis or sedimentation tests, since no suitable rational design procedure is available to evaluate such effects. The settling column may be of any diameter but its length should be equal to the depth of the settling tank to be designed. The settling column usually used in laboratories is 6 in. diameter and about 10 ft deep. It contains sampling ports at 2 ft intervals.

To perform sedimentation tests, a flocculating suspension similar in character to the one to be used in the settling tank is placed in the column in such a way as to ensure a uniform distribution of particles of all sizes from top to bottom. Samples are withdrawn from all the ports at various time intervals and the amount of suspended solids contained in the samples is determined. The percentage removal of suspended solids by sedimentation is then calculated. For example, if a suspension containing 500 mg/L of suspended solids was introduced into the column and the sample showed a concentration of 100 mg/L, then the percentage removed would be 80%.

These percentages are plotted on a depth versus time graph. Points of equal percentage removals are joined and a plot as shown in Fig. 4 is obtained. Overall removals (8) of suspended solids at a certain detention time (t_2) are given by:

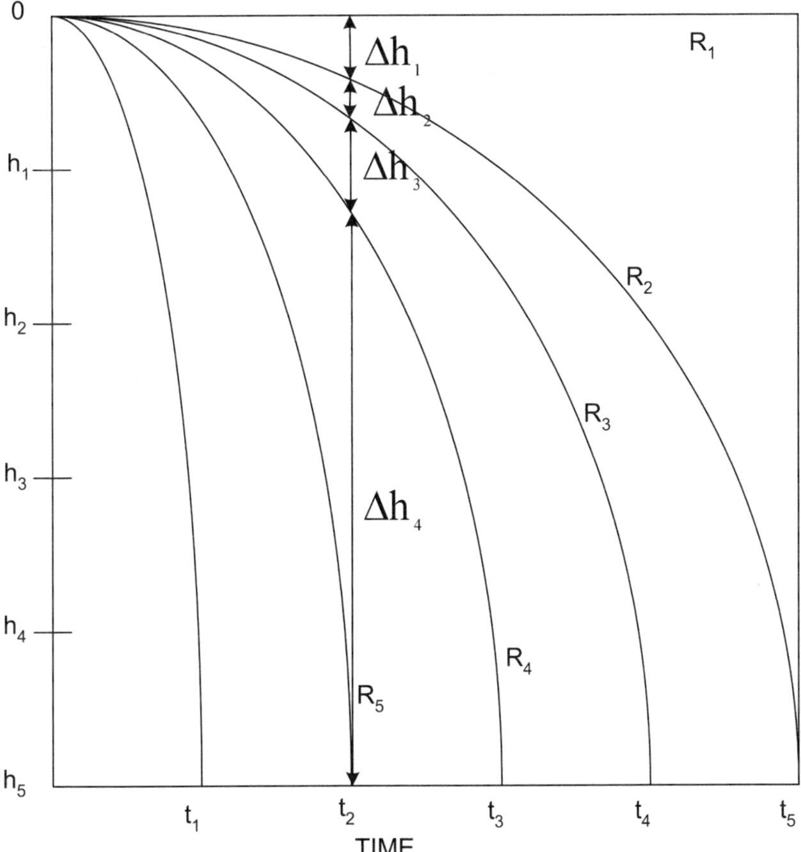

Fig. 4. Equal percent removal curves for clarification of flocculating suspension.

$$R_{t2} = \frac{\Delta h_1}{h_5}\left(\frac{R_1 + R_2}{2}\right) + \frac{\Delta h_2}{h_5}\left(\frac{R_2 + R_3}{2}\right) + \frac{\Delta h_3}{h_5}\left(\frac{R_3 + R_4}{2}\right) + \frac{\Delta h_4}{h_5}\left(\frac{R_4 + R_5}{2}\right) \quad (15)$$

where R_{t2} = the percentage removal at time t_2, R_1, R_2, . . . , R_5 are percentage removals, h_5 = the total depth of water in column, and Δh_1, Δh_2, Δh_3, Δh_4 are depth increments to successive percent removal curves at time t_2.

In order to obtain removals in settling tanks comparable to those indicated by a settling column analysis, Metcalf and Eddy (8) recommend that the detention time be multiplied by 1.75–2.0 and the overflow rate or design settling velocity by 0.65.

3.3. *Zone Settling*

Concentrated suspensions have somewhat different settling characteristics than those of dilute suspensions. Fig. 5 has been plotted to illustrate the effect of a high concentration of suspended solids on settling rates. Curve X is a plot showing settling of a particle in dilute suspensions of different size particles. Initially, the subsidence of the

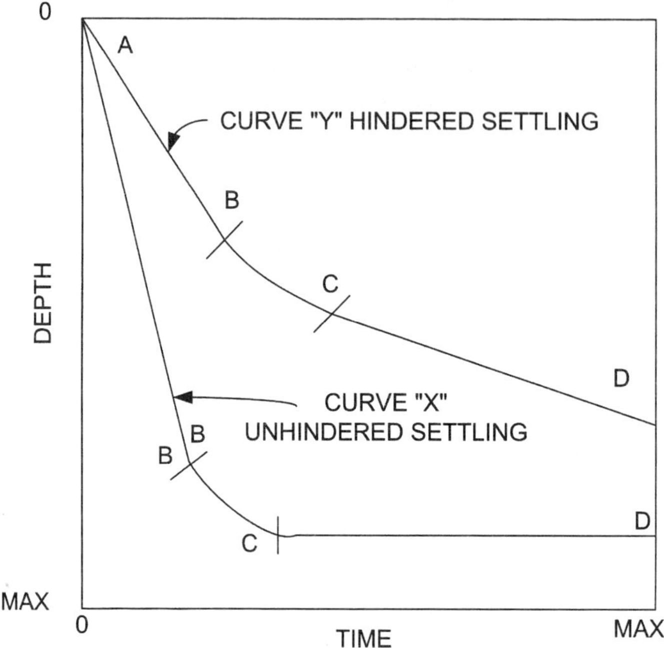

Fig. 5. Effect of high concentration of suspended solids on settling rates.

particle is unhindered and it settles at its own individual velocity of subsidence. The particle accelerates as it moves downward until point B is reached when it starts decelerating. Somewhere between B and C the particle becomes part of the sludge. From C to D the particle is in the compression regime area.

When the concentration of particles is increased, the fast settling particles form a zone at some distance from the start and settle collectively at a reduced velocity. With further increase in the concentration of particles in the suspension, a point is reached where even the initial subsidence is collective. This is shown in curve Y. The particles settle as a zone or "blanket" leaving a relatively clear liquid above the zone-settling region. Some particles are invariably left behind which settle in the relatively clear water as discrete or flocculated particles. A distinct interface is formed between the relatively clear liquid and the zone-settling region.

Settling of particles from a suspension with high concentration of suspended solids usually involves both zone settling and compression settling in addition to free settling. The compression-settling region is formed under the zone-settling region in a settling column. Settling tests are usually required to determine the sedimentation characteristics of suspensions where zone settling and compression settling occur. Talmadge and Fitch (9) developed a method to determine the area required for a solids handling system from the results of settling tests. This method is described below.

A settling test is performed with suspension of solids of uniform concentration (c_0) in a settling column of height (H_0). The position of the interface with time is determined and is plotted on a depth time graph (see Fig. 6).

The critical area for a solid handling system is given by the equation

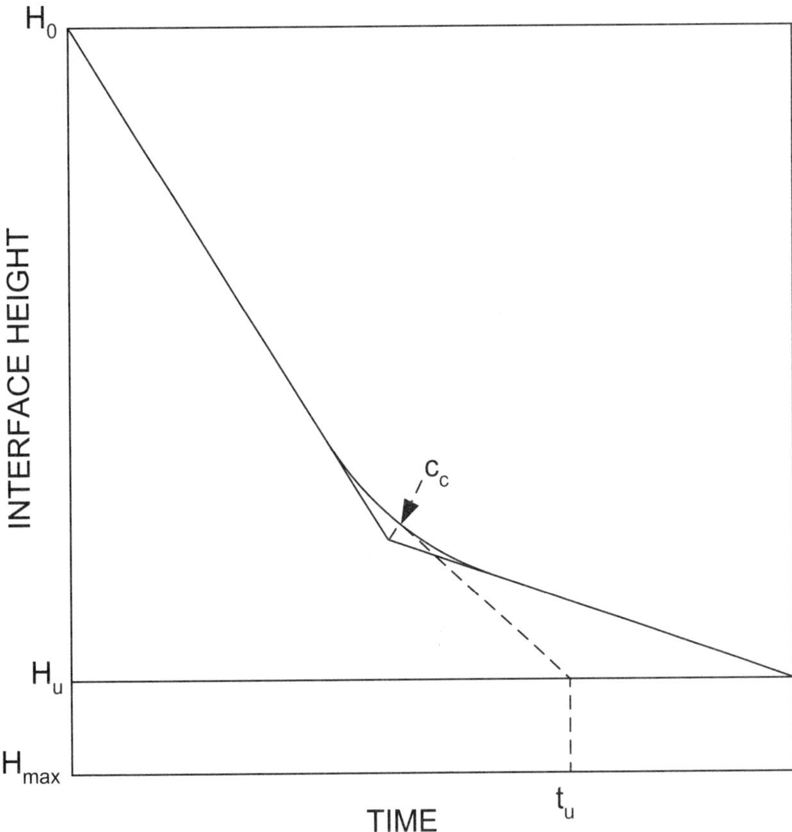

Fig. 6. Graphical determination of t_u, the time required to obtain the desired underflow concentration.

$$A = \frac{Qt_u}{H_0} \tag{16}$$

where A = area required for the solids handling thickener [m^2 (ft^2)], Q = volumetric flow rate into thickener [m^3/s (ft^3/s)], t_u = time required to attain underflow concentration (c_u) (s) and, H_0 = initial column height of the interface in the settling column [m (ft)].

In the above equation Q and H_0 are known and t_u is found graphically from the settling curve. First, the point of critical concentration c_c is determined by bisecting the angle formed by extending the tangents to the hindered settling and the compression settling regions of the settling curve. The bisector cuts the settling curve near the point where compression settling starts. The critical concentration corresponds to the largest cross-sectional area required for a solids-handling system. The value of t_u can be determined by drawing a vertical line to the time axis from the intersection of the tangent at c_c and the horizontal line drawn at depth H_u. H_u is the depth at which all solids are at the desired underflow concentration (c_u):

$$H_u = \frac{c_0 H_0}{c_u} \tag{17}$$

where H_u = the depth at which all solids are at the desired under flow concentration (c_u)[m (ft)], c_u = underflow concentration, and c_0 = initial concentration at depth H_0. Knowing the value of t_u, the area required for solids handling system can be obtained from Eq. (16).

3.4. *Compression Settling*

In the zone of compression, particles come into physical contact with each other and are supported by the layers below. The volume required for sludge in the compression region is also found by settling tests. According to Coulson and Richardson (10), the approximate rate of consolidation is proportional to the difference in sludge depth at time *t* and the final depth of sludge after a long period of settling:

$$-\frac{dHt}{dt} = K\left(H_t - H_\infty\right) \tag{18}$$

where H_t = depth of sludge at any time (t) [m (ft)], H_∞= final depth of sludge after a long period of settling [m (ft)], and K = constant for a given suspension. Integrating Eq. (18) between the limits H_{t1} and H_{t2}, the depths of sludge at time t_1 and t_2, respectively,

$$-\int_{H_{t1}}^{H_{t2}} \frac{dHt}{H_t - H_\infty} = K\int_{t_1}^{t_2} dt$$

$$\text{In}\left(\frac{H_{t1} - H_\infty}{H_{t2} - H_\infty}\right) = K\left(t_1 - t_2\right) \tag{19}$$

$$H_{t1} - H_\infty = \left(H_{t2} - H_\infty\right)e^{K(t_1 - t_2)}$$

H_∞ or the final depth of sludge, after a long period of time, can be calculated from the above relationship.

Gentle stirring generally compacts sludge in the compression region by breaking up the flocs and permitting water to escape. Mechanical rakes are usually employed to provide the required stirring in the sedimentation tank. In settling column tests it may also be proper to provide stirring if the results are to be used in the design of thickeners with mechanical rakes. According to Dick and Ewing (11), stirring also improves the settling efficiency in the zone-settling region.

4. SEDIMENTATION TANKS IN WATER TREATMENT

4.1. *General Consideration*

Most surface water supplies contain large quantities of suspended particles at various times of the year. The suspended particles are generally of an inorganic nature, consisting mostly of sand and clay. A small amount of organic matter and varying numbers of bacteria may also be present. Removal of these impurities is the principal aim of water purification and the technique of sedimentation plays a major role in achieving this goal. The main purpose of sedimentation is to bring about clarification of water by permitting the settling of suspended solids. In water treatment, sedimentation is employed in two principal ways, plain sedimentation and sedimentation following chemical addition for the purpose of coagulation and flocculation or for softening.

Plain sedimentation is used for the removal of settleable solids occurring naturally in surface waters. The detention time required in plain sedimentation usually ranges from 12 h to several days depending upon the settling characteristics of the suspended particles and the clarity desired. Providing a sedimentation tank of such capacity would not be economically justified. Plain sedimentation in water treatment is, therefore, employed only in cases of very turbid waters, occasionally, plain sedimentation tanks of small detention time are provided to lighten the load on coagulation and flocculation processes and, subsequently, on the sedimentation process itself. Cost benefit analyses should be done to determine whether such pre-sedimentation basins are justified.

The major application of the sedimentation operation in water-treatment plants is following coagulation and flocculation. Addition of chemicals renders the solids more settleable and thus facilitates sedimentation. Generally, a detention time of 2–4 h is adequate for the removal of flocculated particles. The effect of sedimentation after coagulation is the reduction in turbidity, color, and also bacteria, which settles with the floc.

A settling basin consists of four different zones: inlet, settling, outlet, and sludge zones. The inlet zone dissipates the kinetic energy of the incoming water and distributes the flow uniformly across the tank. The settling zone provides sufficient time for the settling of suspended particles. The clarified effluent from the settling zone is collected in the outlet zone. The sludge zone provides space for the accumulation of solids. All these zones are important in the proper functioning of a settling basin and need very careful design analysis.

4.2. Inlet and Outlet Control

The inlet to a sedimentation basin is provided to afford a smooth transition from the relatively high velocities in the influent pipe to the very low, uniform velocity distribution desired in the settling zone. The purpose of outlet is the same except that the transition is from the settling zone to the effluent pipe.

The inlet devices should be designed to distribute the flow and suspended matter as uniformly as possible across the full width of the line of travel. If there is more than one tank, the inlet structures should distribute the flow more or less equally into all of the tanks. The velocity of water in the conduits or orifices should be high enough to prevent deposition of suspended matter but should not be so high that the flocs are broken up. A baffle is usually constructed across the sedimentation basin close to the inlet end. This baffle projects several feet below the water surface to dissipate inlet velocities, to provide uniform flow across the basin, and to break up the surface currents. Submerging the inlet pipe can also minimize surface currents.

The outlet devices control the outflow from a sedimentation basin by means of overflow weirs or troughs, sometimes called launders. The weir may be attached to one or both sides of the trough and there may be more than one outlet trough in a single tank, depending on the weir length required. The weir length provided is such as to minimize surges. It has been found by experience that to avoid surges the overflow rate should be less than 372,500 L/d/m (30,000 gpd/ft) of weir (5). In most sedimentation basins the weir is subdivided into a multitude of small V-notches. This helps to prevent non-uniform discharge over the full length of the weir at low flows due to wind currents or slight variations in weir level. The notches are less affected by local differences in weir elevations than a continuous smooth weir would be under the same conditions.

4.3. Tank Geometry

The shape of the basin to be used as a settling tank depends on a number of factors, which include, in addition to the settling characteristics of the suspended matter, the availability of space, the sludge-handling equipment, if required, and the shape of the existing basins, if any. Previous experience of the designer with a particular basin shape may also influence the selection.

Settling tanks used in the field of environmental engineering are basically of three different shapes-rectangular, square, and circular. In water-purification plants, rectangular and square shapes are more common than circular ones. Long, narrow rectangular tanks are generally considered to be more efficient than circular tanks.

The size of a settling tank would depend on the quantity of flow and the number of tanks. Generally in plants with capacity of 1 MGD (million gallons per day) or less, only one settling tank is provided. But where treatment cannot be interrupted, a minimum of two tanks should be provided. In rectangular tanks the length to width ratio generally varies from 2:1 to 4:1 with a maximum limit of 5:1. For mechanically cleaned tanks, dimensions of the available sludge-removal equipment may determine the tank width. A width of 6 m (20 ft) is not unusual. The sludge scrapers may be used in parallel in the same tank and, therefore, the width may be doubled for a two-scraper tank. A width of 25 m (80 ft) is usually the maximum limit. Rectangular tanks of lengths of up to 90 m (300 ft) have been constructed but up to 30 m (100 ft) are more common. Circular tanks are limited in size by the structural requirements of the trusses supporting the bridge. The diameter of circular tanks is generally kept within 30 m (100 ft).

4.4. Short Circuiting

The theoretical detention period of a settling basin is the time required to fill the basin at a given rate of flow. In other words it is equal to the volume of the tank divided by the rate of flow.

In an ideal settling basin the time taken by a small volume of water to pass through the basin at a given rate of flow (flow-through period) is equal to the theoretical detention time at the same rate of flow. Unfortunately, settling basins actually experience a varying degree of short-circuiting due to wind, thermal, and inertial currents and other phenomena occurring in the basin. This results in a portion of the inflow reaching the outlet in less than the theoretical detention time. The amount of short-circuiting can be measured by adding a tracer substance like a fluorescent dye, an electrolyte, etc., at the inlet and measuring the concentration of the tracer reaching the outlet. The tracer substance should be added as a single dose and in sufficient quantity to be easily discernable. Tracer studies are very useful in determining flow pattern in existing tanks.

4.5. Detention Time

According to Maynard (12) a detention time of approx 4 h for alum floc, 3 h for magnesium floc, and 2 h for calcium carbonate particles is adequate. Recommended standards for water works (13) require a minimum of 3 h for presedimentation basins and 4 h for sedimentation basins that follow flocculation. The detention period may be decided upon by performing model studies with the actual surface water to be treated and the coagulant to be used. Model studies are also useful in predicting the removal efficiencies of sedimentation tanks before these are built and could suggest methods of improving

Table 2
Tank Loadings for Common Types of Suspensions

Typical solids	Size of particles	Specific gravity	Settling velocity at 10°C (cm/s)	Surface loading [L/min/m^2 (gpm/ft^2)]	Min. detention time for 3.05 m (10 ft) deep tank
Sand, silt and clay	0.001 cm and larger	2.65	6.9×10^{-3}	5,948 (146)	12.3 h
Alum and Iron flocs	0.1 cm	1.002	8.3×10^{-2}	73,335 (1,800)	1.0 h
Calcium carbonate precipitates	0.1 cm	1.2	4.2×10^{-2}	36,667 (900)	2.0 h
Wastewater organics	0.1 cm and larger	1.001	4.2×10^{-2}	36,667 (900)	2.0 h
Activated sludge organics	0.1 cm and larger	1.005	2×10^{-1}	162,966 (4,000)	0.5 h

Source: Fair et al. (5)

the efficiency of removal without increasing the detention time. Unless it is shown by actual experimental results that reasonable removal efficiency can be obtained in a smaller detention time, the plant would have to be designed for a mandatory detention time as fixed by regulatory standards.

4.6. Tank Design

4.6.1. Theoretical Considerations

The design of a settling tank is based on the settling velocity of the smallest particles to be removed. The settling velocity is determined from Eq. (6). For discrete particles, the variables involved in the equation are the effective particle diameter, the specific gravity of the particles, and the coefficient of drag, which is a function of Reynolds number. Calculation of the settling velocity of discrete particles is, however, not of much help in the design of a sedimentation tank because the sedimentation of discrete particles is virtually non-existent in water-supply practice. The design of sedimentation tanks should, therefore, be based upon the results of settling column analysis. Table 2 lists the size, specific gravity, settling velocities, and detention periods for different particles settling in a 10 ft deep tank (5). It may be noted that the detention times given in the table are much less than actually required in sedimentation tanks. This is because these values do not take into consideration the wind, density, thermal currents, short circuiting, etc., which are always present in actual sedimentation tanks.

4.6.2. Regulatory Standards

Recommended standards for water works, popularly known as Ten State Standards (13), are the most widely used standards for the design of sedimentation tanks. The most important provisions of these standards are listed below:

(a) Plants designed for processing surface waters should provide duplicate units and permit operation of basins in series or parallel.

(b) Presedimentation tanks should be designed for a minimum detention period of 3 h.
(c) Settling tanks that are provided following flocculation and with conventional sedimentation shall have a minimum of 4 h of settling time.
(d) Inlet devices: Inlets shall be designed to distribute the water equally and at uniform velocities. A baffle should be constructed across the basin close to the inlet end and should project several feet below the water surface.
(e) Outlet devices: Outlet devices shall be designed to maintain velocities suitable for settling in the basin and to minimize short circuiting. The use of submerged orifice is recommended.
(f) Overflow rate over the outlet weir shall not exceed 248,400 L/d/m (20,000 gpd/ft). The velocity through settling tanks shall not exceed 0.15 m/min (0.5 ft/min).
(g) Mechanical sludge collection equipment should be provided.

5. SEDIMENTATION TANKS IN WASTEWATER TREATMENT

5.1. General Consideration and Basis of Design

Sedimentation is applied in wastewater treatment in three different kinds of tanks. These are primary sedimentation tanks, secondary or final sedimentation tanks, and gravity thickeners for sludge concentration. Although the theory of sedimentation is the same in both water and wastewater treatments, sedimentation in wastewater treatment differs from that in water treatment in the following aspects:

(a) More suspended solids in wastewater.
(b) Wastewater suspended solids are usually of lower specific gravity.
(c) Size of particles is larger.
(d) Normally no chemicals are employed.
(e) Effluent from wastewater sedimentation tanks usually contains more suspended solids than that from water sedimentation tanks.
(f) Sludge has to be removed continuously to prevent septic condition.

5.1.1. Primary Sedimentation Tanks

Primary sedimentation tanks are provided to remove readily settleable solids and to reduce the load on subsequent biological units. At primary treatment plants where sedimentation is the only treatment, settleable solids are removed to prevent the formation of sludge banks in the receiving waters. The detention time is 120–150 min and the efficiency of removal in such tanks is 50–65% for suspended solids and 25–40% for BOD (biochemical oxygen demand) (8). Basic primary treatment plants are not in common use these days. Such plants do not have the capability to produce the higher quality of effluent mandated by current regulatory standards. In secondary treatment plants, primary sedimentation is always provided as preliminary treatment. These tanks usually have shorter detention times of 60–90 min. Another use of primary sedimentation tanks is as storm water tanks with detention times of only 10–30 min. These tanks remove a substantial portion of the organic and inorganic solids, which would otherwise go into the receiving waters.

5.1.2. Secondary/Final Sedimentation Tanks

Secondary sedimentation tanks are provided to remove the settleable solids that are produced in biological treatment. In secondary treatment plants such tanks are also called final sedimentation tanks. In tertiary treatment plants where phosphorous is removed by chemical precipitation through the addition of chemicals, the settling tanks are called final sedimentation tanks.

For secondary or final sedimentation tanks, a greater efficiency of suspended solids removal is expected than that for primary sedimentation tanks. Secondary or final sedimentation tanks may or may not require scum removal equipment.

Overflow rates allowed for sedimentation tanks vary from 24,500 to 49,000 L/d/m^2 (600 to 1200 gpd/ft^2) for primary tanks and 12,000 to 32,000 L/d/m^2 (300 to 800 gpd/ft^2) for secondary tanks. Weir overflow rates vary from 124,000 to 248,000 L/d/m (10,000 to 20,000 gpd/ft) of weir.

Inlet and outlet hydraulics are similar to those discussed in the case of sedimentation tanks for water treatment. Flow variations in wastewater-treatment plants are much more pronounced than in water-treatment plants. The effect of density current, especially in deep tanks, should also be considered. The tank proportions and the inlet velocity affect density current velocity. Provision of baffles at an inlet velocity of over 0.30 m/s (1 ft/s) minimizes the effect of such density currents.

5.2. *Regulatory Standards*

The following is a list of important provisions described in the Recommended Standards for Sewage Works (14) for the design of sedimentation tanks. These standards are also called Ten State Standards for Sewage Works:

(a) The inlet channel should have a velocity of at least 0.30 m/s (1 ft/s) at one-half the design flow.
(b) The minimum length of the tank should be 3 m (10 ft). The mechanically cleaned settling tank should be as shallow as possible but not less than 2.1 m (7 ft) deep. The final clarifier for activated sludge to be not less than 2.5 m (8 ft) deep.
(c) Provide scum removal facilities in all wastewater sedimentation tanks.
(d) Weir overflow rate should not exceed 10,000 gpd/ft for 1 MGD or smaller plants and 15,000 gpd/foot for larger plants. In metric units this corresponds to 124,000 L/d/m for 3.8 ML/d or smaller plants and 186,000 L/d/m for larger than 3.8 ML/d plants.
(e) The surface overflow rates for primary tanks of capacity greater than 3.8 ML/d (1 MGD) shall not exceed 41,000 L/d/m^2 (1000 gpd/ft^2) at average design flow.
(f) The surface overflow rates for primary tanks of capacity 3.8 ML/d (1 MGD) or less shall not exceed 24,000 L/d/m^2 (600 gpd/ft^2).
(g) Multiple settling tanks are required in all plants of 380,000 L/d (100,000 gpd) or more.
(h) Surface overflow rates for final settling tanks depend on the type of secondary treatment at the plant and the size of the plant. For conventional activated sludge plants the overflow rate shall not exceed 800 gpd/ft^2 for plants of over 1.5 MGD, 700 gpd/ft^2 for plants 0.5–1.5 MGD, and 600 gpd/ft^2 for plants up to 0.5 MGD. In metric units this corresponds to 32,600 L/d/m^2 for plants of over 5.7 ML/d, 28,500 L/d/m^2 for plants 1.9–5.7 ML/d and 24,400 L/d/m^2 for plants up to 1.9 ML/d.
(I) Surface overflow rate for an intermediate settling tank following fixed film processes shall not exceed 61,000 L/d/m^2 (1,500 gpd/ft^2) at peak hourly flow.
(j) The detention time for final sedimentation tanks in conventional activated sludge plants shall be 3 h for up to 0.5 MGD plants, 2.5 h for 0.5–1.5 MGD plants, and 2.0 h for over 1.5 MGD plants. In metric units this corresponds to 3 h for up to 1.9 ML/d plants, 2.5 h for 1.9 to 5.7 ML/d plants, and 2.0 h for over 5.7 ML/d plants.

5.3. *Tank Types*

The sedimentation tanks may be of rectangular, square or circular shape. The type of sedimentation tank in wastewater treatment depends on the following factors:

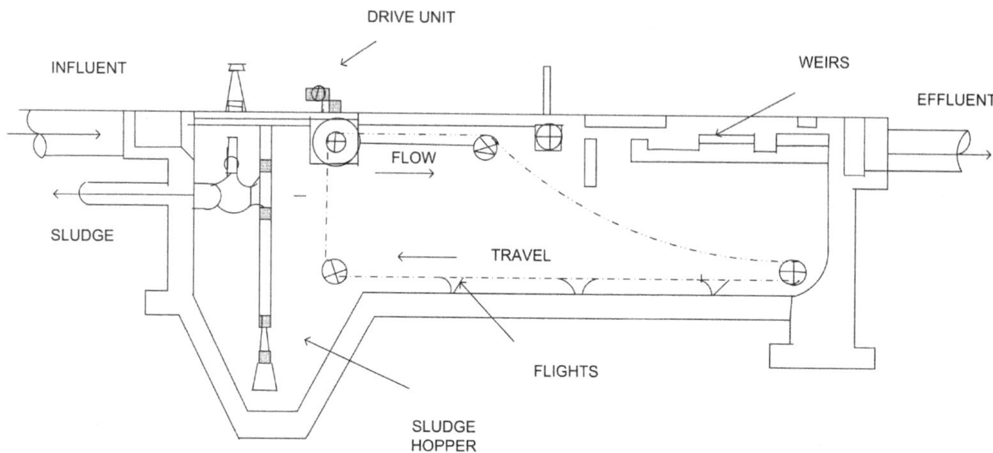

Fig. 7. Schematic diagram of a rectangular setting tank.

(a) Availability of space.
(b) Experience and judgment of the engineer.
(c) Size of the installation.
(d) Rules and regulations of regulatory agencies.
(e) Local site conditions.
(f) Economics involved.

Except at small treatment plants at least two sedimentation tanks are provided. At larger installations the number of tanks depend on the limitation in size of tanks. Sludge scraper mechanisms are now provided in all sedimentation tanks.

5.3.1. Rectangular Tanks

Rectangular tanks are designed with inlets at one end and outlets at the other end (Fig. 7). The sludge hopper is provided at the inlet end of the tank. The sludge removal equipment consists of a pair of endless conveyor chains with 5 cm (2 in.) thick cross pieces of wood attached at 3 m (10 ft) intervals. The conveyor speed is 0.3–0.9 m/min (1–3 ft/min). The lower speed of 0.3 m/min (1 ft/min) is provided in tanks employed for the sedimentation of wastewater from activated sludge units.

Instead of conveyor chains, a bridge-type mechanism may also be used for sludge scraping. The bridge runs on a pair of side rails while spanning the tank width with rakes hanging from it to the tank bottom. The rakes scrape the sludge toward the hopper and are lifted up on the return run. This type of arrangement is not suited for secondary sedimentation tanks, presumably because of longer times required for successive passes of the rakes (15). A sludge hopper about 0.6 m (2 ft) square at the bottom is provided in small tanks. Larger tanks have transverse troughs extending the width of the tank. Sludge is removed from hoppers and troughs with a draw off pipe of a screw conveyor or other transport channels.

The type of inlets provided in rectangular tanks include a perforated baffle, a series of inlet pipes, inlet ports discharging against a baffle, or a return bend discharging against the tank wall. The position of the baffle is usually 0.60–1.0 m (2–3 ft) from the inlet and 0.45–0.60 m (1.5–2 ft) below the water surface as shown in Fig. 7. Scum removal in mechanized rectangular tanks is done during the return travel of the rakes

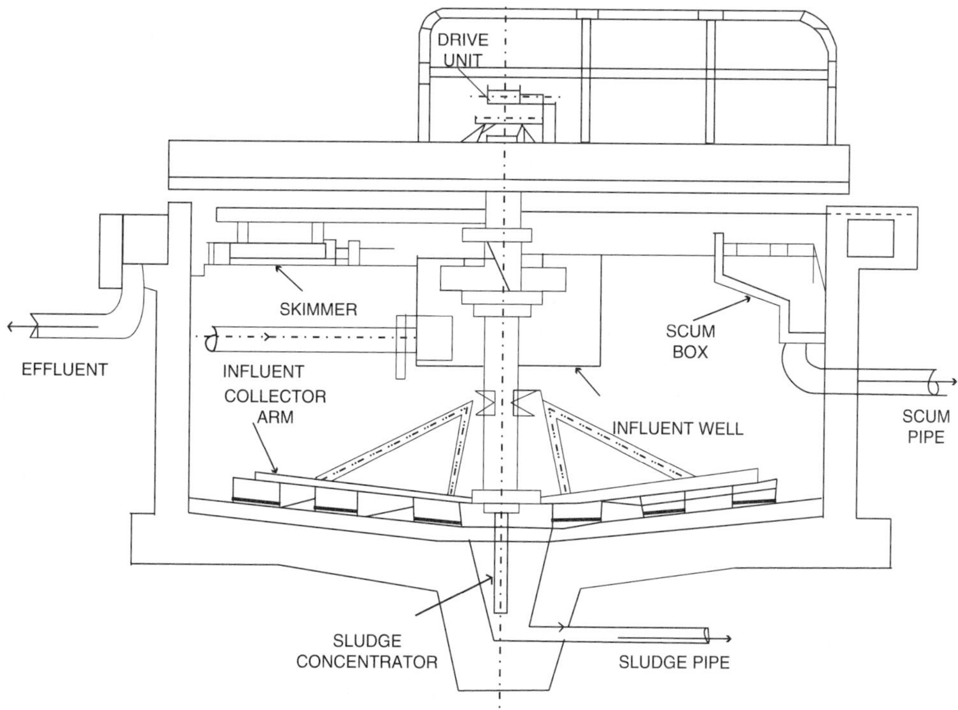

Fig. 8. Schematic diagram of a circular settling tank.

toward the outlet end. The scum is discharged into a scum pipe, which also serves as a scum baffle and the scum collected is pumped into the sludge digestion tank. The outlet weir, located near the effluent end of the tank, is usually provided with a series of 90° V-notches at the crest for uniform distribution of low flow. Multiple outlet troughs are also used in larger tanks.

The limitation on the dimensions of sedimentation tanks as discussed in Section 4 also applies to sedimentation tanks in wastewater treatment.

5.3.2. Circular Tanks

The use of circular tanks for sedimentation of wastewater has increased in recent years. Circular tanks are more adaptable to mechanization than rectangular tanks. The sludge cleaning equipment in circular tanks has fewer moving parts than the chain conveyor equipment in rectangular tanks. The maintenance costs in circular tanks are therefore less.

The circular tank has an influent well in the center (Fig. 8). Wastewater is carried to the influent well in a pipe suspended from a bridge or in an inverted siphon laid beneath the tank floor. The diameter of the influent well varies from 10% to 20% of the tank diameter and the concentric influent baffle making up the influent well extends 0.9–1.8 m (3–6 ft) below the water surface. The influent well distributes the incoming flow equally in all directions. The suspended solids settle to the bottom as wastewater flows radially outward.

The clarified effluent passes over an outlet weir and is collected in an effluent trough or launder near the periphery of the tank. The outlet weir is usually of a V-notch type. A series of adjustable V-notches are attached to the crest of the outlet weir. This ensures a uniform distribution of effluent over the outlet weir even at low flows. The outlet trough is of sufficient capacity to prevent weir flooding at peak flows. A scum baffle is usually provided in front of the outlet weir to retain scum. This baffle is located 20–25 cm (8–10 in.) inward from the outlet weir and extends 20–30 cm (8–12 in.) below the water surface. The scum is removed by a radial arm, which is attached to the sludge-cleaning mechanism and moves over the water surface collecting scum in a radial trough. The scum is then carried to an outside pump from which it is pumped into the sludge digestion system.

Circular tanks have sludge hoppers at the center. The bottom floor is sloped toward the center at a slope of 1 in 12. Typically, a sludge hopper may be 1 m (3 ft) in diameter at the top and 0.30 m (1 ft) in diameter at the bottom, with a depth of 0.75 m (2.5 ft). The sludge cleaning mechanism consists of a revolving shaft to which are attached the radial arms having blades set at an angle. The blades scrape the sludge from the bottom and push it toward the central hopper. The sludge cleaning mechanism revolves at a peripheral speed of 1.5–2.4 m/min (5–8 ft/min) (15).

In some circular tanks the influent enters the tank tangentially near the outer periphery. The wastewater then flows from the annular inlet spirally inward to a central overflow weir. The effluent is collected in a trough and is carried out of the tank in pipes. Circular tanks with this arrangement of flow are not very common.

5.3.3. *Imhoff Tanks*

Imhoff tanks (Fig. 9) are two-story tanks in which sedimentation takes place in the upper chamber and sludge digestion in the lower chamber. Imhoff tanks have the advantage of being simple to operate and are employed at small treatment plants only.

Until recently, Imhoff tanks were designed with no mechanical cleaning or sludge collection equipment. Now mechanically equipped Imhoff tanks are also available.

The settling chamber is designed with an overflow rate of 24,000 L/d/m^2 (600 gpd/ft^2) and a detention time of 3 h is provided. The digester compartment has a sludge storage capacity of about 6 mo.

The settling chamber bottom has a slope of 1.4 vertical to 1.0 horizontal. The settled particles drop into the digestion chamber through a 15 cm (6 in.) wide opening.

Scum is accumulated in the settling chamber and is removed every day. The scum is usually discharged into the digestion chamber through one of the vents. The vents are provided to allow escape of gases produced during digestion of the sludge. Heating of the digestion chamber of an Imhoff tank is not economically possible. The process of sludge digestion is therefore slow.

6. GRIT CHAMBER

6.1. *General*

A grit chamber is the part of a wastewater-treatment plant in which grit is removed to protect subsequent mechanical equipment installations from abrasion and excessive wear as well as to avoid accumulation of grit in sludge digesters, which might then necessitate frequent cleaning. Grit is defined as small, coarse particles of sand, gravel, cinders, or other mineral matter with a specific gravity substantially greater than that of organic matter. At wastewater-treatment plants, the grit may also include bone chips, eggshells, coffee ground, seeds, and similar materials. The grit is characteristically non-putrescible and has settling velocity substantially greater than that of organic

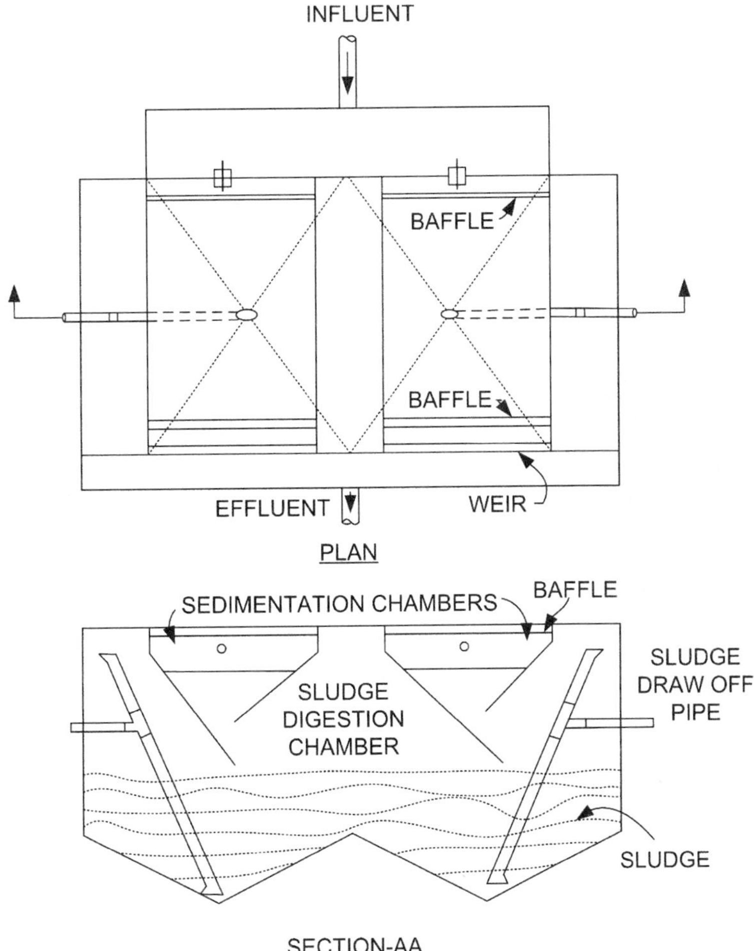

Fig. 9. Schematic diagram of an Imhoff tank.

putrescible solids. The grit chamber in the past has been located prior to the comminutor and bar screen unit. Nowadays there is a tendency of some designers to locate the grit chamber after the comminutor and bar screen device in order to reduce the effect of rags and other gross particles on the mechanical equipment in the grit chamber. In addition to reducing the size of the grit, the comminutor also reduces the size of the organic materials and makes them lighter. This reduces the tendency of the organic matter to settle with the grit in the grit chamber. As a result, the performance of the grit chamber is improved.

6.2. Types of Grit Chambers

Basically two types of grit chambers are installed at wastewater–treatment plants, horizontal flow and circulating flow. A proportional weir or a Parshall flume is used to control the velocity in a horizontal-flow-type chamber. The proportional weir is used for rectangular sections, whereas the Parshall flume is used for parabolic or near parabolic sections. These velocity control devices maintain a substantially constant velocity in the

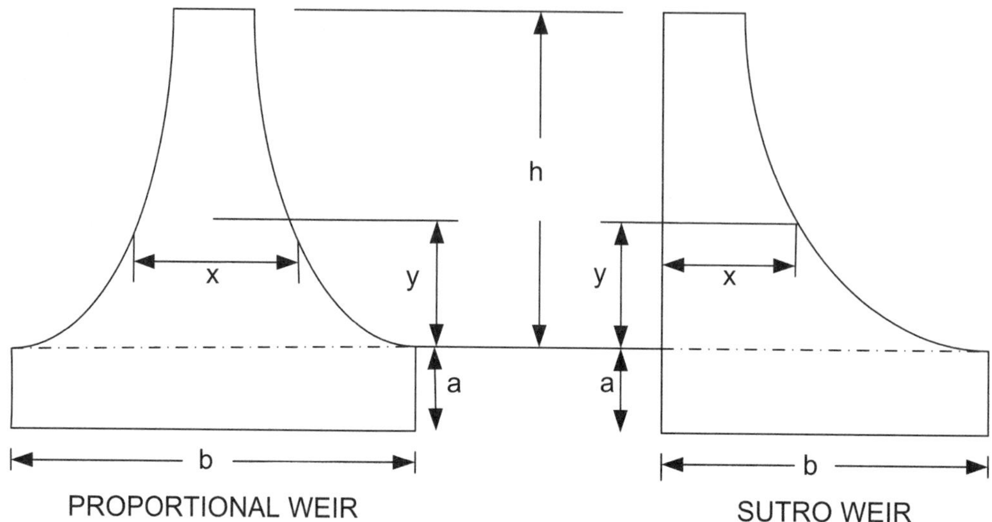

Fig. 10. Proportional and Sutro weirs.

channel at all flows. In circulating flow grit removal devices, the velocity is controlled by means of diffused air, a propeller, or a draft tube. Grit chambers where velocity is controlled by diffused air are also called aerated grit chambers. Grit removal in horizontal flow grit chambers may be manual or mechanical. In the circulating flow type, the grit is always removed mechanically. A majority of grit chambers at plants of over 5.7 ML/d (1.5 MGD) capacities are provided with mechanical cleaning equipment (15). The choice of grit removal methods and equipment would depend upon the quantity and quality of grit and its effect on subsequent units, space requirements and head loss requirements. Head loss in grit chambers could be as little as 6 cm (0.2 ft) where no velocity control section is provided to as much as 60 cm (2 ft) or more where a weir type velocity control section requiring free fall is provided.

6.3. *Velocity Control Devices*

The earliest attempts at velocity control consisted of the use of multiple channels, which were put in or out of operation as the flow increased or decreased. The velocity control devices used currently in grit chambers include a proportional-flow weir, a vertical throat, or a standing wave flume. The vertical throat may be of constant width or adjustable width design. These devices are used at the end of horizontal flow channels to maintain constant velocity in the channel.

Constant velocity is maintained by varying the cross-sectional area of flow in the channel in direct proportion to the flow. Both proportional flow weir and Sutro weir perform this function for rectangular channels. It is, however, essential that water level in the weir should always be in the curved portion and not in the rectangular portion at all flows. Elements of these weirs are shown in Fig. 10. The following equations are used in determining the width and the flow through a proportional or a Sutro weir:

$$x = b\left(1 - \frac{2}{\pi}\tan^{-1}\sqrt{\frac{y}{a}}\right) \tag{20}$$

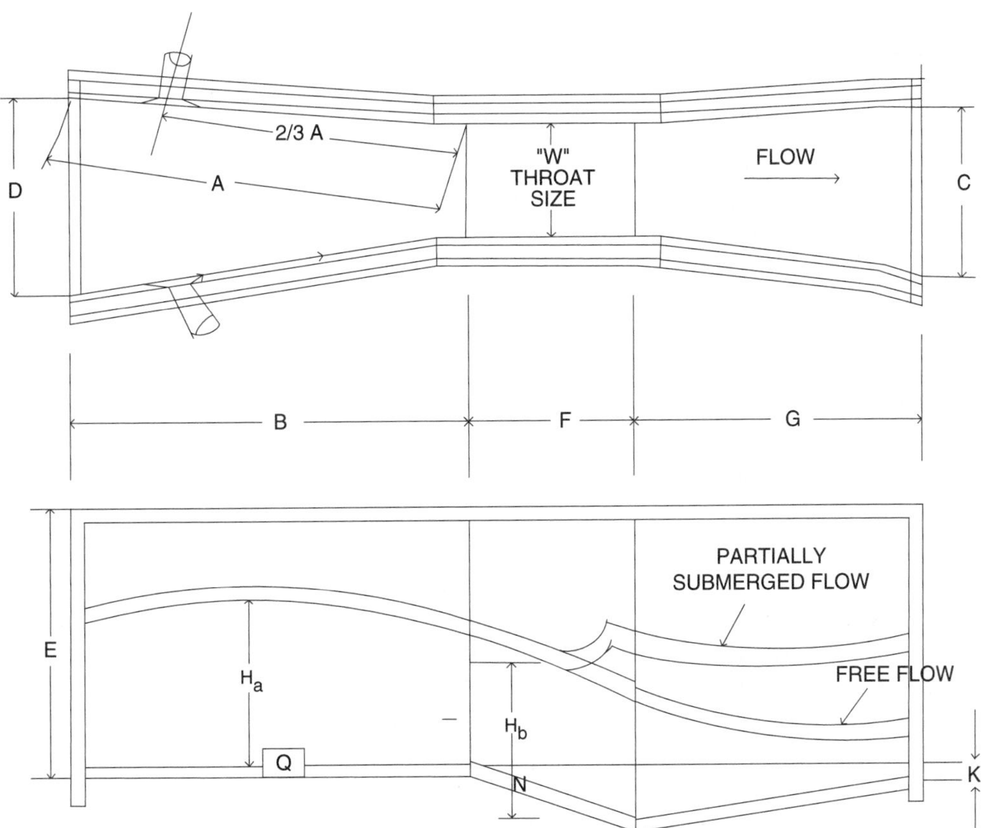

Fig. 11. Schematic diagram of a Parshall flume.

$$Q = cb\sqrt{2ag}\left(h + \frac{2}{3}a\right) \tag{21}$$

$$Q_1 = \frac{2}{3}cb\sqrt{2g}\left[(h+a)^{3/2} - h^{3/2}\right] \tag{22}$$

where Q = total discharge past the weir, Q_1 = discharge through the rectangular part of the weir, C = weir constant, 0.62 for proportional weir and 0.61 for Sutro weir, and a, b, c, h, x, and y are as shown in Fig. 10.

The disadvantage with the proportional or Sutro weir is that both require a lot of head loss. The water level downstream has to be below the crest of the weir. In many situations, enough head loss is not available to provide a proportional weir. In such cases, a rectangular control section is provided to control the velocity.

For a rectangular control section, the cross section of the grit channel must be designed to approach a parabola. Parshall flume is the most widely used rectangular control section. The advantages offered by a Parshall flume are that it is simple in construction, low in cost, and serves as a means of flow measurement with low head loss. It also offers unobstructed flow path so that wastewater solids do not accumulate.

The Parshall flume (Fig. 11) consists of a converging upstream section, a downward sloping throat, and an upward sloping diverging downstream section. Usually it is

Table 3
Parshall Flume Dimensions in Feet and Inches[a]

W	*A*	*B*	*C*	*D*	*E*	*F*	*G*	*K*	*N*
0–3	1–6 ⅜	1–6	0–7	0–10 3/16	2–0	0–6	1–0	0–1	0–2 ¼
0–6	2–0 7/16	2–0	1–3 ⅝	1–3 ⅝	2–0	1–0	2–0	0–3	0–4 ½
0–9	2–10 ⅝	2–10	1–3	1–10 ⅝	2–6	1–0	1–6	0–3	0–4 ½
1–0	4–6	4–4 ⅞	2–0	2–9¼	3–0	2–0	3–0	0–3	0–9
1–6	4–9	4–4 ⅞	2–6	3–4 ⅜	3–0	3–0	2–0	0–3	0–9
2–0	5–0	4–10 ⅞	3–0	3–11½	3–0	2–0	3–0	0–3	0–9
3–0	5–6	5–4 ¾	4–0	5–1 ⅞	3–0	2–0	3–0	0–3	0–9
4–0	6–0	5–10 ⅝	5–0	6–4¼	3–0	2–0	3–0	0–3	0–9

[a]Letters *W*, *A*, *B*, *C*, etc., relate to dimensions of Parshall flume shown in Fig. 11.

constructed of reinforced concrete. Nowadays, at most wasewater treatment plants, stainless steel or reinforced plastic liners are used. The dimensions for Parshall flumes are standardized (Table 3) and the flume is designated by the width of throat only.

If the liquid moves at a high velocity in a thin sheet, conforming closely to the dip at the lower end of the throat, the flow through the Parshall flume is called a free flow. If, however, the backwater raises the water surface causing a ripple, or a standing wave is formed at or just downstream from the end of the throat, the flume is said to operate under partly submerged conditions. There is no retardation in the free flow rate of discharge if the ratio of H_b to H_a (submergence) does not exceed 0.6 for 3–9-in. flumes and 0.7 for 1–8-ft flumes. Under free flow or partly submerged conditions, the discharge through the flume can be obtained from a single upstream measurement of depth. H_b and H_a are as shown in Fig. 11. If submergence exceeds the above values, the flume is said to operate under submerged conditions. In this case the downstream head must also be measured to determine the discharge through the flume. Submerged flow conditions are generally avoided at Parshall flumes employed in grit chambers.

6.4. *Design of Grit Chamber*

The theory of sedimentation discussed in Section 3 is also applicable to the design of grit chambers. The settling velocity of particles to be removed would depend mainly on the size and specific gravity of the grit particles and the temperature of the sewage. Grit chambers are generally designed to remove particles of over 0.2 mm in size. The specific gravity of particles is assumed to be 2.65 and a wastewater temperature of 15.6°C (60°F). It may be noted that under actual operating conditions, some particles having a specific gravity of less than 2.65 are included in the removed grit.

The average settling velocity of particles of 65 mesh (0.21 mm) size and specific gravity of 2.65 is 1.31 m/min (4.3 ft/m). The corresponding overflow rate required for the complete removal of these particles is 1,900 $m^3/d/m^2$ (46, 300 gpd/ft^2) (14). For particles of the same size but with specific gravities of 2 and 1.5 the surface overflow rate is reduced to 1,140 and 570 $m^3/d/m^2$ (28, 000 and 14, 000 gpd/ft^2), respectively. Metcalf and Eddy (8) recommend the use of a settling velocity of 3.7 ft/min for the removal of 65 mesh (0.21 mm) particles and 2.5 ft/min for 100 mesh (0.15 mm) particles, which correspond to 1.13 m/min for 65 mesh and 0.76 m/min for 100 mesh particles.

In the design of grit chambers, particular attention must be given to prevent turbulence at the inlet and outlet ends. Bottom scour velocity is also very important. According to Camp (16) the scouring process itself determines the proper velocity of flow in the unit. The settled particles may not remain at the bottom and may instead be carried into the effluent stream if the velocity is excessive. A velocity of between 0.23 and 0.38 m/s or 0.75 and 1.25 ft/s is usually adequate for the design of a grit chamber and should be as close to 0.30 m/s or 1 ft/s as possible (15). At 0.30 m/s (1 ft/s) most organics remain in suspension while the heavier particles settle out. Another factor to be considered in the design of a grit chamber is the accessibility of the unit and its various components, which may need to be handled frequently. Also a bypass should be provided so that the flow can be diverted into the subsequent units if the grit chamber has to be taken out of service. The length of a horizontal flow grit channel is calculated from the depth as fixed by the settling velocity and the control section, and the cross section required for flow in each channel.

The amount of grit collected depends on the size of the service area, the type of construction and condition of the sewers, whether the system contains separate or combined sewers, type of catch basins, condition of streets, types of soils, and use of household garbage grinders or industrial wastes. According to Fair, et al. (5) the quantity of grit varies from 7.5 to 90 L/ML (1 to 12 ft^3/MG) with an average value of 30 L/ML (4 ft^3/MG). The wastewater-treatment-plant design manual (15) states that the grit quantity to be expected per million gallons of wastewater flow will average between 1/3 ft^3 and 24 ft^3, which correspond to 2.5 to 180 L/ML.

The ultimate disposal of grit depends on the amount and character of the grit and the availability of the disposal site. Unless it is carefully washed, the grit may contain up to 50% by weight of organics. The grit is usually buried or else disposed of in a landfill. It may also be dumped into a sludge lagoon or incinerated.

In an aerated grit chamber the detention time to be provided is about 3 min at the maximum rate of flow (8). The grit hopper is about 1 m (3 ft) deep and the diffusers are 38-60 cm (15-24 in.) above the floor. With proper adjustments of flow of air, grit removal efficiencies of close to 100% could be obtained (8).

Mechanical equipment of grit removal from grit chambers usually consists of a conveyor and a bucket, a plow and scraper or a screw conveyor.

Because velocity through the bottom rectangular portion of the proportional weir does not remain constant as the flow varies, the proportional weir is designed such that the water level in the proportional weir is always in the curved section of the weir.

Points on curved sides of the weir are found by assuming values of y and finding corresponding values of x from Eq. (20). Table 4 is prepared, using Eq. (20), to give values of x/b corresponding to values of y/a. This table will be found to be very useful while designing a proportional or a Sutro weir.

7. GRAVITY THICKENING IN SLUDGE TREATMENT

One of the major problems of any wastewater-treatment plant is the management of sludge or biosolids. It is more economical to increase the concentration of sludge and decrease its volume before it is further treated or finally reused or disposed off. This can be achieved by resettling the sludge in a sludge thickener and stirring it long enough to

Table 4
Values of x/b Corresponding to y/a for Proportional Weir and Sutro Weir[a]

y/a	x/b
0.1	0.805
0.3	0.681
0.5	0.608
1.0	0.500
2.0	0.392
3.0	0.333
4.0	0.295
5.0	0.268
6.0	0.247
7.0	0.230
8.0	0.216
9.0	0.208
10.0	0.195
12.0	0.179
15.0	0.151
20.0	0.140
25.0	0.126
30.0	0.115

[a]Letters a, b, x, and y relate to dimensions of the weir, shown in Fig. 10.

form larger and more rapidly settling aggregates. Some thickening of sludge could be accomplished in the clarifiers themselves if the sludge is not continuously removed from the bottom of the clarifier. However, this will not only decrease the efficiency of clarification in the clarifiers, but also may create anaerobic conditions especially in the secondary clarifiers. Thickening in a separate unit would allow both thickener and clarifier to operate at optimum conditions.

Thickening increases the solids content of sludge by partial removal of liquid. This results in a reduction in the volume of sludge to be handled. For example, a 3% reduction in the water content of a sludge from 99% to 96%, i.e., an increase in solids concentration from 1% to 4%, will result with a thickened sludge that would occupy only 25% of its original volume. Reduction in volume of sludge, in general, reduces the size of subsequent treatment units such as digesters, dewatering equipment, incinerators, etc. Not only is the capital cost reduced, but also there is substantial savings in the operation of subsequent treatment units. This includes a reduction in the amount of heat required for digestion, auxiliary fuel required for heat drying and incineration, and the quantity of chemicals required for sludge conditioning and dewatering. Reduction in sludge volume results in reduction in pipe sizes and pumping costs when sludges are transported long distances such as to sludge lagoons. Sludge thickening also reduces the cost of handling when sludge is to be transported for application on land. Dust (17) made a cost benefit study for providing a thickener at Beaumont, TX. He found that a thickener would reduce the capacity of sludge digestion by half from over 14,200 m^3 (500,000 ft^3) to less that 7,100 m^3 (250,000 ft^3) resulting in saving in overall construction costs.

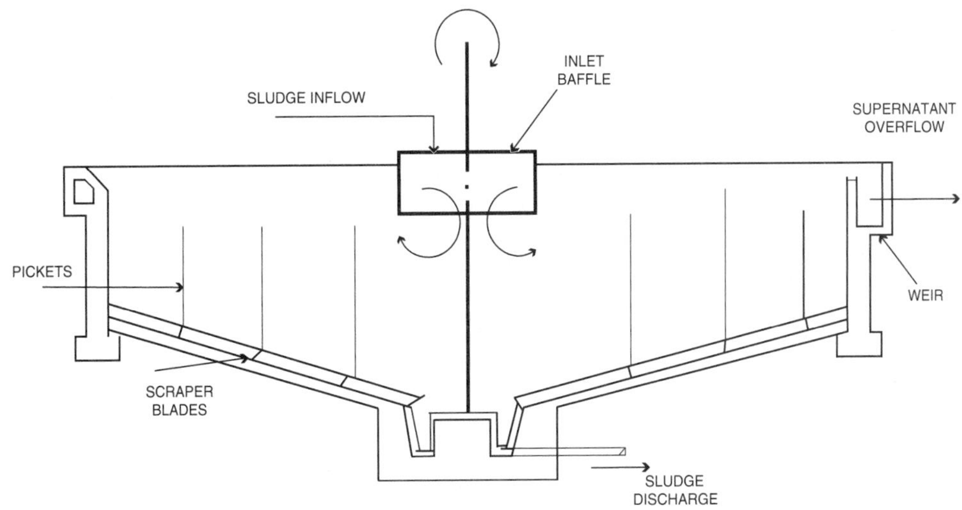

Fig. 12. Schematic diagram of a circular gravity sludge thickener.

Gravity thickening of raw or digested primary sludge is almost always efficient and economical. Primary sludges originally of 2.5–5% solids concentration can be concentrated by gravity thickening to 8–10% (8). In general, sludges from primary clarifiers and from a combination of primary and trickling filter sludges are best thickened by gravity thickeners. Gravity thickening is not very effective for activated sludge. Therefore, activated sludge is not commonly thickened by gravity thickening. According to the EPA (Environment Protection Agency) design manual (18), gravity thickening of a mixture of raw primary and excess air activated sludge is inefficient and hence is rarely used. This, however, is not the case with the oxygen-activated sludge (using pure oxygen instead of air), which has been shown to be more amenable to mixed sludge gravity thickening. Air-activated sludge, on the other hand, has been successfully thickened by dissolved air flotation (DAF).

Gravity thickener (Fig. 12) is basically a settling tank designed primarily for the settling of sludge particles from a dilute suspension of sludge. Sludge enters at the middle of the thickener and the sludge solids settle into a sludge blanket at the bottom. The weight of the overlying solids compacts the sludge as it gathers onto the bottom of the tank. Moving rakes, which dislodge gas bubbles and prevent bridging of the sludge solids, very gently agitates the concentrated sludge. The rakes also provide channels for the released water and scrape the thickened sludge toward a central hopper from where it is withdrawn for further treatment. The supernatant liquid overflows into a peripheral collection weir and is pumped for recycling to the head of the treatment plant.

7.1. Design of Sludge Thickeners

The theory and principles that apply in the design of sludge thickeners are the same as for the design of sedimentation tanks (see Section 3). Empirically, the sludge thickeners are designed on the basis of hydraulic surface loading and solids loading. The solids loadings are given in Table 5 for different kinds of sludge. Typical surface loading rates are 24.4–36.7 $m^3/d/m^2$ (600–900 gpd/ft^2). Torpey (19) studied the thickening

Table 5
Solids Loadings for Thickeners

Type of sludge	Solids loading ($lb/ft^2/d$)	Concentration of solids in unthickened sludge (%)	Concentration of solids in thickened sludge (%)
Separate sludges			
Primary	20–30	2.5–5.5	8–10
Trickling filter	8–10	3.0–6.0	7–9
Air-activated sludge	4–7	0.5–1.2	2–3
Oxygen-activated sludge	5–9	1.0–2.0	2.5–4
Combined sludges			
Primary and trickling filter	12–20	3.0–6.0	7–9
Primary and air-activated sludge	8–15	2.5–4.5	4–8
Primary and oxygen-activated sludge	10–20	1.5–3.5	4–6

of combined primary and activated sludge and indicated that it was desirable to feed relatively large quantities of secondary treated liquor along with the thin sludge to prevent septic conditions in the thickener. A secondary sludge to primary sludge volume ratio of 8 to 1 or greater ensures aerobic conditions (13). According to McCarty (20), surface loadings between 20 and 33 $m^3/d/m^2$ (500 and 800 gpd/ft^2) will ensure aerobic conditions in the thickener. Chlorine can also be used for septicity prevention.

Sludge volume ratio (SVR) is another design factor. SVR is the ratio of the volume of sludge blanket to the daily volume of sludge pumped from the thickener and is a measure of the average retention time of solids in the thickener. SVR is usually kept between 0.5 and 2 d (8). During warm weather, lower values of SVR are preferred. According to Culp and Culp (21) sludge volume index (SVI) of sludge to be thickened is defined as volume in mL occupied by 1 g (dry weight) of activated sludge mixed liquor suspended solids, after 30 min of settling in a 1000 mL graduated cylinder. A low SVI contributes to good thickening (21). Addition of lime, chlorine, or a polymer assists in thickening.

According to the *Design of Municipal Wastewater Treatment Plants, WEF–MOP 8* (15) solids loadings of 39–58.5 $kg/m^2/d$ (8–12 $lb/ft^2/d$) and overflow rates of 16.3–36.7 $m^3/d/m^2$ (400–900 gpd/ft^2) have been used in the design of thickeners.

8. RECENT DEVELOPMENTS

It was realized early in the century by Hazen (22) that the removal of settleable solids was independent of detention time and that it was a function of the overflow rate and the basin depth. He pointed out that the capacity of the sedimentation basin could be increased considerably by inserting horizontal trays in the basin. Camp (7) proposed a design for a settling basin, that would have horizontal trays spaced at 15 cm (6 in.). The detention time for this basin was 10.8 min and its overflow rate over the trays was

27.2 $m^3/d/m^2$ (667 gpd/ft^2). Although shallow depth settling could have minimized the size and cost of water-treatment facilities, the real interest in this concept was not aroused until recently. Hansen et al. (23) pointed out that application of shallow depth settling by inserting trays in conventional sedimentation basins met with only limited success because of two major problems: (1) the unstable hydraulic conditions encountered with very wide, shallow trays and (2) the minimum tray spacing required for mechanical sludge removal equipment. The recent development of tube settlers has essentially overcome these problems.

Two kinds of shallow depth settlers are now commercially available. These are the tube settlers and the Lamella separators. Until recently, both of these settlers had found only limited use in increasing the clarifier capacity at existing overloaded treatment plants. However, currently shallow depth settling is increasingly used in the upgrade of treatment plants, because it has the potential of reducing the size and cost of treatment facilities.

8.1. *Theory of Shallow Depth Settling*

Consider an ideal sedimentation basin for settling of discrete particles of uniform size, shape, and specific gravity. The particles will settle down with a velocity u_t, as determined by Eq. (6). If velocity of flow through the settling basin of length l and depth d is v_1, time t_1 taken by a small parcel of water to pass through the basin is

$$t_1 = \frac{l}{v_1}$$

The time t_2 taken for the discrete suspended particles in that parcel of water to settle through depth d is:

$$t_2 = \frac{d}{u_t}$$

To remove all the discrete particles before water leaves the settling basin t_1 should always be less than t_2. Maximum value of v_1 is obtained by equating t_1 to t_2, i.e.,

$$\frac{l}{v_l}(\max) = \frac{d}{u_t}$$

or

$$v_t(\max) = \frac{lu_t}{d} \tag{23}$$

Also

$$Q = Av_l$$

where Q = flow through the basin [m^3/s (ft^3/s)], A = the cross sectional area of the basin [m^2 (ft^2)], and v_1 = the horizontal velocity [m/s (ft/s)]:

or

$$Q = A\frac{lu_t}{d} \tag{24}$$

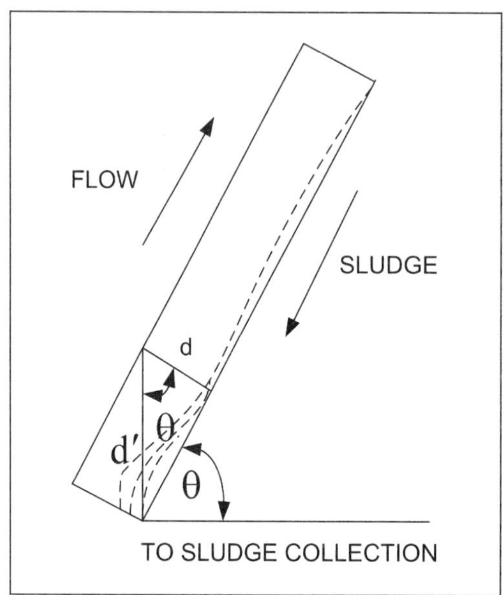

Fig. 13. Flow pattern in an inclined tube for tube settlers.

Equation (24) indicates that for maximum flow through the basin of fixed area, the ratio l/d should be made as large as possible. This can be done by increasing the length of the basin and/or by decreasing the depth of the basin and making it as shallow as possible. The principle of shallow-depth settling is employed in both tube settlers and Lamella separators.

In tube settlers each tube serves as an individual settling basin. For a horizontal tube of size 5cm × 5cm and 60 cm long, l/d ratio is 12, which is not too difficult to obtain in any conventional settling basin. The major advantage of tube settlers is that the tubes are stacked one above the other in a module providing a system with a number of small settling basins occupying the same space. Surface overflow rate through each tube calculated as flow through the tube divided by the surface area of the tube is actually less than that allowed for conventional settling tanks but, because the total surface area exposed to flow in tube settlers is many times more than in the conventional settling tanks, the net result is considerable increase in capacity of the settling basin.

When suspended particles settle in a horizontal tube, the maximum vertical distance traveled by the particles is equal to d, the depth of the tubes. If the tubes are inclined at an angle θ, as shown in Fig. 13, the particles fall through a distance d, such that

$$v_l = \frac{u_t l \cos\theta}{d} \tag{25}$$

where d = vertical distance the particle fall through [m (ft)] and θ = inclination angle of tubes. The above equation is limited to angles between 0° and $\tan^{-1} l/d$.

Laminar flow through the tubes is maintained even at high flow rates by the increased drag force due to a relatively large surface area of the tubes.

8.2. Tube Settlers

Tube settlers consist of numerous open-ended tubes, usually plastic about 5 cm (2 in.) in diameter and 61 cm (24 in.) long, mounted in modules, and placed in a basin. The water that is carrying the suspended solids moves from an influent well, upward through the small tubes, and into an outlet device. The tube cross section may be circular, square, rectangular, hexagonal or any other suitable shape (23). The tube settlers may be of either slightly inclined or steeply inclined design. In a slightly inclined tube settler system the tubes are inclined from the horizontal at an angle of about 5°. Sludge deposits form in the tubes, which need to be backwashed or drained. This system is best suited for use with filters so that the tubes can be back-washed along with the filters without any need for extra water or extra cost. This system of slightly inclined design is used primarily at small plants having a capacity of 3.8 ML/d (1 MGD) or less.

In steeply inclined tube settlers the tubes are inclined at an angle greater than 45°, usually at 60°. Culp et al (24) observed that tube efficiency at 60° was comparable to that obtained at 5°. The sediment does not accumulate in these tubes but continues to settle downward and out of the tubes into the lower influent zone, forming an adsorptive sludge blanket. No backwashing is required in this system for sludge removal from the inside of the tubes. It is necessary, however, as in all sedimentation basins, that means be provided to remove the sludge from the basin.

If tube settler modules were placed in existing basins, the sludge collection method would depend on the original basin and equipment design. At new installations, mechanical cleaning equipment should be provided. Many times the tubes become dirty and unsightly due to some of the suspended solids clinging to the walls or due to biological growth on the wall surface. This may seriously affect the efficiency of the tube settlers. Tube cleaning can be accomplished by providing a grid of diffused air headers beneath the tubes. Culp and Culp (25) have discussed this tube cleaning method. The influent flow is stopped before turning on the air, which scrubs out attached floc, and then a 15–25 min quiescent period is allowed before influent is again released into the tube settlers.

8.2.1. Application of Tube Settlers

White (26) reported construction of tube settlers for settling of power plant ash sluice water. For a flow of 4,540 L/min (1,200 gpm) of ash sluice water containing suspended solids ranging from 1,000 to 8,000 mg/L, conventional settling would have required an area of 290 m^2 (3,000 ft^2). The tube settlers provided had an area of only 70 m^2 (750 ft^2) and were designed at a hydraulic rate of 65 L/min/m^2 (1.6 gpm/ft^2). Bologna (27) reported the successful application of tube settlers to clarification of activated sludge mixed liquor suspended solids.

Culp et al. (24) also reported efficient clarification of both primary and activated sludge with tube settlers. The tubes used were 61 cm (24 in.) long for primary clarification and 122 cm (48 in.) long for activated sludge clarification. The rates of clarification were 120 L/min/m^2 and 82 L/min/m^2 (3 gpm/ft^2 and 2 gpm/ft^2) for primary and activated sludge, respectively. In contrast, Bologna (27) felt that for activated sludge, the upper loading limit was 40.7 L/min/m^2 or 1 gpm/ft^2.

Slechter and Conley (28) reported plant-scale application of tube settlers to primary as well as activated sludge. They observed that tube length was not a significant factor

Table 6
Recommended Overflow Rates for Tube Settlers in Horizontal Flow Tanks (Raw Water Turbidity 0–100)

Desired effluent turbidity (JTU)	Basin overflow rate [L/min/m^2 (gpm/ft^2)]	Tube settler overflow rate [L/min/m^2 (gpm/ft^2)]
Water temperature 4.4°C		
1–3	—	—
1–5	80 (2.0)	100 (2.5)
3–7	80 (2.0)	120 (3.0)
5–10	120 (3.0)	160 (4.0)
Water temperature 10°C		
1–3	80 (2.0)	100 (2.5)
1–5	80 (2.0)	120 (3.0)
3–7	80 (2.0)	160 (4.0)
5–10	—	—

Source: Ref. 25.

in the removal of settleable or suspended solids over settling tube rates of 85.5–134 L/min/m^2 (2.1–3.3 gpm/ft^2) for primary clarifiers. They also pointed out the necessity of cleaning the tubes at least once a week for better performance. With regard to clarification of activated sludge solids, Slechter and Conley (28) point out that a final effluent with suspended solids less than 20 mg/L can be achieved over a wide range of operating conditions. The conditions included tube settling rates of 20.4–93.7 L/min/m^2 (0.5–2.3 gpm/ft^2), solids loading rate of 146.5–415 kg/d/m^2 (30–85 lb/d/ft^2), MLSS of 900–5,000 mg/L, and SVI of 35–135. They, however, pointed out that tube settling rate should not exceed 41 L/min/m^2 (1 gpm/ft^2) using 61 cm (24 in.) tubes, and solids loading should not exceed 171 kg/d/m^2 (35 lb/d/ft^2). Periodic cleaning of tubes was also emphasized.

8.2.2. Design Criteria for Tube Settlers in Water Treatment

The tube settlers are designed on the basis of overflow rates calculated for the horizontal area of sedimentation basins covered by tube settlers. The overflow rates for tube settlers at water-purification plants depend on the raw water turbidity and temperature. Culp and Culp (25) have recommended loading rates for providing tube settlers in existing upflow clarifiers and horizontal flow basins. Typically, for a horizontal flow basin with overflow rate of 82 L/min/m^2 (2 gpm/ft^2) raw water with turbidity of less than 100 JTU (Jackson Turbidity Unit) and 4.4°C temperature, the recommended overflow rates for tube settlers are 102 L/min/m^2 (2.5 gpm/ft^2) for effluent turbidity of 1–5 JTU. Table 6 lists design parameters for tube settlers in a horizontal flow basin. Design of a tube settler employing these design parameters is illustrated in an example at the end of this chapter.

8.3. Lamella Separator

The Lamella separator as developed by the Axel Johnson Institute in Sweden (29) consists of a nest of parallel inclined plates and return tubes (Fig. 14). The plates are 1.5 m wide and 2.5 m long spaced 25–55 mm apart and are inclined at an angle of 25°–45° from the horizontal. The main difference between the Lamella separator and tube settlers

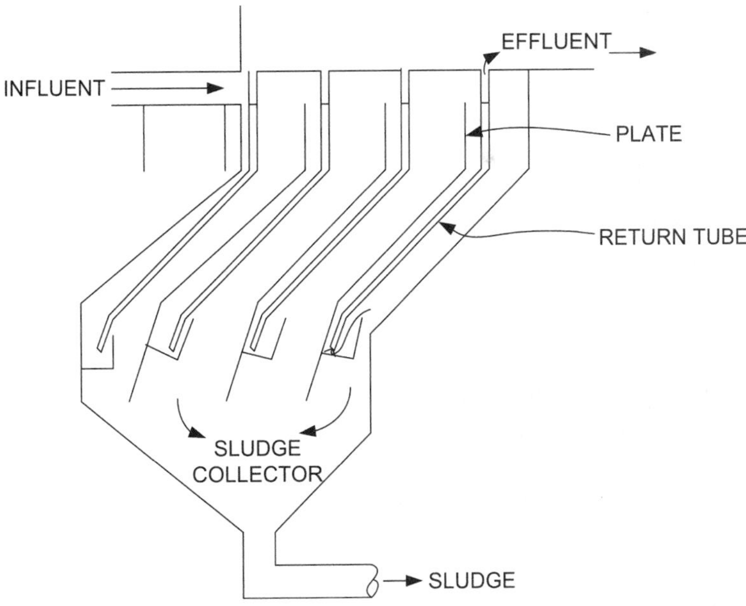

Fig. 14. Lamella separator.

is in the way the influent travels with reference to the solids. In Lamella separators the flow of both influent and solids is concurrent whereas in tube settlers it is countercurrent. The influent in a Lamella separator is introduced at the top of the module. It travels downward with the solids, which settle to the bottom of each plate and are carried by the flow down the incline into the sludge hopper.

A return tube placed at the bottom of each plate carries the effluent back to the top of the unit and into an effluent launder. Miller (29) reported an average solids removal of over 85% at a flow rate of 19 $m^3/d/m^2$ or 470 gpd/ft^2 of projected area. He also pointed out that a conventional sedimentation basin for the same flow at an overflow rate of 32.6 $m^3/d/m^2$, i.e., 800 gpd/ft^2 would require 10 times the space occupied by the Lamella separator.

8.4. Other Improvements

Sedimentation is one of the earliest and most important unit operations in water and wastewater treatment. With time significant improvement has taken place in the process of sedimentation. Particulate matter from the water column can be removed by sedimentation. But now the sedimentation process has been modified and can serve even more purposes. Sedimentation field-flow fractionation (SdFFF) is being used to remove biomass of sediment bacteria (30). The SdFFF is a high-resolution separation technique ideal for characterization of heterogeneous particulate materials. The preliminary experiments in the study showed that the SdFFF method could be applied to complex sediment samples for bacterial biomass measurement. To increase the efficiency of the overall process in sedimentation, significant research has been conducted in the modeling of sedimentation tanks. Models have been developed that are capable of providing useful information such as removal efficiency, size distribution in sludge and in effluent suspension, and thickness of bottom sludge. Based on the model for desired removal efficiency, the length of the tank can be determined (31). Numerical models of gravity sedimentation and thickening were also developed from the governing two-phase flow equations for the liquid and solid phases. The model was calibrated and verified using the data of dynamic porosity profiles of gravity sedimentation and thickening of kaolin suspension in distilled water (32). Batch settling is an important process that finds application for environmental engineering processes such as sedimentation and gravity thickening. The batch settling process has been simulated numerically using a dynamic model for zone settling and compression (33). The test results have been found to be useful in processes such as batch reactors, sedimentation tanks, and gravity thickeners. Theoretical relationships have been established between the initial settling rate and the concentration of suspension in sedimentation process by an analogy between the sedimentation and filtration processes. The calculated results satisfactorily coincided with the reported literature values (34).

Improvement in the modeling approach in sedimentation provided a better understanding of the whole process. This led to modification in the design of sedimentation tanks. The first improvement in the sedimentation tank was to introduce parallel plates that permit solids to reach the bottom after shorter distance of traveling. The horizontal direction of the plates causes them to get filled with solids fairly quickly and at a certain point it also causes the materials to be scoured back into suspension. Inclining the parallel plates can solve this problem to a degree that the sludge can flow in a direction opposite that of the suspended liquid. This led to the development of parallel plate and tube settlers (35). The design developments also include Chevron tube settler and the Pielkenroad separator. The Chevron design has the cross-sectional area of each rhomboidal tube formed in a V shape. The Pielkenroad separator offers some of the advantages of both the Chevron tube settler and the parallel plate Lamella separator. The Pielkenroad separator has corrugated plates that provide for some sludge thickening and appear to be less costly to construct than tube settlers. Gravel bed clarifiers can also provide laminar flow during clarification. A bed of rock serves to establish a zone of laminar flow (35). In compact treatment plant layouts a two-tray sedimentation basin can be built. In such an arrangement the sedimentation basin would have an intermediate suspended floor, with flow moving in one direction in the lower half and flow returning in the opposite direction in the upper half. The plan area required for sedimentation basins can be reduced by about half because the effective settling area of the two-tray basin is doubled (36).

9. SEDIMENTATION IN AIR STREAMS

9.1. *General*

The theory and principles of sedimentation are also employed in the field of air pollution to separate aerosols from gaseous streams. Aerosols may be defined as solid or liquid particles suspended into a gaseous medium. The solid particles are called dusts and the liquid particles, mists. The dusts are of irregular shape but mist particles are of spherical shape. The aerosols in general are finely divided particles of submicron or micron size. Larger particles tend to settle out quickly, but smaller particles remain in suspension and behave like gases. Particulates in air include dusts, flyash, metal oxides, smoke, fumes, pollen grains, fungus spores, mists, and vapors. These particles accumulate in the air from a variety of sources including industrial processes such as pulverizing, sawing, jaw crushing as well as from foundries, cotton gins, power plants, transportation, and solid waste disposal systems. Removal of particulate matter may be achieved by several methods, which include gravity settling, filtration, and electrostatic precipitation. Because this chapter deals with sedimentation, only the gravity settlers will be discussed in some detail.

9.2. Gravity Settlers

Gravity settlers make use of the force of gravity to separate particulates from gas streams. Essentially a gravity settler consists of a long horizontal settling chamber. When a gas stream containing the particulates passes through the chamber, the velocity is reduced and the particulates settle out under the influence of gravity. Equation (6), derived for Class-I clarification of discrete particles, can also be applied to calculate the settling velocity of aerosol particles. It is however observed that most aerosol particles flocculate during settling. Application of Eq. (6) therefore gives only an estimate of the actual settling velocity for such particles. Settling tests should therefore be performed to obtain design parameters for the settling chamber.

In general, the velocity of gas flow in a settling chamber should be kept sufficiently low so that once the particles have settled they do not become re-entrained. However, if the velocity of flow were kept very low, the size of the settling chamber required would become too large and uneconomical. In most cases the velocity of flow is based on test results under actual conditions. According to Rich (37) a velocity of flow less than 3 m/s (10 ft/s) is adequate.

Gravity settlers have low collection efficiencies and require much more space than other dust collection devices and are currently not very widely used. The gravity settlers, however, have the advantage of simplicity in design and maintenance. The friction loss through the gravity settler is very low and occurs mainly at the entrance and the exit. The gravity settlers find application as precleaners installed prior to cyclones to remove large particles. They are also used on natural draft exhausts from kilns and furnaces because of their low-pressure drop and in cotton ginning operations and alfalfa feed mills.

9.2.1. Construction of Gravity Settlers

The gravity settling chambers may consist of a simple enlargement of the air duct, called a balloon duct. The enlargement of air duct reduces the velocity of air, which results in settling of the heavier particles. The gravity settler may also be constructed as a separate settling chamber with inlet and outlet transitions. The settling chamber is usually provided with a dust hopper. The principles of shallow-depth settling are also employed in the design of gravity settlers. Thus, the settling chamber may have multiple collection trays to improve the efficiency of collection. The vertical distance between the trays could be as little as 2.5 cm (1 in.). The cleaning of this chamber may be somewhat difficult.

Special care is to be taken in the design of the inlets to obtain a uniform distribution of air over and across each tray. This is done by the use of gradual transitions, guide vanes, distributor screens, or perforated plates (38). For greater efficiency, the turbulence in the chamber should be minimized. Various forms of gravity settlers are shown in Fig. 15.

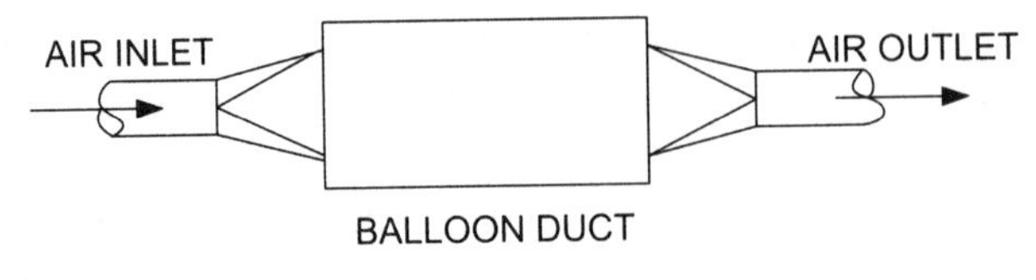

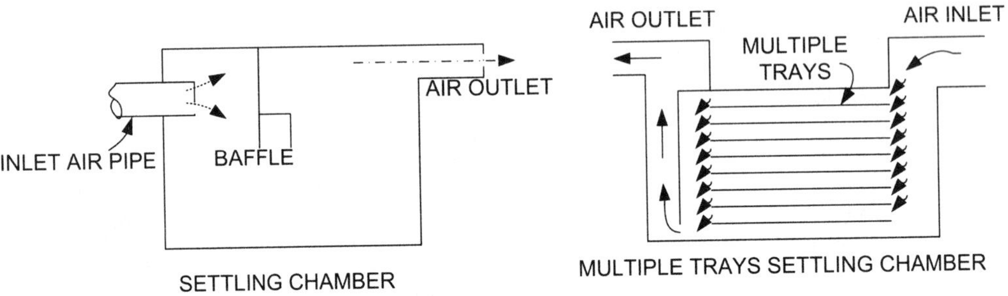

Fig. 15. Balloon duct and gravity settling chambers for settling particulates in air streams.

10. COSTS

10.1. General

Like any other unit of a water- or wastewater-treatment plant, the sedimentation basin involves two kinds of costs: one is the capital cost and the other is the operation and maintenance cost (O&M). The following are the major items to be considered when estimating the capital cost:

(a) Cost of land.
(b) Cost of structure including earthwork, inlet and outlet, bottom slab, walls and walkways.
(c) Cost of sludge collecting mechanism.
(d) Cost of pipes and fittings.
(e) Cost of electrical controls.
(f) Cleanup, site grading, seeding, and sodding.
(g) Engineering and legal fees.
(h) Contractor's profit and overhead.

Operation and maintenance costs should include the following items:

(a) Amortization.
(b) Manpower for normal operation and maintenance.
(c) Supervisory staff.
(d) Annual repairs, spare parts, etc.
(e) Power costs.
(f) Insurance.

10.2. Sedimentation Tanks

Process design manual for suspended solids removal (39) prepared by Hazen and Sawyer, Engineers, for the Technology Transfer Office of the US Environmental Protection Agency (EPA 625/1-75-003a) gives cost curves for sedimentation tanks.

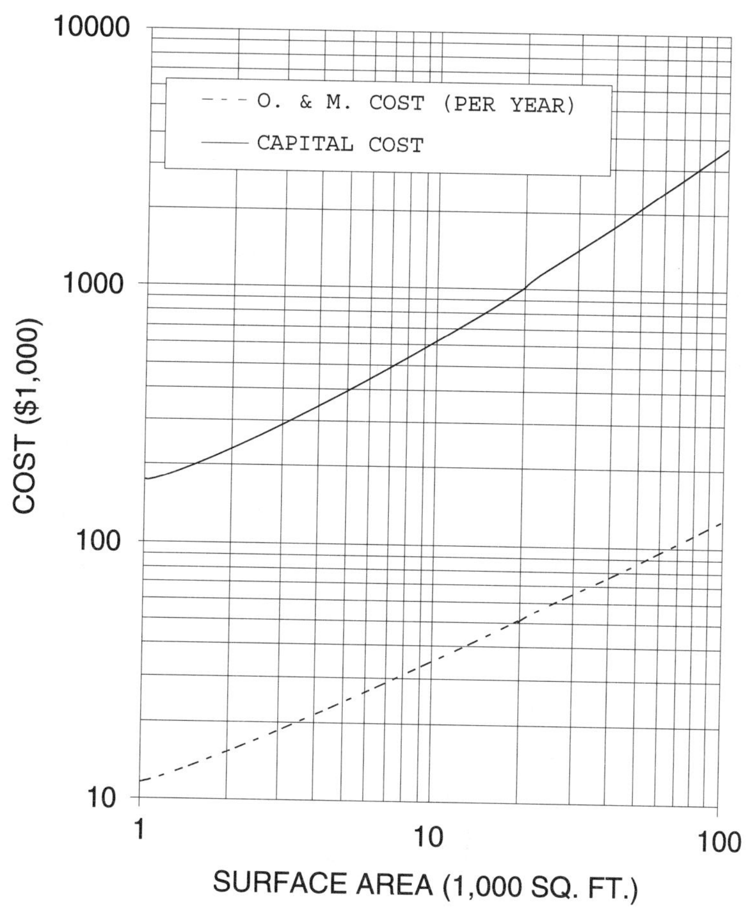

Fig. 16. Cost curves for sedimentation tanks in 2003 US $.

These curves have been updated to the year 2003 using U.S. Army Corps of Engineers Civil Works Construction Cost Index System for utilities, Manual # 1110-2-1304 (40, Appendix), and are presented in Fig. 16.

The curves include the cost of all equipment and controls necessary for a working unit. Included in the capital costs are inlet and outlet appurtenances, sludge-collecting mechanisms, steel or concrete tanks, supports, walkways and sludge draw off, all completely installed. However, the cost of land, building, pumping, sludge disposal, special site conditions, automated or computer controls, and engineering and legal fees are not included. The costs are based on installations using two or more units. The curves provide cost figures for preliminary estimates only and are no substitute for a detailed specific cost analysis.

Operation and maintenance costs include cost of amortization at 7%, manpower for operation and normal maintenance, and power for normal operation.

The cost figures depicted by these curves are applicable to both rectangular and circular tanks.

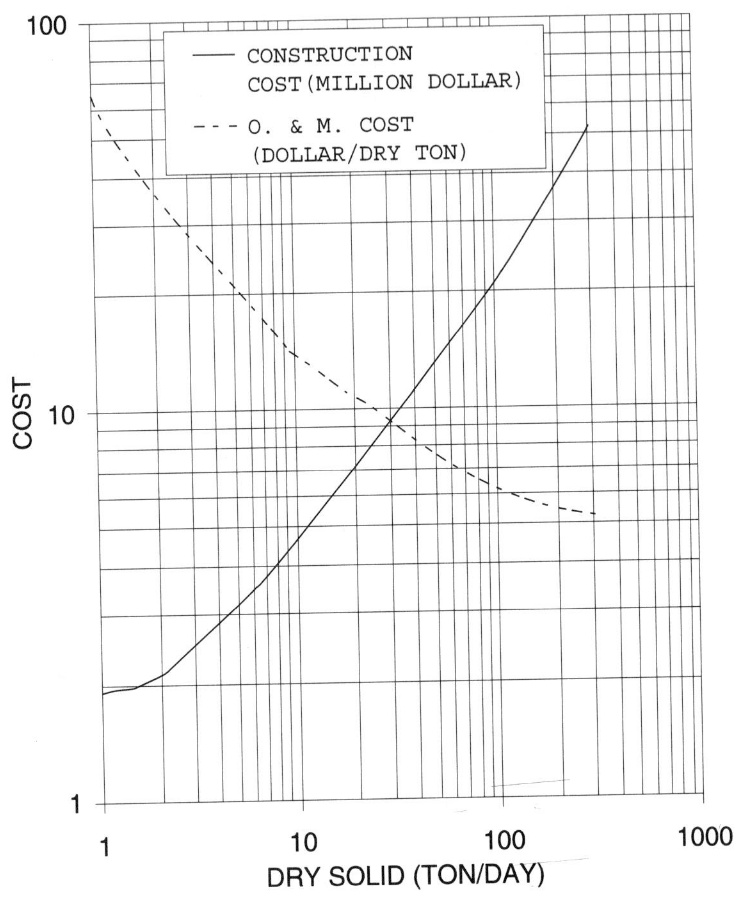

Fig. 17. Cost curves for gravity thickeners in 2003 US $.

10.3. Gravity Thickeners

The cost curves for gravity thickeners are shown in Fig. 17. Both capital costs and operation and maintenance costs are given as a function of daily dry solids production. The influent sludge is assumed to have a solids content of 0.5%. The capital costs include cost of concrete tank, sludge-collecting mechanism, walkway, and electrical controls for normal operation. No sophisticated controls are included nor is the cost of land included. The O&M costs also include amortization at 7% for 20 yr.

The cost figures for gravity thickeners were computed by Stanley Consultants (41) and are taken from US Environmental Protection Agency's *Process Design Manual for Sludge Treatment and Disposal* (18). These costs have been updated to the year 2003 using US Army Corps of Engineers Civil Works Construction Cost Index System for utilities, Manual # 1110-2-1304 (40, Appendix).

10.4. Tube Settlers

Figure 18 presents capital costs for tube settlers in rectangular as well as circular tanks as a function of installed area. The cost figures are taken from US Environmental

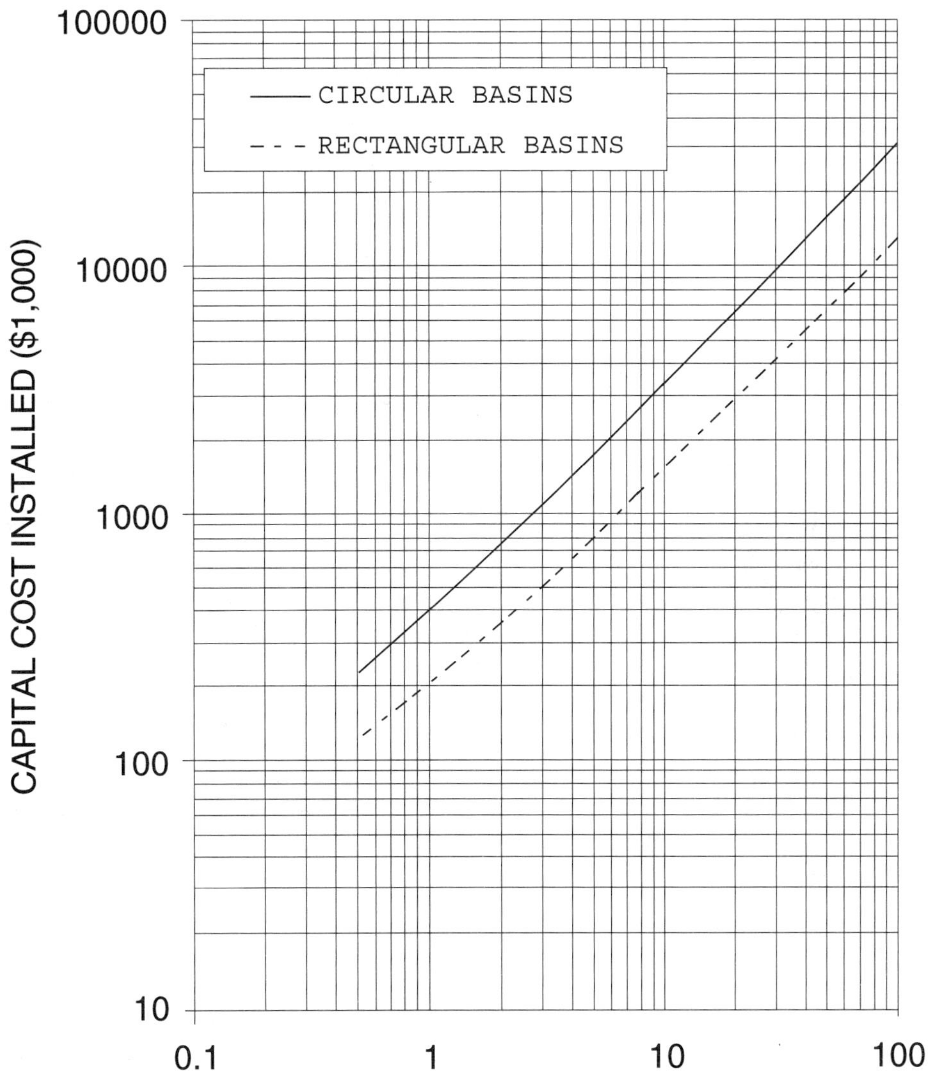

Fig. 18. Cost curves for tube settlers in existing rectangular and circular tanks in 2003 US $.

Protection Agency's *Process Design Manual for Suspended Solids Removal* (39). These curves have been updated to the year 2003 using U.S. Army Corps of Engineers Civil Works Construction Cost Index System for utilities, Manual # 1110-2-1304 (40, Appendix). Capital costs for tube settlers include plastic tubes with 60° inclination and 62 cm (21 in.) deep. Steel supports and additional effluent collector weirs are also included.

The costs of buildings, tanks, air grids for cleaning, sludge disposal, and external piping are not included in the cost figures presented. Engineering and legal fees are also not included.

According to the above design manual the estimated labor required for operation and maintenance of tube settlers is 2 man-hours per basin per week. This appears to be too low. A more reasonable figure would be 4 man-hours per week per basin.

Table 7
Data from Settling Column Analyses for Example 1

Settling time (min)	Fraction remaining	Settling velocity (m/s)
0.5	0.64	5.0×10^{-2}
1.0	0.56	2.5×10^{-2}
2.0	0.44	1.25×10^{-2}
3.0	0.36	0.83×10^{-2}
4.0	0.29	0.63×10^{-2}
6.0	0.17	0.42×10^{-2}
8.0	0.08	0.31×10^{-2}
10.0	0.04	0.25×10^{-2}
12.0	0.02	0.21×10^{-2}

11. DESIGN EXAMPLES

11.1. Example 1

The data shown in Table 7 were obtained from a settling column analysis performed on a dilute suspension of discrete particles. The sampling depth is 1.5 m. Calculate the overall removal of particles at a clarification rate of 0.03 $m^3/s/m^2$

Solution

Draw a settling velocity curve as shown in Fig. 19. The overflow rate given is 0.03 $m^3/s/m^2$, which corresponds to a settling velocity of 0.03 m/s. In other words $u_{t0} = 0.03$ m/s. The horizontal line drawn at the intersection of the curve with the vertical from 0.03 m/s gives x_0, so $x_0 = 0.585$. This means that 58.5% of the particles have velocity less than 0.03 m/s. Particles having velocity more than 0.03 m/s will be completely removed. This fraction is $(1-x_0) = (1-0.585) = 0.415$ or 41.5%.

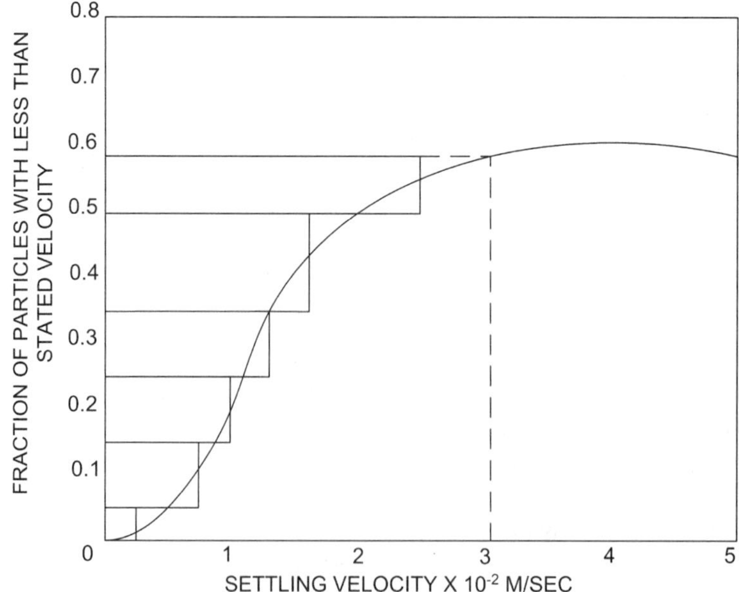

Fig. 19. Settling velocity curve for example 1.

Table 8
Suspended Solids Removal in Percent for Example 2

	Time (min)						
Depth (ft)	10	20	30	45	60	90	120
2	40	58	69	75	80	84	85
4	24	47	60	69	75	81	84
6	16	39	55	65	70	79	82
8	14	33	52	63	68	76	81

Out of 58.5% particles, that have velocity less than 0.03 m/s only a fraction will be removed. This fraction is $\int_0^{x_0} u_t dx$ which is found graphically as follows:

$$(0.166\times0.05+0.33\times0.1+0.5\times0.125+0.75\times0.125+1.37\times0.1+2.3\times0.085)\times10^{-2}$$
$$=0.0053$$

Overall removals

$$x_T=(1-x_0)+\frac{1}{u_{t0}}\int_0^{x_0} u_t dx=(1-0.585)+\frac{0.005}{0.03}$$
$$=0.415+0.177=0.592 \text{ or } 59.2\%$$

11.2. Example 2

The data shown in Table 8 were obtained from a settling column analyses. Find the detention time for a settling tank with 8 ft effective depth to remove 79% of the suspended solids.

Solution

The first step in solving the problem is to plot the data on time-depth graph in the same manner as elevations are plotted on a topographic survey map. Points of equal percentage removals are joined to obtain percentage removal lines as shown in Fig. 20. These lines are drawn similar to contours on a survey map.

To obtain detention time for 75% suspended solids removal, we calculate overall percentage removals, as per Eq. (15), at two different times, one which would give more than 75% removal and the other which gives less than 75% removal. The time required for 75% removal is then obtained by proportion. Examination of the percent removal curves indicates that overall removal at 64 min should be over 75% and that at 41 min should be less than 75%. So calculate overall removals at these two points only.

Overall removal at 64 min equals

$$\frac{5.7}{8}\times\frac{70+80}{2}+\frac{2.3}{8}\times\frac{80+100}{2}=79.3\%$$

Overall removal at 41 min equals

$$\frac{4.9}{8}\times\frac{70+60}{2}+\frac{1.9}{8}\times\frac{80+70}{2}+\frac{1.2}{8}\times\frac{80+100}{2}=71.1\%$$

By proportion 75% removal will be obtained in

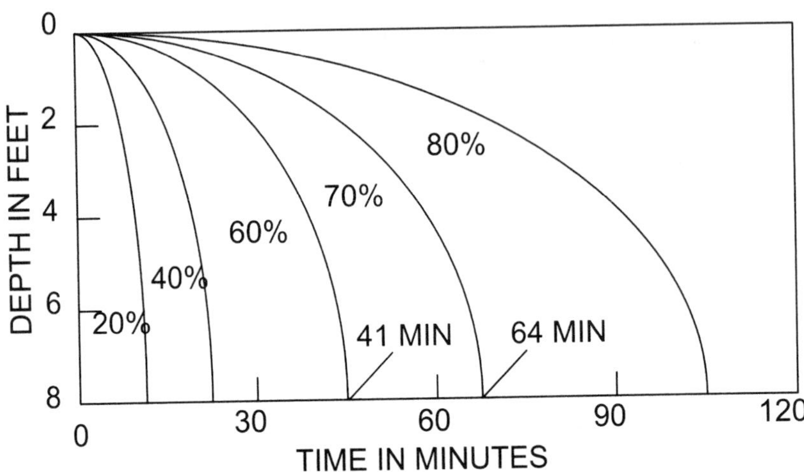

Fig. 20. Equal percentage removal curves for example 2.

$$41 + \frac{3.9}{8.2} \times (64 - 41) = 51.94 \text{ min, say 52 min.}$$

A factor of safety of 2.0 or more is usually added to this theoretical detention time to obtain design detention time for a settling tank. So detention time for the settling tank = 52 × 2 = 104 min.

11.3. Example 3

Settling column analyses performed on a waste sludge with an initial solid concentration of 3,500 mg/L yielded the settling curve plotted in Fig. 21. It is desired to thicken this sludge so that the underflow concentration is 14,000 mg/L. The sludge inflow is 0.02 m^3/s. Determine the area required for thickening, the overflow rate, and the solids loading. H_0 is 80 cm.

Solution

From Eq. (17)

$$H_u = \frac{c_0 H_0}{c_u}$$
$$= \frac{3{,}500 \text{ mg}}{\text{L}} \times 80 \text{ cm} \times \frac{\text{L}}{14{,}000 \text{ mg}} = 20 \text{ cm}$$

Draw tangents to the hindered-settling and the compression-settling regions of the settling curve, as shown in Fig. 21. Bisect the angle formed by these tangents and obtain the point of critical concentration c_c. Determine the value of t_u by drawing a vertical line to the time axis from the intersection of tangent at c_c and the horizontal line drawn at depth H_u (20 cm). As shown in Fig. 21, the value of t_u is 24 min.

From Eq. (16) the area required for thickening is

$$A = \frac{Qt_u}{H_0}$$
$$= \frac{0.02 \text{ m}^3}{\text{s}} \times 24 \text{ min} \times \frac{60 \text{ s}}{1 \text{ min}} \times \frac{1}{80 \text{ cm}} \times \frac{100 \text{ cm}}{1 \text{ m}}$$
$$= 36 \text{ m}^2$$

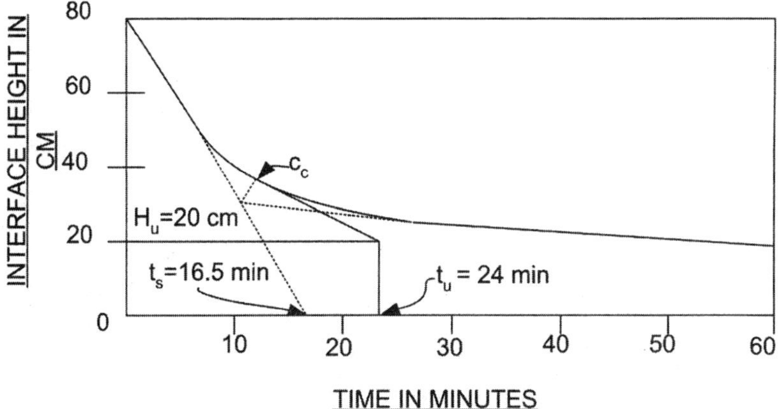

Fig. 21. Settling curve for example 3.

While determining the area required to thicken sludge, it is essential to ensure that this area is adequate for clarification. The settling velocity, v_s, of the sludge is obtained by dividing the depth of the column by the time, t_s, where the tangent to the hindered settling region cuts the time axis:

$$v_s = \frac{80}{16.5 \times 60} = 0.08 \text{ cm/s}$$

The area required for clarification equals

$$F_e = \frac{ma_e}{g_c}$$

which is less than that required for thickening. Therefore the area for thickening governs the design.

Solids loading equals

$$\frac{0.02 \text{ m}^3}{\text{s}} \times \frac{3,500 \text{ mg}}{\text{L}} \times \frac{1}{36 \text{ m}^2} \times \frac{86,400 \text{ s}}{\text{d}} \times \frac{1,000 \text{ L}}{\text{m}^3} \times \frac{\text{g}}{1,000 \text{ mg}}$$
$$= 168,000 \text{ g/m}^2\text{/d} = 168 \text{ kg/m}^2\text{/d}$$

11.4. Example 4

Design sedimentation tanks for a 10 MGD water purification plant. According to Ten State Standards (13) the detention period should be 4 h.

Solution

Capacity of sedimentation tank required:

$$V = 10 \times 10^6 \frac{\text{gal}}{\text{d}} \times \frac{\text{d}}{24 \text{ h}} \times 4 \text{ h} = 29,800 \text{ gal}$$
$$= 29,800 \text{ gal} \times 7.48 = 222,800 \text{ ft}^3$$

Assume an effective depth of 12 ft.

$$\text{Area of tank required} = 222{,}800/12 = 18{,}600\ \text{ft}^2$$

Rectangular tank dimensions:

Assume number of tanks to be provided = 4

The width of the tank should be based on the availability of sludge-collection equipment.

Assume the working width of units to be 40 ft.

$$\text{Length of each tank} = \frac{18{,}600}{4 \times 40} = 116\ \text{ft}$$

Provide four tanks of dimensions 120 ft × 40 ft × 12 ft.

Check for overflow rate:

$$\text{Area of each tank} = 120 \times 40 = 4{,}800\ \text{ft}^2$$

$$\text{Overflow rate} = \frac{2.5 \times 10^6}{4{,}800}$$

$$= 521\ \ \text{gpd/ft}^2, \text{which is acceptable}$$

Check for horizontal flow velocity:

$$\text{Velocity} = \frac{\text{length of tank}}{\text{detention time}}$$

$$\text{Velocity} = \frac{120\ \text{ft}}{4 \times 60\ \text{min}} = 0.5\,\text{ft/min}, \text{which is acceptable}$$

Weir overflow rate:

According to Ten State Standards the maximum weir overflow rate allowed is 20,000 gpd/ft.

So weir length required for each tank

$$= \frac{2.5 \times 10^6}{20{,}000} = 125\ \text{ft}$$

Let n be the number of additional overflow weir troughs, then

$$(2n + 1) \times 40 = 125$$
$$n = 1.06$$

Provide two additional troughs.

$$\text{Weir overflow rate} = \frac{2.5 \times 10^6}{(2 \times 2 + 1) \times 40} = 12{,}500\ \text{gpd/ft}, \quad \text{which is acceptable}$$

11.5. *Example 5*

Design final settling tank(s) for a conventional activated sludge plant of 2 MGD capacity.

Solution

Provide two circular tanks, each will be of 1 MGD capacity. Unless actual experiments show a smaller detention time is justified, the detention time given in the Ten State

Standards should be used. For a conventional activated sludge plant, the detention time is 2.5 h for a flow of 1 MGD.

$$\text{Volume of tank based on detention time} = \frac{1{,}000{,}000 \times 2.5}{24 \times 7.48}$$
$$= 13{,}900 \text{ ft}^3$$

Assume 10 ft side water depth in tank

Area required, A = 1,390 ft^2

Diameter of tank, D = 42.5 ft

The nearest tank diameter for which mechanical equipment is available = 45 ft (Walker Process).

Provide 45 ft diameter tank.

Area of tank provided = 1,590 ft^2

The tank depth could be reduced but we will keep it 10 ft, as assumed, to accommodate variation in flow.

Total volume of tank provided = 1,590 × 10 = 15,900 ft^2

$$\text{Detention time provided based on average design flow} = \frac{15{,}900 \times 7.48 \times 24}{1{,}000{,}000}$$
$$= 2.86 \text{ h}$$

$$\text{Surface overflow rate} = \frac{1{,}000{,}000}{1{,}590}$$
$$= 629 \text{ gpd/ft}^2$$

which is less than 700 gpd/ft^2 required as per Ten State Standards, so it is acceptable.

The length of the outlet weir = 45 × 3.14 = 141.3 ft

Weir overflow rate = 1,000,000/141.3 = 7,077 gpd/ft of weir

which is less than 10,000 gpd/ft required by the Ten State Standards, so it is acceptable.

Inlet well diameter = 10–20% of tank diameter = 4.5–9 ft in diameter,

select inlet well diameter of 8 ft.

11.6. Example 6

Design a grit chamber to remove grit of size 0.2 mm or above at a wastewater-treatment plant with an average flow of 3.2 MGD and a peak flow of 8 MGD. Assume minimum flow to be half the average flow.

Solution

Provide three channels for the grit chamber: two to accommodate peak flow, the third will be a standby unit.

Peak flow in each channel = 8/2

= 4 MGD.

Assume velocity in the channel to be $v = 1$ ft/s.

The cross sectional area required = Q/v = (4 × 1.547)/1 = 6.188 ft^2

Assume a depth of 2.5 ft

$$\text{Width} = \frac{6.188}{2.5} = 2.475 \text{ ft, say } 2.5 \text{ ft}$$

Settling velocity of 0.2 mm particle = 3.7 ft/min.

Theoretical time required for particles of size 0.2 mm to settle to the bottom of the grit chamber

$$t = (2.5 \times 60)/3.7 = 43.2 \text{ s}$$

Minimum length of channel required

$$L = 43.2 \text{ s} \times \frac{1\text{ft}}{\text{s}} = 43.2 \text{ ft}$$

In actual practice the length of channel provided is 25–50 percent more than the theoretical minimum required. Providing 40% more length, then length of channel will be = 43.2 × 1.4 = 60.5, use 60 ft.

Assuming 6 ft^3 of grit per MG, the grit accumulated per day = 6 × 3.2 = 19.2 ft^3.

Provide 4 in. below crest of weir for storage of grit.

$$\text{Storage space in each channel} = 2.5 \times (4/12) \times 60 = 50 \text{ ft}^3.$$

One channel will have to be cleaned every second day.

Also provide 1 ft freeboard. The channels will therefore be 2 ft 6 in. wide and 3 ft 10 inches deep.

The crest of weir will be 4 in. above the bottom of the channel.

The proportional weir is designed as follows:

$$\text{Minimum flow through each grit channel} = 1/2 \times 1/2 \times 3.2 = 0.8 \text{ ft}^3/\text{s}$$

$$\text{Maximum flow through each grit channel} = 1/2 \times 8 = 4 \text{ MGD} = 6.19 \text{ ft}^3/\text{s}$$

$$\text{Assume } a = 0.25 ft$$

$$\text{Therefore } h = 2.5 - 0.25 = 2.25 \text{ ft}$$

$$Q = 6.18 = cb(2 \times 0.25 \times 32.17)^{1/2}\left(2.25 + \frac{0.25 \times 2}{3}\right)$$

$$= 0.62b(16.08)^{1/2} 2.42$$

$$b = 1.03 \text{ ft}$$

Flow through the rectangular section when $h = 0$:

$$Q_1 = \frac{2}{3} cb(2g)^{1/2} a^{3/2}$$

$$Q_1 = \frac{2}{3} \times 0.62 \times 1.03 \times (2 \times 32.17)^{1/2} \times 0.25^{3/2}$$

$$= 0.43 \text{ ft}^3/\text{s}$$

which is less than the minimum flow through the channel. So the water level in the weir will be in the curved section of the weir at all flows.

11.7. Example 7

Design a gravity thickener to thicken sludge from a 2 MGD activated sludge plant. Both primary and activated sludge will be mixed and thickened together. Provide at least two thickeners. Determine the quantity of sludge expected assuming the sludge is undigested. Sludge volume is usually between 7,000 and 10,000 gal/MGD of wastewater flow and solids in the unthickened sludge are between 1,800 to 2,600 lb/MGD of wastewater flow.

Solution

Quantity of sludge per day assuming 10,000 gal/MGD is

$10{,}000 \times 2 = 20{,}000$ gal

Assume an overflow rate of 500 gpd/ft^2

$$\text{Surface area required} = \frac{20{,}000}{500}$$
$$= 40\ \text{ft}^2 \text{ or } 20\ \text{ft}^2 \text{ for each thickener}$$

Quantity of solids per day assuming 2,000 lb of solids/MGD

$= 2{,}000 \times 2 = 4{,}000$ lb

Assume solids loadings of 8 lb/ft^2/d

$$\text{Surface area required} = 4{,}000/8 = 500\ \text{ft}^2,\ \text{or } 250\ \text{ft}^2 \text{ per thickener}$$

Therefore solids loading govern the design.

Diameter of thickeners

$$\text{Area} = 1/4 \times 3.14(\text{diameter})^2 = 250\ \text{ft}^2$$
$$\text{Diameter} = 17.8\ \text{ft}$$

Provide two thickeners of 20 ft diameter each. Because the overflow rate for the anticipated volume of sludge is very low some of the wastewater effluent should be recycled to prevent septic conditions.

11.8. Example 8

Design a tube settler for increasing the capacity of an existing settling tank provided for a water-treatment plant from 3 MGD to 7.5 MGD. The existing tank dimensions are 30 ft × 130 ft × 12 ft deep. Raw water has a turbidity of 20–25 JTU and temperature of up to 40°F. Effluent turbidity desired is 1–5 JTU.

Solution

$$\text{The existing tank over flow rate} = (3 \times 10^6)/(30 \times 130) = 770\ \text{gpd/ft}^2$$

$$\text{The existing tank overflow rate at increased capacity} = (7.5 \times 10^6)/30 \times 130$$
$$= 1{,}923\ \text{gpd/ft}^2$$
$$= 1.34\ \text{gpm/ft}^2$$

Table 6 does not give tube settler rate for 1.34 gpm/ft^2 basin. But for a basin rate of 2 gpm/ft^2 and effluent turbidity 1–5 JTU, the tube settler rate should be 2.5 gpm/ft^2.

At 2.5 gpm/ft^2 design overflow rate, the tube settler area required is

$$A = \frac{7.5 \times 10^6}{2.5 \times 1,440}$$
$$= 2,083 \text{ ft}^2$$

Basin length, which should be covered by tube settlers, is

$$L = \frac{2,083}{30} = 69.4, \text{ use 70 ft}$$

This length is rounded off according to standard module dimensions to allow for an exact number of modules.

The modules are placed for a length of 70 ft from the effluent and extending inward. In order to direct the flow through the module a baffle wall is installed at the inner edge. New effluent launders may also be required to provided uniform flow through the modules. As the basin is quite deep, 4 ft long tubes may be installed.

NOMENCLATURE

a_e acceleration of particle from the external force [m/s^2 (ft/s^2)]
A area required for the solids handling thickener [m^2 (ft^2)]
A cross sectional area of basin [m^2 (ft^2)]
A area of the particle [m^2 (ft^2)]
A_p projected area of particle [m^2 (ft^2)]
a, b, c, h, x, and y are as shown in Fig. 10
c concentration [mg/L]
c_0 initial concentration at depth H_0
c_d coefficient of drag
c_u underflow concentration
C weir constant, 0.62 for proportional weir and 0.61 for Sutro weir
d characteristic diameter of particle [m (ft)]
d vertical distance the particle fall through [m (ft)]
d_p diameter of particle [m (ft)]
F force [dyne (lb-force)]
F_b buoyant force [dyne (lb-force)]
F_d drag force [dyne (lb-force)]
F_e external force [dyne (lb-force)]
g acceleration due to gravity [m/s^2 (ft/s^2)]
g_c Newton's law conversion factor [981 m·g/m·dyne), 32.2 (ft·lb mass/ft·lb-force)]
h_5 total depth of water in column [m(ft)]
Δh_1, Δh_2, Δh_3, Δh_4 depth increments to successive percent removal curves
H depth in column [m (ft)]

H_0	initial column height of the interface in the settling column [m (ft)]
H_t	depth of sludge at any time (t) [m (ft)]
H_u	the depth at which all solids are at the desired under flow concentration (c_u) [m (ft)]
H_∞	final depth of sludge after a long period of settling [m (ft)]
K	constant for a given suspension
L	length [m (ft)]
m	mass of the particle [g (lb-mass)]
n	number
Q	flow through basin [m^3/s (ft^3/s)]
Q	volumetric flow rate into thickener [m^3/s (ft^3/s)]
Q	total discharge past the weir [m^3/s (ft^3/s)]
Q_1	discharge through the rectangular part of the weir [m^3/s (ft^3/s)]
R_e	Reynolds number
$R_1, R_2, \ldots, R_5$	percentage removals (%)
R_{t2}	percentage removal at time t_2 (%)
t	time [h, min, s]
t_u	time required to attain underflow concentration (c_u)
u	linear velocity of particle [m/s (ft/s)]
u_t	terminal velocity [m/s (ft/s)]
v_l	horizontal velocity [m/s (ft/s)]
V	volume [m^3 (ft^3)]
x	fraction of particles removed
θ	inclination angle of tubes (deg)
μ	kinematic viscosity [m^2/s (ft^2/s)]
ψ	sphericity of particle

REFERENCES

1. H. E. Babbit and J. J. Doland, *Water Supply Engineering*, McGraw-Hill, New York, 1949.
2. AWWA, *Water Treatment Plant Design*. American Water Works Association, 1997.
3. E. B. Fitch, Anaerobic digestion and solids-liquid separation, in *Biological Treatment of Sewage and Industrial Wastes*, J. McCabe and W.W. Eckenfelder Jr. (eds.), Reinhold Publishing Corporation, New York, 1958, Vol. 2, pp. 159–170.
4. T. R. Camp, *Trans. Am. Soc. Civil Engineers*, **103**, 897 (1946).
5. G. M. Fair, J. C. Geyer, and A.A. Okun, *Water and Wastewater Engineering*, Willey, New York, 1968, Vol.2.
6. J. S. McNown and J. Malailea, *Trans. Am. Geophys. Union,* **31**, 74, (1950).
7. T. R. Camp, *Trans. Am. Soc. Civil Engineers*, **111**, 895–936 (1946).
8. X. Metcalf and Y. Eddy, *Wastewater Engineering—Treatment, Disposal and Reuse*, McGraw-Hill, New York (1991).
9. W. P. Talmadge and E. B. Fitch, *Industrial and Engineering Chemistry*, **47**, 38 (1955).
10. J. M. Coulsan and J. F. Richardson, *Chemical Engineering* McGraw-Hill, New York, 1955, Vol. 2.
11. R. I. Dick and B. B. Ewing, *J. San Eng., Am. Soc. Civil Engineers*, **93**, 9–29 (1967).
12. S. B. Maynard, *Water and Sewage Works*, Reference Number, pages R202–204, (1961).

13. Great Lakes-Upper Mississippi River Board of State Sanitary Engineering, *Recommended Standards for Water Works*, Albany, New York, 1968; addeda, 1972.
14. Great Lakes-Upper Mississippi River Board of State Sanitary Engineer, *Recommended Standards for Sewage Works*, Albany, New York, 1997.
15. WEF, *Design of municipal Wastewater Treatment Plant MOP-8*, Water Environment Federation, Alexandria, VA, 1998.
16. T. R. Camp, *Sewage Works J*, **14**, 368 (1942).
17. J. Dust, *Civil Eng.* **247**, 68–72, (1956).
18. US EPA, *Process Design Manual for Sludge Treatment and Disposal* US Environmental Protection Agency, EPA 625/1-79-011, 1979.
19. W. N. Torpey, *Proc. Am. Soc. Civil Engineers*, **80**, 443–1 (1954).
20. P. L. McCarty, *J. Water Pollut. Control Fed.*, **38**, 493–507 (1966).
21. R. L. Culp and G. L. Culp, *Advanced Wastewater Treatment*, Van Nostrand Reinhold Company, New York, 1971.
22. A. Hazen, *Trans. Am. Soc. Civil Engineers*, **53**, 63 (1904).
23. S. P. Hansen, G. H. Richardson, and A. Hsiung, *Chemical Engineering Progress Symposium Series*, **65**, 97, 207–217 (1969).
24. G. L. Culp, S. P. Hansen, and G. H. Richardson, *J. Amer. Water Works Assoc.* **60**, 681 (1968).
25. G. L. Culp and R. L. Culp, *New Concepts in Water Purification*, Van Nostrand Reinhold Company, New York, 1974.
26. J. E. White, *Public Works*, November, pp. 71–73 (1974).
27. A. E. Bologna, *J. Water Pollut. Control Assoc. of Pennsylvania*, September–October, p. 14 (1969).
28. A. F. Slechter and W. R. Conley, *J. Water Pollut. Control Fed.*, **43**, 1724 (1971).
29. R. Miller, *Water Waste Eng.*, September, pp. 42–44 (1974).
30. A. Khoshmanesh, R. Sharma, and R. Beckett, *J. Environ. Eng, Am. Soc. Civil Engineers*, **127**, 19–31 (2001).
31. Y .C. Jin, Q. C. Guo, and T. Viraghavan, *J. Environ. Eng, Am. Soc. Civil Engineers*, **126**, 754–760 (2000).
32. J. R. Karl and S. A. Wells, *J. Environ. Eng, Am. Soc. Civil Engineers*, **125**, 792–805 (1999).
33. Y. Zheng and D.,M. Bagley, *J. Environ. Eng, Am. Soc. Civil Engineers*, **125**, 1007–1013 (1999).
34. M. A. Islam and R. D. Karamisheva, *J. Environ. Eng, Am. Soc. Civil Engineers*, **124**, 39–50 (1998).
35. JMM Consulting Engineers, *Water Treatment Principles and Design,* A Wiley-Interscience Publication, New York, 1985.
36. D. J. Corbin, D. G. Monk, C. J. Hoffman, and S. F. Crumb, *J. Amer. Water Works Assoc.*, **84**, 36–42 (1992).
37. L. G. Rich, *Unit Operation of Sanitary Engineering*, Wiley, New York, 1961.
38. US Department of Health Education and Welfare, *Control Techniques for Particulates Air Pollutants*, NAPCA, Publication No. AP-51, 1969.
39. US EPA, *Process Design Manual for Suspended Solids Removal*, US Environmental Protection Agency, EPA 625/1-75-003a, Washington, DC, 1975.
40. US ACE, *Civil Works Construction Cost Index System Manual*, # 1110-2-1304, U.S. Army Corps of Engineers, Washington, DC, USA, PP 44 (PDF file is available on the Internet at http://www.nww.usace.army.mil/cost), 2000 (Revised 2003).
41. Stanley Consultants, *Sludge Handling and Disposal Report*, Muscatine, IA, November 15, 1972.

Appendix
US Yearly Average Cost Index for Utilities[a]

Year	Index	Year	Index
1967	100	1986	347.33
1968	104.83	1987	353.35
1969	112.17	1988	369.45
1970	119.75	1989	383.14
1971	131.73	1990	386.75
1972	141.94	1991	392.35
1973	149.36	1992	399.07
1974	170.45	1993	410.63
1975	190.49	1994	424.91
1976	202.61	1995	439.72
1977	215.84	1996	445.58
1978	235.78	1997	454.99
1979	257.20	1998	459.40
1980	277.60	1999	460.16
1981	302.25	2000	468.05
1982	320.13	2001	472.18
1983	330.82	2002	484.41
1984	341.06	2003	495.72
1985	346.12		

[a]Extracted from US ACE 2000 *Civil Works Construction Cost Index System Manual*, # 1110-2-1304, U.S. Army Corps of Engineers, Washington, DC, USA, PP 44 (PDF file is available on the Internet at http://www.nww.usace.army.mil/cost), (Tables Revised 31 March 2003).

12
Dissolved Air Flotation

Lawrence K. Wang, Edward M. Fahey, and Zucheng Wu

CONTENTS

1. INTRODUCTION

1.1. Adsorptive Bubble Separation Processes

Adsorptive bubble separation processes make use of the selective adsorption of impurities at the gas/liquid or gas/solid interfaces of rising bubbles. The adsorbed impurities, which can be in soluble or insoluble form, are carried to the top of the bubble separation reactor, where they can be removed from the aqueous system (1,131,141,144). Today, the adsorptive bubble separation processes are used for a variety of solute/liquid and solid/liquid separation applications (2–4,12–79) and many analytical and control methods have been developed for control and monitoring of the processes (5–11,14,53, 63,102–1–3,131,134). The process applications include water purification (12–18,

From: *Handbook of Environmental Engineering, Volume 3: Physicochemical Treatment Processes*
Edited by: L. K. Wang, Y.-T. Hung, and N. K. Shammas © The Humana Press Inc., Totowa, NJ

Table 1
Classification of Adsorptive Bubble Separation Processes (141,144,180–183)

I. Classification according to the technique used for generating fine gas bubbles:
  1. Dissolved gas system (Example: Dissolved air flotation)
  2. Dispersed gas system (Example: Dispersed air flotation)
  3. Vaccum system (Example: Vaccum air flotation)
  4. Electrolysis system (Example: Electroflotation)
  5. Biological system (Example: Biological flotation)

II. Classification according to the technique used for separating impurities or pollutants:
  1. Foam separation:
    A. Foam fractionation
    B. Froth flotation
      B1. Precipitate flotation
      B2. Ion flotation
      B3. Molecular flotation
      B4. Microflotation and colloid flotation
      B5. Macroflotation and ore flotation
      B6. Adsorption flotation
      B7. Adsorbing colloid flotation
  2. Nonfoaming adsorptive bubble separation
    A. Bubble fractionation
    B. Solvent sublation
    C. Nonfoaming flotation
      C1. Nonfoaming precipitate flotation
      C2. Nonfoaming adsorption flotation
      C3. Nonfoaming flotation thickening

28–36,55,56,78–83,89,92,101,106,107,111,112,118–124,130,138,142–147), storm water runoff treatment (19,61,104,105,149), toxic substance removal (20,21,31–34,57, 63,107), algae separation (22,23,48,87), odor substance stripping (24,107), bacteria separation (25,26,55,56), groundwater purification (27,109–111,133,150), industrial waste treatment (49–54,57,58,60–78,81,82,93,96–101,113–117,119,127,129,132,137,144), municipal primary and secondary clarification (40–47,85,86,91–95,108,126,135,136, 139,140,146), tertiary clarification (88,89,115), sludge thickening (3,13,37–39,59,90, 125, 141), deinking operation, mineral compounds separation (31,57,84,154), and so forth. While there are various adsorptive bubble separation processes technically available, dissolved air flotation is the most commonly used flotation process in industry and municipalities today. In particular, dissolved air flotation is gradually replacing conventional sedimentation processes for clarification. Recently, new sequencing batch reactors (SBR) involving the use of dissolved air flotation (DAF) instead of sedimentation have been developed (133,151–153). The new SBR-DAF can be either a biological process or a physicochemical process.

Adsorptive bubble separation processes can be classified by the technique used to generate the gas bubbles or by the technique used to separate impurities, as indicated in Table 1 (141,144,180–183). Each process system is briefly introduced here.

All adsorptive bubble separation processes in Table 1 involve the use of gas bubbles for separating substances from water. The gases can be air, nitrogen, carbon dioxide, and so on. The substances to be separated can be waste materials or useful industrial products and can be in soluble or insoluble form. Commonly, the processes are classified into five categories according to the techniques used for the production of selective gas bubbles: (a) dissolved gas system (such as dissolved air flotation if dissolved air is used): gas bubbles are generated from a supersaturated solution of a pressurized gas/liquid mixture by pressure release; (b) dispersed gas system (such as dispersed air flotation if air is dispersed): gas bubbles are generated by the diffusion of gas through porous media or the mixing/shearing action of propellers at atmospheric pressure; (c) vacuum system (such as vacuum air flotation) : gas bubbles are produced by saturating the water or wastewater with gas at atmospheric pressure, followed by application of a vacuum to the liquid; (d) electrolysis system (such as electroflotation): the gas used in this system usually consists of fine hydrogen bubbles and oxygen bubbles produced by the electrolysis of water; and (e) biological system (such as biological flotation): the gas bubbles, such as nitrogen and carbon dioxide, are produced in a biological nitrification and denitrification system. The selection of a feasible bubble—producing system depends on the cost of such bubble reactor and the characteristics of water or wastewater to be treated. Each system is fully described elsewhere (1,141,144,180–183). This chapter presents mainly the dissolved air flotation process.

It should be noted that if the density of the substances to be separated by flotation is less than that of water, no gas bubbles need to be generated in the separation reactor. In such a case, the process is termed "gravity flotation." Typical examples include oil/water separation, wax/water separation, and sticky/water separation by gravity flotation process (2,141,144).

Adsorptive bubble separation processes can also be classified into two main categories according to the techniques used for separating impurities or pollutants: (a) foam separation: the target floated substances are separated from bulk water in a foam phase; and (b) nonfoaming adsorptive bubble separation: the target floated substances are separated from bulk water near the water surface without foam. Foam separation can be further divided into two subcategories: (a) foam fractionation, in which the bulk water is a homogeneous solution containing negligible amounts of suspended solids; and (b) froth flotation, in which the bulk water is a heterogeneous water mixture containing significant amounts of suspended solids. Nonfoaming adsorptive bubble separation processes can be divided into three subcategories: (a) bubble fractionation, (b) solvent sublation, and (c) nonfoaming flotation. (144)

For potable water treatment using dissolved air bubbles, the process is termed dissolved air flotation in accordance with the techniques used for production of air bubbles, or termed nonfoaming flotation in accordance with the techniques used for separation of impurities.

It should be noted that a process system using dissolved air for solid/liquid separation can be "dissolved air flotation" according to the technique for generating fine gas bubbles, or "nonfoaming flotation" according to the technique for separating impurities. Similarly, a process system using dispersed air for solute/liquid separation can be "dispersed air flotation" as well as "froth flotation."

The process units can be round, square, or rectangular. In addition, gases other than air can be used. The petroleum industry has used nitrogen, with closed vessels, to reduce the possibilities of fire. Other gases used in dissolved gas flotation process include carbon dioxide and ozone. Therefore, there can be a "dissolved nitrogen flotation" process when nitrogen bubbles are used, a "dissolved carbon dioxide flotation" process when carbon dioxide bubbles are used, or a "dissolved air-ozone flotation" process when both air and ozone bubbles are involved.

In general, grease, light solids, grit, heavy solids, colloidal substances, and dissolved solutes can be removed all in one adsorptive bubble separation unit. The separation unit usually has high overflow rates and short detention periods, which mean smaller tank sizes resulting in less space requirements and possible savings in construction costs. Odor nuisance is minimized because of the short detention periods and the presence of dissolved oxygen in the clarified effluent (i.e., dissolved air containing oxygen is the most commonly used gas in the bubble separation process). In many cases, the recovered materials may be reusable as a source of fuel or recovered paper stock. For total odor control a flotation unit can be enclosed on the top with a sealed cover.

In a flotation system for solid/liquid separation, there are at least two methods by which gas bubbles can be used to increase the buoyancy of suspended solids: (a) entrapment of the bubbles in the particle structure; and (b) adhesion of the bubbles to the particle surface (see Fig. 1). In the former case, as the gas bubbles rise toward the surface, the controlled turbulence in the inlet compartment causes contact between the solids. The floc, formed by the natural floc-forming properties of the materials or by the chemicals that have been added, increases in size because of more contact with other solids. Eventually, a structure is formed that does not permit rising gas bubbles to pass through or around it. The buoyancy of the floc is continually increased as more gas bubbles are entrapped (and the floc grows), and ultimately the flocs are floated to the surface.

Adhesion of gas bubbles to the surface of the solids for solid/liquid separation results from interfacial tension arising from intramolecular forces that exist at an interface between two phases. At some point, the buoyant force of the suspended particles and the adhered gas bubbles is sufficient to float the particles.

In a foaming system or a fractionation system for solute/liquid separation, the gas bubbles are used to provide gas/liquid interfaces. The solute to be separated can be surface-active dissolved matter, or colloidal matter. A surface-active substance (i.e., surfactant) containing both hydrophilic and hydrophobic groups, reduces the surface tension at a liquid surface, and orients itself between two interfaces (such as a water–air interface or a water–solid interface) in such a way that it brings them into more intimate contact. The strong adsorption of surface-active solutes at surfaces or interfaces is in the form of an oriented monolayer. The solute to be removed is adsorbed on the water–air interfaces of the rising bubbles and carried upward to the top of the bubble reactor where the bubbles become foam, or break. Subsequent removal of the concentrated liquid layer from the top then allows effective removal of the solute. In summary, the solute/liquid separation in a bubble reactor is accomplished by surface adsorption, not by the buoyancy, in this particular case. How hydrophilic and hydrophobic solids will attach themselves to air bubbles in both dissolved air flotation and dispersed air flotation is shown in Fig. 2.

For water or wastewater containing both suspended solids and surface-active solutes, all the aforementioned mechanisms may be involved in an adsorptive bubble separation reactor.

1.2. Content and Objectives

This chapter briefly presents the historical development, principles, design, and applications of various adsorptive bubble separation processes, with emphasis on the

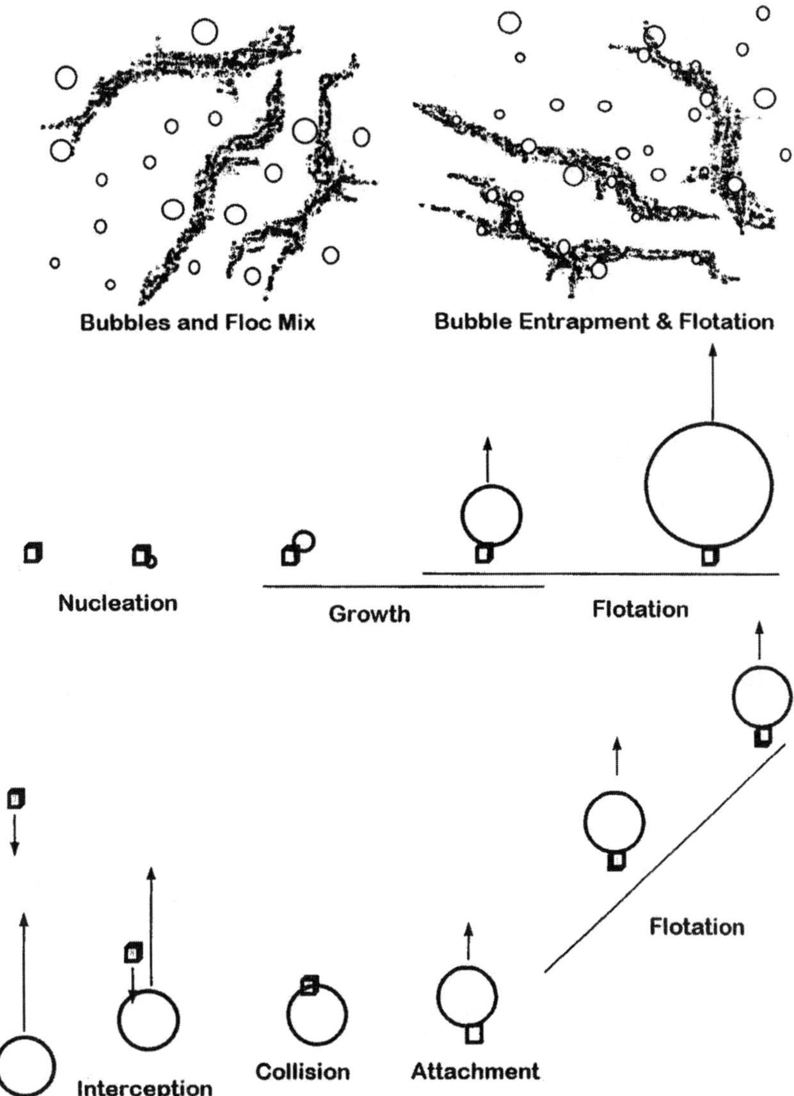

Fig. 1. Bubbles and flocs mixing, entrapment, and flotation (144).

most common dissolved air flotation systems. Bench-scale laboratory tests are useful for the treatability studies, troubleshooting, and determination of chemical dosages or flow parameters. Some test procedures and apparatus are recommended for researchers and practicing engineers. The industrial applications, site selection, cost estimation, installation, troubleshooting, maintenance, and pilot-plant operations of the well-established dissolved air flotation process units are introduced in detail.

2. HISTORICAL DEVELOPMENT OF CLARIFICATION PROCESSES

2.1. Conventional Sedimentation Clarifiers

The progress made during the last 50 yr in design, operation, and performance of dissolved air flotation clarifiers and sedimentation clarifiers are presented in this section

The Contact Angle (θ) is a measure of the relative size of the Cohesive Forces (F_C) that are pulling interfacial water molecules at the edge into the water and the Adhesive Forces (F_A) that are pulling these molecules towards the solid surface. The Resultant of the two forces will be perpendicular to the water/air interface.

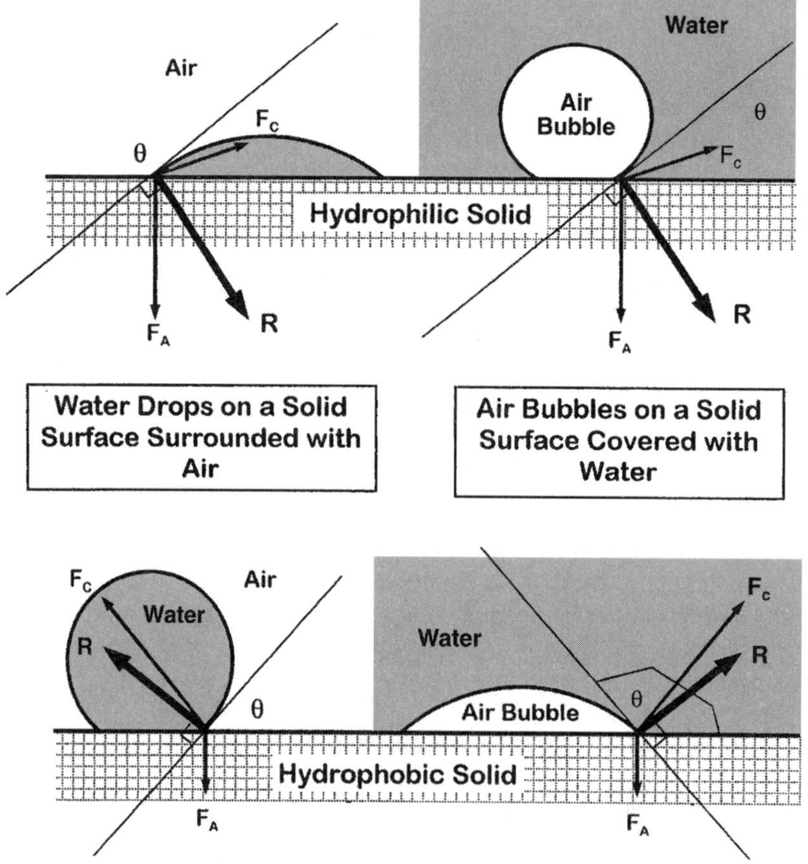

Fig. 2. Hydrophilic and hydrophobic solids and their attachment to bubbles (144).

for the purpose of process comparison. The development of sedimentation clarifiers has been very limited. Table 2 shows development of sedimentation clarifiers during the last 50 yr. From the very start, sedimentation operated with the maximum theoretical specific clarification. The normal settling velocity is approx 0.8 in./min (2 cm/min), which corresponds to 0.5 gpm/ft^2 (20 LPM/m^2). The specifications of modern settlers are still limited to overflow rate at 0.34–0.61 gpm/ft^2 and retention time at 60–120 min.

Because the settled sludge does not compact readily, the sludge bed must be kept rather deep, 39 in. (1 m) or more and this restricts the design of shallow sedimentation clarifiers. Improvements made are:

(a) Use of flocculating chemicals to improve clarification and compacting of settled sludge.

Table 2
Sedimentation Clarifiers—Brief History of Development (141,144)

Year	Type	Maximum capacity (gpm or MGD)	Rate (gpm/ft²)	Retention time (min)	Sludge consistency (%)
1900–Present	Round Central Shaft Scraper Drive	7,000 gpm 10 MGD	0.5	100	1–5
	Round With Bridge Scraper Drive	12,000 gpm 17 MGD	0.5	100	1–5
	Rectangular	500 gpm 0.7 MGD	0.5	100	1–5
	Rectangular High Volume	1800 gpm	0.5	400	3–10
	VOITH Kratzer	2.6 MGD			
	Settling Cone	410 gpm 0.6 MGD	0.5	130	1–5
1960	Settling Cone with bottom thickening	1050 gpm	0.5	130	3–10
	VOITH Purgator KROFTA Sedimentator	1.5 MGD			
1965	Lamella clarifiers A. Parkson B. Krofta Laminar Settler	1260 gpm 1.8 MGD	8.3	8	A. 1–5 B. 3–10
1970	Lamella top insert for conventional settling clarifiers	10,500 gpm 15 MGD	0.75	50	1–5

(b) In settling cones, bottom stirrers are installed for better thickening of the settled sludge or for preventing packing of the settled sludge, particularly in coating effluent clarification.
(c) Inclined lamella settling clarifiers have been developed, allowing a substantial specific load increase, but still requiring rather deep sludge hoppers. In clarification of fresh water with very low amounts of settled material these units are successful.
(d) Lamella packages can be mounted on the free surface of a settling tank with central shaft bottom scraper drive. In this way the specific clarification per surface can be increased by 50–100%. However, the settled sludge may require a still deeper sludge level. These surface inserts are applicable when the suspended solids in the raw water are low.

The above improvements in settling clarifiers are of rather marginal importance and the fact remains that sedimentation operates close to the theoretical limit without the possibility of improvement in space requirement, retention time, and sludge thickening.

2.2. *Innovative Flotation Clarifiers*

The general expression "flotation" is used for both dissolved air flotation and dispersed air flotation. Because this often creates confusion, Fig. 3 shows a brief comparison of the two flotation clarifiers and the sedimentation clarifier:

(a) Dissolved air flotation with air addition at 1% of water flow.
(b) Dispersed air (foaming) flotation with air addition at 400% of water flow.
(c) Sedimentation without air addition.

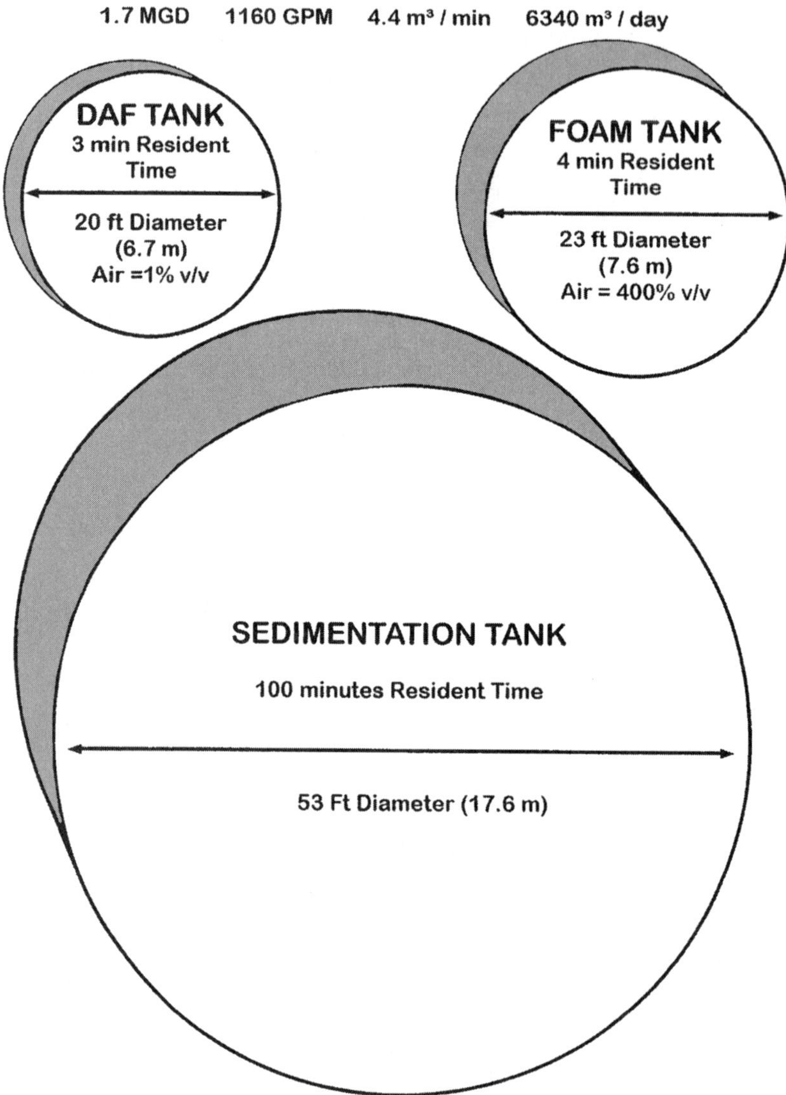

Fig. 3. Size comparison among dissolved air flotation, foam separation, and sedimentation (144).

Dispersed air flotation is successfully used in mineral separation techniques and in waste paper deinking processes for separation of ink from paper slurry. The applications of dispersed air flotation are different from that of dissolved air flotation, only the latter is gradually replacing conventional sedimentation processes for solid/liquid separation. Accordingly, this section emphasizes the comparison between dissolved air flotation clarifiers and sedimentation clarifiers (180–184).

Table 3 shows the evolution of dissolved air flotation (DAF) clarifiers during the last 50 yr. The following progress has been made:

(a) Specific clarification load increased from 1.5 gpm/ft^2 (60 LPM/m^2) to 5 gpm/ft^2 (210 LPM/m^2) and for triple stacked unit to 10 gpm/ft^2 (420 LPM/m^2)

Table 3
Dissolved Air Flotation Clarifiers—Brief History of Development (141,144)

Year	Type	Maximum capacity (gpm or MGD)	Rate (gpm/ ft^2)	Retention time (min)	Dissolved air type	Air dissolving tank retention time (s)
1920	SVEEN PETERSON	790 gpm 1.1 MGD	2.0	25	Full	60
1930–1935	ADKA SAVALLA	600 gpm 0.85 MGD	2.0	20	Vacuum	—
1948	KROFTA Unifloat ADKA Simplex KOMLINE SANDERSON	2500 gpm 3.8 MGD	2.0	20	Full Partial	60
1955	KROFTA Flotator	2800 gpm 4.0 MGD	4.0	20	Full	60
1965	KROFTA Sedifloat ADKA Standard INFILCO Carborundum	4700 gpm 6.6 MGD	2.0	40	Partial	60
1970	PERMUTIT Erpac	4000 gpm 5.8 MGD	3.0	12	Full Partial	60
1975	KROFTA Supracell	8,000 gpm 11.5 MGD	3.5	3	Partial Recycle	10
1993	KROFTA Sandfloat BP	20,000 gpm 28.8 MGD	5.0	5	Partial Recycle	10

(b) The retention time of water in the flotation clarifier decreased from 30 min to 3 min.
(c) The largest unit size increased from 260 gpm (1000 LPM) to 7900 gpm (30,000 LPM) and for triple stacked units to 23,700 gpm (90,000 LPM).
(d) The size of modern DAF units is much smaller. It allows construction predominantly in stainless steel, prefabricated for easy erection.
(e) The smaller size and weight 120 lb/ft^2 (60 kg/m^2) allows installation on posts leaving free passage under the unit. It is easier to find available space for indoor installation and to construct inexpensive housing.
(f) Air dissolving is improved and now requires only 10 s retention time in the air dissolving tube instead of the previous 60 s. This reduction in retention time results in smaller air-dissolving tubes, which are predominantly built from stainless steel.
(g) Availability of excellent flocculating chemicals gives a high stability of operation and high degree of clarification.

Presently the DAF clarifiers have not yet reached the theoretical limit of specific clarification. Normal flotation velocity is 12 in./min (30 cm/mm), corresponding to 7.5 gpm/ft^2 (300 LPM/m^2). The highest present specific clarification is 5 gpm/ft^2 (210 LPM/m^2) for normal operation.

In summary, modern dissolved air flotation (DAF) units with only 3 minutes of retention time can treat water and wastewater at an overflow rate of up to 5 gpm/ft^2 for a single unit, and up to 10.5 gpm/ft^2 for triple stacked units. The comparison between a DAF clarifier and a settler shows that (a) DAF floor space requirement is only 15 % of the settler; (b) DAF volume requirement is only 5% of the settler; (c) the degrees of

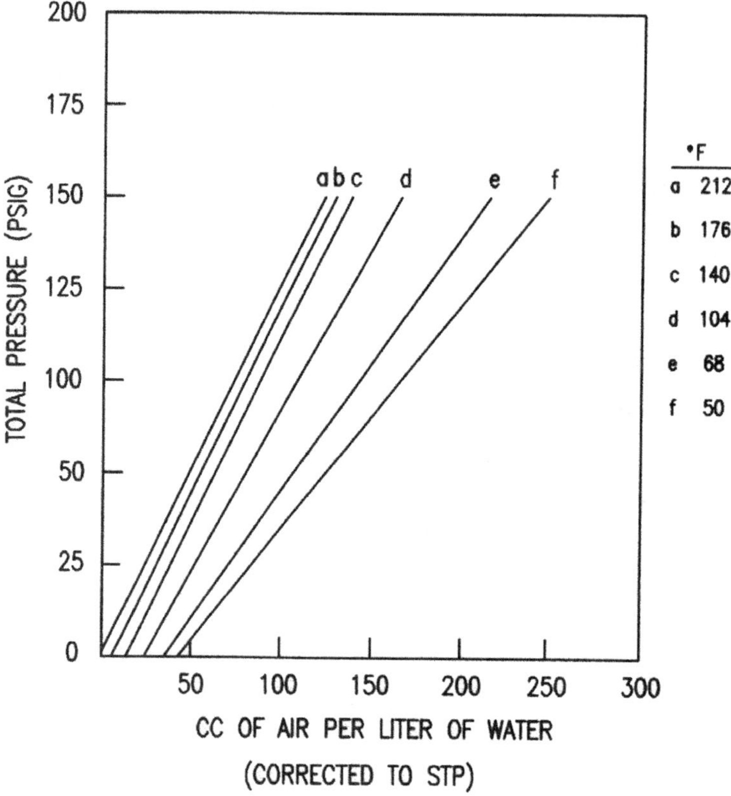

Fig. 4. Total pressure versus dissolved air (*Source:* TAPPI).

clarification of both clarifiers are the same with the same flocculating chemical addition; (d) the operational cost of the DAF clarifier is slightly higher than that for the settler, but this is offset by considerably lower cost of the installation's financing; and (e) DAF clarifiers are mainly prefabricated in stainless steel for erection cost reduction, corrosion control, better construction flexibility, and possible future changes, contrary to *in situ* constructed heavy concrete sedimentation tanks. Typical examples of industrial application of DAF clarifiers for treatment of industrial wastewater and municipal wastewater are presented later.

3. DISSOLVED AIR FLOTATION PROCESS

3.1. *Process Description*

Dissolved air flotation is used mainly to remove suspended and colloidal solids by flotation (rising) by decreasing their apparent density. The influent feed liquid can be raw water, wastewater, or liquid sludge. The flotation system consists of four major components: air supply, pressurizing pump, retention tank, and flotation chamber. According to Henry's law, the solubility of gas (such as air) in an aqueous solution increases with increasing the pressure (Fig. 4). The influent feed stream can be saturated at several times atmospheric pressure (25–90 psig) by a pressurizing pump. The pressurized feed stream is held at this high pressure for about 0.5–3.0 min in a retention tank (i.e., a pressure vessel) designed to provide sufficient time for dissolution of air into the stream to be treated. From the retention tank, the stream is released back to atmospheric pressure in the flotation chamber. Most of the pressure drop occurs after a pressure-reducing valve and in the transfer line between the retention tank and flotation chamber so that the turbulent effects of the depressurization can be minimized. The sudden reduction in pressure in the flotation chamber results in the release of microscopic air bubbles (average diameter 80 μm or smaller), which attach themselves to suspended or colloidal particles in the process water in the flotation chamber. This results in agglomerations, which, owing to the entrained air, give a net combined specific gravity less than that of water, or cause the flotation phenomenon. The vertical rising rate of air bubbles ranges between 0.5 and 2.0 ft/min. The floated materials rise to the surface of the flotation chamber to form a floated layer. Specially designed flight scrapers or other skimming devices continuously remove the floated material. The surface sludge layer can in certain cases attain a thickness of many inches and can be relatively stable for a short period. The layer thickens with time, but undue delays in removal will cause a release of particulates back to the liquid. Clarified water (effluent) is usually drawn off from the bottom of the flotation chamber and either recovered for reuse or discharged. Figures 5A–5C illustrate three dissolved air flotation systems.

The retention time in the flotation chambers is usually about 3–60 min depending on the characteristics of process water and the performance of a flotation unit. The process effectiveness depends on the attachment of air bubbles to the particles to be removed from the process water. The attraction between the air bubbles and particles is primarily a result of the particle surface charges and bubble-size distribution. The more uniform the distribution of water and microbubbles, the shallower the flotation unit can be. Generally, the depth of effective flotation units is between 1 and 9 ft.

3.2. *Process Configurations*

The three common flotation system configurations are (a) full flow pressurization, (b) partial flow pressurization without effluent recycle, and (c) recycle flow pressurization, which are graphically illustrated in Figs. 5A, 5B, and 5C, respectively.

In the full flow pressurization system (Fig. 5A), the entire influent feed stream is pressurized by a pressurizing pump and held in the retention tank. The system is usually applicable to the feed stream with suspended solids exceeding 800 mg/L in concentration, and not susceptible to the shearing effects caused by the pressurizing pump and the high pressure drop at the pressure release valve. It is occasionally used for separating some discrete fibers and particles, which require a high volume of air bubbles. It is particularly feasible for solid–water separation where suspended solids will flocculate rapidly with the addition of chemical coagulants in the inlet compartment in the presence of the released air. The air bubbles may become entrapped within the floc particles resulting in a strong air to solids bond, thus in a highly efficient separation process.

In the partial flow pressurization without effluent recycle system (Fig. 5B), only about 30–50% of the influent feed stream is pressurized by a high-pressure pump and held in the retention tank. The remaining portion of influent stream is fed by gravity or a low-pressure pump to the inlet compartment of the flotation chamber where it mixes with the pressurized portion of the influent stream. Materials with low specific gravity can be removed with the partial flow pressurization system. This system is not

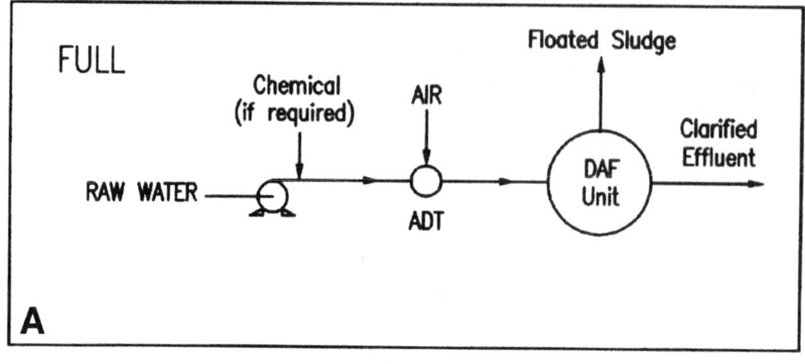

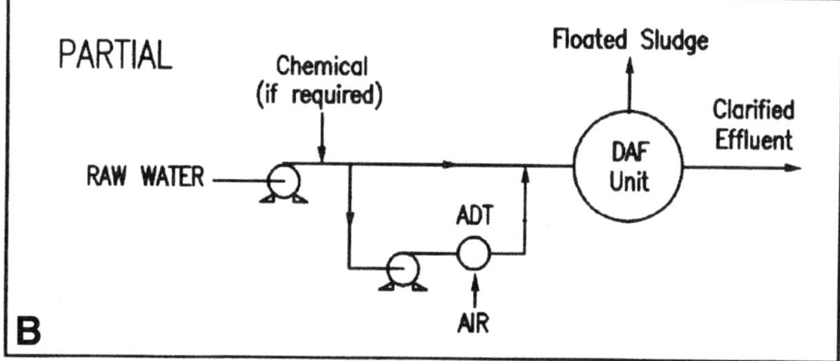

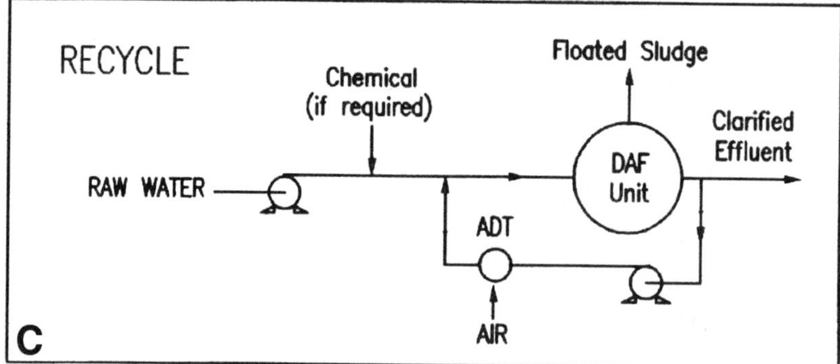

Fig. 5. Operational modes of dissolved air flotation (141).

recommended for use when the suspended solids are susceptible to the shearing effects of the pressurizing pump and the high-pressure drop at the pressure release valve. It is generally employed in applications where the suspended solids concentrations are low, resulting in lower air requirement, and, in turn, lower operation and maintenance costs.

In the recycle flow pressurization system (Fig. 5C), a portion (15–50%) of the clarified effluent from the flotation chamber is recycled, pressurized, and semisaturated with air in the retention tank. The recycled flow is mixed with the unpressurized main influent stream just before admission to the flotation chamber, with the result that the air bubbles come out of the aqueous phase in contact with suspended particulate matter at the inlet compartment of the flotation chamber. The system is usually employed in applications where preliminary chemical addition and flocculation are necessary ahead of flotation. It eliminates the problems with shearing the floc particles as only clarified effluent passes through the pressurizing pump and the pressure release valve. It should be noted, however, that the increased hydraulic flow on the flotation chamber due to the flow recirculation must be taken into account in the flotation chamber design.

While all the aforementioned three system configurations can be used for sludge (or fiber) separation, only the recycle flow pressurization system and partial flow pressurization system are recommended for water purification or wastewater treatment.

3.3. *Factors Affecting Dissolved Air Flotation*

There are many factors affecting a dissolved air flotation system. Some factors have been discussed in Sections 3.1 and 3.2, and some will be presented in detail in the following sections. This particular section briefly summarizes the most important DAF process factors.

3.3.1. *Nature of the Particles*

The specific gravity is a characteristic of the particle or liquid to be abated or separated. It can easily be accepted that sand, for example, cannot be floated while voluminous material, such as activated sludge, or a water immiscible liquid such as oil, can be floated. Surface-active substances can be separated by foam separation. Bubbles can strip off volatile substances.

3.3.2. *Size of Particles*

Minutely small particles, particularly of the granular high-specific-gravity nature, cannot be floated. Generally, floatability increases with the size of the particle. In many cases, the size of particles can be increased by flocculation with various chemical coagulants.

3.3.3. *Dispersing Agents*

Certain wastewaters and liquids contain unusual concentrations of various chemicals, resulting in specific flotation problems or advantages. Surfactants, such as detergents, tend to alter the physical properties of the sludge particle surface to be floated. The quantity and type of surfactant present in the influent may cause a variation (either positive or negative) in flotation results.

3.3.4. *Composition and Nature of the Influent*

The composition and nature of the influent is most important. Equalization of composition and flow improves the performance of the flotation unit.

3.3.5. *Liquid Currents*

The liquid currents are governed by the physical design and hydraulics of the flotation unit. This becomes a consideration in the design of the tank and hydraulic loadings of the flotation unit.

3.3.6. *Air to Solids (A/S) Ratio*

The amount of air and the method of mixing the air with the material to be floated are functions of the design of a particular flotation unit. For a specific application, a definite amount of air is necessary for flotation. In thickening applications it has been shown that increased performance is obtained at higher A/S ratios.

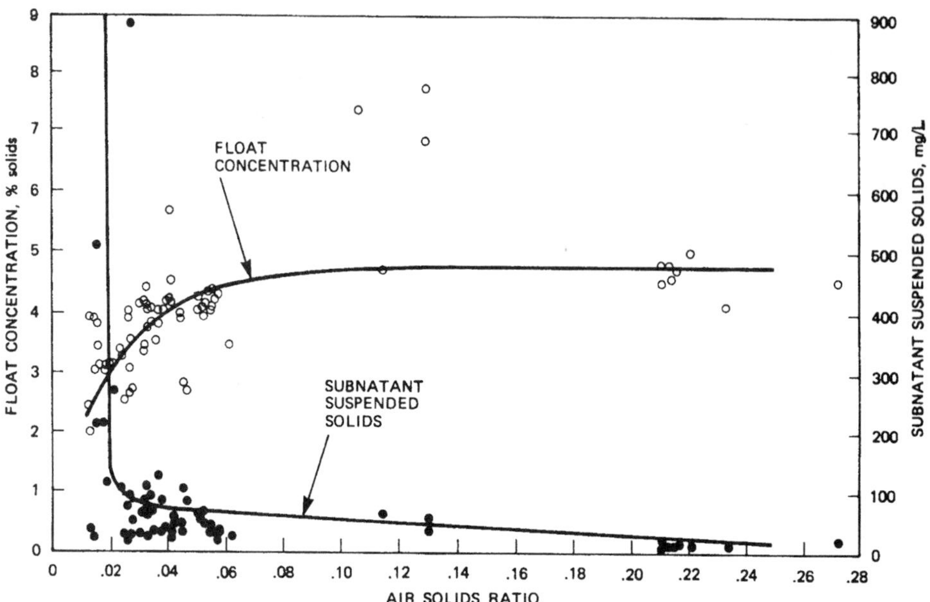

Fig. 6. Effects of air–solids ratio on float concentration and subnatant suspended solids (*Source:* US EPA).

3.3.7. *Float Removal*

A float-removal mechanism must be designed to have adequate capacity to remove water carryover. Various items to be considered in this design are the depth of submergence of the scooping mechanism and the speed of scoop operation.

4. DISSOLVED AIR FLOTATION THEORY

4.1. *Gas-to-Solids Ratio of Full Flow Pressurization System*

The performance of a dissolved air flotation (DAF) unit will be mainly dependent on the ratio of the amount of gas to the amount of suspended solids applied to the unit. Figure 6 shows the effect of the air-to-solids ratios (A/S) on the percentage concentration of floated solids and the effluent (subnatant) suspended solid concentration. It is clear that increasing the A/S ratio beyond an optimum value results in no significant increase in the performance efficiency of a DAF unit. Figure 7 shows the effect of hydraulic loading rate on subnatant chemical oxygen demand (COD) (131,183).

The mass ratio of gas to solids can be derived from gas concentrations, solid concentrations, pressure, and flow parameters. In the case of full flow pressurization system (Fig. 5A), the mass flow rate of dissolved gas entering the flotation chamber is

$$G_{\text{in}} = QC_r \tag{1}$$

where G_{in} = mass flow rate of dissolved gas entering the flotation chamber (mg/s), Q = flow rate of influent feed stream (L/s), and C_r = solubility of gas in water in the pressurized retention tank (mg/L). The dissolved gas in the liquid leaving the flotation chamber is

$$G_{\text{out}} = QC_e \tag{2}$$

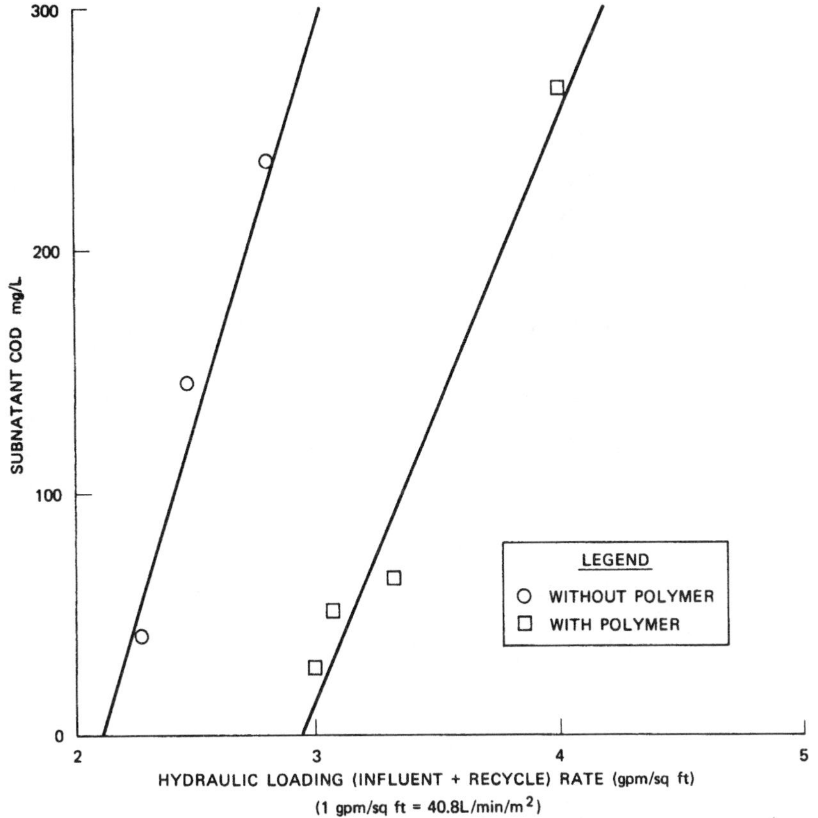

Fig. 7. Effect of hydraulic loading rate on subnatant COD (*Source:* US EPA).

where G_{out} = mass flow rate of dissolved gas leaving the flotation chamber (mg/s) and C_e = solubility of gas in the flotation effluent (mg/L). The gas released for flotation of suspended solids (G, mg/s) is given by

$$\begin{aligned} G &= G_{in} - G_{out} \\ &= Q(C_r - C_e) \end{aligned} \tag{3}$$

At low dissolved gas concentrations where Henry's law is valid, the dissolved gas concentrations at saturation are proportional to gas pressures (131):

$$C_r / C_e = P_r / P_e = P/1 \tag{4}$$

where P_r = gas pressure in the retention tank, atmosphere, assigned to be P, and P_e = gas pressure in the flotation chamber's effluent compartment, atmosphere, taken to be 1 atm. The released gas can be expressed in terms of pressure as

$$G = QC_e(P-1) \tag{5}$$

A correction factor, F, can be applied to the pressure term because complete gas saturation of liquid is often not achieved in a pressurized retention tank:

$$G = QC_e(FP-1) \tag{6}$$

$$G = QC_e(f)(P-1) \tag{6a}$$

where f = factor of gas dissolution at pressure P (where P is any pressure above normal atmospheric pressure of 1 atm), fraction, usually 0.167–1.0, and F = Fraction of gas dissolution at pressure P (where P is any pressure higher than 2 atm), fraction, usually 0.5–1.0.

Equation (6) has successfully been used by design engineers since 1970s for high pressure systems, such as DAF, in which P is higher than 2 atm (102,103,131,141,144). Equation (6) , however, is invalid for a lower pressure system in which P is below 2 atm. Recently Dr. William Selke and his colleagues (179) have suggested a new factor (f) and Eq. (6a) to cover any pressure ranges above normal atmospheric pressure of 1 atm.

The mass flow rate of suspended solids entering the flotation system is

$$S = QX \tag{7}$$

where S = mass flow rate of suspended solids entering the flotation system (mg/s) and X = suspended solid concentration of influent feed stream (mg/L). The gas to solids ratio is thus given by

$$G/S = C_e(FP-1)/X \tag{8}$$

$$G/S = C_e(f)(P-1)/X \tag{8a}$$

If the gas to be pressurized is air, the air-to-solids ratio for the full flow pressurization system can be derived as

$$A/S = 1.3\ a(FP-1)/X \tag{9}$$

$$A/S = 1.3\ a\ (f)(P-1)/X \tag{9a}$$

where A/S = air-to-solids ratio, A = mass flow rate of air released for flotation of suspended solids (mg/s), S = mass flow rate of suspended solids entering the flotation system (mg/s), a = air solubility in effluent at 1 atm pressure (mL/L), 1.3 = weight in milligrams of 1 mL of air, 1 = 1 atm of air remaining in solution after depressurization, f = factor of gas dissolution at pressure P (where P is any pressure above normal atmospheric pressure of 1 atm), fraction, usually 0.167–1.0, and F = fraction of gas dissolution at pressure P (where P is any pressure higher than 2 atm), fraction, usually 0.5–1.0.

4.2. Gas-to-Solids Ratio of Partial Flow Pressurization System

In the case of partial flow pressurization system (Fig. 5B), the dissolved gas entering the flotation chamber is

$$G_{\text{in}} = Q_p C_r + Q_n C_f \tag{10}$$

where Q_p = portion of influent feed stream which is pressurized (L/s), Q_n = portion of influent feed stream which is not pressurized (L/s), and C_f = dissolved gas concentration in the raw influent feed stream, mg/L

The dissolved gas out of the flotation chamber is

$$G_{\text{out}} = (Q_p + Q_n)C_e = QC_e \tag{11}$$

Assuming that at 1 atm

$$C_f = C_e \tag{12}$$

The gas released to float suspended solids is

$$\begin{aligned} G &= G_{\text{in}} - G_{\text{out}} \\ &= Q_p\left(C_r - C_e\right) \end{aligned} \tag{13}$$

Assuming dissolved gas concentrations are low and Henry's law is valid, then applying the correction factor, F, the mass flow rate of released gas is

$$G = Q_p C_e (FP - 1) \tag{14}$$

$$G = Q_p C_e (f)(P - 1) \tag{14a}$$

The gas-to-solids ratio for the partial flow pressurization system (Fig. 5B) is

$$G/S = Q_p C_e (FP - 1)/QX \tag{15}$$

$$G/S = Q_p C_e (f)(P - 1)/QX \tag{15a}$$

The Q_p/Q ratio ranges between 0.3 and 0.5. (131).

If the gas to be pressurized is air, the air-to-solids ratio for the partial flow pressurization system becomes

$$A/S = 1.3\,a\,Q_p (FP - 1)/QX \tag{16}$$

$$A/S = 1.3\,a\,Q_p (f)(P - 1)/QX \tag{16a}$$

4.3. *Gas-to-Solids Ratio of Recycle Flow Pressurization*

In the case of recycle flow pressurization system (Fig. 5C), the dissolved gas entering the flotation chamber is

$$G_{\text{in}} = QRC_r + QC_f \tag{17}$$

where R = recirculation ratio = Q_r/Q, Q_r = recycle flow (L/s), and C_f = dissolved gas concentration in influent feed stream (mg/L)

The dissolved gas out of the flotation chamber is

$$G_{\text{out}} = Q(1 + R)C_e \tag{18}$$

Assuming that at 1 atm

$$C_f = C_e \tag{19}$$

The gas released to float suspended solids is

$$G = G_{\text{in}} - G_{\text{out}} \tag{20}$$

Assuming dissolved gas concentrations at saturation are proportional to pressure, or follow Henry's law (131), and applying the correction factor, F or f, the mass flow rate of release gas is

$$G = QRC_e (FP - 1) \tag{21}$$

$$G = QRC_e (f)(P - 1) \tag{21a}$$

Table 4
Air Characteristics and Solubilities[a] (141, 144)

Temperature		Volume Solubility		Weight Solubility		Density	
ºC	ºF	mL/L	ft^3/1000 gal	mg/L	lb/1000 gal	g/L	lb/ft^3
0	32	28.8	3.86	37.2	0.311	1.293	0.0808
10	50	23.5	3.15	29.3	0.245	1.249	0.0779
20	68	20.1	2.70	24.3	0.203	1.206	0.0752
30	86	17.9	2.40	20.9	0.175	1.166	0.0727
40	104	16.4	2.20	18.5	0.155	1.130	0.0704
50	122	15.6	2.09	17.0	0.142	1.093	0.0682
60	140	15.0	2.01	15.9	0.133	1.061	0.0662
70	158	14.9	2.00	15.3	0.128	1.030	0.0643
80	176	15.0	2.01	15.0	0.125	1.000	0.0625
90	194	15.3	2.05	14.9	0.124	0.974	0.0607
100	212	15.9	2.13	15.0	0.125	0.949	0.0591

[a]Values presented in absence of water vapor and at 14.7 psia pressure.

The mass flow rate of suspended solids (S) is again given by QX, the gas to solids ratio with effluent recycle is

$$G/S = RC_e(FP-1)/X \tag{22}$$

$$G/S = RC_e(f)(P-1)/X \tag{22a}$$

If the gas to be pressurized is air, the air-to-solids ratio for the recycle flow pressurization system can then be derived as:

$$A/S = 1.3a\,R(FP-1)/X \tag{23}$$

$$A/S = 1.3a\,R(f)(P-1)/X \tag{23a}$$

$$A/S = 1.3a(Q_r/Q)(FP-1)/X \tag{23b}$$

$$A/S = 1.3a(Q_r/Q)(f)(P-1)/X \tag{23c}$$

4.4. *Air Solubility in Water at 1 Atm*

The air solubility is a function of water temperature, and can be determined according to the following table (131, 144):

Temperature (ºC)	Air solubility (mL/L)
0	28.8
10	23.5
20	20.1
30	17.9

More detailed air solubility data can be found from Table 4, which was prepared by Boyd and Shell (4).

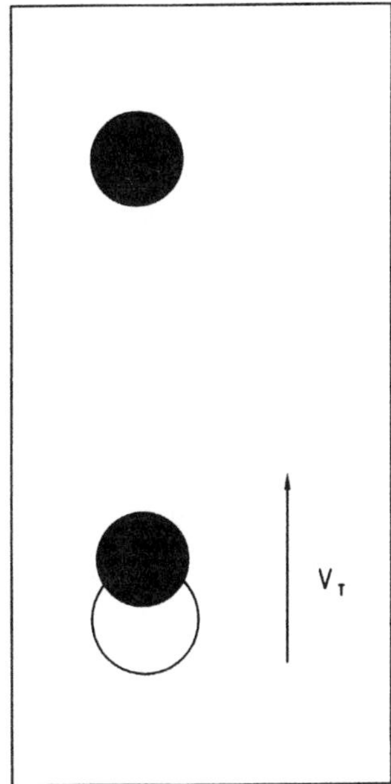

Fig. 8. Separation of particle from wastewater by DAF (*Source*: US EPA).

4.5. *Pressure Calculations*

The gage pressure of the pressurized retention tank can be controlled and readily read. The following are two equations for converting the gage pressure to the atmospheric pressure.

$$P_{at} = (P + 14.7)/14.7 \tag{24}$$

$$P_{at} = (P' + 101.35)/101.35 \tag{25}$$

where P_{at} = pressure (atm), P = gage pressure (psig), and P' = gage pressure (kPa).

4.6. *Hydraulic Loading Rate*

As discussed previously, dissolved air flotation is a process for removing suspended matter from a water, wastewater, or sludge stream by means of minute air bubbles, that upon attachment to a discrete particle reduce the effective specific gravity of the aggregate particle to less than that of water. Reduction of the specific gravity for the aggregate particle causes separation from the carrying liquid in an upward direction. As Fig. 8 suggests, the particle to be removed may have a natural tendency either to rise or to settle. Attachment of the air bubble to the particle induces a vertical rate of rise (m/s) noted as V_T.

Figure 9 illustrates the basic design considerations of the flotation unit. The measurement of V_T will be discussed later. Because the influent feed stream must pass through the flotation chamber, the particle to be removed will have a horizontal velocity. Certain criteria have been established for limits of the parameter V_H, which sets the width and depth of the flotation chamber:

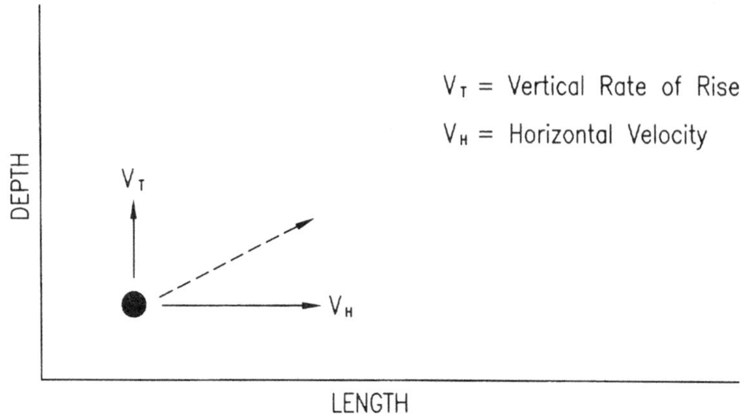

Fig. 9. Basic design concept of flotation unit (*Source*: US EPA).

$$V_H = Q/A_c \tag{26}$$

where V_H = horizontal velocity (m/s), Q = influent flow rate (m^3/s), and A_c = cross-sectional area of a flotation chamber (m^2).

Figure 9 shows that the effective length (L) of the flotation chamber is directly proportional to the horizontal velocity and depth and inversely proportional to the vertical rate of rise of the particle to be removed. In the design of a flotation chamber, the usual procedure is to select the target suspended solids to be removed with a rise rate of V_T', and design the chamber so that all suspended solids that have a rise rate equal to or greater than V_T will be separated. The suspended solids must have sufficient rise velocity to travel the effective depth (the distance from the bottom to the water surface of the flotation chamber) within the detention time in order to be floated (131,183). That is, the rise rate V_T must be at least equal to the effective depth divided by the detention time, or equal to the flow divided by the surface area:

$$V_T = D/T = Q/A_s \tag{27}$$

where V_T = vertical rise rate of suspended solids (m/s), D = effective depth of the flotation chamber (m), T = detention time (s), Q = influent flow rate (m^3/s), and A_S = surface area of flotation chamber (m^2)

The ratio of Q/A_s is also defined as the hydraulic loading rate, which is another very important design parameter. Theoretically, any particles having a rise rate equal to or greater than the hydraulic loading rate will be removed in an ideal flotation chamber. Figure 7 shows that practically the separation efficiency of COD in a dissolved air flotation unit is a function of hydraulic loading rate. Figure 10 shows the effects of solids loading rate on float concentration and subnatant suspended solids.

In practical design, the rise rate (V_T) of suspended solids to be floated can be measured in the laboratory or in the field, and the influent feed rate Q is generally known. The minimum required surface area (A_s) of a flotation chamber can then be determined according to Equation (27). The effects of short circuitry and turbulence in the flotation chamber, which interfere with the suspended solids rising through the water, should also be considered. Assuming the flotation chamber is rectangular in shape, the width (W)

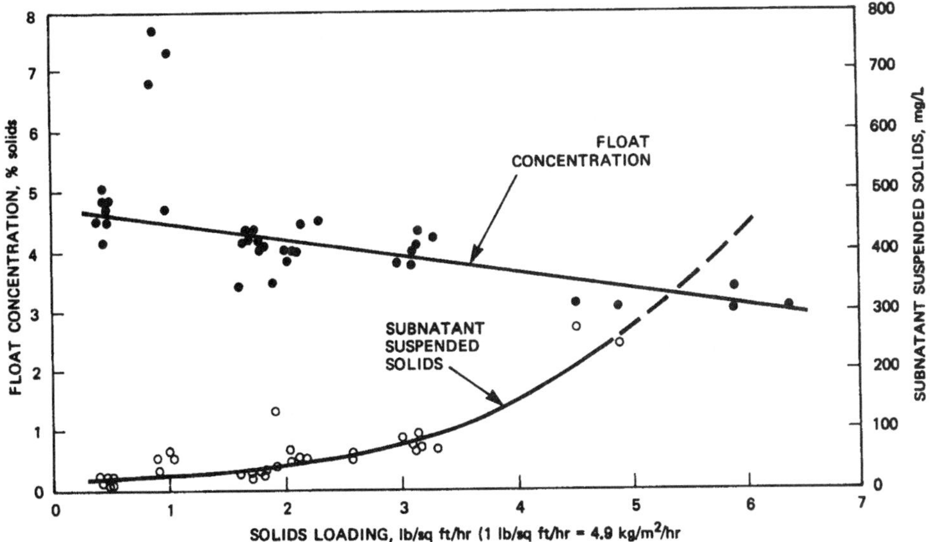

Fig. 10. Effects of solids loading rate on float concentration and subnatant suspended solids (*Source*: US EPA).

and effective length (L) of a flotation chamber can then be determined by Eq. (28) and (29), respectively. The D/W ratio is usually between 0.3 and 0.5 (131,144):

$$W = A_c / D = A_s / L \tag{28}$$

$$L = (A_s / W)F' = (V_H / V_T)F'D \tag{29}$$

where W = width of flotation chamber (in.), L = effective length of flotation chamber (in.), and F' = factor for short circuiting and turbulence, assumed as 1.4.

4.7. Solids Loading Rate

In the flotation systems that contain low concentrations of suspended solids, only "free flotation" occurs. In free flotation, the suspended solids near the bottom of a flotation chamber rise freely toward the surface, and the floated suspended solids near the surface will not continue to compress with time. In "compression flotation," the suspended solids concentration usually is high. Initially the suspended solids also rise freely toward the surface forming a scum layer. As flotation continues, the floated suspended solids near the surface accumulate and the scum layer continues to compress with time therefore, compression flotation permits the production of a high scum (or float) concentration for ease of handling and disposal. Figure 11 shows that there are three different kinds of technologies for dissolving air into water: (a) deep-shaft air dissolving technology (94,95); (b) conventional air dissolving technology in a pressure tank (144); and (c) innovative air dissolving technology in an air-dissolving tube using aspirated air (141,144,152).

The flotation phenomenon that occurs when a concentrated suspension, initially of uniform concentration throughout, is placed in a graduated cylinder, may be observed as shown in Fig. 12.

Because of the hydraulic characteristics of flow around the particles and other interparticle forces, the particles float as a zone, maintaining the same relative position with

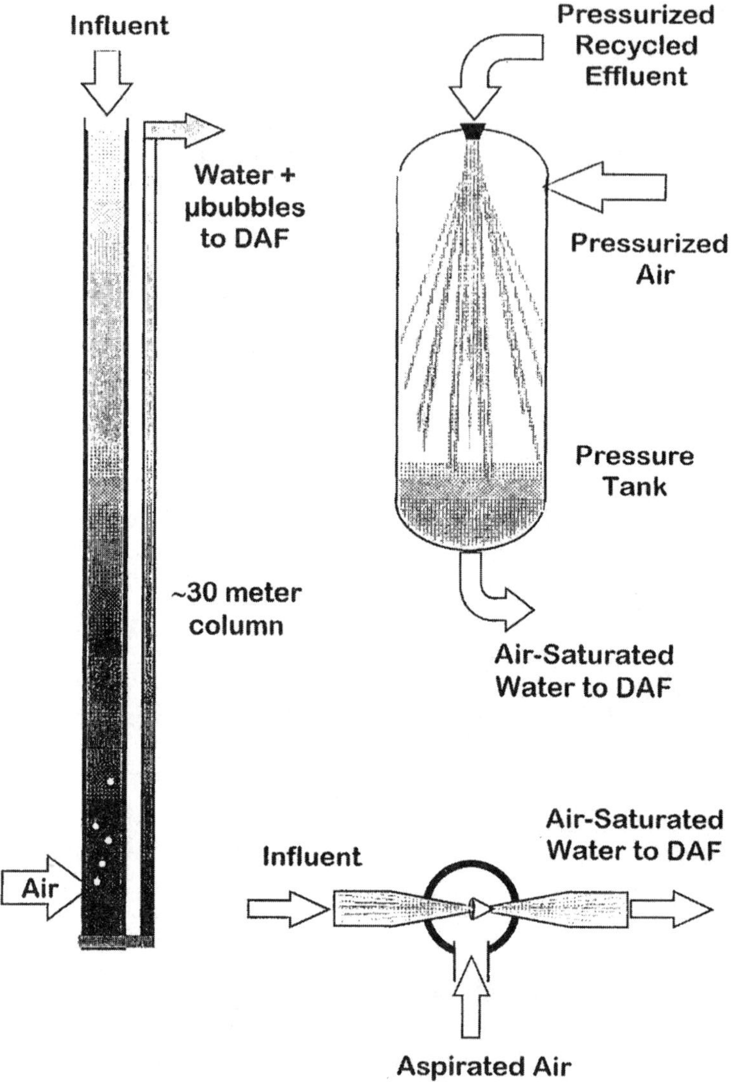

Fig. 11. Three technologies for dissolving air into water (144).

respect to each other. As this region floats, a volume of relatively clear water is produced below the zone flotation region. Particles remaining in this region float as discrete or flocculated particles. A distinct interface exists between the discrete flotation region and the hindered-flotation region shown in Fig. 12. The rate of flotation in the hindered-flotation region is a function of the concentration of solids and their condition. As flotation continues, a compressed layer of particles begins to form on the top of the cylinder in the compression-flotation region. The particles in this region now apparently form a structure in which there is physical contact between particles. The forces of physical interaction between the particles are especially strong.

In design of a flotation system for sludge thickening, the overflow rate determination should be based on three factors: (a) the area needed for free flotation in the discrete

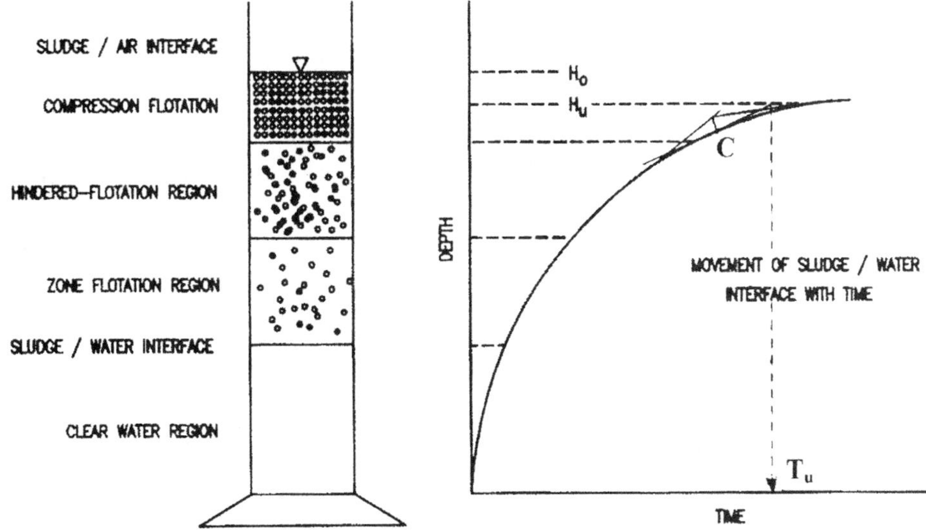

Fig. 12. Schematic of flotation regions for solid/liquid separation.

flotation region; (b) the area needed on the basis of the rate of floating of the interface between water and sludge; and (c) the rate of floated sludge withdrawal from the compression region.

The area requirement for compression flotation may be calculated by the following method. Initially the flotation of a suspension is observed, and the position of the solid/liquid interface versus time is plotted as shown on Fig. 12. The rate at which the interface rises is then equal to the slope of a tangent to the curve at that point in time. According to the procedure, the critical area for flotation thickening is given by the following equation:

$$A_s = Q(T_u)/H_0 \tag{30}$$

where Q = influent flow rate (m^3/s), A_s = surface area required to reach desired sludge consistency (m^2), T_u = time to reach desired sludge consistency (s), and H_0 = height of floated sludge (float) in test cylinder (m).

The critical concentration controlling the sludge thickening capacity of dissolved air flotation is point *C* on Fig. 12. This point is determined by extending the tangents to the free and compression regions of the curve, intersecting the tangents, and bisecting the angle formed. The time T_u can be determined as follows: (a) construct a horizontal line at the height that corresponds to the desired float consistency; (b) construct a tangent to the curve at point *C*; and (c) construct a vertical line from the point of intersection of those two lines to the time axis.

5. DESIGN, OPERATION, AND PERFORMANCE

5.1. *Operational Parameters*

Dissolved air flotation units are typically designed based on air-to-solids ratio, hydraulic loading rate (or overflow rate), air pressure, effluent recycle, solids loading, etc. The air-to-solids ratio can be determined according to Section 4 and is defined as the ratio of air feed to dry suspended solids feed by weight. The density of air is approx 0.08 lb/ft^3 (see Table 4). The air-to-solids ratio is known when both air flow rate and influent suspended solids load are known (141):

$$A/S = 0.08Q_a/w = dQ_a/w \tag{31}$$

where d = air density (lb/ft^3) (Table 4), Q_a = air flow rate (cfm), and w = influent suspended solids load (lb/min).

The air-to-solids ratio is important because it affects the sludge rise rate. The air-to-solids ratio needed for a particular application is a function primarily of the sludge's characteristics such as sludge volume index (SVI). The most common ratio used for design of a waste activated sludge thickener is 0.03.

Hydraulic loading rate (overflow rate) is the flow rate through the flotation chamber divided by the liquid surface area normally expressed in GPD/ft^2; and must be adequately controlled. Air pressure in flotation is important because it determines air saturation or the size of the air bubbles formed. It influences the degree of solids concentration and the subnatant (separated water) quality. In general, an increase in either pressure or air flow produces greater float (solids) concentrations and a lower effluent suspended solids concentration. There is an upper limit, however, as too much air will tend to break up floc particles.

Additional recycle of clarified effluent does two things: (a) it allows a larger quantity of air to be dissolved because there is more liquid; and (b) it dilutes the feed sludge. Dilution reduces the effect of particle interference on the rate of separation. The concentration of sludge increases and the effluent suspended solids decrease as the sludge blanket detention period increases.

The solid loading is the dry weight of suspended solids per unit time per square foot of flotation surface area. This is normally expressed as lb dry suspended solids per h per square foot of surface area. The gas or air released at atmospheric pressure at one atmospheric pressure can be estimated from Eqs. (32) and (33):

$$C_g = C_{gs}(FP - 14.7)/14.7 \tag{32}$$

$$C_g = C_{gs}(f)(P - 14.7)/14.7 \tag{32a}$$

$$C_a = C_{as}(FP - 14.7)/14.7 \tag{33}$$

$$C_a = C_{as}(f)(P - 14.7)/14.7 \tag{33b}$$

where C_g = gas released at one atmospheric pressure (mg/L), C_{gs} = gas saturation concentration at atmospheric conditions (mg/L), C_a = air released at 1 atm pressure (mg/L), C_{as} = air saturation concentration at atmospheric conditions (mg/L), P = absolute pressure in air dissolving tank (psig), F = system gas dissolving efficiency, fraction, usually 0.5–1.0, and f = system gas dissolving efficiency, fraction, usually 0.167–1.0.

Table 4 lists the solubility of air in water with varying temperature. The values listed may be used in Eq. (33) for the design of a dissolved air flotation system. Similar values of C_{as} for other gases are available in standard textbooks. It should be noted that an appropriate correction factor should be used when applying the data of Table 4 to wastewater or sea water containing a high salt concentration. For instance, less than 80% of the air soluble in pure water is soluble in sea water.

Table 5
Dissolved Air Flotation Operation and Performance (141,144,183)

Operation Parameter	Range
Solids loading, lb dry solids/hr/ft^2 of surface	
with chemicals	2–5
without chemicals	1–2
detention time (min)	3–60
Air-to-solids ratio	0.01–0.1
Blanket thickness (in.)	1–24
Retention tank pressure (psi)	25–70
Recycle ratio (% of influent flow)	5–50
Hydraulic loading rate	
gpd/ft^2	2880–7200
gpm/ft^2	2–5
Solids removal efficiency	
with flotation aid (% removal)	95
without flotation aid (% removal)	50–80
Float solids concentration (%)	2–10
Oil and grease removal efficiency	
with flotation aid (% removal)	85–99
without flotation aid (% removal)	60–80

5.2. *Performance and Reliability*

Typical operation and performance parameters are shown in Table 5 and were developed from governmental and manufacturer's operation manuals. Operating parameters for some actual flotation plants are shown in Table 6. Dissolved air flotation systems are very reliable from a mechanical standpoint. Variations in influent feed characteristics can affect process reliability, and may require operator's attention. Chemical pretreatment is essential.

6. CHEMICAL TREATMENT

To obtain optimum treatment with some raw water or wastes, it is necessary to use chemical pretreatment before dissolved air flotation. The necessity for use of chemical conditioning is normally associated with (a) a high degree of emulsification of oil or grease matter in waste stream flow, thus a requirement to break the emulsion and form a floc to absorb the oil or grease; (b) a high concentration of stable colloids in raw water, thus a requirement to destabilize the colloids for floc generation; and (c) a finely divided small suspended solids in a liquid sludge stream, thus a requirement to increase particle size or to render the particles more hydrophobic. It has been known (141) that increasing the particle size increases the rate of separation. Flocculation as a means of promoting particle growth preceding flotation contributes to the effectiveness of the flotation process where chemical conditioning is used. The points of chemical injection and the possible use of flocculation associated with the three methods of air injection are shown in Figs. 5A–5C.

Common chemicals used as flotation aids include alum, ferric chloride, inorganic polymers, and organic polymers. They are added for promoting flocculation, in turn, inducing the development of buoyant forces by the methods described in Section 1.

Table 6
DAF Operating Conditions and Performance for Thickening of Biological Settled Sludge (SS)

Location	Feed	Influent SS (mg/L)	Subnatant SS (mg/L)	Removal (% SS)	Float (% solids)	Loading (lb/hr/ft^2)	Flow (gpm/ft^2)	Remarks
Bernardsville, NJ	M.L.[a]	3,600	200	94.5	3.8	2.16	1.2	Standard[c]
Bernardsville, NJ	R.S.[b]	17,000	196	98.8	4.3	4.25	0.5	Standard
Abington, PA	R.S.	5,000	188	96.2	2.8	3.0	1.2	Flotation aid[d]
					6.0			After 12 hrs
Hatboro, PA	R.S.	7,300	300	96.0	4.0	2.95	0.8	Flotation aid
Morristown, NJ	R.S.	6,800	200	97.0	3.5	1.70	0.5	Standard
Omaha, NB	R.S.	19,660	118	99.8	5.9	7.66	0.8	Flotation aid
					8.8			After 24 hrs
Omaha, NB	M.L.	7,910	50	99.4	6.8	3.1	0.8	Flotation aid
Belleville, IL	R.S.	18,372	233	98.7	5.7	3.83	0.4	Flotation aid
Indianapolis, IN	R.S.	2,960	144	95.0	5.0	2.1	1.47	Flotation aid
					7.8			After 12 hrs
Warren, MI	R.S.	6,000	350	95.0	6-9	5.2	1.75	Flotation aid
Frankenmuth, MI	M.L.	9,000	80	99.1	6-8	6.5	1.3	Flotation aid
Oakmont, PA	R.S.	6,250	80	98.7	8.0	3.0	1.0	Flotation aid
Columbus, OH	R.S.	6,800	40	99.5	5.0	3.3	1.0	Flotation aid
Levittown, PA	R.S.	5,700	31	99.4	5.5	2.9	1.0	Flotation aid
Nassau Co., NY	R.S.	8,100	36	99.6	4.4	4.9	1.2	Flotation aid
Bay Park S.T.P.	R.S.	7,600	460	94.0	3.3	1.3	0.33	Standard
Nashville, TN	R.S.	15,400	44	99.6	12.4	5.1	0.66	Flotation aid

Souce: US EPA
[a]M.L.: Mixed liquor from aeration tanks.
[b]R.S.: Return sludge.
[c]Standard: Indicates no flotation aid / holding before sampling.
[d]Flotation aid: Indicates use of coagulant, flotation aid.

Increasing floc size facilitates the entrapment of rising air bubbles. Certain organic chemicals promote flotation of suspended solids in the influent stream by altering the surface properties of the three phases involved, these being the solid phase (suspended matter), gas phase (air), and liquid phase (water or wastewater). These alterations of surface properties result in changes to one or more of the interfacial tensions existing and subsequent improvement in adhesion of the suspended matter and adsorption of the solutes to the air bubbles. Chemicals suitable for this purpose include amyl alcohol, pine oil, cresylic acid, and surfactants.

7. SAMPLING, TESTS, AND MONITORING

7.1. *Sampling*

Sampling should be performed carefully. These samples may be obtained through valves provided in the respective flotation piping. If sampling points are not provided, they should be installed to facilitate operation and control of the process. Samples of the supernatant can be obtained at the overflow weir.

7.2. *Laboratory and Field Tests*

Samples should be analyzed according to procedures specified in Standard Methods (APHA, AWWA, WPCF, "Standard Methods for the Examination of Water and Wastewater," American Public Health Association, Washington DC, 2000, or later) and, in addition, should be visually observed.

Operating experience will allow most operators to judge the performance of the flotation system. A suspended solids rise test is useful to visually check operating results. On most units, a sampling valve is provided on the inlet mixing chamber. When the flotation unit is in operation, a liter jar sample is withdrawn, placed in a graduated cylinder, and the time for the water–solid interface to rise is recorded. A clear separation between suspended solids and liquid can be observed. Normal rise times are 10–25 s, and experience will indicate an average time for each particular plant. The relative depth of the blanket, the subnatant clarity, and the general appearance of the flocculated solid particles are also good visual indicators.

If polymer surfactant or inorganic coagulants are used as flotation aid(s), quantitative determinations of their initial and residual concentrations are important for flotation process control. Anionic surfactants can be measured by the Methylene Blue Method (5), the Azure A Method (6), or the CDBAC Titration Method (7). Cationic surfactants can be measured by the Methyl Orange Method (8) or the STPM Titration Method (9). Wang and his co-workers have developed an analytical method for the quantitative determination of both cationic and anionic polyelectrolytes (10,11).

The optimal chemical dosage for the influent feed stream should be determined at the start of each shift using jar test procedures described below.

8. PROCEDURES AND APPARATUS FOR CHEMICAL COAGULATION EXPERIMENTS

Chemical coagulation and flocculation are an important part of water and wastewater clarification. Coagulation or destabilization of a colloidal suspension results in joining of minute particles by physical and chemical processes. Flocculation results in formation of larger, floatable/settleable flocs by bridging. These are commonly the first processes in a water- or wastewater-treatment sequence to remove either suspended matter or color. Adsorption of ionic forms also occurs to varying degrees depending on the type of ion involved and the presence and amounts of other chemical constituents in the water or wastewater.

Inorganic coagulants (aluminum, iron, magnesium salts) may be used to coagulate particles and to form floatable or settleable flocs composed of the hydrous metal oxide precipitates and impurities. Alkalinity and pH controls are extremely important whenever the inorganic coagulants are used. Wang and Wang (134) present several recommended laboratory experiments involving the use of inorganic coagulants.

Polyelectrolytes are high-molecular-weight polymeric substances used in water purification and waste treatment to aid in the clarification of turbid suspensions or the dewatering of sludges. These compounds consist of a long-chain organic "backbone" with various types of ionic (cationic or anionic) or non-ionic solubilizing groups. Because of the extremely long chain lengths, one end or segment of the polymer molecule is capable of reacting independently of the other end or segments. The individual segments are adsorbed onto the surfaces of the dispersed particles and bridging occurs between the normally stable (unsettleable or unfloatable) solid particles under proper conditions of time, temperature, concentration, and mixing, this bridging leads to a floatable, settleable, or filterable floc. This mechanism of destabilization by polymers is commonly known as chemical coagulation. Although electrostatic interactions between oppositely charged polyelectrolyte and particle are important, it has been observed that anionic (negative) polymers will also destabilize negative solids. Polyelectrolytes may be obtained commercially with various molecular weights and compositions, as well as charges. They may be natural products, such as some starches and gums, or synthetically produced. Not all of these substances are acceptable for use with drinking waters. At the present time there is no analytical way to predict the behavior or applicability of a given polymer with a particular water or waste. Polymers are less sensitive to pH variations than metal coagulants, however, the dose required for optimum clarification varies over much wider ranges. Alkalinity control is not important when an organic polymer is used alone as the sole coagulant for floc formation or enhancement. Wang and Wang (134) and Krofta and Wang (141) have recommended a laboratory experiment involving the evaluation of a polymer as a coagulant.

When both an inorganic salt (such as alum) and an organic polymer are used as coagulants and coagulant aids, respectively, in a treatment system, alkalinity and pH controls become important. In this case, the laboratory experiments recommended by Wang and Wang (134) may be followed for process optimization.

In general, chemical coagulation (or flocculation) experiments can be conducted in a Standard Jar Test Apparatus for determining the optimum chemical dosage. The apparatus has six motorized stirrers which can be turned at the same speed. Jar tests are conducted with various coagulants and coagulant aids at different dosages and pH conditions. Usually, 1 L of test sample and desired chemicals are placed in a beaker and rapidly mixed by the stirrer at 100 rpm for 1 min, then slowly flocculated at 15–30 rpm for 10 min or longer, and finally settled at 0 rpm for 30 min. The settled supernatant liquid in the beaker is sampled for water quality analysis. Both the supernatant and the settled sludge are also visually observed and their conditions recorded.

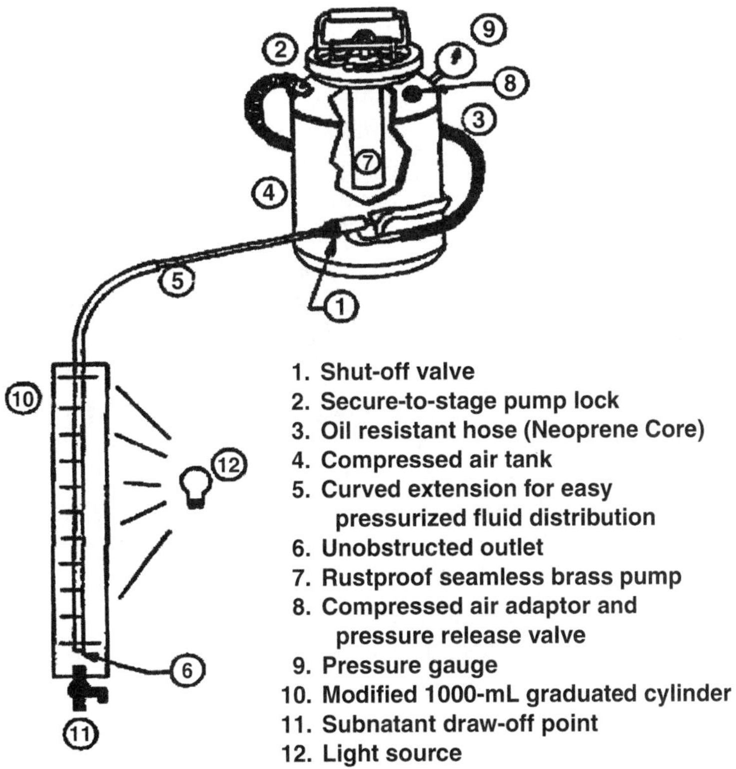

Fig. 13. Laboratory dissolved air flotation apparatus (15,141).

The chemical coagulation and flocculation tests determine the optimum chemical dosages for removal of turbidity and color, the pH adjustment requirements, and the necessity for the supplemental use of activated carbon. Jar tests yield a wealth of qualitative information on the rate of agglomeration as a function of energy input (paddle speed), the settleability of the floc formed, the floatability of the floc formed, and the clarity of the treated water (which might impact the subsequent length of filter run).

9. PROCEDURES AND APPARATUS FOR LABORATORY DISSOLVED AIR FLOTATION EXPERIMENTS

9.1. *Full Flow Pressurization System*

A commercially available compressed air tank is modified by removing the nozzle on its hose extension and fitting a pressure gauge into the tank. A 1000-mL plastic graduated cylinder is fitted with a valve on the bottom to draw off subnatants. Both the modified compressed air tank and cylinder are shown in Fig. 13.

The laboratory-scale dissolved air flotation (DAF) unit can be operated under the following three conditions: (a) full flow pressurization (Fig. 5A), (b) partial flow pressurization (Fig. 5B), and (c) recycle flow pressurization (Fig. 5C).

To start a DAF full flow pressurization experiment, the compressed air tank, shown in Fig. 13, is filled half full with raw or pretreated liquid sample (i.e., influent) that has been adjusted to process temperature. The compressed air tank is then pumped to 45–65 psig and manually shaken for 2 min to allow the air to dissolve in the sample. The first portion of pressurized sample (approx 100 mL) is released into the sink to allow any bound air to escape and to clear the outlet.

The remaining pressurized sample (1000 mL) is then carefully and slowly released into the modified 1000-mL graduated cylinder (see Fig. 13) by putting the outlet tube all the way to the bottom of the cylinder and raising the tube upward with the upward flow of the filled sample. (If the proper dose of chemical is added at the same time when the 1000 mL of pressurized sample is released into the 1000-mL cylinder, it is suggested that the filled cylinder be capped and inverted once to ensure proper mixing of chemicals.) Note when a line of demarcation (line of water–sludge separation) first appears between the clarified liquid at the bottom and the solids layer at the top, it is time to estimate the rising velocity of the sludge blanket by noting the level of the interface at regular time intervals. The rising velocity (in./min) of the sludge–water interface should be estimated accurately with the aid of a stop watch. A light source behind the cylinder may enhance the visibility of the particles. After 2–3 min the mL of floating sludge, mL of settled sludge if any, and the physical characteristics of the subnatant are recorded. Finally, at least 200 mL of subnatant (i.e., effluent) are drawn off from the cylinder bottom for testing suspended solids and other water quality parameters.

Equation (9) can be used to calculate the air to solids (*A/S*) ratio for the full flow pressurization system based on the laboratory experimental results. The flotation efficiency in terms of suspended solids removal can be calculated by the following formula:

$$E = 100\left(1 - X_e / X\right) \tag{34}$$

where E = percentage of suspended solids removal, X = suspended solids of influent (mg/L), and X_e = suspended solids of effluent (mg/L).

9.2. *Partial Flow Pressurization System*

The experimental apparatus for partial flow pressurization of DAF (Fig. 5B) is identical to that for full flow pressurization, shown in Fig. 13. To start a DAF partial flow pressurization experiment, the 1000-mL graduated cylinder (Fig. 13) is initially filled with the volume (V_i) of raw or pretreated liquid sample (i.e., influent not to be pressurized), which is adjusted to the process temperature. The value of V_i is decided based on the desired percentage of partial flow usually brought to 1 L:

Percentage of partial flow	Unpressurized influent volume (V_i)	Pressurized influent volume (V_p)
10%	900 mL	100 mL
20%	800 mL	200 mL
30%	700 mL	300 mL
40%	600 mL	400 mL
50%	500 mL	500 mL

The compressed air tank (Fig. 13) is then filled approximately half full with influent adjusted to the process temperature. The compressed air tank is subsequently pumped to 45–65 psig and shaken for 2 min to allow air to dissolve in the pressurized influent. The first portion of pressurized water (approx 100 mL) is released into the sink to allow any bound air to escape and to clear the outlet. The remaining pressurized water (V_p) is then released into the 1000-mL graduated cylinder (which is initially filled with V_i mL of the influent) by putting the outlet tube all the way to the cylinder bottom and raising the tube upward with the upward flow of the sample ($V_i + V_p$).

The rising velocity (in./min) of floating flocs is timed with a stopwatch by observation. After 2–3 min important physical characteristics of floating sludge (float), settled sludge, and subnatant are recorded. At least 200 mL of subnatant is drawn off from the cylinder bottom for testing suspended solid and other water quality parameters.

Equation (34) can be used for calculation of flotation efficiency.

9.3. *Recycle Flow Pressurization System*

The recycle flow pressurization system and the experimental apparatus for the system are shown in Fig. 5C and Fig. 13, respectively.

To start a DAF recycle flow pressurization experiment, the 1000-mL graduated cylinder (Fig. 13) is initially filled with the desired volume (V_i) of raw or pretreated liquid sample (i.e., influent), which is adjusted to the process temperature. The volume of liquid is determined based on the desired percentage of recycle flow as follows:

Percentage of recycle flow	Unpressurized influent volume (V_i)	Pressurized recycle water volume (V_r)
11.1%	900 mL	100 mL
25.0%	800 mL	200 mL
42.9%	700 mL	300 mL
66.7%	600 mL	400 mL
100.0%	500 mL	500 mL

The compressed air tank (Fig. 13) is then filled approximately half full with the recycle water (e.g., clarified effluent or other sources of clean water) adjusted to the process temperature. Suspended solids in the recycle water (*X*) are measured to provide a correction factor in the final calculation. The compressed air tank is subsequently pumped to 45–65 psig and shaken for 2 min to allow air to dissolve in the clean water. The first portion of pressurized water (approx 100 mL) is released to allow any bound air to escape and to clear the recycle outlet. The remaining pressurized water (V_r) is then released into the 1000-mL graduated cylinder (which is initially filled with the calculated volume of the influent) by putting the outlet tube all the way to the cylinder bottom and raising the tube with the upward flow of the sample ($V_i + V_r$).

The rising velocity (in./min) of floating sludges or flocs is timed with a stop watch by observation. After 2–3 min important physical characteristics of floating sludge, settled sludge, and subnatant are recorded. At least 200 mL of subnatant are drawn off from the cylinder bottom for testing suspended solids and other water quality parameters. The following material balance equations should be used for recycle flow correction:

$$X_e(V_i) + X_r(V_r) = X_e(1000 - V_r) + X_f V_f \tag{35}$$

$$V_i = 1000 - V_r \tag{35a}$$

where X_i = influent suspended solids (including chemicals, when required) (mg/L), V_i = volume of influent used (mL), X_r = suspended solids of recycle water (mg/L), V_r = volume of recycle water used, (mL), X_e = suspended solids of the clarified effluent or subnatant (mg/L), X_f = suspended solids of float (mg/L), and V_f = volume of float (mL).

The value of X_e calculated in Eq. (35) is then used in Eq. (34) for determination of flotation efficiency in terms of suspended solids: $X_r = X_e$.

Equation (36) should be used for calculation of the air-to-solids ratio based on the laboratory experimental results:

$$A/S = 1.3aV_r(FP-1)/V_iX \tag{36}$$

The quantity of air which will theoretically be released from solution following pressure reduction can be computed from:

$$a_r = a\left(FP_r - P_e\right)/14.7 \tag{37}$$

where a_r = air released at atmospheric pressure at 100% saturation (mL/L liquid) (air volume at standard conditions, e.g., 0°C + atm. absolute), a = air saturation at one atmosphere and pressure [mL/L (std. conditions), Section 4.4], P_r = pressure before release [lb/in.^{2} (absolute)], and P_e = pressure after release [lb/in.^{2} (absolute)].

Equation (37) is valid for all dissolved air flotation systems (Figs. 5A–5C). The actual quantity of air released will depend on how close the equilibrium solubility at P_r is attained and on the turbulent mixing conditions at the point of pressure reduction. The closeness to equilibrium solubility will depend on the time of retention. The closeness to equilibrium solubility will depend on the time of retention under pressure, on the mass transfer contact surface between air and water, and on the degree of mixing. Conventional static holding tanks can usually yield up to 50% saturation in the normal retention times. The use of packing or mixing can produce 90% saturation in conventional retention times. This can be taken care of in the calculations by multiplying P by a factor, F, where F is the fraction of saturation attained in the retention tank, and is equal to 1, for water fully saturated with air.

It should be noted that the operation of the pressure cell closely simulates the recirculation of effluent as used in the full-scale flotation systems. The returned effluent (recycle water) may be developed by repeated flotation of several different portions of raw waste. After the recycle water has been developed and used in the flotation tests, samples may then be withdrawn for chemical analysis.

Wang and Wang (134) recommend a laboratory experiment for the recycle flow pressurization system. The volume of gas bubbles released into the water after the pressure is released can be determinated according to the method presented in Fig. 14.

10. NORMAL OPERATING PROCEDURES

10.1. Physical Control

Typically the flow through the flotation unit is continuous and should be set for as constant a rate as possible. The drive mechanism normally turns continuously and contains a torque monitor or shear pins that will shut down the drive (and sound an alarm) if the drive mechanism is overloaded.

Flow meters are normally provided for the flow through the flotation chamber, the recycle flow, and the air flow. A control is normally provided on the retention tank to automatically control the liquid level by blowing off excess accumulations of air within the tank.

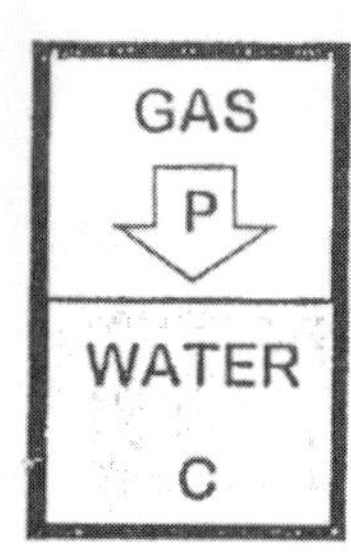

P = Pressure of the the gas

Note: 1 atm = 1.01 Bar = 14.7 psi

C = Concentration of the gas in a saturated solution

mL(STP) / 100 mL water = % v/v gas / water

Solubilities of some gasses, P = 1 atm

Temp °C	Air	Oxygen	Nitrogen	Hydrogen	CO_2
4	2.63	4.40	2.14	0.206	14.7
20	1.87	3.10	1.54	0.182	8.78
50	1.30	2.09	1.09	0.161	4.36

Air is dissolved in water at 20°C and 80 psi. What will be the % volume of air bubbles released when the pressure is reduced to 14.7 psi ?

$$80 \text{ psi} \times \frac{1 \text{ atm}}{14.7 \text{ psi}} = 5.4 \text{ atm}$$

solubility of air at 5.4 atm = 1.87 x 5.4 / 1 = 10.10

% volume of bubbles released 10.10 - 1.87 = 8.23 % v/v

Fig. 14. Volume of bubbles released when pressure is released (144).

The discharge pressure of the air compressor is normally regulated by a pressure reducing valve and is typically set at 75 psi.

10.2. Startup

The following are the startup procedures recommended by the US Environmental Protection Agency.

(a) Close drain valves as required.
(b) Open appropriate valves on the recycle water system.
(b-1) If the unit has been drained, open the necessary valves to the auxiliary water supply.
(b-2) If the unit has not been drained, do not open the auxiliary water supply valves.
(c) Start the recirculation pump. If the unit has been drained, wait until it is full and the auxiliary water supply valve has been closed before proceeding to Step d.
(d) Start the air feed and adjust to the required flow.

(e) Allow unit to run 10–15 min before starting influent (raw water, raw wastewater, or sludge) feed. This will charge the unit with chemical and aerated water.

10.3. Routine Operations

A check on the following unit operations at least twice per shift is recommended by the US Environmental Protection Agency (US EPA):

(a) Visual check for proper chemical conditioning and operation. For example, large floc carrying over into recycle water indicates a problem with the reaeration system. A very turbid effluent with no floc development shows deficiency or overloading of the unit.
(b) Flow.
(c) Skimmer speed setting.
(d) Recycle rate.
(e) Chemical supply.
(f) Obtain and analyze samples as required.
(g) Chemical V—notch weir setting (dosing).
(h) Retention tank air cushion.

A mechanical check should be made on the following units at 2 h intervals.

(a) Pumps: chemical feed, recycle, reaeration, and sludge sumps.
(b) Air manometer operation.
(c) Retention tank pressure.
(d) Sludge pit mixers.

10.4. Shutdown

(a) Shut off influent feed.
(b) Shut off chemical supply.
(c) If possible, allow unit to operate for 30 min before shutting down the sludge removal system (skimmer flights or equivalent). This serves to clear the unit of suspensions and the sludge removal system clears the water surface of sludge. The unit can then be shut down with the flotation retention tank filled with practically clean water and the flotation unit primed for startup.
(d) Shut off air supply.
(e) Turn off reaeration pump.
(f) Turn off recirculation pump.
(g) Turn off sludge mechanism drive motor(s).
(h) Shut off chain oilers.
(i) If no other units are operating to the same pit, shut off sludge pit mixer and pump.
(j) If the unit is to be shut down for an extended period or for internal maintenance, it must be drained.
(j-1) Open drain valves on air flotation unit and retention tank.
(j-2) Flush the unit, flights, beaching plate, baffles with the high pressure hose.

11. EMERGENCY OPERATING PROCEDURES

11.1. Loss of Power

In the event of loss of power, the dissolved air flotation unit should be shut down unless emergency electrical generation is available. After power is restored, a normal start up should be performed and the unit placed back in operation.

11.2. Loss of Other Treatment Units

Loss of chemical feed to the flotation unit will generally affect performance. If this occurs, operating parameters such as recycle ratio may require readjustment to obtain the best possible performance. Best performance without chemical feed will generally be very inferior to performance with chemical feed.

12. OPERATION AND MAINTENANCE

12.1. Troubleshooting

Common design shortcomings and possible solutions are presented in Table 7. A troubleshooting guide recommended by the U.S. Environmental Protection Agency for solid/liquid separation by dissolved air flotation is introduced in Table 8. The DAF solids loading rate for thickening various sludges can be found from Table 9.

12.2. Labor Requirements

Table 10 indicates the expected labor requirements for flotation operation and maintenance.

12.3. Construction and O & M Costs

The 2003 capital cost could be approximated by Eq. (38):

$$C = 6.867 \times 10^4 \times Q_w^{1.14} \tag{38}$$

where C = 2003 capital cost of DAF thickening proces ($) and Q_w = WWT plant design flow (MGD).

The associated costs include those for excavation, process piping, equipment, concrete and steel. In addition, such cost as those for administrating and engineering are equal to 0.2264 times Eq. (38).

Chemical feeding, effluent storage, pumping, and sludge disposal costs are not included. Cost adjustment from past to present can be done by using either the Chemical Engineering (CE) Fabricated Equipment cost index, or the Engineering News Record (ENR) cost index (147).

12.4. Energy Consumption

Electrical energy requirements of a recycle flow pressurization system for treatment of water and wastewater are presented below as a function of influent flows 1 MGD = 1 million gallons per day = 3.785 million liters per day = 3.785 MLD):

Plant flow (MGD)	Electrical energy required (kwh/yr)
0.1	4.5×10^4
1.0	2.0×10^5
10.0	2.0×10^6

The aforementioned energy consumption includes that required for recycle flow, air injection and chemical feed pumping

Table 7
Common Design Shortcomings and Solutions

Shortcoming	Solution
1. Excessive wear in sludge mechanism chains and gears	1. Install automatic oilers
2. Poor results in mixing chemicals (polymer)	2a. Install automatic feed system or an aspirator wetting system to insure initial wetting of polymer (powders) 2b. Prepare a less concentrated mixture of 0.25–0.50 percent of weight by polymer to water
3. Early failure of pressure gauges and controls	3. Install sensitive equipment on panels isolated from equipment vibration
4. Sludge feed pumps run on–off cycle causing pulsed feed to DAF unit	4. Install a flow meter and flow control system to provide consistent, controllable inflow rate
5. Only primary effluent available for auxiliary recycle	5. Install line so that secondary effluent can be used for recycle during periods when primary effluent has more than 200 mg/L solids or contains unusual amount of stringy material
6. Wide variation in feed solids concentration occur due to direct feed of DAF from final clarifier	6. Install a mixed storage tank to minimize the fluctuations

Source: US EPA.

When a recycle flow pressurization DAF system is used for sludge/fiber thickening or concentration, more detailed electrical energy consumption data are presented below as a function of flotation chamber's surface area:

Surface area (ft^2)	Electrical energy (kwh/yr)
100	1×10^5
1,000	7×10^5
10,000	5×10^6

The aforementioned energy consumption data for sludge/fiber thickening are estimated based on typical solids loading rates: (a) 8–20 lb/ft^2/d without polymer addition, and (b) 24–60 lb/ft^2/d with polymer addition. The float solids concentration will be in the range of 2.5–10%.

12.5. Maintenance Considerations

A good preventive maintenance program will reduce breakdowns, which could be not only costly, but also very unpleasant for operating personnel. The following are the major elements that should be inspected semiannually for wear, corrosion, and proper adjustment: (a) drives and gear reducers, (b) chains and sprockets, (c) guide rails, (d) shaft bearings and bores, (e) bearing brackets, (f) baffle boards, (g) flights and skimming units, (h) suction lines and sumps, (i) pumps, and (j) compressors.

Table 8
Troubleshooting Guide for Dissolved Air Flotation Operation

Observations	Probable cause	Check or monitor	Solutions
1. Floated sludge layer too thin	1a. Flight speed too high		1a. Adjust flight speed as required
	1b. Unit overloaded	1b. Proper operation	1b. Turn off inlet and calibration of flow to allow polymer pumps unit to clear
	1.c Polymer dose too low	1c. Proper operation and calibration of polymer pumps	1c. Adjust as required
	1d. Excessive air/solids ratio	1d. Float appearance	1d. Reduce air flow (very frothy) to pressurization system
	1.e Low dissolved air		1e. See Item 2
2. Low dissolved air	2a. Reaeration pump off, clogged or malfunctioning	2a. Pump condition	2a. Clean as required
	2b. Eductor clogged		2b. Clean eductor
	2c. Air supply malfunction	2c. Compressor and supply lines	2c. Repair as required
3. Effluent solids too high	3a. Unit overloaded	3a. See Item 1b	
	3b. Polymer dosage too low	3b. See Item 1c	
	3c. Skimmer off or too slow	3c. Skimmer operation	3c. Adjust speed
	3d. Low air/solids ratio	3d. Poor float formation	3d. Increase air flow
	3e. Improper recycle flow	3e. Recycle pump flow rate	3e. Adjust flow
4. Skimmer blade leaking on beaching plate	4a. Skimmer wiper not adjusted properly		4a. Adjust
	4b. Hold down tracks too high		
5. Skimmer blades binding on beaching plate	5. Skimmer wiper not adjusted properly		5. Adjust
6. High water level in retention tank	6a. Air supply pressure low	6a. Compressor and air lines	6a. Repair
	6b. Level control system bleeding continuously	6b. Level control system	6b. Repair bleed system
	6c. Insufficient air injection	6c. Compressor and air lines	6c. Increase air flow
7. Low water level in retention tank	7a. Recirculation pump not operating or clogged	7a. Pump operation	7a. Inspect and clean
	7b. Level control system not bleeding air properly	7b. Level control	7b. Repair
8. Low recirculation pump capacity	8. High retention tank pressure	8. Recirculation flow rate	8. Increase recirculation flow

Source: US EPA.

Table 9
Solids Loading Rate for Thickening Various Sludges Using DAF

Type of sludge	Solids loading rate (lb/ft^2/h)[a] No chemical addition	Optimum chemical addition
Primary only	0.83–1.25	up to 2.5
Waste activated sludge (WAS) Air	0.42	up to 2.0
Oxygen	0.6–0.8	up to 2.2
Trickling filter	0.6–0.8	up to 2.0
Primary + WAS (air)	0.6–1.25	up to 2.0
Primary + trickling filter	0.83–1.25	up to 2.5

Source: US EPA.
[a]1 lb/ft^2/h = 4.9 kg/m^2/h

12.6. *Environmental Impact and Safety Considerations*

Dissolved air flotation requires very little use of land. The air released in the unit is unlikely to strip volatile organic material into the air. The air compressors will need silencers to control the noise generated. The sludge generated will need methods for disposal. This sludge will contain high levels of chemical coagulants used.

Although the dissolved air flotation equipment presents no special hazards, general safety considerations should apply. At least two persons should be present when working in areas not protected by handrails. Walkways and work areas should be kept free of grease, oil, leaves, and snow. Protective guards must be in place unless mechanical/electrical equipment is locked out of operation.

The retention tank is a hydropneumatic tank and should not be pressurized beyond the working pressure rating. The tank should have a functional relief valve and should be inspected on a regular basis for excessive corrosion.

13. RECENT DEVELOPMENTS IN DISSOLVED AIR FLOTATION TECHNOLOGY

13.1. *General Recent Developments*

Dissolved air flotation (DAF) is a promising process for removing suspended solids from water by utilizing the lifting force induced when bubble–floc aggregates of low

Table 10
Flotation Clarifier Labor Requirements

Flotation chamber Surface area(ft^2)	Operation labor (h/yr)	Maintenance labor (h/yr)	Total labor (h/yr)
11	215	260	475
21	320	350	670
53	550	540	1090
105	840	750	1590
210	1300	1050	2350
520	2200	1600	3800

Source: US EPA.

density are produced (49–51). The scale-up or full-scale plants utilizing DAF technique are discussed elsewhere (12,13,16,25,29,52,53,60,69,70,79–82,97,104–114, 121–130). DAF has successfully been used in wastewater-treatment processes for separating oil from aqueous dispersion (54), removing biosolids and bacteria (55,56), reducing metals (57), and thickening activated sludge (58–60). DAF was found effective for BOD, COD, and turbidity reduction with chemical precipitation (61, 62). Dissolved air flotation was found to be a suitable treatment method for soaking basin overflow of a plywood mill using birch as raw material. According to DAF pilot treatment studies, over 90% reductions of suspended solids are possible with a hydraulic surface load of $m^3/(m^2h)$. In subsequent experience in full scale, the following reductions have been achieved: suspended solids 93%, BOD 50%, COD 57%, P 92%, and N 52%. Malley found that the removal of dissolved organic carbon (DOC) was independent of the separation process but depended on the pretreatment process (63). Bunker et al. suggested that DOC removal improved with increasing the time allowed for flocculation (64).

The electrical charges on particles and bubbles are important in DAF (65) for treatment of oil refinery wastewater (66) and oil in emulsions from refinery wastewater that was chemically pretreated. It is clear that chemical pretreatment is an essential requirement for high efficiency in DAF. Generally, high shear rates are necessary for chemical dosing and lower shearing rates are needed to promote the aggregation of fine particles through flocculation. Large-diameter oil droplets in the presence of inorganic salts are essential for the rapid separation of emulsified oils. However, it is still not clear what the conditions producing optimal flocculation for DAF might be. Some workers indicated that prolonged flocculation times are not needed in DAF as good separations could be achieved with short residence times, typically 5–15 min for fine flocs (67), while Fukushi et al. (68) remarked that large flocs are more desirable because the larger number of bubbles attached to them will make the separation more efficient. Klute et al. (69) indicated that relatively high mixing intensities in the flocculation stage produce larger flocs corresponding to increased removal of suspended solids. However, it was suggested that small velocity gradients in the flocculation stage produce good floc size distributions for efficient removal by DAF (70). Ho and Ahmad (71) noted the effect of ionic additives on the stability or electrokinetic properties of vegetable oil emulsions stabilized by non-ionic emulsifiers. Al-Shamrani et al. (54) reported on the DAF separation of mineral oil–water emulsions stabilized by non-ionic surfactants and flocculated by inorganic salts. They investigated the role of aluminum and ferric sulfates as destabilizing agents for oil droplets stabilized by a non-ionic surfactant, and the use of the electrokinetic properties of the emulsion to interpret the experimental results.

The DAF process takes place in two separated zones, each with a different mechanism (49). The contact zone provides for the contact and attachment of bubbles and flocs (aggregation), while the separation zone separates the generated aggregates from the water phase. Previous research established that the contact zone configuration could be of great importance for the removal efficiency, but lacked quantitative guidelines. Lundh et al. (49) presented the quantifying design criteria and analysis of an experimental study of the flow structure in a traditionally designed rectangular DAF pilot plant and the influence of varying the internal geometry, meaning the variation in height, length, and inclination of the contact zone shaft wall. Studies showed that the air content influences the flow structure in the separation zone (72,73), and to a lesser extent in the contact zone. The flow structure in the separation zone has been modeled (74) and measured experimentally in pilot plants and full-scale plants.

To better understand the mechanism of bubble–floc interaction in DAF Shawwa, et al. determined the hydrodynamic characteristics in the contact zone, in terms of contact time and degree of mixing, as a function of hydraulic loading rate and recycle ratio. Another approach to understanding the mechanisms in the dissolved air flotation process involved the formation of bubbles obtained by the release of pressurized water through a nozzle, and a further examination was to study the parameters influencing the size of gas bubbles (76,77).

The agglomeration (coagulation/flocculation) phase has been indicated as essential for determining the downstream process efficiency, which is a prerequisite for process improvement. Vlaski et al. (70) promoted a kinetic modeling addressing particle (floc) size–density relationship by using cyanobacteria as a surrogate for the process removal efficiency assessment of particles. The process efficiency was assessed as a function of the preceding agglomeration phase and the obtained particle (floc) size distributions, including the influence of coagulant dose, coagulation pH, flocculation time, energy input, single stage versus tapered flocculation and application of cationic polymer as coagulant aid.

A combination technique of flotation and filtration, countercurrent dissolved air flotation filtration, born out of a review of the technology of flotation, was developed in the UK (78). The patented process (74) has been designed as a compact water-treatment process to overcome operational problems with seasonal blooms of filter-blocking algae. One advantage of the technology is its operational flexibility, as it can be turned off to ensure it can maintain maximum capacity during the worst algal blooms. The difference from the standard DAF design is that the recycle water is introduced after the flocculated water inlet structure (but above the filter media), which generates an even bubble blanket field in the flotation tank through which all the flocculated water must pass. Flocculated water inlet entered the middle of the tank and the dissolved air recycle inlet was set below the water inlet through a nozzle. The countercurrent water was filtered through a sand bed for subsequent treatment, while sludge was removed by a hydraulic desludging blanket. The advantage of this design of moving the recycle inlet away from the flocculated water inlet is that the potential for floc damage (shear) by the recycle is eliminated.

13.2. Physicochemical SBR-DAF Process for Industrial and Municipal Applications

A combined physicochemical Sequencing Batch Reactor (SBR) and Dissolved Air flotation (DAF) process (i.e. physicochemical SBR-DAF process) has been developed by Wang, Kurylko, and Wang (151) and demonstrated in pilot scale for treatment of an electroplating effluent containing hazardous heavy metals and volatile organic compounds (VOCs).

The same physicochemical SBR-DAF process (133,151–154) is also feasible for treating potable water, contaminated groundwater, sewage, and other industrial effluents.

The process equipment and process steps of physicochemical SBR-DAF are very similar to that of physicochemical SBR, except that the former adopts DAF clarification, and the later adopts sedimentation clarification.

This process equipment (physicochemical SBR-DAF) may also be further equipped with an enclosure and air purification means when it is used for treating hazardous and odorous wastewater. For portable water purification and industrial effluent treatment, addition of PAC may enhance final polishing efficiency.

13.3. Adsorption Flotation Processes

There are at least five kinds of powdered activated carbon (PAC) adsorption systems: (a) conventional continuous PAC system involving continuous powdered activated carbon feeding, mixing, adsorption, coagulation, and sedimentation; (b) sequencing batch reactor (SBR) PAC system involving batch powdered activated carbon feeding, mixing, adsorption, coagulation, and batch sedimentation; (c) continuous adsorption flotation system involving continuous powdered activated carbon feeding, mixing, adsorption, coagulation, and continuous dissolved air flotation; (d) sequencing batch reactor (SBR) adsorption flotation system involving batch powdered activated carbon feeding, mixing, adsorption, coagulation, and batch dissolved air flotation; and (e) precoat carbon filtration system. Filtration is usually used as an advanced treatment step after the above process systems (a), (b), (c), and (d) for final polishing (141,144,151,153,159–163).

Both a continuous adsorption flotation system and a SBR adsorption flotation system were developed by Wang (9,176) and further demonstrated by Krofta and Wang (18,80,141,144) in pilot and full-scale water-treatment plants mainly for removal of taste, odor, and toxic organic substances.

Precoat carbon filtration is very similar to conventional precoat filtration system, except that coarse powdered activated carbons and/or fine granular activated carbons are used as the precoat medium (170,171).

13.4. Dissolved Gas Flotation

Theoretically any kind of gas can be used for production of fine gas bubbles to be used in dissolved gas flotation (DGF) process system. Practically, however, air is readily available free of charge, and is usually used almost in all dissolved gas flotation systems. Other types of gases are used in DGF system only when there is a need. The use of various gases for DGF treatment of industrial/municipal water and wastewater is an important technological development (141,144,164–169,173).

For instance, air bubbles (containing oxygen) are not suitable for DGF treatment of military explosive wastes (174), therefore, nitrogen bubbles (instead of air bubbles) can be generated for flotation treatment. Under this process condition, the DGF process is also called Dissolved Nitrogen Flotation (DNF).

Carbon dioxide gas is a waste air stream from many industrial plants, and it can be recovered for pH adjustment, recarbonation, and, of course, DGF treatment of an industrial wastewater (144,175), or an industrial water aiming at hardness removal (110,111, 130,173). Under this process condition, the DGF process is also called Dissolved Carbon Dioxide Flotation (DCDF).

Both ozonation and DAF processes are popular among environmental engineers for treatment of potable water, cooling waster, and industrial effluent (49,98,99,143,144). Wang and his coworkers (144,164–169) have proven that ozonation and DAF can be combined into one process, known as Dissolved Air–Ozone Flotation, for space saving and cost saving.

Figure 14 shows the volumes of various gas bubbles released (air, oxygen, nitrogen, hydrogen, carbon dioxide) when pressure is released to 1 atm (144). A typical example is also provided in Fig. 14 for a step-by-step calculation of percentage gas volume versus water volume.

13.5. *Combined Sedimentation and Flotation*

Combined sedimentation and flotation is another important recent process development. Since the process equipment has been used extensively for upgrading the existing sedimentation process, it is introduced as a practical example in Section 14, Application and Design Examples .

Additional recent development in potable water treatment by DAFF is reported by Krofta and Wang (144). Section 14 further introduces the first five potable water DAFF plants built in the US, the first European DAFF plant , and many new DAFF plants under design or construction.

14. APPLICATION AND DESIGN EXAMPLES

14.1. *Example 1: First Five Potable Water Flotation-Filtration (DAFF) Plants Built in America and Their Common Special Features*

This chapter is written by the authors in memory of late Dr. Milos Krofta, the father of flotation technology, who designed and built the first five full-scale potable flotation-filtration (DAFF) plants in America during the period of 1982–1988: (a) Lenox Water Treatment Plant, MA; (b) Coxsackie Water Treatment Plant, NY; (c) Pittsfield Water Treatment Plant, MA; (d) Nanty Glo Water Treatment Plant, PA; and (e) Howell Water Treatment Plant, NJ. Dr. Krofta passed away on August 22, 2002, shortly after he celebrated his 90th birthday. Dr. Krofta had over 50 US and foreign patents on flotation technologies. His organizations (Lenox Institute of Water Technology, Krofta Engineering Corporation, Krofta Waters, Inc., KWI, Inc., and Krofta Technologies Corporation) had developed, designed, and built over 3000 flotation units for potable water treatment and industrial effluent treatment around the world. Both late Dr. Milos Krofta and Dr. Lawrence K. Wang received the Pollution Engineering magazine's PE Five-Star Award and Korean Society of Water Pollution Research and Control's Outstanding Engineering Award, owing to their outstanding design and the five DAFF plants' excellent performance (144).

The flotation/filtration (DAFF) process is a modern development for treatment of potable water by use of DAF combined with dual media (sand/anthracite) filtration. In the DAFF process, DAF replaces sedimentation in conventional systems in order to improve operation and reduce space requirements. The DAFF process is designed and built for five installations in the eastern US, with some particular advantages of each system in terms of contaminant removal and space savings. The five flotation/filtration (DAFF) units supply 1.1–37.5 MGD (million gallons per day) to municipal water systems (144).

Dissolved air flotation (DAF) is widely used for industrial effluent treatment (49–54,81, 82,96–101,113–117,119,127,129,132,137,144), while flotation/filtration (DAFF) has been specifically developed for treatment of drinking water. DAFF combines DAF and filtra-

The influent raw water or waste water enters the inlet at the center near the bottom (A), and flows through a hydraulic rotary joint (B), and an inlet distributor (C), into the rapid mixing section (D), of the slowly moving carriage. The entire moving carriage consists of rapid mixer (D), hydraulic static flocculator (F), air dissolving tube (G), backwash pumps (H) and sludge discharging spiral scoop (I). To flock out colliods and suspended solids, alum is added at (Q) at the inlet (A). For additional improvement of flocculation, polyelectrolytes can be added at the same inlet (A).

At the outlet of the flocculator on the carriage, pressurized water with dissolved air is added (D). At the bottom of the carriage (L), a small volume of the water preclarified by dissolved air flotation is taken by a pressure pump (M), that feeds an Air Dissolving Tube (G), where compressed air is added from a separate compressor (N), riding with the pressure pump, and Air Dissolving Tube (G), on the carriage air is dissolved under pressure in the water and mixed with the flocculated raw water at the outlet of the flocculator.

The flocks and suspended solids are floated to the water surface. The floating scum or sludge accumulated on the water surface is scooped off by a sludge discharging spiral scoop (I), and discharged into the center sludge collector (O), where there is a sludge outlet (P) to an appropriate sluge treatment facility.

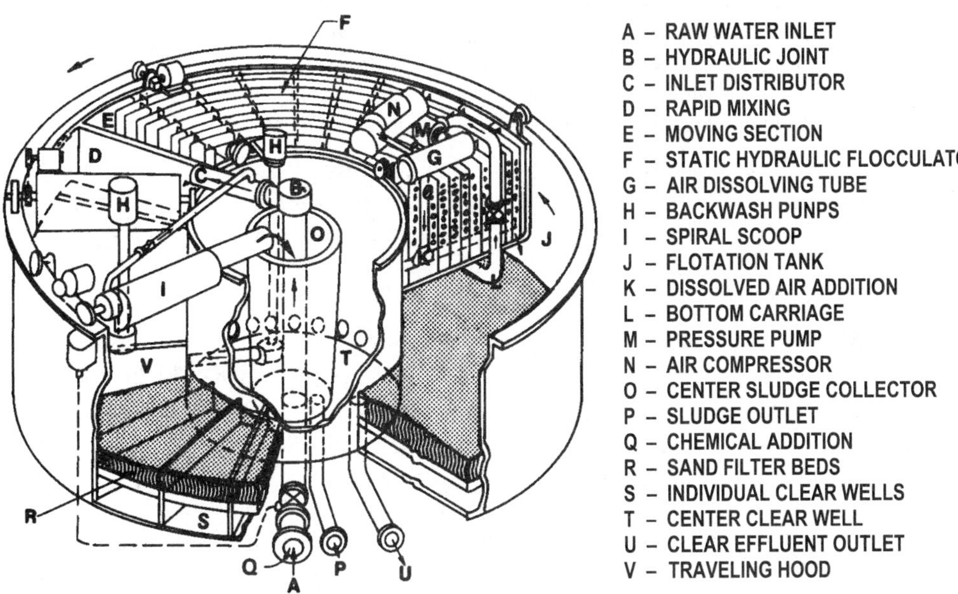

Fig. 15. Lenox water treatment plant—a typical DAFF plant.

tion in a single tank. (12,13,29,78–80,121,155,156). Figure 15 illustrates the layout of a typical DAFF plant (the Lenox Water Treatment Plant). All first five American DAFF plants (Lenox, Coxsackie, Pittsfield, Nanty Glo, and Howell) designed and built by Dr. Krofta and his associates are similar to each other. The following paragraphs briefly describe how a typical potable DAFF plant works (12,13).

The influent raw water or wastewater enters the inlet at the center near the bottom, and flows through a hydraulic rotary joint and an inlet distributor into the rapid mixing section of the slowly moving carriage. The entire moving carriage consists of rapid mixer, flocculator, air-dissolving tube, backwash pump, sludge discharge scoop, and sludge recycle scoop.

From the rapid mixing section, the water enters the hydraulic flocculator for gradually building up the flocs by gentle mixing. The flocculated water moves from the flocculator into the flotation tank clockwise with the same velocity as the entire carriage including flocculator moving counterclockwise compensated by the opposite velocity of the moving carriage, resulting in a "zero" horizontal velocity of the flotation tank influent. The flocculated water thus stands still in the flotation tank for optimum clarification.

At the outlet of the flocculator, clean water with microscopic air bubbles is added to the flotation tank in order to float the insoluble flocs and suspended matters to the water surface. The floating scum or sludge accumulated on the water surface is scooped off by a sludge discharge scoop and discharged into the center sludge collector where there is a sludge outlet to an appropriate sludge treatment facility.

The bottom of the DAFF unit is composed of multiple sections of sand filter and clear well. The clarified flotation effluent passes through the sand filter downward and enters the clear well. Through the circular hole underneath each sand filter section, the filter effluent enters the center portion of the clear well where there is an outlet for the filter effluent.

The backwash hood is also a portion of the moving carriage. A filter section is to be backwashed only when the filter top is covered by the backwash hood. A semicontinuous filter backwashing takes place by pumping the clarified water through one section of the sand filter upwards with a backwash pump, and sucking the washwater out from the filter top covered by the backwash hood with another backwash pump. Backwash water is discharged into the main distribution inlet header for reprocessing.

The floating scum or sludge is removed by the sludge discharge scoop and wasted. A separate skimming device, called the "sludge recycle scoop," lifts the remaining floating sludge back into the main rapid mix inlet for additional sludge contact with the newly formed flocs. Under optimum operational conditions, the practice of sludge recycle reduces the chemical dosage for water or wastewater treatment.

The DAFF plant can be manually operated, or completely automated with the level control that operates the inlet flow valve. Filter backwashing is also automated by time and/or head loss control.

As stated above, the overall flow diagram of the DAFF system is also similar to that of the conventional system, except that DAF replaces sedimentation in the conventional system in order to improve operation and reduce the process equipment's foot print. Flotation takes place in the top layer of the tank, and filtration in the bottom layer of the same tank. In addition, a static circular flocculation tank is built into the center (inlet) area. The segmented sand filter uses dual media (sand and anthracite) and features continuous backwashing. Backwash water from the sand filter is recycled directly back to the flocculator. "First Filtrate" from each sand section can be isolated and is also recycled (in the aeration system).

The five DAFF installations have much in common in terms of application, location, and climate. Some of the common benefits of the DAFF system experienced in these systems are discussed here along with the application in general to surface water treatment in cold-weather climates. Advantages of this type of system include space savings, reduced sludge volume, and simplified recirculation of backwash and first filtrate. There is also evidence that the use of DAF results in lower chemical requirements for flocculation, and greater removal of certain contaminants including *Giardia* and *Cryptosporidia* cysts.

Many of the advantages found in the DAFF installations described in the later sections have general application to many northern US and Canadian locations. Particular features illustrated in the mentioned installations include:

(a) Compact construction (clarifier and filter in one tank) allows economical installation inside a single building. In Pittsfield, MA-for instance, 24 MGD capacity can be fit into a building space 70 ft × 240 ft, with room inside to add bulk chemical storage, laboratories, and parked vehicles.

(b) Chemical flocculation requirements are generally less for flotation compared to sedimentation. This is due to flotation's ability to utilize a smaller flocculated particle size. All of the above examples utilize alum or aluminum chloride flocculation, with minimal or no polymer required to build particle size.
(c) Flotation has advantages in removal of organic color and cysts. Bulky or light weight flocculated materials can be more efficiently removed than with settling. Good removal of bacteria and low particle counts are experienced.
(d) Flotation/filtration (DAFF) is successful in treatment of cold water amid water with low turbidity. All of the mentioned cases experienced incoming turbidities below 2 NTU and water temperatures below 35°F.
(e) First filtrate separation and backwashing is accomplished with no waste of water. Only the Lenox plant is now operating with first filtrate separation, but all of the mentioned examples recycle filter backwash back into the flocculator section before the flotation clarifier. Newer models now feature dual media filtration, and complete first filtrate separation as required by some new regulations, but with no extraneous discharge of wastes other than the floated sludge.
(f) Greater sludge consistencies and more concentrated reject are obtained compared to conventional sedimentation systems. The only waste discharge is thickened sludge at a solids content of over 2%. This is a tremendous advantage in considering loading on a sewer system (Pittsfield, Howell), or the volume of sludge to be stored and dewatered (Lenox, Coxsackie, Nanty-Glo). The high consistency sludge reduces the size and increases the effectiveness of subsequent dewatering beds or sludge presses.

Each of the first five DAFF plants built in the US is presented in the subsequent examples for more detailed illustration.

14.2. Example 2: Lenox Water Treatment Plant, MA, USA—The First Full Scale Flotation-Filtration Plant for Potable Water Treatment in America

Lenox is a small town located in the Berkshire Hills of Western Massachusetts. The town's population is about 6500 in regular seasons and usually reaches about 10,000 during the Boston Symphony Orchestra Festival at Tanglewood. The town's main water sources consist of two small surface reservoirs—(a) Upper Root Reservoir and (b) Lower Root Reservoir—which together supply approximately 1 MGD to the town. Lenox was the site of the first operating full-scale flotation/filter installation in North America; a 1.5-MGD, 22-ft-diameter plant installed "temporarily" in 1981. Habitual water shortages due to occasional drought conditions and high seasonal summer demand made the water-saving feature of 100% backwash water recycle attractive in this case. A new permanent system was installed in 1992, which includes separation and reuse of the "first filtrate." The new Lenox Plant has adopted two DAFF units (two Krofta package plants model SASF-18), each has a diameter of 18 ft, capable of treating 1.5 MGD (Fig. 15).

The incoming water from this reservoir has low turbidity (0.6–10 NTU) and moderate color (3–15 color units) similar to the situation in nearby city of Pittsfield, MA. The turbidity and color in the filtered water (i.e., DAFF effluent) are below 0.15 NTU and below 1 color unit, respectively. About 11 mg/L of alum is used for water treatment.

The Lenox treatment systems are not near city sewer lines, and therefore have to deal with sludge disposal. The floated sludge at 2% solids content is the only waste discharged from the plant. The sludge is held in dewatering ponds (i.e., sludge drying beds) and allowed to freeze over the winter. The freeze/thaw cycle aids in dewatering of the sludge, which is eventually dug out and disposed of in a landfill. The Lenox system operates for 24 h/d without on-site supervision. It has a computer-monitoring system in plant, and alarms are remotely monitored in the water department offices 4 miles away. The original "temporary" DAFF system has been dismantled and is being reinstalled in a different location in a unique experiment in removal of phosphorus from a lake water with recreational use. The new system was recognized by the Massachusetts DEP as one of the best water systems in Massachusetts in 1995.

Special emphasis of the Lenox Plant is placed on water conservation and cost saving. The water loss of a conventional flocculation, sedimentation, filtration plant is about 9% due to the fact that its filter backwash wastewater is totally wasted. A comparable DAFF (including flocculation, flotation, and filtration) recycles its filter backwash wastewater and chemical flocs for reproduction of drinking water, thus its water loss is only about 0.7% in the floated sludge. The rates of water treatment by the two plants can be estimated as follows.

The water consumption rate of a conventional plant is calculated by the following two equations:

$$\mathrm{WPR} = \mathrm{PF} \times 0.91 \tag{39}$$

$$\mathrm{PF} = 1.1\mathrm{WPR} \tag{40}$$

where WPR = water production rate, assuming = water consumption rate (MGD) and PF = plant flow (MGD).

The water consumption rate of a DAFF plant, such as at Lenox, is calculated by the following two equations:

$$\mathrm{WPR} = \mathrm{PF} \times 0.993 \tag{41}$$

$$\mathrm{PF} = 1.007\ \mathrm{WPR} \tag{42}$$

Assuming the coagulant dosages (mg/L) for both conventional and innovative plants are identical, the conventional plant requires about 10% more coagulants by weight (ton/d) because the conventional plant must treat about 10% more water (i.e., factor 1.1 vs factor 1.007) in order to supply the same water consumption rate to the community. For the same reason, the plant size, construction cost, and building's heating cost of a conventional water purification plant will be comparatively much higher than that of a potable DAFF plant.

Both Dr. Milos Krofta and Dr. Lawrence K. Wang received the Pollution Engineering magazine's Five-Star Engineering award in 1983 for innovative design and excellent performance of the Lenox Plant.

14.3. *Example 3: Coxsackie Water Treatment Plant, NY, USA—The First Full-Scale Flotation-Filtration Plant for Potable Water Treatment in the State of New York*

The 1.1-MGD DAFF potable water system was installed at the village of Coxsackie, NY in 1985 for treatment of surface reservoir water. Three 12-ft-diameter DAFF units were retrofitted into an existing building that housed poorly operating diatomaceous earth filters. The filters were replaced one at a time by adding prefabricated DAFF units in their place. These new DAFF systems are remotely operated during part of each day and weekends (29).

Alum is the only chemical used for flocculation. The Coxsackie raw water quality is 5–70 color units and 2–20 NTU, respectively. The treated DAFF effluent quality is excellent, always meeting the US Federal Drinking Water Standards (29).

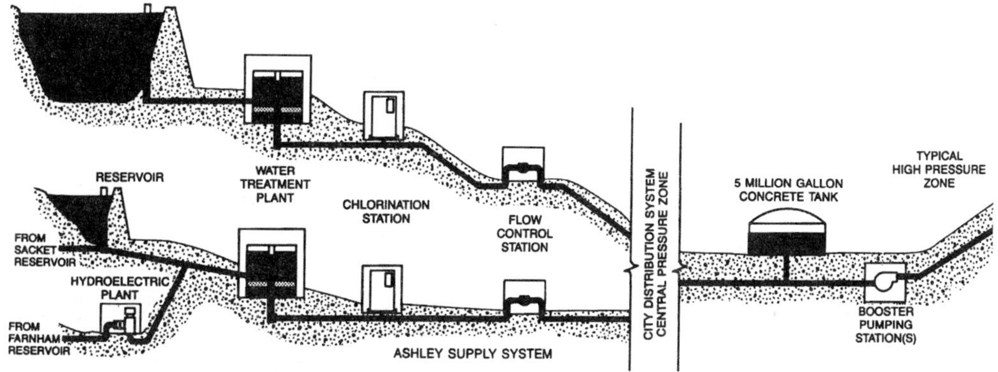

Fig. 16. Pittsfield water treatment system.

The backwash wastewater is 100% recycled for water conservation, and the only discharge is 2% floated sludge. The sludge is dewatered in a freeze/thaw dewatering pond (i.e., sludge drying beds).

The Coxsackie plant was chosen as the "Water System of the Year for 1990" by the New York Rural Water Association (144).

14.4. *Example 4 Pittsfield Water Treatment Plant, MA, US—Once the World's Largest Flotation-Filtration Plant for Potable Water Treatment)*

The city of Pittsfield (population 55,000) in the Commonwealth of Massachusetts is, like many New England communities, dependent on surface reservoirs for its water supply. In the late 1970s, the city recognized that it would need to do more than just chlorinating its drinking water and launched a study of its water system that, 10 yr later, resulted in the start up of two unique water treatment plants that provide Surface Water Treatment Rule (SWTR) performance (16, 130).

Once the world's largest potable DAFF system (Fig. 16) with 37.5 MGD total capacity was built in Pittsfield, MA in August 1986. The city of Pittsfield has two plants: (a) Ashley Water Treatment Plant (Ashley WTP) and Cleveland Water Treatment Plant (Cleveland WTP) (155, 156).

Each city's plant relies on an innovative package treatment process using coagulation, dissolved air flotation, and sand filtration (DAFF) that requires minimal space and operator attention, yet produces top quality water. Both Krofta Engineering Corporation (KEC) and the Lenox Institute of Water Technology (LIWT; formerly the Lenox Institute for Research) designed the DAFF package plants together. The late Dr. Milos Krofta and Dr. Lawrence K. Wang were the President and the Assistant to the President, respectively, of KEC. Dr. Krofta and Dr. Wang were also the President and the Director, respectively, of LIWT, when the Pittsfield plant was designed and built. The O'Brien & Gere Engineers and other consulting engineering firms, designed the buildings and related distribution system improvements.

Pittsfield has two main reservoirs, each serviced by a separate DAFF installation. Water from these reservoirs is typical of New England impoundment water: soft, low alkalinity water with relatively low turbidity. The Cleveland Plant which treats mainly Cleveland Reservoir water has four 49-ft-diameter DAFF units, producing up to 24+ MGD total of filtered water. The Ashley Plant, which treats mainly Ashley Reservoir water, has two

49-ft-diameter DAFF units with total capacity of 12+ MGD. Both Ashley and Cleveland Plants discharge their product water directly into the city's water supply system, which has filtered water storage in separate water storage towers located miles from the treatment plants. Start-up of individual filters can be accomplished in less than 15 min, if necessary, so that two or three DAFF units are left off line in the larger plant until demand or equipment rotation requires their startup.

Furthermore, water from the high elevation reservoir system is brought into the Ashley WTP through a low-flow, high-head hydroelectric power plant. This hydroelectric power generator was incorporated into the treatment plant design to remove 320 ft of excess head while providing the tangible benefit of reduced electric bills. The Byron-Jackson multistage turbine generator produces about 70 kW in normal operation (2.5 MGD), but can generate up to 225 kW at full flow. Power not required for the treatment plant is sold to the Northeast Utilities power grid.

Total detention time of the innovative DAFF plant is less than 20 min in comparison with a conventional sedimentation-filtration plant requiring 7–9 h of detention time.

Excellent performance results have been obtained on color and turbidity removal. Low turbidity, low alkalinity waters are difficult to flocculate and treat properly with conventional water-treatment plants using sedimentation clarification. Pittsfield raw water turbidity is in the range of 0.4–2 NTU, and the Pittsfield DAFF system has routinely produced filtered water with low turbidities in the range of 0.01–0.1 NTU. Raw water color ranging 8–70 CU has been routinely reduced to less than 1 CU. Pittsfield experienced a *Giardia* outbreak just in the year prior to the installation of the DAFF system, giving ample first-hand evidence of the need for filtration of even low turbidity surface water sources.

Both the Ashley and Cleveland plants directly discharge their floated sludge into the city sewer system for phosphorus removal, so no sludge thickening on site is required.

On behalf of all performing organizations for the success of the Pittsfield Water Treatment System, Dr. Lawrence K. Wang received the Korean Society of Water Pollution Research and Control's Outstanding Engineering Award, for innovative design and excellent performance of the world's largest DAFF plant in Pittsfield.

14.5. Example 5: Nanty Glo Water Treatment Plant, PA, USA—The First Full-Scale Flotation-Filtration Plant for Potable Water Treatment in the State of Pennsylvania

The Nanty Glo Water Treatment Plant was installed in 1986, and has a capacity of 2 MGD. There are two 20-ft-diameter flotation-filtration (DAFF) units, followed by a large on-site water storage tank, and water distribution to the town of Nanty Glo.

The source is an upland reservoir, which has experienced algal bloom problem in the summer. Incoming raw water turbidity (2–20 NTU) and color (5–20 CU) are normally low. Outgoing DAFF effluent turbidity is generally below 0.1 NTU, and effluent color is always below 1 CU. The plant is manned only during part of the day-shift, and is remotely monitored during the remainder of the time (144).

14.6. Example 6: Howell Water Treatment Plant, NJ, USA—The First Full-Scale Flotation-Filtration Plant for Potable Water Treatment in the State of New Jersey

The 8-MGD Howell Township's regional water supply system uses flotation-filtration (DAFF) units, and was installed in 1988. The raw water is from a river that is influenced by industrial and municipal effluent discharge. Summer algal growth is common, and manganese content is a problem. The incoming raw water quality is very poor with raw water turbidity ranging from 2 to 200 NTU.

This system also includes granular activated carbon (GAC) treatment following the DAFF treatment. Polymer only is used for flocculation, and the water is filtered by DAFF to less than 0.4 NTU turbidity prior to final polishing by the GAC. Color and algae removal capability was the major factor in choosing the compact DAFF units (144).

14.7. *Example 7: Millwood Water Treatment Plant, NY, USA—The First American Plant Using European DAF Technology*

The 7.5 MGD Millwood Water Treatment Plant takes water from New York City's Catskill Reservoir system and, as an alternative supply, the Croton Reservoir system. Although both sources normally have relatively low turbidity and color during spring runoff , turbidity levels can increase dramatically. The Millwood plant was the first in the US to use European dissolved air flotation (DAF) technology (instead of American DAF technology developed by Lenox Institute of Water Technology and Krofta Engineering Corporation). The Millwood Plant is owned by the Town of New Castle, and operated by a contract operations firm.

The treatment process includes rapid mixing, three-stage flocculation, dissolved air flotation clarification with surface skimmers, ozone disinfection using air-fed generators, and dual-media declining-rate filtration.

Thickened sludge from the skimmed flotation tanks is pumped to nearby lagoons, where it dewaters naturally by freezing and thawing during the winter months. Waste backwash water is treated using an inclined plate settler/thickener, with supernatant being recycled to the head of the plant and thickened sludge passing to the lagoons. The fully automatic Millwood water treatment plant (WTP) has been operational since 1993 and has consistently produced a filtered water quality of about 0.03 NTU or better (157).

14.8. *Example 8: Lee Water Treatment Plant, MA, USA—A New Generation of Flotation-Filtration Plant for Potable Water Treatment at 5 gpm/ft^2 Overflow Rate*

The Lee DAFF plant with a design capacity of 2.0 MGD (7570 m^3/day) was commissioned in December, 1998 to serve a population of approximately 6400 residents. The Lee DAFF plant utilizes the following unit processes: chemical addition/mixing, oxidation, coagulation, dissolved air flotation (DAF), automatic backwash dual media filtration (ABF), disinfection, and corrosion control.

To comply with the required filtering of three surface water sources, the town chose to install an innovative DAFF system as the best and most economical answer to their needs. In the design for 2.0 MGD flow, two dissolved air flotation-filtration (DAFF) clarifiers (Krofta Sandfloat SAF BP-24) were utilized as the main treatment system in the plant. The new generation DAFF plant (Fig. 17) has a overflow rate of 5 gpm/ft^2, representing a significant improvement in flotation technology development (79, 144).

Two main surface water sources for potable purposes are utilized, Leahey and Schoolhouse Reservoirs, which supply the approx 1.2 MGD (4542 m^3/day) to the town. The center of the Lee DAF facility is a package plant consisting of chemical pretreatment, coagulation, dissolved air flotation, and automatic backwash filtration. The following are the design criteria for the Lee DAFF Clarifiers:

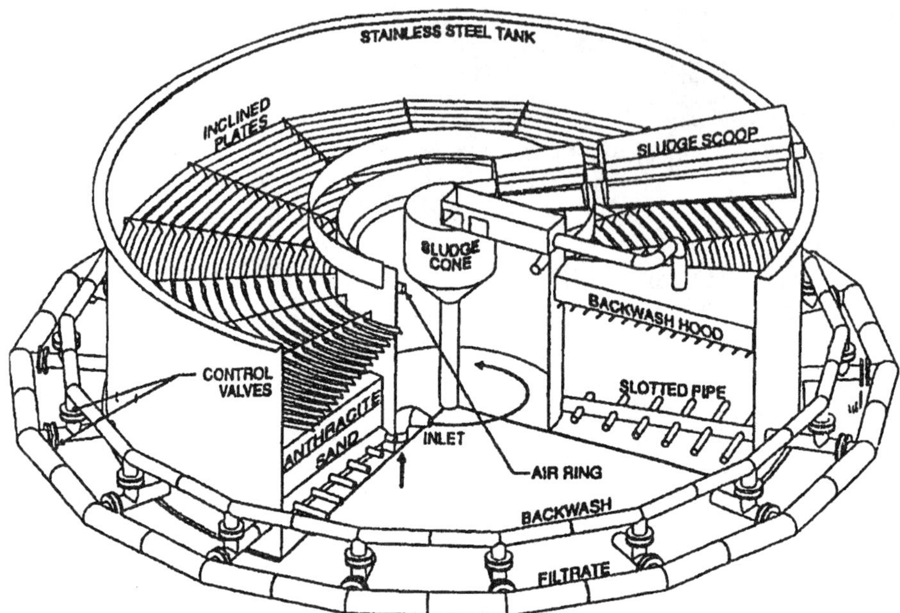

Fig. 17. Lee water treatment plant.

(a) Design flow = 2.0 MGD (7570 m^3/day).
(b) Clarifier diameter = 24 ft (7.3 m).
(c) Clarifier retention time = 16 min.
(d) Total filter area = 400 ft^2 (37.2 m^2).
(e) Design hydraulic loading = 3.5 gpm/ft^2 (0.141 m^3/min/m^2).
(f) Number of filter cells = 17.
(g) Filter cell area = 23.5 ft^2 (2.19 m^2).
(h) Filtration rate = 3.5 gpm/ft^2 (0.141 m^3/min/m^2).
(i) Filter media = 12 in. (30.5 cm) layer of 1.1 mm anthracite, and 12 in. (30.5 cm) layer of 0.35 mm sand.
(j) Filter backwash = 20 gpm/ft^2 (0.812 m^3/min/m^2) full flow; 16 gpm/ft^2 (0.650 m^3/min/m^2) partial flow.

The Town of Lee currently employs two full time operators for the treatment plant. Complete automation includes continuous monitoring of pH, turbidity, chlorine residual and flow, with alarms to alert the operator of any malfunction.

Typical chemical dosage for coagulation of Leahey Reservoir consists of a low dose of sodium hydroxide (to raise the pH to 6.5–7) followed by 0.5 mg/L of sodium aluminate, and 12–15 mg/L of aluminum sulfate [as $Al_2(SO_4)_3$]. Estimated cost for this level of chemical pretreatment is approx $0.01/1000 gal ($0.01 / 3.785 m^3) treated.

Based on the operation and performance data generated to date, the innovative Lee, MA DAFF plant consisting of chemical pretreatment, oxidation, coagulation, dissolved air flotation, automatic backwash dual media filtration, disinfection, and corrosion control has proven to be a feasible system for water purification since 1998 (79,144).

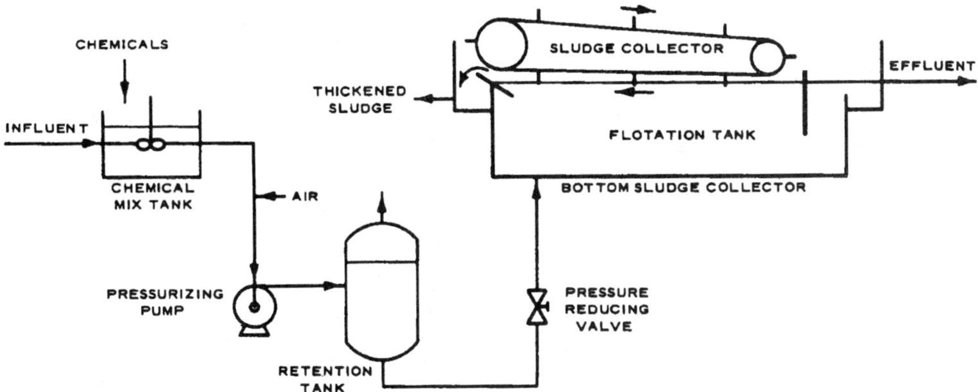

Fig. 18. Typical rectangular dissolved air flotation full-flow pressurization system (*Source*: US EPA).

14.9. Example 9: Applications and Future of Dissolved Air Flotation Technology

14.9.1. Old and New Applications of Flotation Technology

Dissolved air flotation (DAF) is traditionally used for sludge thickening in biological wastewater treatment plants (WWTP), and its sludge thickening application will continuously grow with time. A typical rectangular DAF unit is shown in Fig. 18.

New applications of dissolved air flotation (DAF) and flotation-filtration (DAFF) include (a) municipal potable water treatment; (b) municipal wastewater primary treatment (primary flotation clarification); (c) municipal wastewater secondary treatment (secondary flotation clarification); (d) municipal wastewater tertiary treatment (tertiary flotation-filtration, or tertiary DAFF); (e) storm water treatment; (f) industrial water treatment; (g) industrial effluent physicochemical treatment; (h) industrial biological treatment; (i) groundwater decontamination; and (j) wastewater recycle. Both rectangular DAF (Fig. 18) and circular DAF (Fig. 19) are equally effective for the above applications, if both are properly designed and built.

In case a combined dissolved air flotation and filtration process system (DAFF) is needed for treating a specific water or wastewater, such a combined process system may be designed and installed with two options: (a) a combined DAF-filtration (DAFF) package plant similar to the Lenox Plant, Pittsfield Plant, Coxsackie Plant, and others designed by Krofta Engineering Corporation, the Lenox Institute of Water Technology, Dongshin EnTech, etc. (144, 172); or (b) a combined treatment system consisting of individual DAF and filtration units. Both option (a) and option (b) discussed above in this section will be equally effective for water or waste treatment.

Applications for DAF and DAFF will continuously expand in both industrial and municipal applications around the world as more demonstration projects show that the processes are both technically and economically feasible for treatment of potable water, industrial effluents, and sludges.

Existing DAF and DAFF plants and new preliminary engineering designs will both confirm that the DAF and DAFF capital and O&M costs are competitive with other

Fig. 19. Typical circular dissolved air flotation system.

high-rate water, wastewater, and sludge treatment processes. When both DAF and DAFF are connected together for a two stage treatment, DAF can be placed on the top of DAFF for space saving.

There are now competing equipment manufacturers, so that DAF and DAFF are no longer considered proprietary processes, and many consulting engineering firms are capable of designing all components of DAF systems, allowing any competent contractors to install DAF and DAFF plants. The progress of DAF and DAFF technology development around the world is presented below with special emphasis on US applications.

14.9.2. *Municipal Applications*

14.9.2.1. Municipal Potable Water Treatment

Currently DAF and DAFF technology is mainly adopted in the US as smaller potable water treatment plants (WTP) where filtration has yet to be implemented, and perhaps one or two huge metropolitan areas such as New York City or Boston. For smaller WTP, DAF and DAFF have historically not been marketed as aggressively as high-rate packaged roughing filter clarifiers, which have dominated the small plant market on previously unfiltered sources. In addition, a significant upgrade and retrofit market could become available for large to medium sized plants. Several large capacity DAF and DAFF potable water treatment plants will come on line over the next few years. These plants will much enhance the reputation of DAF and DAFF as mainstream water clarification processes. The following is a partial list of potable DAF and DAFF built in the US since 1982:

(a) Ashley WTP and Cleveland WTP: 37.5 MGD (138 MLD); City of Pittsfield, MA in service since 1986 (144,155,156).
(b) Howell WTP: 8 MGD (30 MLD); New Jersey American Water Works, Howell, NJ in service since 1990 (144).
(c) Mt. Vernon WTP: 8 MGD (30 MLD); Mt. Vernon Water Works, Mt. Vernon, IN in service since 1993 (144).

(d) Millwood WTP: 7.5 MGD (28 MLD); Town of New Castle, NY in service since 1993 (144).
(e) Lee WTP: 2.0 MGD (7.5 MLD); Town of Lee, MA USA; in service since 1998 (74).
(f) Nanty-Glo WTP: 2 MGD (7.6 MLD); Town of Nanty-Glo, PA in service since 1991.
(g) Westmoreland WTP: 1.8 MGD (6.8 MLD); Westmoreland County Water Authority, Westmoreland, PA in service since 1995.
(h) Lenox WTP: 1.2 MGD (6 MLD); Town of Lenox, MA in service since 1982 (12,13,121).
(i) Coxsackie WTP: 1.1 MGD (4.2 MLD); Coxsackie, NY in service since 1990 (29,144).
(j) Lakeville WTP: 0.75 MGD (2.8 MLD); Bridgeport Hydraulic Company, Lakeville, CT in service since 1996.
(k) Lake Vangum WTP: 0.75 MGD (2.8 MLD); Bridgeport Hydraulic Company, Lakeville, CT in service since 1996.
(l) Greenville Water System: 75 MGD (280 MJD); Greenville Water System, Greenville, SC in service since 2000 (177,178).
(m) West Nyack WTP: 20 MGD (T5 MLD); West Nyack, NY in service since 2003 (184).

It is encouraging to note that more new DAF and DAFF potable water treatment plants are being designed or under construction. The following is a partial list of potential new DAF and DAFF water treatment plants to be built in the US:

(a) Boston WTP: 450 MGD (1700 MLD); Massachusetts Water Resources Authority, Boston, MA.
(b) Hemlocks WTP: 50 MGD (190 MLD); Bridgeport Hydraulic Company, CT.
(c) Cambridge WTP: 24 MGD (91 MLD); City of Cambridge, MA.
(d) Montrose WTP: 7 MGD (26 MLD); Northern Westchester Joint Water Works, NY.
(e) Danbury WTP: 5.5 MGD (21 MLD); City of Danbury, CT.

Internationally DAF and DAFF technology has been adopted for potable water treatment in the entire nation of South Korea, including the Seoul Water Treatment Plant (80,124,172). Under an assignment of the US Department of State, Dr. Lawrence K. Wang served as the Senior Advisor of the United Nations Industrial Development Organization (UNIDO), Vienna, Austria. DAF and DAFF technology was inspected and recommended by UNIDO for adoption by all developing countries that urgently need cost-effective water and wastewater treatment systems (129,130,154,175). Figure 20 shows a high-rate DAFF plant (Source: Dongshin, Seoul, Korea) for potable water treatment in Korea (172). This Korean potable water DAFF plant (Fig. 20) is very similar to the Krofta/LIWT potable water DAFF plant (Fig. 17) installed in Lee, MA.

14.9.2.2. Municipal Waste Treatment

Sludge thickening by DAF technology is considered a mainstream process. In general, the municipalities adopt conventional gravity thickeners for treatment of primary sludges, but adopt innovative DAF thickeners for treatment of secondary sludges from biological waste-treatment plants.

New municipal waste-treatment applications of DAF and DAFF include: (a) storm runoff treatment (19,61,104,105,144,149); (b) primary flotation clarification; (c) secondary flotation clarification (40–47,85–86,91–95,108,126,136,139–141,144–146); and (d) tertiary flotation-filtration (DAFF) clarification (88, 89,115,144). Primary clarification and/or secondary clarification can be replaced by DAF for cost saving. Figures 21 and 22 show the positions of clarifiers where DAF can be used for clarification in

Fig. 20. DAFF plants (each 10,000 m^3/day capacity; Model SDF-27-0210) installed in Misa-ri, Hanam-city, and Gyeonggi-do, Korea, for the boat racing arenas (Source: Dongshin, Seoul, Korea).

activated sludge process systems and trickling filter systems, respectively. Similarly DAF can be applied to nitrification and denitrification process systems for clarification, as shown in Fig. 23 (144). The use of DAF as primary, secondary, or tertiary flotation clarification is fully demonstrated for full-scale industrial operations. It will take time for municipalities to adopt industrial applications for their own applications. DAFF generally is only suitable for secondary clarification or tertiary clarification.

14.9.3. *Industrial Applications*

14.9.3.1. Industrial Water Treatment

The quantity of water consumed by industries is huge. For instance, about 33% of the total water produced by the City of Pittsfield Water Treatment Plant, MA, is consumed by the General Electric Co. Many industrial water-treatment plants using DAF and DAFF technology have been built around the world due to the low cost of DAF and DAFF clarifiers. It is unquestionable that the DAF and DAFF technology used in municipalities can be readily adopted for industrial water treatment. It is easy for industry to adopt a municipal technology as long as the technology is technically and economically feasible (130,155,156,172).

14.9.3.2. Industrial Waste Treatment

With technological collaboration with the Lenox Institute of Water Technology (LIWT), Krofta Engineering Corporation (KEC) alone has designed and installed over 3000 DAF and DAFF clarifiers for industrial wastewater treatment around the world. Over 90% of all DAF and DAFF units built by all flotation manufacturers are used for industrial waste treatment (144,172). It has been known that DAF and DAFF are technically and economically feasible for treating most industrial wastewaters and other

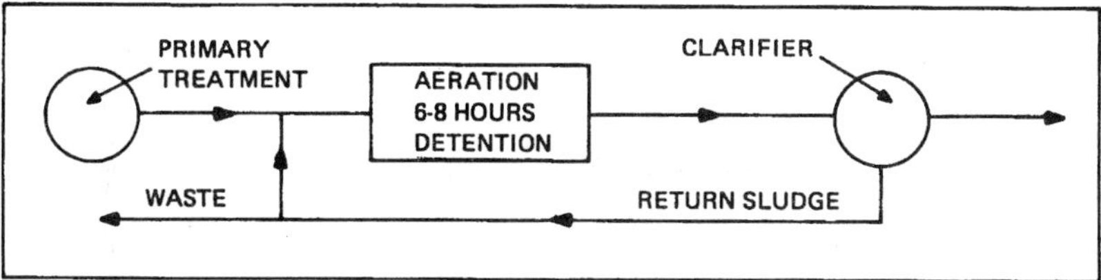

Conventional activated-sludge process.

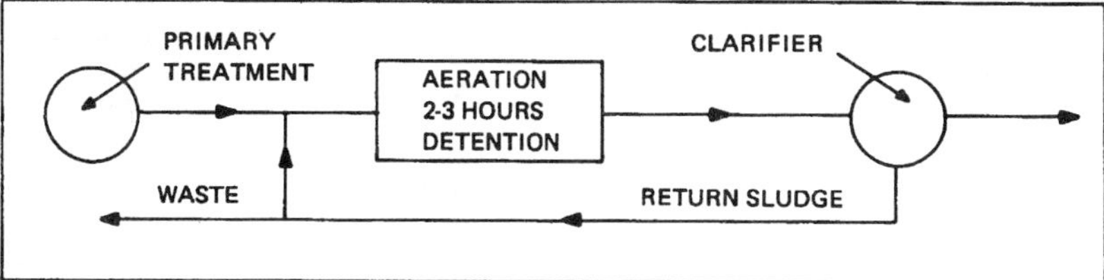

High-rate activated-sludge process.

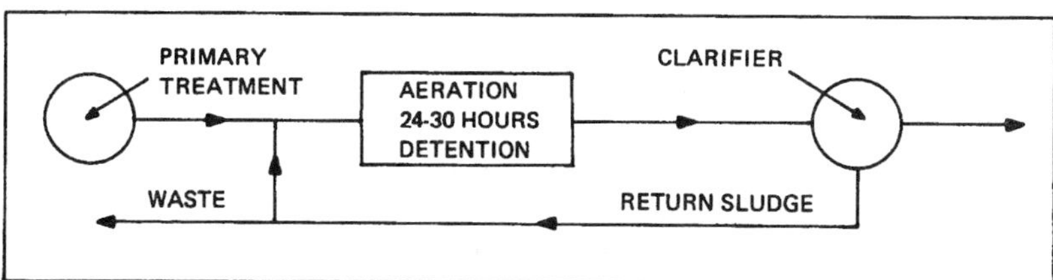

Extended-aeration process.

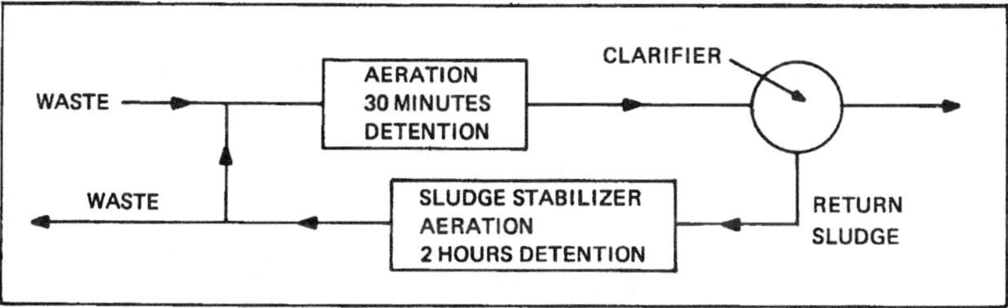

Contact-stabilization process.

Fig. 21. Activated sludge process systems in which DAF can be used for clarification (Source: US EPA).

polluted waters, such as: (a) paper and pulp mills; (b) tannery factories; (c) can food manufacturing; (d) seafood processing; (e) livestock processing; (f) petroleum refineries; (g) gas stations; (h) foundries; (i) metal finishing; (j) iron and steel manufacturing; (k) textiles; (l) steam electric power plants; (m) inorganic chemicals manufacturing; (n) ore mining and dressing; (o) porcelain enameling; (p) paint and ink formulation; (q) coil coating; (r) nonferrous metals manufacturing; (s) aluminum forming; (t) battery

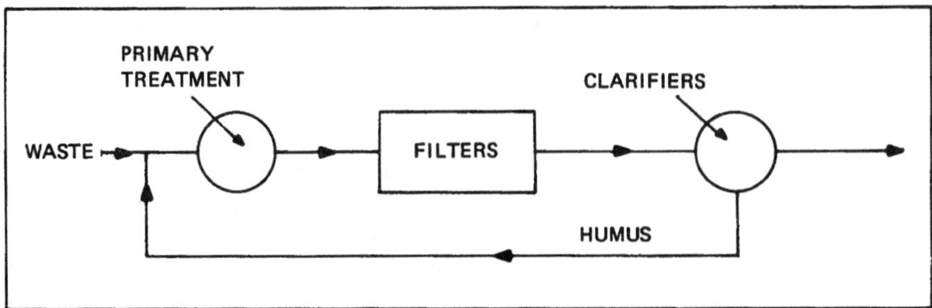

Standard-rate trickling filters.

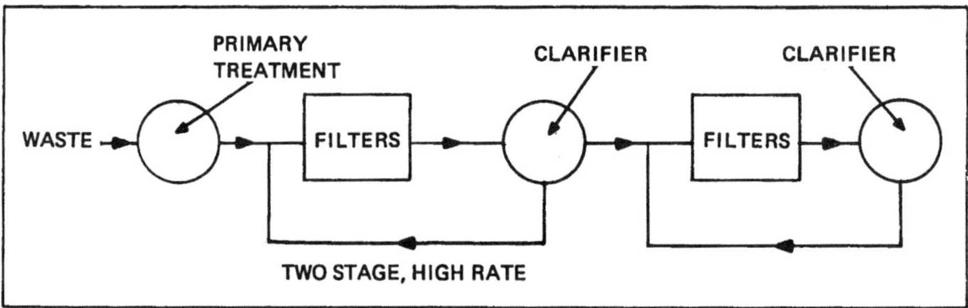

High-rate trickling filters.

Fig. 22. Trickling filter systems in which DAF can be used for clarification (Source: US EPA).

manufacturing; (u) electrical and electronic components; (v) copper coating; (w) organic and inorganic wastes; (x) auto and other laundries; (y) contaminated groundwater; (z) combined industrial and municipal wastewater.

Tables 11 and 12 document the removal data from the US Environmental Protection Agency, for treatment of auto and laundry wastes and petroleum refining wastes, respectively, by a combined flotation and filtration system (158). Either a combined DAFF package plant or a combined treatment system consisting of separate DAF and filtration units will be equally effective for the intended industrial waste treatment.

The use of DAF and DAFF for treating various industrial water or wastewater can be found from the literature (49–54,81,82,93,96–101,113–117,119,127,129,132,137,144, 154,175). Figure 24 shows a newly developed activated sludge contact stabilization system using DAF for both primary clarification and secondary clarification in a complete tannery wastewater treatment system (175).

It should be noted that of over 3000 DAF and DAFF units designed and installed by Krofta Engineering Corporation (KEC) and the Lenox Institute of Water Technology (LIWT) around the world, over 95% of installed DAF and DAFF units were for industrial applications (industrial water, wastewater, and sludge treatment) (144). Municipal applications of DAF and DAFF have been well demonstrated, but the municipal market remains a new frontier to be further developed.

14.10. Example 10: Combined Sedimentation and Flotation Process

Sedimentation clarification and flotation clarification, each has its own advantages and disadvantages. Under certain situations, a combination of the two will give the best

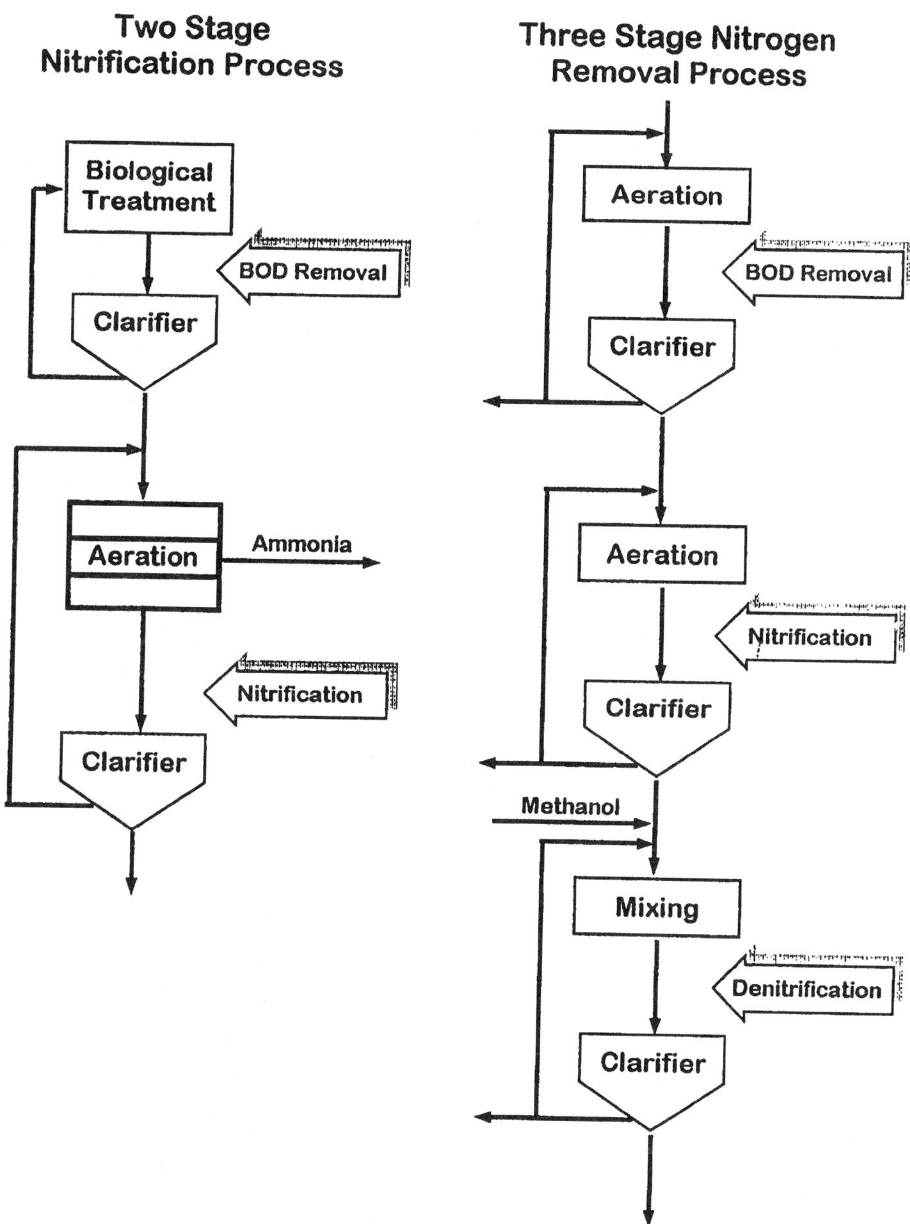

Fig. 23. Nitrification and denitrification process systems using DAF for clarification (Source: Lenox Institute of Water Technology, Lenox, MA).

performance. Krofta organizations (including the Lenox Institute of Water Technology, Krofta Engineering Corporation, Krofta Waters, Inc., KWI, Krofta Technology Corporation) in Lenox–Pittsfield, MA, USA, have developed a combined sedimentation-flotation clarifier, commercially known as Sedifloat, which is circular in shape.

Dongshin in Seoul, Korea, has developed a rectangular combined sedimentation-flotation clarifier, shown in Fig. 25. While both circular and rectangular combined

Table 11
Treatment of Auto and Laundry Wastes by DAF and Filtration

	Removal Data[a] Sampling: 2 d composite and grab		
	Concentration		
Pollutant/parameter	Influent	Effluent	Percentage removal
Classical pollutants (mg/L)			
BOD_5	180	93	48
COD	2100	1100	48
TOC	540	290	46
TSS	740	71	90
Oil and grease	76	47	38
Total phenol	0.094	0.076	19
Total phosphorus	12	2.1	83
Toxic pollutants (μg/L)			
Antimony	2300	1800	22
Arsenic	3.5	BDL	86*
Cadmium	40	9.5	76
Chromium	360	200	44
Copper	660	350	47
Cyanide	<10	13	NM
Lead	1000	180	82
Mercury	1.0	<1.0	>0
Nickel	270	<220	19
Selenium	BDL	BDL	NM
Silver	66	53	20
Zinc	2300	1200	48
Bis(2-ethylhexyl) phthalate	90	98	NM
Butyl benzyl phthalate	41	ND	>99
Di-*n*-butyl phthalate	300	270	10
Di-*n*-octyl phthalate	11	ND	>99
Phenol	28	19	32
Ethylbenzene	3.0	2.0	33
Toluene	4.5	5.0	NM
Anthracene/Phenanthrene	10	3.5	65
2-Chloronapthalene	17	17	0
Carbon tetrachloride	210	15	93
Chloroform	10	20	NM
Methylene chloride	8	110	NM
Tetrachloroethylene	ND	0.5	NM
Mapthalene	11	1.5	86
1,1,1-Trichloroethane	860	55	94
Trichlorofluoromethane	ND	6	NM
Acrolein	360	ND	>99

Source: US EPA.

[a]Blanks indicate data not available. BDL, below detection limit. ND, not detected. NM, not meaningful.

*Approximate value.

Table 12
Treatment of Petroleum Refining Wastes by DAF and Filtration

Removal data[a]
Sampling: Average of three daily samples analysis

	Concentration		
Pollutant/parameter	Influent	Effluent	Percentage removal
Classical pollutants (mg/L)			
COD	140	56	59
TOC	43	22	49
TSS	50	4	92
Oil and grease	35	6	83
Total phenol	0.024	0.023	4
Toxic pollutants (µg/L)			
Chromium	200	34	83
Copper	28	7	75
Mercury	0.8	<0.5	>37
Zinc	200	92	55
Silver	<3	<3	NM
Beryllium	<2	<2	NM
Cadmium	<1.5	<1.5	NM
Nickel	<10	<10	NM
Lead	<18	<18	NM
Arsenic	<20	<20	NM
Antimony	<25	<25	NM
Thallium	<15	<15	NM

Source: US EPA
[a]Blanks indicate data not available. NM, not meaningful.

sedimentation-flotation clarifiers perform equally well, the rectangular combined sedimentation-flotation clarifiers (Fig. 25) are popular due to the fact that many existing sedimentation clarifiers have been easily upgraded to a combined sedimentation-flotation clarifier for increasing both performance efficiency and hydraulic capacity.

14.11. Example 11: Greenville Water System, SC, USA—The Largest Potable Water DAF Plant in the USA in 2003

The City of Greenville, in the foothills of the Blue Ridge Mountains, obtains its drinking water from three surface water supplies: Table Rock Reservoir, North Saluda Reservoir, and Lake Keowee (177, 178). Table Rock Reservoir in Greenville County is a man-made lake, created in 1930, along the South Saluda River. The watershed to the reservoir covers approx 9000 acres and is completely owned by the Greenville Water System (GWS), which is run by the commissioners of Greenville's Public Works Division. The North Saluda Reservoir, located on the North Saluda River, is a man-made reservoir that was brought on line in 1961. It has a watershed area of approx 17,000 acres, which is also entirely owned by the Greenville Water System. In July of 2000, the Greenville Water System began using a new filtration plant that uses state-of-the-art

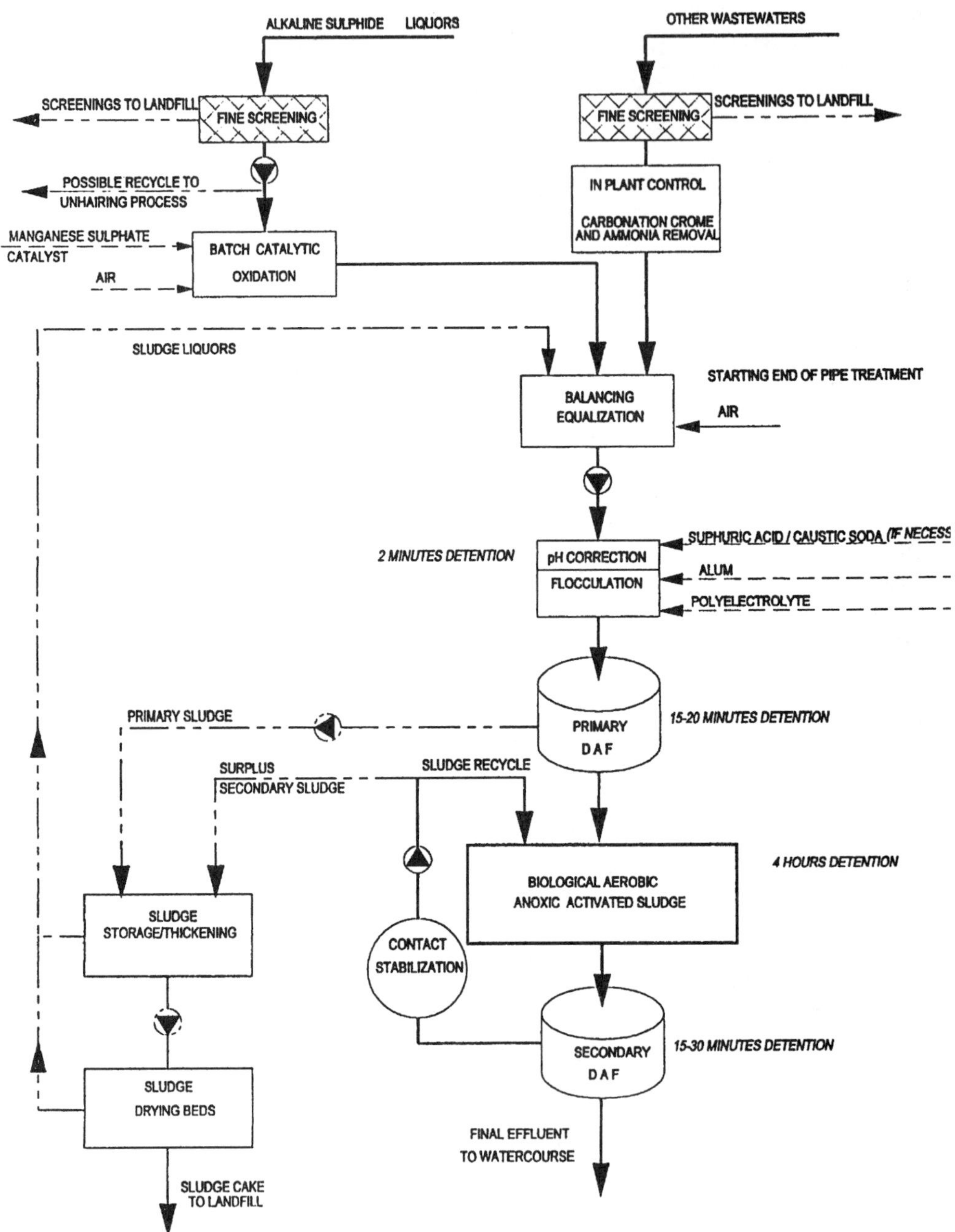

Fig. 24. An activated sludge contact stabilization process using primary DAF clarification and secondary DAF clarification for tannery wastewater treatment (Source: Lenox Institute of Water Technology, Lenox, MA).

DAF technology to clarify and filter the raw water from Table Rock and North Saluda. Before this new DAF plant was brought into service, the water from these two reservoirs was unfiltered. The new 75-MGD DAF facility is the largest plant of its type in the United States (177, 178). At the Table Rock and North Saluda treatment plants, the raw water is filtered, chlorinated, and fluoridated. Small amounts of caustic soda are added for pH control,

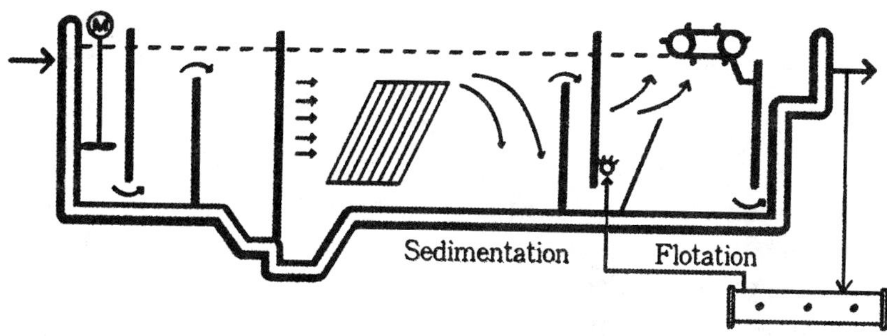

Fig. 25. Combined sedimentation and flotation process (Source: Dongshin, Seoul, Korea).

and a corrosion control agent is added to protect the integrity of the pipes in the distribution system.

Because the Greenville Water System owns all the land within the watersheds of the Table Rock and North Saluda Reservoirs, it maintains control over the activities that occur within the vicinity of the lakes. The watersheds are completely undeveloped, and the GWS employs full-time staff to patrol the watersheds and ensure that there are no activities occurring that could contaminate the public water supply. Public recreation is prohibited within the watershed. The commissioners of the Public Works Division signed a conservation easement with The Nature Conservancy in 1995 to offer even greater protection to the watersheds and the public water supply.

Lake Keowee is designated as a No Discharge Zone by the US EPA. This determination was published in the Federal Register on March 4, 1999, after the state of South Carolina requested that the USEPA determine that "adequate and reasonable" pump-out facilities exist for boaters on several lakes, including Lake Keowee. The No Discharge Zone designation prohibits the direct discharge of sewage from any vessel into the lake.

NOMENCLATURE

- a — Air solubility or saturation at 1 atmospheric pressure (std. conditions) (mL/L)
- A — Mass flow rate of air released for flotation of suspended solids (mg/s)
- A_c — Cross-sectional area of a flotation chamber (m^2)
- a_r — Air released at atmospheric pressure at 100% saturation in liquid (mL/L) (air volume at standard conditions, e.g., 0°C + atm. absolute)
- A_s — Surface area required to reach desired sludge consistency (m^2)
- C — 2003 capital cost of DAF thickening proces ($)
- C_a — Air released at 1 atmospheric pressure (mg/L)
- C_{as} — Air saturation concentration at atmospheric conditions (mg/L)
- C_e — Solubility of gas in the flotation effluent (mg/L)
- C_f — Dissolved gas concentration in the raw influent feed stream (mg/L)
- C_g — Gas released at 1 atmospheric pressure (mg/L)
- C_{gs} — Gas saturation concentration at atmospheric conditions (mg/L)
- C_r — Solubility of gas in water in the pressurized retention tank (mg/L)
- d — Air density (lb/ft^3)
- D — Effective depth of the flotation chamber (m)
- E — Percentage of suspended solids removal

f	Factor of gas dissolution at pressure P (where P is any pressure above normal atmospheric pressure of 1 atm), fraction, usually 0.167–1.0
F	Fraction of gas dissolution at pressure P (where P is any pressure higher than 2 atm), fraction, usually 0.5–1.0
F'	Factor for short circuiting and turbulence, assumed as 1.4
G_{in}	Mass flow rate of dissolved gas entering the flotation chamber (mg/sec)
G_{out}	Mass flow rate of dissolved gas leaving the flotation chamber (mg/sec)
H_o	Height of floated sludge (float)
L	Effective length of flotation chamber (in.)
P	Gage pressure, psig or absolute pressure in air dissolving tank, psig
P'	Gage pressure (kPa)
P_{at}	Pressure (atm)
P_e	Gas pressure in the flotation chamber's effluent compartment, atmosphere taken to be 1 atm
PF	Plant flow (MGD)
P_r	Gas pressure in the retention tank before release [lb/in$^{2.}$ (absolute) or atmosphere]
Q	Influent flow rate (m^3/sec)
Q_a	Air flow rate (cfm)
Q_n	Portion of influent feed stream which is not pressurized (L/s)
Q_r	Recycle flow (L/s)
Q_w	WWT plant design flow (MGD)
R	Recirculation ratio Q_r/Q
S	mass flow rate of suspended solids entering the flotation system (mg/s)
T	Detention time (s)
V_f	Volume of float (mL)
V_H	Horizontal velosity (m/s)
V_i	Volume of influent used (mL)
V_r	Volume of recycle water used (mL) = 1000 mL – V_i
w	Influent suspended solids load (lb/min)
W	Width of flotation chamber (in.)
WPR	Water production rate (MGD)
X	Suspended solids concentration of influent feed stream (mg/L)
X_e	Suspended solids of the clarified effluent or subnatant (mg/L) = X_r
X_f	Suspended solids of float (mg/L)
X_i	Influent suspended solids (including chemicals, when required) (mg/L)
X_r	Suspended solids of recycle water (mg/L)

ACKNOWLEDGMENTS

This chapter is written by the authors in memory of the late Dr. Milos Krofta, the father of flotation technology, who designed and built the first five full-scale potable flotation-filtration (DAFF) plants in America in the period of 1982–1988: (a) Lenox Water Treatment Plant, MA; (b) Coxsackie Water Treatment Plant, NY (c) Pittsfield Water Treatment Plant, MA; (d) Nanty Glo Water Treatment Plant, PA; and (e) Howell Water Treatment Plant, NJ. Dr. Krofta passed away on August 22, 2002, shortly after he celebrated his 90th birthday. Dr. Krofta had over 50 US and foreign patents on flotation technologies. His organizations (Lenox Institute of Water Technology, Krofta Engineering Corporation, Krofta Waters, Inc, KWI, Inc., and Krofta Technologies Corporation) had developed, designed, and built over 3000 flotation units for potable water treatment and industrial effluent treatment around the world. Both late Dr. Milos Krofta and Dr. Lawrence K. Wang received the Pollution Engineering magazine's PE Five-Star Award and Korean Society of Water Pollution Research and Control's Outstanding Engineering Award, owing to their outstanding design and the five DAFF plants' excellent performance. The late Dr. Milos Krofta was the President of the Lenox Institute of Water Technology (LIWT) and all of his Krofta organizations. In Dr. Krofta's final days, he established the International Association of Flotation Technology. This chapter is one of many publications written by the former LIWT faculty, students and friends in memory of the late Dr. Krofta for his lifelong accomplishment in development and promotion of dissolved air flotation for water, wastewater and sludge treatment. At the time of this study, Dr. Lawrence K. Wang was the Dean, Director, and Professor of LIWT and the "Assistant to the President" of Krofta Engineering Corporation. Mr. Edward Fahey was an Environmental Engineer as well as a part-time Graduate Student at LIWT, from where Mr. Fahey received his Master of Engineering degree.

REFERENCES

1. L. K. Wang, *Environmental Engineering Glossary*, Veridian Engineering (formerly Calspan Corporation), Buffalo, New York, 420 pp. 1974.
2. L. K. Wang, *Theory and Applications of Flotation Processes*. US Department of Commerce, National Technical Information Service. Springfield, VA, PB 86-194198/AS, 1985.
3. I. Kumar and W. E. Eustance, *Flotation Processes*, Lenox Institute of Water Technology (formerly Lenox Institute for Research), Lenox, MA. Technical Report No. LIR 03-88-284, 1988.
4. L. Boyd and C. L. Shell, *Proceedings of the 27th Industrial Waste Conference*, 1972, pp. 705–713.
5. L. K. Wang, *J. American Water Works Assoc.*, **67**, 19–21; 182–184 (1975).
6. L. K. Wang and R. C. Ross, *International J. Environ. Anal. Chem.*, **4**, 285–300 (1976).
7. L. K. Wang, *J. Am. Oil Chem. Soc.* **52**, 340–346 (1975).
8. L. K. Wang and D. F. Langley, *Industrial Engineering & Chemistry, Product Research and Development* **14**, 210–212 (1975).
9. L. K. Wang, *J. Appl. Chem. Biotech.* **25**, 475–490 (1975).
10. L. K. Wang and W. W. Shuster, *Industrial Engineering & Chemistry, Product Research and Development*, **14**, 312–314 (1975).
11. L. K. Wang, M. H. S. Wang, and J. F. Kao, *Water, Air and Soil Pollution* **9**, 337–348 (1978).
12. M. Krofta and L. K. Wang, *J. New England Water Works Assoc.* 249–264 (1985).
13. M. Krofta and L. K. Wang, *J. New England Water Works Assoc.* 265–284 (1985).
14. L. K. Wang, *Engineering Manual for Operation, Testing, and Monitoring of Hillcrest Wastewater Treatment Facilities*. Lenox Institute of Water Technology (formerly Lenox Institute for Research), Lenox, MA. Technical Report No. LIR 08/87–255, 1987.
15. M. Krofta and L. K. Wang, *J. American Water Works Assoc.* **74**, 304–310 (1982).
16. M. Krofta and L. K. Wang, *Proceedings of the 1987 Joint Conference of American Water Works Association and Water Pollution Control Federation*, Cheyenne, WY, 1987.
17. L. K. Wang, D. Barns, P. Milne, B. C. Wu, and J. Hollen, *Removal of Extremely High Color from Water Containing Trihalomethane Precursor by Flotation and Filtration*, US

Department of Commerce, National Technical Information Service, Springfield, VA. Technical Report No. PB83-240374, 11, 1983.

18. M. Krofta, L. K. Wang, and M. Boutroy, *Development of a New Treatment System Consisting of Adsorption Flotation and Filtration.* US Department of Commerce, National Technical Information Service, Springfield, VA. Report No.PB85-20940l/AS, 1984.
19. M. Krofta, L. K. Wang, R. Robinson, and W. Mahoney, *Flotation Treatment of Contaminated Storm Run Off Water.* Lenox Institute of Water Technology (formerly Lenox Institute for Research), Lenox, MA. Technical Report No. LIR/03-85-124, 1985.
20. L. K. Wang, B. C. Wu, A. Meier, et al., *Removal of Arsenic from Water and Wastewater.* Lenox Institute of Water Technology (formerly Lenox Institute for Research), Lenox, MA. Technical Report No. LIR/10-84/6; US Department of Commerce, National Technical Information Service, Springfield, VA. Report No. PB86-169299, 1984.
21. M. Krofta, L. K. Wang, B. C. Wu, and R. Foote, *Treatment of West Springfield Raw Water for Ethylene Dibromide Removal*, Lenox Institute of Water Technology (formerly Lenox Institute for Research), Lenox, MA. Technical Report No. LIR/05-85/138, 1985.
22. M. Krofta, L. K. Wang, R. L. Spencer, and J. Weber, *Proceedings of the Water Quality and Public Health Conference*, Worcester Polytechnic Institute, Worcester, MA, pp. 103–110; US Department of Commerce, National Technical Information Service, Springfield, VA. Report No. PB83-219550, 1983.
23. Y. Nurdogan, *An Advanced Dissolved Air Flotation System for Microalgae Separation.*, University of California, Berkeley, CA, 1987.
24. L. K. Wang, *Odor Pollution Control by Chemical Treatment*, Lenox Institute of Water Technology (formerly Lenox Institute for Research), Lenox, MA. Technical Report No. LIR/10-85/156, 1985.
25. L. K. Wang and P. J. Kolodziej, *Proceedings of the Water Quality and Public Health Conference.* Worcester Polytechnic Institute, Worcester, MA, pp. 17–29; US Department of Commerce, National Technical Information Service, Springfield, VA. Report No. PB83-244053, 1983.
26. M. Krofta and L. K. Wang, *American Institute of Chemical Engineers National Conference Proceedings*, Houston, TX; US Dept of Commerce, National Technical Information Service, Springfield, VA. Report No. PB83-232843, 1983.
27. L. K. Wang and B. C. Wu, *OCEESA J.* **1**, 15–18 (1984); US Department of Commerce, National Technical Information Service, Springfield, VA. Report No. PB85-167229/AS.
28. L. K. Wang and B. C. Wu, *Treatment of Watervliet Reservoir Water*, Lenox Institute of Water Technology (formerly Lenox Institute for Research), Lenox, MA. Technical Report No. LIR/05-87/262, 1987.
29. L. K. Wang and C. Gaetani, *Investigation of Potable Water Supply at the Village of Coxsackie, New York*, Lenox Institute of Water Technology (formerly Lenox Institute for Research), Lenox, MA. Technical Report No. LIR/09-86/204, 1986.
30. L. K. Wang and B. C. Wu, *Investigation of a Sandfloat Pilot Plant for the City of Beacon, New York.* Lenox Institute of Water Technology (formerly Lenox Institute for Research), Lenox, MA. Technical Report No. LIR/07-87/263, 1987.
31. C. J. Bien, *A Study of Lead Content and Proposed New Flotation Technique of Lead Removal in Drinking Water.* Lenox Institute of Water Technology (formerly Lenox Institute for Research), Lenox, MA. Master's Thesis No.LIR/MS-l982/Bien (M. Krofta and L. K. Wang, advisors), 1982.
32. B. C. Wu, *Separation of Organics and Inorganics from Water by Bubble Separation Process.* Lenox Institute of Water Technology (formerly Lenox Institute for Research), Lenox, MA. Master's thesis No. LIR/MS-1982/Wu. (M. Krofta and L. K. Wang, advisors), 1982.

33. M. Krofta and L. K. Wang, *Treatment of Raw Water from Black River, NY by Flotation-Filtration Process*, Lenox Institute of Water Technology (formerly Lenox Institute for Research), Lenox, MA. Technical Report No. LIR/06-84/5, 1984.
34. M. Krofta and L. K. Wang, *Proceedings of the American Water Works Association, Water Reuse Symposium III*, San Diego, CA, Vol. 3, pp. 1238–1250, 1984.
35. M. Krofta and L. K. Wang, *Proceedings of the American Water Works Association, Water Reuse Symposium II*, San Diego, CA, Vol. 3, pp. 1251–1264, 1984.
36. T. F. Zabel and J. D. Melbourne, *Development in Water Treatment.* Water Research Center, Medmenham, Bucks, UK, 1980.
37. M. Krofta and L. K. Wang, *Drying*, Hemisphere Publishing Corp., Harper & Row Publishers, New York, 1986, Vol. 2, pp. 765–771.
38. M. Krofta and L. K. Wang, *Drying*, Hemisphere Publishing Corp., Harper & Row Publishers, New York, 1986, Vol. 2, pp. 772–780.
39. US EPA, *Process Design Manual for Sludge Treatment and Disposal.* US Environmental Protection Agency, Cincinnati, OH. Technical Report No. 625/1-74-006, 1974.
40. M. Krofta, D. B. Guss, and L. K. Wang, *Civil Engineering for Practicing & Design Engineers*, Pergamon Press, New York, 1983, Vol. 2, pp. 307–324.
41. C. P. C. Poon, L. K. Wang, and M. H. S. Wang, In: *Handbook of Environmental Engineering,* L. K. Wang and N. C. Pereira (eds.), Humana Press, Totowa, NJ, 1986, Vol. 3, pp. 229–303.
42. M. Krofta and L. K. Wang, *TAPPI J.* **70**, 92–96 (1987).
43. M. V. Childers, *Proceedings of National Council for Air and Improvement Southern Regional Meeting*, 1984.
44. L. K. Wang, M. H. S. Wang, and C. P. C. Poon, Trickling Filters. In: *Handbook of Environmental Engineering,* L. K. Wang and N. C. Pereira (eds.), Humana Press, Totowa, NJ, 1986, Vol. 3, pp. 361–426.
45. M. Krofta and L. K. Wang, *Proceedings of the 41st Industrial Waste Conference*, Lewis Publishers Inc., Chelsea, MI, 1987, pp. 67–72.
46. L. K. Wang, M. H. S. Wang, and C. P. C. Poon, *Effluent and Water Treatment Journa*l. **24**, 9–95 (1984).
47. C. P. C. Poon, L. K. Wang, and M. H. S. Wang, Waste Stabilization Ponds and Lagoons. In: *Handbook of Environmental Engineering*, L. K. Wang and N. C. Pereira (eds.), Humana Press, Totowa, NJ, 1986, Vol. 3, pp. 305–360.
48. M. Krofta, L. K. Wang, R. L. Spencer, and J. Weber, *Algae Separation by Dissolved Air Flotation.* US Department of Commerce, National Technical Information Service, Springfield, VA. Technical Report No. PB83-219550, 1983.
49. M. Lundh, L. Jonsson, and J. Dahlquist, *Water Research* **36**, 1585–1595 (2002).
50. A. R. Shawwa and D. W. Smith, *Water Science Technology* **38**, 245–252 (1998).
51. F. W. Pontius, *Water Quality and Treatment*, 4th ed., McGraw-Hill, New York, 1990, pp. 367–453.
52. Y. Chung, Y. C. Choi, Y. H. Choi, and H. S. Kang, *Water Research* **34**, 817–824 (2000).
53. S. Steinbach and J. Haarhoff, *Water Science Technology* **38**, 303–310 (1998).
54. A. A. Al-Shamrani, A. James, and H. Xiao, *Water Research* **36**, 1503–1512 (2002).
55. J. K. Edzwald and M. B. Kelley, *Water Science Technology* **37**, 1–8 (1998).
56. K. French, R. K. Guest, G. R. Finch, and C. N. Haas, *Water Research* **34**, 4116–4119 (2000).
57. L. Stoica, M. Dinculescu, and C. G. Plapcianu, *Water Research* **32**, 3021–3030 (1998).
58. G. Offringa, *Water Science Technology* **31**, 159–72 (1995).
59. T. H. Chung and D. Y. Kim, *Water Science Technology* **36**, 223–230 (1997).
60. N. T. Manjunath, I. Mehrotra, and R. P. Mathur, *Water Research* **34**, 1930–1936 (2000).
61. S. Laine, T. Poujol, S. Dufay, J. Baron, and P. Robert, *Water Science Technology* **38**, 99–105 (1998).

62. P. Jokela and P. Keskitalo, *Water Science Technology* **40**, 33–41 (1999).
63. J. P. Malley, Jr., *Environmental Technology* **11**, 1161–1168 (1991).
64. D. Q. Bunker, Jr., J. K. Edzwald, J. Dahlquist, and L. Gillberg, *Water Science Technology* **31**, 63–71 (1995).
65. D. M. Leppinen, *J. Water Supply: Research And Technology-Aqua* **49**, 258–259 (2000).
66. D. Miskovic, B. Dalmacija, Z. Zivanov, E. Karlovic, Z. Hain, and S. Maric, *Water Science Technology* **18**, 105–114 (1986).
67. J. K. Edzwald, J. P. Walsh, G. S. Kaminski, and H. J. Dunn, *J. Am. Water Works Assoc.*, **84**, 92–100 (1992).
68. K. Fukushi, Y. Matsui, and N. Tambo, *J. Water Supply: Research And Technology-Aqua.* **47**, 76–86 (1998).
69. R. Klute, S. Langer, and R. Pfeifer, *Water Science Technology* **31**, 59–62 (1995).
70. A. Vlaski, A. N. Van Breemen, and G. J. Alaerts, *J. Water Supply: Research And Technology-Aqua.* **45**, 53–261 (1996).
71. C. C. Ho and K. Ahmad, *J. Colloid Interface Sci.* **216**, 25–33 (1999).
72. M. Lundh, L. Jonsson, and J. Dahlquist, *Water Research* **34**, 21–30 (2000).
73. M. Lundh, L. Jonsson, and J. Dahlquist, *Water Science Technology* **43**, 185–194 (2001).
74. C. T. Ta, J. Beckley, and A. Eades,*Water Science Technology* **43**, 145–152 (2001).
75. A. R. Shawwa and D. W. Smith, *Water Science Technology* **38**, 245–252 (1998).
76. V. Dupre, M. Ponasse Y. Aurelle, and A. Secq, *Water Research* **32**, 2491–2497 (1998).
77. M. Krofta and L. K. Wang, *Development of An Innovative and Cost Effective Municipal-Industrial Waste Treatment System.* US Department of Commerce, National Technical Information Service, Springfield, VA. PB88-168109/AS, 1985.
78. R. J. Scriven, S. K. Ouki, A. S. Doggart, and M. J. Bauer, *Water Science Technology*, **39**, 211–215 (1999).
79. E. Fahey, *Pilot Scale Demonstrations and Full Scale Operation of Potable Water Flotation-Filtration Plants.* Lenox Institute of Water Technology, Lenox, MA. Master's thesis. (L. K. Wang and D. B. Aulenbach, advisors), 2001.
80. M. Krofta and L. K. Wang, *Treatment of Potable Water from Seoul, Korea by Flotation, Filtration and Adsorption.* US Department of Commerce, National Technical Information Service, Springfield, VA. PB88-200530/AS, 1985.
81. M. Krofta and L. K. Wang, *Proceedings of the 1985 Powder and Bulk Solids Conference*, Chicago, IL, 28 pp. 1985.
82. L. K. Wang, B. C. Wu, R. Fat, and F. Rogalla, *Treatment of Scallop Processing Wastewater by Flotation, Adsorption and Ion Exchange.* US Department of Commerce, National Technical Information Service, Springfield, VA. PB89-143556/AS, 1985.
83. G. M. Huntley, L. K. Wang, and L. W. Layer, *Evaluation of Sodium Aluminate as a Coagulant for Cost Savings at Water Treatment Plants.* US Department of Commerce, National Technical Information Service, Springfield, VA. PB88-168075/AS, 1985.
84. M. Krofta, L. K. Wang, and R. Foote, *Separation of High Grade Sulfur from an Ore by Flotation.* Lenox Institute of Water Technology (formerly Lenox Institute for Research), Lenox, MA., Technical Report No. LIR/09-85/155, 1985.
85. L. K. Wang, *Secondary Treatment of Bangor Primary Effluent by Supracell Clarifier.* Lenox Institute of Water Technology (formerly Lenox Institute for Research), Lenox, MA. Technical Report No. LIR/11-85/160, 1985.
86. M. Krofta and L. K. Wang, *Investigation of Municipal Wastewater Treatment by a Compact Innovative System.* Lenox Institute of Water Technology (formerly Lenox Institute for Research), Lenox, MA. Technical Report No. LIR/01-86/165, 1986.
87. L. K. Wang, *Removal of Algae from Lagoon Effluent.* Lenox Institute of Water Technology (formerly Lenox Institute for Research), Lenox, MA. Technical Report No. LIR/01-88/167, 1988.

88. M. Krofta and L. K. Wang, *Tertiary Wastewater Treatment.* US Department of Commerce, National Technical Information Service, Springfield, VA. Technical Report No. PB88-168133/AS, 1986.
89. L. K. Wang, *Symposium on Environmental Technology and Management*, US Department of Commerce, National Technical Information Service, Springfield, VA. Technical Report No PB88-200589/AS, 1986.
90. L. K. Wang, *A Promising and Affordable Solution to Sludge Treatment.* US Department of Commerce, National Technical Information Service, Springfield, VA. Technical Report No. PB88-168398/AS, 1986.
91. M. Krofta, and L. K. Wang, *Municipal Waste Treatment by Supracell Flotation, Chemical Oxidation and Star System*, US Department of Commerce, National Technical Information Service, Springfield, VA. Technical Report No. PB88-200548/AS, 1986.
92. M. Krofta, and L. K. Wang, *Dissolved Air Flotation Processes.* US Department of Commerce, National Technical Information Service, Springfield, VA. Technical Report No. PB88-168448/AS, 1986.
93. M. Krofta, D. Guss, and L. K. Wang, *Proc. 42nd Industrial Waste Conference*, Purdue Univ., Lewis Publishers, Chelsea, MI, 1988, pp. 185–195.
94. L. K. Wang and P. G. Daly, *Preliminary Design Report of a 10-MGD Deep Shaft Flotation Plant for the City of Bangor.* US Department of Commerce, National Technical Information Service, Springfield, VA. Technical Report No. PB88-200597/AS, 1987.
95. L. K. Wang and P. G. Daly, *Preliminary Design Report of a 10-MGD Deep Shaft Flotation Plant for the City of Bangor*, US Department of Commerce, National Technical Information Service, Springfield, VA. Technical Report No. PB88-200605/AS, 1987.
96. M. Krofta, L. K. Wang, and C. D. Pollman, *Procedings of the 43rd Industrial Waste Conference*, Lewis Publishers, Chelsea, MI, pp. 535. 1989.
97. M. Krofta and L. K. Wang, *Proceedings of the 43rd Industrial Waste Conference*, pp. 673. 1989.
98. L. K. Wang, *Recent Development in Cooling Water Treatment with Ozone.* Lenox Institute of Water Technology, Lenox, MA. LIR/03-88-285, 1988.
99. L. K. Wang, *Treatment of Cooling Tower Water with Ozone*, Lenox Institute of Water Technology, Lenox, MA. LIR/05-88/303, 1988.
100. M. Krofta, D. Guss, and L. K. Wang, *Proceedings of the 1988 Food Processing Waste Conference*, GTI, Atlanta, GA, 1988.
101. M. Krofta, and L. K. Wang, *Proceedings of National Water Supply Improvement Association Conference* , San Diego, CA, 1988.
102. L. K. Wang and M. H. S. Wang, *Proceedings of the 44th Industrial Waste Conference*, 1990, pp. 493–504.
103. M. Krofta, and L. K. Wang, *Proceedings of the 44th Industrial Waste Conference*, 1990, pp. 505–515.
104. L. K. Wang and W. J. Mahoney, *Water Treatment* **9**, 223–233 (1994).
105. L. K. Wang, M. H. S. Wang, and W. J. Mahoney, *Proceedings of the 44th Industrial Waste Conference*, 1990, pp. 667–673.
106. L. K. Wang, *Using Air Flotation and Filtration in Removal of Color, Trihalomethane Precursors and Giardia Cysts.* NYSDOH workshop on water treatment chemicals and filtration, American Slow Sand Association Annual Meeting, 1989.
107. L. K. Wang, M. H. S. Wang, and F. M. Hoagland, *Water Treatment* **7**, 1–16 (1992).
108. L. K. Wang and J. P. VanDyke, *Proceedings of Annual Meeting of Engineering Foundation*, Palm Coast, FL, Dec. 1989.
109. L. K. Wang, *Great Lakes 90 Conference Proceedings*, Hazardous Materials Control Research Institute, Silver Spring, MD, Sept. 1990.

110. L. K. Wang, *Proceedings of New York–New Jersey Environmental Exposition*, NYNJE, Belmont, MA, Oct. 1990.
111. L. K. Wang, *Proceedings of Modern Engineering Technology Seminar*, Taipei, Taiwan, ROC, Dec. 1990.
112. T. Pieterse and R. Kfir, *Water Quality Int.*, **4**, 31 (1991).
113. B. Rusten, B. Eikebrokk, and G. Thorvaldsen, *Water Science Technology*, **22**, 108 (1990).
114. P. Keskitaol and I. Sundholm, *Proceedings of the 6th International Symposium on Agricultural and Food Processing Wastes*, Chicago, IL, Dec. 1990.
115. H. J. Kiuru, *Water Science Technology*, **22**,139 (1990).
116. D. Guss and R. Hebert, *1990 TAPPI Nonwovens Conference & Technology Exposition*, 1990.
117. D. Guss and D. Brown, *1991 TAPPI Environmental Conference*, April, 1991.
118. J. A. Kollajtis *Proceedings of the Annual AWWA Conference, Water Quality for the New Decade*, Philadelphia, PA, 1991, pp. 433–448.
119. L. K. Wang, *Proceedings of the 1991 Annual Conference of the Korea Society of Water Pollution Research and Control*, Seoul, Korea, Feb. 1991.
120. J. P. Malley and J. K. Edzwald, *J. Water SRT-Aqua* **40**, 7–17 (1991).
121. L. K. Wang, M. H. S. Wang, and P. Kolodzicj, *Water Treatment* **7**, 387–406 (1992).
122. L. Mahony, *Proceedings TAPPI Contaminents Seminar*, 1992.
123. K. A. Graham and M. Venkatesh, *Proceedings of the 47th Industrial Waste Conference*, Lewis Publishers, Chelsea, MI, 1992.
124. L. K. Wang, and C. S. Hwang, *Proceedings of the 1991 Annual Conference of the Korean Society of Water Pollution Research and Control, Seoul Korea, Water Treatment* **8**, 7–16, 1993.
125. S. Cizinska, V. Matejo, C. Wase, Y. Klasson, J. Krejci, and G. Dalhammar, *Water Research*, **26**, 139 (1992).
126. N. Dewitt and N. K. Shammas, *Proceedings Water Environment Federation 65th Annual Conference & Expo.* , New Orleans, LA, Sept. 1992.
127. M. Viitasaari, *Environ. & Safety Tech.*, pp. 49, 51, 53, 55 (1993).
128. B. Pascual, B. Tansel, and R. Shalewitz, *Proceedings of the 49th Industrial Waste Conference*, Lewis Publishers, Chelsea, MI, 1994.
129. L. K. Wang and M. Cheryan, *Application of Membrane Technology in Food Industry for Cleaner Production*. United Nations Industrial Development Organization (UNIDO) Technical Paper No. 8-6-95, 1995.
130. L. K. Wang, *The State of the Art Technologies for Water Treatment and Management*. United Nations Industrial Development Organization (UNIDO) Technical Paper No. 8-8-95, 1995.
131. L. K. Wang, *Water Treatment* **10**, 41–54 (1995).
132. M. Viitasaari, P. Jokela, and J. Heinanen, *Water Science Technology*, **31**, 299–313 (1995).
133. L. K. Wang, P. Wang, and N. L. Clesceri, *Water Treatment* **10**, 121–134 (1995).
134. L. K. Wang and M. H. S. Wang, *Water Treatment* **10**, 261–282 (1995).
135. R. Gnirss and A. Peter-Frohlich, *Water Science Technology* **34**, 257–265 (1996).
136. M. Krofta, D. Miskovic, D. Burgess, and E. Fahey, *Water Science Technology* **33**, 171–179 (1996).
137. R. E. Carawan and E.G. Valentine, *Dissolved Air Flotation Systems for Bakeries*, North Carolina Cooperative Extension Service, Tchnical Report No. CD-43, March, 1986.
138. L. K. Wang, *OCEESA J.*, **13**, 12–16 (1996).
139. M. Krofta and D. Burgess, *Proceedings of the First International Conference on Environmental Restoration*, Ljubljana, Slovenia, July, 1987.
140. D. Guss and R. L. Klaer, *Water Environ. Fed. Conf.*, Chicago, IL, Oct. 1997.
141. M. Krofta and L. K. Wang, *Flotation and Related Adsorptive Bubble Separation Processes*. 4th ed. Lenox Institute of Water Technology, Lenox, MA. Technical Manual No. Lenox 7-25-1999/348, 1999.

142. L. K. Wang and S. Kopko, *City of Cape Coral Reverse Osmosis Water Treatment Facility,*. US Deparment of Commerce, National Technical Information Service, Springfield, VA. 1987; *Proceedings of Advances in Filtration and Separation Technology, American Filtration and Separation Society*, Vol. 11, pp. 499–506. http://www.afssociety.org/publications/Contents/vol11.shtml.
143. C. C. Yannoni, Q. Zhu, and S. D. Clark, *J. New England Water Works Assoc.*, **113**, 115–127 (1999).
144. M. Krofta and L. K. Wang, *Flotation Engineering*. 1st ed. Lenox Institute of Water Technology, Lenox, MA. Technical Manual No. Lenox/1-06-2000/368, Jan. 2000.
145. T. Hedberg, J. Dahlquist, D. Karlsson, and L. O. Sorman, *Water Science and Technology*, **37**, 81–88 (1998).
146. H. Kiuru and R. Vahala, *Proceedings of the 4th International Conference on DAF in Water and Wastewater Treatment*, IWA Publishing (International Water Association); 2001.
147. Editor, *Engineering News Records* (ENR), p. 55 May 19, 2003.
148. Pan America Environmental, *Dissolved Air Flotation Operational Theory,*. Pan America Environmental, Wauconda, IL, http://www.panamenv.com/dissolved-air-flotation-theory.html, 2003.
149. L. K. Wang, and M. H. S. Wang, *Handbook of Industrial Waste Treatment,* 1st ed., Marcel Dekker, New York, pp. 61–125, 1992.
150. L. K. Wang, Site Remediation and Groundwater Decontamination. In: *Handbook of Industrial and Hazardous Wastes Treatment*, 2nd ed., L. K. Wang, Y. T. Hung, H. H. Lo, and C. Yapijakis (eds.), Marcel Dekker, New York, pp. 923–969, 2004.
151. L. K. Wang, L. Kurylko and M. H. S. Wang, *Sequencing Batch Liquid Treatment*. US Patent No. 5354458. US Patent & Trademark Office, Washington, DC, 1996.
152. D. Nolasco, D. Irvine, and M. Monoharan, *Water Environment & Technology*. **10**, 91–96 (1998).
153. L. K. Wang, and Y. Li, Sequencing Batch Reactors. In: *Biological Treatment Processes*, L. K. Wang, N. C. Pereira, and Y. T. Hung (eds.), Humana Press, Totowa, NJ, 2005.
154. L. K. Wang, J. V. Krouzek, and U. Kounitson, *Case Studies of Cleaner Production and Site Remediation*. United Nations Industrial Development Organization (UNIDO), Vienna, Austria. Training Manual No. DTT-5-4-95, 1995.
155. L. K. Wang, *Water Treatment*, **6**, 127–146 (1991).
156. W. L. Forestell and L. K. Wang, *J. New England Water Works Assoc.* **99**, 249–284 (1991).
157. D. Nickols and I. A. Grossley, *The Current Status of Dissolved Air Flotation in the USA*. Technical Report. Hazen and Sawyer, New York, 2003.
158. US EPA, Personal communications with Dr. Lawrence K. Wang of Lenox Institute of Water Technology, Lenox, MA, 2002.
159. L. K. Wang, *Feasibility Study of Treating Field Military Wastewater by a Process System Including Powdered Carbon Adsorption, Polymer Coagulation, and Diatomite Filtration*, U.S. Defense Technical Information Center, Alexandria, VA, ADA077198, 1973.
160. L. K. Wang, *Investigation of Methods for Determining Optimum Powdered Carbon and Polyeletrolyte Dosages in Military Wastewater Treatment Systems*, U.S. Defense Technical Information Center, Alexandria, VA, ADA082506, 1973.
161. L. K. Wang, *J of Applied Chemistry and Biotech*, **25**, 491–503 (1975).
162. L. K. Wang, *Water and Sewage Works*, **123**, 42–47, 1976.
163. L. K. Wang, *Water and Sewage Works* **124**, 32–36 (1977).
164. L. K. Wang, *Gas Dissolving System and Method*, U . S. Patent No. 5049320, September 17 1991. US Patent & Trademark Office, Washington, DC, 1991.
165. L. K. Wang, *Gas Dissolving and Releasing Liquid Treatment System*. U. S. Patent No. 5167806, December 1, 1992. US Patent & Trademark Office, Washington, DC, 1992.
166. L. K. Wang, *Water and Wastewater Treatment*. U.S. Patent No. 5240600, August 31, 1993. US Patent & Trademark Office, Washington, DC, 1993.

167. L. K. Wang, *Improved Method and Apparatus for Liquid Treatment.* U.S. Patent No. 5256299, October 26, 1993, US Patent & Trademark Office, Washington, DC, 1993.
168. L. K. Wang, *Combined Coarse and Fine Bubbles Separation System.* U.S. Patent No. 5275732, January, 4, 1994. *US* Patent & Trademark Office, Washington, DC, 1994.
169. M. Krofta and L. K. Wang, *Proceedings of the Seventh Mid-Atlantic Industrial Waste Conference*, 1985, pp. 207–216.
170. L. K. Wang, *Micro/Ultra Filtration System*, U.S. Patent No. 4,973,404, Nov. 1990. US Patent & Trademark Office, Washington, DC, 1990.
171. L. K. Wang, *Reduction of Color, Odor, Humic Acid and Toxic Substances by Adsorption, Flotation and Filtration*, Annual Meeting of American Institute of Chemical Engineers, Symposium on Design of Adsorption Systems for Pollution Control, Philadelphia, PA (P926-08-89-20), 1989.
172. Dongshin, *Dissolved Air Flotation.* Technical Report. Dongshin, Seoul, Korea. September 26, 2003.
173. L. K. Wang, J. S. Wu, N. K. Shammas, and D. A. Vaccari, Recarbonation and Softening. In: *Physicochemical Treatment Processes,* L. K. Wang, Y. T. Hung, and N. K. Sahmmas (eds.). Humana Press, Totowa, NJ, pp. 199–228, 2005.
174. L. K. Wang, *Canad. J. Chem. Engin.* **60**, 116–122 (1982) (NTIS-A131138).
175. L. K. Wang, *Flotation and Best Available Technologies for Tannery Waste Treatment.* Manual No. Lenox/10-18-1999/362. Lenox Institute of Water Technology, Lenox, MA, USA. Training Manual No. UNIDO-DTT-1-15-96. United Nations Industrial Development Organization, Vienna, Austria, 1999.
176. L. K. Wang, *1974 Earth Environment and Resources Conference Digest of Technical Papers*, **1**(74), 56–57 (1974).
177. J. M. Wong, and F. Y. Chang, *Application of High-Rate Clarification Processes (DAF, AquaDAF, and Actiflo) to Optimize Drinking Water Treatment*, Technical paper presented at the California-Nevada Section AWWA Fall Conference in San Diego, CA, October 6–9, 2003.
178. L. B. Stovall, *Source Water Protection: Greenville, South Carolina.* US Environmental Protection Agency, Washington, DC, March 12, 2003.
179. W. A. Selke, L. K. Wang, N. K. Shammas, and D. A. Aulenbach, *Correction Factor of Gas Dissolution Under Pressure for Flotation System Design*, Technical Note, International Association of Flotation Technology, Newtonville, NY, Nov. 2003.
180. M. H. S. Wang, Separation of Lignin from Aqueous Solution by Adsorptive Bubble Separation Processes. Ph.D. thesis. 241 pp. Rutgers University, New Brunswick, NJ. 1972.
181. L. K. Wang, Continuous Bubble Fractionation Process. Ph.D. thesis. 171 pp. Rutgers University, New Brunswick NJ. 1972.
182. L. K. Wang, Y. T. Hung, H. H. Lo and C. Yapijakis (eds.). *Handbook of Industrial and Hazardous Wastes Treatment.* Marcel Dekker Inc., NYC, NY. pp. 873–921, 2004.
183. L. K. Wang, N. K. Shammas and Y. T. Hung (eds.). *Advanced Physicochemical Treatment Processes.* Humana Press, Totowa, NJ, 2005.
184. R. M. Manamy. *Public Works.* pp. 24–28, Dec. (2003).

13
Gravity Filtration

J. Paul Chen, Shoou-Yuh Chang, Jerry Y. C. Huang, E. Robert Bauman, and Yung-Tse Hung

CONTENTS

1. INTRODUCTION

The purpose of filtration is to remove the particulates suspended in water by passing the water through a layer of porous material. Larger particulates are retained by straining and sedimentation, while colloidal matter is retained by adsorption, or coagulation and sedimentation. Biological interactions occur only when the water passes very slowly through the porous mass.

There are three basic terms used to describe the method of applying the motive force used in filtration systems—vacuum filtration, pressure filtration, and gravity filtration. In vacuum filtration, the filter is located on the suction side of a pump and the pressure drop across the filter is limited to the suction lift differential that can be generated by the pump, usually *5.5–6.7 m (18–22 ft)* of water. The filter itself is generally operated under a pressure less than atmospheric pressure. In pressure filtration, the filter is located on the discharge side of the pump, and the pressure drop across the filter can be any differential that can be generated by the characteristics of the pump, usually 3–12 m (10–40 ft) of water. If the discharge of the filter is not directly to a storage tank at atmospheric pressure, the pressure within the filter is usually at a pressure greater than atmospheric pressure. In gravity filtration, the water flows through the filter media under the force of gravity. In other words, a gravity filter is a special type of pressure filter in which the water is delivered to the filter and the water on the influent side of the filter is at atmospheric pressure. Because no pump is directly involved, the difference

From: *Handbook of Environmental Engineering, Volume 3: Physicochemical Treatment Processes*
Edited by: L. K. Wang, Y.-T. Hung, and N. K. Shammas © The Humana Press Inc., Totowa, NJ

in elevation of the water level between the influent and the effluent side of the filter is made large enough that the total motive force desired is available by gravity flow through the filter. If the effluent water discharge elevation is higher than the filter media surface elevation, the pressure in the media will always be greater than atmospheric pressure. If the effluent water discharge elevation is below the media surface elevation, the pressure within the media can be less than atmospheric pressure and a condition of "negative-head" can develop. Gravity filters in which the granular media depth ranges from 0.3 m (12 in.) to as much as 2.1 m (84 in.) are frequently referred to as deep-bed gravity filters.

This chapter focuses on gravity filtration and does not attempt to present a detailed discussion of gravity filter design, because it has been well presented in various textbooks (1–3). The emphasis is on the principles and current advances in gravity filtration.

2. PHYSICAL NATURE OF GRAVITY FILTRATION

The removal of suspended matter achieved by the porous media is generally considered to be composed of two steps: transport and attachment. Detachment might also take place during filtration, but mostly during the backwashing cycle. Transport mechanisms move a particle into and through a filter pore so that it comes very close to the surface of the filter media or existing deposits where attachment mechanisms serve to retain the suspended particle in contact with the media surface or with previously deposited solids. Detachment mechanisms occur due to the action of hydrodynamic forces of the flow such that a certain part of the previously adhered particles, less strongly linked to the others, is detached from filter media or previous deposits and carried further into or through the filter.

2.1. Transport Mechanism

The flow patterns in a filter bed of randomly packed grains are too complex to analyze in a precise geometric way. However, in the grain size of interest in rapid granular media filtration (0.4–2 mm), at the filtration rate of interest (2–8 gpm/ft^2 or 81.4–325.6 L/min · m^2) with a water temperature between 0°C and 30°C (viscosity 0.018 and 0.008 poises), laminar flow conditions exist. Experimental results reported by Ives (4), who used dyed streams to visualize the flow around 5 mm grains, showed no disturbance of the streamline flow even at rates considerably higher than those encountered in practice. This means transport mechanisms are responsible for providing forces to move particles out of their flow streamlines into the vicinity of the grain surface.

Transport mechanisms are straining, interception, inertial forces, sedimentation, diffusion, and hydrodynamic forces (see Fig. 1). The suspended solids removal efficiency and the type of dominating transport mechanisms depend on the sizes of the particles and their size distribution in the filtering water. The study by Craft shows that straining is important for particulates with diameters greater than about 20% that of the grains through which they are filtered (5). It should be noted that filtration systems in which straining is dominant are not operating under optimal conditions, because clogging is very fast and thus frequent backwashing is required. According to Herzig et al., sedimentation is important for particulates with sizes >30 μm, and negligible for particulates

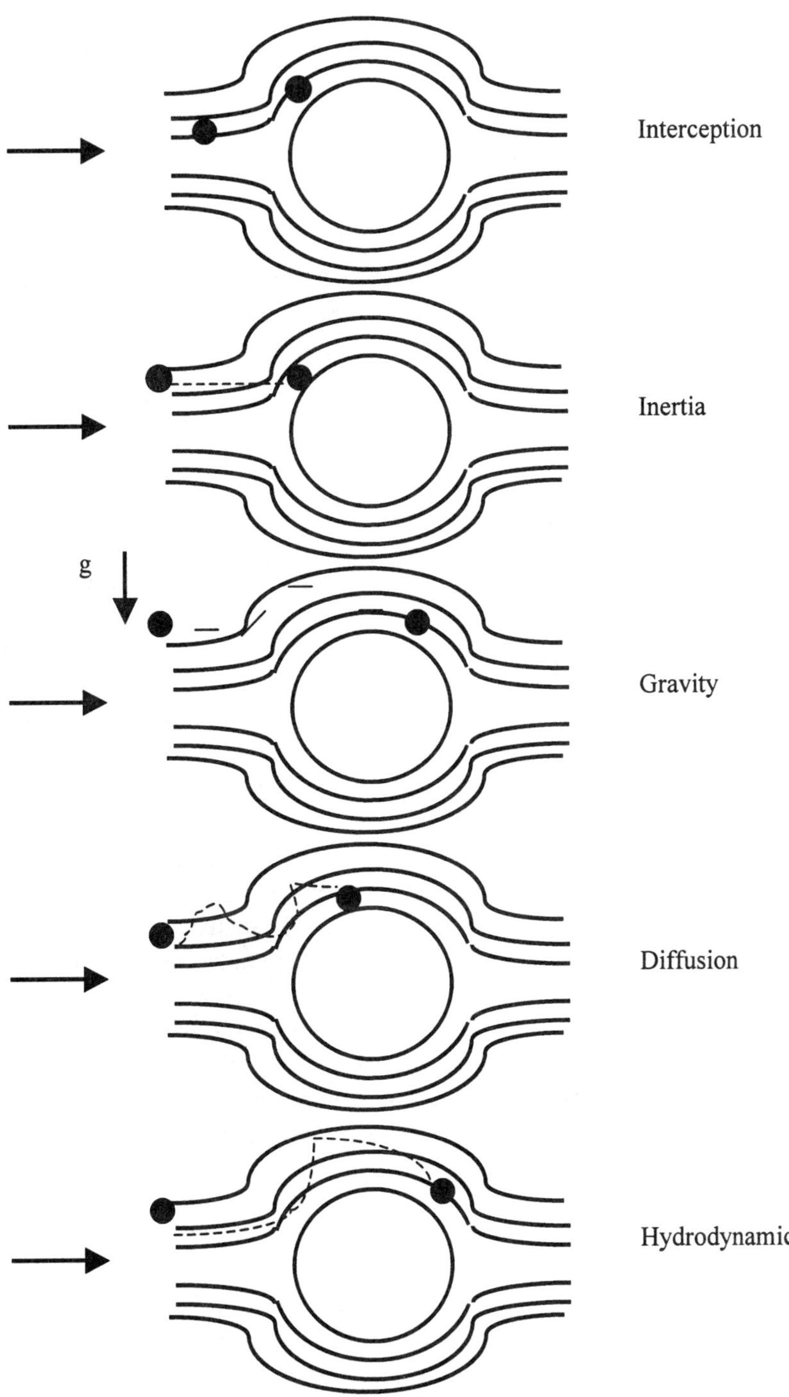

Fig. 1. Simplified diagrams of particle transport mechanisms.

with a diameter <3 μm (6). Interception has to be considered for any size, while inertial and hydrodynamic forces can be neglected. Diffusion is important only for colloidal particles (less than 0.1 μm size). The study by Yao et al. demonstrates that the removal is negligible when a size of suspended solids is less than a critical value of 1 μm (7). The experimental results by Ghosh et al. confirm this finding (8).

2.2. *Attachment Mechanisms*

Attachment of particles to grain surfaces or existing deposits of particles has generally been attributed to four kinds of forces: axial pressure of the fluid, friction, surface forces (van der Waals and electrical), and chemical forces. Precoating of the media grains with a cationic polymer increases attachment of negatively charged colloidal particles (7). According to these authors there is an optimum polymer dosage that can be determined from jar tests and electrophoretic mobility measurements. Use of polymer addition can improve the performance of existing filtration plants, which fail to meet quality standards (9,10).

2.3. *Detachment Mechanisms*

Accumulated deposits have an unequally strong structure. Under the action of hydrodynamic forces due to the flow of water through the media, which increase with increasing head loss, this structure is partially destroyed. A certain portion of previously adhered particles less strongly linked to the others is detached from the grains. Consequently, as the deposits accumulate they become unstable and parts of them are torn away by the flow, to go back into suspension in the pores. The detachment is observed when flow rate is suddenly increased (11,12). Moreover, the detachment occurs even at constant flow rate (13,14).

Another group of researchers such as Ives, Lerk, Mackrle, and Mackrle opposed this detachment mechanism (15,16). They considered that, as the interstitial velocity increases, and as the surface available in the filter pores and the amount of divergence and convergence of flow diminish due to the deposits accumulating in the pores, there is a reduction in the probability of particles being brought to a surface for adherence. Stanley (17) observed that, even in the presence of a continuously flowing suspension, radioactively labeled iron floc was not detached from its original place of deposition in the filter.

Nevertheless, from a macroscopic point of view, it is impossible to determine if deteriorating effluent quality is caused by either solids detachment or decreasing solids attachment efficiency as the filter becomes clogged.

3. MATHEMATICAL MODELS

Mathematical models used in filtration are intended to represent both the solids clarification process and the pressure drop increase caused by the deposited material. Two different kinds of models are reported in the literature. A theoretical approach to the removal mechanisms has led one group of research workers to develop “idealized models,” which combine theoretical and empirical results. Another group of workers has adopted a “black box” approach to the filtration process, developing several “purely empirical models,” orientated toward the design of efficient filtration systems. The following paragraphs highlight the most relevant achievements in this area. A detailed study of this subject can be found in the cited references.

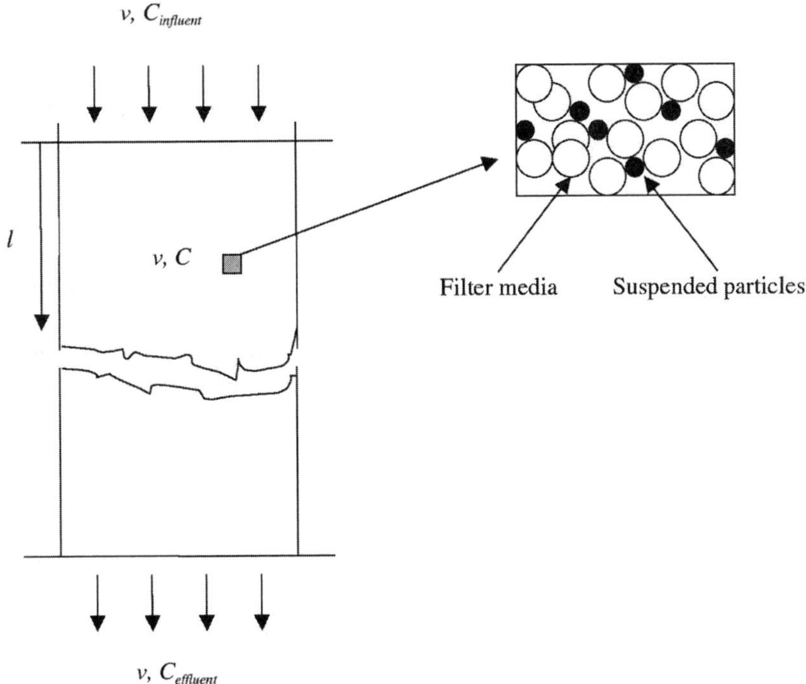

Fig. 2. Schematic representation of filtration process.

3.1. Idealized Models

3.1.1. Clarification Equation

The general mass balance equation has been developed by considering that the rate of accumulation of solids within a unit filter volume equals the rate of inflow of solids minus the rate of outflow of solids (3). Figure 2 shows a typical filtration process unit operation. The simplified mass balance equation can be presented as:

$$\frac{\partial \sigma}{\partial t} = -v \frac{\partial C}{\partial l} \tag{1}$$

where σ = specific deposit (volume of deposits per unit of filter volume), C = volumetric concentration (volume of suspended particles per unit of suspension volume), t = filtration time from the beginning of the run, v = approach velocity of the suspension, l = filter media depth.

Equation (1) can be simplified using to the following assumptions: (1) the diffusion is always negligible when particle size is larger than 1 μm; (2) the porosity of the clogged filter bed does not differ significantly from that of a clean bed; and (3) the volume of the moving particles can be neglected with respect to the volume of the retained particles (6,18). This simplification is reasonable if the suspension concentration (C) is small compared to the specific deposit (σ). Although the simplification is best for large values of σ, it can be used throughout the run maintaining the error below 10% (6).

The mass balance equation alone is not enough to solve the problem, given that two unknowns (σ and C) are present. A second equation is needed to introduce the relationship between the suspended particles and the porous medium.

The basic theoretical hypothesis, first introduced by Iwasaki (19), is that the change in the concentration with respect to depth in the filter is proportional to the concentration:

$$-\frac{\partial C}{\partial l} = \lambda C \tag{2}$$

where λ = filter coefficient, which is an empirical measure of the interaction between the suspended particles and the filter medium.

Because the solids removal efficiency is affected by the changes in the pores due to deposits, which are both time and depth dependent, the filter coefficient (λ) is not a constant. Equation (2) is not easily solved; thus, analytical expressions for σ and C can not be readily available.

Iwasaki (19) assumed that λ increased linearly with σ because of the increased surface area available for adhesion. Stein (20) realized that a careful experimental study made a few years previously by Eliassen (21) keyed in with Iwasaki's mathematics. Stein modified the equation so that λ increases linearly, then decreases nonlinearly with the amount of clogging.

Alternatively, Mintz (13) proposed an approach based on the physical consideration that a change of suspended solids concentration at each individual layer of the filter bed is the overall result of two opposing processes: (1) extraction of particles from the water and their attachment to filter media under the action of adhesive forces and (2) detachment of previously adhering particles under the action of the hydrodynamic forces of the stream. That is expressed as

$$-\frac{\partial C}{\partial l} = \lambda C - \frac{q\sigma}{v} \tag{3}$$

where q = detachment or scour coefficient.

Ornatskii et al. reported an extensive study on the clogging of sand beds by clay suspensions (22). They assumed that the filter coefficient (λ) decreases linearly with the specific deposit (σ) being inversely proportional to interstitial velocity. Ives reconciled the two conflicting assumptions of Stein and Ornatskii by suggesting that λ first increases with σ because of the increased surface area, then decreases because of the increased interstitial velocity and the smoothing of flow paths (23,24).

Ives and Sholji found that filter efficiency increases with solids retention initially, perhaps due to an increase in specific surface of the filter media, and then decreases (25). These findings were confirmed by the experimental results of Fox and Cleasby (26).

Ives (27) proposed a general expression as

$$\frac{\lambda}{\lambda_0} = \left(1 + \frac{\beta\sigma}{1-\varepsilon_0}\right)^y \left(1 - \frac{\beta\sigma}{\varepsilon_0}\right)^z \left(1 - \frac{\sigma}{\sigma_u}\right)^x \tag{4}$$

where λ_0 = initial filter coefficient, β = bulking factor, [= $1/(1 - \varepsilon_d)$] where ε_d is the porosity of the deposited solids, ε_0= initial porosity of the filter bed, σ_u= ultimate or maximum specific deposit, and x, y, z = empirical constants.

The third term on the right-hand side of Eq. (4) does not have any theoretical explanation. It was introduced in order to take into account the fact that retention stops ($\lambda = 0$) well before the bulk specific deposit $\beta\sigma$ equals the initial porosity ε_0. The value of the porosity of the deposited solids ε_d is difficult to determine. However, presumably it is highly dependent on flow conditions. In addition, one still has to determine the empirical constants x, y, and z, as well as the ultimate specific deposit δ_u. As a matter of fact, most of the mathematical models proposed for the variation of λ with σ can be expressed by Eq. (4), with proper selection of values for x, y, and z.

Equation (4) is theoretically correct but does not reflect the observed initial increase of λ unless a very high value is used for the exponent y. The experimental results by several investigators indicate that the filter coefficient attains a maximum of two to four times its initial value (28–30). Assuming a twofold increase, this kind of behavior can be represented by an equation of the form (31,32):

$$\lambda = \lambda_0 \left[1 + \left(\frac{S}{S'} \right)^{\frac{1}{3}} \right] \qquad \text{if } S \le S' \tag{5a}$$

$$\lambda = \lambda_0 2 \left[\frac{(S'' - S)}{(S'' - S')} \right]^{\frac{1}{3}} \qquad \text{if } S > S' \tag{5b}$$

where S = the volume of void space occupied or blocked by the deposits per unit filter volume, and S' = the value of S at which λ is maximum; and S'' = the value of S at which is zero.

The exponent 1/3 is purely empirical. It is intended to provide a smooth variation of λ in accordance with experimental results (26). The parameter S represents the fraction of the void space that is no longer available for the flow path. If the particles are relatively small compared to the grains, a uniform coating will form and $S = \beta\sigma$. However, if the particles are large, they will tend to block the channels and will cause a much larger reduction in the zone available for the flow to pass through. Thus, the bulking factor will be a higher value. At present no analytical approach is available to determine the values of β for different relative size groups d/D (d = filtered particle diameter, D = grain diameter). Thus, β has to be determined experimentally.

The initial filter coefficient λ_0 is difficult to determine. The extensive experimental studies were carried out in this area (6); however, the empirical equations developed by various investigators have very limited practical use. Another group of research workers have followed a more theoretical approach, studying trajectories of particles in suspension as they approach a filter grain. Using numerical analysis to solve the trajectory equations and then correlating the results empirically with dimensionless parameters, Rajagopalan and Tien (33) proposed the following equations:

$$\lambda_0 = \left(\frac{1}{D} \right) \left[\frac{\left(\frac{\pi}{6} \right)}{(1 - \varepsilon_0)} \right]^{-\frac{1}{3}} \ln \left[\frac{1}{(1 - n_0)} \right] \tag{6a}$$

and

$$n_0 = A_S N_{Lo}^{1/8} N_R^{15/8} + 0.00338 A_S N_G^{1.2} N_R^{-0.4} + 4 A_S^{1/3} N_{Pe}^{-2/3} \tag{6b}$$

where n_0 = initial collection efficiency, $A_s = 2(1 - p^5)/w$, $p = (1 - \varepsilon_0)^{1/3}$, $w = 2 - 3p + 3p^5 - 3p^6$, N_{Lo} = London group = $4H_a / (9\pi\mu d^2 v)$, N_R = Relative size group = d/D, N_G = Gravity group = $v_s/v = d^2(\rho_p - \rho)g/(18\mu v)$, N_{Pe} = Peclet number = $3\pi\mu v dD/(k_B T)$, k_B = Boltzmann's constant (1.38048×10^{-16} erg/°K or 1.38048×10^{-23} J/°K), H_a = Hamaker constant (10^{-13} erg or 1×10^{-20} J), T = temperature, in degrees Kelvin, ρ = density of water, ρ_p = density of the particles, v_s = settling velocity of the particles, g = accelaration due to gravity.

The only restrictions that apply are that the relative size group (*d/D*) be less than 0.18 and that the London group (N_{Lo}) be greater than zero (attractive net surface forces). An excellent review of the research done in this area can be found in Tien and Payatakes (34).

3.1.2. Head Loss Equation

Porous media present a resistance to flow, even for a clean liquid, which can be calculated using the Carman–Kozeny equation (1,35):

$$\left(\frac{\partial H}{\partial l}\right)_0 = \frac{f}{\phi}\frac{1-\varepsilon_0}{\varepsilon_0^3}\frac{v^2}{dg} \tag{7}$$

where $\left(\frac{\partial H}{\partial l}\right)_0$ = initial head loss gradient, f = friction factor = $150\left[(1-\varepsilon_0)/N_{Re}\right]+1.75$

ϕ = particle shape factor (dimensionless): $\phi = 1$ for spherical particles, N_{Re} = Reynolds number = $\rho v d/\mu$ (dimensionless), and μ = absolute viscosity of water. Other equations that can also be used are those proposed by Fair and Hatch (36) and Rose (37).

Particle deposition causes head loss to increase. Two distinct types of deposition can occur, which can be referred to as "coating" and "blocking." Coating takes place when the deposits uniformly cover the pore walls, reducing its diameter. The pore is never blocked because increasing velocity stops deposition at a certain level. Blocking is caused when particles are large with respect to pore size. One particle or a number of them could block a pore and hinder or even prevent water from flowing through it.

Most of the theories proposed to evaluate this head loss increase during filtration assume that it is caused by coating of the pores (38). If MacLaurin series expansion is used (assuming $\sigma \ll \varepsilon_0$), all the models show an increased head loss per unit depth proportional to the specific deposit, that is,

$$\left(\frac{\partial H}{\partial l}\right) \Big/ \left(\frac{\partial H}{\partial l}\right)_0 = 1 + \text{constant} \times \sigma + \cdots \tag{8}$$

According to Ives (27), when this formulation is combined with the simple analytical model for σ proposed by Heertjes and Lerk (39) (linear decrease of λ with σ from λ_0 when $\sigma = 0$ to 0 when $\sigma = \varepsilon_0$), it yields a head loss increase linear with time. When this type of equation is used to simulate the behavior of real filters, measured

Table 1
Bulking Factors and Head Loss Constants for Various Suspensions

	Bulking factor β			Head loss constants K		
Type of suspensions	Average	Range	Sample No.	Average	Range	Sample No.
Clay/alum suspensions	5	4–7	25	280	200–400	25
Ferric Chloride suspensions	17	12–20	20	330	200–500	20
Clay/polymer suspensions	7	5–12	20	7800	6000–9000	20

head loss is usually much higher than predicted head loss, particularly when large particles are present in the influent. It is possible that this is because partial blocking of the pores takes place. Because it is practically impossible to study blocking from a theoretical point of view, an empirical constant relating the volume of the deposits no longer available to the flow has to be introduced. Herzig et al. discussed this subject in depth (6).

Maroudas and Eisenklam found that the most significant parameter is the ratio *d/D* (40). If $d/D > 0.15$, the porous medium is irreversibly blocked and a filter cake is formed. If $d/D < 0.065$, the retention is always low, and for intermediate values a partial blocking of the porous bed may occur.

Garcia-Maura proposed a general approach (31). The influent is assumed to be composed of particles of *m* different sizes. In order to take into account the different blocking capacity of particles of different sizes, the volume, which no longer is available for the flow path, is expressed as

$$S = \sum_{j=1}^{m} \beta_j \sigma_j \tag{9}$$

where β_j = expansion factor for particle size *j* and σ_j = specific deposit of particles of size *j*.

The expansion factors β_j correspond to bulking factors if deposition is in coating form, but are much higher if blocking takes place and a few particles are able to divert the flow from a large zone. The head loss increase is proposed as follows,

$$\frac{\partial H}{\partial l} = \left(\frac{\partial H}{\partial l}\right)_0 \left(1 + KS^{1.5}\right) \tag{10}$$

where K = an empirical constant and the exponent 1.5 has been adjusted according to experimental results (41,42).

The value of *K* depends on the influent suspension characteristics, but not on the filter operating characteristics such as flow rate and grain size. The experimentally determined bulking factor (β) and head loss constant (*K*) for various types of suspensions are presented in Table 1.

3.2. *Empirical Models*

Acknowledging the fact that complex and unpredictable effects on filter performance are caused by various characteristics of influent suspended solids in addition to the fil-

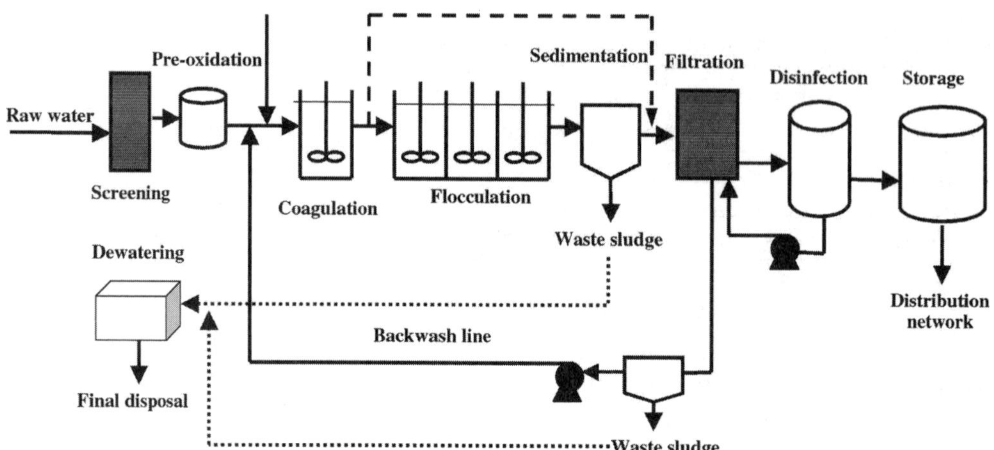

Fig. 3. Illustration of a typical water treatment system.

ter physical characteristics and operating conditions, several investigators have tried to find a way to correlate measurable properties of suspended matter with the basic filter design criteria. One approach to develope such a correlation has been to determine the filtrability of a water source. Filtrability has been defined as the ease with which water can be passed through a given filter and the effectiveness with which the solids are removed in the filter.

Considerable work has been done on the use of the membrane filtration test for defining quantitatively the filter-clogging properties of natural waters. Works of Hudson (43,44), Gamet and Rademacher (45), Hsiung and Cleasby (46), Cleasby (47), and Hsiung (48) present some ways to predict the performance of granular filters.

Conley and Hsiung extended the empirical model to the design and application of multimedia filters (49). They proposed that the variables that affect filtration efficiency and head loss, such as flow rate, media size, filter depth, and amount of suspension in the influent, can be arranged in grouped terms. Experimental data are then used to establish the proper exponentials to be used with these grouped terms.

4. DESIGN CONSIDERATIONS OF GRAVITY FILTERS

In water treatment, coagulation, sedimentation, and filtration processes are important in quality of final effluent. Figure 3 illustrates a typical water treatment system. These multiple-barrier unit operations, if properly functioning, can improve the taste and appearance of water, remove sorbed metals, trace organics, and pesticides, and inactivate (or disinfect) pathogens: bacteria, viruses, and protozoa. The design of gravity filter units involves the determination of the optimum combination of the variables, which affect both plant performance and plant operating cost.

4.1. Water Variables

4.1.1. Influent Water Characteristics

The quality of the incoming water varies with time in three major ways:

- In the concentration of the suspended solids.
- In the physicochemical properties of the suspended solids (surface characteristics, organic versus inorganic, etc.).
- In the particle size distribution of the suspended solids.

As the quality of raw or pretreated water (e.g., after coagulation-flocculation-sedimentation) varies, the following are expected to occur:

- Variation in the proportion of the total solids, which are to be removed by surface removal versus depth removal mechanisms.
- Variation in the proportion of the total solids where removal is more affected by transport phenomena versus those where removal is more affected by attachment phenomena and/or detachment phenomena.

One difference observed in the filtration of potable water as compared to direct filtration of wastewater involves the pressure drop that can be built up across a filter prior to solids breakthrough. In water filtration, solids breakthrough can be observed even though the pressure drop does not exceed 1 m (3.28 ft) of water. In direct filtration of secondary effluents, however, pressure drops of 6–10 m (20–32 ft) of water have been encountered without substantial increase in solids passing.

4.1.2. Desired Effluent Water Quality

According to the US National Primary Drinking Water Regulations, at no time can turbidity (cloudiness of water) go above 5 nephelolometric turbidity units (NTU); systems that filter must ensure that the turbidity goes no higher than 1 NTU (0.5 NTU for conventional or direct filtration) in at least 95% of the daily samples in any month (1). As of January 1, 2002, turbidity may never exceed 1 NTU, and must not exceed 0.3 NTU in 95% of daily samples in any month. At present, there is no standard quality expected as a result of filtration of wastewaters. Direct filtration of effluents from trickling filter plant can be expected to produce an effluent turbidity of 2–5 NTU, 5–10 mg/L of suspended solids (SS), and 5–10 mg/L of BOD (biochemical oxygen demand). Direct filtration of activated sludge plant effluents can be expected to produce an effluent turbidity of 1–3 NTU, 3–7 mg/L of SS, and 1–6 mg/L of BOD. Improved performance of the filtration process can be achieved by using chemical treatment.

4.1.3. Water Temperature

In most of water-treatment plants, the water temperature depends largely on the source of water. Ground water normally has a relatively small temperature variation (3–5°C) throughout the year and is generally between 10 and 15°C (higher in southern states like Florida). Surface waters, on the other hand, vary in temperature from a low of about 1–2 °C in winter to a high of 30–35 °C in summer.

Filtration hydraulic theory indicates that, all other conditions being equal, the head loss through a filter is higher with cold water than with warm water due principally to the viscosity effects of temperature. In a year-round study of the filtration of Lake Erie water, however, Dostal and Robeck found that during winter operation the colder water resulted in a higher initial head loss in accordance with theory, but that the colder water also resulted in a deeper floc penetration and a lower rate of head loss buildup during the same filter run (50). Thus, the net effect of the colder temperature was to produce a longer run under equivalent influent–effluent quality conditions.

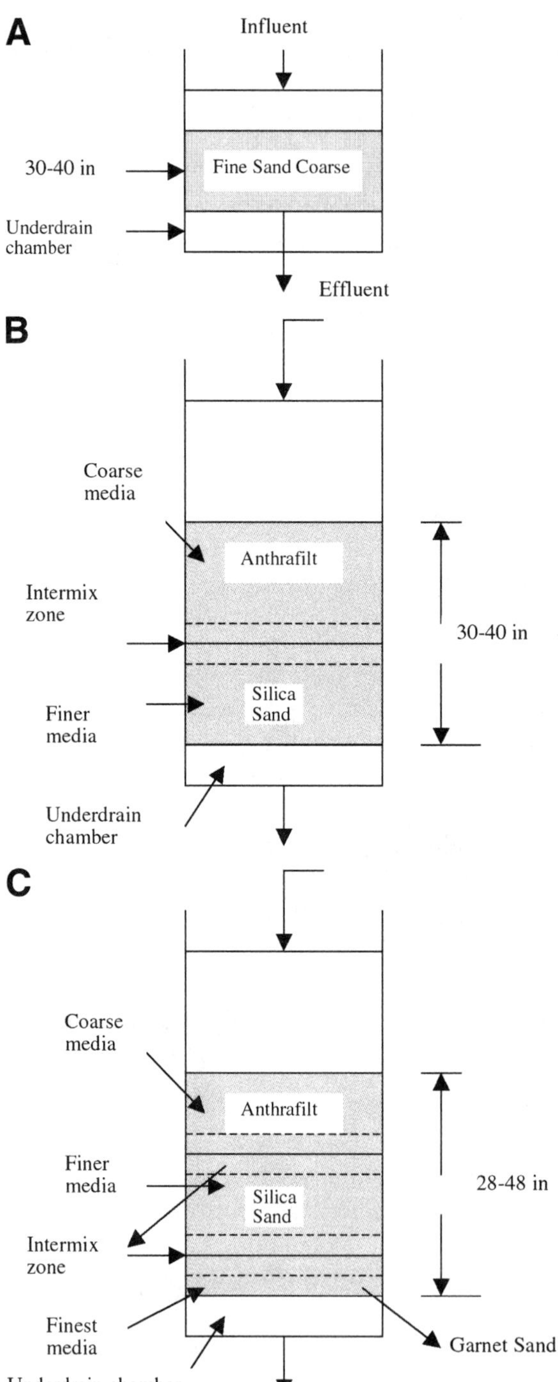

Fig. 4. Filter configurations **(A)** single media conventional filter; **(B)** dual media filter; and **(C)** multimedia filter.

4.2. Filter Physical Variables

4.2.1. Filter Configuration and Media Types, Sizes, and Depths

Figure. 4 shows several filter configurations, which have been used extensively in gravity filtration. Granular filter media commonly used in water and wastewater filtration includes silica sand, garnet sand, and anthracite coal. These media can be purchased in a broad range of effective sizes and uniformly coefficients (1,2). The media have specific gravities approximately as follows:

Anthracite coal	1.35–1.75
Silica sand	2.65
Garnet sand	4.0–4.2

Significant changes are taking place in the type, size, and gradation of filter media used in filter design. Whereas a number of years ago the typical rapid sand filter used a graded sand with an effective size between 0.4 and 0.55 mm with a uniformity coefficient between 1.35 and 1.75, recent designs tend to use coarse media. In addition, current design is placing emphasis on the use of dual- or multimedia filters for potable water filtration. Typical design parameters can be found from Table 2 (51).

The literature describing filtration of secondary effluents makes it clear that some of the biological floc carryover to the filters tends to be removed at the top surface and in the upper layers of the media, causing rapid head loss development, short filter runs, and excessive backwash requirements (42). The detrimental effects of this strong surface-removal tendency can be corrected by use of one of several design modifications, such as:

- A coarser top media size.
- Dual-or mixed (triple)-media filters to protect the filtrate quality when using coarse top sized media.
- Deep beds of near unisized course filter media (unstratified).
- Up-flow filtration (coarse-to-fine media in the flow direction) in a hydraulically stratified media filter.
- Higher terminal head loss capability to achieve acceptable run length.

These design alternatives are not all mutually compatible and have secondary implications that must be considered. Each media layer will consist of fine media at the top and coarser media at the bottom. A coarse top media size in the anthracite layer means a still coarser bottom size. The coarse bottom size dictates the minimum backwash rate required to fluidize the bed. Use of too coarse media may require a backwash rate that is abnormally high, causing extra costs for backwash pumps, water storage, piping, and appurtenances. The coarse bottom size can be reduced by specifying a more uniform media, which may also have some minor benefit to filtrate quality. Such specification, however, increases the media cost.

Media type, size, and depth are three interrelated variables that affect the solids distribution across the filter. Figure 5 shows the suspended solids removal from a trickling filter final effluent through the depth of the filter at 3 and 5 h from the beginning of the run. The removal efficiency at a depth of 0.6 m (24 in.) is nearly the same from all filters—single- and dual-media filters. However, a difference in removal efficiency is shown in the upper layers. Layers of anthracite media show a superior removal efficiency to that of sand media. It is interesting to observe that the same media size but

Table 2
Typical Media Design Values for Various Filters

Parameter	Single-medium filters	Dual-media filters	Mixed-media filters
Antracite layer			
Effective size (mm)	0.50–1.5	0.7–2.0	1.0–2.0
Uniformity coefficient	1.2–1.7	1.3–1.8	1.4–1.8
Depth (cm)	50–150	30–60	50–130
Sand layer			
Effective size (mm)	0.45–1.0	0.45–0.60	0.40–0.80
Uniformity coefficient	1.2–1.7	1.2–1.7	1.2–1.7
Depth (cm)	50–150	20–40	20–40
Garnet layer			
Effective size (mm)			0.2–0.8
Uniformity coefficient			1.5–1.8
Depth (cm)			5–15

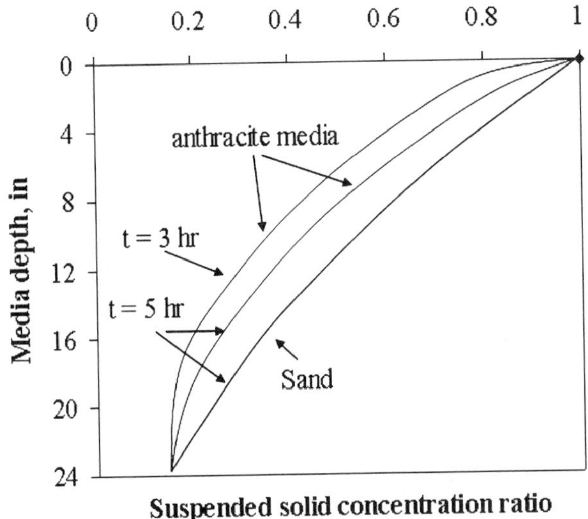

Fig. 5. Effect of media type, size, and depth on removal efficiency of suspended solids. Note that both sand and anthracite medium have size of 1.09 mm.

different grain shape and porosity (1.09 mm sand versus 1.09 mm anthracite) results in different removal efficiency.

Because dual-media filters are the common method in tertiary wastewater filtration, and because several of the other filter design alternatives are at least partly proprietary in nature, specific recommendations will be limited to the design of dual-media filters.

In tertiary wastewater treatment applications, the top size of the coal should be greater than 1 mm. An effective coal size at least 1–1.2 mm should be specified. After placement in the filter, the coal should be backwashed two or three times, and 2.5 cm (1 in.) layers of the fine surface material should be skimmed off after each wash to remove unwanted fine coal.

Nearly uniform coal as practicable (low uniformity coefficient) should be specified to minimize the bottom coal size and the backwash rate required. The minimum fluidization velocities (V_{mf}) are dependent on the size of filtration media (e.g., coal, sand, and garnet sand) and temperature (52). The minimum uniformity coefficient commercially available is appropriately 1.3. A reasonably uniform media can be achieved by specifying that all coal lie between alternative adjacent sieve sizes (e.g., −12+16, passing 12 mesh and retained on 16 mesh). However, it will usually be neccesary to permit 10% of the total media to fall outside of this range on both the coarse and fine sides of the range.

A sand specification compatible with the specified coal must be selected. The bottom sand size (e.g., the 90% finer size) should have the same minimum fluidization velocity as the bottom coal to ensure that the entire bed fluidizes at about the same backwash rate. The effective size of sand should be such as to achieve the goal of coarse-to-fine filtration without causing excessive media intermixing. A ratio of the 90% finer coal size to the 10% finer sand size equal to about 3 will result in a few inches of media intermixing at the interface (41,42,53). A larger ratio, say 4, will increase the degree of media intermixing, a smaller ratio, say 1, will cause a sharp interface to be present between the two media. After selecting the 10% finer sand size, the uniformity coefficient (or other specification) should be determined graphically to ensure that the desired top and bottom media sizes could be achieved.

After the sand is installed in the filter, the media should be backwashed and skimmed once or twice to remove any unwanted excessively fine material before installing the coal.

In determining the depth of media, there is no reasonable method other than pilot-plant operation that can be used to determine the optimum depth of filter media. Huang established that, for filtration of trickling filter final effluent, a depth of at least 38 cm (15 in.) of unisized 1.84 mm anthracite and 30.4–38 cm (12–15 in.) of unisized 0.55 mm sand was needed (42). Baumann and Cleasby recommend coal depths of 38–50 cm (15–20 in.) and sand depths of 30–38 cm (12–15 in.) (52).

4.2.2. *Head Loss Development*

As filtration progresses, the suspended solids are retained in the filter pores and the flow resistance increases which, in turn, decreases the pressure (driving force) available for maintaining the flow rate. Head loss development in a gravity filter is affected by many variables. The effects of two profound variables (media type and size, and filtration rate) are illustrated in Fig. 6. Finer media size and higher filtration rate result in higher head loss development than that which would develop from coarser media and lower flow rates.

The shape of head loss curves depends on the nature of the solids removal mechanisms, which take place within the filter. Figure 6 shows typical head loss curves when solids removal occurs only within the filter bed. Increasing the filtration rate increases the initial head loss. Because the head loss curves are essentially parallel, increasing the filtration rate slightly decreases water production to any particular terminal head loss. Depth removal of this type may be experienced using larger-sized surface media and the various filter designs providing coarse-to-fine filter media gradation. Nearly linear head loss curves have been observed in filtration of alum-treated secondary effluent (54).

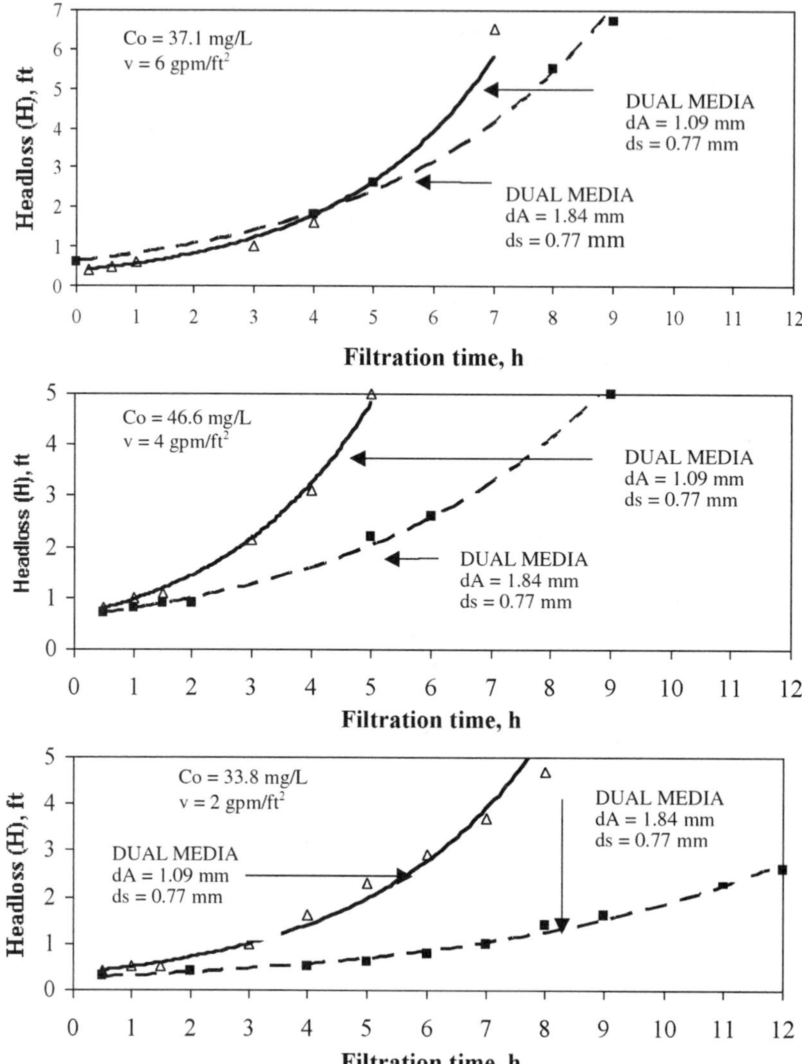

Fig. 6. Comparison of head losses of various anthracite sizes at different filtration rates through the 24 in. full depth filter.

When the solids are removed partly on the surface and partly in the depth of the filter, surface removal will predominate at low filtration rates. With higher filtration rates, the solids are carried deeper into the filter bed and more filtrate is produced before the surface cake forms. Hence, there exists an optimum filtrate production rate. If the filtration rate is increased to a level such that surface removal does not take place and all the solids are carried into the bed, depth removal will predominate.

Filtration of secondary effluent from a trickling filter plant tends to involve both surface and depth removal resulting in the filter head loss (41,42). Therefore, plots of head loss versus time or volume of filtrate can, especially in the operation of tertiary wastewater filters, yield valuable information on the design of the filter media or the selection of an optimum filtration rate.

4.3. Filter Operating Variables

4.3.1. Filtration rate

Filters can be classified to few types according to filtration rate:

1. Slow sand filters, which have a hydraulic application rate less than 10 $m^3/m^2 \cdot d$ (0.17 gpm/ft^2), are more extensively used in Europe. Kiely (9) suggests that this type of filter may become more popular because of its excellent abilities in improving the microbiological water quality (removal efficiencies of up to 99.9 % for *Giardia* and *Cryptospiridium* cysts). According to Ray (10), the filter builds up a layer of filtered contaminants on the surface, which becomes the active filtering medium. This active filtration layer is termed a *schmutzdecke*. The use of slow sand filters has declined in the past because of high cost of construction, large filter area needed, and unsuitability for processing waters with high-turbidity or polluted waters requiring chemical coagulation.
2. Rapid filters, which have a hydraulic application rate of about 120 $m^3/m^2 \cdot d$ (2 gpm/ft^2), are more extensively used in the Unites States.
3. High rate filters, which have a hydraulic application rate greater than 240 $m^3/m^2 \cdot d$ (4 gpm/ft^2), are also more extensively used in the United States.

4.3.2. Mode of Operating Filtration Cycle

There are four basic modes of operating filters that differ primarily in the way that the driving force is applied across the filter. They are referred to as constant-pressure filtration, constant-rate filtration with effluent rate control or with influent flow-splitting rate control, and variable declining-rate filtration (52,56).

In constant-pressure filtration, the total available driving force is applied across the filter throughout the filter run. At the beginning of the filter run, the filter media are clean, the filter resistance is low, and the flow rate is therefore high. As the filter retains the solids, filter resistance increases and flow rate decreases. The constant pressure mode of filtration is seldom used with water or wastewater gravity filters because it requires a relatively large volume of water storage on the upstream and the downstream sides of the filter.

In constant-rate filtration with effluent rate control, a constant pressure is applied (by constant water level) and the flow rate is held constant by the action of a manually operated or automatic effluent flow control valve. At the beginning of the filter run, the effluent control valve is nearly closed to provide the additional resistance needed to maintain the desired flow rate. As filtration progresses, the filter media becomes clogged with suspended solids, and the flow control valve opens slowly. When the flow control valve is fully open, the run must be terminated, since any further increase in filter resistance will not be balanced by a corresponding decrease in the resistance of the flow.

The disadvantages of this system include: (1) the initial and operating costs of a fairly complex control system are high; (2) the filtrate quality is not as good as that obtained by a filter of the influent flow-splitting rate control or of declining rate filtration modes described later. In addition, the valves used as rate controllers frequently require an excessive amount of maintenance, so much so that in many plants they never operate properly.

The influent flow-splitting constant-rate filter system splits the flow nearly equally to all the operating systems, usually by means of an influent weir box on each filter. A typical filter arrangement and its operating characteristics are depicted in Fig. 7.

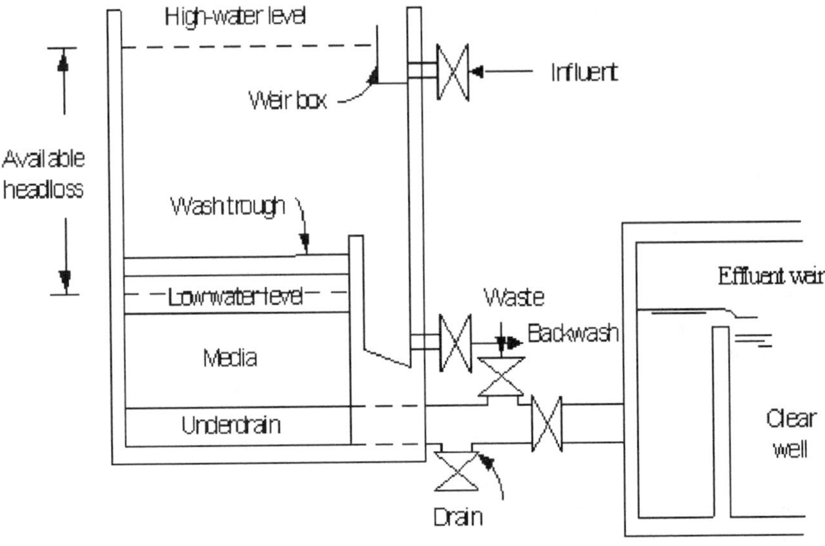

Fig. 7. Typical filter and clear well arrangement in influent flow splitting filtration.

The advantages of this system include: (1) constant-rate filtration is achieved without rate controllers as long as the total plant inflow remains constant; (2) when a filter is taken out of service for backwashing or returned to service after backwashing, the water level on the filter gradually rises or lowers in the operating filters until sufficient head is achieved to handle the flow. The changes in filtration rate are made slowly and smoothly without the abrupt effects associated with automatic or manual control equipment, thus providing the least harmful effect to filtrate quality (12); (3) the head loss for a particular filter is evidenced by the water level in the filter box. When the water reaches a desired terminal level above the media, backwashing of that filter is initiated.

The effluent weir must be higher than the level of filter media to prevent accidental dewatering of the filter bed; this arrangement eliminates the possibility of negative head in the filter. Thus, the undesirable problems that sometimes result from negative head are eliminated. The only disadvantage is the additional depth of filter box required. The depth must be enough to provide a reasonable available head loss across the filter media.

Variable declining-rate filtration is based on the operation of a filter between constant-pressure and constant-rate controls and has all of the advantages of constant-rate operation by influent flow splitting plus some additional advantages. The principal differences between declining-rate and constant-rate with influent-flow splitting is the location and type of influent arrangement and the provision of less available head. As illustrated in Fig. 8, the influent enters below the low water level of the filters through a relatively large influent header serving all the filters and a relatively large influent valve to each individual filter. Thus, the head losses in the header and influent valves are small and do not restrict the flow to a filter so the water level is essentially the same in all operating filters at all times.

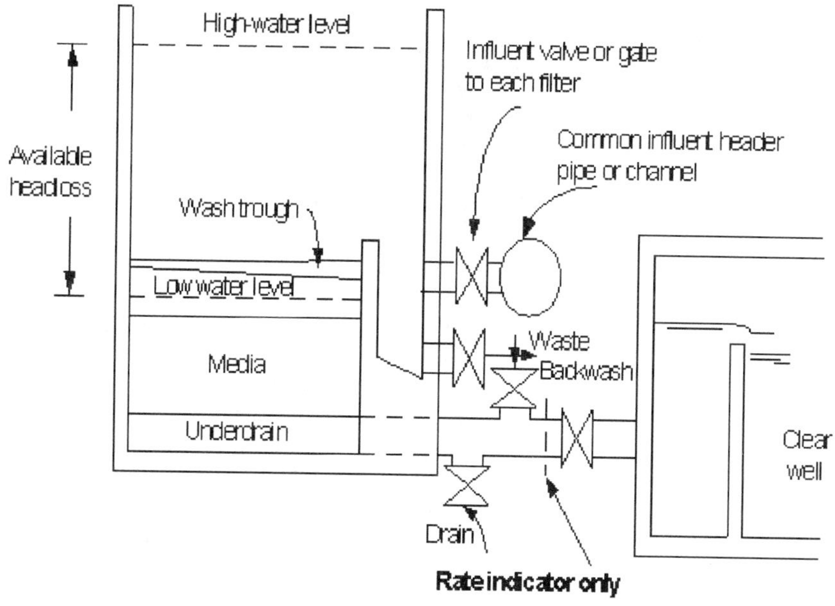

Fig. 8. Typical filter and clear well arrangement in variable declining-rate filtration.

As the filters served by a common influent header get dirty, the flow through the dirty filters decreases, automatically causing the cleaner filters to pick up the capacity lost by the dirtier filters. The water level in all filters rises slightly and this provides the additional head to the cleaner filters as they pick up the flow diverted to them by the dirtier filters. Rate changes occur gradually and smoothly without any automatic control equipment. As the dirty filter gets dirtier, the rate of flow through the filter decreases. The loss of head through the media then decreases linearly (laminar flow) with the decrease in the flow rate. However, the head loss through the underdrain systems usually decreases exponentially (turbulent flow) with the decrease in the flow rate. In this way, the variable declining filtration rate decreases the head loss through the underdrain system with the decreasing filtration rate and makes that head available for greater head loss through the dirty filter media. This accounts for the provision of smaller total available heads with variable declining rate filters (Fig. 8) as compared to influent-flow-splitting constant-rate filters (Fig. 7). The head loss provided for is generally in the range of 2.4–3.0 m (8–10 ft) of water with constant-rate filters and 1.5–2.4 m (5–8 ft) of water with declining-rate filters where potable water filtration is involved. In the case of wastewater filtration, Baumann and Cleasby (52) observed that regardless of filtration rate, the filtration of trickling filter effluents produced a head loss of 1.8–2.1 cm (0.06–0.07 ft) of water per pound of suspended solids removed. Table 3 provides a guideline for troubleshooting in filtration systems (51).

4.3.3. *Mode of Operating Backwashing Cycle*

The purpose of backwashing is to restore the solids retaining capacity of the filter by removing the solids collected during the filter cycle. Gravity filters are periodically cleansed *in situ* by an upflow of clean water or with a combination of air scouring

Table 3
Troubleshooting Guide for Filtration Systems

Conditions	Possible Causes
1. High head loss through a filter unit, or filter runs too short.	Filter bed in need of backwashing Air binding Mud balls in the filter bed Improper rate-of-flow controller operation Clogged underdrains Improper media design: too small, or too deep Floc strength too strong - will not penetrate media
2. High turbidity	Filter bed in need of backwashing Rate of flow too high Improper rate-of-flow controller operation Disturbed filter bed Mud balls in the filter bed Air binding Inappropriate media size or depth Low media depth (caused by loss during backwash) Floc too small or too weak caused by improper chemical pretreatment

when the effluent quality becomes unacceptable or the terminal head loss reaches the actual head available. In some cases, however, filters are backwashed on a regular time cycle based on experience with the two criteria above, say once each day. Amirtharajah found that cleaning granular filters by merely fluidizing the filter bed through water backwash alone is inherently a weak cleaning method due to the limitations in particle collisions (57). In fact, the primary cleaning mechanism is a result of the hydrodynamic shear forces, and the maximum hydrodynamic shear occurs at a porosity of 0.68–0.71 for the media size normally used in filtration. For graded sand this optimum porosity occurs in the top layers of the fluidized bed at the bed expansions of 40–50%. For effective cleaning, air scour and surface wash that promote interparticle abrasions during backwash are indispensable.

A study on air dynamics through filter media during air scour has been conducted by Hewitt and Amirtharajah (58). They concluded that:

1. The effectiveness of backwashing filters with air scour is related to the abrasion between sand grains. The intensity of abrasion is related to (a) the effective stresses between the grains; and (b) the magnitude of their relative movement. These two effects are contradictory with increasing water backwashing flow rate and thus there exists an optimum air–water flow rate combination for cleaning effectiveness.
2. At water backwashing flow rate of about 20–45% of minimum fluidization, depending on the air flow rate, the air moves through the bed in a pulsing action in which the media moves in a general downward and inward direction toward the air pockets that form and collapse. This is the point at which simultaneous air scour and subfluidization water backwash becomes most effective at cleaning. An empirical design equation was developed to predict the optimum backwashing air/water flow rates combination.
3. Media loss during air scour occurs due to the media being carried above the static level of the sand bed by the turbulent wake of air bubbles. Larger bubbles and higher water

Table 4
List of Equations for Determination of Backwash Operational Parameters

Filter media	Equation*	Applicable range of Q_a
Sand	$0.8\ Q_a^{\ 2} + \%(V/V_{\mathrm{mf}}) = 43.5$	1.8 to 4.6 scfm/ft^2
Anthracite	$1.7\ Q_a^{\ 2} + \%(V/V_{\mathrm{mf}}) = 43.0$	1.5 to 4.2 scfm/ft^2
Dual media	$1.7\ Q_a^{\ 2} + \%(V/V_{\mathrm{mf}}) = 39.5$	0.8 to 2.4 scfm/ft^2
Granular activated carbon	$3.3\ Q_a^{\ 2} + \%(V/V_{\mathrm{mf}}) = 26.6$	$Q_a < 2.7$ scfm/ft^2
Granular activated carbon-sand	$3.0\ Q_a^{\ 2} + \%(V/V_{\mathrm{mf}}) = 27.2$	$Q_a < 2.0$ scfm/ft^2

*V_{mf} is based on $d_{90\%}$ size.

flow rate tend to carry media higher. A minimum in the media loss also occurs at the same combination of air and water flow rates that result in most effective cleaning. Media loss will be excessive when fine or light media are washed with air and water simultaneously. The best practical approach to eliminate media loss is to increase the clearance of the wastewater gutter above the media surface and terminate the air–water wash before overflow.

Amirtharajah and coworkers (1,57,58) developed the so-called "collapse-pulsing" theory. An empirical equation relating airflow rate, fluidization velocity and backwash water flow rate was presented below:

$$aQ_a^{\ 2} + \%(V/V_{\mathrm{mf}}) = b \tag{11}$$

where a,b = coefficients for a given media; Q_{a} = airflow rate; $\%(V/V_{\mathrm{mf}})$ = percentage of minimum fludization water flow (the ratio of superficial water velocity devided by the minimum fluidization velocity based a specified size of the medium).

The coefficients and applicable range of airflow rate for commonly used filter are given in Table 4.

Backwashing of wastewater filters encounters a unique condition substantially different from water filters. The wastewater filter receives heavier and more variable influent suspended-solids loads, and the solids tend to stick more tenaciously to the filter media. Thus, effective media cleaning during backwashing is essential for wastewater filtration. The findings of Cleasby and Lorence (59) and Young (60) can be used as guidelines for successful backwashing. They are

1. Backwash procedures must provide adequate scouring action to clean the media grains completely during each backwash and to remove the loosened solids from the filter. The most effective backwash procedure is to use air and water simultaneously throughout all but the last 1 or 2 min of the backwash sequence.
2. Adequate volume of backwash water must be used to effectively remove solids from the water zone between the media bed and the backwash trough. This volume may vary with filter design, but 3–4 m^3/m^2 (75–100 gal/ft^2) per backwash seems to be a minimum for most filter designs.
3. Sudden surges in backwash water should be avoided, especially if gravel layers are used for media support. In all cases, slow opening and closing backwash values are recommended. Adequate freeboard to overflow must be provided and the rates of air and water flow must be selected to ensure that loss of media will be minimized.
4. Some means of measuring backwash head loss and some method of relieving excessive underdrain pressure during backwash should be included in filter designs.

5. APPLICATIONS

5.1. *Potable Water Filtration*

The most commonly used unit process in a water-treatment plant is the gravity granular media filter. Such filters have been used in potable water production for the removal of solids present in surface waters pretreated by coagulation and sedimentation, for the removal of precipitates resulting from lime or lime/soda-ash softening, and for the removal of iron or manganese found in many underground water supplies.

Numerous technical solutions may be available for any particular water-filtration problem. The desired quality of water may be obtained by filtering the water in a conventional rapid sand filter plant, in a dual or multimedia filter plant, or in a diatomite filter plant. Until about 1960, most of the rapid sand filter plants were designed with a single media (silica sand, effective size of 0.4–0.6 mm, depth of 30–36 in.) and operated at a low flow rate (commonly 2 gpm/ft^2) to a relatively low head loss (8–10 ft of water). This design usually provided a good quality of effluent, but no consideration was given to finding the extent of chemical pretreatment and the proper combination of flow rate, total head loss allowance, sand size, sand depth, and run length that could produce the desired quality of effluent at the low cost.

During the last decades, a significant advancement has been made in understanding the mechanism of filtration and in developing applications in potable water filtration. Application of dual or multimedia filtration has been advocated. Also, for high-quality raw waters, the use of chemical coagulation-flocculation and filtration, without sedimentation preceding filtration, has gained popularity. This new scheme is referred to as contact coagulation filtration or direct filtration.

The contact coagulation filtration process eliminates sedimentation units, and a coagulant is applied before the flocculator. Flocculation continues to occur within the filter bed as the mixture of coagulant and water travel downward. Provisions are usually made to feed both a coagulant and a filter aid or polyelectrolyte. It is expected that either one or both are necessary to obtain an acceptable water quality. This depends on the characteristics of each individual water supply. In some waters having low turbidities and low temperatures that need excessive dosage of coagulant to form a settleable floc, contact coagulation filtration usually requires much less coagulant (1,2,61–63). All of the above results in some reduction in the first cost of the water-treatment plant (1,2,64,65).

5.2. *Reclamation of Wastewater*

Global water shortage is an issue that is increasing in magnitude, severity, and urgency. Escalating water needs is the outcome of world population growth of almost 1 billion per decade, with almost four-fifths in urban areas. Take Singapore as an example, the population has tripled since 1950 accompanied by an eightfold increase in daily water consumption. In 1999, the daily water consumption was 1.3 million cubic meters. As a land-scarce country and ranked sixth as the most water-deficient country in the world, Singapore is unable to develop any new large-scale surface water collecting schemes to support its growing population and expanding economy (66).

The global search for alternative water sources began some decades ago and has found two promising technologies: seawater desalination and wastewater reclamation. The desalination option has been proposed to provide additional water. However, although there has been rapid development in desalination, its operation is still costly because of high energy consumption. There should be a parallel effort to explore the wastewater reclamation option.

Wastewater reclamation had arisen from a need to provide the world's spiraling population with potable water. The partial solution to an inadequate fresh water supply is to reuse water by reclaiming wastewater and augmenting the water supply with such recycled water. In many countries, wastewater undergoes secondary treatment before discharge to the nearby waters. Further treatment of the secondary effluent to a reusable standard is technologically feasible (65–68). Municipal wastewater has been traditionally selected as the starting point of wastewater reclamation. However, as the world's communities now emphasize sustainable development, industrial wastewater effluent will be in the limelight. Although not recommended for potable water applications, industrial wastewater reclamation is a viable option capable of producing high-quality water suitable for indirect potable use (IPU) such as cooling water and feed water for the demineralization unit for production of boiler feed water (69).

Wastewater reclamation processes can be divided into the following two processes: traditional treatment and membrane-based treatment. Traditional treatment used in water reclamation includes coagulation, sedimentation, filtration, carbon adsorption, hardness removal if necessary, and disinfection. The successful example is the Water Factory 21, Orange County, CA, USA. Developments in membrane technology have resulted in increased applications in many areas including wastewater reclamation. The Water Factory 21 in the USA and the NEWATER project in Singapore have provided successful examples for the IPU (66,68,69). It has been stressed repeatedly by researchers and professionals in field of membrane technology that proper feed pretreatment and well-developed cleaning protocols are imperative in ensuring success of any membrane processes. Gravity filtration can be a promising treatment for both traditional treatment and membrane-based treatment.

5.2.1. *Traditional Treatment*

In the traditional treatment for water reclamation shown in Fig. 9, filtration plays an important role:

1. Removal of residual biological flocs in settled effluent from secondary treatment by trickling filters or activated-sludge processes.
2. Removal of precipitates resulting from alum, iron, or lime precipitation of phosphates in secondary effluents from trickling filters or activated-sludge processes. The suspended solids to be filtered can be substantially different from those in normal secondary effluent as discussed above.
3. Removal of solids remaining after the chemical coagulation of raw wastewaters in physical–chemical waste treatment processes. Again, the solids to be filtered can be substantially different from normal secondary effluent solids as discussed above.

Much of the design and operation of wastewater filters has been based on the experience from potable water filtration. However, several differences need to be noted. With built-in raw and filtered water-storage capacity, potable water filters can be and generally are operated at constant filtration rates for long periods, and steady operating conditions will prevail. Thus, plant design can be based on the maximum daily demand, not on peak hourly demand. In wastewater filtration, however, the plant must be

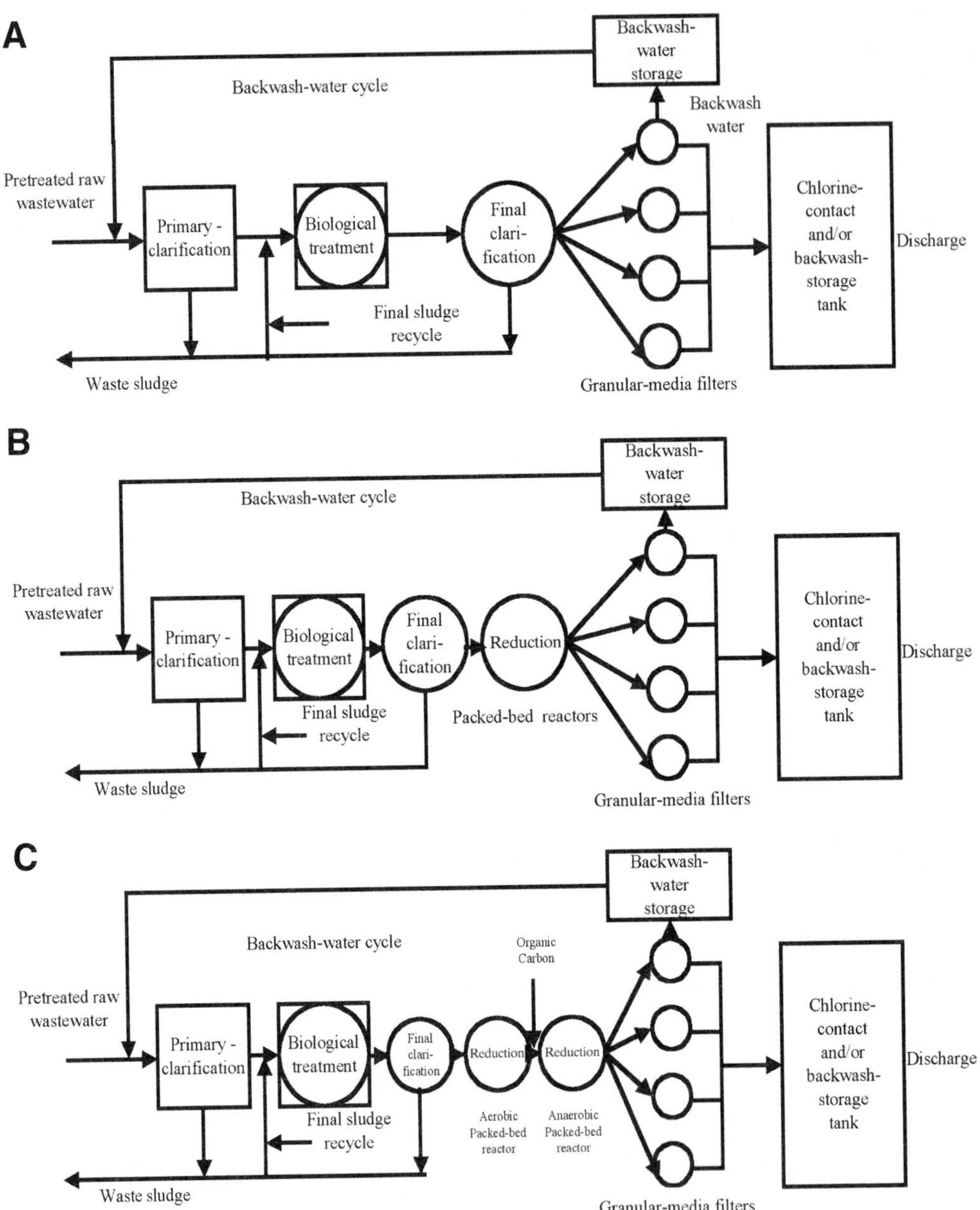

Fig. 9. Granular filters for tertiary treatment: **(A)** following biological secondary treatment; **(B)** following biological secondary and biological nitrification treatments; **(C)** following biological secondary and biological nitrification denitrification treatments.

designed to handle a continuously varying rate of flow with variations from a nighttime dry weather flow to peak hourly flows in storm water runoff periods. Therefore, the potential effects of peak flow rates must be considered in the design.

In potable water filtration, the water is much more consistent in both the level of solids and in their filtering characteristics. Even with pretreated surface water, the solids to be removed consist of low levels of floc carryover, with some attached colloidal solids that contributed to the original raw water turbidity. The filtering characteristics of solids that are mainly organic are more predictable than the filtering characteristics of the inorganic–organic solids found in typical wastewaters to be treated. Considerable data are available to demonstrate that in raw wastewater the suspended solids levels will vary directly with the flow. Because all wastewater-treatment processes are least efficient under their peak-load operating conditions, higher suspended solids will be carried over to tertiary filters during peak flows. Thus, applied to filters, the wastewater presents its highest solids concentration to be removed during the highest flow-rate periods. Such high loadings contribute to high head loss and, consequently, to potential short filter runs. Therefore, the critical design condition to be considered occurs under the peak-flow operating conditions. In addition, the solids in wastewater are much stickier than water-plant solids, and are much more difficult to remove effectively during backwashing.

The problems encountered in designing filters for wastewater treatment require that the following be considered (51,52,66,69):

1. A completely mixed flow equalization basin ahead of gravity filters should be considered. Fifteen to twenty percent of mean daily flow-storage capacity would permit constant flow and nearly constant solids loads to the filters.
2. The higher solids loadings to wastewater filters require better distribution of solids throughout the filter bed. This improvement can be accomplished by:
   - using coarse top media, requiring a higher backwash rate,
   - using dual- or triple-media or upflow filters to achieve coarse-to-fine filtration,
   - using coarse, deep bed and nearly unisize media filters, and
   - providing higher terminal head losses.

5.2.2. *Membrane-Based Treatment*

In order to obtain high-grade reclaimed water, reverse osmosis (RO) membrane filtration must be used as one of the last processes in the water reclamation. However, the RO membrane is not able to tolerate the influent feed water with higher organic contents as well as suspended solids. Controlling RO membrane fouling continues to be a major challenge in desalination and wastewater reclamation. Proper pretreatment has been stressed repeatedly as the first line of defense in controlling membrane fouling and ensuring success for the RO operation. The feed must pass through a series of pretreatments before entering the RO filtration units. The pretreatments can be classified into two groups: membrane-based treatment and conventional treatment. The former includes microfiltration (MF) and ultrafiltration (UF). The details of the membrane-based technology for the pretreatment can be found in this book.

A typical conventional pretreatment configuration would include flocculation, lime or alum clarification, recarbonation, settling, filtration, and activated carbon adsorption (66). Biological activity is controlled by chlorination. The Water Factory 21 in the US reported that 26% of TOC was removed by lime clarification. Concentration of inorganic constituents such as calcium, magnesium, iron, fluoride, and silica were significantly reduced; over 99% of coliform bacteria were removed. Multimedia filtration produced an average effluent turbidity of 0.14–0.16 NTU. Granular activated carbon (GAC) removed 30–50% of the organics. The average TOC concentration of GAC effluent was 5.5 mg/L (70).

Alum, alum–polyelectrolyte, and polyelectrolyte in-line flocculation filtration systems were compared by Cikurel et al. (71). Results showed that high-molecular-weight (2×10^6) branched-chain polyacrylamide at dosage as low as 0.5 mg/L was more effective at turbidity removal than alum dosed at 10–20 mg/L. Results were consistent at both low (6–12 NTU) and high (20-27 NTU) feed turbidity. The performance of polyacrylamide as primary flocculant was comparable to the alum–polyelectrolyte system. Various polyelectrolytes of different molecular weights (MW) were evaluated. High-MW polyacrylamide was more effective than medium- and low-MW polyamides at the same dosage of 0.5 mg/L. Effectiveness of medium-MW polyamide increased when dosage was increased above 5 mg/L. Low-MW polyamide was not effective in reducing turbidity even at a dosage of 7 mg/L. The charge density of the polymers did not play a significant role in particle removal. It was postulated that the main mechanism of high-MW polymer is bridging in contrast to adsorption and charge-neutralization action of alum.

Studies on in-line filtration showed that effluent turbidity of less than 2 NTU is achievable with alum dosage of 0–8 mg/L and cationic polymer dosage of 0–0.5 mg/L (72). However, the performance of this system is dependent on feed water quality, hydraulic loading rate of the granular media column, and good control over the dosage of alum and the cationic polymer. High feed water turbidity and high polymer dosage have adverse effects on the duration of filtration cycle.

Kim and co-workers studied the performance of an RO system that was used to treat secondary effluent from a wastewater treatment plant that receives 70% of its influent as industrial wastewater to a standard suitable for industrial reuse (66). A comparison study was performed to compare three different pretreatment schemes for the RO unit. The three pretreatment systems investigated included:

1. System I (UF).
2. System II (dual-media filtration and granular-activated-carbon adsorption with addition of an organic flocculant),
3. System III (dual-media filtration and granular-activated-carbon adsorption without addition of an organic flocculant).

Table 5 summarizes the quality of the water before and after pretreatment by the three systems (66). Although the feed quality varied for each experiment, such variations were not considered significant. A good pretreatment system should be able to handle feed variations of such degree. These three systems provide efficient turbidity removal. Effective removal of color, TOC, COD, and BOD_5 is observed when Systems II and III are used. The quality of the treated water after the pretreatments and the RO is summarized in Table 6 together with the cooling water requirement and the typical water quality from a local reservoir. The turbidity of the treated water from all these three treatment systems was much lower than both the cooling water requirement and the local reservoir water. Nonetheless, the effluent from these three pretreatment systems and RO may be considered for re-use in some industrial operations (e.g., cooling water). If the RO membrane fouling is taken into consideration, the UF membrane provides the best pretreatment amongst the three systems. However, Systems II and III are more competitive because of the maturity of the technologies.

The current trend is that more membrane filtration system replace traditional filtration systems. Take Singapore as an example. Two factories with the capacity of 22,500

Table 5
Water Quality of Feed after Pretreatment (before RO) in Wastewater Reclamation[a]

	System I		System II		System III	
Parameter	Feed	After Treatment	Feed	After Treatment	Feed	After Treatment
Turbidity (NTU)	9.23–10.96	0.62–1.15	9.64–10.80	2.74–3.32	11.80–16.70	1.80–2.25
Color (LU)	80–125	30–50	70–80	< 20	80–90	< 20
TOC	26.7–44.4	14.2–34.3	12.2–21.7	3.5–5.4	12.6–20.7	6.0–7.4
COD	52–68	45–55	42–81	6–20	46–64	11–19
BOD_5	10–12	8–10	12–14	6–8	12–18	8
Nitrate	3.0–4.1	2.4–3.3	3.7–5.0	1.7–3.1	3.8–4.7	1.2–2.1
Phosphate	9.2–10.0	6.8–9.2	9.4–12.3	8.3–9.8	10.7–15.4	7.8–14.7

[a]All units are in mg/L.

m^3/day each have been commenced in January 2003 to supply NEWATER to the wafer fab industry and IPU (73). The Singapore government has planned to increase NEWATER use up to 15% of total demand by 2008.

6. DESIGN EXAMPLES

6.1. *Example 1*

Outline an engineering procedure for a step-by-step design of a filtration system.

Solution

The design of filters depends on the influent water or wastewater characteristics, process and hydraulic loading, method and intensity of cleaning, nature, size, and depth of the filtration material, media and multimedia filters are more effective and easier and less expensive to operate than single media sand filters for the treatment of water or wastewaters; therefore, they are more widely accepted. The design of a multimedia filter is illustrated below. However, the procedure can be applied to the design of a rapid sand filter also.

Step 1. Input Data Collection:

(a) Wastewater characteristics.
(a_1) Average daily flow (MGD).
(a_2) Peak flow (MGD).
(b) Suspended solid (mg/L).
(c) Suspended solids characteristics.
(c_1) Biological floc.
(c_2) Chemical floc: alum dose, lime dose, ferric chloride.
(d) Temperature (°F).
(e) Alkalinity.
(f) pH.

Step 2. Design Parameters Assumptions and Decisions: Engineering assumptions and decisions shall be made based on the specifications and regulations established by the US EPA, the local governments, and professional associations.

Table 6
Water Quality of Final Effluent in Wastewater Reclamation[a]

Parameter	Feed	System I & RO	System II & RO	System III & RO	Cooling Water	Reservoir water[c]
pH	7.6–7.9	6.3–6.8	6.5–6.9	6.6–7.0	6.0–8.3	7.5-8.5
Turbidity (NTU)	9.2–16.7	< 0.7	< 0.8	1.1	—[b]	< 2.5
Color (LU)	70–125	< 10	< 10	< 10	—	< 5
TDS	1100–1400	134–225	243–381	340–880	500	—
Alkalinity (as $CaCO_3$)	263–338	34–50	43–66	58–74	350	30
Hardness (as $CaCO_3$)	186–371	37–82	27–76	68–146	650	130
TOC	12.1–44.4	2.6–13.4	2.9–5.0	4.3–5.3	—	2.8-5.3
COD	42–81	< 12	< 7	< 17	5	5
BOD_5	10–18	3–4	4–5	4–5	—	< 5
Chloride	320–533	88–173	80–171	93–211	500	200
Fluoride	3.0–7.2	1.9–2.6	1.3–2.3	1.6–3.5	—	—
Nitrate (as N)	3.0–5.0	0.9–1.2	1.1–2.9	1.4–1.9	—	—
Ammonia (as N)	27.5–39.7	3.0–15.0	6.0–16.0	8.0–20.3	—	< 0.15
Phosphate (as P)	3.0–5.0	0.3–1.0	0.3–1.9	0.8–2.2	—	< 0.15
Calcium	78.4–111.0	9.4–21.6	5.5–17.1	11.5–25.1	50	—
Magnesium	13.9–24.9	3.3–6.9	3.2–8.2	4.8–9.6	—	—
Potassium	29.0–53.9	6.0–14.1	6.0–15.1	9.6–20.4	—	—
Sodium	300–505	69.7–123	80.3–133.0	127–205	—	—
Copper	0.07–0.33	0.11–0.28	0.09–0.22	0.14–0.28	—	< 0.005
Chromium	0–0.064	0–0.015	0.005–0.018	0.010–0.050	—	—
Iron	0–0.116	ND	ND	ND	0.5	< 0.1
Lead	ND	ND	ND	ND	—	—
Manganese	0.06–0.34	0.05–0.09	0.01–0.04	0.03–0.06	0.5	—
Nickel	0.42–0.87	0.17–0.20	0.10–0.18	0.11–0.22	—	—
Zinc	0.10–0.56	0.03–0.28	0.19–0.30	0.07–0.44	—	—

[a]All units are in mg/L.
[b]Not specified/measured.
[c]From a local reservoir.

(a) Rate of wastewater application (gal/min/ft^2).
(b) Size distribution of filter media and mean diameter (mm).
(c) Approach velocity (fps).
(d) Number of layers.
(e) Shape factor for each layer.
(f) Porosity of unexpanded bed depth.
(g) Specific gravities of filter medium and of water.
(h) Depth of each filter medium (ft).
(i) Permeability of each layer.
(j) Size distribution of gravel.
(k) Depth of gravel medium (ft).
(l) Desired degree of bed expansion (20–50%).
(m) Type of underdrain system.
(n) Head loss in underdrain system (ft).
(o) Diameter of 60 percentile particles d^{60} (mm).
(p) Specific weight of sand (lb/ft^3).
(q) Density of water (g/cm^3).
(r) Absolute viscosity of water (centipoises).
(s) Porosity of expanded bed.
(t) Number of troughs.
(u) Width of trough (ft).
(v) Depth of underdrain (ft).
(w) Operating depth of water above sand (ft).
(x) Freeboard (ft).
(y) Time of backwash (min).
(z) Distance from top of trough to underdrain (ft).

Step 3. Design Procedure:

(a) Select a loading rate (filtration rate) and calculate filter surface area (from the federal or state regulations and professional associations' specifications):

$$SA = \frac{Q_{\text{avg}}\left(10^6\right)}{LR(24)(60)}$$

where SA = surface area (ft^2), Q_{avg} = average flow (MGD), and LR = loading rate (gal/min·ft^2).

(b) Select filter medium and evaluate size distribution and depth of each layer.
(c) Calculate terminal head loss through filter using Kozeny equation:

$$\frac{h_f}{L} = K\left(\frac{\nu}{g}\right)\left[\frac{(1-\varepsilon)^2}{\varepsilon^3}\right](v)\left(\frac{\sigma_s}{d_p}\right)^2\left(\frac{1}{3.28\times10^{-3}}\right)^2$$

where h_f = loss of head in depth of bed (ft), L = length (ft), K = coefficient of permeability (≈ 6), ν = kinematic viscosity (ft^2/s), g = gravitational acceleration (ft/s^2), ε = porosity of layer, v = approach velocity (fps), σ_s = shape factor (6 for spherical, 8.5 for crushed granules), d_p = mean particle size (mm), 3.28×10^{-3} = conversion factor (mm to ft).

(d) Calculate unit head loss through each expanded media:

$$\Delta p = D(1-\varepsilon)\left(G_{s,m} - G_{s,w}\right)$$

where Δp = pressure drop across fluidized bed (ft), D = unexpanded bed depth (ft), ε = porosity of unexpanded meida, $G_{s,m}$ = specific gravity of filter medium, and $G_{s,w}$ = specific gravity of water.

(e) Calculate total head loss through expanded media.

$$(\Delta p)_T = \sum \Delta p$$

where $(\Delta p)_T$ = total pressure drop across fluidized bed (ft).

(f) Calculate rate of filter backwashing for any desired expansion as follows:
(f_1) Calculate minimum fluidization velocity.

$$v_f = \frac{0.00381(d_{60})^{1.82}\left[W_s(W_m - W_s)\right]^{0.94}}{\mu^{0.88}}$$

where v_f = minimum fluidization velocity (gpm/ft^2), d_{60} = 60% finer size of the sand (mm, ≈ 0.75), W_s = specific weight of water (lb/ft^3), W_m = specific weight of sand (lb/ft^3), μ = absolute viscosity of water (centipoises).

(f_2) Calculate Reynolds number corresponding to the minimum fluidization velocity.

$$(R_n)_f = \frac{\rho_l v_f d_{60}(3.28\times10^{-3})}{\mu(7.48)(60)(2.09\times10^{-5})}$$

where $(R_n)_f$ = Reynolds number, ρ_l= density of water (lb-sec^2/ft^4), v_f = minimum fluidization velocity (gpm/ft^2), d_{60} = 60% finer size of the sand (mm, ≈ 0.75), 3.28×10^{-3} = conversion factor (mm to ft), μ= absolute viscosity of water (centipoise), 7.48 = conversion factor (ft^3 to gal), 60 = conversion factor (s to min), and 2.09×10^5 = conversion factor (centipoise to lb-s/ft^2).

(f_3) If $(R_n)_f > 10$, apply a correction factor to the calculated value of v_f as follows:

$$K_R = 1.775(R_n)_f^{-0.272}$$

where K_R = correction factor.

(f_4) Calculate the unhindered settling velocity as follows:

$$v_s = 8.4v_f$$

where v_s = unhindered settling velocity (gpm/ft^2) and v_f = minimum fluidization velocity (gpm/ft^2).

(f_5) Calculate Reynolds number based on the unhindered settling velocity:

$$(R_n)_s = \frac{\rho_l v_s d_{60}(3.28\times10^{-3})}{\mu(7.48)(60)(2.09\times10^{-5})}$$

where $(R_n)_s$ = Reynolds number, ρ_1 = density of water (lb-s^2/ft^4) v_s = unhindered settling velocity (gpm/ft^2), d_{60} = 60% finer size of the sand (mm, ≈ 0.75), and μ = absolute viscosity of water (centipoise).

(f_6) Calculate expansion coefficient as follows:

$$n_e = 4.45(R_n)_s^{-0.1}$$

where n_e = expansion coefficient.

(f_7) Using v_f and n_e, calculate the constant K_e for the system.

$$v_f = K_e(\varepsilon)^{ne}$$

where v_f = minimum fluidization velocity (gpm/ft^2), ε = porosity of unexpanded media, and n_e = expansion coefficient.

(f_8) Calculate the desired porosity at the desired bed expansion:

$$\frac{D_e}{D} = \frac{1-\varepsilon}{1-\bar{\varepsilon}}$$

where D_e = depth of expanded bed (ft) D = depth of unexpanded bed (ft), $\bar{\varepsilon}$ = porosity of unexpanded bed, and ε = porosity of expanded bed.

(f_9) Calculate the backwash rate:

$$BR = K_e(\bar{\varepsilon})^{ne}$$

where BR = backwash rate (gpm/ft^2), K_e = system constant, $\bar{\varepsilon}$ = porosity of expanded bed, and n_e = expansion coefficient.

(g) Select wash water troughs arrangement and calculate depth of wash water troughs:

$$Q = 2.49bh_0^{3/2}$$

where Q = total trough flow (cfs), b = trough width (ft), h_0 = trough minimum depth (ft).

(h) Select an underdrain system and calculate the minimum total filter depth:

$$TD = UD + GD + MD + MOD + FB$$

where TD = total filter depth (ft), UD = depth of underdrain (ft), GD = gravel depth (ft), MD = media depth (ft), MOD = maximum operating depth of water above sand, 3–5 ft, and FB = freeboard (ft).

(i) Calculate total head necessary to backwash water:

$$TH = HLUD + HLG + HLM + DWT$$

where TH = total head necessary for backwash, $HULD$ = head loss in underdrain (manufacturer's requirement), HLG = head loss in gravel (c above), HLM = head loss through fluidized bed (d above), DWT = total depth from top of wash trough to underdrain (ft).

(j) Assume backwash time and calculate total backwash water needed:

$$BWW = (BR)(BWT)(SA)$$

where BWW = total backwash water (gal), BR = backwash rate (gpm/ft^2), BWT = backwash time (min), and SA = surface area (ft^2).

Step 4. Output Data:

(a)

Layer	Depth (ft)	Diameter (ft)	Shape factor	Specific gravity
1	xx.x	xx.x	xx.x	xx.x
2	xx.x	xx.x	xx.x	xx.x
3	xx.x	xx.x	xx.x	xx.x
4	xx.x	xx.x	xx.x	xx.x

(b) Average loading rate (gpm/ft²).
(c) Surface area (ft²).
(d) Underdrain head loss (ft).
(e) Wash water gutter width (ft).
(f) Wash water gutter depth (ft).
(g) Terminal head loss through bed (ft).
(h) Maximum head for backwashing (ft).
(i) Total filter depth (ft).
(j) Wash water needed (gal).
(k) Backwash rate (gpm/ft²).

6.2. Example 2

Select a loading rate and calculate filter surface area, assuming the average flow is 1 MGD.

Solution

$$SA = \frac{Q_{\text{avg}}\left(10^6\right)}{LR(24)(60)}$$

where SA = surface area (ft²), Q_{avg}= average flow (1 MGD), LR = loading rate (6 gal/min·ft²), 24 = h/d, 60 = min/h:

$$SA = \frac{1\times 10^6}{6(24)(60)}$$

$$SA = 116 \text{ ft}^2$$

6.3. Example 3

Continue Example 2; select filter media and evaluate size distribution and depth of each layer. Calculate the terminal loss through the filter.

Solution

Step 1. Anthracite is selected:

Depth = 18 in.
Effective size = 1.2 mm
Uniformity coefficient = 1.5

Sand is also selected:

Depth = 12 in.
Effective size = 0.5 mm
Uniformity coefficient = 1.4

Step 2. Calculate terminal head loss through filter using Kozeny equation:

$$\frac{h_f}{L} = K\left(\frac{\nu}{g}\right)\left[\frac{(1-\varepsilon)^2}{\varepsilon^3}\right](v)\left(\frac{\sigma_s}{d_p}\right)^2\left(\frac{1}{3.28\times 10^{-3}}\right)^2$$

where h_f= loss of head in depth of bed (ft), L = length (1.5 ft anthracite, 1.0 ft sand), K = coefficient of permeability (6), ν = kinematic viscosity (1.088×10^{-5} ft²/s), g = gravitational acceleration (32.2 ft/s²), ε = porosity (0.50 anthracite, 0.40 sand), v = approach velocity

(0.04 fps), σ_s = shape factor (8.5), d_p = mean particle size (1.2 mm anthracite, 0.5 mm sand), and 3.28×10^{-3} = conversion factor (mm to ft).

For anthracite:

$$\frac{h_f}{1.5} = 6\left(\frac{1.088\times10^{-5}}{32.2}\right)\left[\frac{(1-0.5)^2}{0.5^3}\right](0.04)\left(\frac{8.5}{1.2}\right)^2\left(\frac{1}{3.28\times10^{-3}}\right)^2$$

$$h_f = 1.13 \text{ ft}$$

For sand:

$$\frac{h_f}{1.0} = 6\left(\frac{1.088\times10^{-5}}{32.2}\right)\left[\frac{(1-0.4)^2}{0.4^3}\right](0.04)\left(\frac{8.5}{0.5}\right)^2\left(\frac{1}{3.28\times10^{-3}}\right)^2$$

$$h_f = 5.45 \text{ ft}$$

Total

$$h_f = 6.58 \text{ ft}$$

6.4. *Example 4*

Continue Example 3, calculate (a) total head loss through expanded media and (b) calculate the rate of filter backwash.

Solution

Step 1. Calculate total head loss through expanded media:

$$(\Delta p)_T = \sum \Delta p$$

$$(\Delta p)_T = 0.50 + 0.99$$

$$(\Delta p)_T = 1.49 \text{ ft}$$

Step 2. Calculate rate of filter backwashing:

(a) Calculate minimum fluidization velocity:

$$v_f = \frac{0.00381(d_{60})^{1.82}\left[W_s(W_m - W_s)\right]^{0.94}}{\mu^{0.88}}$$

where v_f = minimum fluidization velocity (gpm/ft^2), d_{60} = 60% finer size of the sand (mm, 0.75 mm), W_s = specific weight of water (62.4 lb/ft^3), W_m = specific weight of sand (165 lb/ft^3), and μ = absolute viscosity of water (1.009 centipoise):

$$v_f = \frac{0.00381(0.75)^{1.82}\left\{62.4\left[2.65(62.4) - 62.4\right]\right\}^{0.94}}{1.009^{0.88}}$$

$$v_f = 8.68 \text{ gpm/ft}^2 = 0.0193 \text{ fps}$$

(b) Calculate the Reynolds number corresponding to the minimum fluidization velocity:

$$(R_n)_f = \frac{\rho_l v_f d_{60}(3.28\times10^{-3})}{\mu(7.48)(60)(2.09\times10^{-5})}$$

where $(R_n)_f$ = Reynolds number, ρ_l = density of water (1.94 lb-s²/ft⁴), v_f = minimum fluidization velocity (8.68 gpm/ft²), d_{60} = 60% finer size of the sand (mm, 0.75 mm), μ = absolute viscosity of water (1.009 centipoise) 3.28×10^{-3} = conversion factor (mm to ft), 7.48 = conversion factor (ft³ to gal), 60 = conversion factor (sec to min), and 2.09×10^5 = conversion factor (centipoise to lb-s/ft²):

$$(R_n)_f = \frac{1.94(8.68)(0.75)(3.28\times10^{-3})}{1.009(7.48)(60)(2.09\times10^{-5})}$$

$$(R_n)_f = 4.4$$

(c) Because $(R_n)_f < 10$, no correction factor is needed.

(d) Calculate the unhindered settling velocity as follows:

$$v_s = 8.45 v_f$$

where v_s = unhindered settling velocity (gpm/ft²) and v_f = minimum fluidization velocity (8.68 gpm/ft²):

$$v_s = 8.45(8.68)$$

$$v_s = 73.3\,\text{gpm/ft}^2$$

(e) Calculate the Reynolds numbers based on the unhindered settling velocity:

$$(R_n)_s = \frac{\rho_l v_s d_{60}(3.28\times10^{-3})}{\mu(7.48)(60)(2.09\times10^{-5})} = 8.45(R_n)_f$$

where v_s = unhindered settling velocity (73.3 gpm/ft²), $(R_n)_f$ = Reynolds number for minimum fluidization velocity (4.4), and $(R_n)_s$ = Reynolds number for unhindered settling velocity:

$$(R_n)_s = 8.45(4.4)$$

$$(R_n)_s = 37.2$$

(f) Calculate expansion coefficient:

$$n_e = 4.45(R_n)_s^{-0.1}$$

where n_e = expansion coefficient and $(R_n)_s$ = Reynolds number for unhindered settling velocity (37.2):

$$n_e = 4.45(37.2)^{-0.1}$$

$$n_e = 3.1$$

(g) Calculate K_e for the system:

$$v_f = K_e(\varepsilon)^{n_e}$$

where v_f = minimum fluidization velocity (8.68 gpm/ft²), K_e = constant for system, ε = porosity of unexpanded media (0.40), and n_e = expansion coefficient (3.1):

$$8.68 = K_e(0.40)^{3.1}$$

$$K_e = 149\,\text{gpm/ft}^2$$

(h) Calculate the desired porosity at the desired bed expansion:

$$\frac{D_e}{D} = \frac{1-\varepsilon}{1-\bar{\varepsilon}}$$

where D_e = depth of expanded bed (1.2 ft), D = depth of unexpanded bed (1.0 ft), ε = porosity of unexpanded bed (0.40), and $\bar{\varepsilon}$ = porosity of expanded bed:

$$\frac{1.2}{1.0} = \frac{1-0.40}{1-\bar{\varepsilon}}$$

$$\bar{\varepsilon} = 0.50$$

(i) Calculate the backwash rate:

$$BR = K_e(\bar{\varepsilon})^{n_e}$$

where BR = backwash rate (gpm/ft^2), K_e = system constant (149 gpm/ft^2), $\bar{\varepsilon}$ = porosity of expanded bed (0.50), and n_e = expansion coefficient (3.1):

$$BR = 149(0.50)^{3.1}$$

$$BR = 17.4\ \text{gpm/ft}^2$$

6.5. Example 5

Continue Example 4. Calculate the unit head loss through each expanded media.

Solution:

The following design equation is used:

$$\Delta p = D(1-\varepsilon)(G_{s,m} - G_{s,w})$$

where Δp = pressure drop across fluidized bed (ft), D = unexpanded bed depth (1.5 ft anthracite, 1.0 ft sand), ε = porosity of unexpanded media (0.50 anthracite, 0.40 sand), $G_{s,m}$ = specific gravity media (1.67 anthracite, 2.65 sand), and $G_{s,w}$ = specific gravity water (1.0):

$$\Delta p \text{ anthracite} = 1.5(1-0.5)(1.67-1.0)$$
$$\Delta p = 0.50 \text{ ft}$$
$$\Delta p \text{ sand} = 1.0(1-0.4)(2.65-1.0)$$
$$\Delta p = 0.99 \text{ ft}$$

6.6. Example 6

This is an example for illustration of a 4-Cell Gravity Filter (41,42). The design of a wastewater filtration plant is based principally on the characteristics and flow rate of the wastewater. Not only are the typical characteristics and the average plant flow rates important, but the hourly, daily, and monthly variations in these parameters result in significant variations in operational requirements and plant performance. The essential information required consists of:

1. Projected wastewater flow rate during plant design life:

	Flow rate	
Parameter	MGD	m^3/d
Maximum hourly flow (peak 4-h)	18.1	68,508
Maximum daily flow	11.3	42,770
Mean annual flow	8.82	33,384
Minimum 4-h flow in low-flow month	3.82	14,459

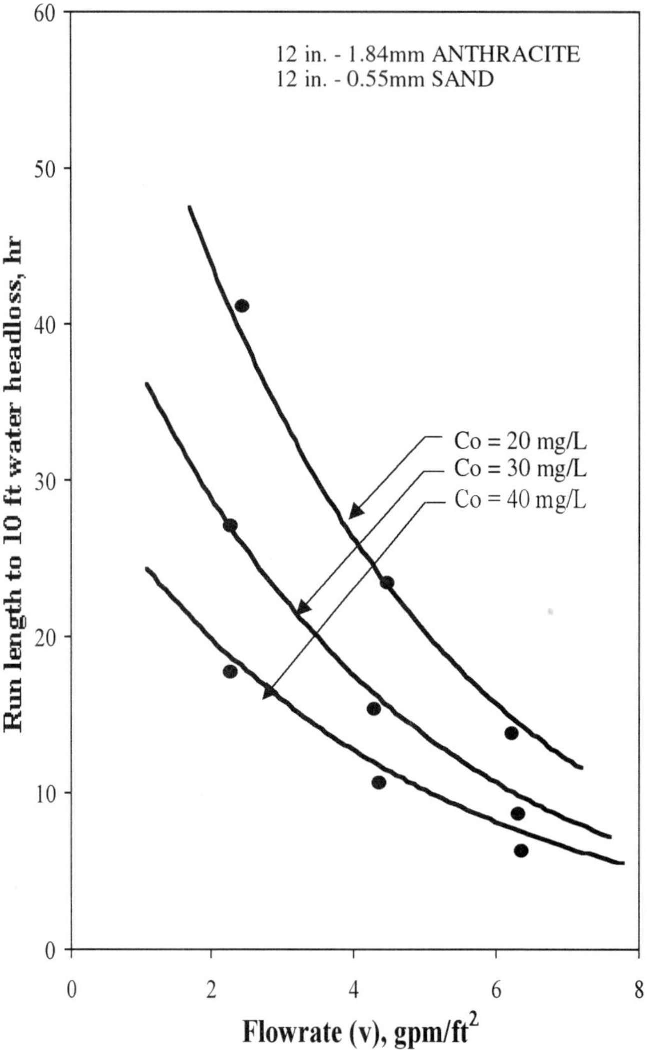

Fig. 10. Run length vs flow rate at various influent SS concentrations.

2. Results from pilot-plant operation (41,42):
   a. Filter media type, size, and depth:

Media	Size (mm)	Depth (in. (cm))
Anthracite (top media)	1.84 (10/12)*	15 (38.10)
Sand (bottom media)	0.55 (30/35)*	15 (38.10)

*Standard sieve number, passed/retained

   b. Backwash requirements: 3 min air scouring at a rate of 0.9–1.5 $m^3/min/m^2$ (3–5 $scfm/ft^2$), followed by 5 min water wash at a rate of 841 $L/min/m^2$ (20 gpm/ft^2).
   c. Curves relating run length to filtration rate at various influent SS (suspended solids) concentrations (Fig. 10).
   d. Curves relating net water production from the filters to filtration rate under various run length conditions (Fig. 11).

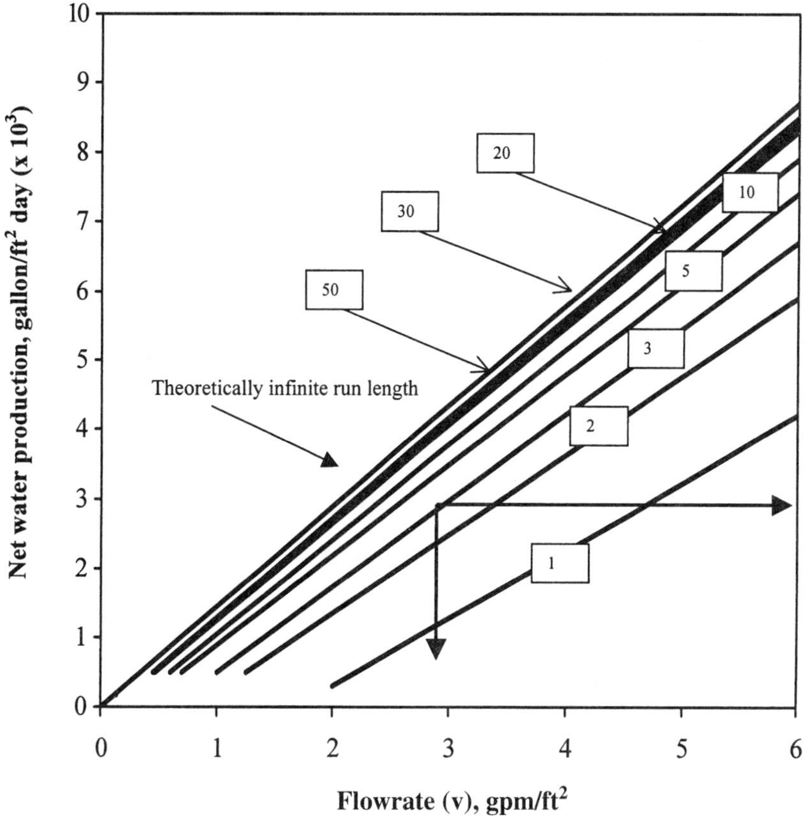

Fig. 11. Net water production vs flow rate at various run lengths.

The first step in the calculation is to determine the nominal design filtration rate based on the maximum 4-h peak flow rate at the worst SS concentration in the wastewater expected to occur during the design life. Table 7 shows such a calculation; it is assumed that the maximum 4-h peak flow is 18.1 MGD (68,508 m^3/day) and the worst SS concentration in the wastewater is 40 mg/L. A part of the filtrate is used for backwashing and will be equalized and recycled to the final clarifier for further settling. Because the water used for backwashing will reach the filter again, the filter capacity should be designed on the basis of the maximum 4-h peak flow plus the estimated amount of backwashing water. Ten percent of the flow is assumed to be required for backwashing in this example. This backwashing water will increase the amount of wastewater that will have to be filtered.

The determination of the nominal filter design rate proceeds by trial and error. As shown in Table 7, the nominal maximum filtration design rate was first assumed to be 7.0 gpm/ft^2 (284.9 L/min · m^2) in the first alternative. Thus, the nominal filter area required was determined as follows:

$$\text{Nominal filter area} = \frac{\text{Filter capacity required}}{\text{Assumed nominal filter rate}}$$

$$= \frac{19.91\ \text{MGD}\left(\dfrac{694\text{gpm}}{\text{MGD}}\right)}{\dfrac{7.0\ \text{gpm}}{\text{ft}^2}}$$

$$= 1980\ \text{ft}^2\left(183.9\ \text{m}^2\right)$$

Table 7
Determination of Nominal Design Filtration Rate Based on max. 4-h Peak Flow at Worst SS Concentration:

Trial	Flow Rate (MGD)	Filter Capacity (MGD)	Nominal filter rate (gpm/ft²)	Nominala run length (hr)	Nominal filter area (ft²)	Actual[b] filter area (ft²)	Actual filter rate (gpm/ft²)	Actual[a] run length (hr)	Net[c] water prod. (gpd/ft²)	Net[c] water prod. (MGD)	Remarks
1	18.1	19.91	7.0	6.5	1980	2640	5.25*	9.00*	6,900*	18.2* > 18.1	Sufficient
							7.0**	6.5**	8,950**	17.7** < 18.1	Not sufficient
2	18.1	19.91	6.5	7.5	2125	2830	4.89*	10*	6,500*	18.4* > 18.1	Sufficient
							6.50*	7.5**	8,500**	18.05** < 18.1	Not sufficient
3	18.1	19.91	6.25	8.0	2210	2940	4.70*	10.5*	6,250*	18.4* > 18.1	Sufficient
							6.25**	8.0**	8,250**	18.2** > 18.1	Sufficient
4	18.1	19.91	6.0	8.5	2310	3070	4.60*	10.6*	6,100*	18.7* >18.1	Sufficient
							6.0**	8.5**	7,900**	18.3** > 18.1	Sufficient
5	18.1	19.91	5.0	9.8	2760	3680	3.76*	12.7*	5,050*	18.6* > 18.1	Sufficient
							5.0**	9.8**	6,590**	18.4** > 18.1	Sufficient

* From four filter cells in operation.
** One of the filter cells down for repairs.
[a] From Fig. 10.
[b] 1.33 times nominal filter area.
[c] From Fig. 11.
[d]Maximum 4-h peak = 18.1 MGD, 10% backwash water = 1.81 MGD, Co = 40 mg/L.

As a safety factor, 1.33 times this area, is provided so that one of the filter cells can be out of service at any time for repairs or backwashing. Thus, under normal conditions, the actual filtration rate would be 5.25 gpm/ft^2 (213.7 L/min · m^2) instead of 7.0 gpm/ft^2 (284.9 L/min·m^2). However, a filtration rate of 7.0 gpm/ft^2 (284.9 L/min · m^2) will be the filtration rate when one of the filter cells is out of service. Actual run lengths can be determined by using the actual filtration rate and SS in the wastewater from Fig. 10. The actual run lengths would be 9.0 and 6.5 h when the actual filtration rates are 5.25 and 7.0 gpm/ft^2 (213.7 and 284.9 L/min·m^2), respectively. After the actual filtration rate and run length are determined, net water production under either condition (four filter cells in operation or one cell down for repairs or backwashing) can be estimated from Fig. 11. The result is included in Table 7. This result indicates that under the assumed nominal filtration rate of 7.0 gpm/ft^2 (284.9 L/min·m^2) the filter design is sufficient to meet the maximum 4-h peak flow operating condition when all four filter cells are in operation. However, it is not sufficient to meet the maximum 4-h peak flow when one of the filter cells is out of service. Thus, the assumed nominal filtration rate of 7 gpm/ft^2 (284.9 L/min · m^2) is too high. Another trial is required.

The third trial, with an assumed nominal filtration rate of 6.25 gpm/ft^2 (254.4 L/min · m^2), shows that the filter design is sufficient to meet the flow as designed for both conditions. So do the fourth and fifth alternatives. Consequently, the highest nominal filtration rate that provides a filter design sufficient to meet both conditions is between 6.25 and 6.5 gpm/ft^2 (254.4 and 264.6 L/min · m^2). For practical purposes, it appears that the nominal filtration rate of 6.25 gpm/ft^2 (254.4 L/min · m^2) is adequate for this design. A filter design based on the maximum 4-h peak flow condition has to be checked to see if it is sufficient to produce filtered water under all other flow conditions, such as maximum daily, mean annual, and minimum 4-h in the low-flow month. In this case, the filtration rate is satisfactory for all levels of flow and SS.

NOMENCLATURE

A_s	parameter in Eq. (6)
C	concentration of suspension vol/vol
d	filtered particle diameter
D	grain diameter
f	friction factor in Eq. (7)
g	gravitational acceleration
H	head loss
H_a	Hamaker constant in Eq. (6)
H_0	initial head loss
k_B	Boltzmann's constant in Eq. (6)
K	head loss constant
l	filter depth
m	number of sizes in the influent particle size distribution
N_G	gravity group in Eq. (6)
N_{Lo}	London group in Eq. (6)
N_{Pe}	Peclet number in Eq. (6)
N_R	relative size group in Eq. (6)
N_{Re}	Reynolds number in Eq. (7)
p	parameter in Eq. (6)

q detachment or scour coefficient
S volume of void space occupied or blocked by the deposits per unit filter volume
S' value of S for which γ is maximum
S'' value of S for which γ is zero
t time
T absolute temperature
v approach velocity (Q/A)
V_S settling velocity in Eq. (6)
w parameter in Eq. (6)
x exponent of velocity term in Eq. (4)
y exponent of spherical specific surface term in Eq. (4)
z exponent of the capillary specific surface term in Eq. (4)
β bulking factor
β_j expansion or bulking factor for the deposits of size j
ε_d porosity of the deposits
ε_0 porosity of the clean filter medium
γ filter coefficient
γ_0 initial filter coefficient
μ dynamic viscosity of the filtering liquid
η_0 initial collection efficiency
ρ density of filtering liquid
ρ_p density of filtered particles
σ specific deposit, vol/unit filter volume
σ_U saturation value of the specific deposit
ϕ particle shape factor

REFERENCES

1. R. D. Letterman, (ed.), *Water Qunality and Treatment, A Handbook of Community Water Supplies*, 5th Ed., McGraw-Hill, New York, 1999.
2. A. Rushton, (ed.), *Solid-liquid filtration and Separation Technology*, 2nd ed., Wiley-VCH, New York, 2000.
3. C. Tien, *Granular Filtration of Aerosols and Hydrosols,* Butterworths Publishers, Boston, 1989.
4. K. J. Ives, *Capture Mechanisms in Filtration. Scientific Basis of Filtration*, NATO Advanced Study Institute, Series E, Vol. 2, Noordhoff International Publishing Co., Netherlands, 1975.
5. T. F. Craft, *Radiotracer Study of Rapid Sand Filters*, Ph.D. Dissertation, Georgia Institute of Technology, Atlanta, Georgia, 1969.
6. J. P. Herzig, D. M. Leclerc, and P. LeGoff, Flow of suspensions through porous media—application to deep bed filtration. *Industrial & Engineering Chem.* **62**, 8–35 (1970).
7. K. M. Yao, M. T. Habiban, and C. R. O'Melia, Water and waste water filtration:concepts and application. *Environmental Sci. &Technol.* **5**, 1105–1112 (1971).
8. M. M. Ghosh, T. A. Jordan, and R. L. Porter, Physicochemical approach to water and wastewater filtration. *Environ. Eng. Div.*, ASCE **101**, 71–86 (1975).
9. G. Kiely, *Environmental Engineering*, McGraw-Hill, UK, 1997
10. B. T. Ray, *Environmental Engineering*, PWS Series in Engineering, Boston, MA, 1995.
11. J. L. Tuepker and C. A. Buescher Jr., Operation and maintenance of rapid sand and mixed media filters in a lime softening plant. *J. Amer. Water Works Assn.*, **60**, 1377–1388 (1968).

12. J. L. Cleasby, M. M. Williamson, and E. R. Baumann, Effect of filtration rate changes on quality. *J. Amer. Water Works Assn.* **55**, 869–878 (1968).
13. D. M. Mintz, Kinetics of filtration. *Dokl. Ak. Nauk*, **78**, 12 (1951).
14. D. M. Mintz, *Modern Theory of Filtration*, Special Subject No. 10, International Water Supply Congress and Exhibition, London, 1966.
15. C. F. Lerk, *Some Aspects of the Deferrisation of Groundwater*, Thesis, Technical University, the Netherlands 1965.
16. V. Mackrle and S. Mackrle, *Adhesion in filters. J. Sanit. Eng. Div., ASCE.* **87**, 17–32 (1961).
17. D. R.Stanley, Sand filtration studied with radiotracers. *Amer. Soc. Civil Eng. Proceedings*, **81**, 1–23 (1955).
18. K. J. Ives and R. M. W. Horner, Radial filtration. *Proceedings of the Institution of Civil Engineers*, London, **55**, 229–245 (1973).
19. T. Iwasaki, Some notes on sand filtration. *J. Amer. Water Works Assn.* **29**, 1591–1597 (1937).
20. P. C. Stein, *A Study of the Theory of Rapid Filtration of Water Through Sand*, Sc. D. Thesis, MIT, Boston, MA, (1940).
21. R. Eliassen, *An Experimental and Theoretical Investigation of the Clogging of a Rapid Sand Filter*, Sc. D. Thesis, MIT, Boston, MA, 1935.
22. N. V. Ornatskii, E. M. Sergeev, and Y. M. Shekhtman, *Investigation of the Process of Clogging of Sands*, University of Moscow, 1955.
23. K. J. Ives, Deep filters. 61st *National Meeting, Amer. Inst. of Chemical Eng.*, Houston, TX, 1967.
24. K. J. Ives, Optimization of deep bed filtration, Paper presented at the 1st Pacific Chemical Engineering Congress, Kyoto, Japan, October, 1972.
25. K. J. Ives and I. Sholji, Research on variables affecting filtration. *J. Sani. Engr.* Div., ASCE **91**, 1–18 (1965).
26. D. M. Fox and J. L. Cleasby, Experimental evaluation of sand filtration theory. *J. Sani. Engr. Div., ASCE* **92**, 61–82 (1966).
27. K. J. Ives, Mathematical models of deep bed filtration, In: *Scientific Basis of Filtration*, Nato Advanced Study Institute, Series E, Volume 2, Noordhoff International Publishing Co., Netherlands, 1975.
28. T. R. Camp, Theory of water filtration. *J. Sani Engr. Div., ASCE* **90**, 1–30 (1964).
29. A. K. Deb, Theory of sand filtration. *J. Sani. Engr. Div., ASCE* **95**, 399–422 (1969).
30. K. J. Ives, Filtration using radioactive algae. *J. Sani. Engr. Div., ASCE* **87**, 23–37 (1961).
31. F. Garcia-Maura, *Filtration of Polydispersed Particles through Multimedia Filters*, M. S. Thesis, University of Wisconsin, Milwaukee, WI, 1984.
32. J. D. Logan, *Transport Modeling in Hydrogeochemical Systems*. Springer, New York, 2001.
33. R. Rajagopalan and C. Tien, Trajectory analysis of deep bed filtration using the sphere-in-cell porous media model. *AIChE.* **22**, 523–533, (1976).
34. C. Tien and A. C. Payatakes, Advances in deep bed filtration. *AIChE.* **25**, 737–759 (1979).
35. J. Kozeny, On capillary conduction of water in the soil, *Sitzungsber Akad. Wiss.*, Vienna, Abt. IIIa, 136, 276 (1927).
36. G. M. Fair and L. P. Hatch, Fundamental factors governing the streamline flow of water through sand. *J. Amer. Water Works Assn.* **25**, 551–1565 (1933).
37. H. E. Rose, On the resistance coefficient -reynolds number relationship for fluid flow through a bed of granular materials, *Inst. Mechanical Engr. Proceedings*, **153**, 141–148 (1945).
38. R. Sakthivadivel, V. Thanikachalam, and S. Seetharaman, Head-loss theories in filtration. *J. Amer. Water Works Assn.* **64**, 233–238 (1972).
39. P. M. Heertjes and C. F. Lerk, The function of deep-bed filters Part I: the filtration of colloidal solutions. *Trans. Instn. Chem. Engrs.* **45**, T129–T145 (1967).
40. A. Maroudas, and P. Eisenklam, Clarification of suspensions: a study of particle deposition in granular media. *Chemical Engineering Science.* **20**, 867–888 (1965).

41. E. R. Baumann and J. Y. C. Huang, Granular filters for tertiary wastewater treatment. *J. Water Pollution Control Fed.* **46**, 1958–1973, 1974.
42. J. Y. C. Huang, *Granular Filters for Tertiary Wastewater Treatment.* PhD dissertation, Iowa State University of Science and Technology, Ames, IA, (1972).
43. H. E. Hudson Jr., A theory of the functioning of filters. *J. Amer. Water Works Assn.* **40**, 868–872 (1948).
44. H. E. Hudson Jr., Factors affecting filtration rates. *J. Amer. Water Works Assn.* **48**, 1138–1154 (1956).
45. M. B. Gamet and J. M. Rademacher, Measuring filter performance. *Water Works Engineering* **112**, 117–118 (1959).
46. K. Y. Hsiung and J. L. Cleasby, Prediction of Filter Performance. *J. Sanit. Eng. Div., ASCE* **94**, 1043–1069 (1968).
47. J. L. Cleasby, Approaches to a filtrability index for granular filters. *J. Amer. Water Works Assn.* **61**, 372–381 (1969).
48. K. Y. Hsiung, Filtrability study on secondary effluent filtration. *J. Sanit. Eng. Div., ASCE* **98**, 505–513 (1972).
49. W. R. Conley and K. Y. Hsiung, Design and application of multimedia Filter. *J. Amer. Water Works Assn.* **61**, 97–101 (1969).
50. K. A. Dostal and G. G. Robeck, Studies of modifications in treatment of Lake Erie water. *J. Amer. Water Works Assn.* **58**, 1489–1504 (1966).
51. S. R. Qasim, E. M. Motley, and G. Zhu, *Water Works Engineering: Planning, Design and Operation*, Prentice-Hall, New Jersey, 2000.
52. J. L. Cleasby and E. R. Baumann, *Wastewater Filtration—Design Considerations.* EPA Technology Transfer Seminar Publication, July, (1974).
53. T. R. Camp, Discussion—Experience with anthracite-sand filters, *J. Amer. Water Works Assn.* **53**, 1478–1483 (1961).
54. J. L. Cleasby, A. M. Malik, and E. W. Stangl, Optimum backwash of granular filters. 46th Annual Conference, Water Pollution Control Federation, Cleveland, OH, 1973.
55. J. L. Cleasby and E. R. Baumann, (1962) Selection of sand filtration rates. *J. Amer. Water Works Assn.* **54**, 579–602 (1962).
56. J. J. Cleasby, Filter rate control without rate controllers, *J. Amer. Water Works Assn.* **61**, 181–185 (1968).
57. A. Amirtharajah, Optimum backwashing of sand filters. *J. Environ. Engr. Div., ASCE,* **104**, 917–932 (1978).
58. S. R. Hewitt and A. Amirtharajah, Air dynamics through filter media during air scour. *J. Environ. Engr. Div., ASCE* **110**, 591–606 (1984).
59. J. L. Cleasby and J. C. Lorence, Effectiveness of backwashing for wastewater filters. *J. Environ. Engr. Div., ASCE* **104**, 749–765 (1978).
60. J. C. Young, Operating problems with wastewater filters. *J. Water Pollution Control Fed.* **57**, 22–29 (1985).
61. G. G. Robeck, K. A. Dostal, and R. L. Woodward, Studies of modifications in water filtration. *J. Amer. Water Works Assn.* **56**, 198–213 (1964).
62. T. V. Garel, Depth filters do double duty. *Wat. Works & Wastes Engr,* **2**, 34–36 (1965).
63. W. R. Conley and R. H. Evers, Coagulation control. *J. Amer. Water Works Assn.* **60**, 165–174 (1968).
64. A. Adin and M. Rebhum, High-rate contact flocculation-filtration with cationic polyelectrolyte, *J. Amer. Water Works Assn.* **66**, 109–117 (1974).
65. Metcalf and Eddy, Inc. (ed.) *Wastewater Engineering: Treatment Disposal and Reuse*, 4th ed., McGraw-Hill, New York, 2002.
66. S. L. Kim, J. P. Chen, and Y. P. Ting, Study on feed pretreatment for membrane filtration of secondary effluent. *Separation & Purification Technol.* **29**, 171–179 (2002).

67. M. J. Hammer and M. J. Hammer, Jr., (eds.), *Water and Wastewater Technology*, 3rd ed., Prentice-Hall, Inc., New Jersey, 1996.
68. J. P. Chen, S. L. Kim, and Y. P. Ting, Optimization of feed pretreatment for membrane filtration of secondary effluent. *J. Membrane Sci.* **219**, 27–45 (2003).
69. Singapore Public Utilities Board *Singapore Water Reclamation Study, Expert Panel Review and Findings*. Singapore, 2002.
70. W. R. Mills, Jr., S. M. Bradford, M. Rigby, and M. P. Wehner, *Wastewater Reclamation and Reuse*. Technomic Publishing, Lancaster, PA, (1998).
71. H. Cikurel, M. Rebhun, A. Amirtharajah and A. Adin, Wastewater effluent reuse by in-line flocculation filtration process. *Wat. Sci. & Technol.* **33**, 203–211 (1996).
72. D. Jolis, R. Campana, R. A. Hirano, P. Pitt, and B. Mariñas, Desalination of municipal wastewater for horticultural reuse: process description and evaluation. *Desalination.* **103**, 1–10 (1995).
73. J. J. Qin, M. H. Oo, H. Lee, and R. Kolkman, Dead-end ultrafiltration for pretreatment of RO in reclamation of municipal wastewater effluent. *J. Membrane Sci.* **243,** 107–113 (2004).

14
Polymeric Adsorption and Regenerant Distillation

Lawrence K. Wang, Chein-Chi Chang, and Nazih K. Shammas

CONTENTS

1. INTRODUCTION

One of the most significant advances in ion exchange resin and adsorbent technology has been the development of the macroreticular pore structure (1–24). Various synthetic routes have been developed for preparing both ion exchange resins and polymeric adsorbents of high surface area and pore volume. Furthermore, the synthesis has been developed to the degree that the surface area and pore parameters can be varied over a wide range. Several of these macroreticular polymers based on the crosslinked styrene and acrylate systems are now available commercially. A polymeric adsorbent is defined as a macroporous or macroreticular polymeric material that has similar properties to ion-exchange resin, but has no functional ionic group (11,21). These polymeric adsorbents are hard, durable, insoluble spheres of high surface area and porosity. They are also available in a variety of polarities.

In general, adsorbents are solids that possess high specific surfaces, usually well above 5 m^2 of exposed surface area per gram of solid. Adsorbents fall into two major physical classes, porous and non-porous. The porous adsorbents consist of particles that are usually large (greater than 50 mesh) and the high surface area is a result of pores of varying diameters that "permeate" the particles. The diameters of these pores are larger than molecular distances. Non-porous adsorbents are usually finely divided solids (less than 10 μm), and the high surface area of such materials is due to the fine state of subdivision that is achieved by various techniques such as grinding, precipitation, etc. The specific surfaces of several commercial porous and non-porous adsorbents are given in Table 1.

From: *Handbook of Environmental Engineering, Volume 3: Physicochemical Treatment Processes*
Edited by: L. K. Wang, Y.-T. Hung, and N. K. Shammas © The Humana Press Inc., Totowa, NJ

Table 1
Specific Surfaces of Typical Porous and Non-porous Adsorbents

Porous		Non-porous	
Adsorbent	Specific surface (m^2/g)	Adsorbent	Specific surface (m^2/g)
Granular carbons	500–2000	Carbon black	100
Silica gel	600	TiO_2 pigment	70–80
Bone char	60–80	ZnO pigment	1–10
Soils	10–100		
Asbestos	17		
Polymeric, macroreticular	100–600		

The selection by the authors of the terms "macroreticular" and "microreticular" is to characterize the physical pore structure of the new ion exchange resins and polymeric adsorption terminology. The terms microporous and macroporous, usually used in adsorption terminology, refer to those pores less than 20 Å and greater than 200 Å, respectively.

Pores of diameters between 20 Å and 200 Å are referred to as transitional pores and polymeric adsorbents. This classification could, of course, also be applied to the macroreticular ion exchange resins; however, the terminology would not distinguish those pores that are part of the organic gel structure of the macroreticular ion exchange resins and polymeric adsorbents.

In essence, the new macroreticular ion exchange resins have both a microreticular as well as a macroreticular pore structure. The former refers to the distances between the chains and crosslinks of the swollen gel structure and the latter to the pores that are not part of the actual chemical structure. The macroreticular portion of structure may actually consist of micro-pores, macro-pores, and transitional pores depending on the pore size distribution. Confusing as this terminology may appear, the terms defined above are necessary to distinguish the various structures from one another. The use of the terms "microporous" and "macroporous" for distinguishing between the gel-type standard ion exchange resins and the "macroreticular" ion exchange resins is considerably more confusing and, unfortunately, has no direct relationship to the terms "micropores" and "macropores" as normally defined for absorbents. All too often the term "macroporous" has been used for materials that cannot be distinguished from ordinary gel-type materials by any of the available physical methods.

The macroreticular, polymeric absorbents constitute a new and unique class of absorbents because of the wide range of pore structures that one can develop within the framework of a particular framework system. For example, for the styrene-divinylbenzene class of polymeric absorbents having surface areas ranging from 7 to 600 m^2/g and average pore diameters ranging from 60 to 1,000,000 Å, pore volumes vary from 10 to 90%. Furthermore the surface characteristics are also quite well defined. This flexibility is also possible with other polymer systems such as those based on the acrylates, the vinylpyridines, and the phenol formaldehyde condensate polymers. The classical absorbents such as the silicas, aluminas, and carbons do not offer such flexibility.

The polymeric adsorption process is very similar to granular activated carbon process in theory and principles (11). The former uses granular, porous resins to be the adsorbents, while the later uses granular activated carbon to be adsorbents. Detailed adsorption theory and principles can be found in Chapter 15 and from the literature (1,2).

2. POLYMERIC ADSORPTION PROCESS DESCRIPTION

2.1. *Process System*

Polymeric adsorption, also referred to as resin adsorption, is an adsorption process involving the use of granular polymeric adsorbents (GPA) to extract and, in some cases, recover dissolved organic solutes from aqueous water and wastewater (3–12). A polymeric adsorption system is very similar to a granular activated carbon (GAC) adsorption system in terms of adsorption theory, principles, and process equipment. The former uses GPA as the granular adsorbents, while the latter uses GAC (11).

A polymeric adsorption system is also very similar to an ion exchange process system in terms of process equipment and polymeric resins used in the process. However, the former (polymeric adsorption) is an adsorption process using nonionic polymeric resins to be the adsorbents for removing pollutants, while the latter (ion exchange) is an ion exchange process using ionic polymeric resins (cationic ion exchange resins or anionic exchange resins) to be the ion exchange agents for removing pollutants (11,12). The two processes, GAC adsorption process (i.e., another granular adsorption process) and ion exchange process (i.e., another resin process), are fully covered in other chapters. A schematic of a polymeric adsorption system used for the removal and recovery of phenol from water is shown in Fig. 1.

2.2. *Process Steps*

Waste treatment by resin adsorption involves two basic steps:

1. Contacting the liquid waste stream with the resins and allowing the resins to adsorb the solutes from solution.
2. Subsequently regenerating the resins by removing the adsorbed chemicals, often effected by simply washing with the proper solvent.

Commonly, a typical system for treating low-volume waste streams will consist of two fixed beds of resin. One bed will be on stream for adsorption, while the second is being regenerated. In cases where the adsorption time is very much longer than the regeneration time (as might be when solute concentrations are very low), one resin bed plus a hold-up storage tank could suffice.

2.3. *Regeneration Issues*

Solvent regeneration will be required unless the solute-laden solvent can be used as a feed stream in some industrial process at the plant, or the cost of the solvent is low enough so that it may be disposed of after one use. Solvent recovery, usually by distillation (Section 6), is most common when organic solvents are used. Distillation will allow solute recovery for reuse if such is desired.

Resin lifetimes may vary considerably depending on the nature of the feed and regenerant streams. Regeneration with caustic is estimated to cause a loss of 0.1–1% of

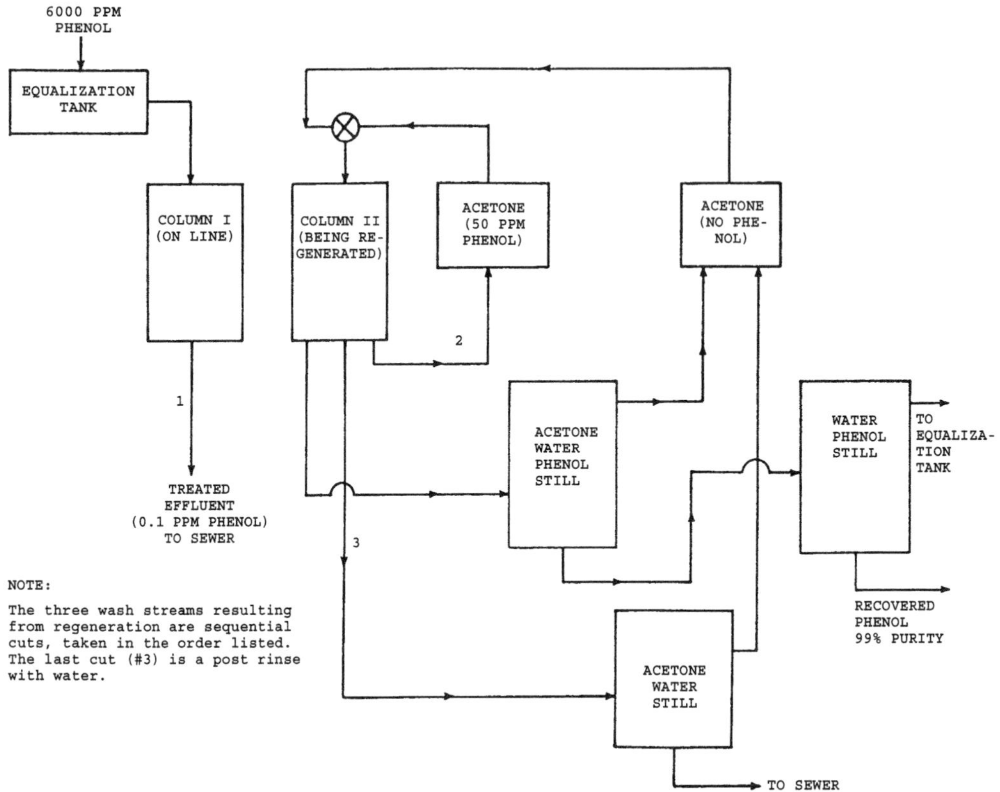

Fig. 1. Schematic of a resin adsorption system for the removal of phenol from water.

the resin per cycle; replacement of resins at such installations may be necessary every 2–5 yr. Regeneration is accomplished with hot water or steam.

3. POLYMERIC ADSORPTION APPLICATIONS AND EVALUATION

3.1. Applications

3.1.1. Case Studies

Polymeric adsorption process is adopted for water or wastewater treatment when

(a) Selective adsorption is desired.
(b) Low leakage rates are essential.
(c) Carbon regeneration is not practical.

Several current applications of resin adsorption for which some information is available are discussed below. A dual resin adsorption system is being used to remove color associated with metal complexes and other organics from 10^6 L/d (300,000 gpd) waste stream from a dyestuff production plant. The system also removes copper and chromium present in the influent waste stream, both as salts and as organic chelates.

Two large systems currently operating in Sweden and Japan remove colored pollutants (derived from lignin) from paper mill bleach plant effluents. The Swedish plant, which produces 70 Mg (300 tons) of pulp/d, uses the resin adsorption system and is reported to remove 92–96% of the color, 80–90% of the chemical oxygen demand (COD), and 40–60% of the 5-d biochemical oxygen demand (BOD_5) from the effluent of the caustic extraction stage in the bleach plant. The system consists of three resin columns, each containing about 20 m^3 (700 ft^3) of resin. The system in Japan is for a 420 Mg/d (760 ton/d) pulp plant and consists of four resin columns, each with about 30 m^3 (1060 ft^3) of resin. In both cases, the resins are regenerated with a caustic wash followed by reactivation with an acid stream (e.g., sulfuric acid).

Some resin adsorption units in operation are used to remove color from water supply systems; others are used to decolorize sugar, glycerol, wines, milk whey, pharmaceuticals, and similar products. One plant in Louisiana, which removes color from an organic product stream, is said to be in operation for 8 yr now without replacement of the initial resin.

Another plant in Indiana has used a resin system to recover phenol from a waste stream. A dual resin system is currently being installed at a coal liquefaction plant in West Virginia to remove phenol and high-molecular-weight polycyclic hydrocarbons from a 38 L/min (10 gpm) waste stream. Methanol will be used as the regenerant for the primary resin adsorbent. One resin adsorption system, in operation for 5 yr, is removing fat from the wastewaters of a meat production plant.

Other applications include the recovery of antibiotics from a fermentation broth, the removal of organics from brine, and the removal of drugs from urine for subsequent analysis. Adsorbent resins are also currently being used on a commercial scale for screening out organic foulants prior to deionization in the production of extremely high purity water.

3.1.2. Process Adoption

Advantages cited for the use of polymeric adsorbents include efficient removal of both polar and non-polar molecules from wastewater, ability to tailor-make an adsorbent for a particular contaminant, and small energy inputs for regeneration when compared to carbon. The systems are relatively compact and thus require little space. High levels of total dissolved solids (particularly inorganic salts) do not interfere with the action of resin adsorbents on organic solutes. There are clear indications that some organic chemicals are more easily removed from solutions of high concentrations of dissolved salts than from salt-free solutions (in some cases of high salt content, the adsorbent may have to be prerinsed before regeneration).

Polymeric adsorption is similar in nature to activated carbon adsorption, making the two processes competitive in many applications. The most significant difference between carbon and resin adsorption is that resins are always chemically regenerated (through the use of caustic or organic solvents), while carbons, because the adsorption forces are stronger, must usually be thermally regenerated, eliminating the possibility of material recovery. On the other hand, resins generally have a lower adsorption capacity than carbons. Polymeric adsorption is not likely to be competitive with carbon for the treatment of high-volume waste streams containing moderate to high concentrations of mixed wastes with no recovery value. However, a combination of the two processes may be attractive.

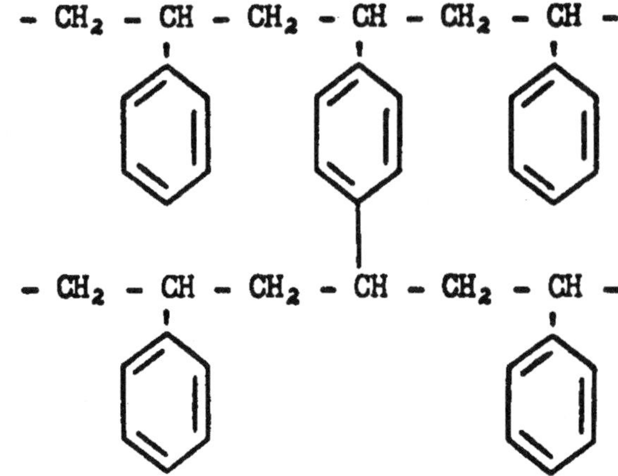

Fig. 2. Structure of amberlite XAD-2 and amberlite XAD-4 (23).

3.2. *Process Evaluation*

3.2.1. *Limitations*

Among its limitations, resin adsorption generally has lower adsorption capacity than activated carbon, and it also has a relatively high cost when the two are compared. It is necessary to keep the amount of suspended solids in the influent low enough to prevent clogging of the bed (no higher than 50 ppm and in some cases below 10 ppm). Another disadvantage is the susceptibility of the process to certain poisons such as oxidants or organic foulants that are not efficiently removed by the regenerant. Resin adsorption may be used over a wide pH range; some resins have been able to operate at as low as pH 1–2 and at as high as pH 11–12. However, in many cases, adsorption will be pH dependent, and will thus require pH control. Temperature may also vary significantly. Resins have been used in applications where the influent temperature was as high as 80°C (176°F). Adsorption will, however, be more efficient at lower temperatures. Conversely, higher temperatures will aid regeneration.

3.2.2. *Reliability*

Reliability is still under evaluation for this technology.

3.2.3. *Chemicals Required*

Adsorbents are commercially available (13–15). Regenerants that are in use include basic, acidic, and salt solutions or regenerable non-aqueous solvents.

4. POLYMERIC ADSORBENTS

4.1. *Chemical Structure*

In contrast to the macroreticular ion exchange resins, the polymeric adsorbents are truly non-ionic and their properties are totally dependent on their surface characteristics. Figures 2–4 describe the chemical structures of two of the more interesting polymeric

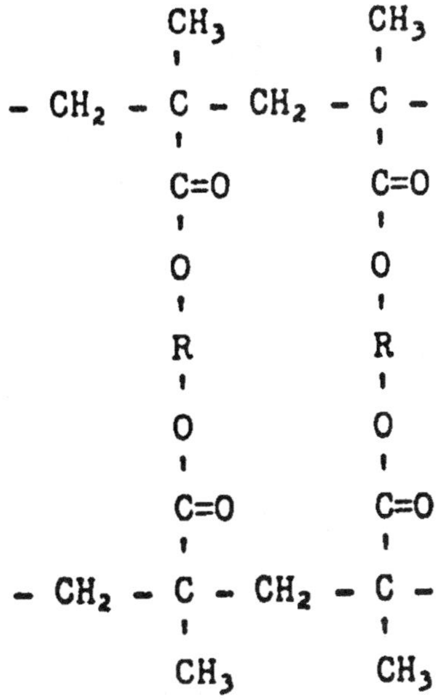

```
          CH3          CH3
           |            |
 - CH2 -  C  - CH2 -  C  -
           |            |
          C=O          C=O
           |            |
           O            O
           |            |
           R            R
           |            |
           O            O
           |            |
          C=O          C=O
           |            |
 - CH2 -  C  - CH2 -  C  -
           |            |
          CH3          CH3
```

Fig. 3. Structure of amberlite XAD-7 (23).

adsorbent classes. One (Fig. 2) is based on a crosslinked styrene–divinlybenzene polymer that is hydrophobic. The other (Figs. 3 and 4) is a crosslinked polymethacrylate structure, which is a considerably more hydrophilic structure. These macroreticular polymeric adsorbents are dimensionally and chemically quite stable. They are also quite insoluble.

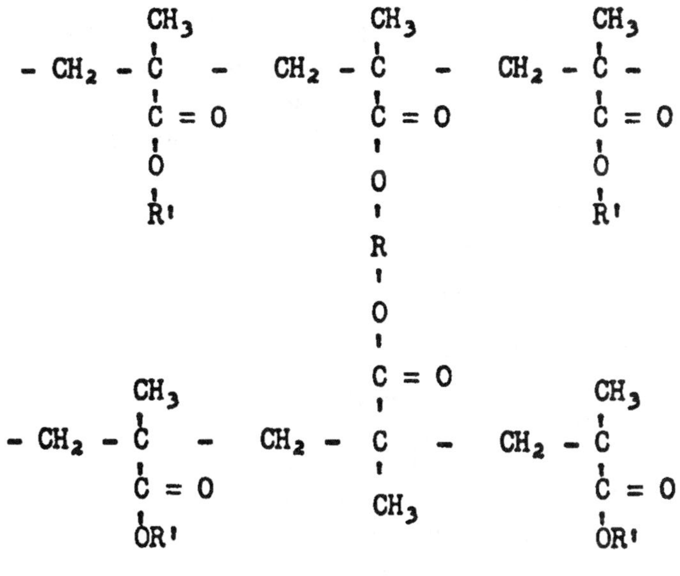

```
          CH3               CH3               CH3
           |                 |                 |
 - CH2 -  C    -   CH2 -   C    -   CH2 -   C  -
           |                 |                 |
           C = O             C = O             C = O
           |                 |                 |
           O                 O                 O
           |                 |                 |
           R'                R                 R'
                             |
                             O
                             |
                             C = O
          CH3                |                CH3
           |                 |                 |
 - CH2 -  C    -   CH2 -   C    -   CH2 -   C
           |                 |                 |
           C = O            CH3                C = O
           |                                   |
           OR'                                 OR'
```

Fig. 4. Structure of amberlite XAD-8 (23).

Table 2
Typical Properties of Amberlite Polymeric Adsorbents

Polymeric adsorbents	Chemical nature	Helium porosity volume (%)	Helium porosity (mL/g)	Surface area (m^2/g)	Average pore diameter (Å)	Skeletal density (g/mL)	Nominal mesh size
			Nonpolar				
XAD-1	Polystyrene	37	0.69	100	200	1.06	20–50
XAD-2	Polystyrene	42	0.69	330	90	1.06	20–50
XAD-4	Polystyrene	51	0.99	750	50	1.09	20–50
			Intermediate polarity				
XAD-7	Acryllc ester	55	1.06	450	80	1.25	20–50
XAD-8	Acryllc ester	52	0.82	140	250	1.26	25–50

4.2. *Physical properties*

The physical properties of the above-described macroreticular polymeric adsorbents are described in Table 2 and Fig. 5. In appearance, the particles are white, hard, and spherically shaped (4). Figure 6 describes the pore structures of polymeric adsorbents as measured by mercury penetration.

4.3. *Adsorption Properties*

Many of the adsorptive properties of these polymeric adsorbents have been described previously. In essence, their adsorptive properties may be predicted from their theoretical solubility parameters and the solubility characteristics of the adsorbate. For example, the less soluble the solute, the more readily it is absorbed. Furthermore, the aromatic-based Amberlite XAD-2 and Amberlite XAD-4 resins are more selective for aromatic solutes. Figure 7 and Table 3 describe the effect of the solubility of the solute. The curves illustrate the adsorption capacity of a series of chlorinated phenols. It is clear that as the degree of chlorine substitution of the phenol increases, the adsorptive capacity of the aromatic polymeric adsorbent increases. Elution of the adsorbed solutes is normally achieved with solvents. Elution or desorption of the adsorbed solutes may also be predicted by the solubility parameter of the solvent. As the solubility parameter increases, the elution efficiency decreases.

5. DESIGN CONSIDERATIONS

5.1. *Adsorption Bed, Adsorbents, and Regenerants*

The equipment for resin adsorption systems consists of two or more steel tanks (stainless or rubber-lined) with associated piping and pumps, and solvent (and perhaps solute) recovery equipment (e.g., a still). Up to three stills may be required in some systems.

Materials needed include a regenerant solution (e.g., aqueous caustic solution or organic solvent) and resin. In one full-scale installation for the removal of organic dye wastes from water, two different resins are employed. The waste stream is first contracted with a normal polymeric adsorbent and then with an anion exchange resin.

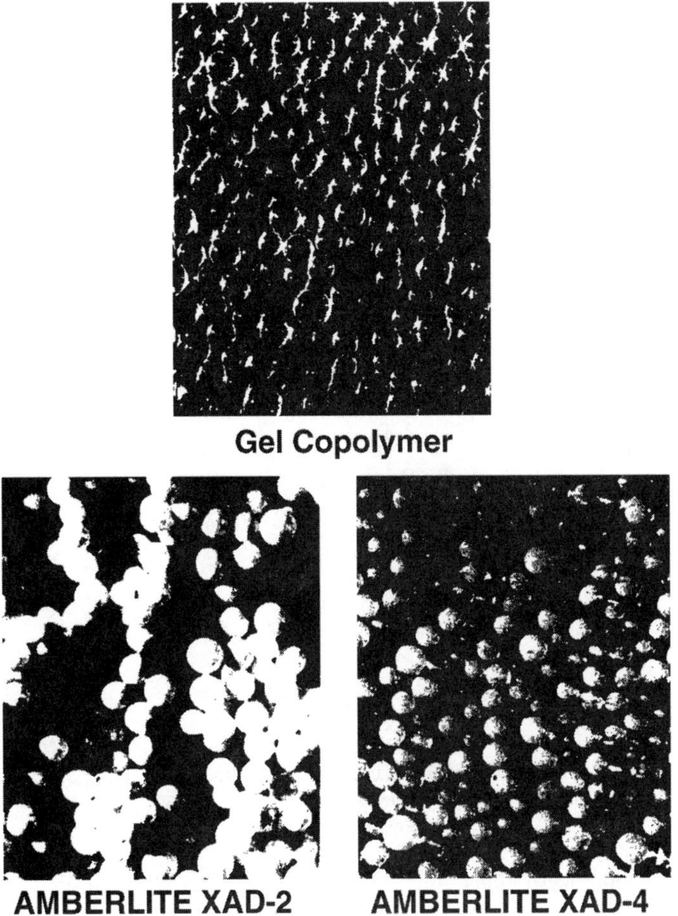

Fig. 5. Photomicrographs of macroreticular polymeric adsorbents.

The adsorption bed is usually fed downflow at flow rates in the range of 0.6–5 L/s/m^3 of resin (0.25 to 2 gpm/ft^3 of resin). This is equivalent to 2–16 bed volumes (BV) per hour, and thus contact times are in the range of 3–30 min. Surface hydraulic loading rates range from 2 to 22 L/s/m^2 (1 to 10 gpm/ft^2). Adsorption is stopped when the bed is fully exhusted and/or the concentration in the effluent rises above a certain level.

Properties of a few currently available resin adsorbents are shown in Table 4. Surface areas per unit weight of resin adsorbents are generally in the range of 100–700 m^2/g (490,000–3,400,000 ft^2/lb). These figures are below the typical range for activated carbons, 800–1200 m^2/g (3,900,000–5,900,000 ft^2/lb) and, in general, indicate lower absorptive capacities, although the chemical nature and pore structure of the resin may be more important factors. This has been demonstrated in one application relating to color removal.

Tests should be run on several resins when evaluating a new application. Important properties are the degree of hydrophilicity and polarity, particle shape (granular versus spherical), size, porosity, and surface area.

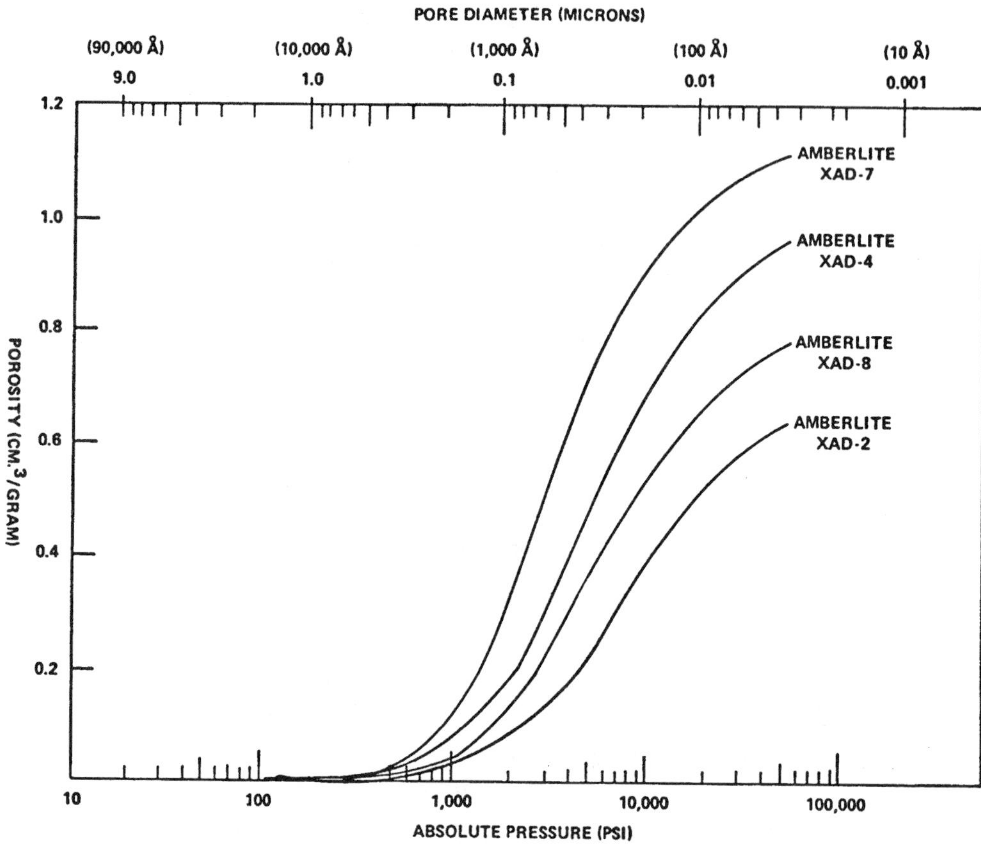

Fig. 6. Pore distributions of some amberlite adsorbents as determined by mercury intrusion.

It is frequently possible to "tailor" a resin for specific applications because much greater control over the chemical and surface nature can be achieved in resin production than in activated carbon manufacture. The cost of developing a totally new resin could be prohibitive for most applications; however, minor modifications of currently available resins are feasible and would be cost effective.

Almost always, menthol is a good regenerant. It washes off the adsorbents easily, and is about the most available and least expensive solvent. Solubility of the sorbate in the regenerant is also important. Not only must van der Waal's attractive forces binding the sorbate to the adsorbent be overcome, but the solubility must be high enough to permit rapid dissolution after the solvent diffuses to the adsorption site.

Careful selection of the solvent often allows the recycling of the regenerant stream laden with the adsorbed organic compound. Thus, what would have been a non-productive pollution control step is transformed into a closed-loop materials recovery process. Even when recycling the regenerant stream is not feasible, solvent and sorbate can be separated by common liquid–liquid separation techniques. These adsorptive methods make good sense today when shortages and high prices of raw materials

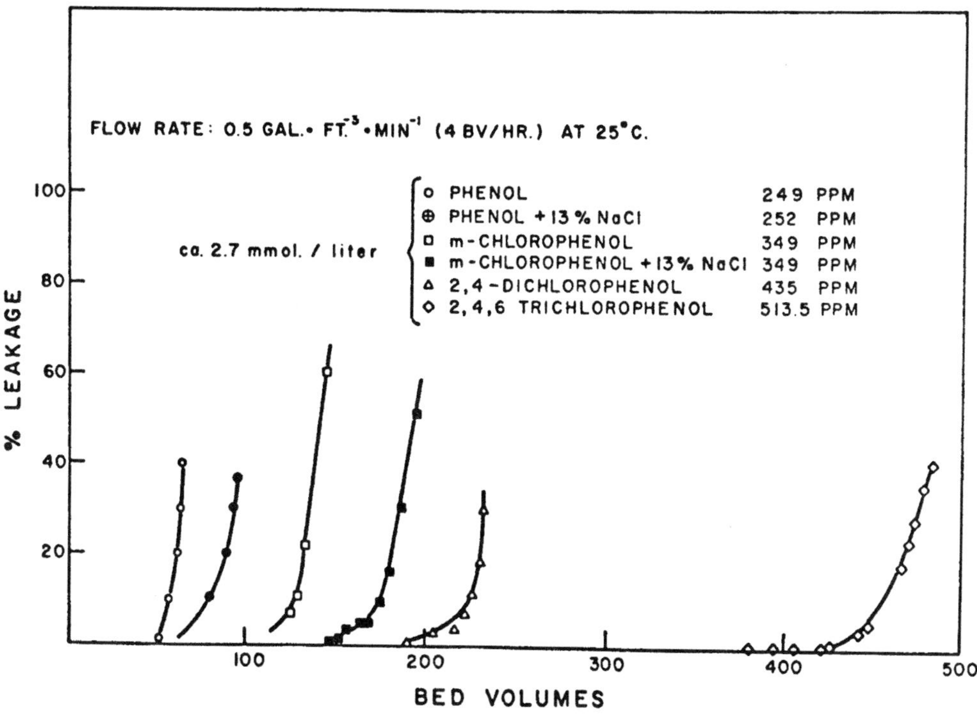

Fig. 7. Column adsorption of amberlite XAD-4 from aqueous solutions.

plague the chemical industry. A distillator system required for recovery of regenerant (solvent) is available commercially (16,17).

5.2. Generated Residuals

The used regenerant solution and/or extracted solutes would require disposal if they are not recycled. For example, when highly colored wastewaters are treated, the used regenerant solution (containing 2–4% caustic plus the eluted wastes) if not recycled, it must be disposed off either by evaporation or incineration. A second example is the removal of pesticides from water, with regeneration being effected by an organic solvent.

Table 3
Adsorption of Phenol and Substituted Chlorophenols on Aberlite XAD-4, 25°C

Solute	Solubility in water (ppm)	Solute in influent (ppm)	Solute in influent mol/L	Solute adsorbed (lb/ft^3). Zero leakage	Solute adsorbed (lb/ft^3). 10 ppm leakage
Phenol	82,000	250	2.7	0.78	0.83
m-Chlorophenol	26,000	350	2.7	2.40	2.53
2,4-Dichlorophenol	4,500	430	2.7	5.09	5.49
2,4,6-Trichlorophenol	900	510	2.6	12.02	13.81

Note: Flow rate = 0.5 gal/min/ft^3.

Table 4
Properties of Currently Available Resin Adsorbents

Name[a]	Base	Specific gravity (wet)	Void volume (%)	Particle size mesh	Bulk density (kg/m^3)	Surface area (m^2/g)	Average pore size (Å)
XAD-1		1.02	37	20–50	—	100	200
XAD-2	Styrene–divinylbenzene	1.02	42	20–50	640–740	300	90
XAD-4		1.02	51	20–50	625	780	50
XAD-7		1.05	55	20–50	655	450	90
XAD-8	Acrylic ester	1.09	52	20–50	690	140	235
Dow MFS 4256[b]	Styrene–divinylbenzene	—	40	+10	430	400	110
Dow MFS 4022		—	35	20–50	—	100	200
Dow MFS 4257		—	40	20–50	—	400	110
Duolite S-30		1.11	35	16–50	460	128	—
Duolite S-37		1.12	35–40	16–50	640	—	—
Duolite ES-561	Phenol–formaldehyde[c]	1.12	35–40	18–50	640–720	—	—
Duolite A-7D		—	—	—	—	24	—
Duolite A-7		1.12	35–40	16–50	640	—	—

[a]XAD resins manufactured by Rohm and Haas Company; Dow MFs resins manufactured by Dow Chemical USA; Duolite resins manufactured by Diamond Shamrock Chemical Company. [b]Resins designed for use in vapor phase adsorption applications. [c]Functional groups such as phenolic hydroxyl groups, secondary and tertiary amines are present on the basic phenol – formaldehyde structure; physical form of these resins is granular as opposed to a bead form for the other brands.

Source:US EPA.

6. DISTILLATION

6.1. Distillation Process Description

Distillation (Fig. 8) is a unit process usually employed to separate volatile components of a waste stream or to purify liquid organic product streams. Figure 9 shows how distillation is used for the recovery of solvent (regenerant) in the polymeric adsorption process. The process involves boiling a liquid solution and collecting and condensing the vapor, thus separating the components of the solution. The process relies on the differences in vapor pressure exhibited by materials at various temperatures. If one component of a mixture has a higher vapor pressure than the others at a certain temperature, then boiling the mixture at this temperature will concentrate the more volatile components in the vapor phase. The vapor is collected in a vessel (accumulator) where it is condensed, resulting in a separation of materials in the feed stream into two streams of different composition. If there are only two components in the liquid, one concentrates in the condensed vapor (condensate) and the other in the residue liquid (bottoms). If there are more than two components, the less volatile components concentrate in the residual liquid and the more volatile in the vapor or vapor condensate. If the vapor is condensed and then reboiled, a vapor stream with a different concentration in its composition may be obtained, allowing further separation of the material. This is the basis for multistage distillation operations (e.g., packed columns or tray distillation).

The ease with which a component is vaporized is called its volatility, and the relative volatilities of the components determine their vapor–liquid equilibrium relationships. If one of the two components in a mixture is more volatile than the other, it will be more concentrated in the vapor phase and leaner in the liquid phase. The degree to which the separation will take place, under a given set of equilibrium conditions, depends on the extent of variance in the volatilities of the components. If the volatilities of two components are the same, the mixture is azeotropic (i.e., there will be no difference in the composition between the liquid and vapor at equilibrium) and it cannot be separated by ordinary distillation methods.

6.2. Distillation Types and Modifications

There are five general types of distillation described below (9,10,16,17):

6.2.1. Batch Distillation

The simplest form of distillation is a single equilibrium (vapor/liquid contact) stage operation carried out in a "still." The liquid is heated to a sufficient temperature to volatilize the lower boiling material with the vapor condensed and collected in an accumulator. If the residual liquid is the product, then the operation continues until the desired purity of the liquid phase has been obtained. Batch distillation is the most common process used for industrial waste and particularly for solvent recovery. A typical solvent recovery distillation unit consists of a boiling chamber into which contaminated solvent is pumped, steam jacket and boiler to supply the heat, vapor collector and condensing unit, and instrumentation.

6.2.2. Continuous Fractional Distillation

The fractional distillation process is used when the liquid feed is to be separated into more than one product or when a nearly pure product is required. The process

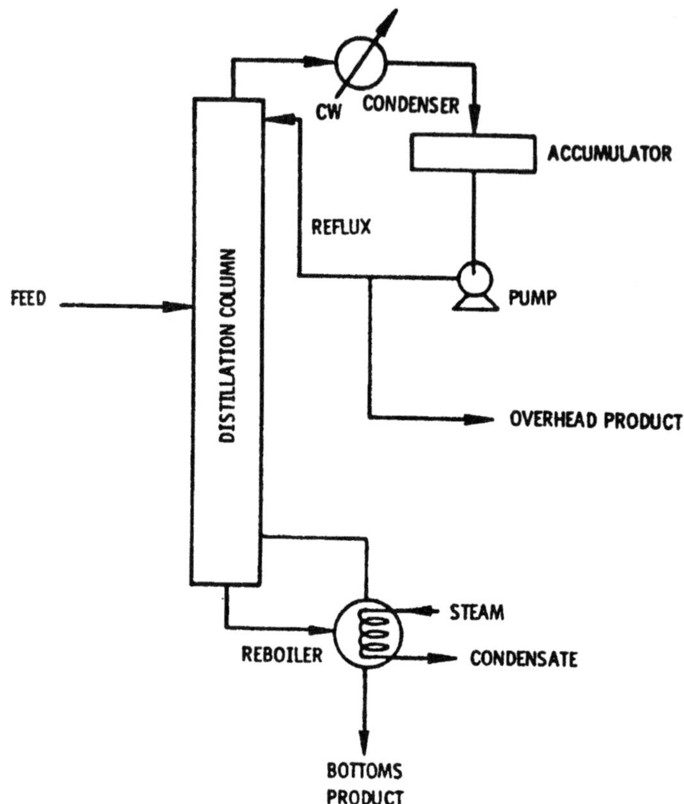

Fig. 8. Schematic of a continuous fractional distillation column.

consists of returning part of the condensate or overhead product back to the distillation process. The returned condensate liquid or reflux is introduced at the top of the column and as it flows down it is brought into intimate contact with the rising vapor stream and results in nearly pure overhead product. The intimate contact between vapor and reflux is achieved by having a number of perforated plates or trays, or packing material in the column. The bottom product can also be purified by introducing the feed in a central position of the column rather than to the still. The feed material flows down through the column and some of the volatile components are stripped before it reaches the still. The stripped feed is further boiled in the still, also referred to as reboilers, and is continuously withdrawn as a liquid bottom product. Figure 8 illustrates a schematic diagram of a continuous fractional distillation column.

6.2.3. Azeotropic Distillation

An azeotrope is a liquid mixture whose components have the same volatility and thus produces a vapor phase of the same composition as that of the liquid. Separation of an azeotrope is often achieved by adding an additive to the mixture to form a new boiling-point azeotrope with one of the original constituents. The volatility of the new azeotrope

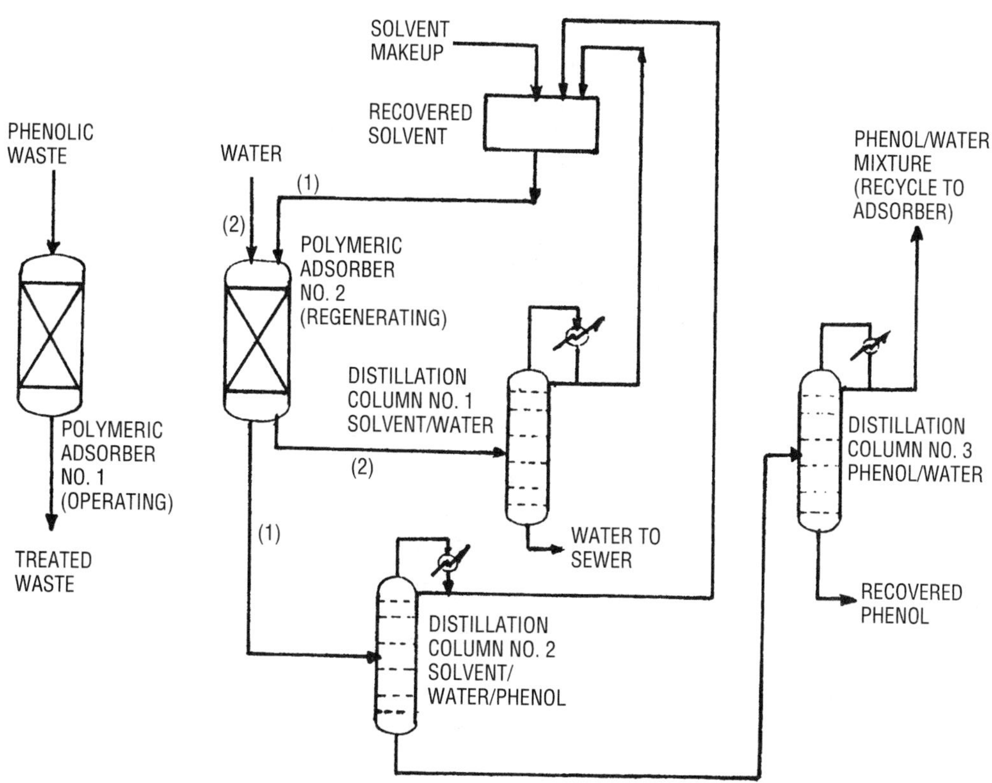

Fig. 9. Phenol removal and recovery system with solvent regeneration of amberlite adsorbent.

is such that it may be easily separated from the other original constituents. Azeotrope distillation does not find extensive use because of the difficulty of finding a solvent that forms a new azeotrope with the required properties (e.g., to form a compound with the right volatility, to be easily recovered from the new azeotrope, relatively inexpensive, non-toxic, non-corrosive, and non-reactive).

6.2.4. Extractive Distillation

This is a distillation process where a non-volatile separating agent is added to a mixture that is difficult or impossible to separate by ordinary means due to the relative volatility of the components of the mixture. The solvent alters the relative volatility of the original constituents, thus permitting separation. The added solvent is of low volatility and is not appreciably vaporized. The solvent and the component with the reduced volatility are removed in the liquid stream. Further treatment of the liquid stream is required to separate the agent from the liquid for reuse.

6.2.5. Molecular Distillation

Molecular distillation is a form of a very low-pressure distillation conducted at absolute pressures in the order of 0.003 mm of mercury. The process is useful when a heat-sensitive material is involved or when the volatility of the material is very low.

6.3. *Distillation Process Evaluation*

6.3.1. *Technology Status and Reliability*

Distillation is well developed for processing applications. This process is highly reliable for proven applications when properly operated and maintained.

6.3.2. *Advantages and Limitations*

Distillation can recover materials that otherwise would be destroyed by waste treatment. It can separate, segregate, or purity to high-quality standards. This could mean that the recovered solvent may be directly recyclable or salable.

Distillation does have several limitations:

(a) The equipment is expensive, and is often complex, requiring operation by highly skilled personnel.
(b) Recovery is energy-intensive.
(c) Its application to feed is limited in that it will handle only liquid solutions that are relatively "clean" and of a consistent composition.
(d) Materials being distilled should not contain appreciable quantities of solids or non-volatile materials.
(e) Feeds that tend to polymerize should be avoided.
(f) Still bottoms sometimes contain tars and sludge, which must be disposed by landfilling or incineration.

6.3.3. *Distillation Design Considerations*

The design criteria for a specific application will be dependent on the physical properties of the waste stream and the required effectiveness of the separation. The key properties will relate to the relative volatilities of the pollutants and the stream matrix (e.g., water or a recoverable solvent). Solvents may be required in some distillation processes. The condensate stream or the liquid bottoms stream will contain the concentrated pollutants, which require subsequent handling.

7. DESIGN AND APPLICATION EXAMPLES

Three polymeric adsorbents are currently being used for a host of applications in the pharmaceutical industry (recovery and purification of antibiotics and vitamins), as an analytical tool (analysis of drugs and other natural products), and for the treatment of industrial wastes. The applications and case studies discussed in the following examples will be mainly devoted to the treatment of industrial wastes.

7.1. *Example 1*

This example introduces a case study of phenolic waste treatment using polymer adsorption. The treatment of wastes containing phenol is a universal problem because of the widespread usage of this chemical. It poses various problems during waste treatment and, if not removed, presents serious problems to municipal water-treatment plants. Many industrial wastes contain thousands of parts per million of phenol, which not only present pollution problems but also constitute economic losses of a valuable raw material. The ideal waste-treatment process would be one that would remove the phenol from the waste and would recover the phenol in a usable form. Discuss how this has been accomplished and demonstrate using the polymeric adsorbent, Amberlite XAD-4.

Table 5
Removal of Phenol from Waste Effluent Using Amberlite XAD-4[a]

Bed volumes (BV) throughput	Phenol leakage (ppm)
Test A: Initial phenol concentration = 6700 ppm	
0	0
2	0
4	0
8	0
12	0
13	0
13.3	10
13.6	25
13.8	40
Test B: Initial phenol concentration = 3000 ppm	
0	0
8	0
16	0
20	0
24	0
25	23
26	47

[a]Phenol/XAD-4; pH = 6.4–7.0; flow rate = 2–4 BV/h.

Solution

Phenol can be readily adsorbed by Amberlite XAD-4 from solutions containing several thousands of parts per million (ppm) of this pollutant and yield effluents containing less than 1 ppm of phenol. Tables 3 and 5 illustrate this capability of Amberlite XAD-4. The unusual property of Amberlite XAD-4 is the fact that various options are available for eluting and recovering the phenol in a usable form. This property distinguishes it from the classical carbonaceous adsorbents that have been used for phenol removal but which had to be regenerated thermally, destroying the valuable phenol. The various options open for eluting and recovering phenol are summarized as follows.

(a) Regeneration with Dilute Caustic: This system will find the greatest application in those situations where direct recycle of a dilute sodium phenolate stream is desirable, where waste caustic is available, and where raw waste phenol concentrations are below 5000 ppm. This limitation on phenol concentration in waste streams is a result of less efficient stoichiometric regeneration resulting in an increasingly more dilute sodium phenolate stream, until no concentration effect is observed.

In order to recover the phenol from the sodium salt, a sulfuric acid treatment is recommended. If on-site recovery is not attractive, the sodium phenolate can be sold to firms specializing in recovery of this material.

(b) Regeneration with Solvent and Recycle of the Phenol–Solvent Mixture: When recycle of the phenol in a water-wet solvent is acceptable or where waste solvents are available, this system will find the greatest application. The solvents found to be most effective in phenol regeneration from Amberlite XAD-4 are methanol and acetone. Two bed volumes

Table 6
Acetone Regeneration of Amberlite XAD4 Exhausted with Phenol[a]

Bed volumes (BV) regenerant	Phenol concentration (%)
0	0
0.15	0.2
0.38	0.4
0.60	0.5
0.75	8.0
0.90	15.0
1.10	13.0
1.30	8.6
1.40	5.4
1.70	0.1
1.90	0
2.00	0
3.00	0

[a]Phenol/XAD-4; acetone; flow rate = 2 BV/h.

of either solvent are usually effective in removing virtually all of the adsorbed phenol. Phenol loadings vary with the concentration from 1–10% in the solvent. Note that water entrained in the bed is unavoidably present in the phenol solvent mixture but generally can be held to less than 25%.

(c) Solvent Regeneration, Recovery of the Solvent and Recycle of an Aqueous Phenol Stream: Again two bed volumes of solvent are used to regenerate the resin. The spent methanol or acetone regenerant is fed to a distillation unit where the solvent is recovered for subsequent regenerations. A phenol-rich water stream is taken off as a bottom product. If desired, a second distillation step to recover solvent washed from the resin bed after the regeneration will increase solvent recovery.

(d) Solvent Regeneration, Recovery of the Solvent and Recycle of a Concentrated Phenol Stream: System (c) is further modified by adding a distillation column to dehydrate the phenol rich water stream to yield 99% phenol. A small 10% phenol and 90% water azeotrope stream is recycled to the influent of the adsorption train. The continuous distillation operation could also be conducted batchwise.

(e) If the phenol is to be recovered from the waste effluent of a plant producing phenol formaldehyde resins (Bakelite), it may be eluted with 37% formaldehyde and the elute recycled. Some elution data are illustrated in Table 6 for a system in which acetone is used as the solvent from which pure phenol is recovered. The entire process is outlined in Fig. 9. The value of phenol recovered in 2 yr is approximately equal to the installed cost of the phenol-recovery system.

7.2. *Example 2*

This example introduces a case study of treating trinitrotoluene (TNT) waste using polymeric adsorption. Experimental results are presented in Table 7.

Solution

Table 7 describes the performance of Amberlite XAD-4 for the treatment of the waste effluent from a munitions plant. The data support the excellent performance of the polymeric

Table 7
Adsorption of TNT by Amberlite XAD4

Total column throughput (BV)	Influent TNT concentration (mg/L)	Effluent TNT concentration (mg/L)
0	118	0
100	112	0
200	125	0
300	117	0
400	108	0
500	118	0
600	92	0.4
700	94	0.6
800	95	0.9*

*Note: Below effluent TNT standard = 1.0 mg/L.

adsorbent, Amberlite XAD-4. The use of carbon is clearly not indicated as desorption can only be achieved thermally and such practices are clearly unsafe for a carbon saturated with an explosive such as TNT.

7.3. *Example 3*

This example introduces a case study of removing noxious compounds from water using polymeric adsorption. Regardless of the efficacy of industrial waste control systems, traces (parts per billion, ppb) of many noxious compounds are entering into drinking water supplies. There has been much concern that has been expressed over this problem and several studies have been initiated on the use of adsorbents to remove these noxious compounds even though there has been no concrete evidence concerning the harmful nature at these compounds that are present at such trace concentrations. The macroreticular polymeric adsorbents have shown considerable effectiveness in removing trace levels of such compounds, and therefore they were adopted for the treatment of water supplies.

Solution

A recent publication by the Department of Chemistry of Iowa State University, Amos, Iowa summarizes the results of a comprehensive study on the use of Amberlite XAD-2 and Amberlite XAD-7 for identifying and removing a host of typical organic pollutants from water. Using both Amberlite XAD-2 and Amberlite XAD-7, Prof. Fritz and his associates have developed a method for extracting trace organic containments from potable water. They have demonstrated that Amberlite XAD-2 and Amberlite XAD-7 are capable of adsorbing weak organic acids and bases and neutral organic compounds quantitatively from water containing parts per billion to parts per million concentrations of the compounds listed in below:

Acenaphthylene
1-Methylnaphthalene
Methylindene
Indene
Acenaphthene
2-2-Benzothiophene
Isopropylbenzene
Ethyl benzene

Naphthalone
2,3-Dihydroindene
Alkyl-2,3-dihydroindene
Alkyl benzothiophenes
Alkyl naphthalenes
Methyl isobutyl ketone
n-Hexanol
Ethyl butyrate
Benzene
Naphthalene
Benzoic zcid
Phenylenediamine
Phenol
2,4-Dimethylphenol
p-nitrophenol
2-Methylphenol
Aniline
O-Cresol

Although the study was analytically oriented, it does point to the potential use of the Amerlite macroreticular polymeric adsorbents for treating potable water contaminated with various noxious organic compounds. Many other investigators in the United States are continuing these studies.

7.4. *Example 4*

This example introduces a case study of removing toxin from blood using polymeric adsorption. Although this is not generally considered a waste-treatment process, the removal of toxins from blood is the most important treatment process to man. When industrial pollution control and abatement systems do not perform efficiently or are not instituted, man must depend on his own pollution control system, the kidney, because these toxins may enter in man's blood system. The kidney, however, does not have at all times the capacity or ability to effectively treat and cleanse the blood of certain toxins and some device such as an artificial kidney must be employed.

Many drugs commonly implicated in intentional or accidental overdoses are removed by hemodialysis (artificial kidneys). This technique is cumbersome, slow, and requires large volumes of solutions and highly trained personnel. The various studies that have been discussed previously encouraged several investigators to experiment with ion exchange resins and adsorbents for removing these toxic materials by directly treating the blood of a patient and thereby avoiding the need for the hemodialysis procedure. If one considers the low rate of diffusion across membranes, the large surface area of membranes required for hemodialysis to be practical, and compare these factors with the rapid rate of adsorption and high surface area of ion exchange resins, it is obvious that treating blood directly over ion exchange resins and polymeric adsorbents (hemoperfusion) has many advantages over the process of dialyzing blood through membranes (hemodialysis). The past 25 yr of research have resulted in techniques for preparing resins and adsorbents (hemoperfusion) that have many advantages over the process of dialyzing blood through membranes (hemodialysis). The research has also resulted in techniques for preparing resins and adsorbents that are sterile and free of pyrogen reactions. Furthermore one may now choose from a host of products to select an optimum product for the hemoperfusion of particular toxins that may occur in the bloodstream of patients.

Solutions

Although much of the above-described effort on hemoperfusion has been devoted to the use of ion exchange resins, the availability of macroreticular polymeric adsorbents has aroused much interest because of their

(a) High surface areas
(b) Inertness
(c) Ability to adsorb a spectrum of toxic drugs from the bloodstream without altering the ionic composition or pH of the blood.

The Albert Einstein Medical Center (Philadelphia, PA) has been quite successful in removing toxins from blood by hemoperfusion through columns of the macroreticular Amberlite polymeric adsorbents. The center has demonstrated through an exhaustive study on animals and on a number of humans that one can safely and readily remove toxins such as barbiturates and glutethimide from the bloodstream of comatose patients by hemoperfusion of the blood over Amberlite XAD-2 and Amberlite XAD-4 polymeric adsorbents. The center also compared the procedure with hemodialysis (artificial kidney) and found the hemoperfusion technique using the polymeric adsorbent to be less complicated and faster. Although carbons have also been used in hemoperfusion, their use has been found to be inherently troublesome due to the instability of the carbon particles.

The studies by the center performed with Amberlite polymeric adsorbents were conducted with resin that had carefully been treated to remove any potentially harmful impurities, microorganisms, and pyrogens. When the hemoperfusion tests were performed on several patients intoxicated with various barbiturates and glutethimide, comparative tests were made using hemodialysis. Extracorporeal Medical Specialties, Inc. of King of Prussia, PA, is manufacturing cartilage containing polymeric adsorbents specially prepared for hemoperfusion. The patients responded well and the toxins were cleared from the patients much faster using the hemoperfusion techniques with the Amberlite XAD-2 resin.

The overall promise of hemoperfusion with ion exchange resins and polymeric adsorbents for the treatment of drug intoxication can best be summarized with the following quotation by the Center:

> The Amberlite XAD-2 resin hemoperfusion systems appears to be clinically superior to hemodialysis in the treatment of drug intoxication....It results in higher clearance rates of intoxicants and is mechanically simpler and less expensive. In patients with overwhelming, life-threatening intoxication, hemoperfusion therapy may be of value in reducing coma time and the occurrence of residual complications, particularly pneumonitis. Moreover, the potential range for effective adsorption of toxins by the resin column has not been fully explored and, with the use of combinations of lipophilic, hydrophilic, anion, and cation exchange resins, may be broader than for hemodialysis.

Although the work of the Center represents the beginning of a new era in life-saving hemoperfusion techniques, it also culminates a quarter of a century of studies by others in the medical profession. It must be noted, however, that hemoperfusion using resins does not and cannot replace hemodialysis. Whereas it can take over the kidney and liver functions for a temporary period as in the case of drug intoxication, it cannot replace the hemodialysis (artificial kidney machine) procedure for those who have lost their kidney function permanently. In the future, however, hemoperfusion using resins may be a useful adjunct to hemodialysis in such cases. It is the authors' opinion that this application represents the ultimate use of polymeric adsorbents for the treatment of wastes.

7.5. Example 5

Distillation is a required supplemental process for recovery of solvent (regenerant) used in a polymeric adsorption system. Discuss other environmental and industrial applications of distillation process.

Solution

Treatment of wastes by distillation is not widespread, perhaps because of the cost of energy requirements. The distillation process is currently being used to recover solvents and chemicals from industrial wastes, where such recovery is economical. The use of distillation for treatment is increasing, as regulations for discharge become stricter making the cost of by-products recovery through distillation a more competitive means of waste solvent recovery. Other means of reclamation competitive to distillation include steam stripping and evaporation, which are presented in other chapters.

Typical industrial wastes that can be handled by distillation include the following:

(a) Plating wastes containing an organic component (usually the solvents are evaporated and the organic vapors distilled).
(b) Organic effluents from printed circuit boards are adsorbed on activated carbon. Regeneration of the activated carbon gives a liquid, which is distillable for recovery of the organic component.
(c) Methylene chloride that contains contaminates is a disposal problem, but it can be salvaged for industrial application by distilling.
(d) Methylene chloride can be recovered from polyurethane waste.
(e) The separation of ethylbenzene from styrene and recovery of both.
(f) Waste solvents for reuse in cleaning industrial equipment; this is usually a mixture of acetone, ketones, or alcohols, and some aromatics.
(g) Recovery of acetone from a waste stream that was created by the regeneration of a carbon adsorption bed used to remove acetone vapor from the off gas in plastic filter products.
(h) The production of antibiotics (e.g., penicillin) results in the generation of large quantities of wastes containing butyl acetate. The waste is distilled, and a portion of the butyl acetate can be recycled. The still bottoms, however, are hazardous wastes, which contain 50% butyl acetate and 50% dissolved organics (fats and protein). These are disposed of by incineration.
(i) Waste motor oil from local service stations and from industrial locations can be re-refined to produce regenerated lube oil or fuel oil with the aid of distillation.

7.6. Example 6

Introduce the prefabricated polymeric adsorption process equipment, which are commercially available.

Solution

Polymeric adsorption is similar in nature to granular activated carbon adsorption (GAC adsorption). The major difference between polymeric adsorption and GAC adsorption is that polymeric adsorbents are always chemically regenerated (by such means as distillation), while GAC are usually thermally regenerated.

Both the prefabricated GAC and non-GAC adsorption process equipment and the adsorbents used for polymeric adsorption are commercially available:

(a) Editor, Activated carbon adsorbers and non-carbon adsorbers. *Pollution Engineering* **32**, 22–23 (2000).

(b) Editor, Absorbents/adsorbents and absorption/adsorption equipment. *Water Environment Federation* **14**, 72–73 (2002).
(c) Editor, Adsorbents and adsorption systems. *Chemical Engineering* **107**, 470–478 (2000).

7.7. Example 7

Introduce the prefabricated distillators, which are commercially available.

Solution

Many prefabricated distillation equipment are available commercially:

(a) Editor, Distillation equipment. *Environmental Protection* **14**, 124 (2003).
(b) Editor, Distillation columns. *Chemical Engineering* **107**, 406 (2000).

7.8. Example 8

The theory and principles of polymeric adsorption are similar to that of carbon adsorption (18). Introduce the advanced activated carbon test methods that can also be applied to testing polymeric adsorbents.

Solution

Below are method titles and brief descriptions of advanced adsorption/adsorbent test methods available (1):

(a) Full Adsorption Pore Characterization: Measures the statistical distribution of the numbers of sites present over the range of adsorption forces. These data are fit into a polynomial for use in performance prediction programs.
(b) Heat of Adsorption: When adsorbent is immersed into a solvent like mineral oil, heat is given off which can be measured with a thermometer in the solvent. The amount of heat is directly related to the available adsorption space.
(c) Heterogeneity: Test designed to separate material from the outside to the inside of adsorbent granules (gem stone polishing) or differences among particles, based on size or with density separation. Once separated, the materials are tested using the Heat of Adsorption method to provide the heterogeneity profile of the material.
(d) Adsorption Determination: Test method designed to determine number of high-energy adsorption sites per gram of sorbent. Some adsorbate(s) require these sites for their removal from water such as MTBE and other water-soluble materials.
(e) Adsorbate(s) Determination: Provides a precise and accurate sample for the subsequent instrumental analysis of adsorbate(s) for specific use on new adsorbents.
(f) Microscale Activation: Instead of activating kilograms of raw material to evaluate material source, this method allows only milligrams to be used. Three thermogravimetric analysis (TGA) runs provide what is possible to be used from a particular raw source material, such as nut or fruit shell, waste material or agricultural matter.
(g) Accelerated Miniadsorption Column Evaluation: Miniadsorption columns are used to simulate pilot- and large-scale commercial systems. A useful and quick method to evaluate and compare sorbent performance.

7.9. Example 9

Polymeric adsorption has a new application for purification of contaminated groundwater. Please introduce the new technology (12,18,24).

Solution

The Ambersorb 563 adsorbent is a regenerable polymeric adsorbent that treats groundwater contaminated with hazardous organics (see Fig. 10). Ambersorb® is a registered

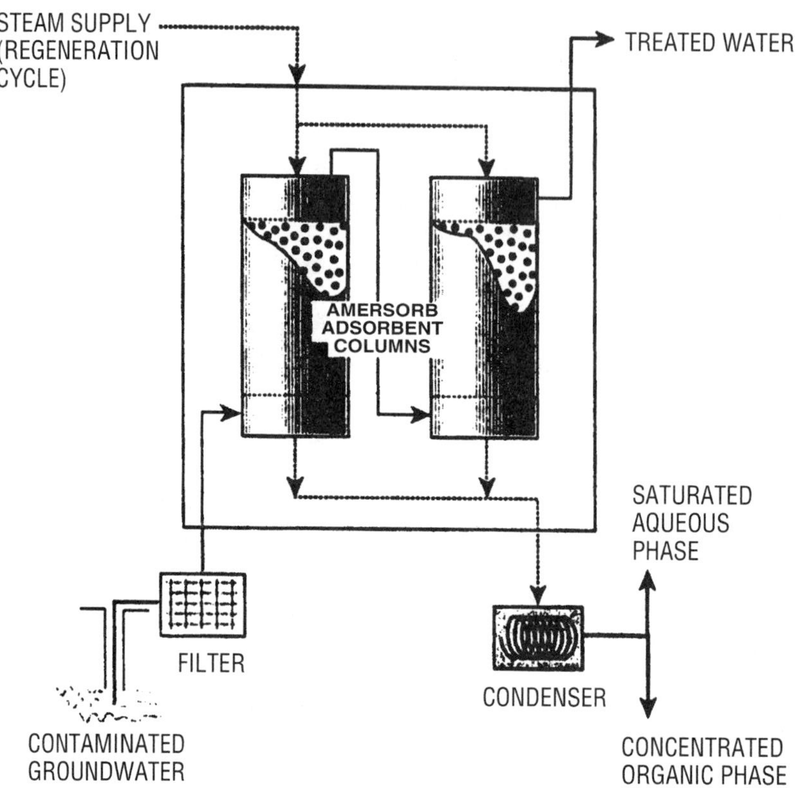

Fig. 10. Application of polymeric adsorption (Ambersorb 563 adsorbent) for groundwater decontamination.

trademark of the Rohm and Haas Company (727 Norristown Road, P.O. Box 904, Spring House, PA 19477).

The groundwater decontamination system includes a pump (not shown), a filter, a pair of polymeric adsorption columns (i.e., Ambersorb adsorbent columns), a steam supply facility, and a condenser, as shown in Fig. 10. In actual process system operation, the contaminated groundwater is pumped to a filter and the two Ambersorb adsorbent columns for treatment. The treated water is discharged from the top of the second adsorbent column. The two Ambersorb adsorbent columns are connected in series. After the Ambersorb adsorbent is exhausted, the process system is switched to regeneration mode, and the influent pump stops pumping. In the regeneration cycle, steam is supplied from the top of the column for regeneration of the Ambersorb adsorbent, and VOCs (volatile organic compounds), i.e., the contaminants, are collected by a condenser, from which there will be a saturated aqueous phase and a concentrated organic phase.

Ambersorb 563 adsorbent has 5–10 times the capacity of granular activated carbon (GAC) for low concentrations of VOCs. Current GAC adsorption techniques are well established for groundwater remediation, but require either disposal or thermal regeneration of the spent carbon. In these cases, the GAC must be removed from the site and shipped as a hazardous material to the disposal or regeneration facility. Ambersorb 563 adsorbent has unique properties that result in several key performance benefits:

(a) Ambersorb 563 adsorbent can be regenerated on site using steam, thus eliminating the liability and cost of off-site regeneration or disposal associated with GAC treatment. Condensed contaminants are recovered through phase separation.
(b) Because Ambersorb 563 adsorbent has a much higher capacity for volatile organics than GAC (at low concentrations), the process can operate for significantly longer service cycle times before regeneration is required.
(c) Ambersorb 563 adsorbent can operate at higher flow rate loadings compared with GAC, which translates into a smaller, more compact system.
(d) Ambersorb 563 adsorbents are hard, nondusting, spherical beads with excellent physical integrity, eliminating handling problems and attrition losses typically associated with GAC.
(e) Ambersorb 563 adsorbent is not prone to bacterial fouling
(f) Ambersorb 563 adsorbents have extremely low ash levels.

In addition, the Ambersorb 563 carbonaceous adsorbent-based remediation process could eliminate the need to dispose of by-products. Organics can be recovered in a form potentially suitable for immediate reuse. For example, removed organics could be burned for energy in a power plant. Reclamation of waste organics is an important benefit, as recovered materials could be used as resources instead of disposed of as wastes. This combination of benefits may result in a more cost-effective alternative to currently available treatment technologies for low-level VOC-contaminated groundwater.

Ambersorb 563 adsorbent is applicable to any water stream containing contaminants that can be treated with GAC, such as 1,2-dichloroethane, 1,1, 1-trichloroethane, tetrachloroethene, vinyl chloride, xylene, toluene, and other VOCs. This technology was accepted by the US Environmental Protection Agency into the SITE Emerging Technology Program in 1993.

7.10. Example 10

A combined air stripping and polymeric adsorption system has been developed by Purus, Inc., San Jose, CA, for site remediation . Please introduce the system and discuss its feasibility for decontamination of both soil and groundwater.

Solution

The combined air stripping and polymeric adsorption system (12) developed by Purus, Inc. (Purus), is shown in Fig. 11, and is for vapor treatment. The process system purifies air streams contaminated with volatile organic compounds (VOC). It works directly from soil extraction wells or from groundwater (or wastewater) air strippers.

The process system (Fig. 11) traps the contaminants using filter beds that contain a proprietary polymeric adsorption resin (polymeric adsorbent). This regenerative adsorption method involves one on-line treatment bed for influent air, while another bed undergoes a desorption cycle (see Fig. 11). An on-board controller system automatically switches between adsorption and desorption cycles. The desorption cycle uses a combination of temperature, pressure, and purge gas (N_2) to desorb VOCs trapped in the adsorbent bed. The contaminants are removed, condensed, and transferred as a liquid to a storage tank. Thus, the recovered material can be easily reclaimed.

Historically, granular activated carbon (GAC) has been the principal medium for separating volatile or semivolatile organic compounds from an air stream. However, because the GAC beds are difficult to regenerate on site, most treatment technologies use a passive GAC system that requires hauling the spent GAC off site for disposal or treatment. Another problem with GAC is decreased treatment efficiency resulting from moisture in

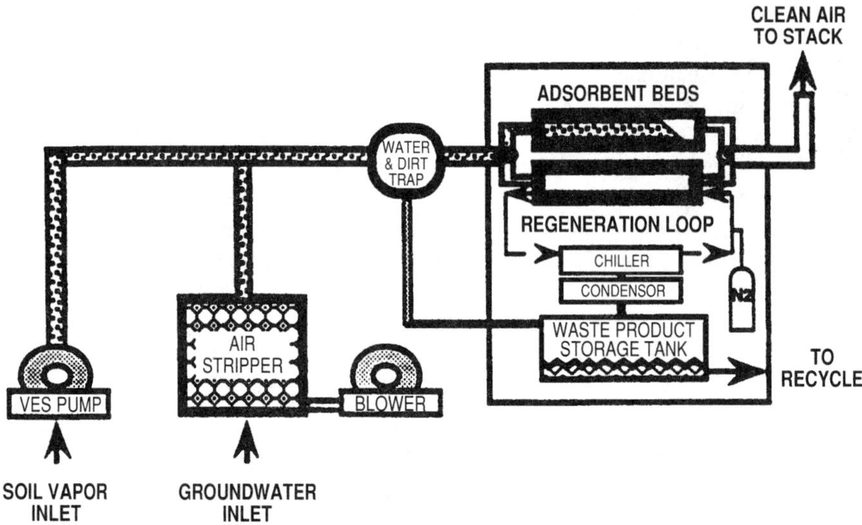

Fig. 11. Application of polymeric adsorption and air stripping for site remediation.

the waste stream. Moisture in humid contaminated air dramatically reduces the GAC's ability to adsorb organic contaminants; GAC treatment efficiency declines to 30% of original efficiency as the relative humidity (RH) exceeds 75%.

Polymeric adsorbent beds used in the combined air stripping and polymeric adsorption system have been recycled on a test stand more than 1000 times with no measurable loss of adsorption capacity. In addition, the polymeric adsorption resin (i.e., polymeric adsorbent) has a relatively high tolerance for water vapor, allowing efficient treatment of air streams with an RH greater than 90%. These two capabilities make on-site treatment of VOCs possible with substantially lower operating costs.

The combined air stripping and polymeric adsorption system (Fig. 11) controls VOC emissions at site remediation projects, industrial wastewater facilities, and industrial air processing sites. Site remediation usually involves vacuum extraction of solvents or fuels from soils, as well as the pumping and treatment of groundwater by air stripping. The newly developed process system has also treated industrial waste containing solvents using an emission-free, closed-loop air stripping process.

Under a US Environmental Protection Agency demonstration project, the process system (Fig. 11) has simultaneously treated vapors from soil vacuum extraction wells and a groundwater air stripper.

ACKNOWLEDGMENTS

This chapter, Polymeric Adsorption and Regenerant Distillation, has been written in honor of Dr. Robert Kunin who was the former Research Manager of Rohm and Haas Company, Philadelphia, PA. Dr. Kunin has published over 30 scientific papers and research reports, and invented several ion exchange resins and polymeric adsorbents. As an internationally well-known researcher, he has been a mentor and advisor to many professional engineers and scientists around the world during his lifelong research career at Rohm and Haas. Dr. Lawrence K. Wang is among those researchers, who learned ion exchange and polymeric adsorption processes from Dr. Kunin. This book chapter was completed under Dr. Kunin's guidance. The authors of this chapter thank Dr. Kunin for his review, comments, and approval of the manuscript, and salute him for his dedication of entire professional career to development and improvement of ion exchange and adsorption processes.

REFERENCES

1. M. Greenbank, H. Nowicki, H. Yuto, and B. Sherman, *Water Conditioning and Purification* **45**, 98–103 (2003).
2. D. Vidic, T. Suidan, and C. Brenner , *Wat. Res.* **28**, 263–268 (1994).
3. J. S. Fritz, *Anal. Chem.* **44**, 139 (1972).
4. K. A. Kun, R. Kunin, and J. Polym, *Science* Part C, No. 16,1457 (1967).
5. J. Rosenbaum, Artificial Organs *Trans. Amer. Soc.* **16**, 134 (1970).
6. J. Rosenbaum, *Clinical Toxicology* **5**, 331 (1972).
7. US EPA, *Technical Review of the Best Available Technology, Best Demonstrated Technology, and Pretreatment Technology for the Gum and Wood Chemicals Point Source Category*. US Environmental Protection Agency, Washington DC, 1979.
8. US EPA, *Development Document for Effluent Limitations, Guidelines, and Standards for the Pulp, Paper, and Paperboard and the Builders Paper and Board Mills Point Source Categories*. EPA-440/1-80/025-b. U. S. Environmental Protection Agency, Washington DC, 1980.
9. US EPA, *Physical, Chemical, and Biological Treatment Techniques for Industrial Wastes*. Volume II. NTIS- PB 275 287; U. S. Environmental Protection Agency, Washington DC, 1977.
10. US EPA, *Development Document for Effluent Limitations, Guidelines, and Standards for the Leather Tanning and Finishing Point Source Category*. EPA-440/1-79/016. US Environmental Protection Agency, Washington DC, 1979.
11. L. K. Wang, *Environmental Engineering Glossary*, Calspan Corporation, Buffalo, NY, 1974.
12. L. K. Wang, J. V. Krouzek, and U. Kounitson, *Case Studies of Cleaner Production and Site Remediation*. UNIDO Manual DTT-5-4-95. United Nations Industrial Development Organization, Vienna, Austria, 1995.
13. Editor, *Pollution Engineering* **32**, 22–23 (2000).
14.. Editor, *Water Environment Federation* **14**, 72–73 (2002).
15. Editor, *Chemical Engineering* **107**, 470–478 (2000).
16. Editor, *Environmental Protection* **14**, 127 (2000).
17. Editor, *Chemical Engineering* **107**, p. 406 (2000).
18. L. K. Wang, *J Appl. Chem. Biotech.* **25**, 475–490 (1975).
19. Y.S. Lipatov, *Polymer Reinforcement*, Chem Tec Publishing, Toronto, Ontario, Canada, 1995.
20. J. Toth, *Adsorption: Theory, Modeling and Analysis*. Marcel Dekker, New York, 2002.
21. Rohm Hass, *Polymeric Adsorbent with an Acrylic Ester Matrix*. www.rohmhaas.com, 2003
22. L. K. Wang, Y. T. Hung and N. K. Shammas (eds.) *Advanced Physicochemical Treatment Process*. Humana Press, Totowa, NJ, 2005.
23. R. Kunin, *Polymeric Adsorption*. Lenox Institute of Water Technology (formerly Lenox Institute for Research), Lenox, MA. LIR/02-87-223B, 1987.
24. L. K. Wang, Y. T. Hung, H. H. Lo, and C. Yapijakis (eds.), *Handbook of Industrial and Hazardous Wastes Treatment*. Marcel Dekker, NYC, NY, 2004.

15

Granular Activated Carbon Adsorption

Yung-Tse Hung, Howard H. Lo, Lawrence K. Wang, Jerry R. Taricska, and Kathleen Hung Li

CONTENTS

1. INTRODUCTION

Adsorption is the process of binding and removing certain substances from a solution through the use of an adsorbent. Activated carbon is the most commonly used adsorbent in the treatment of water, municipal wastewater, and organic industrial wastewaters, because of its ability to adsorb a wide variety of organic compounds, as well as the economic feasibility of use. In water treatment, activated carbon is used to remove organic compounds that cause objectionable taste, odor, and color. In advanced wastewater treatment, carbon is used to adsorb organic compounds, and in industrial wastewater treatment, it is used to adsorb toxic organic compounds. It is usually used in the granular form in the carbon adsorption column application in water and wastewater purification, but it is also used in the powdered form in the powder-activated, carbon-activated sludge process for wastewater treatment.

Adsorption may be classified as chemical adsorption or physical adsorption. In chemical adsorption, a chemical reaction occurs between the solid and the adsorbed solute,

From: *Handbook of Environmental Engineering, Volume 3: Physiochemical Treatment Processes*
Edited by: L. K. Wang, Y.-T. Hung, and N. K. Shammas © The Humana Press Inc., Totowa, NJ

Table 1
Several Activated Carbon Removal Applications

Acetaldehyde	Gasoline
Acetic Acid	Glycol
Acetone	Herbicides
Activated-sludge effluent	Hydrogen sulfide
Air-purification scrubbing solutions	Hypochlorous acid
Alcohol	Insecticides
Amines	Iodine
Ammonia	Isopropyl acetate and alcohol
Amyl acetate and alcohol	Ketones
Antifreeze	Lactic acid
Benzine	Mercaptans
Biochemical Agents	Methyl acetate and alcohol
Bleach solutions	Methyl-ethyle-ketone
Butyl acetate and alcohol	Naphtha
Calcium hypochlorite	Nitrobenzene
Can and drum washing	Nitrotoluene
Chemical tank wash water	Odors
Chloral	Organic compounds
Chloramine	Phenol
Chlorine	Potassium permanganate
Chlorobenzene	Sodium hypochlorite
Chlorophenol	Solvents
Chlorophyl	Sulfonated oils
Cresol	Tastes (organic)
Dairy process wash water	Toluene
Decayed organic matter	Trichlorethylene
Defoilants	Trickling-filter effluent
Detergents	Turpentine
Dissolved oil	Vinegar
Dyes	Well water
Ethyl acetate and alcohol	Xylene

and the reaction is rarely used in water and wastewater treatment. Physical adsorption is widely used in water and wastewater treatment. Physical adsorption is primarily due to van der Waals' forces and is a reversible process. When the molecular forces of attraction between the solute and the adsorbant are greater than the forces of attraction between the solute and the solvent, the solute will be adsorbed onto the surface of adsorbent materials. Adsorption by activated carbon is a typical example of physical adsorption. Activated carbon has a large number of capillaries within the carbon particles. The total surface available for adsorption of solute includes the surfaces of the pores as well as the external surface of the particles. In fact, the pore surface area is much larger than the external surface area of the particles and most of the adsorption occurs on the pores' surfaces. The ratio of the total surface area to the mass of activated carbon is very large.

In its initial applications, the granular activated carbon (GAC) adsorption was used in a tertiary treatment process for the removal of organic pollutants in the secondary effluent from biological wastewater-treatment systems. Recently, GAC adsorption has been used as a secondary treatment process in the physical–chemical treatment (PCT)

Table 2
Carbon Adsorption Technology Milestones

1550 BC	Early recorded use of charcoal
1811 AD	Bone char used for sugar processing
1828	First char regeneration instituted
1852	First granular charcoal filter (Elizabeth, NJ)
1889	Hershoff multiple hearth furnace introduced
1906	First commercial production of activated carbon (Eponite, Europe)
1910	First application of GAC in drinking water treatment (Reading, England)
1913	First commercial production of activated carbon in the US
1928	First use of powdered activated carbon (PAC) for taste and odor control (Chicago meat packers)
1929	First GAC filter installed (Hamni, Germany)
1930	First municipal use of powdered activated carbon for taste and odor control (Hackensack Water Company, United Water, NJ)
1961	GAC filters installed at Hopewell, VA water-treatment plant
1965	First advanced wastewater-treatment plant incorporating GAC (South Lake Tahoe, CA)
1974	First PAC used in combination with dissolved air flotation (DAF) (Cornell Univ. Aeronautical Lab., NY) (60)
1978	First fluidized bed GAC regeneration furnace installed in the U.S.
1978	First municipal physicochemical wastewater plant using GAC filter and GAC regeneration (85-MGD Niagara Falls Wastewater Treatment Plant, NY) (20–22, 59)
1981–1989	First physicochemical fluidized bed GAC processs First biological fluidized bed GAC process First physicochemical GAC-SBR process First biological GAC-SBR process First physicochemical PAC-SBR process First biological PAC-SBR process First physicochemical PAC-DAF-SBR process First biological PAC-DAF-SBR process (Lenox Institute of Water Technology and Zorex Corporation, MA) (30)
1989	First precoat GAC filtration process (Lenox Institute of Water Technology, Krofta Engineering Corporation, and Zorex Corporation, MA) (26)
1996	First biological GAC filtration plant for water treatment. (Saskatoon, Canada) (42)
2000	First full-scale GAC fluidized bed biological GAC plant for groundwater decontamination (US Filter, USA) (53)
2003	First dual-stage biological GAC filtration plant for potable water treatment (Ngau Tam Mei Water Works, Hong Kong, China) (49)

plants, because activated carbon can remove biodegradable as well as refractory organics present in the raw wastewaters. The physical–chemical treatment consists of chemical coagulation and precipitation of raw municipal or industrial wastewater, followed by adsorption on activated carbon for removal of soluble and insoluble organic pollutants. The GAC anaerobic filter process has been used in the treatment of industrial wastewaters containing slowly biodegradable or toxic organics. Table 1 lists several applications for activated carbon's adsorption abilities (1). The milestones for the GAC adsorption technology are shown in Table 2 (2).

Table 3
Properties of Granular Activated Carbons[a]

	ICI America Hydrodarco 3000 (Lignite)	Calgon Filtrasorb 300 (8 × 30) (Bituminous)	Westvaco Nuchar WV-L (8 × 30) (Bituminous)	Witco 517 (12 × 30) (Bituminous)
Physical properties				
Surface area, m^2/g (BET)	600–650	950–1050	1000	1050
Apparent density, g/cm^3	0.43	0.48	0.48	0.48
Density, baskwashed and drained, lb/ft^3	22	26	26	30
Real density g/cm^3	2.0	2.1	2.1	2.1
Particle density g/cm^3	1.4–1.5	1.3–1.4	1.4	0.92
Effective size, mm	0.8–0.9	0.8–0.9	0.85–1.05	0.89
Uniform coefficient	1.7	1.9 or less	1.8 or less	1.44
Pore volume g/cm^3	0.95	0.85	0.85	0.60
Mean particle diameter, mm	1.6	1.5–1.7	1.5–1.7	1.2
Specification				
Sieve size (US standard rises)				
Larger than No. 8 (max %)	8	8	8	*c*
Larger than No. 12 (max %)	*c*	*c*	*c*	5
Smaller than No. 30 (max %)	5	5	5	5
Smaller than No. 40 (max %)	*c*	*c*	*c*	*c*
Iodine Np.	650	900	950	1000
Abrassion No., minimum	*b*	70	70	85
Ash (%)	*b*	8	7.5	0.5
Moisture as packed (max %)	*b*	2	2	1

[a]Other sizes of carbon are available on request from the manufacturers.
[b]no available data from the manufacturer.
[c]Not applicable to this size of carbon.

The manufacture of activated carbons and properties of activated carbons are described in the chapter on powdered activated carbon (PAC) in this handbook series. The GAC are particles that are larger than US Sieve Series No. 50, while the PAC is smaller. Table 3 shows the properties of several commercially available granular activated carbons (3).

2. PROCESS FLOW DIAGRAMS FOR GAC PROCESS

Two alternatives exist for the use of GAC in municipal and industrial-wastewater treatment. One alternative is to use the GAC process in the tertiary treatment following conventional primary and biological secondary treatment. Tertiary treatment processes consisting of activated carbon adsorption process range from treatment of the secondary treated effluent with only activated carbon addition to systems with chemical clarifications, nutrient removal, filtration, carbon adsorption, and disinfection. The second alternative uses activated carbon in a physical–chemical treatment (PCT) process in which raw wastewater is treated in a primary settling tank with chemical addition prior to GAC adsorption. Filtration and disinfection may also be included in PCT, but biological processes are not used.

Figures 1 and 2 show the flow diagrams for alternative treatment schemes for tertiary treatment and PCT systems (4). If biological treatment and efficient filtration are used upstream of carbon treatment, then several benefits could be obtained: (1) the applied loads of biochemical oxygen demand (BOD), chemical oxygen demand (COD), and other organics are reduced allowing either the production of a higher-quality carbon column, effluent at a given contact time, or the production of equal water quality at a shorter contact period; (2) the applied loads of suspended and colloidal solids are less, thus reducing head loss through the bed of GAC column to minimize physical plugging, ash buildup, and progressive loss of adsorptive capacity in the carbon particles after several cycles of regeneration; (3) biological growth, septic condition in the bed, and hydrogen sulfide production may be minimized due to the reduction of the supply of bacterial food and oxygen demanding substances applied to the carbon.

Figure 3 shows the typical schematics of the biological activated carbon (BAC) process for drinking water treatment (5). In the BAC process, the biological activity in the GAC column for removal of organics present in the filter effluent will increase the GAC column life.

The PCT process maximizes the use of GAC by extending its removing refractory dissolved organics to adsorption of biodegradable organics and, in some cases, by using the GAC bed as a filter to remove suspended and colloidal materials, which will shorten column life. With this approach, the GAC is loaded as heavily as possible within the limits of effluent water quality criteria. PCT process has a lower capital cost but a higher operation and maintenance cost than biological treatment, followed by tertiary treatment. However, it should be noted that in several cases the effluent quality can be optimized only when biological oxidation, chemical coagulation, filtration, and adsorption are operated in series as separate processes. The capabilities of PCT must be evaluated regarding specific effluent quality requirements to determine its applicability to a given problem.

3. ADSORPTION COLUMN MODELS

In the granular activated carbon (GAC) adsorption column, the feed wastewaters containing adsorbates or contaminants are passed through a layer of GAC particles. Three types of GAC adsorption column processes in common use consist of downflow-gravity fixed-bed process, upflow-pressure fixed-bed process, and upflow fluidized-bed process. All three processes are considered as plug flow processes. During the adsorption process, the GAC particle density increases significantly as adsorption progresses. The denser GAC granules that have been saturated with adsorbates will migrate to the bottom of the

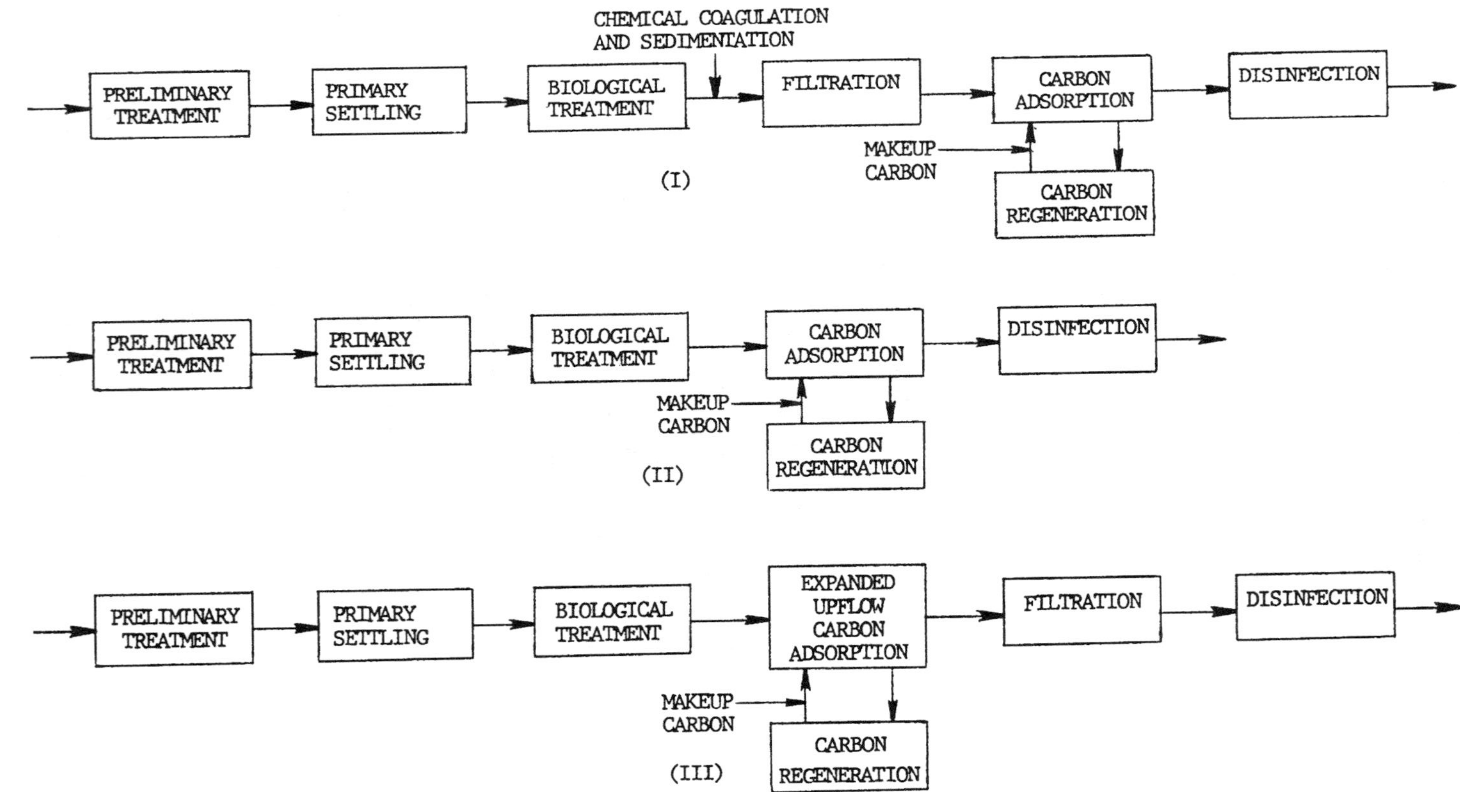

Fig. 1. Wastewater-treatment process with carbon adsorption as a tertiary treatment step.

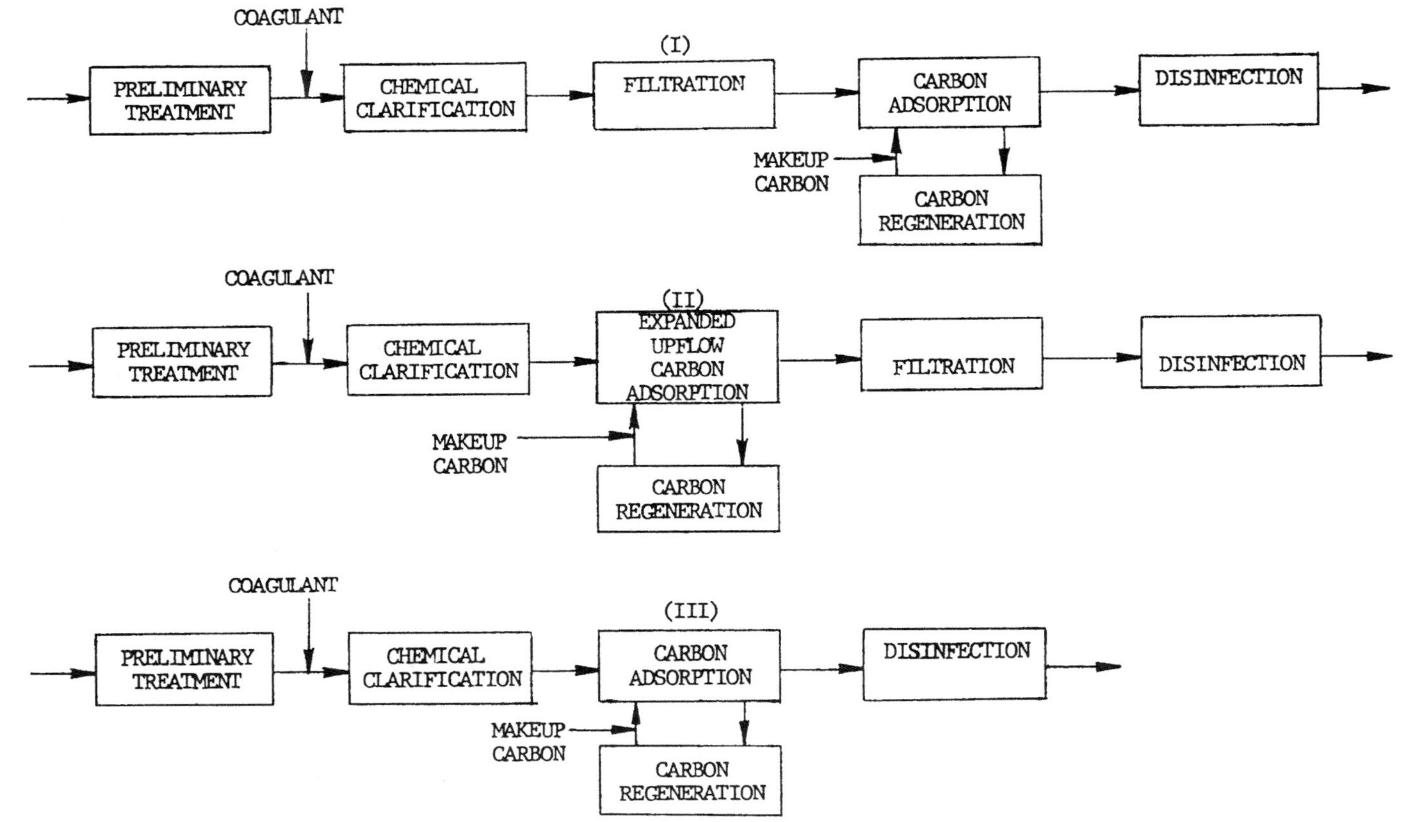

Fig. 2. Physicochemical treatment process.

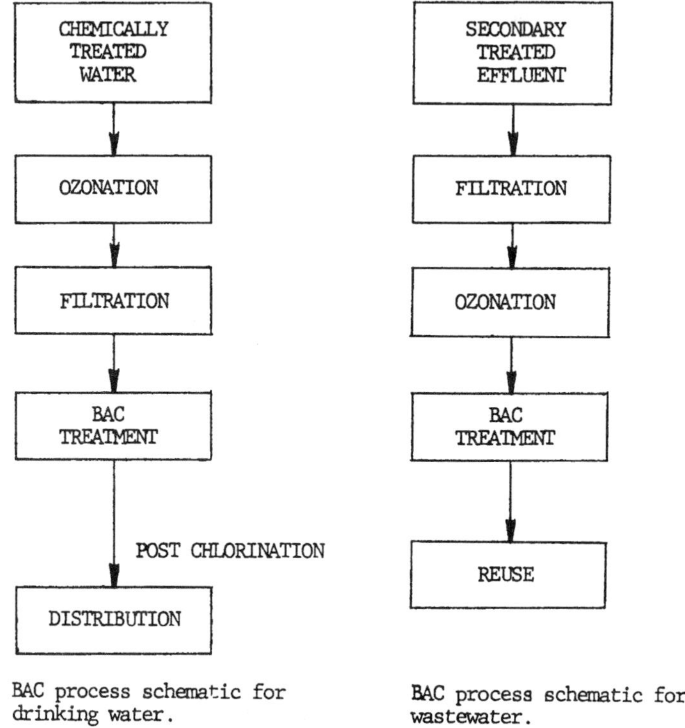

Fig. 3. BAC process schematic.

carbon columns. In all three processes, the influent wastewater has contact with GAC that has adsorbed the greatest amount of adsorbates or organic pollutants.

The three GAC adsorption systems vary in operating characteristics, especially with regard to clogging. Down flow gravity adsorption columns are the most susceptible to clogging and are used only for wastewaters of low turbidity. Upflow pressure adsorption columns have fewer clogging problems due to the direction of flow. For a fluidized bed with adsorption process, clogging of GAC columns are not a concern. However, the abrasion and GAC particle breakup are problems that affect the fluidized-bed process.

Figure 4 depicts an idealized GAC breakthrough curve. The GAC breakthrough curve is developed based on data collected from the pilot plant study in treating the wastewater of concern and is useful in the design of GAC adsorption columns. The curve relates the effluent adsorbate or organic concentration to the volume of wastewater treated by the GAC columns. In Fig. 4, C is the solute concentration and V is the volume of liquid passed through the column.

The breakthrough is related to the formation and movement of the active adsorption zone and the saturated zone in the GAC adsorption columns. When feed wastewater is introduced at the top of a clean bed of activated carbon, most adsorbate removal initially occurs at the top of the column. The saturated zone moves through the bed preceded by the active adsorption zone with depth δ. For simplicity's sake, it is assumed that all the adsorption occurs in the active adsorption zone and that the zone behind the active zone is completely saturated. When the front of the active adsorption zone reaches the

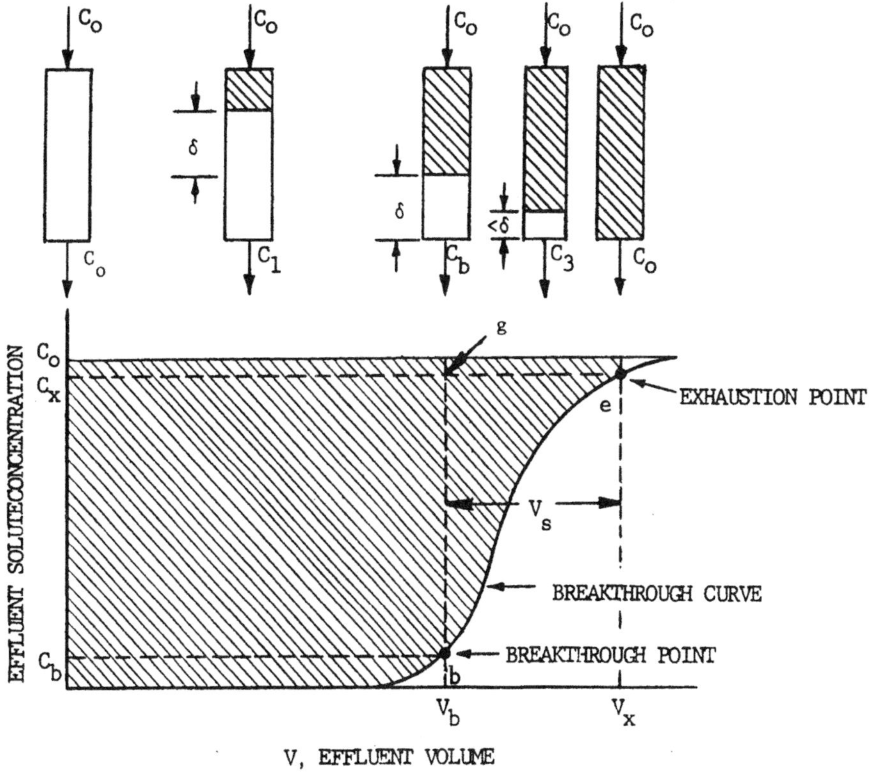

Fig. 4. Idealized breakthrough curve for GAC adsorption column.

bottom of the bed, breakthrough at column occurs, and the effluent solute concentration increases rapidly. Finally, the GAC will be completely saturated and the effluent solute concentration at exhaustion, C_x, will approach C_0, the influent solute concentration. In practice, breakthrough is defined as the effluent solute concentration reaches either $0.05C_0$ or the effluent quality requirement. Exhaustion is usually defined as $C_x = 0.95C_0$.

The point on the S-shaped breakthrough curve at which the effluent solute concentration reaches its maximum allowable value is called the breakthrough point. The point where the effluent solute concentration reaches 95% of its influent value is referred to as the exhaustion point.

The mathematical equations dealing with the formation and movement of the adsorption zone has been described by Hutchins (6). For the active adsorption zone calculation the time to exhaustion is defined as t_x, while the time of passage through the active adsorption zone as t_δ:

$$t_x = V_x/Q = V_x/(F_m A) \tag{1}$$

$$t_\delta = (V_x - V_b)/Q = (V_x - V_b)/(F_m A) \tag{2}$$

where V_b = volumes of liquid passed through column at breakthrough, V_x = volume of liquid passed through column at exhaustion, Q = volumetric flow rate, F_m = mass liquid flux, and A = cross-sectional area of column.

The velocity at which the active adsorption zone moves through the column is assumed to be constant except during the period it is being formed. This velocity defines the adsorption-zone height δ:

$$\delta = U_{\delta} t_{\delta} = t_{\delta} L / (t_x - t_f) \tag{3}$$

where L = column length, t_f = active adsorption zone formation time, U_{δ} = active adsorption zone velocity, and t_{δ} = time of pass the adsorption-zone height.

The formation time at the active adsorption zone can be estimated as t_f:

$$t_f = (1 - f) t_{\delta} \tag{4}$$

$$f = P_s / P_{tc} \tag{5}$$

where f = fractional capacity of the active adsorption zone, P_s = quantity of solute adsorbed in the active adsorption zone from breakthrough to exhaustion; and P_{tc} = total capacity of carbon in the active adsorption zone:

$$P_s = \int_{V_b}^{V_x} (C_0 - C)\, dV \tag{6}$$

$$P_{tc} = (V_x - V_b) C_0 \tag{6a}$$

Then

$$f = \frac{P_s}{P_{tc}} = \frac{\int_{V_b}^{V_x} (C_0 - C)\, dV}{(V_x - V_b) C_0} \tag{7}$$

$$f = \int_0^1 \left(1 - \frac{C}{C_o}\right) \frac{d(V - V_b)}{(V_x - V_b)} \tag{8}$$

P_s is equal to the cross-hatched area bounded by points b, δ, and e in Fig. 4. C is the solute concentration, C_0 is influent solute concentration, C_x is the effluent solute concentration at exhaustion, V is the volume of liquid passing through column, and dV is differential volume. As f approaches 1, t_f approaches zero, and ideal plug flow conditions are approximated. When channeling or mass-transfer limitations prevail, the breakthrough curve rises very rapidly following breakthrough until it approaches C_x, the exhaustion effluent solvent concentration. Under these conditions, P_s is small and the fractional capacity, f, is approaching zero. The depth of the active adsorption zone, δ, can now be expressed in terms of fractional capacity, f:

$$\delta = \frac{L(V_x - V_b)}{V_b + f(V_x - V_b)} \tag{9}$$

When column breakthrough occurs, the only portion of the adsorption column bed that has not been saturated, exhausted, or in equilibrium with the influent solute concentration C is the final portion of depth δ in the column. Total adsorption capacity to breakthrough can then be defined as:

$$S_b = \left(\frac{x}{m}\right)_{C0} \rho_p\left[(L-\delta)+(f\,\delta)\right] \tag{10}$$

where S_b = mass of solute adsorbed at equilibrium per unit of cross-sectional area of adsorption bed, $\left(\frac{x}{m}\right)_{C0}$ = mass of organic adsorbed per unit mass of carbon at the equilibrium concentration C_0, ρ_p = apparent packed density of the carbon adsorbent, and L = bed depth.

The effective adsorption capacity of a carbon column depends on the fractional capacity f. As fractional capacity (f) decreases, adsorption-zone height (δ) increases in size and mass of solute absorbed per cross section area (S_b) decreases in magnitude. Therefore, it is very important to maximize the fractional capacity (f) of the active adsorption zone.

By lumping all the mass transfer effects together into a simple mass transfer rate term, the following mass transfer model is presented:

$$F_m = \frac{dC}{dy} = ka\left(C - C_e\right) \tag{11}$$

where dy = differential column length, k = mass transfer rate coefficient, a = external area of the adsorbent per unit volume, C_e = equilibrium concentration for the amount of organic adsorbed, y = column length, and F_m = mass liquid flow rate.

$C - C_e$ represents the difference in concentration between the actual and the equilibrium value of any point in the adsorber and is the driving force for adsorption. Rearranging Eq. (11) gives:

$$ka/F_m\,dy = dC/\left(C - C_e\right) \tag{12}$$

Then integrating Eq. (12) over the adsorption zone for constant k gives:

$$\int_0^y \left(ka/F_m\,dy\right) = \int_{C_b}^{C_x} \frac{dC}{\left(C - C_e\right)} \tag{12a}$$

$$\frac{ka\delta}{F_m} = \int_{C_b}^{C_x} \frac{dC}{\left(C - C_e\right)} \tag{13}$$

Any value of column length (y) less than the adsorption-zone height (δ) corresponds to concentrations between C_b and C_x, then Eq. (13) becomes:

$$\frac{kay}{F_m} = \int_{C_b}^{C} \frac{dC}{\left(C - C_e\right)} \tag{14}$$

Combining Eqs. (13) and (14) yields the following relationship:

$$y/\delta = \frac{V - V_b}{V_x - V_b} = \frac{\displaystyle\int_{C_b}^{C} \frac{dC}{\left(C - C_e\right)}}{\displaystyle\int_{C_b}^{C_e} \frac{dC}{\left(C - C_e\right)}} \tag{15}$$

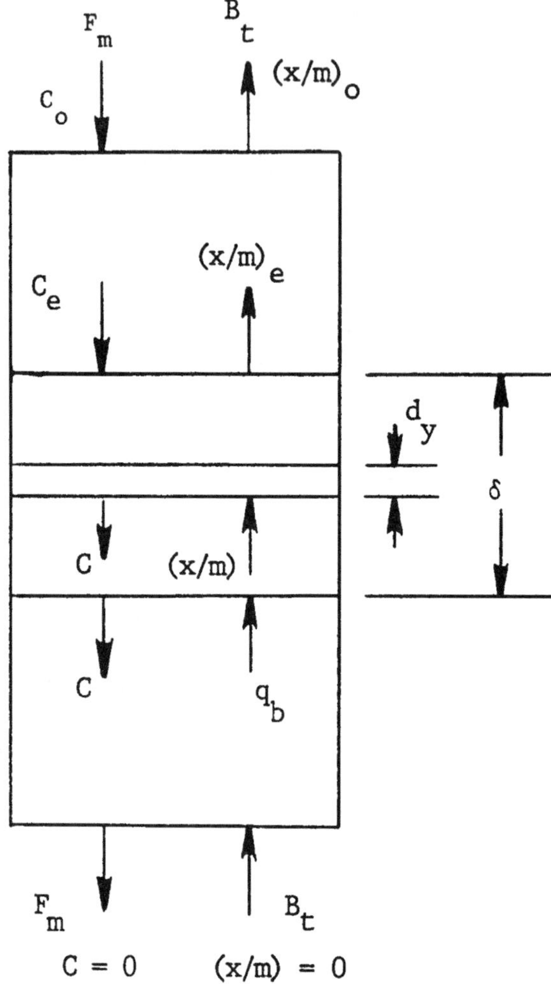

Fig. 5. Schematic diagram of steady state adsorption column.

In Eq. (15) the two expressions can be either graphically or numerically integrated.

Over the entire carbon adsorption column, the adsorption rate can be defined as:

$$F_m C_0 = P_s \left(\frac{x}{m} \right)_{C0} \tag{16}$$

where P_s = superficial or apparent rate of saturation.

Equation (16) provides one point on the operating curve of the GAC adsorber. Since the curve is above linear for *x/m* versus *C*, and, in most cases, the curve passes very close to the origin. Therefore, Eq. (14) can be expressed as

$$F_m C_0 = P_s \left(\frac{x}{m} \right) \tag{17}$$

A material balance around the entire column as shown in Fig. 5 yields

$$F_m(C_0 - 0) = B_t\left(\left(\frac{x}{m}\right)_0 - 0\right) \tag{18}$$

where F_m = liquid mass flow rate to GAC adsorption column, C_0 = influent solute concentration, B_t = mass flow rate of solid to maintain a stationary adsorption zone, and $\left(\frac{x}{m}\right)_0 = \delta_0$ mass solute/mass adsorbent at equilibrium with influent concentration, C_0.

A material balance between the lower end of the column and some arbitrary cross section of the column is

$$F_m(C_0 - 0) = B_t\left(\left(\frac{x}{m}\right) - 0\right) \tag{19}$$

From Eqs. (18) and (19),

$$\frac{B_t}{F_m} = \frac{C}{\left(\frac{x}{m}\right)} = \frac{C_0}{\left(\frac{x}{m}\right)_0} \tag{20}$$

Equation (20) is plotted as a straight line on Fig. 6. The straight line passes through the origin and the equilibrium isotherm curve of influent solute concentration, C_0, is called an operating line. Each point on the operating line represents the compositions of the liquid and solid streams passing each other at some cross section of the column.

A vertical line drawn between the isotherm and the operating line gives, $C - C_e$, the difference in concentration between the actual and the equilibrium value of any point, is the driving force for adsorption. By assuming that the operating curve goes through the origin, the curve is defined by the choice of C_0 and the characteristic isotherm. This information can then be used for graphical integration to determine the column characteristics.

The shape of the breakthrough curve will depend on the nature of the feed wastewater. For single-component adsorbate, the adsorption zone will be short and the breakthrough curve will be steep like curve *a* in Fig. 7.

On the other hand, for multicomponent adsorbates having different adsorbabilities, the adsorption zone will be long and the breakthrough curve will be gradual as in curve *b* in Fig. 7. The breakthrough curve for various wastes is shown in Fig. 8.

4. DESIGN OF GRANULAR ACTIVATED CARBON COLUMNS

4.1. *Design of GAC Columns*

Granular activated carbon adsorption columns are most suitable for the treatment of industrial wastewaters containing high concentrations of organics to be removed because (a) separation of the spent carbon from the wastewater after treatment is not necessary and (b) the concentration of the adsorbed solute is in equilibrium with the influent solute concentrations rather than the effluent solute concentration, which provides a greater flexibility of operation.

Several modes of GAC column operation are available depending on the results desired.

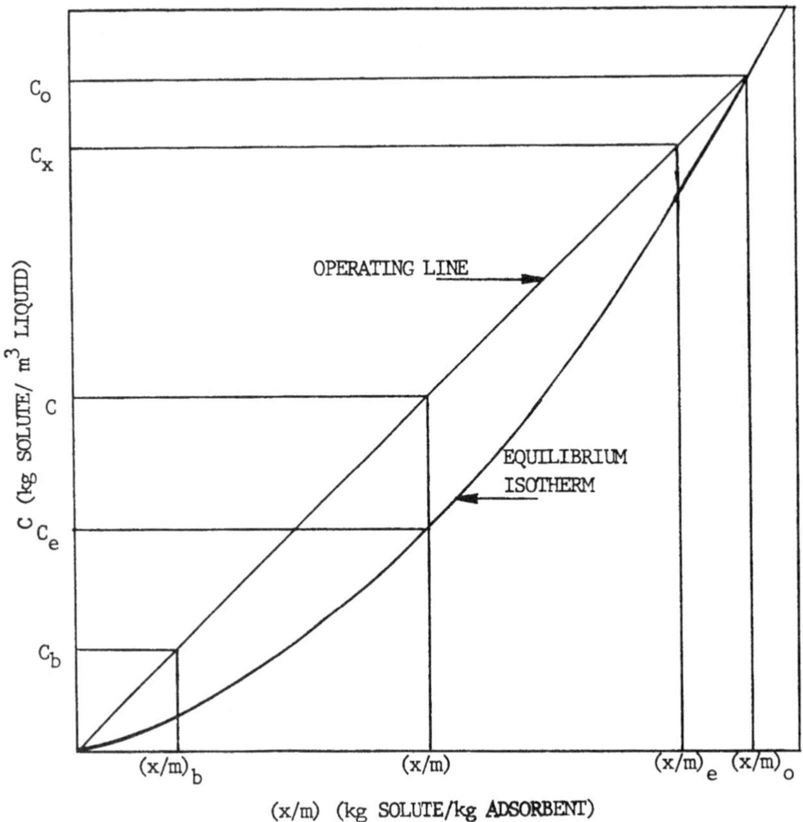

Fig. 6. Operating line for steady state adsorption column.

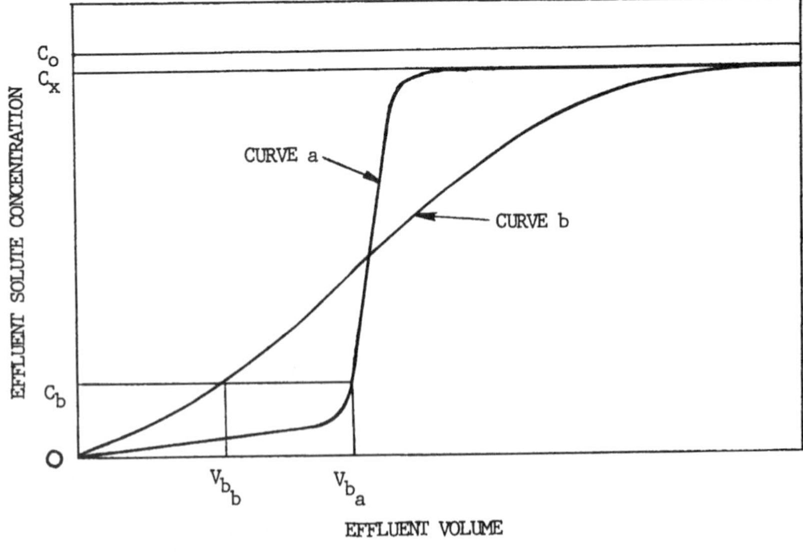

Fig. 7. Typical adsorption breakthrough curves.

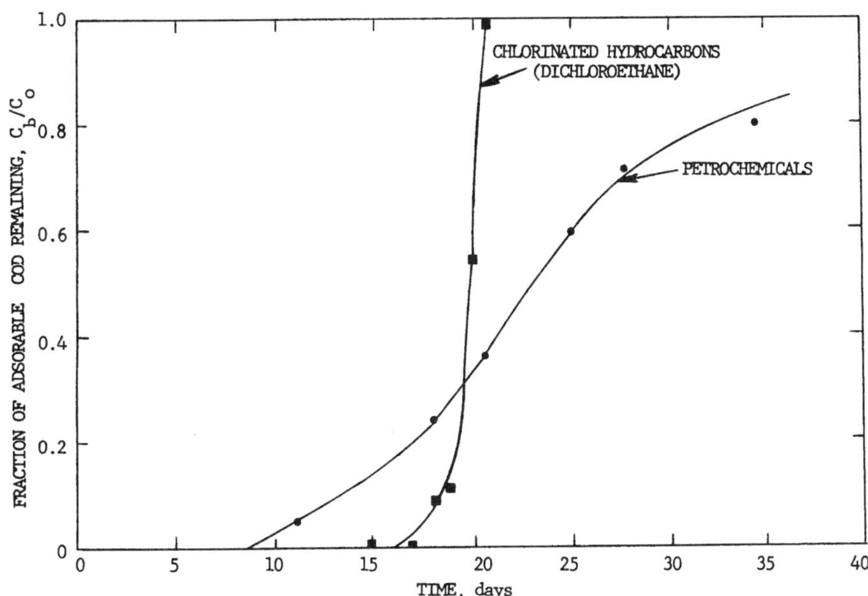

Fig. 8. Breakthrough curves for various wastes.

1. Downflow, fixed beds in series. When breakthrough occurs in the last column, the first column is in equilibrium with the influent concentration in order to achieve a maximum carbon adsorption capacity. After replacement of carbon in the first column, it becomes the last column in a series.
2. Multiple units, operated in parallel. The effluent is blended to achieve the final desired quality. The effluent from a column ready for regeneration or replacement, which is high in COD, is blended with the other effluents from fresh carbon columns to achieve the desired quality. This mode of operation is most adaptable to waters in which the capacity at breakthrough/capacity at exhaustion is great (near 1.0).
3. Upflow, expanded beds. These are used when influent has suspended solids or when biological action occurs in the GAC columns.
4. Continuous counterflow, column or pulsed beds. The spent carbon from the bottom is sent to regeneration with the regenerated and makeup carbon fed to top of the reactor.

The maximum economy would require that spent carbon be in equilibrium with the influent wastewater. The depth of carbon removed for regeneration and therefore the depth of the total carbon system will depend on the depth of the active adsorption zone.

In designing a carbon adsorption system, it will be necessary to conduct pilot column tests under operating conditions similar to those expected in the full-scale plants and using the same wastewaters. Because these conditions may not be known at the time of pilot-plant study, a method for extrapolating the experimental data to various operating conditions is required to reduce the number of experiments. Hutchins and co-workers (6, 8) have used the Bohart–Adams (7) equation in the form of bed-depth service time (BDST) for interpretation of column data and process design, including data extrapolation to conditions other than tested.

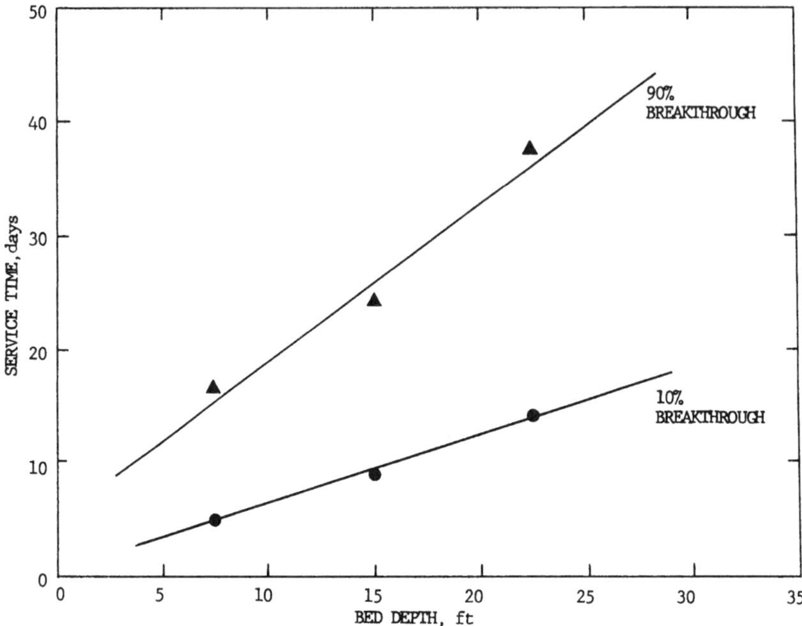

Fig. 9. Bed depth-service time curve for a single stage GAC column.

For developing a BDST correlation, a number of pilot columns of equal depth are operated in series and breakthrough curves are plotted. These data are then used for producing a BDST correlation by recording the operating time required to reach a certain removal at each depth. Plots of BDST correlations for 10% and 90% breakthrough are shown in Fig. 9. These correlation lines correspond to 90% and 10% removals, respectively. The Bohart–Adams equation for the BDST is shown below:

$$t = \left[\left(N_0/C_0 v\right)X\right] - \left(1/KC_0\right)\ln\left[\left(C_0/C_b\right) - 1\right] \tag{21}$$

where t = service time (h), X = bed depth (ft), v = hydraulic loading, or linear velocity of fluid (ft/h), C_0 = concentration of impurity in influent (lb/ft^3), C_b = concentration of impurity in effluent (lb/ft^3), N_0 = adsorption efficiency (lb/ft^3), and K = adsorption rate constant (ft^3/1b-h).

In Fig. 9, the slope of the BDST line is equal to the reciprocal velocity of the adsorption zone, and the x-intercept, the critical depth, is the minimum bed depth required for obtaining the desired effluent quality at time zero. Hutchins has shown that N_0 is dependent on C_b as can also be seen from the variation in slope for different breakthrough values in figure.

If the adsorption zone is arbitrarily chosen as the carbon layer through which the solute concentration ranges from 10% to 90% of the feed concentration, the depth of this zone is given by the horizontal distance between these two lines in the BDST graph. The depth of the adsorption zone increases with time or with depth of bed.

It should be noted that the BDST equation is only valid for a single-stage column in which the feed is always applied to the head end of the bed, and the entire bed is in

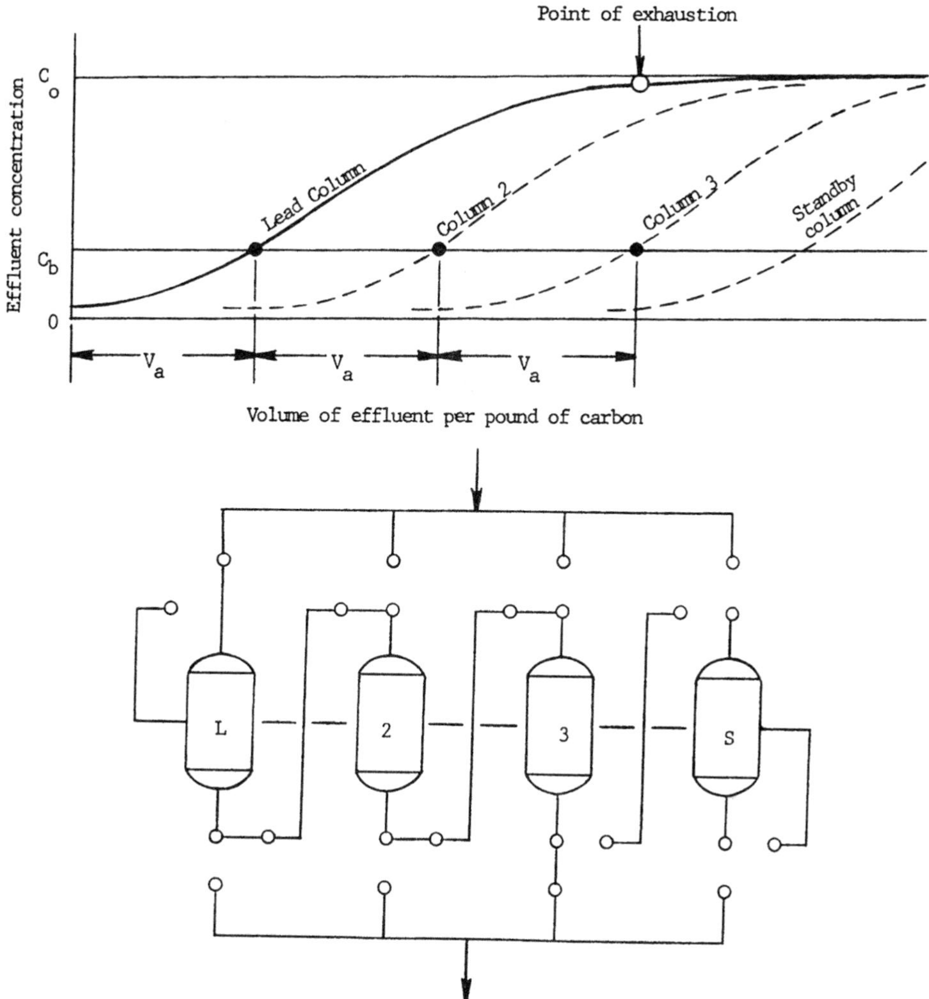

Fig. 10. Graphical approach to the design of adsorption column series.

service throughout the run. These conditions do not apply for a continuous operation of a multistage system, because the head section of the column for this type of system is removed and a fresh section is added to the tail end when the effluent quality is no longer acceptable. Figure 10 shows a multistage column operation. One testing approach is to simulate a multistage operation by repeatedly removing the head column when the concentration C_b in the last column exceeds the acceptable level and adding a new column containing fresh or regenerated carbon at the tail end. This will ensure a steady-state operation and a reliable design. The results of such an experiment can be presented in a BDST-type plot as shown in Fig. 11. In this plot the lines start at different slopes indicating a different velocity of the adsorption zone for different breakthrough level. However, after several cycles a steady-state condition is approached and the forward velocity of the adsorption zone is the same for all levels of breakthrough.

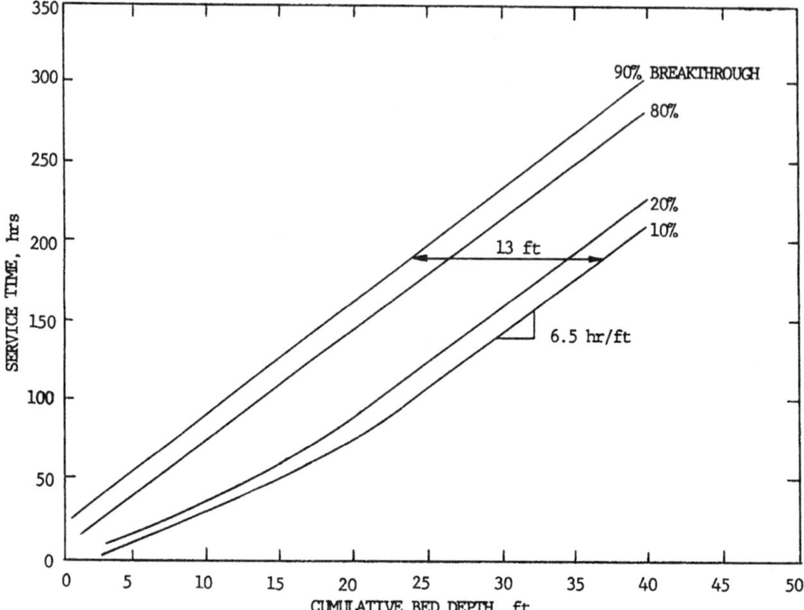

Fig. 11. Hypothetical BDST curves for a multistage GAC columns operation.

Deviation from this behavior can be expected if the carbon adsorption capacity changes on successive regenerations or by using different fresh carbons.

4.2. Pilot Plant and Laboratory Column Tests

Laboratory evaluation of GAC columns pilot-plant GAC column tests should be performed to obtain design data for full-scale plant construction. The pilot-plant studies are designed to obtain the following information:

1. Compare the performance of carbons under the same operating conditions.
2. Determine the minimum contact time required to achieve the desired effluent quality.
3. Determine headloss at various flow rates through different bed depths.
4. Check the backwash flow rate necessary for carbon bed expansion for cleaning purposes.
5. Determine the carbon dosage required.
6. Determine the effects of various methods of pretreatment of influent on carbon column performance, carbon dosage, and overall plant costs.

Pilot column tests are also described in the US EPA Manual (3). Procedures for laboratory GAC column studies are listed below:

1. Select carbons with suitable physical and chemical characteristics.
2. Conduct a preliminary batch adsorption study or isotherm study to determine the degree of treatment attainable by the various carbons and to select the carbon that is most effective.
3. Conduct a laboratory column test to obtain data for use in system design.
4. Analyze the laboratory data, using an appropriate design model.

Laboratory column tests should be conducted with the same GAC as in the full-scale plant. Care must be taken when filling the test columns with carbon to avoid trapping air that would produce channeling of flow and reduce contact between the carbon and liquid. To avoid air entrapment, slurring the carbon with boiling water and feeding the hot slurry to the column will be needed (9).

Laboratory GAC columns should be about the same height as the full-scale unit. Short columns need to be arranged in series to achieve the desired depth. Furthermore, the ratio of laboratory column diameter to carbon particle size shall be at least 25:1 in order to prevent wall effects from influencing test results (10). A column diameter is to be at least 4 in. to simulate the packing characteristics of a full-scale column and to minimize wall effects (9).

Laboratory column units should be operated at application rates that will yield empty-volume detention times of 25–50 min. Minimum application rates should not be less than 1 gpm/ft^2, are typical of full-scale units, and should be used in laboratory studies if column length is sufficient to provide the necessary detention time. Adsorption depends more on the detention time than the application rate. The detention times should be maintained constant during scale-up (9).

5. REGENERATION

In the regeneration process, the objective is to remove from the carbon porous structure the previously adsorbed materials, thus reinstituting its ability to adsorb impurities. Several modes of regeneration exist for granular activated carbon: steam treatment, solvent extraction, acid or base treatment, chemical oxidation, and thermal regeneration. Of these, thermal regeneration is most widely used in wastewater treatment.

Thermal regeneration of activated carbon proceeds by three major steps: vaporization of water near 100°C, baking of the adsorbate at temperature up to 800°C, and activation between temperatures of 800 and 950°C. During the baking step, the organic adsorbates are converted to a volatile fraction and free carbon residues on the surface. In the activation step, the carbon residues are oxidized by steam, carbon monoxide, or oxygen from the gas phase. The properties of the regenerated carbon are most affected by the activation step.

Multiple hearth furnaces are presently the most commonly used equipment for regenerating granular activated carbon from wastewater treatment. The spent carbon is fed to the top hearth and is then moved across each hearth by rotating arms. Hot gases generated by combustion at the lower hearths flow upward through the column countercurrent to the carbon. Steam is usually introduced at the lower hearths to serve as one of the oxidants for the activation step. Carbon losses during multiple hearth regeneration range from 5% to 10% of the feed. The details of multiple hearth GAC regeneration are well covered in the literature (11, 12).

Quality control of regenerated carbon is provided by measuring the apparent density (AD) of the spent and regenerated carbon. The AD of virgin carbon is about 0.48 g/cm^3, it may increase to 0.50 or 0.52 g/cm^3 or more as carbon becomes saturated with adsorbed organics. During the regeneration of spent carbon, the adsorbed organics are removed, and the weight of carbon decreases. If properly regenerated, the AD will return to 0.48 g/cm^3. The AD could be easily and rapidly determined by weighing a known volume of carbon. The AD of the regenerated carbon can be controlled by several parameters. The AD can be decreased by increasing the furnace hearth temperatures, decreasing the spent carbon feed rate, increasing the use of steam, or decreasing the rabble arm speed. The AD can be increased by decreasing the furnace hearth temperatures, increasing the spent carbon feed rate, increasing the use of steam, or increasing the rabble arm speed.

6. FACTORS AFFECTING GAC ADSORPTION

Adsorption of pollutants on granular activated carbon (GAC) is influenced by the characteristics of both the adsorbent and the adsorbate (solute). Factors that affect these characteristics are discussed below.

6.1. Adsorbent Characteristics

6.1.1. Surface Area and Particle Size

Surface area is one of the most important characteristics affecting the adsorptive capacity of an adsorbent. The adsorptive capacity of solid adsorbents generally is directly proportional to the specific surface area, i.e., adsorption increases with an increase of surface area. The surface area of a nonporous adsorbent increases rapidly as the particle size decreases, but for a highly porous adsorbent, such as activated carbon, much surface area resides in the internal micropore structure; the total surface area is relatively independent of particle size.

The pore size distribution in activated carbon indicates that micropores contribute a major portion of the specific surface, and many of these are in molecular dimensions. The surface area accessible to the adsorbate (solute) is determined by the size of solute molecules, i.e., large solute molecules will be excluded from the micropores. Therefore, the large volume of pores in the smaller size range (micropores) possesses relatively large surface area for adsorbing small molecules, while those in the larger size range (e.g., macropores) affects the adsorption of large molecules.

6.1.2. Chemistry of Surface

The presence of specific functional groups on the surface of the adsorbent also affects the adsorption process. The formation of polar surface groups of chemisorbed oxygen in activated carbon during the activation process affects the adsorptive capacity of many solutes. Surface oxides consisting of acidic functional groups reduce the adsorbing capacity of carbon for many organic solutes; on the other hand, surface oxides of carbonyl groups enhances adsorption of aromatic solutes such as phenol and naphthalene.

The different functional groups formed during carbon activation process are very much determined by the temperature. For instance, the presence of phenolic and lactone functional groups on carbon surface could result if carbon was activated to 400°C temperatures. On the other hand, carbons that preferentially adsorb acid usually are activated at much higher temperature, near 1000°C.

6.2. Adsorbate Characteristics

6.2.1. Solubility

Adsorption by activated carbon in aqueous solutions is influenced by several physiochemical properties of the organic solutes. Solubility probably is the most important property affecting adsorptive capacity. In general, a high solubility indicates a strong solute–solvent affinity, so the extent of adsorption will be low because of the necessity of breaking the strong solute–solvent bond before adsorption. As the molecule of solute becomes larger, its solubility decreases and the extent of adsorption increases as long as the molecule can enter the pores of adsorbent.

6.2.2. Chain Length

The solubility of organic compounds in water normally decreases with increasing chain length, thus resulting in an increase in adsorptive capacity. This is because the expulsion of increasing long-chain molecules permits an increasing number of water–water bonds to reform, which help the adsorption of the solute from the solvent (water). The relationship between adsorptive capacity and chain length is not linear; for example, a one-carbon increase for a hydrophilic parent molecule produces a 50% increase in adsorptive capacity, whereas a two-carbon increase in chain length from C6 to C8 produces only a net 10% increase.

6.2.3. Molecular Weight

The adsorptive capacity of activated carbon tends to increase with increasing molecular weight and size of the adsorbate molecule. This is particularly true for the cases where the rate of adsorption is primarily controlled by intraparticle transport. The rate dependence of adsorption on molecular weight and size is expected for rapidly agitated batch reactors.

6.2.4. Polarity

A major portion of the carbon surface is nonpolar or hydrophobic. In aqueous systems, nonpolar solutes are more rapidly and strongly adsorbed to activated carbon than polar solutes. A general rule for predicting the effect of solute polarity on adsorption is that a polar solute will prefer the phase which is more polar; namely, a polar solute will be strongly adsorbed from a nonpolar solvent by a polar adsorbent, but will much prefer a polar solvent (e.g., water) to a nonpolar adsorbent (e.g., activated carbon). It therefore follows that adsorption decreases as polarity of the solutes increases.

6.2.5. Degree of Ionization

Decreasing adsorption with increasing ionization has been observed for many types of organic acids. Adsorption for a series of substituted, benzoic acids appeared to vary inversely with the dissociation constant. For an amphoteric compound that has the capacity to be an acid or a base, the maximum adsorption occurs at the isoelectric point, where the compound is ionized with a net charge of zero. Therefore, adsorption is at a maximum for neutral species and at a minimum for charged species.

7. PERFORMANCE AND CASE STUDIES

Activated carbon columns have been applied for tertiary treatment following biological treatment and for the treatment of sewage and industrial wastewaters following a chemical precipitation step in a PCT process. In the former case, biochemical oxygen demand (BOD) has substantially been removed and little or no biological activity occurs in the carbon column. In the latter case, as soluble BOD passes through the columns, biological activity (aerobic or anaerobic) occurs in the columns and the BOD is lowered. The performance of PCT systems is shown in Table 4 (14).

Table 4
Results of Adsorption Isotherm Tests on Different Industrial Wastes

	Effluent COD			Effluent TOC			Effluent BOD		
	Raw COD	Removal		Raw TOC	Removal		Raw COD	Removal	
	(mg/L)	(mg/L)	(%)	(mg/L)	(mg/L)	(%)	(mg/L)	(mg/L)	(%)
Blue Plains Pilot Plant	320	16	95	100	8	92	150	6	96
Owosso, MI	250–350	24–30	≈ 91	—	—	—	140	8	84
Pomona, CA	321	19	94	—	—	—	120*	7.8	78.5**
Rosemount, MN (first year)	—	—	—	—	—	—	230	23	90
Rosemount, MN (last 3–4 mos)									
Battelle Pilot Plant at Westerly	527	42	92	—	—	—	240	26	89
CRSD Pilot Plant at Westerly	437	56	87	90	21	77	206	32	84

*Estimated based on BOD similar to COD removals across clarifier.
**Just around carbon columns.

Biological action in the carbon columns provides biological regeneration of the carbon, thus increasing the apparent capacity of the carbon. It was found that anaerobic activity could be inhibited and some nonsorbable compounds converted to sorbable compounds by ozonation prior to the carbon columns. The ozone increases the dissolved oxygen level in the carbon column influent thereby reducing the possibility of sulfide production and anaerobic biological activity. Ozone dosages at the Cleveland Westerly Plant range from 4 to 9 mg/L. The results of adsorption tests on various industrial wastewaters are shown in Table 5 (14).

A large number of large-scale systems are employed for industrial/municipal wastewater treatment, several of which will be listed and briefly described below.

1. Tertiary Treatment Plants (see Table 6): Carbon adsorption, when applied to well-treated secondary effluent, is capable of reducing COD to less than 10 mg/L and the BOD to under 2 mg/L. Removal efficiency may be in the range of 30–90% and varies with flow variations and different bed loadings. It should be noted that all of the plants listed in Table 6 treat large flows (1–100 MGD), require contact time of around 30 min, and presumably use thermal reactivation (3).
2. Physical Chemical Treatment (PCT) Plants (see Table 7) (3): PCT acts on the effluent of a primary treatment system to remove BOD, COD, suspended solids, and some color. Carbon adsorption in conjunction with lime clarification has been shown to be capable of achieving over 90% BOD and suspended solid reduction. Carbon loadings for PCT plants are in the range of 0.4–0.6 1b of COD/lb of carbon and 0.15–0.3 1b of TOC/lb of carbon. It is clear that both tertiary and PCT plants have biological activity taking place in the carbon beds, which increases the apparent carbon capacity.
3. Treatment of Industrial Waste Streams (see Table 8) (15): Quite frequently, segregated industrial waste streams are treated with activated carbon. Contaminants removed include BOD, TOC, phenol, color, cresol, polyethers, cyanide, acetic acid, and other, mostly organic

Table 5
Summary of PCT Pilot-Plant and Full-Scale Plant Performances

Type of Industry	Initial TOC (or phenol) (mg/L)	Initial color (OD)	Average reduction (%)	Carbon exhaustion rate (lb/1000 gal)
Food and kindred products	25–5300	—	90	0.8–345
Tobacco manufacturers	1030	—	97	58
Textile mill products	9–4670	—	93	1–246
	—	0.1–5.4	97	0.1–83
Apparels and allied products	390–875	—	75	12–43
Paper and allied	100–3500	—	90	3.2–156
products	—	1.4	94	3.7
printing, publishing and allied industries	34–170	—	98	4.3–4.6
Chemicals and allied	19–75500	—	85	0.7–2905
products	(0.1–5325)	—	99	1.7–185
	—	0.7–275	98	1.2–1328
Petroleum refinig and	36–4400	—	92	1.1–141
related industries	(7–270)		99	6–24
Rubber and miscellaneous plastic products	120–8375	—	95	5.2–164
Leather and leather products	115–9000	—	95	3–315
Stone, clay and glass products	12–8300	—	87	2.8–300
Primary metal industries	11–23000	—	90	0.5–1857
Fabricated metal products	73000	—	25	606

chemicals. The flows being treated are generally small in comparison with tertiary or PCT systems. Several systems treat less than 20,000 gpd, the lowest being 5,000 gpd. As indicated in Tab1e 8, thermal reactivation of the carbon is used on some systems with flows as low as roughly 10,000 gpd, although this method of regeneration is commonly used for the flows above about 60,000 gpd.

4. Carbon Treatment at Waste-Management Facilities: Two different waste-management facilities use carbon as part of their waste treatment facilities. Chem-Tro1 Pollution Services, Inc. (Model City, NY) uses a granular carbon contacting system to remove BOD and COD associated with higher-molecular-weight organic chemicals. Flows are around 100,000 gpd and influent contains several thousand ppm of TOC. At Hyon Waste Management Services, Inc. (Chicago, IL) carbon is added to the aqueous biological treatment system to assist in pollutant removal and to protect the system against shock from sudden increases in certain toxic pollutants.

8. ECONOMICS OF GRANULAR ACTIVATED CARBON SYSTEM

Operating costs of granular carbon waste-treatment systems reflect two major items of expense: carbon make-up and equipment amortization. Carbon make-up is necessary

Table 6
Tertiary Treatment Plants for Municipal Wastewaters

Site	Average plant capacity (MGD)	contractor type	Number of contractors in series	Contract Time (min)*	Hydraulic loading (gpm/ft^2)	Total carbon depth (ft)	Carbon size	Effluent requirements (oxygene demand) (mg/L)
1. Arlington, VA	30	Downflow gravity	1	38	2.9	15	8 × 30	BOD < 3
2. Colorado Spring, CO	3	Downflow	2	30	5	20	8 × 30	BOD < 2
3. Dallas, TX BOD<5 (by 1980)	100	Upflow packed	1	10	8	10	8 × 30	BOD < 10
4. Fairfax Cty, VA	36	Downflow gravity	1	36	3	15	8 × 30	BOD < 3
5. Los Angeles, CA	5**	Downflow gravity	2	50	4	26	8 × 30	COD < 12
6. Montgomery Cty, MD	60	Upflow packed	1	30	6.5	26	8 × 30	BOD < 1 COD < 10
7. Occoquan Cty, CA	18	Upflow packed	1	30	5.8	24	8 × 30	BOD < 1 COD < 10
8. Orange Cty, CA	15	Upflow Packed	1	30	5.8	24	8 × 30	COD < 30
9. Piscataway, MD	5	Downflow pressure	2	37	6.5	32	8 × 30	BOD < 5
10. St. Charles. MO	5.5	Downflow	1	30	3.7	15	8 × 30	
11. South Lake Tahoe, CA	7.5	Upflow packed	1	17	6.2	14	8 × 30	BOD < 5 COD < 30
12. Windhoek, South Africa	1.3	Downflow Pressure	2	30	3.8	15	12 × 40	COD < 10

* Empty bed (superficial) contact time for average plant flow.

Table 7
Physical Chemical Treatment Plants for Municipal Wastewaters

Site	Average plant capacity (MGD)	contractor type	Number of contractors in series	Contract Time (min)*	Hydraulic loading (gpm/ft^2)	Total carbon depth (ft)	Carbon size	Effluent requirements (oxygene demand) (mg/L)
1. Cortland, NY	10	Downflow pressure	1 or 2	30	4.3	17	8 × 30	TOD < 35
2. Cleveland Westerly, OH	50	Downflow pressure	1	35	3.7	17	8 × 30	BOD < 15
3. Fitchburg, MA	15	Downflow pressure	1	35	3.3	15.5	8 × 30	BOD < 10
4. Garland, TX	30**	Upflow downflow	2	30	2.5	10	8 × 30	BOD < 10
5. LeRoy, NY	1	Downflow pressure	2	27	7.3	26.8	12 × 40	BOD < 10
6. Niagra Falls, NY	48	Downflow gravity	1	20	3.3	9	8 × 30	COD < 112
7. Owosso, MI	6	Upflow packed	2	36	6.2	30	12 × 40	BOD < 7
8. Rosemount, MN	0.6	Upflow downflow	3 (max)	66 (max)	4.2	36 (max)	12 × 40	BOD < 10
9. Rocky River, OH	10	Downflow pressure	1	26	4.3	15	8 × 30	BOD < 15
10. Vallejo, CA	13	Upflow expanded	1	26	4.6	16	12 × 40	BOD < 45 (90% of time)

* Empty bed (superficial) contact time for average plant flow.
** 90 MGD ultimate capacity.

Table 8
Industrial Wastewater Treatment Using Granular Activated Carbon Process

Industry Location	Design Flow Rate (1000 gpd)	Organic contaninants	Pretreatment	Contact time (min)	Adsorption System: Adsorber type	Adsorption System: Carbon reactivation
1. Carpet Mill British Columbia	50	Dyes	Screens	—	Moving bed	None
2. Textile Mill, VA	60	Dyes	Filtration	57	Moving bed	None
3. Oil Refinery, CA	4200	COD	Equalization Oil Flotation	60	Gravity beds in parallel	Multiple hearth furnace
4. Oil Refinery, PA	2200	BOD	Equalization Oil Flotation Filtration	—	Moving bed	Multiple hearth furnace
5. Detergent, NJ	15	Xylene Alcbhols TOC	None	540	Downflow beds in series	Multiple hearth
6. Chemicals, AL	500	Phenolics Resin	Chemical Clarification	173	Moving beds	Multiple hearth
7. Resins, NY	22	Xylene Phenolics	Chemical Clarification	30	Downflow beds in series	Rotary kiln
8. Herbicide, OR	150	Chlorophenols Cresol	None	105	Upflow beds in series	Multiple hearth
9. Chemicals, NY	15	Phenol COD	Equalization	200	Downflow beds in series	None
10. Chemicals, TX	1500	Nitrated Aromatics	Activated Sludge Filtration	40	Moving beds	Rotary kiln
11. Chemicals, NJ	100	Polyols	Equalization Clarification		Moving bed	Multiple hearth furnace
12. Explosives Switzerland	5	Nltrated Phenols	Equalization	150	Downflow beds in series	None
13. Pharmaceuticals Switzerland	25	Phenol	Equalization pH adjusted Settling	90	Downflow beds in series	None
14. Insecticide England		Chlorophenol	Equalization Clarification		Downflow beds in series	Rotary kiln
15. Wood Chemicals, MS	3000	TOC	pH adjustment	50	Moving beds	Multiple hearth furnace
16. Dyestuffs, PA	1500	Color TOC	Equalization Clarification Filtration	50	Moving beds	Multiple hearth furnace

Table 9
Optirmzation of Investment and Operatmg Costs, Granular Carbon Treatment Mumcipal Secondary Waste (10 MGD)

	Adsorber configuration			
	Single stage	Two-stage series	Three-stage series	Four-stage series
Basis				
Residence time, min.	36	36	36	36
Carbon, lb/MG	1,350	950	870	830
Adsorber diameter, ft	50	40	40	40
Bed depth, ft	16	27	27	24
Flow rate, (gal/min)/ft^2	4	4	4	4
Capital Cost, thousands of dollars				
Adsorbers	128	102	140	170
Reactivation unit	120	120	105	105
Pumps, piping	222	280	371	432
Electrical, instruments	75	88	101	114
Land, building, off-sites	55	65	73	80
40% Contingencies and profit	240	266	316	360
Subtotal	840	931	1,106	1,261
Granular activated carbon	566	424	375	353
TOTAL CAPITAL COST, $1000	1,406	1,355	1,481	1,614
Operating cost, thousands of dollars/yr				
Amortization, 7.5%	105	102	111	121
Carbon make-up, 5%	69	49	45	42
Labor	27	27	27	27
Fuel	6	5	4	4
Power	18	31	44	57
Supervision, maintenance, analytical, supplies	31	33	38	42
TOTAL OPERATING COST	256	247	269	293
Cost, ¢/1000 gal	7.0	6.8	7.4	8.1

because approx 5% of the granular activated carbon is lost during each cycle of use and reactivation. To achieve longer contact time and greater countercurrent stages, additional equipment (adsorber, pumps, piping) and carbon inventory are required. Increased pumping costs would also be expected.

System optimization therefore requires an analysis of all factors as they vary with system design. An analysis of the various capital and operating cost variables for several absorber systems is shown in Table 9 for a 10 MGD municipal waste-treatment plant. It can be seen that a two-stage series system offers somewhat lower total capital and operating costs than the other systems studied. This lower cost system was then applied to 1 and 100 MGD plants. These results are shown along with the 10 MGD plant in Table 10 (16). A considerable size effect is indicated in both capital and operating costs.

Table 10
Summary of Investment and Operatmg Cost, Granular Carbon Treatment (Two-Stage Systems) Municipal Secondary Waste

	Million Gallons Per Day (MGD)		
	100	10	1
Basis			
Residence time, min	36	36	36
Carbon, lb/MG	950	950	950
Adsorber diameter, ft	72	40	12
Bed depth, ft	27	27	27
Flow rate, (gal/min)/sq.ft^2	4	4	4
Capital Cost, thousands of dollars			
Adsorbers	600	102	28
Reactivation unit	330	102	60
Pumps, piping	775	280	83
Electrical, Instruments	175	88	30
Land, buildings, off-sites	120	65	20
40% Contingency and profit	800	266	88
Subtotal	2,800	931	309
Granular activated carbon	3,410	424	43
TOTAL CAPITAL COST, $1000	6,210	1,355	352
Cost, $1000/MGD	62	135	352
Operating Costs, thousands of dollars/yr			
Amortization, 7.5%	465	102	26
Carbon make-up, 5%	486	49	5
Labor	60	27	9
Fuel	30	5	1
Power	150	31	2
Supervision, maintenance, analytical	105	33	13
TOTAL OPERATING COST	1,296	247	56
Cost, ¢/1,000 gal.	3.5	6.8	15.3

Figures 12–27 present cost data on components of granular carbon systems as developed by Culp/Wesner/Culp Consulting Engineers under USEPA contract 68-03-2516 (61). Costs are based on *Engineering News-Record* (ENR) Construction Cost Index (CCI) for October 1978 of 265.38. In 2002, the average CCI for United States had an average value of 608.65 (62). To update the construction costs use the following equation:

$$\text{Update cost} = \text{Total construction cost from curve}\left(\frac{\text{Current CCI}}{265.38}\right)$$

In 2002, the cost multiplier would be:

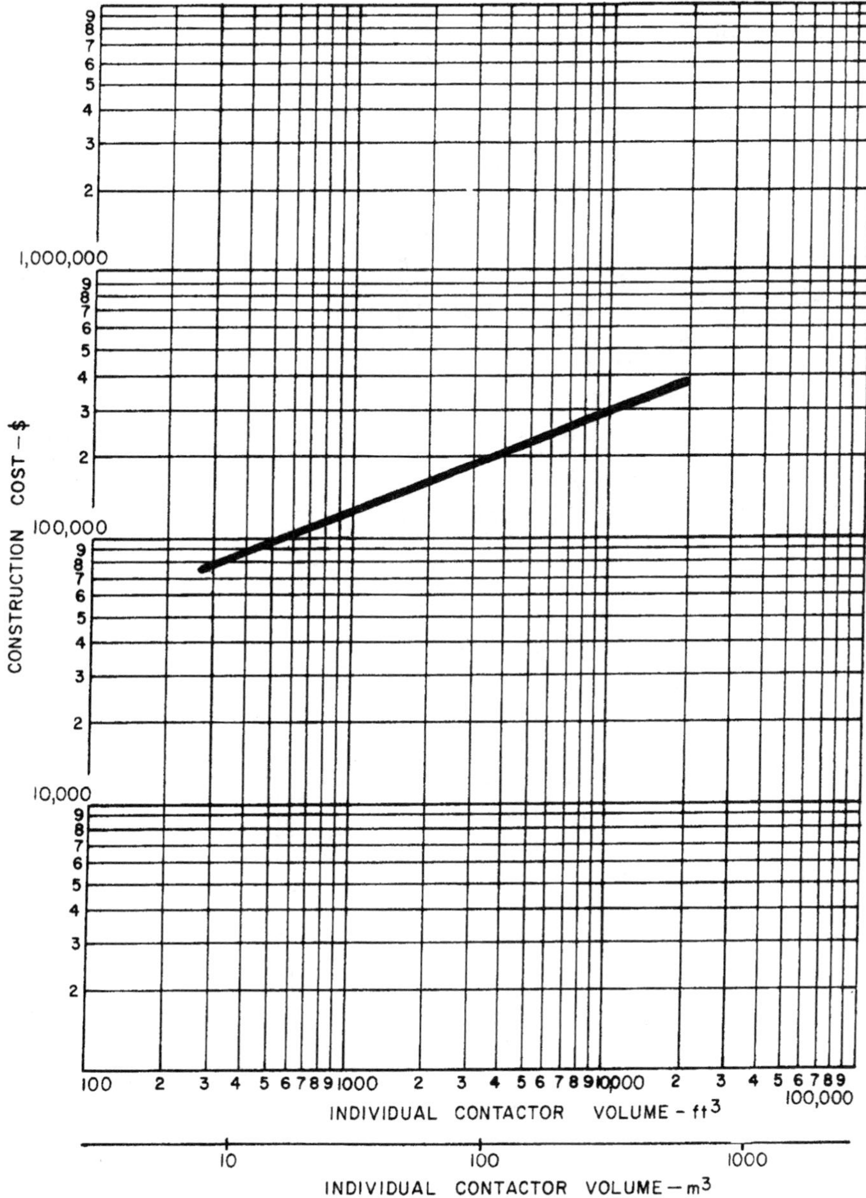

Fig. 12. Construction cost for gravity carbon contactor—concrete construction.

$$\text{Updated cost} = \text{Total construction cost from curve}\left(\frac{608.65}{265.38}\right)$$

$$\text{Updated cost} = \text{Total construction cost from curve} \times 2.29$$

Energy costs were based on \$0.03/kW-h for electricity, \$0.0013/scf for natural gas and \$0.45/gal for diesel fuel. More detailed cost data for granular carbon system are available in the US EPA Technology Transfer Manual on carbon adsorption.

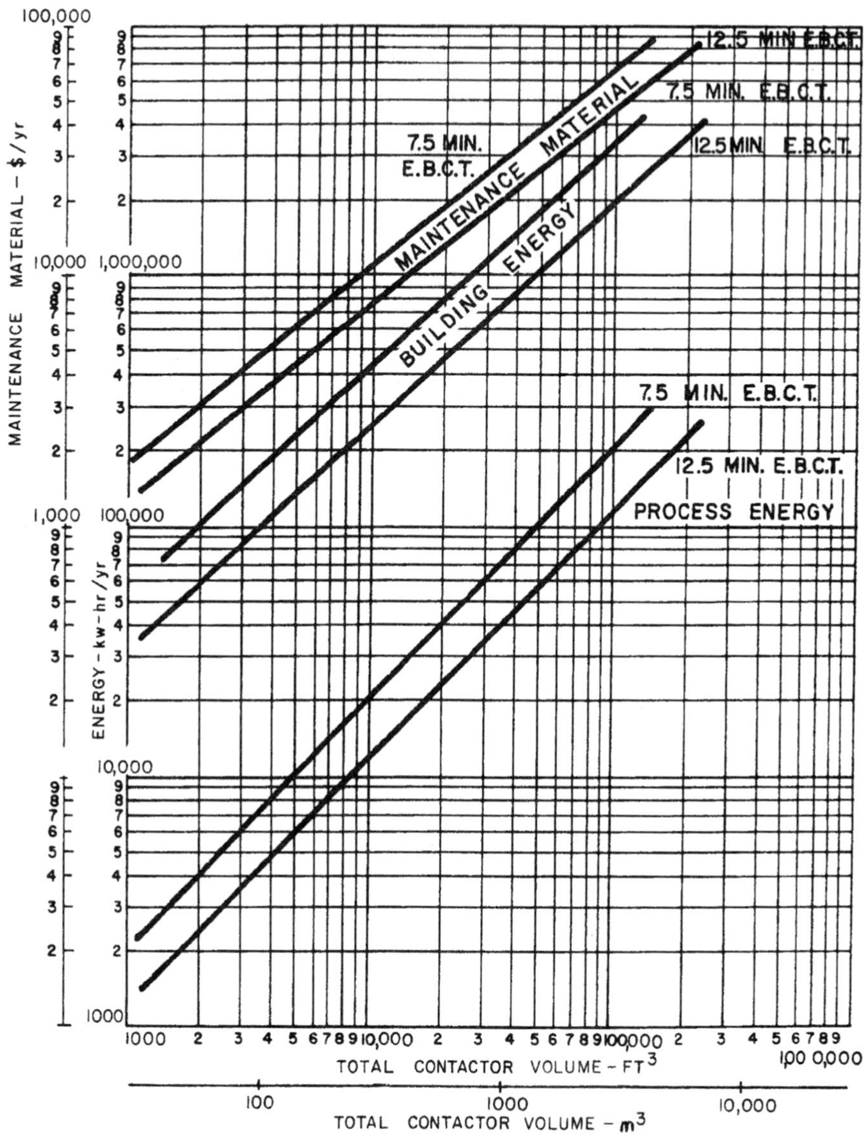

Fig. 13. Operation and maintenance requirements for concrete gravity carbon contactors—building energy, process energy, and maintenance material needed for 7.5 and 12.5 min empty bed contact times.

9. DESIGN EXAMPLES

9.1. *Example 1*

A laboratory GAC adsorption column, 4 in. in diameter and 12 ft deep, is operated at a flow rate of 20 gal/h. Calculate the following:

1. The application rate in gpm/ft^2.
2. The detention time, t, in the column.
3. The volumetric flow rate, Q, in bed volumes per hour, at this detention time.

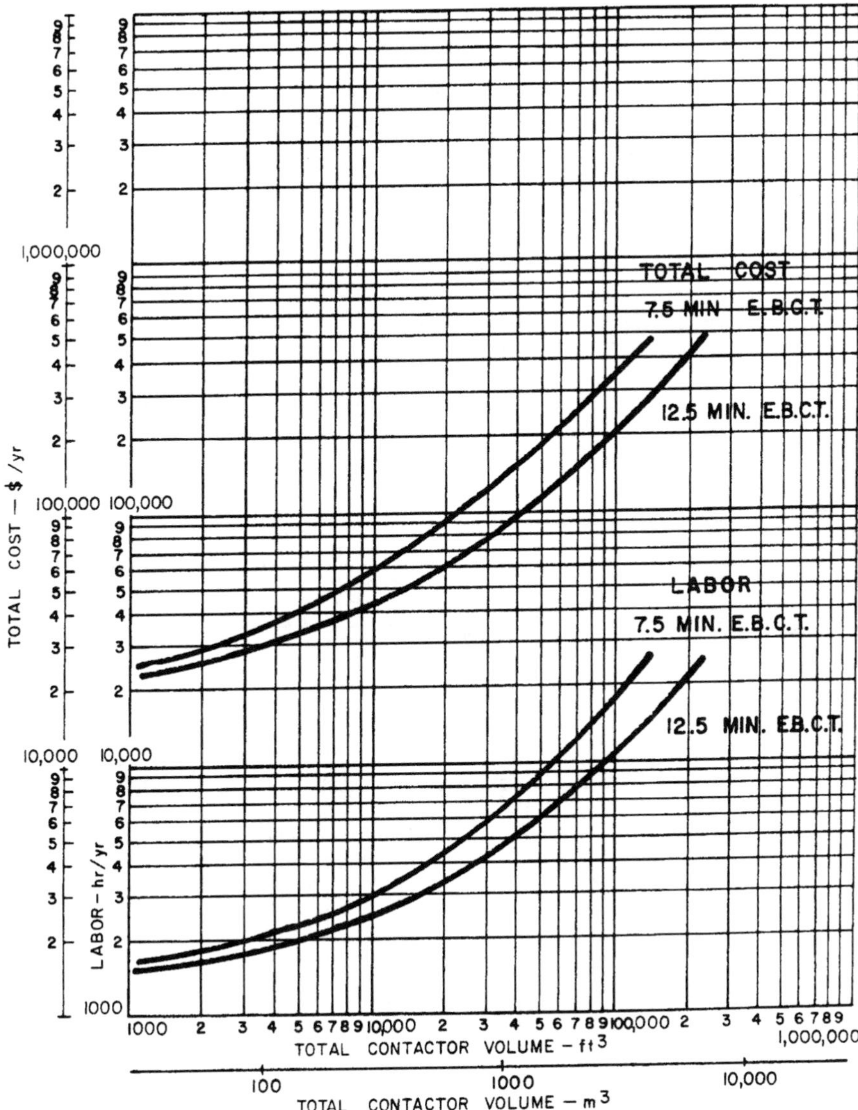

Fig. 14. Operation and maintenance requirements for concrete gravity carbon contactors—labor and total cost needed for 7.5 and 12.5 min empty bed contact time.

4. The application rate that would yield the same detention time in a production column that was 20 ft in diameter and 30 ft tall.

Solution

1. Compute application rate:

$$\text{column area} = \pi(4)^2/4 = 12.57\ \text{in.}^2 = 0.087\ \text{ft}^2$$

$$\text{application rate} = 20\ \text{gal/h}/\left[(60\ \text{min/h})(0.087\ \text{ft}^2)\right] = 3.83\ \text{gpm/ft}^2$$

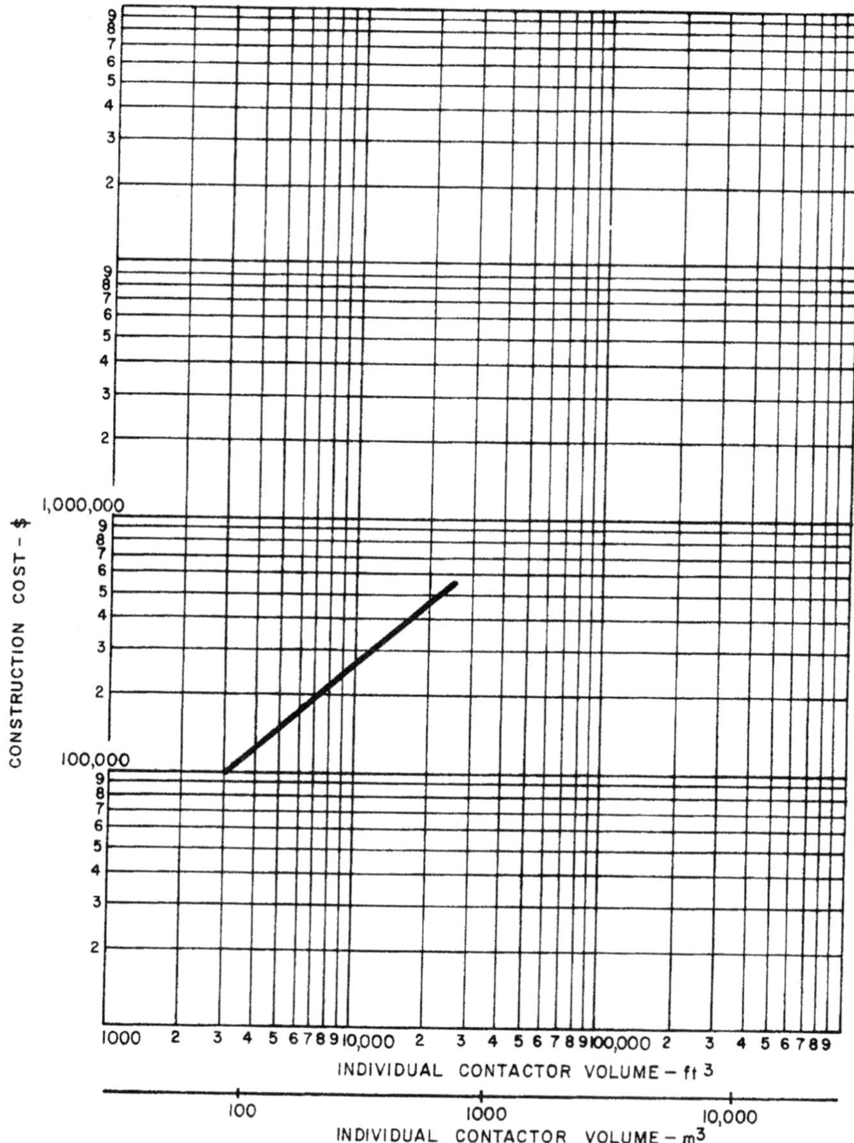

Fig. 15. Construction cost for steel gravity carbon contactor.

2. Compute detention time, t, in the column:

$$\text{linear velocity} = 3.83\,\text{gpm/ft}^2/7.48\ \text{gal/ft}^3 = 0.512\ \text{ft/min}$$

$$t = 12\ \text{ft}/0.512\ \text{ft/min} = 23.44\ \text{min}$$

3. Compute volumetric flow rate, Q (bed volumes/hour). Fractional void volumes for granular carbon columns normally range from 0.40 to 0.55 and is assumed to average 0.50 (Fornwalt and Hutchin (8)). Therefore,

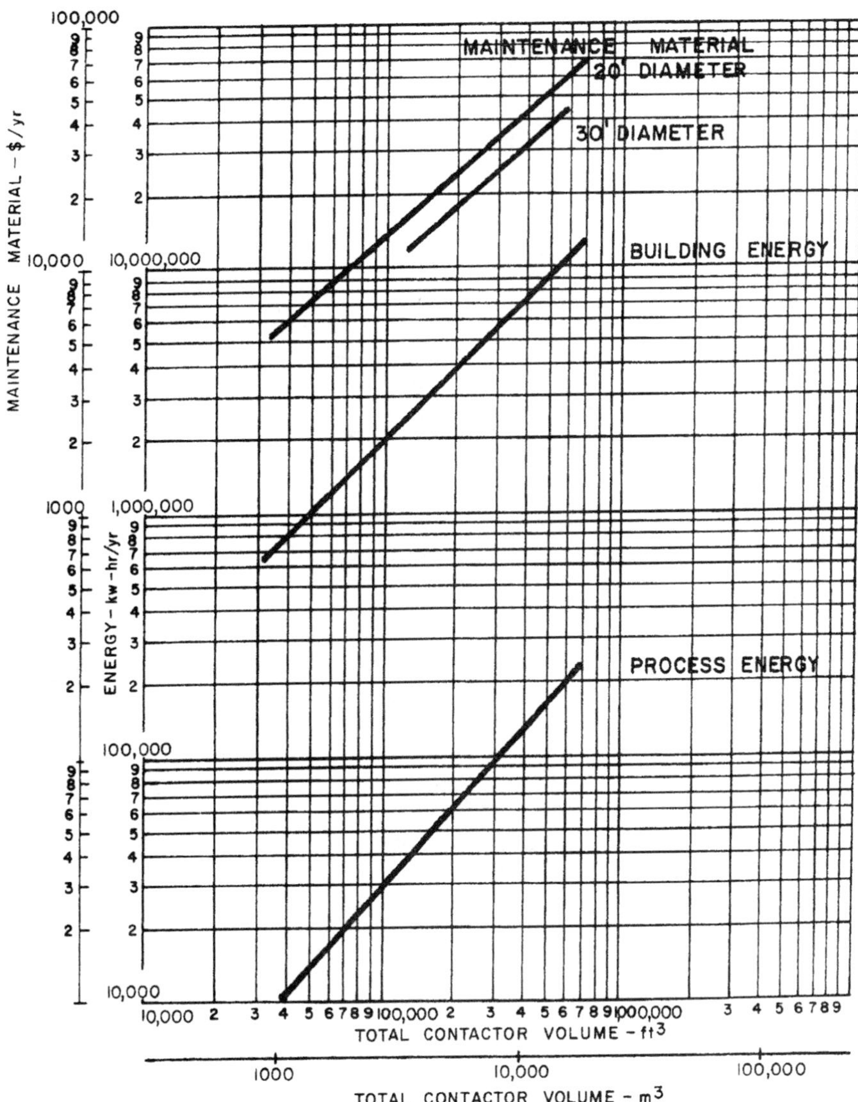

Fig. 16. Operation and maintenance requirement for steel gravity carbon contactors—building energy, process energy, and maintenance material.

$$Q = 0.50/t$$

$$Q = 0.50/(23.44 \text{ min}/60 \text{ min/h}) = 1.28 \text{ bed volumes/h}$$

4. Compute application rate for a 20 ft diameter by 30 ft deep column having same detention time as in the laboratory column:

$$\text{required linear velocity} = 30 \text{ ft}/23.44 \text{ min} = 1.28 \text{ ft/min}$$

$$\text{application rate} = (1.28 \text{ ft/min})(7.48 \text{ gal/ft}^3) = 9.57 \text{ gpm/ft}^2$$

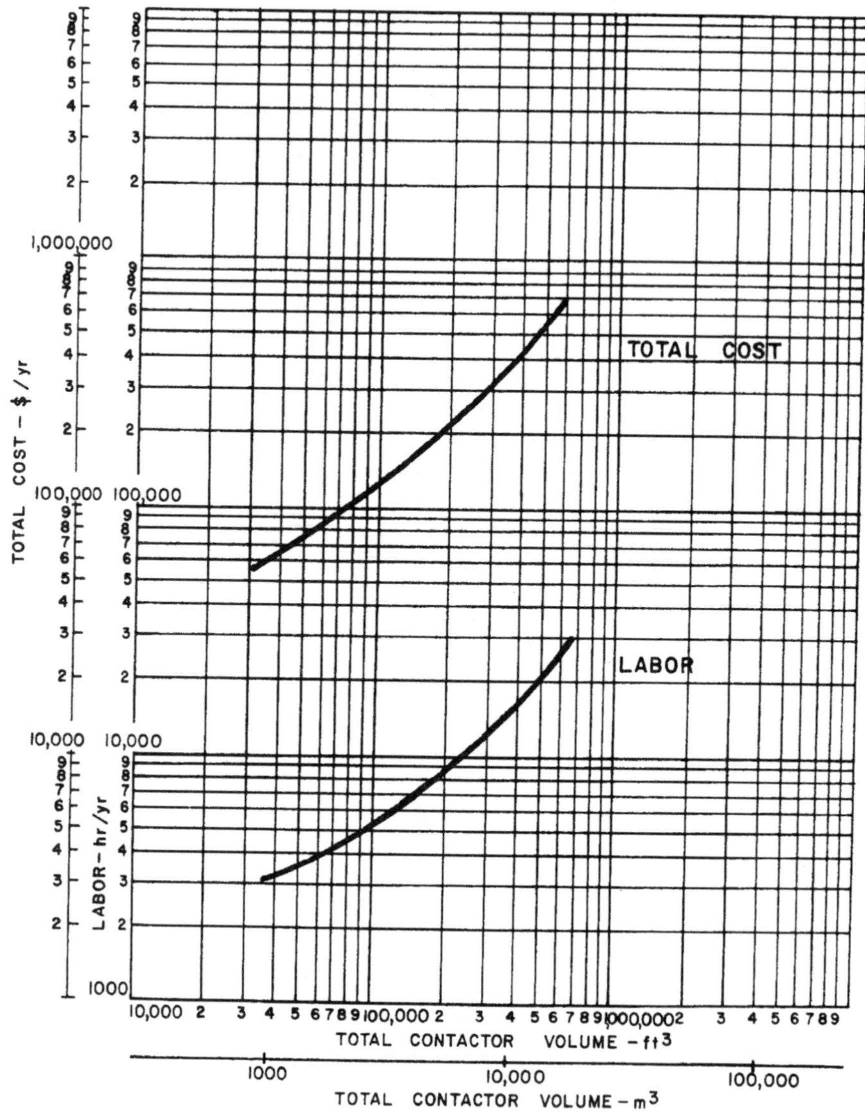

Fig. 17. Operation and maintenance requirements for steel gravity carbon contactors—labor and total cost.

9.2. *Example 2*

Wastewater containing 20 mg/L phenol is fed to a fixed bed of activated carbon at a superficial flow rate (F_m) of 0.2 m^3/m^2 (min). The height of carbon in the column is 4 m. The breakthrough point and exhaustion point concentrations are taken as 2 mg/L phenol and 18 mg/L phenol, respectively. The bulk density of carbon in the bed is 0.6 g/mL. Assume that the mass transfer $ka = 15\ min^{-1}$ and that the equilibrium isotherm curve is shown in Fig. 28. It was determined that $V_x - V_b$ is equal to 16.30 m^3.

Calculate:

1. The shape of the breakthrough curve.

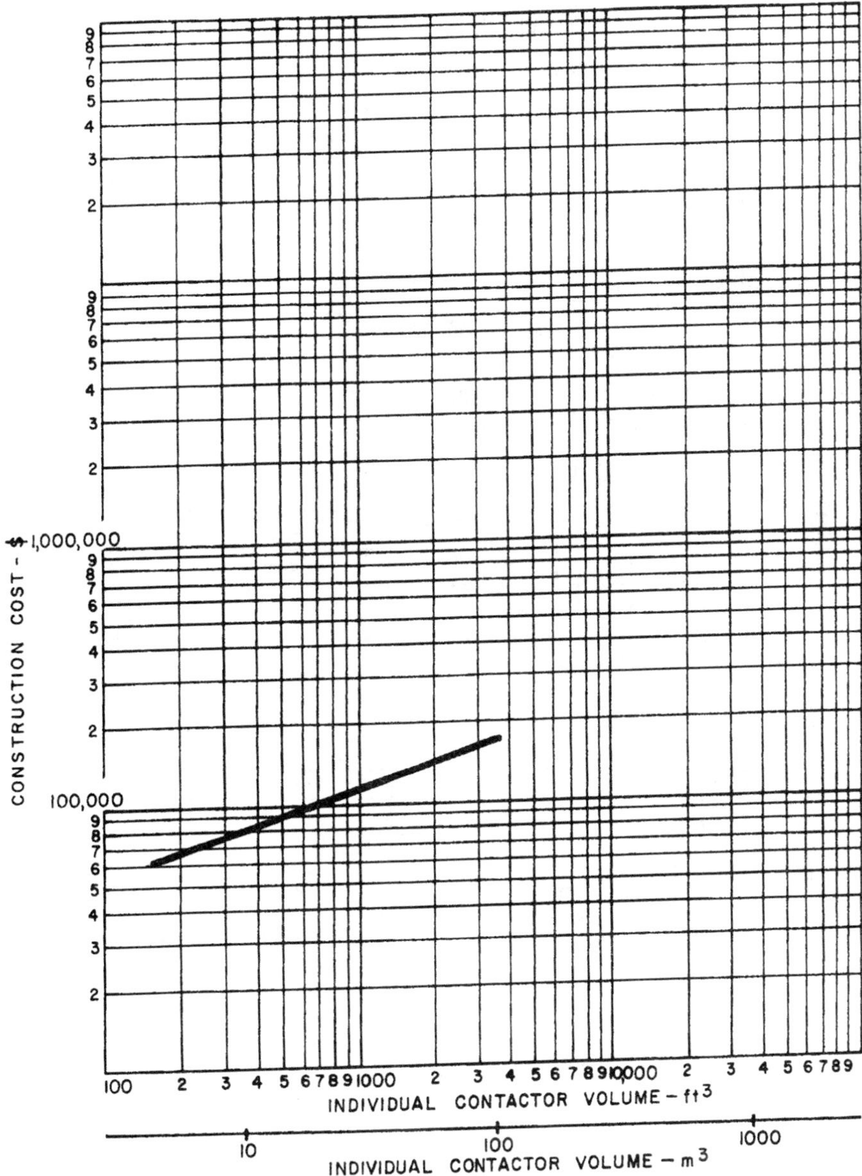

Fig. 18. Construction cost for pressure carbon contactors.

2. The degree of saturation at the breakthrough point.
3. The time and volume of water treated at breakthrough point.

Solution

1. The fresh carbon contains no phenol and the initial effluent also contains no phenol. The operating line thus passes from the origin to the equilibrium curve at 20 mg/L phenol as shown in Fig. 29. The following table presents the calculations to develop the breakthrough curve.

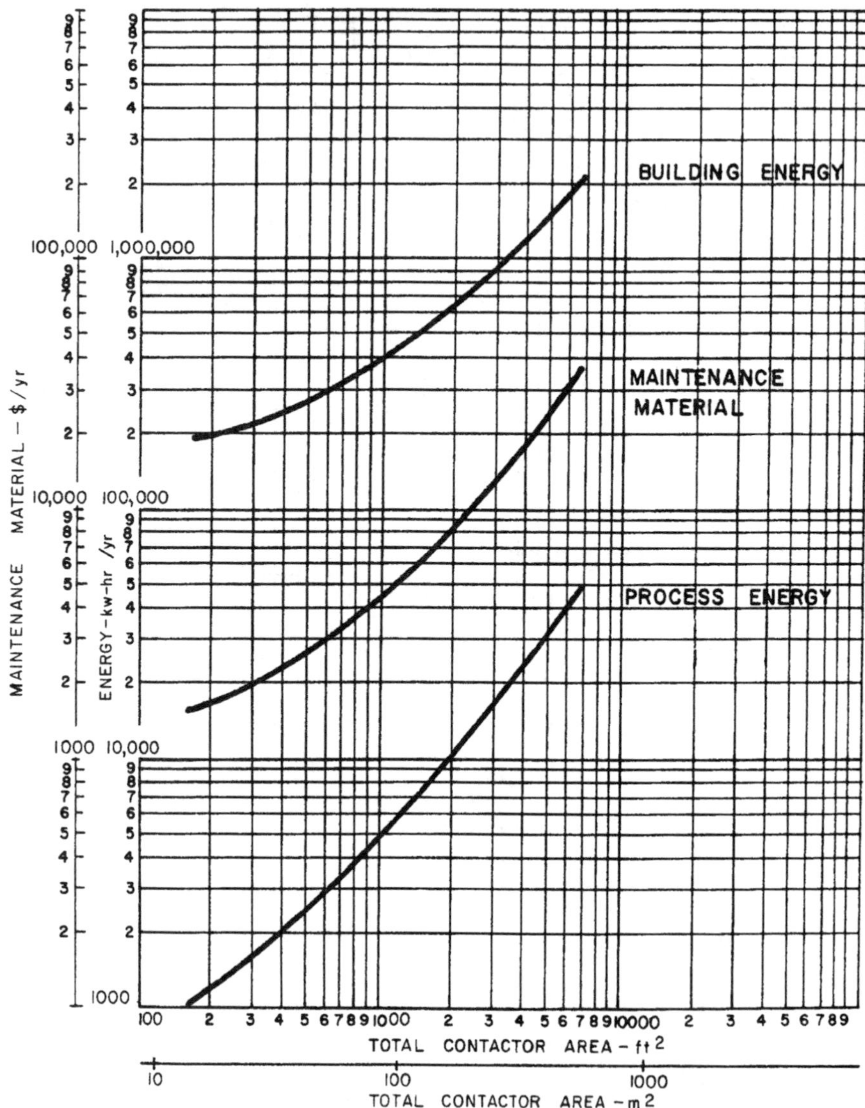

Fig. 19. Operation and maintenance requirements for pressure carbon contactors—building, energy, and maintenance material.

C (1)	C_e (2)	$\dfrac{1}{C-C_e}$ (3)	$\int_2^C \dfrac{dC}{C-C_e}$ (4)	$\dfrac{V-V_b}{V_x-V_b}$ (5)	$\dfrac{C}{C_0}$ (6)
2	1.4	1.67	0	0	0.10
4	2.9	0.909	2.40	0.147	0.20
6	4.6	0.714	4.05	0.248	0.30
8	6.6	0.714	5.45	0.334	0.40
10	8.6	0.714	6.85	0.420	0.50
12	10.7	0.769	8.35	0.511	0.60
14	13.0	1.00	10.15	0.623	0.70
16	15.3	1.43	12.55	0.770	0.80
18	17.6	2.50	16.30	1.00	0.90

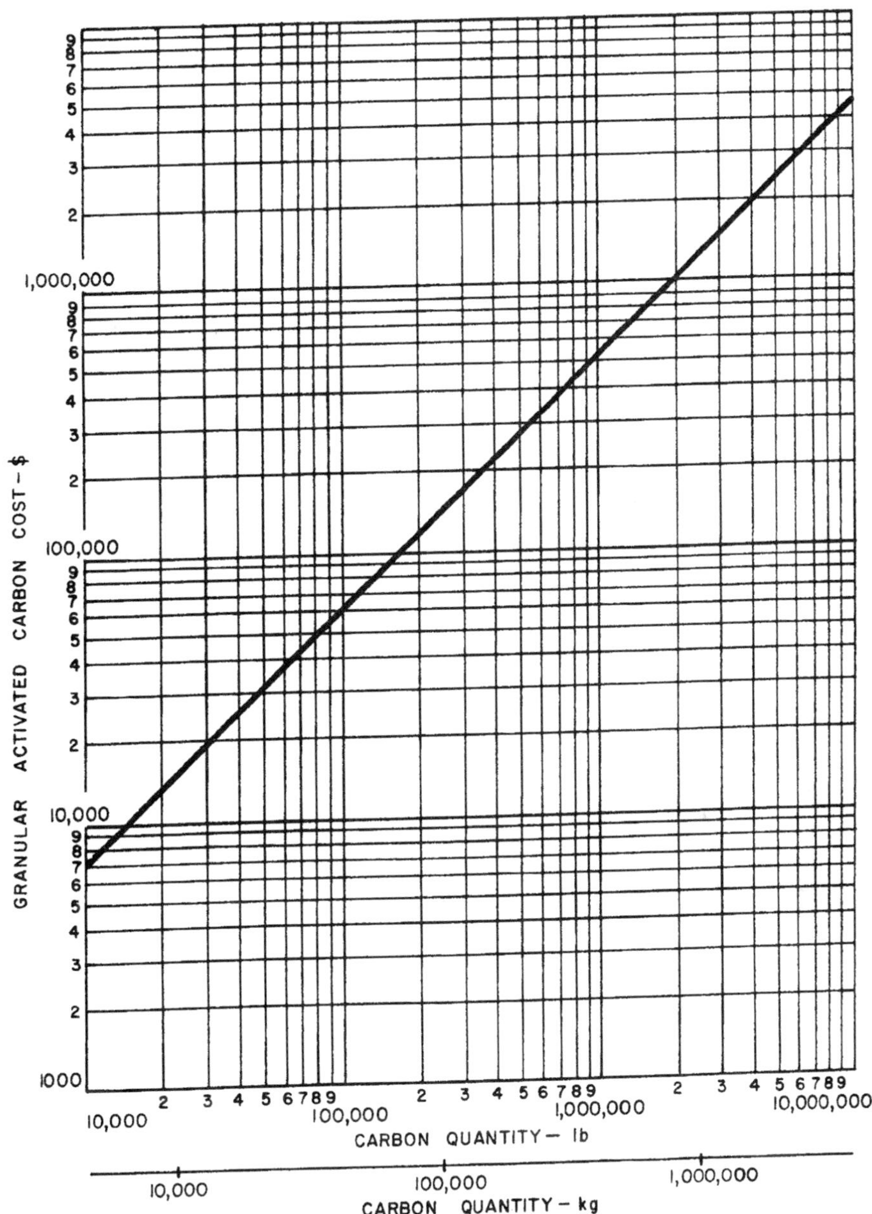

Fig. 20. Operation and maintenance requirements for pressure carbon contactors—labor and total cost.

Column 1 lists values of C between C_b and C_x. Column 2 lists corresponding values of C obtained from Fig. 30 from the equilibrium isotherm curve at the same value of $\frac{x}{m}$. Column 4 is obtained by plotting Column 3 vs Column 1 and determine graphically the area beteen $C_b = 2$ and each value of C. Column 5 is the ratio of each value in column 4 to 16.30 or $(V_x - V_b)$. The breakthrough curve is a plot of Column 6 vs Column 5 and is shown in Fig. 30.

2. The fractional capacity of the carbon at the breakthrough point is determine from Eqs. (7) and (8)

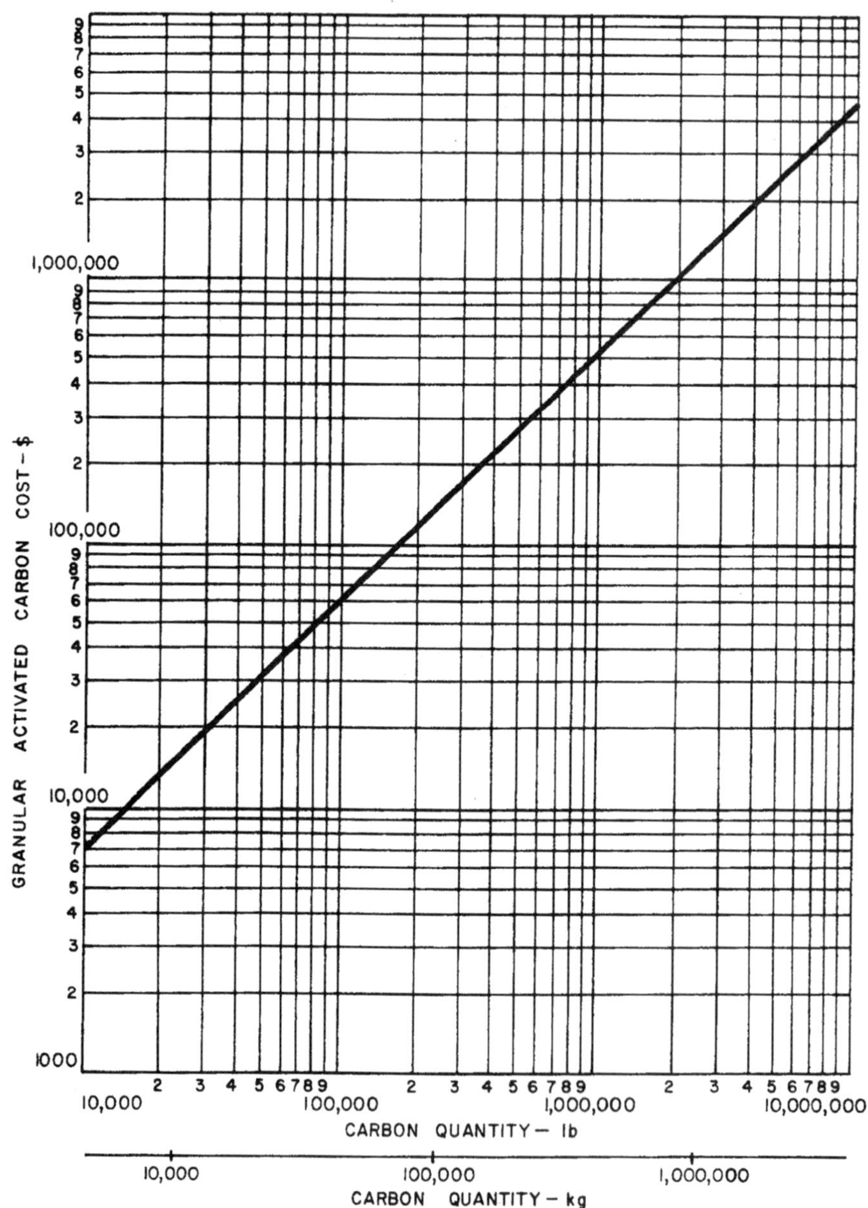

Fig. 21. Material cost for granular activated carbon, including cost for purchase, delivery, and placement.

$$f = \frac{P_s}{P_{tc}} = \frac{\int_{V_b}^{V_x} (C_0 - C) dV}{(V_x - V_b) C_0} \tag{7}$$

$$f = \int_0^1 \left(1 - \frac{C}{C_0}\right) d \frac{(V - V_b)}{(V_x - V_b)} \tag{8}$$

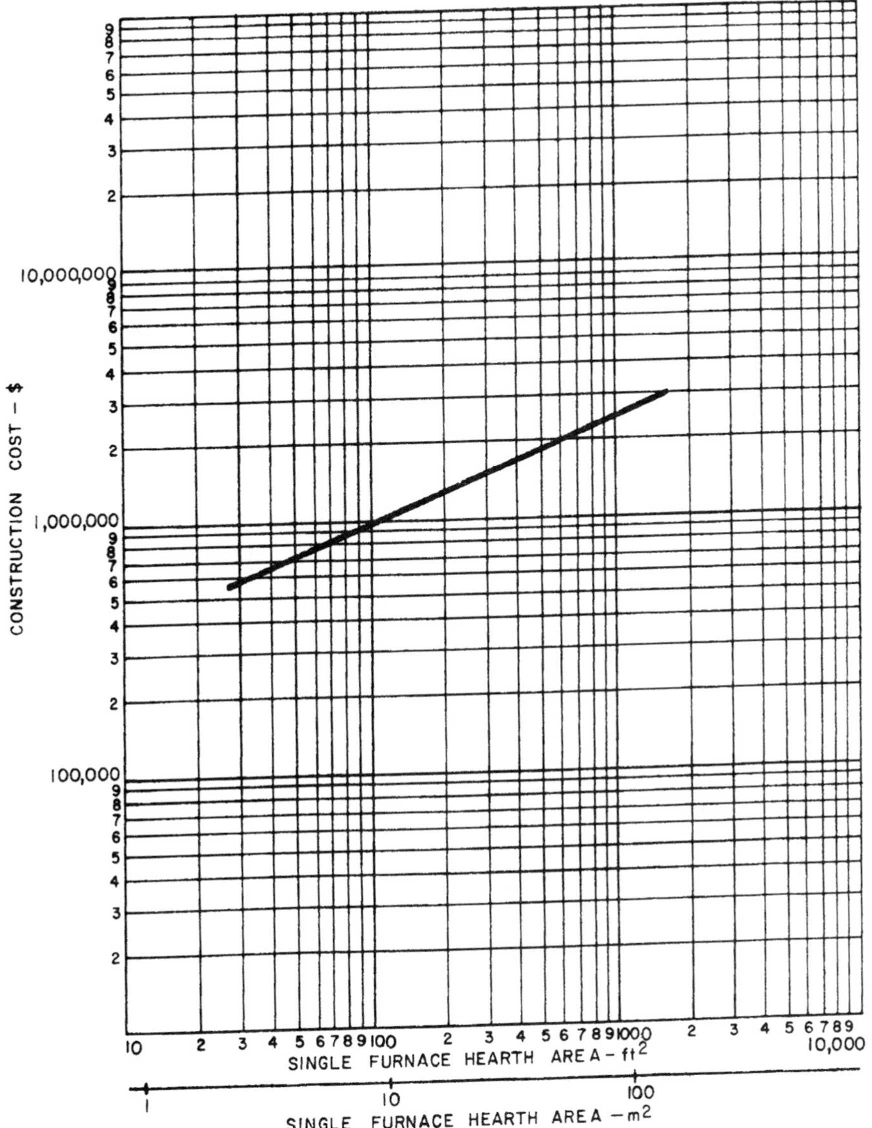

Fig. 22. Construction cost for multiple hearth granular carbon regeneration.

The integral is the area between the breakthrough curve and $C/C_0 = 1$ in Fig. 30. From graphical integration, the area is found to be

$$f = 0.46$$

The depth of the active adsorption zone is calculated from Eq. (13)

$$\frac{kay}{F_m} = \int_{C_b}^{C} \frac{dC}{C - C_e} \tag{14}$$

$$\frac{kay}{F_m} = \int_{2}^{18} \frac{dC}{C - C_e} = 16.3 \tag{14}$$

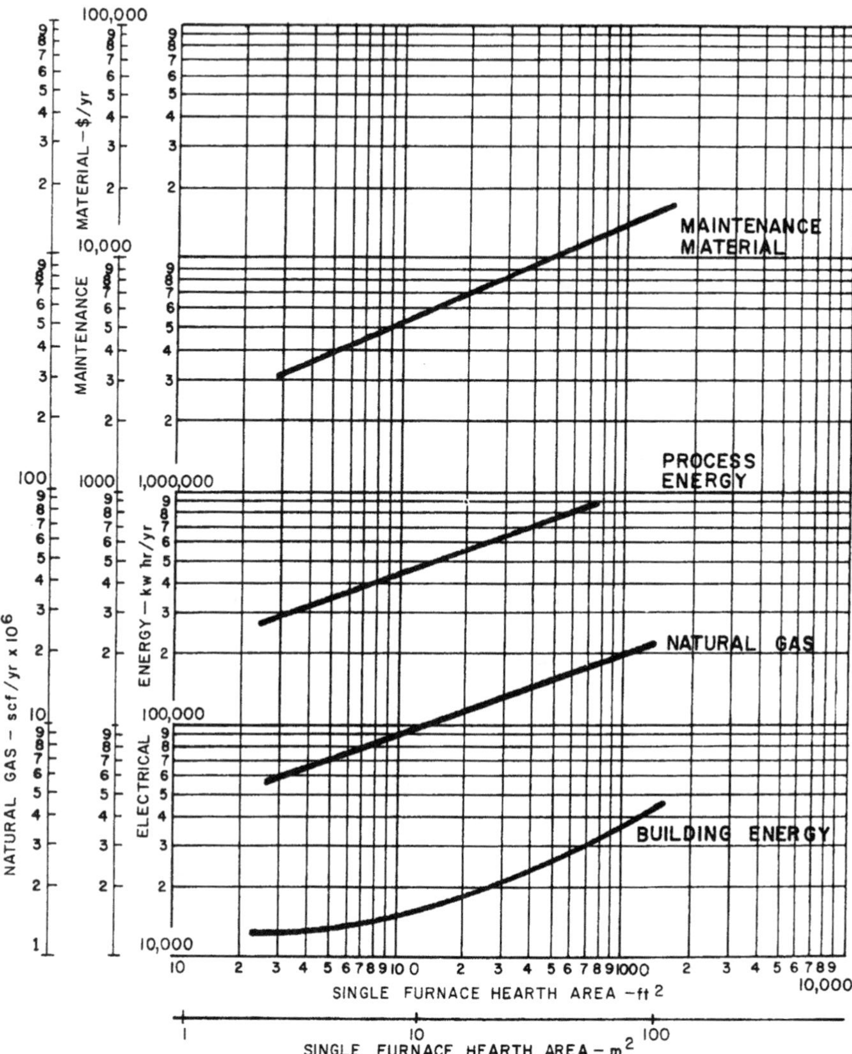

Fig. 23. Operation and maintenance requirements for multiple hearth granular carbon regeneration—building energy, process energy, natural gas, and maintenance material.

Since

$$F_m = 0.\ 0.2\ \text{m}^3\ /\ \text{m}^2\ (\text{min}) = 0.2\ \text{m/min}$$

$$ka = 15\ \text{min}^{-1}$$

therefore

$$\delta = 16.3(0.2)/15 = 0.217\ \text{m}$$

The degree of saturation at the breakthrough point can be evaluated by

$$M/M_s = \left[y - (f\delta)\right]/y$$

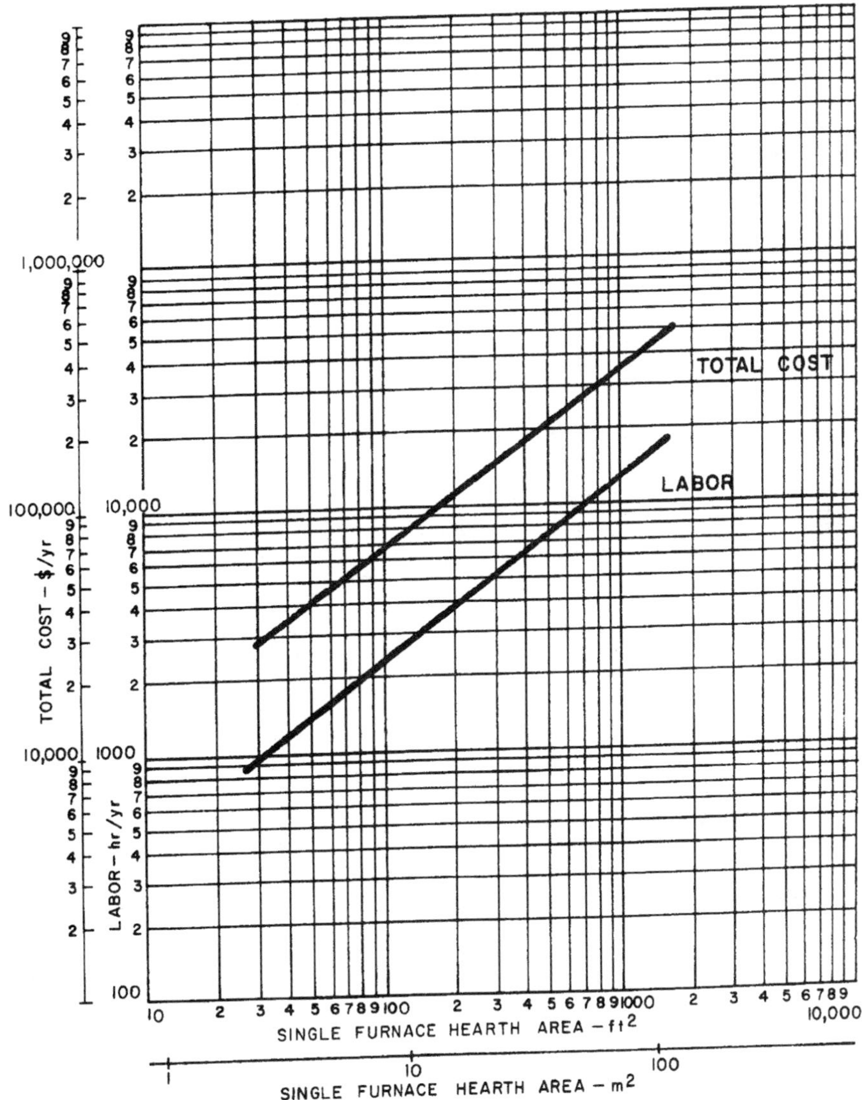

Fig. 24. Operation and maintenance requirements for multiple hearth granular carbon regeneration—labor and total cost.

where M = total amount of solute adsorbed at breakthrough point, M_s = total amount of solute adsorbed at exhaustion, y = total column depth, f = fractional capacity of adsorbent, δ= depth of active adsorption zone, $M/M_s = [4-(0.46)(0.217)]/4$, and $M/M_s = 0.975 = 97.5\%$.

3. Weight of carbon in column per m^2

$$\rho = (0.6 \text{ g/mL})(1000 \text{ kg mL/g m}^3) = 600 \text{ kg/m}^3$$

$$\rho\ y = (600)(4) = 2400 \text{ kg/m}^2$$

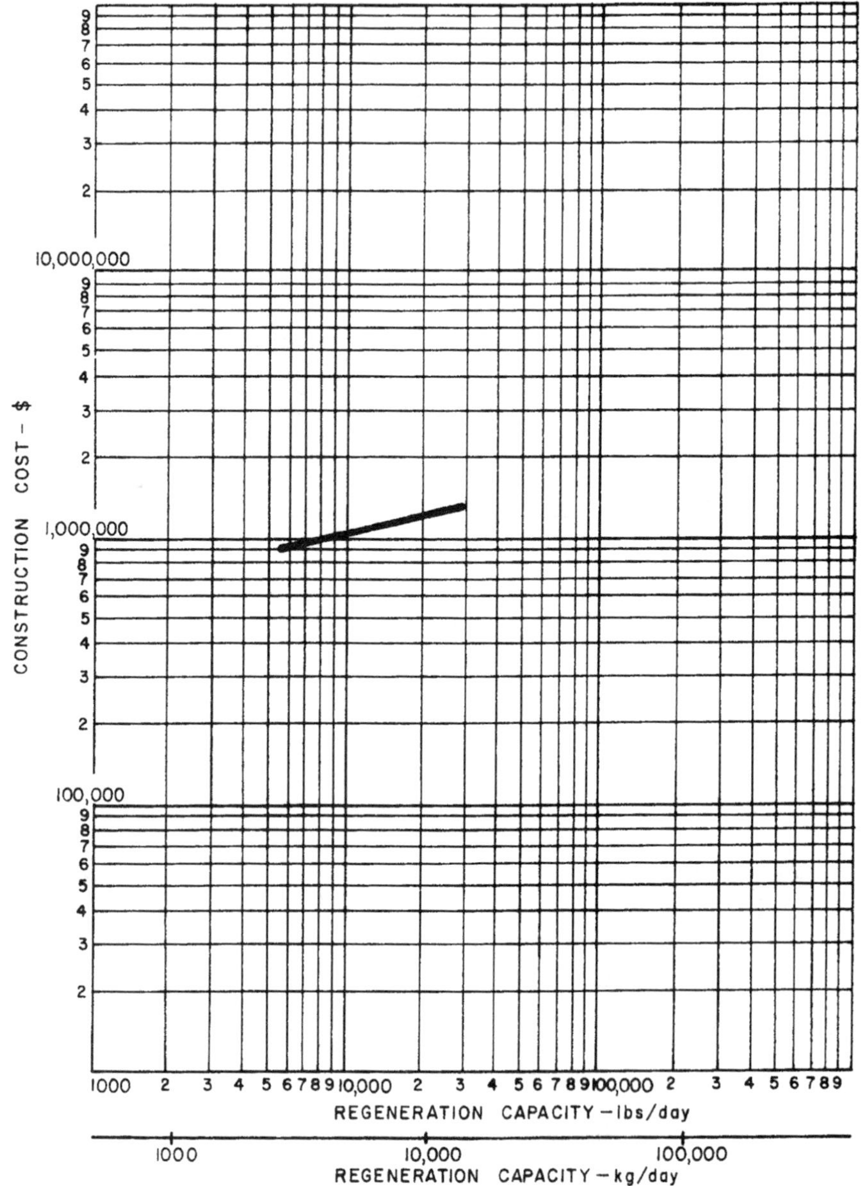

Fig. 25. Construction cost for granular carbon regeneration—fluid bed process.

At influent concentration of phenol equal to

$$C_0 = 20\,\text{mg/L}$$

$$q_0 = 0.098 \text{ mg phenol/mg carbon}$$

Carbon is 97.5% saturated at the breakpoint.
Thus, phenol adsorbed at breakthrough point

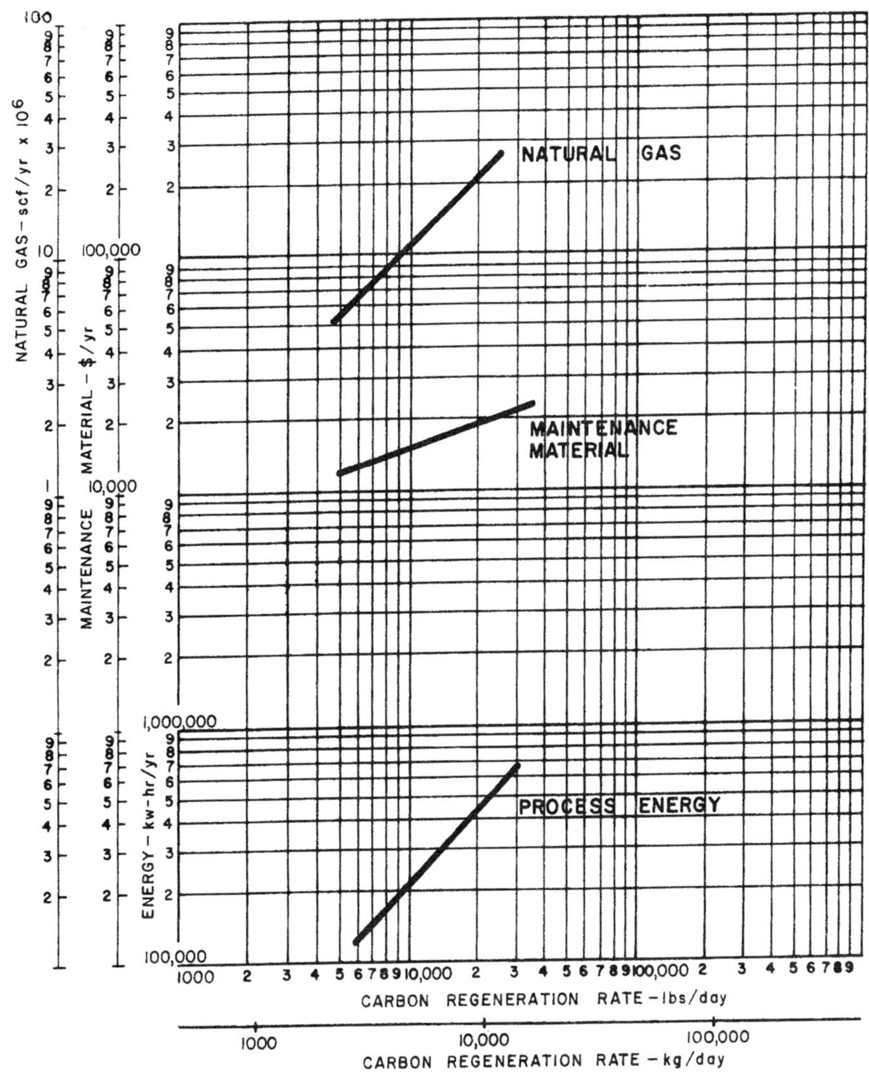

Fig. 26. Operation and maintenance requirements for granular carbon regeneration, fluid bed process—natural gas, process energy and maintenance material.

$$\text{Phenol absorbed} = \text{kg phenol/m}^2 = 2400(0.098)(0.975)$$

$$\text{Phenol absorbed} = 229 \text{ kg/m}^2$$

$$\text{Water flow rate} = \left[0.2 \text{ m}^3/\text{m}^2(\text{min})\right]\left[1000 \text{ kg/m}^3\right] = 200 \text{ kg/m}^2 (\text{min})$$

$$\text{Influent phenol} = 20 \text{ mg/L} = 20 \text{ ppm}$$

$$\text{Influent phenol} = 20 \times 10^6 \text{ kg/kg water}$$

Thus, flow rate of phenol to column:

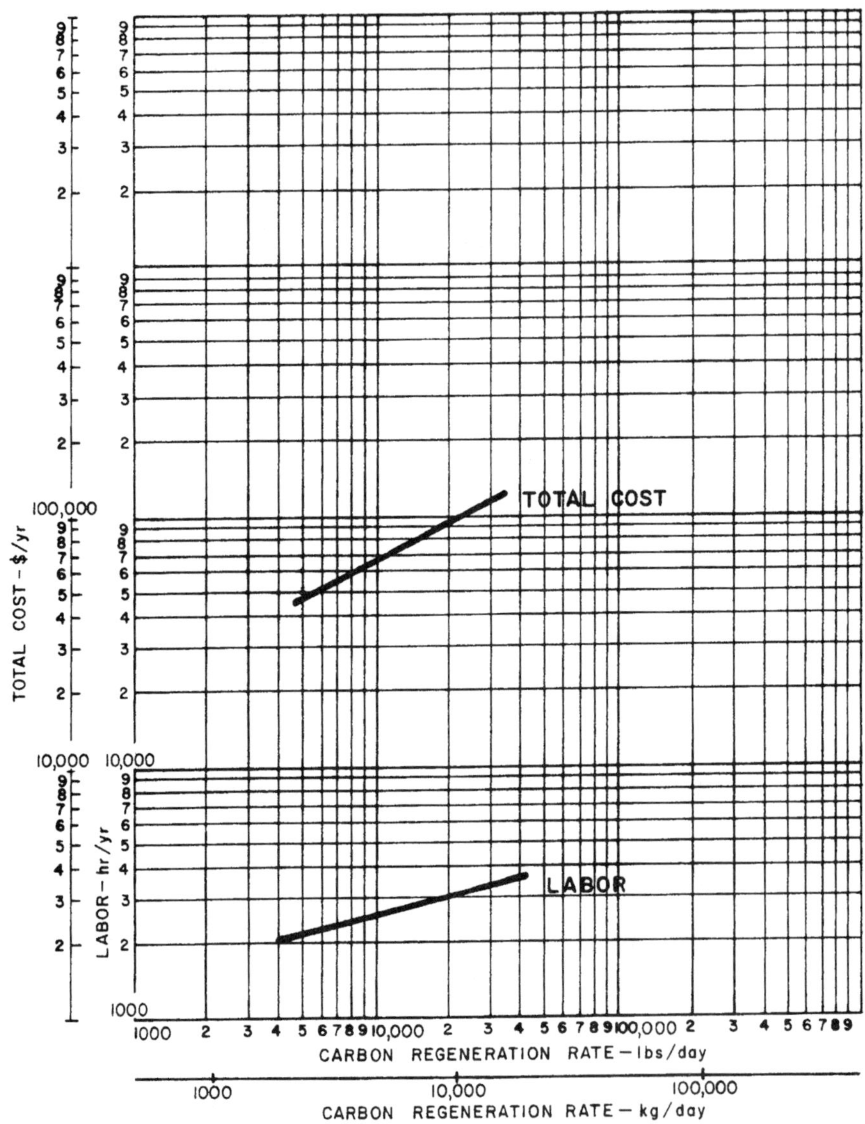

Fig. 27. Operation and maintenance requirements for granular carbon regeneration, fluid bed process—labor and total cost.

$$= 200 \times 20 \times 10^{-6}$$
$$= 4 \times 10^{-3} \text{ kg phenol/m}^2 \text{(min)}$$

$$\text{Breakthrough time} = \text{phenol adsorbed/phenol flow r}$$
$$= 229/4 \times 10^{-3}$$
$$= 5.7 \times 10^{4} \text{ min}$$
$$= 40 \text{ d}$$

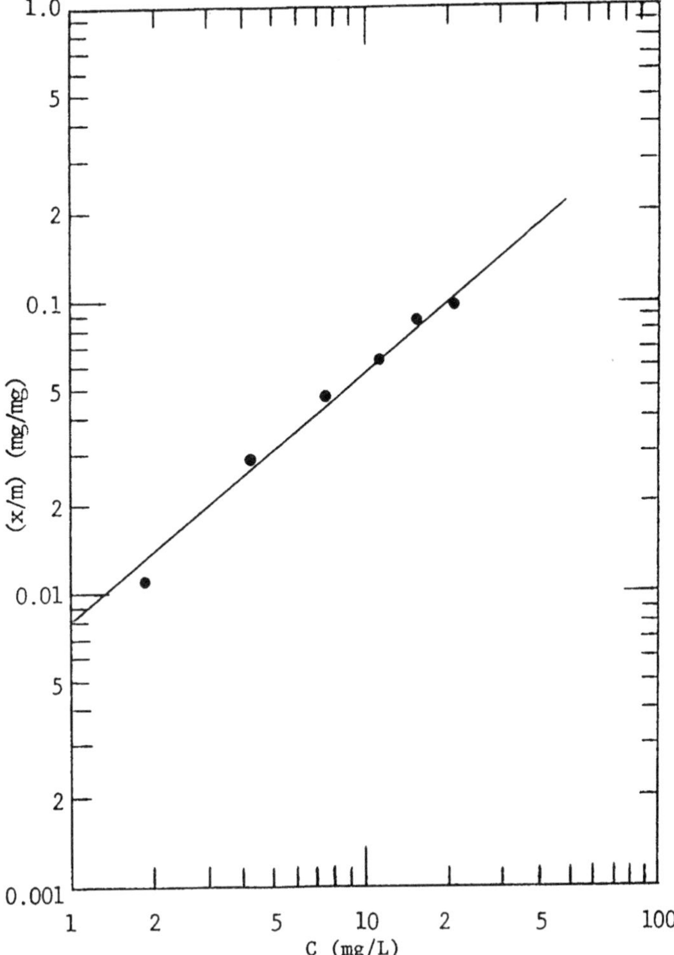

Fig. 28. Determination of Freundlich constant.

Volume of water treated at breakthrough point

$$= \left[0.2 \text{ m}^3/\text{m}^2 \text{ (min)}\right]\left[5.7 \times 10^4 \text{ min}\right]$$
$$= 1.14 \times 10^4 \text{ m}^3/\text{m}^2$$

Therefore, each square meter of cross-sectional area of adsorber can treat 11,400 m^3 of wastewater before the breakthrough point at 40 d. The concentration in the effluent will then rise above 2 mg/L phenol and the carbon in the column should be regenerated.

9.3. *Example 3*

A series of continuous bench-scale carbon column studies was conducted in the laboratory, using 1-in. diameter columns. The wastewater contains 12 mg/L of ABS synthetic detergent and is to be reduced to a final concentration of 0.5 mg/L by carbon adsorption. Laboratory data shown below include bed depths, flow rates, the throughput volume and time associated with a breakthrough concentration of 0.5 mg/L.

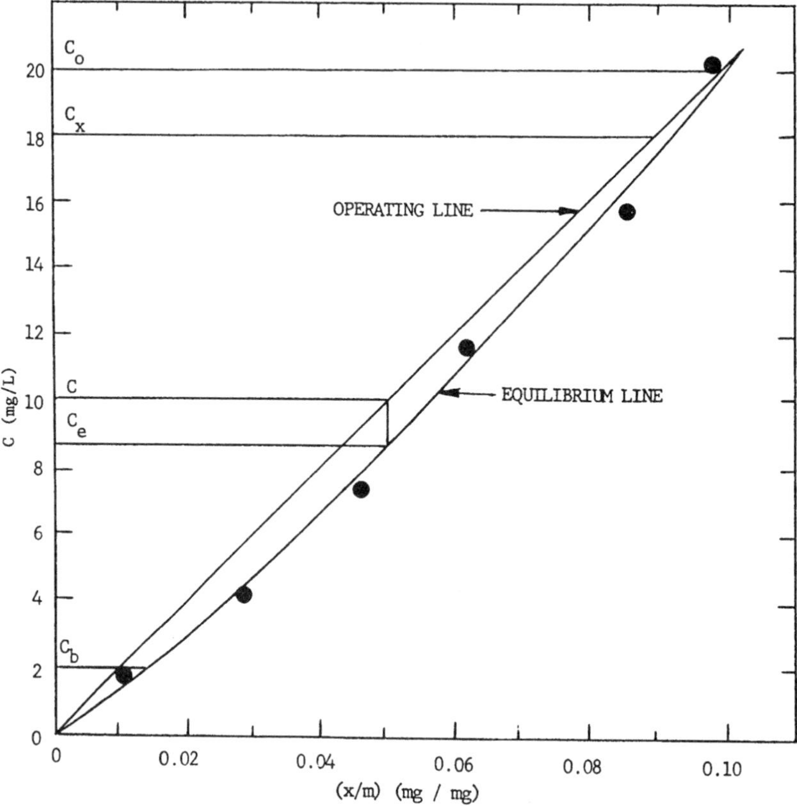

Fig. 29. Operating and equilibrium curves.

Determine values for the Bohart–Adams constants K and N_0, and find the value of x_0 for each flow rate.

Flow rate (gpm/ft^2)	Bed depth (ft)	Throughput volume (gal)	Time (h)
2.5	3.0	820	1,000
2.5	5.0	1,810	2,215
2.5	7.0	2,790	3,410
4.5	3.0	590	400
4.5	5.0	1,450	990
4.5	9.0	3,180	2,160
8.0	5.0	1,145	440
8.0	9.0	2,775	1,060
8.0	12.0	3,990	1,525

Solution

1. The Bohart–Adams equation is shown below:

$$t = \left(\frac{N_0}{C_0 V}\right)X - \frac{1}{KC_0}\ln\left(\frac{C_0}{C_b} - 1\right)$$

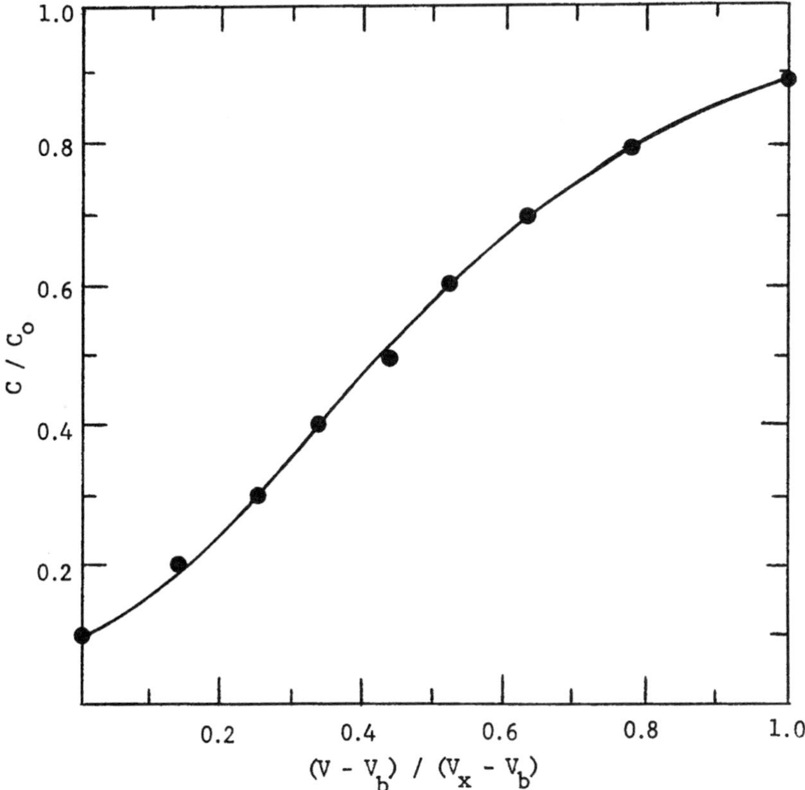

Fig. 30. Breakthrough curve.

where C_0 = initial concentration of solute (lb/ft^3), C_b= desired concentration of solute at breakthrough (lb/ft^3), K = rate constant (ft^3 liquid/lb carbon-h), N_0 = adsorptive capacity of carbon (lb/ft^3), x = depth of carbon bed (ft), V = linear flow of velocity of feed to carbon bed (ft/h, gpm/ft^2), and t = service time of column under above conditions (h). This equation is of the form $y = mx + b$, a straight line on arithmetic paper. A plot of service time (t) to breakthrough vs bed depth should yield a line with a slope, m, equal to $\left(\frac{N_0}{C_0 V}\right)$ and an intercept, b, of

$$-\frac{1}{KC_0}\ln\left(\frac{C_0}{C_b}-1\right)$$

The values of N_0 and K can be determined from the graph.

2. The data are plotted as in Fig. 31 and the slope and intercept of each line is determined.
3. Compute N_0 , using the slopes of the lines.

$$\text{Slope} = \frac{N_0}{C_0 V}$$

where V = linear velocity of flow and C = initial ABS concentration (lb/ft^3).
For 2.5 gpm/ft^2:

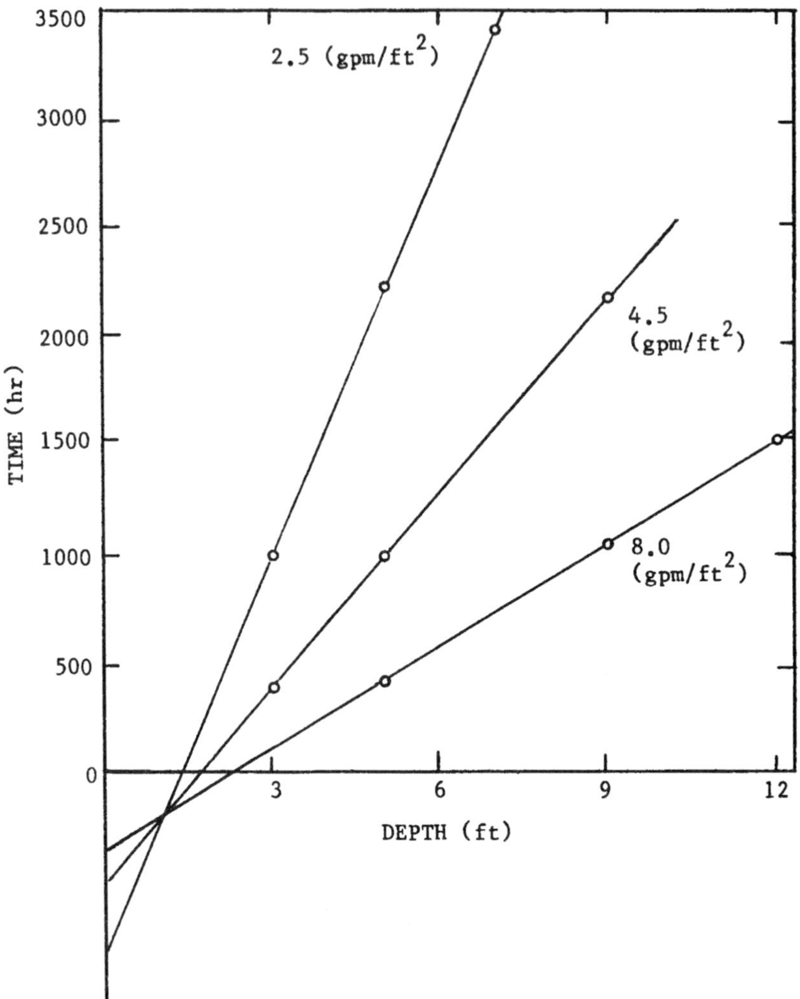

Fig. 31. Plot to determine slope and intercept for Bohart—Adams equation.

$$v = \left(2.5\,\frac{\text{gpm}}{\text{ft}^2}\right)\left(\frac{\text{ft}^3}{7.48\text{ gal}}\right)\left(60\,\frac{\text{min}}{\text{h}}\right) = 19.23$$

$$C_0 = (12\text{ mg/L})\left(10^{-3}\,\frac{\text{g}}{\text{mg}}\right)\left(2.205\times10^{-3}\,\frac{\text{lb}}{\text{g}}\right)\left(28.32\,\frac{1}{\text{ft}^3}\right)$$

$$C_0 = 7.49\times10^{-4}\text{ lb/ft}^3$$

$$\text{Slope} = 603\text{ h/ft} = \frac{N_0}{C_0 V}$$

$$N_0 = \left(603\,\frac{\text{h}}{\text{ft}}\right)(C_0)(V)$$

$$N_0 = \left(603\,\frac{\text{h}}{\text{ft}}\right)\left(7.49\times10^{-4}\,\frac{\text{lb}}{\text{ft}^3}\right)\left(19.23\,\frac{\text{ft}}{\text{h}}\right)$$

$$N_0 = 8.68\text{ lb/ft}^3$$

4. Compute K, using the intercepts of the lines.

$$\text{Intercept} = -\frac{1}{C_0 K}\ln\left(\frac{C_0}{C_b}-1\right)$$

For 2.5 gpm/ft²

$$\text{Intercept} = -800\text{ h} = -\frac{1}{7.49\times 10^{-4}\text{ lb/ft}^3(K)}\ln\left(\frac{12\text{ mg/L}}{0.5\text{ mg/L}}-1\right)$$

$$K = -\frac{1}{7.49\times 10^{-4}\text{ lb/ft}^3(-800\text{ h})}\ln(24-1)$$

$$K = 5.23\frac{\text{ft}^3}{\text{lb-h}}$$

5. In the same manner, K values can be calculated. Calculate x_0 for the various flow rates. x_0 is the width of the exchange zone.

$$x_0 = \frac{V}{KN_0}\ln\left(\frac{C_0}{C_b}-1\right)$$

For 2.5 gpm/ft²:

$$x_0 = \frac{19.23\text{ ft/h}}{(5.23\text{ ft}^3/\text{lb - h})(8.68\text{ lb/ft}^3)}\ln\left(\frac{12}{0.5}-1\right)$$

$$x_0 = 1.33\text{ ft}$$

6. The Bohart-Adams constants are summarized below:

Flow rate (gpm/ft²)	V (ft/hr)	Slope (h/ft)	Intercept (h)	N_0 (lb/ft³)	K (ft³/lb-h)	x_0 (ft)
2.5	19.23	603	-800	8.68	5.23	1.33
4.5	36.09	294	-488	7.95	8.58	1.66
8.0	64.17	155	-300	7.45	12.38	2.18

9.4. *Example 4*

Using the data from Example 3, design an adsorption column to treat a waste flow of 20,000 gal/d containing 12 mg/L ABS. The required effluent concentration is 0.05 mg/L, and the column should operate for 120 d before it reaches exhaustion. Operation is to be 16 h/d, 5 d a week.

Solution

1. N_0 and K values from Example 3 are plotted as shown in Figure 32. This depicts the variation in these parameters with flow rate.
2. Compute the wastewater flow rate for 16 h/day operation.

$$Q(\text{gpm}) = \frac{20{,}000\text{ gal/d}}{16\frac{\text{h}}{\text{d}}\times 60\frac{\text{min}}{\text{h}}} = 20.83\text{ gpm}$$

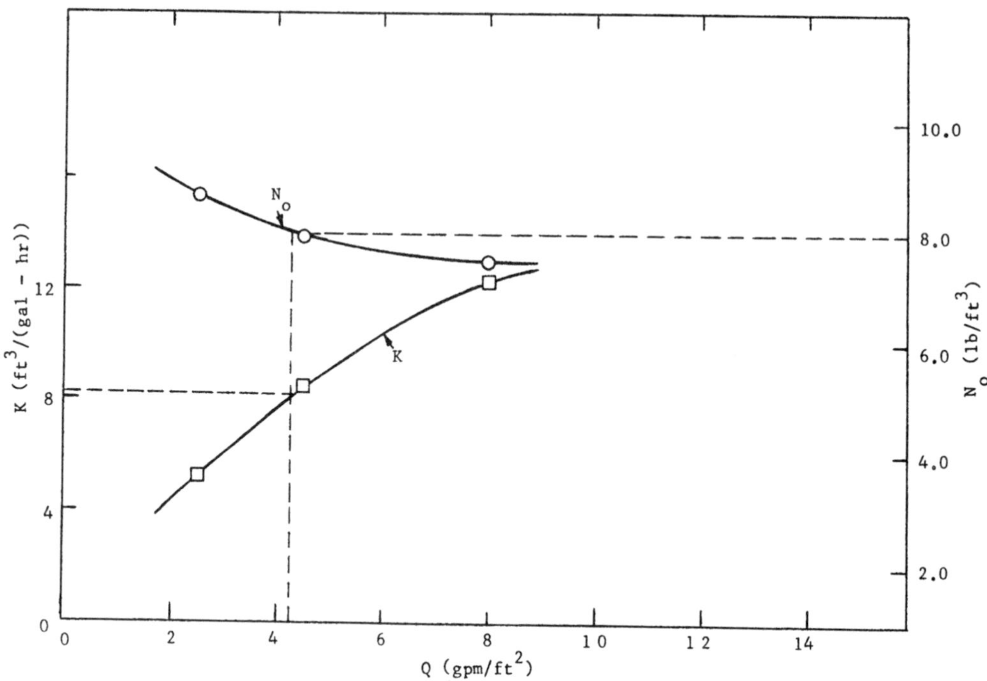

Fig. 32. Effects of Q on K and N_0.

3. Choose a column diameter of 2.5 ft, and compute the corresponding hydraulic loading rate:

$$\text{Area of column} = \frac{\pi(2.5)^2}{4} = 4.90 \text{ ft}^2$$

$$\text{Hydraulic loading of column} = \frac{20.83 \text{ gpm}}{490 \text{ ft}^2} = 4.25 \text{ gpm/ft}^2$$

4. From Fig. 32 read N_0 and K for this hydraulic loading, then:

$$N_0 = 8.05 \text{ lb/ft}^3$$
$$K = 8.20 \text{ ft}^3/\text{gal-h}$$

5. Calculate the depth of bed required for 120 d of operation. Solve the following equation:

$$x = \frac{C_0 V}{N_0}\left[t + \frac{1}{C_0 K} \ln\left(\frac{C_0}{C_b} - 1\right)\right]$$

$$x = \frac{C_0 V t}{N_0} + \frac{V}{N_0 K} \ln\left(\frac{C_0}{C_b} - 1\right)$$

where

$$C = 12 \text{ mg/L} = 7.49 \times 10^{-4} \text{ lb/ft}^3$$
$$C_b = 0.5 \text{ mg / L}$$

$$V = 4.25\,\frac{\text{gpm}}{\text{ft}^2} \times \frac{\text{ft}^3}{7.48\text{ gal}} \times 60\,\frac{\text{min}}{\text{h}} = 34.09\,\frac{\text{ft}}{\text{h}}$$

$$t = 120\text{ d} \times 8\text{ h/d} = 960\text{ h}$$

$$x = \frac{(960\text{ h})\left(7.49\times10^{-4}\,\frac{\text{lb}}{\text{ft}^3}\right)\left(34.09\,\frac{\text{ft}}{\text{h}}\right)}{8.05\,\frac{\text{lb}}{\text{ft}^3}} + \frac{34.09\,\frac{\text{ft}}{\text{h}}}{\left(8.05\,\frac{\text{lb}}{\text{ft}^3}\right)\left(8.20\,\frac{\text{ft}^3}{\text{gal - h}}\right)}\ \ln\left(\frac{12}{0.5}-1\right)$$

$$x = 3.04 + 1.62 = 4.66\text{ ft}$$

6. Compute the volume of carbon required to fill the bed:

$$\text{Volume of carbon required} = 4.90\text{ ft}^2 \times 4.46\text{ ft} = 21.85\text{ ft}^3$$

7. Compute the amount of carbon required on an annual basis with no regeneration:

$$\frac{\text{Number of carbon changes}}{\text{yr}} = \frac{52\,\text{wks} \times 5\,\frac{\text{d}}{\text{wk}}}{120\,\frac{\text{d}}{\text{change}}} = 2.16$$

$$\text{Annual volume of carbon required} = 2.16\text{ changes} \times 21.85\,\frac{\text{ft}^3}{\text{change}} = 47.19\text{ ft}^3$$

10. HISTORICAL AND RECENT DEVELOPMENTS IN GRANULAR ACTIVATED CARBON ADSORPTION

10.1. Adsorption Technology Milestones

Adsorbents include (a) granular-activated carbon (GAC); (b) polymeric adsorbent; and (c) powdered-activated carbon (PAC). All are important adsorbents for the adsorption processes (1–53). The polymeric adsorbent and PAC are discussed in detail in separate chapters of this handbook series.

Both GAC and PAC can remove synthetic organic chemicals (2,15–17,31,37,50), arsenic (18,23), heavy metals (19–23,25,29–32), color (24–27,32), odor (24,32), trihalomethane (27), humic acid (32), residual chlorine (25, 29–31,33), fluoride (29), toxic substances (37–43,45–48,53) from the aqueous phase.

The silver-impregnated GAC can have a disinfection effect during the adsorption process (33,34). Recently, GAC has been applied to air pollution control for removal of pollutants from air emissions (39,41, 46,50).

Table 2 presents the GAC and PAC adsorption technology milestones. Only recent developments (1970–2003) and GAC applications are introduced below.

The first and largest municipal physicochemical wastewater treatment plant to employ GAC filters for tertiary treatment is the Niagara Falls Wastewater Treatment Plant, NY, which was built in 1978 with a 85-MGD peak capacity (20–22). Figure 33 shows the flow diagram of the Niagara Falls plant.

The first precoat GAC filtration process was developed by Wang (26) on a pilot scale in 1989. A large amount of research was conducted on GAC/PAC technology developments at both Zorex Corporation and the Lenox Institute of Water Technology

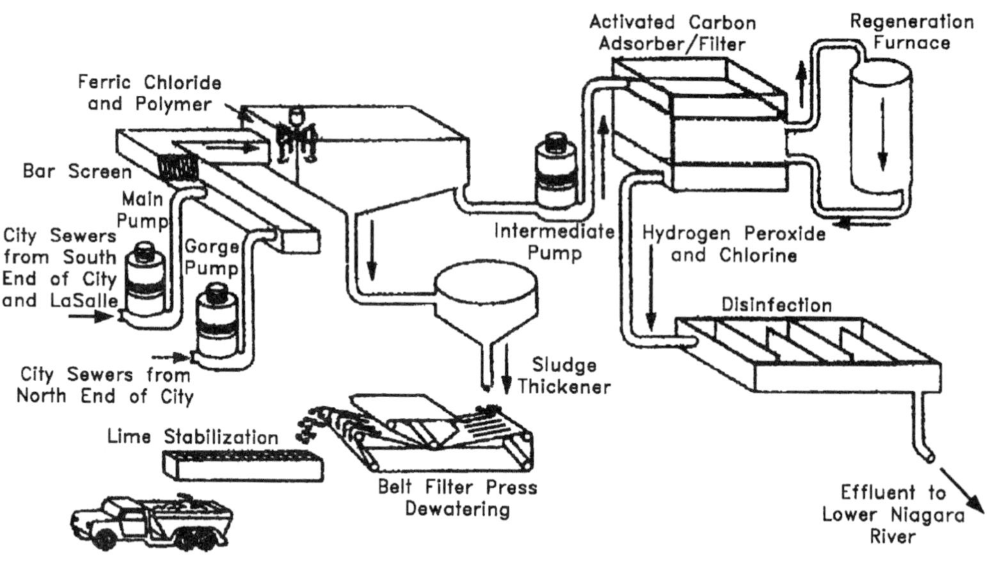

Fig. 33. Niagara Falls Physicochemical Wastewater Treatment Plant (Source: *Water Engineering and Management*, ref. 59).

from 1981 to 1989 (30). Their research findings and engineering designs of GAC/PAC processes were partially introduced at the 20th Annual Meeting of the fine Particle Society Symposium on Activated Carbon Technology, Boston, MA in 1989 (30). These processes were developed in laboratory experiments and verified in pilot studies in 1980s; they became popular only recently: (a) first physicochemical fluidized bed GAC process; (b) first biological fluidized bed GAC process; (c) first physicochemical GAC sequencing batch reactor (SBR); (d) first biological GAC-SBR; (e) first combined dissolved air flotation (DAF) and GAC process; (f) first DAF-PAC process; (g) first physicochemical PAC-SBR process; (h) first biological PAC-SBR process; (i) first physicochemical PAC-DAF-SBR process; (h) first biological PAC-DAF-SBR process; (i) first ion exchange SBR process; (j) first physicochemical SBR process; and (k) first regenerable gas phase GAC system.

Owing to the importance of the above technologies, many US patents for GAC/PAC in combination with SBR, DAF, and precoat filtration were filed by and granted to Wang and his co-workers (35,41,54,55). For details about GAC- and PAC-related SBR processes, please refer to another handbook in this series (51). The gas-phase GAC system for air emission control is also described in the literature (41,52,57).

The biological GAC filtration process was introduced as a competitive process to DAF-GAC (27) in 1989. As shown in Table 2, Mainstream Bio-Manipulation Systems Ltd. adapted both the slow sand filtration and biological GAC filtration processes in 1996 for drinking water production (42). In 2003, the 230-MLD (230-million liters per day) Ngau Tam Mei Water Works, Hong Kong, China became the first dual-stage biological GAC filtration plant (49).

Table 2 further shows that in 2000 the first biological fluidized bed GAC system was built by both Envirogen and US Filter (53) for groundwater decontamination. The well-established innovative GAC systems are further discussed in later sections of this chapter.

10.2. Downflow Conventional Biological GAC Systems

10.2.1. General Introduction

Granular-activated carbon (GAC) adsorption system can remove many adsorbable organics and inorganics (30,63,64), but not non-adsorbable pollutants such as dimethyl-nitrosamine, acetone cyanohydrin, butylamine, cyclohexylamine, diethylene glycol, ethylenediamine, triethanolamine, and ethanol. Biological processes, on the other hand, can remove biodegradable pollutants, but not any non-biodegradable pollutants. Combination of both GAC processes will solve many traditionally unsolvable environmental pollution control problems.

Environmental engineers recognize that biological activity plays a major role in removal of organics by activated carbon. When granular activated carbons are used as the filter media as well as the growth media in an attached growth biological oxidation-adsorption system, is the result is a biological GAC adsorption system.

The conventional biological GAC process consists of a fixed bed of GAC media over which wastewater is applied for aerobic biological and adsorption treatment aiming at toxic organic substances removal. Biological slimes form on the GAC media, which assimilate and oxidize substances in the wastewater. The bed is dosed by a distributor system, and the treated wastewater is collected by an underdrain system.

Organic material present in the wastewater is degraded by population of microorganisms attached to the GAC media and partially adsorbed by GAC macropores and micropores. The thickness of the slime layer increases, as the microorganisms grow during bio-oxidation. The macropores and micropores of GAC are also gradually saturated by the target organic pollutants during adsorption. Microorganisms are also partially responsible for continuous GAC regeneration and prolonged adsorption. Periodically, the GAC bed must be backwashed and regenerated for reuse.

Both the downflow pressurized biological GAC system and downflow gravity biological GAC system are technically feasible for water and wastewater treatment as long as oxygen is available for bio-oxidation (43,44).

10.2.2. Saskatchewan-Canada Biological GAC Filtration Plant for Biological Treatment of Drinking Water

Slow sand filtration (water moves through such filters 10–20 times slower than rapid sand filters) relies on the formation of a biological layer at the top of the filter. The filter does not become effective until this layer has formed (27,42). In addition to removing dissolved organics, slow sand filtration can be very effective in particle removal. The American Water Works Association (AWWA) states: "The slow sand filtration process is expected to remove such biological particles as cysts, algae, bacteria, viruses, parasite eggs, nematode eggs, and amorphous organic debris at 100-fold to 10,000-fold levels when the filter is biologically mature." As effective as sand filtration can be, it is possible to maintain much greater numbers of microorganisms if the support material is GAC instead of sand. It is therefore preferable to use GAC for the removal of dissolved organics (42).

Mainstream Bio-Manipulation Systems Ltd., Canada, has worked on adapting both the slow sand filtration and biological GAC filtration processes with the support of the National Research Council. Such treatment systems have been installed at three different sites across Saskatchewan. One site has been in operation since 1996 and removal rates of turbidity, dissolved organic carbon, and color have been good for both the sand filter and the biological GAC filter. Both have provided high-quality household water with no color or odor (removal rates of turbidity, dissolved organic carbon, and color are consistently above 50%). For drinking water purposes, the water is polished by a reverse osmosis unit. At this site, all of the household water was hauled before installation of the biological treatment system. Based on successes like this one, it is anticipated that biological treatment will become one of the most common future treatment tools for dealing with surface waters on the Canadian prairie (42).

10.2.3. Ngau Tam Mei Water Works, Hong Kong, China

When the Northwestern New Territories of Hong Kong faced projected shortfalls of potable water in 1994, the Water Supplies Department initiated new facilities for treatment, conveyance, and storage of water from its major supply, the Dongjiang River in Guangdong Province, People's Republic of China, via the Western Aqueduct.

In 2000, the Ngau Tam Mei water treatment works was commissioned, officially opening on December 2. It became the first water treatment plant worldwide to use dual-stage biological filtration with granular activated carbon (GAC) to remove ammonia, replacing break-point chlorination (49).

The HK$1.8 billion (US$227 million) project treats raw water from the Dongjiang River, which is contaminated by wastewater. CDM designed the plant with an initial capacity of 230 million liters/day (MLD), expandable to 450 MLD.

The innovative plant meets or surpasses water quality goals given a challenging raw water source with (a) 4 preozone contact tanks with a design detention time 5 min; (b) 12 triple-deck sedimentation basins with a designed surface loading rate of 1.3 m/h; (c) intermediate ozone contact tanks with a design retention time of 15 min for achieving 1-log inactivation of *Cryptosporidium*; (d) 12 first-stage GAC (1.5-m depth) filters with minimum filter fun times of 24 h and filtration rate of 12 m/h followed by 12 second-stage GAC (1.8-m depth) filters with filtration rates of 8 m/h; and (e) ozone peak dosage of 5 mg/L, ozone production rates of 1150 kg/d, and ozone concentration of 7.5%. The plant reduces operating and maintenance costs by (a) generating high-quality oxygen on site, eliminating more costly truck-delivered liquid oxygen, (b) using dual-stage GAC filters to remove ammonia and replace breakpoint chlorination, (c) providing flexibility for operating in direct-filtration mode during periods of acceptable raw water quality to reduce coagulant chemical doses and sludge production, and (d) reducing labor cost and improving plant management through a supervisory control and data acquisition (SCADA) system (49).

The process train combines two advanced technologies: (a) dual-stage ozonation for preoxidation and primary disinfection and (b) dual-stage GAC biological filtration for nitrification—a first-of-its-kind application in drinking water treatment.

Special features of the largest biological GAC filtration plant include (49):

(a) *Dual-stage biological GAC filtration (first application worldwide)*: First-stage filters remove turbidity, particles, biodegradable organic carbon, and taste- and odor-causing compounds. Second-stage filters remove ammonia, eliminating breakpoint chlorination and associated high chlorine doses. Results since commissioning show complete removal of ammonia (<0.02 mg/L).

(b) *Ozonation for primary disinfection (first in Hong Kong)*. This inactivates *Giardia* and *Cryptosporidium* and reduces chlorine usage, helping to eliminate formation of chlorinated byproducts (THMs) and enhancing downstream biological filtration by oxygenating water and increasing formation of biodegradable organic carbon.

(c) *Ozone injection dissolution system (first in Hong Kong)*: Developed and patented by CDM, the sidestream Venturi injection with downflow tube (SVI-DT) system improves ozone dissolution and oxidation efficiency and reduces operational costs.

(d) *Onsite oxygen generating systems (first in Hong Kong)*: System uses vacuum pressure swing adsorption (VPSA) technology to produce high-quality gaseous oxygen instead of truck-delivered liquid oxygen, reducing chemical costs.

(e) *Ozonation for manganese removal*: Process uses preozone for oxidation of reduced manganese to its insoluble form (manganese dioxide) for subsequent removal by coagulation and settling, followed by intermediate ozone, which oxidizes remaining manganese in the settled water to permanganate for subsequent catalytic removal by first-stage GAC filters.

(f) *Sludge treatment facilities*: CDM identified plate and frame filter presses as the only equipment capable of achieving minimum required sludge solids content (30%).

10.3. Upflow Fluidized Bed Biological GAC System

The upflow fluidized bed biological GAC system has fewer clogging problems than the two downflow biological GAC systems introduced in Section 10.2. Accordingly, the downflow biological GAC filtration process is mainly used for potable water treatment, while the upflow fluidized bed biological GAC system may be used for both water and wastewater treatment (27,30). Many researchers are studying the upflow fluidized bed biological GAC systems (30,36,47,53,57,60). While the first fluidized bed biological GAC system was designed and built by Envirogen and US Filter (53) for groundwater decontamination in 2000, the authors of this chapter chose Hydroxyl Systems' FBB-GAC (Fluidized Bed Bioreactor) line of treatment systems for the purpose of illustration (58).

The FBB-GAC system (Fig. 34) can be used in aerobic, anoxic, or anaerobic conditions and can accommodate a variety of granular and other media. When adsorbent media such as GAC is used, the FBB combines the benefits of adsorption and bio-oxidation—contaminants are adsorbed onto the media surface and oxidized by biofilm also present on the surface. Unlike other biological treatment systems, the requirement for operator attention is minimal and unattended operation is practical. A major advantage of the FBB-GAC is that treatment detention times are typically minutes instead of hours.

The FBB-GAC system is supplied either as a single skid or module of shippable height, incorporating a low profile reactor, or as a two-piece unit with a detachable tall cylindrical reactor. This system can use aerobic, anoxic, or anaerobic treatment of waterborne biodegradable matter, particularly adsorbable contaminants in low ppm concentrations. Typical applications include treatment of groundwater contaminated with BTEX and as a complement to Advanced Oxidation Technologies for complete mineralization of

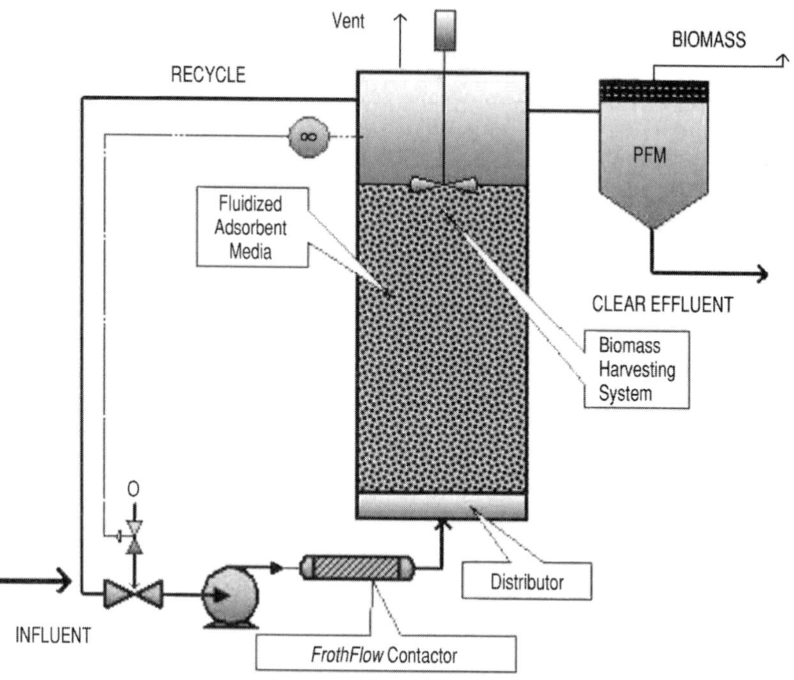

Fig. 34. Fluidized bed biological (FBB) granular-activated carbon plant (Source: Hydroxyl System Inc.).

biorefractory contaminants. As an anaerobic reactor, the FBB-GAC system can be used to treat high-strength wastewaters. Typical contaminants are BTEX, glycol, MTBE, soluble oil and grease, and organic solvents. The FBB-GAC system has the following special features: (a) fast bio-oxidation; (b) fully automated with PLC control; (c) weatherproof container (optional); (d) no plugging or sludge bulking; (e) no post-clarification required; (f) compact and portable size; (g) unattended operation possible; and (h) no off-gas (58).

A typical configuration for aerobic applications (Fig. 34) is depicted below to illustrate the operation process. Influent enters a recirculation loop operating at several times the flow rate of the influent. Enough oxygen is added and thoroughly dissolved in the stream to maintain adequate dissolved oxygen in the reactor. The flow is equally distributed across the reactor, resulting in even fluidization (expansion) of the media bed. An extremely high concentration of biomass develops because of the huge media surface area, abundant oxygen, and optimized mass transfer conditions. Excess biomass is periodically and automatically removed by extracting media, shearing the biomass and returning the cleaned biomass to the reactor. The effluent from the FBB-GAC is typically very low in suspended solids, allowing discharge without further treatment (58).

NOMENCLATURE

a	external area of the adsorbent per unit volume
A	cross-sectional area of column
AD	apparent density of the spent and regenerated carbon

B_t	mass flow rate of solid to maintain a stationary adsorption zone
BAC	biological activated carbon
BDST	bed-depth service time
BOD	biochemical oxygen demand
BTEX	butyl, toluene, ethylbenzene, and xylene
C	actual concentration
C_b	concentration of impurity in effluent (lb/ft^3)
C_b	desired concentration of solute (lb/ft^3)
C_e	equilibrium concentration for the amount of organic adsorbed.
C_{eff}	effluent solute concentration
C_0	concentration of impurity in influent (lb/ft^3)
C_0	influent solute concentration
C_0	initial concentration of solute (lb/ft^3)
C_x	exhaustion effluent solute concentration
CCI	construction cost index
COD	chemical oxygen demand
dC	differential concentration
dy	differential column length
DAF	dissolved air floatation
ENR	*Engineering News-Record*
f	fractional capacity of the active adsorption zone
F_m	mass liquid flux
F_m	mass liquid flow rate
ft	feet
ft^2	square feet
FBB	fluidized bed bioreactor
g	grams
GAC	granular activated carbon
gal	gallon
gpd	gallons per day
gpm	gallons per minute
h	hour
k	mass transfer rate coefficient
K	adsorption rate constant (ft^3/lb-h)
kg	kilograms
kW	kilowatt
L	bed depth
L	column length
m	meters
m^2	square meters
m^3	cubic meters
M	total amount of solute adsorbed at breakthrough point
M_s	total amount of solute adsorbed at exhaustion
min	minute
mg/L	milligrams per liter

mL	milliliter
MGD	million gallons per day
MLD	million liters per day
MTBE	methyl tertiary-butyl ether
N_0	adsorption efficiency (lb/ft^3)
P_s	quantity of solute adsorbed in the active adsorption zone from breakthrough to exhaustion
P_{tc}	total capacity of carbon in the active adsorption zone
PAC	powdered-activated carbon
PCT	physical–chemical treatment
Q	volumetric flow rate
S_b	mass of solute adsorbed at equilibrium per unit of crosssectional area of adsorption bed
SBR	sequencing batch reactor
scf	standard cubic feet
t	service time (h)
t_f	active adsorption zone formation time
t_x	time of exhaustion
t_δ	time of passage through the active zone
TOC	total organic carbon
U_δ	active adsorption zone velocity
V	volume of liquid passed through column
V_b	volumes of liquid passed through column at breakthrough
v	hydraulic loading, or linear velocity of fluid (ft/hr)
V_x	volume of liquid passed through column at exhaustion
X	bed depth (ft)
$\left(\frac{x}{m}\right)_{C_0}$	mass of organic adsorbed per unit mass of carbon at the equilibrium concentration C_0
$\left(\frac{x}{m}\right)_0$	mass solute/mass adsorbent at equilibrium with influent concentration C_0
y	column length
y	total column depth
δ_0	mass solute/mass adsorbent at equilibrium with influent concentration C_0
δ	adsorption-zone height
ρ_p	apparent packed density of the carbon adsorbent
π	pi (3.14)
ρ	carbon density (kg/m^3)

REFERENCES

1. P. N. Cheremisinoff and A. C. Morresi, *Poll Eng* **6,** pp. 66–69 (1974).
2. B. H. Kornegay, *Seminar on Control of Organic Chemical Contaminants in Drinking Water*, Los Angeles, CA, US Environmental Protection Agency, Office of Drinking Water, Washington, DC, November, 1978.
3. US EPA, *Process Design Manual for Carbon Adsorption*, Technology Transfer, US Environmental Protection Agency, Washington, DC, October, 1973.

4. J. J. McElhaney, et al., *Proceedings AWWA 6th Water Quality Technology Conference*, 1978.
5. A. S. Michaels, *Industrial and Engineering Chemistry* **44**, 1922 (1952).
6. R. A. Hutchins, *Chemical Engineering*, **80**, 133 (1973).
7. G. S. Bohart and E. Q. Adams, *J. Am Chem Soc* **42**, 523 (1920).
8. H. J. Fornwalt and R. A. Hutchins, *Chemical Engineering* **73**, 179 (1966).
9. R. A. Conway and R. D. Ross, *Handbook of Industrial Waste Disposal*, Van Nostrand Reinhold Company, New York, 1980.
10. H. Sontheimer, *J Am Water Works Assoc* **71**, 618–627 (1979).
11. P. N. Cheremisinoff and A. C. Morresi, *Carbon Adsorption Handbook*, P.N. Cheremisinoff and F. Ellerbusch (eds). Ann Arbor Science Publisher, Ann Arbor MI, 1978, pp. 1–53.
12. W. W. Eckenfelder, Jr., *Principles of Water Quality Management*, CBI Publishing Company, 1980.
13. D. G. Hager, Industrial wastewater treatment by granular activated carbon, *Industrial Water Engineering*, pp. 14–28, January/February (1974).
14. J. C. Cooper and D. G. Hager, *Chemical Engineering Progress* **62**, 85–90 (1966).
15. L. K. Wang, *J. Appl Chem and Biotechn* **25**, 491–503 (1975).
16. L. K. Wang, *Water and Sewage Works* **123**, 42–47 (1976).
17. L. K. Wang, *Water and Sewage Works* **124**, 32–36 (1977).
18. L. K. Wang, *Removal of Arsenic from Water by Continuous Granular Carbon Adsorption Process*, US Department of Commerce, National Technical Information Service, Springfield, VA, PB86-197597/AS, 1985.
19. L. K. Wang, *Treatment of Scallop Processing Wastewater by Flotation, Adsorption and Ion Exchange,* Lenox Institute of Water Technology, Lenox, MA, Technical Report #LIR/05-85/139, 1985.
20. L. K. Wang, M. H. S. Wang, and J. Wang, *Design, Operation and Maintenance of the Nation's Largest Physicochemical Waste Treatment Plant, Volume 1,* Lenox Institute of Water Technology, Lenox, MA, Report #LIR/03-87-248, March, 1989.
21. L. K. Wang, M. H. S. Wang, and J. Wang, *Design Operation and Maintenance of the Nation's Largest Physicochemical Waste Treatment Plant, Volume 2* Lenox Institute of Water Technology, Lenox, MA, Report #LIR/03-87/249, March, 1987.
22. L. K. Wang, M. H. S. Wang, and J. Wang, *Design Operation and Maintenance of the Nation's Largest Physicochemical Waste Treatment Plant, Volume 3* Lenox Institute of Water Technology, Lenox, MA, Report #LIR/03-87/250, March, 1987.
23. L. K. Wang, *Removal of Arsenic and Other Contaminants from Storm Run-off Water by Flotation, Filtration, Adsorption and Ion Exchange*, US Department of Commerce, National Technical Information Service, Springfield, VA, PB88-200613/AS, 1988.
24. L. K. Wang, *Treatment of Potable Water from Seoul, Korea by Flotation, Filtration and Adsorption*, US Department of Commerce, National Technical Information Service, Springfield, VA, PB88-200530/AS, 1988.
25. L. K. Wang, *Procedures for Evaluation of the Scott Fetzer Adsorption Filter in Removing Lead and Other Contaminants from Drinking Water Zorex Corporation., Pittsfield, MA,* Tech. Report #P917-4-89-3, 1989.
26. L. K. Wang, *Advanced Precoat Filtration and Competitive Processes for Water Purification*, Harvard University, The Harvard Club, Boston, MA, Jan, 1989.
27. L. K. Wang, *Using Air Flotation and Filtration in Removal of Color, Trihalomethane Precursors and Giardia Cysts*, the NY State Department of Health Workshop on Water Treatment Chemicals and Filtration & the 1989 American Slow Sand Association Annual Meeting, 1989 (No. P904-8-89-23).
28. L. K. Wang, *Manufacturers and Distributors of Activated Carbons and Adsorption Filters*, Zorex Corporation, Pittsfield, MA, Technical Report #P917-5-89-7, 1989.
29. L. K. Wang, *Reduction of Chlorine, Fluoride, Lead and Organics by Adsorption Filters*, Zorex Corporation, Pittsfield, MA, Technical Report No. P.917-6-89-14, June, 1989.

30. L. K. Wang, *New Dawn in Development of Adsorption Technologies*, the 20th Annual Meeting of the Fine Particle Society Symposium on Activated Carbon Technology, Boston, MA, Aug, 1989.
31. L. K. Wang, *Removal of Heavy Metals, Chlorine and Synthetic Organic Chemicals by Adsorption*, Zorex Corporation, Pittsfield, MA, Tech. Report #P917-5-89-8, 1989.
32. L. K. Wang, *Reduction of Color, Odor, Humic Acid and Toxic Substances by Adsorption, Flotation and Filtration*, Annual Meeting of American Institute of Chemical Engineers, Symposium on Design of Adsorption Systems for Pollution Control, Philadelphia, PA, 1989.
33. L. K. Wang, *Dechlorination with Innovative Metal Filter Medium and Silver Impregnated Granular Activated Carbon,* Zorex Corporation, Pittsfield, MA, Tech. Report #P931-10-89-30, 1989.
34. L. K. Wang, *Critical Review of Brass KDF Medium and Silver Impregnated Granular Activated Carbon,* Zorex Corporation, Pittsfield, MA, Tech. Report #P931-10-89-31, 1989.
35. L. K. Wang, L. Kurylko, and M. H. S. Wang, *Improved Method and Apparatus for Liquid Treatment,* US Patent and Trademark Office, Washington, D.C., Patent No. 5256299, Oct 26, 1993.
36. S. S. Cheng, *Water Science and Technology*, **30**, 131–142 (1994).
37. L. K. Wang, *The State-of-the-art Technologies for Water Treatment and Management*, United Nations Industrial Development Organization (UNIDO), Vienna, Austria, UNIDO Training Manual No. 8-8-95, 1995.
38. L. K. Wang, *Management of Hazardous Substances at Industrial Sites*, United Nations Industrial Development Organization (UNIDO), Vienna, Austria, UNIDO Technical Manual No. 4-4-95, 1995.
39. L. K. Wang, *Case Studies of Cleaner Production and Site Remediation*, United Nations Industrial Development Organization (UNIDO), Vienna, Austria, UNIDO Training Manual No. 5-4-95, 1995.
40. L. K. Wang, *Identification, Transfer, Acquisition and Implementation of Environmental Technologies Suitable for Small and Medium Size Enterprises*, United Nations Industrial Development Organization (UNIDO), Vienna, Austria, UNIDO Technical Paper No. 9-9-95, 1995.
41. L. K. Wang, L. Kurylko, and O. Hyrcyk, *Site Remediation Technology*, US Patent and Trademark Office, Washington, DC, Patent No. 5552051, 1996.
42. H. Peterson, *Critical Reviews in Environmental Control*, 2 (www.quantumlynx.com/water'back/vol5no2/story3.html), 1996.
43. N. Davis, L. Erickson, and R. Hayter, *Centerpoint*, Georgia Institute of Technology, GA, **4** (2) (1998).
44. O. Griffini, M. Bao, D. Burrini, D. Santianni, C. Barbieri, and F. Pantani, *Journal Water SRT–Aqua* **48**, 177–185 (1999).
45. L. K. Wang, Site remediation and groundwater decontamination in the US. *The Encyclopedia of Life Support Systems*, Eolss Publishers, Co., Ltd., UK. UNESCO, World Summit, Johannesburg, 2002.
46. L. K. Wang, Hazardous waste management. *The Encyclopedia of Life Support Systems*, Eolss Publishers, Co., Ltd., UK. UNESCO, World Summit, Johannesburg, 2002.
47. G. J. Wilson, *Water Science and Technology* **38**, 9–17, (2002).
48. L. K. Wang, Industrial ecology. *The Encyclopedia of Life Support Systems*, Eolss Publishers, Co., Ltd., U.K., UNESCO, World Summit, Johannesburg, 2002.
49. P. Y. Chung, *Ngau Tam Mei Water Works, Hong Kong*, Camp Dresser & McKee Inc., NY, 2003.
50. L. K. Wang, Y. T. Hung, H. H. Lo, and C. Yapijakis, (eds.), *Handbook of Industrial and Hazardous Wastes Treatment*, Marcel Dekker, New York, 2004.
51. L. K. Wang, N. Pereira, and Y. T. Hung, (eds.), *Biological Treatment Processes*, Humana Press, Inc., Totowa, NJ, 2005.

52. L. K. Wang, N. Pereira, and Y. T. Hung, (eds.), *Air Pollution Control Engineering*, Humana Press, Inc., Totowa, NJ, 2004.
53. US Filter, *Implementation of GAC Fluidized - Bed for Treatment of Petroleum Hydrocarbons in Groundwater at Two BP Oil Distribution Terminals, Pilot and Full-Scale.* US Filter (www.usfilter.com), 2000.
54. L. K. Wang and L. Kurylko, *Contamination Removal System Employing Filtration , Plural Ultraviolet & Chemical Treatment Steps & Treatment Controller*, US Patent No. 5,190,659; *Method and Apparatus for Filtration with Plural Ultraviolet Treatment Stages*, US Patent No. 5,236,595. US Patent and Trademark Office, Washington, DC, 1993.
55. L. K. Wang, L. Kurylko, and M. H. S. Wang, *Sequencing Batch Liquid Treatment*, US Patent and Trademark Office, Washington, D.C., U. S. Patent No. 5,354,458, 1994.
56. T. C. Voice, X. Zhao, J. Shi, and R. F. Hickey, *Biological Unit Processes for Hazardous Waste Treatment—1995 Third International In Situ and On-Site Bioremediation Symposium*, Battelle Press, Columbus, OH, 1995.
57. V. F. Medina, J. S. Devinny, and M. Ramaratnam, *Biological Unit Processes for Hazardous Waste Treatment—1995 Third International In Situ and On-Site Bioremediation Symposium.* Battelle Press, Columbus, OH, 1995.
58. Hydroxyl System, Fluidized Biological Reactor. Hydroxyl System Inc., 9800 McDonald Park Road, Sidney, BC V8L 5W5 Canada (www.hydroxyl.com/products/biological/hydroxyl-fbb.html. info@hydroxyl.com), 2003.
59. I. Lisk, *Water Engineering & Management*, p. 6, Feb. (1995).
60. L. K. Wang, *1974 Earth Environment and Resources Conference Digest of Technical Papers* **1**, 56–57 (1974).
61. US EPA, *Estimating Water Treatment Costs, Volume 2. Cost Curves Applicable to 1 to 200 mgd Treatment Plants*, Research and Development, EPA-600/2-79-162b, U.S. Environmental Protection Agency, Washington, DC, 1979.
62. J. L. Tuchman (ed.), *Engineering News-Record*, November 3, (ENR.com) p. 39. (2003).
63. M. H. S. Wang (formerly M. H. Sung). Adsorption of Nitrogen and Phosphorus Using a Continuous Mixing Tank with Activated Carbon. Master thesis. Univ. of Rhode Island, Kingston, RI, 1968.
64. L. K. Wang, Y. T. Hung, H. H. Lo, J. R. Taricska and K. Hung Li. *J. OCEESA,* Dec. (2004).

16
Physicochemical Treatment Processes for Water Reuse

Saravanamuthu Vigneswaran, Huu Hao Ngo, Durgananda Singh Chaudhary, and Yung Tse Hung

CONTENTS

1. INTRODUCTION

The continuing processes of industrialization and urbanization coupled with uncontrolled population growth, deforestation, and water pollution are exerting pressure on the planet's limited freshwater resources. The need to recycle and reuse wastewater has been more and more realized, as the global supplies of clean water diminish and demand for water rises. Advanced wastewater treatment is becoming an international focus for the rational use of scarce water resources, and as means of safeguarding aquatic environments from the harm caused by wastewater disposal. Conventionally, wastewater was discharged into the environment after removing the majority of suspended solids in primary treatment and biodegradable organic substance in secondary treatment. These treatments are not sufficient to produce effluent of reusable quality. Now the trend is changing toward the total water recycle approach, which promotes ecological sustainability by managing the treated wastewater as a resource instead of a waste and, at the same time, reducing the demand for water from the existing water resources. Tertiary wastewater treatment is therefore required to remove most of the remaining organics, solids, and pathogenic microorganisms.

The advanced treatment processes utilized in the polishing wastewater-treatment system are physicochemical in nature. Coagulation–flocculation, filtration, and sedimentation followed by chlorination are the standard form of advanced treatment scheme that is mostly utilized in practice. In coagulation–flocculation processes, chemicals such as alum, ferric chloride, and some advanced forms of flocculant are added to neutralize the charge on particles and to agglomerate the colloidal particles into settleable and filterable particles. In these processes, most of the organic pollutants are destabilized and are

From: *Handbook of Environmental Engineering, Volume 3: Physicochemical Treatment Processes*
Edited by: L. K. Wang, Y.-T. Hung, and N. K. Shammas © The Humana Press Inc., Totowa, NJ

subsequently removed in sedimentation and/or filtration processes. In the treatment plant, a mechanical mixer can be provided to disperse coagulant chemicals uniformly (by rapid mixing) and bring the destabilized colloidal particles together to form flocs (by slow mixing). Sedimentation is the solid–liquid separation process that makes use of the gravitational settling principle. The sedimentation tank is provided to collect the settleable particles destabilized and agglomerated in coagulation–flocculation processes and to pass relatively clear effluent to the filtration process. Filtration is the process whereby the impurities are removed by a combination of different processes such as sedimentation, interception, adsorption, and straining. Straining removes those suspended particles that are too large to pass through the pores of the filter bed. Sedimentation removes fine suspended solids as they are deposited onto the surface of the filter media grains. Adsorption occurs as a result of electrostatic attraction of particles toward the filter media particles of the filter bed. The organics substances accumulated on the filter medium due to all these actions may undergo biochemical and bacterial activity and thus organics are biodegradated.

When the filtrate contains an excessive amount of dissolved organic substances, then it is further treated by adsorption and ion-exchange processes. The effluent is finally disinfected with chlorine or UV and discharged into waterways for various reuse purposes.

With technological advances and the ever-increasing stringency of water-quality criteria, membrane processes are becoming a more attractive solution to the challenge of water reuse. The use of membrane technology, particularly in wastewater treatment and reuse, has received increased attention since early 1990s. Membrane technologies currently being used in different industries include microfiltration (MF), ultrafiltration (UF), nanofiltration (NF), reverse osmosis (RO), pervaporation, dialysis, and electrodialysis. Although membrane processes such as reverse osmosis and nanofiltration could in theory remove all pollutants, including dissolved organics, their operational costs are high because of high-energy requirements and membrane fouling. Micro- and ultrafiltration are cost-effective options, but they cannot remove dissolved organic matter due to their relatively larger pore sizes. Therefore, in water-reuse applications, ultrafiltration or microfiltration needs to be combined with biological processes. For example, in a water-mining project in Canberra, Australia, biological filtration is combined with continuous microfiltration. Microfiltration is preferable choice in wastewater treatment and reuse applications because it can be operated at very low pressure (1 bar) compared to ultrafiltration (100–500 kPa) and reverse osmosis (2000–8000 kPa).

2. CONVENTIONAL PHYSICOCHEMICAL TREATMENT PROCESSES

2.1. *Principle*

2.1.1. *Coagulation–Flocculation*

The principal use of coagulation and flocculation is to agglomerate particles into settleable or filterable flocs prior to sedimentation or filtration. Coagulation consists of adding chemicals to the colloidal suspensions. This results in particle destabilization by a reduction in the repulsive forces, which tend to keep particles apart. The colloidal particles then agglomerate to form settleable or filterable solids by the process of flocculation.

Coagulation involves the reduction of surface charges and the formation of complex hydrous oxides. The process forms either flocculant suspensions of compounds, which entrap desired pollutants and carry them out of solution, or insoluble precipitates of the pollutants themselves. The coagulation phase is practically instantaneous and the particles are usually submicroscopic in size. The chemistry involved is very complex and the known major interactions that occur are (a) the reduction of the zeta potential to a degree where the attractive van der Waals' forces and the agitation provided cause the particles to coalesce; (b) the aggregation of particles by interparticulate bridging between reactive groups of colloids; and (c) the enmeshment of particles in the precipitate floc that is formed. The interparticulate forces acting on a colloidal particle are repulsive forces, due to the electrostatic zeta potential, and attractive forces, due to van der Waals' forces acting between the particles. The net resultant force is attractive to a certain distance. Beyond this distance, the net resultant is repulsive. When a coagulant salt is added to a water, it dissociates, and the metallic ion undergoes hydrolysis and creates positive hydroxometallic ion complexes. There are a number of species of hydroxometallic complexes formed because the complexes, which are hydrolysis products, tend to polymerize. These complexes are polyvalent with high positive charges, and are adsorbed onto the surface of the negative colloids. This results in a reduction of the zeta potential to a level where the colloids are destabilized. The coagulation of colloids by organic polymers occurs by a chemical interaction or bridging. The polymers have ignitable groups such as carboxyl, amino, and sulfonic, and these groups bind with reactive sites or groups on the surfaces of the colloids. In this manner, several colloids are bound to a simple polymer molecule to form a bridging structure. Bridging between particles is optimum when the colloids are about one-half covered with adsorbed segments of the polymers.

Collisions of destabilized particles lead to agglomeration. The collisions can occur by three separate mechanisms:

(a) Aggregation resulting from random Brownian movement of fluid molecules (perikinetic flocculation). When submicron particles move in water under Brownian motion, they collide with other particles. On contact, they form large particles and continue to do so until they become too large to be affected by Brownian motion. Perikinetic flocculation is predominant for submicron particles. The higher the initial concentration of particles in the suspension, the faster is the floc formation, because the opportunity for collision is higher.

(b) Aggregation induced by velocity gradient in the fluid (orthokinetic flocculation). Orthokinetic flocculation that involves particle movement with gentle motion of water considers that particles will agglomerate if they collide and become close enough to be within a zone of influence of one another. It also considers that particles have negligible settling velocity; hence, there is a need for agitation of the water, or a velocity gradient to promote the collisions. The rate of flocculation is proportional to the velocity (shear) gradient, the volume of the zone of influence, and the concentration of particles.

(c) Differential settling, where flocculation is due to the different rates of settling of particles of different sizes. Larger particles settle faster than smaller particles, which makes the relative velocities between the particles different. This also helps in orthokinetic flocculation as because velocity gradients are produced, causing further agglomeration.

The two main modes of process operations used in flocculation (i.e., creation of velocity gradients) are hydraulic flocculator and mechanical flocculator. Mechanical

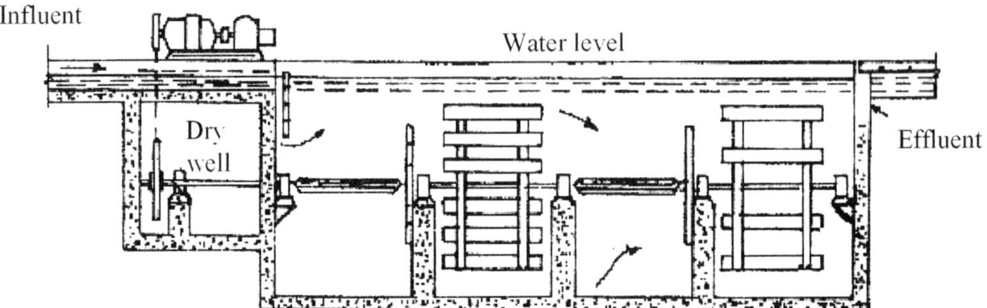

Fig. 1. Longitudinal flow flocculator.

flocculators use mechanical mixing devices such as paddles, turbines, or propellers. Mechanical flocculators are used widely. Paddle flocculator is one of the commonly used types of mechanical flocculator. There are vertical flocculators and longitudinal-flow flocculators (Figs. 1 and 2). The design criteria are shown in Table 1.

In large plants, it is desirable to provide more than one compartment in series to lesson the effect of short circuiting. The paddle is driven by electric motors. The direction of flow is usually horizontal moving parallel or at right angles to the paddles shafts. The shaft of the container also affects the process of flocculation. For the same volume and height of water in the containers of several shapes such as circular, square, pentagonal, and hexagonal, it was observed that the pentagonal shape gave the best performance.

Electrocoagulation is a relatively new technology that has been tried in water and wastewater treatment. In this process, the coagulating agent (i.e., aluminum or ferrous ion) is introduced as a result of anodic dissolution of the electrode (i.e., aluminum or ferrous ion) and the process permits a careful control of the dosage of the reagent. The process has advantages over conventional chemical coagulation (2). First, the coagulation process is enhanced due to local attraction of pollutants by the electrode; second,

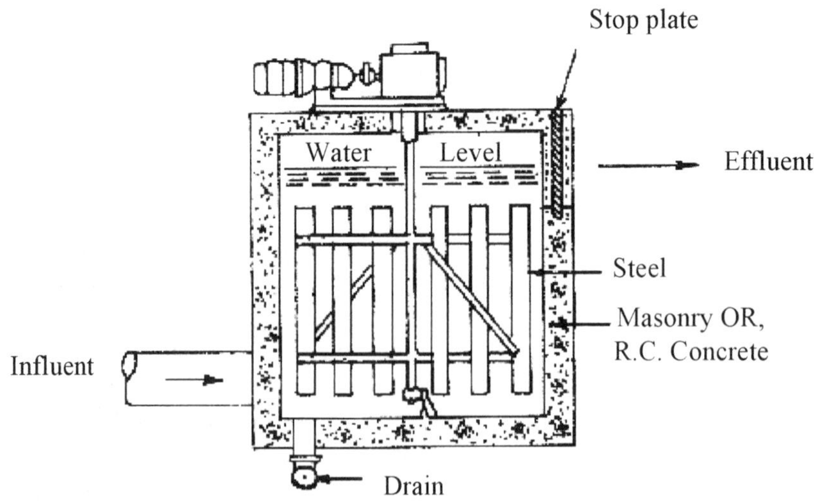

Fig. 2. Vertical flocculator.

Table 1
Typical Paddle Flocculator Parameters

Parameter	Value
Velocity of flow (m/s)	0.2–0.8
Depth of tank (m)	3–4.5
Detention time (min)	10–40
Velocity gradient (1/s)	10–75
Velocity of blades (m/s)	0.2–0.6
Outlet velocity (m/s)	0.15–0.25
Power consumption (kW/MLD)	10–36

the cathode generates hydrogen bubbles promoting the growth of precipitate flocs for easier separation; and third, the strength of the flocs formed during electrocoagulation is higher than that formed during chemical coagulation.

2.1.2. Sedimentation

Sedimentation is a solid–liquid process, which is based on the gravitational settling principle. In wastewater-treatment plants, sedimentation is used to remove settleable solids. The efficiency of the sedimentation process is related to various factors such as loading rate, water quality, particle/floc size and weight, tank geometry, and so on. A sedimentation tank can be designed for optimum efficiency (90–95% floc removal) or can be designed to operate at lower efficiencies, allowing the filters to remove most of the remaining solids. Usually the latter approach leads to a total plant optimization. Particles settle from suspension in different ways, depending on the characteristics and concentration of the particles. Four distinct types of sedimentation have been classified, reflecting the influence of the concentration of the suspension and flocculating properties of the particles:

(a) First-class clarification: Settling of dilute suspensions that have little or no tendency to flocculate.
(b) Second-class clarification: Settling of dilute suspensions with flocculation taking place during the settling process.
(c) Zone settling: Particles settle as a mass and not as discrete particles. Interparticle forces hold the particles (which are sufficiently close) in a fixed position, so that the settlement takes place in a zone.
(d) Compression settling: settlement takes place over the resistance provided by the compacting mass resulting from particles that are in contact with each other.

In a conventional sedimentation tank, the flow is usually horizontal. Circular or rectangular configurations are common for sedimentation tanks with horizontal flow (Figs. 3 and 4). The main design criteria for a sedimentation tank with horizontal flow are the surface loading rate, adequate depth and detention time for settling, and suitable horizontal flow velocity and weir loading rate to minimize turbulence. Typical design values are shown in Table 2.

Various features must be incorporated into the design to obtain an efficient sedimentation process. The inlet to the tank must provide uniform distribution of flow across the tank. If more than one tank exists, the inlet must provide equal flow to each tank. Baffle

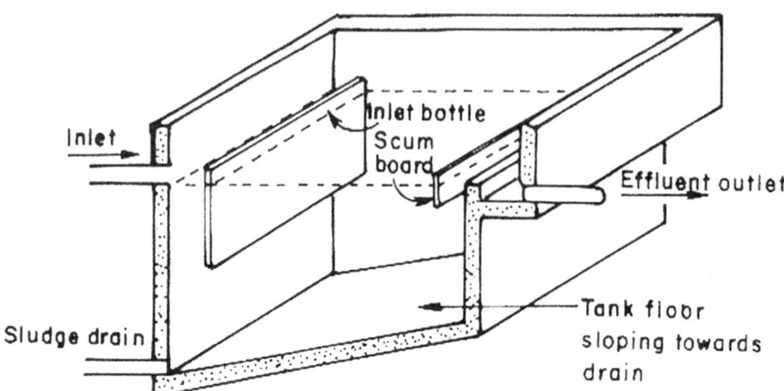

Fig. 3. Schematic diagram of a rectangular sedimentation tank (3).

walls are often placed at the inlet to distribute even flow, by use of 100–200-mm-diameter holes evenly spaced across the width of the wall.

The solid contact clarifier incorporates both flocculation and sedimentation in one unit, thereby reducing the plant size. It is also known as a sludge blanket clarifier or clariflocculator, depending on the design. In a clariflocculator, the flocculation is achieved by mixing the flocculent with turbid water at the central zone of the clarifier and the settling at the outer zone. The mixing of flocculent is usually carried out mechanically. Sludge is periodically removed from the sludge storage provided at the bottom. When the reactor is started up after some stoppage, it takes time for the blanket to form. Various shapes and different designs of sludge blanket clarifiers are available. A simple design of a solid contact clarifier is shown in Fig. 5.

2.1.3. *Rapid Filtration*

Rapid filtration is employed in water and tertiary wastewater treatments to remove various sizes of particles. These particles range from 0.1 to 100 μm in size, including microorganisms such as bacteria. Removal of these particles by rapid filtration involves

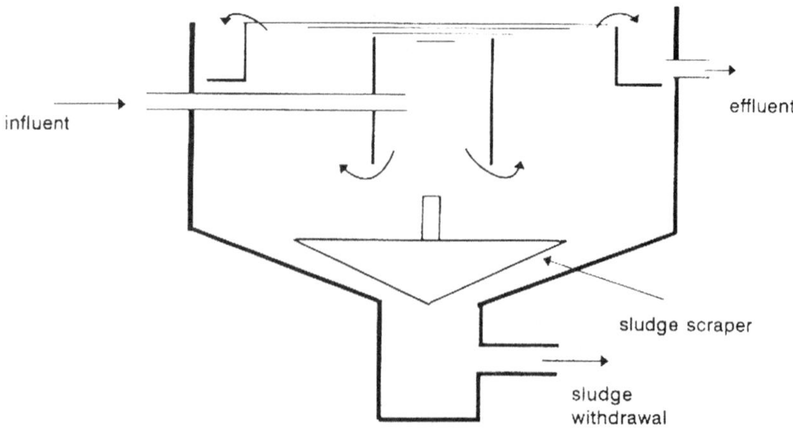

Fig. 4. Schematic diagram of a circular sedimentation tank (3).

Table 2
Basic Design Criteria for Horizontal Flow Sedimentation Tanks

Parameter	Design value
Surface loading rate ($m^3/m^2{\cdot}d$)	20–60
Mean horizontal velocity (m/min)	0.15–0.90
Water depth (m)	3–5
Detention time (min)	120–240
Weir loading rate ($m^3/m{\cdot}d$)	100–200

complex mechanisms. First, particles in suspension are transported near filter grains by mechanisms such as sedimentation, interception, diffusion, inertia, and the hydrodynamic effect, and then are attached to the filter grains or to the particles already attached on the filter grains. Removal mechanisms involved in rapid filtration not only depend on the physical and chemical properties of particles and filter medium but also on their surface characteristics. Particles that are brought near to filter grains by different transport mechanisms are attached to the surfaces of filter grains subsequently. The attachment is due to the surface forces, which act between the particles and the filter grains. These surface forces include van der Waals' attractive force, electric double-layer force, Born repulsive force and hydration force. Thus, the effective removal of particles in rapid filtration depends on the surface forces. This in turn depends on the surface chemistry of the particles as well as the filter grains. The surface chemistry depends on several other factors such as ionic strength of the suspension, which affects the electric double-layer force by altering the diffuse layer thicknesses of particles as well as filter grains.

Furthermore particles in suspension and filter grains will have surface charges. If the charges of particles and filter grains are opposite, then the condition of filtration is said to be favorable as particles will have attractive interaction with filter grains. If the charges of particles and filter grains are similar, the condition of filtration is termed as unfavorable as particles will have repulsive interaction with filter grains. Generally, filter media such as sand, glass beads, and particles in water or wastewater possess negative

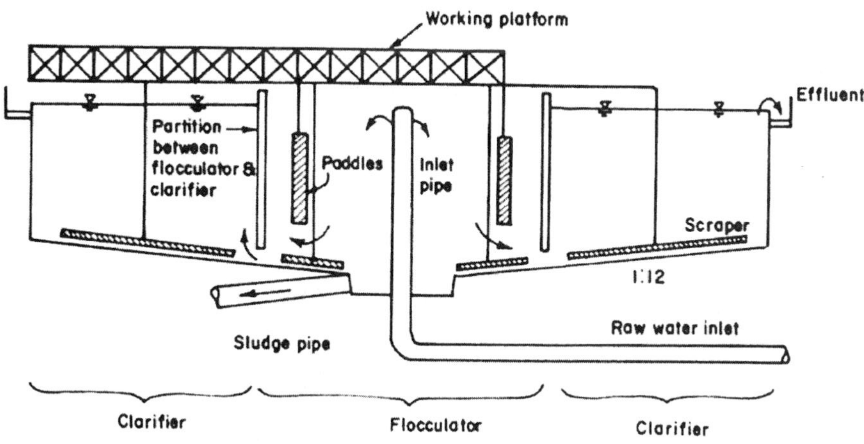

Fig. 5. Sludge contact clarifier (3).

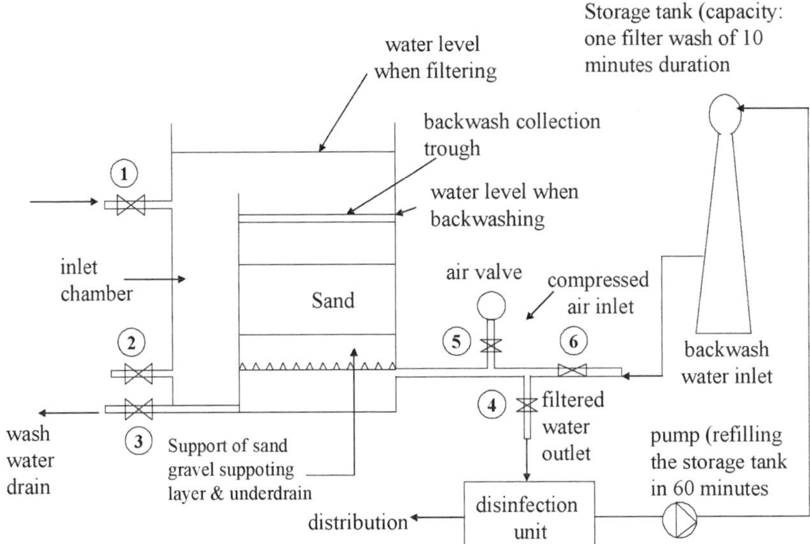

Fig. 6. Schematic diagram of a rapid filter.

surface charge; thus, most of the time filtration will occur under unfavorable conditions, if no chemicals are added to alter the conditions

Rapid filtration is generally placed in a treatment train after coagulation, flocculation, and sedimentation. The raw water to be treated by this train generally will have a turbidity of more than 10–20 NTU. Rapid filter can be either gravity type or pressure type. There is no biological action taking place in a rapid filter; however, some nitrification can occur (3). A typical rapid filter is shown in Fig. 6.

2.1.3.1. Dual–Media Filtration

Traditionally, only sand is used for the filter bed, with a filter depth of 0.6–1.0 m and a grain size of 0.4–1.2 mm. In this single-medium sand filter, stratification takes place during the backwashing, resulting in the very fine size of the medium accumulating at the top of the bed and the coarse particles remaining at the bottom. As a result of this, a major portion of the suspended matter is removed at the top layer (10–15 cm) of the bed. The particles that escape this top few centimeters of the bed tend to pass through the rest of the filter. Such conventional single-medium filters are normally restricted to a flow of 5–10 $m^3/m^2 \cdot h$ and an applied turbidity of 10 NTU or less. This shortcoming can be overcome if the arrangement of different sizes of grain is reversed, that is, if the arrangement is made from coarse to fine in the flow direction. In order to have this arrangement, media of different sizes and different specific gravity should be used: the lighter and coarser material at the top and heavier and finer material at the bottom. This arrangement is known as dual-media filter. Generally, a coarse medium of low density, such as anthracite (specific gravity = 1.35–1.70), over a fine but heavier medium, like sand (specific gravity = 2.65–2.70), is used. On backwashing in an upward direction, the coarse and lighter grains remain over the fine and heavier medium. This makes the penetration, and removal of impurities take place throughout the entire bed.

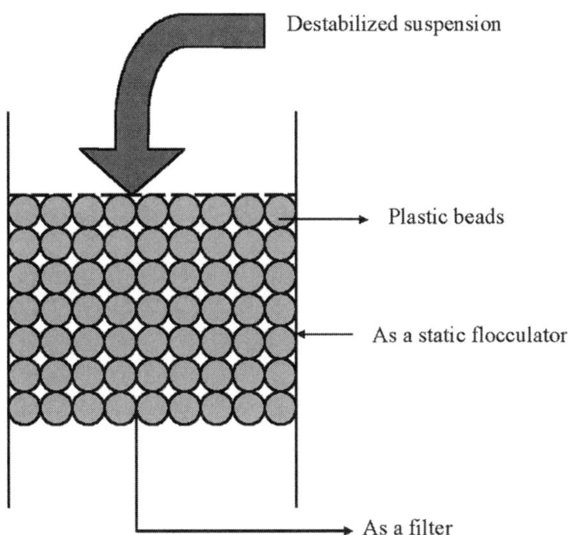

Fig. 7. Schematic diagram of floating-medium flocculator/filter.

2.1.3.2. FLOATING–MEDIUM FILTER

Filtration technology has been used for centuries in water treatment. Its use is becoming increasingly important as water reuse is envisaged. There have been a number of modifications made during the last two decades. The two main objectives to upgrade this technology are (a) to make the system compact and energy efficient to reduce the capital and operating costs and (b) to meet stringent water-quality standards for water reuse. There have been many designs, namely, high-rate multimedia filter, mobile-bed filter. The major problem of these designs is the high energy requirement for cleaning of the filter. The floating-medium filter developed overcomes these short comings. Generally, a large quantity of water for backwash is required to fluidize the highly dense filter media such as sand on anthracite. Backwash requirements become very high in the case of direct filtration (more specifically in contact flocculation/filtration). In direct filtration, the entire solid–liquid separation is within the filter bed itself. With the objective of reducing the backwash requirement, the use of synthetic buoyant filter materials (less dense than water) such as polypropylene, polystyrene, etc., are used. A saturated/unsaturated downflow buoyant-medium packed-bed filtration system with in-line flocculation arrangement has been successfully developed for water and tertiary wastewater treatment (4,5).

The floating medium filter is a direct filter that operates in a downflow mode, under constant head pressure, and incorporates in-line flocculation (Fig. 7). Coagulated wastewater enters the filter at the top where it comes into contact with the floating medium. Flocculation then occurs in the floating medium due to the promotion of interparticle contacts by the water flow around individual grains of media. This is followed by the separation of particles and flocs by floating-filter medium. Thus, it has a dual function of flocculation and solid–liquid separation. The floating-medium filter can be operated at very high velocity of 30–50 m/h. The backwashing of filters of this type can be achieved with a small quantity of water at a much smaller backwash velocity. Apart from this advantage, floating-filter media are also found to have a high solids retention capacity and low head loss development.

Ngo and Vigneswaran (6) found that the most suitable backwash method for a floating-medium filter was a combination air (30 kPa) and water (20–30 $m^3/m^2 \cdot h$) in the upward direction for 30 s followed with upward flow of water (20–30 $m^3/m^2 \cdot h$) for another 30 s. The backwash frequency of 90–120 min was found to be suitable for a long filter run. The consumption of backwash water was only 1.2–1.8% of filtered water production. In this backwash method, no air is necessary and the water requirement can be reduced by another 35%. An option of mechanical backwash using rotating paddles, followed by downflow water backwash could be the cheapest backwash method.

2.1.4. Adsorption

Adsorption is a surface phenomenon by which molecules of pollutants (adsorbates) are attracted to the surface of adsorbent by intermolecular forces of attraction. It takes place when atoms of surface functional groups of adsorbent (activated carbon) donate electrons to the adsorbate molecules (usually organic pollutants). The position of the functional groups (which are generated during activation process) of the adsorbent determines the type of adsorbent–adsorbate bond, and thus the type of adsorption. The physical adsorption is mainly caused by van der Waals' and electrostatic bonds between the adsorbate molecules and the atoms of the functional groups. The process is reversible, and thus desorption of the adsorbed solute can occur. The physical adsorption takes place at lower temperature (in the neighborhood of room temperature), and it is not site-specific. The adsorption can occur over the entire surface of the adsorbent at multilayers. On the other hand, the chemical adsorption involves ionic or covalent bond formation between the adsorbate molecules and the atoms of the functional groups of the adsorbent. The chemical adsorption is irreversible, and the heat of adsorption is typically high. The chemical adsorption process is site-specific and it occurs only at certain sites of the adsorbent at only one layer (monolayer).

Because the wastewater contains a large amount of organic and inorganic substances, it is possible that both physical and chemical adsorption takes place when it comes into contact with an adsorbent (usually activated carbon). However, for simplicity, only the physical adsorption process is discussed, as most of the adsorption-separation processes depend on physical adsorption.

The adsorption process with wastewater is competitive in nature. The extent of competition depends on the strength of adsorption of the competing molecules, the concentrations of these molecules, and the characteristics of the adsorbent (activated carbon). In a competitive adsorption environment, desorption of a compound may takes place by displacement by other compounds, as the adsorption process is reversible in nature. It sometimes results in an effluent concentration of an adsorbate greater than the influent concentration (7).

There are basically four steps an adsorbate passes through to get adsorbed onto the porous adsorbent. First, the adsorbate must be transported from bulk solution to the boundary layer of the wastewater surrounding the adsorbent (bulk solution transport). The transport occurs by diffusion if the adsorbent is in a quiescent state. In the fixed-bed or in the turbulent mixing batch reactors, the bulk solution transport occurs by turbulent mixing. Second, the adsorbate must be transported by molecular diffusion through the boundary layer surrounding the adsorbent particles (film diffusion transport). Third, after passing through the boundary layer, the adsorbate must be transported through the pores of the adsorbent to the available adsorption sites (pore transport). The intraparticle transport may occur by molecular diffusion through the wastewater solution in the pores (pore diffusion) of by diffusion along the surface of the adsorbent (surface diffusion). Finally, when the adsorbate reaches the adsorption site, the adsorption bond is formed between the adsorbate and the adsorbent. This step is very rapid for physical adsorption (8). Thus, it is either the bulk solution transport or film diffusion transport or pore transport that controls the rate of organic removal from the wastewater. In turbulent mixing condition (in fixed-bed or in batch reactor), it is most likely that a combination of film diffusion and pore diffusion controls the rate of adsorption of organics. At the initial stage, the film diffusion may control the adsorption rate but after the accumulation of adsorbates within the pore of the adsorbent, it is possible that the adsorption rate is controlled by the pore transport.

2.1.4.1. Design of Adsorption Systems

The adsorption processes used in practice are either batch mode or fixed bed mode depending on the characteristics of the adsorbent. In the batch mode, adsorbent is added to the tank containing wastewater. The pollutants such as persisting organics, pesticides, herbicides, and heavy metals are adsorbed onto the adsorbent surface and are subsequently removed by sedimentation–filtration processes. In fixed–bed mode, adsorbents are packed in a column, and the wastewater is passed through the column either from the top or from the bottom (fluidized mode). The pollutants are adsorbed on the adsorbent surface and thus the effluent of better quality is achieved.

Activated carbons, both granular activated carbon (GAC) and powdered activated carbon (PAC), are the oldest and most widely used adsorbents commercially as well as in the laboratory. They can be used in wastewater effluent treatment, potable water treatment, solvent recovery, air treatment, decolorizing, and many more other applications. The GAC is used as a fixed filter bed whereas the PAC is used directly in the aeration tank.

2.1.4.1.1. Batch Adsorption System. The batch adsorption system is usually used for the treatment of small volumes of wastewater. In the batch adsorption system, the adsorbent is mixed with the wastewater to be treated in an agitated contacting tank for a period of time. The slurry is then filtered to separate the adsorbent from wastewater. It can be performed in the single-stage or multistage system depending on the characteristics of adsorbate and adsorbent (Fig. 8).

2.1.4.1.2. Fixed-Bed Adsorption System. Depending on the characteristics of the wastewater and the adsorbent, the fixed-bed adsorption column can be operated in single or multiple units. The operation can be upflow or downflow. In the downflow operation, the filtration process is more effective. However, it suffers more pressure drop compared to the upflow operation. When a highly purified effluent is required, the fixed-bed adsorption columns are operated in series. The fixed-bed column operation in series and parallel are shown in Figs. 9 and 10.

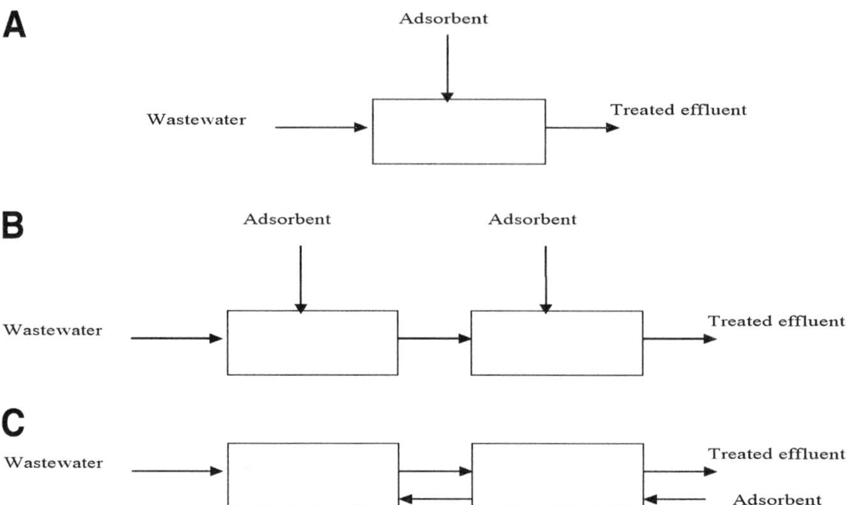

Fig. 8. Batch adsorption systems; (a) single-stage design, (b) two-stage crosscurrent design, and (c) two-stage countercurrent design.

2.1.4.1.3. Pulsed-Bed Adsorption System. In the pulsed-bed adsorption system, the adsorbent is removed at regular intervals from the bottom of the column and replaced by the fresh adsorbent from the top. The column is normally packed full of adsorbent so that there is no freeboard for bed expansion during operation (Fig. 11).

2.1.4.1.4. Fluidized-Bed Adsorption System. In upflow operation, the adsorption bed is completely fluidized and hence expanded. When the adsorbent particle size is small, it is advantageous to use the fluidized-bed adsorption system. It reduces the excessive head due to the fixed bed clogging with particulate matter often experienced in downflow adsorption system.

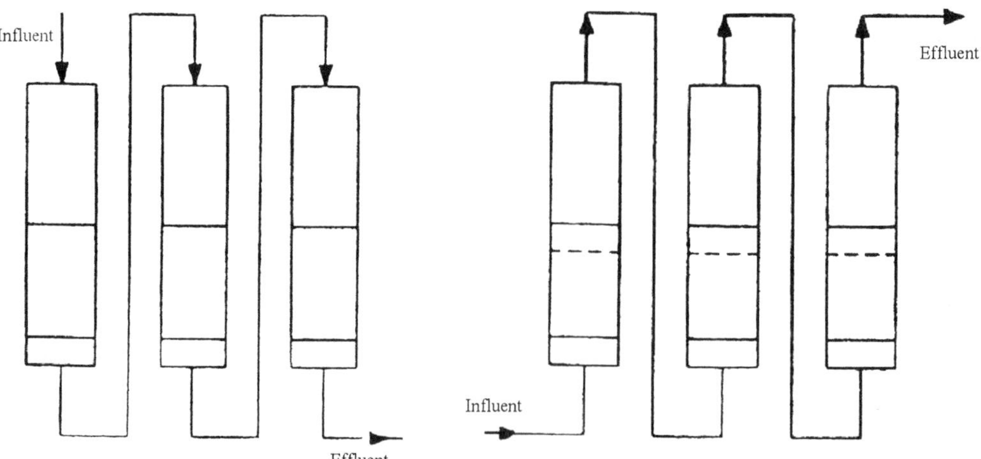

Fig. 9. Fixed bed adsorption columns in series downflow and upflow operational mode.

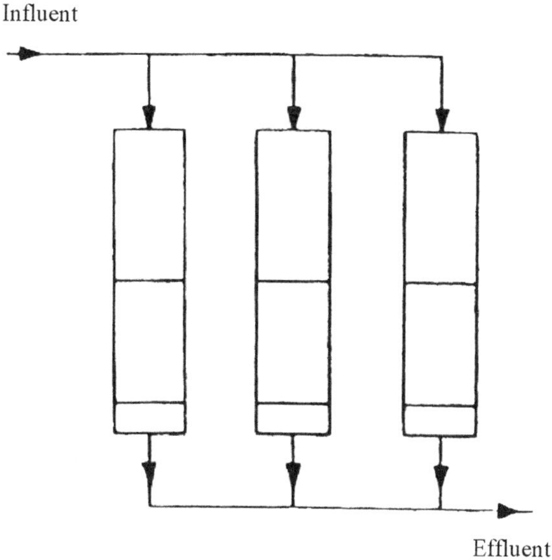

Fig. 10. Fixed bed adsorption columns in parallel operational mode.

2.1.4.1.5. Powdered Activated Carbon Treatment (PACT). The performance of the aerobic or anaerobic biological treatment process can be improved by adding powdered activated carbon (PAC) to the process. The PAC particles help in reducing the problems of bulking of sludge or foaming associated with the activated-sludge process. The PAC particles enhance the biological assimilation of organics. During the process, the adsorption capacity of the PAC is also partially renewed by concurrent microbial degradation of adsorbed organic substances (9). The primary advantages of using PAC are (a)

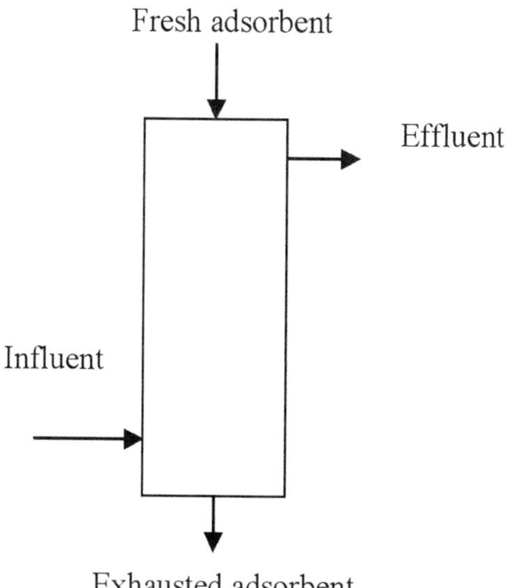

Fig. 11. Pulsed bed adsorption system.

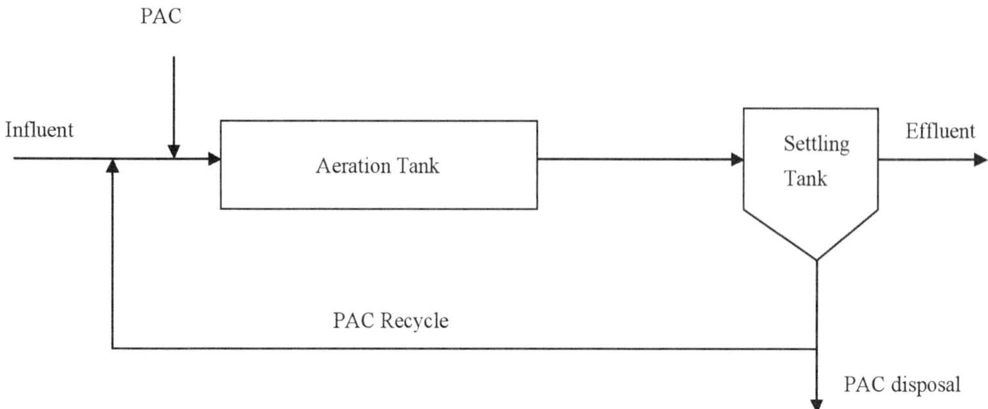

Fig. 12. Schematic diagram of PACT adsorption system.

low capital investment cost and (b) possibility of changing the PAC dose as the water quality changes. The main disadvantages of use of PAC are (a) high operating cost if high PAC dose is required, (b) low TOC removal, (c) inability to regenerate, and (d) difficulty of sludge disposal. However, the use of PAC can enhance the performance of the existing biological treatment system by removing dissolved substances, forming settleable flocs, and stabilizing the system against toxicity and shock loadings (10). The general PACT is shown in Fig.12.

2.1.5. Flotation

Flotation is used to separate solids or dispersed liquids from a liquid phase. The separation is effected by introducing fine gas bubbles, usually air, into the system. The added fine air bubbles either adhere to or are trapped in the particle's structure, making the particles buoyant and bringing them to the surface. Even particles with a density greater than that of the liquid phase can be separated by flotation. Surface properties (rather than size or relative density) of particles play a predominant role during flotation.

Two types of flotation in use are dispersed and dissolved air flotation. In dispersed air flotation, air is directly introduced into the liquid through diffusers, whereas, in the case of dissolved air flotation, air bubbles are produced by precipitation from a solution supersaturated with air. Production of air bubbles can be achieved by dissolved air pressure flotation. Here the influent to the flotation unit is pressurized and then released in the unit to produce air bubbles.

Dissolved air pressure flotation can be used (a) to remove solids from industrial wastewater and (b) to separate and concentrate biomass after the biological treatment of the municipal wastewater. A typical flow diagram of a dissolved air pressure flotation system is shown in Fig. 13. The important factors in designing the unit are the influent solids concentration, the quantity of air expressed as air-to-solid ratio, and the overflow rate. For better design of the flotation unit, laboratory tests of the preliminary design and pilot-plant studies are always recommended.

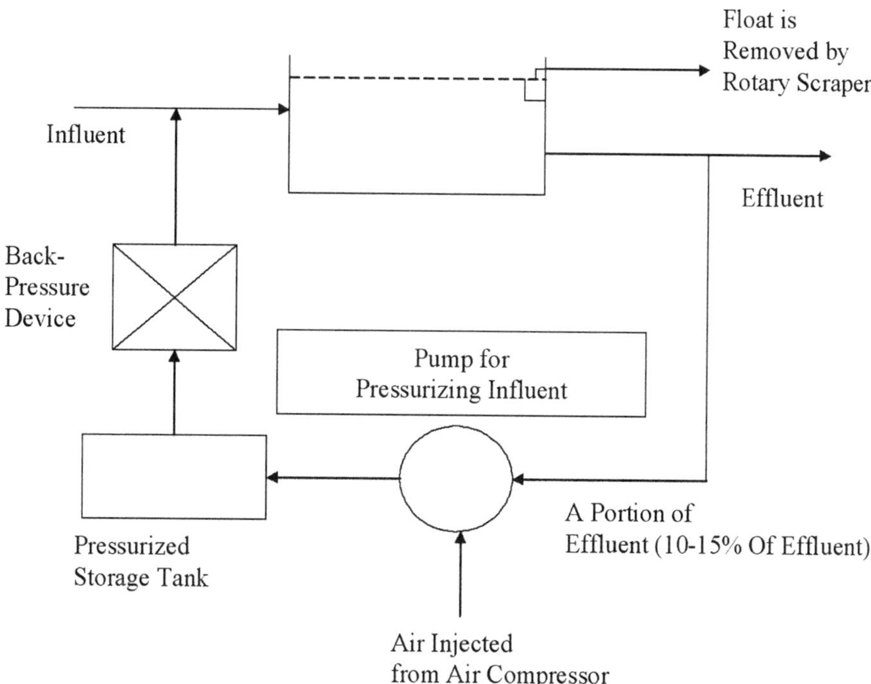

Fig. 13. Flow diagram of dissolved air pressure flotation system.

2.1.6. *Ion Exchange*

Ion exchange is defined as a process where an insoluble substance (resin) removes ions of positive or negative charge from an electrolytic solution and releases other ions of like charge in a chemically equivalent amount. The ions in solution diffuse into the molecular network of the resin where exchange occurs without any structural change of the resin. The ion exchange proceeds until ion exchange equilibrium is established.

The general reaction of the exchange of ions A and B on a cation exchange resin can be written as follows:

$$nR^{-}A^{+} + B^{n+} \Leftrightarrow Rn^{-}B^{n+} + nA^{+}$$

(Resin) (Solution) (Resin) (Solution)

where R^- is an anionic group attached to the ion exchange resin.

There are two types of resins: (a) cation exchange resins, which contain exchangeable cations, and (b) anion exchange resins with exchangeable anions. Cation and anion resins are also known as acid exchange and base exchange resins, respectively. Based on the ion exchange capacity (which is a measure of the total quantity of ions that can be theoretically exchanged per unit mass or per unit volume of resin), resin can be further classified as strong and weak cation/anion resins. In general, ions of high valency are preferred over ions of low valency. For example, Fe^{3+} ion is more easily removed than Na^+ ion:

$$Fe^{3+} > Mg^{2+} > Na^{+}; \qquad PO^{3-} > SO^{2-} > NO^{-}_{3}$$

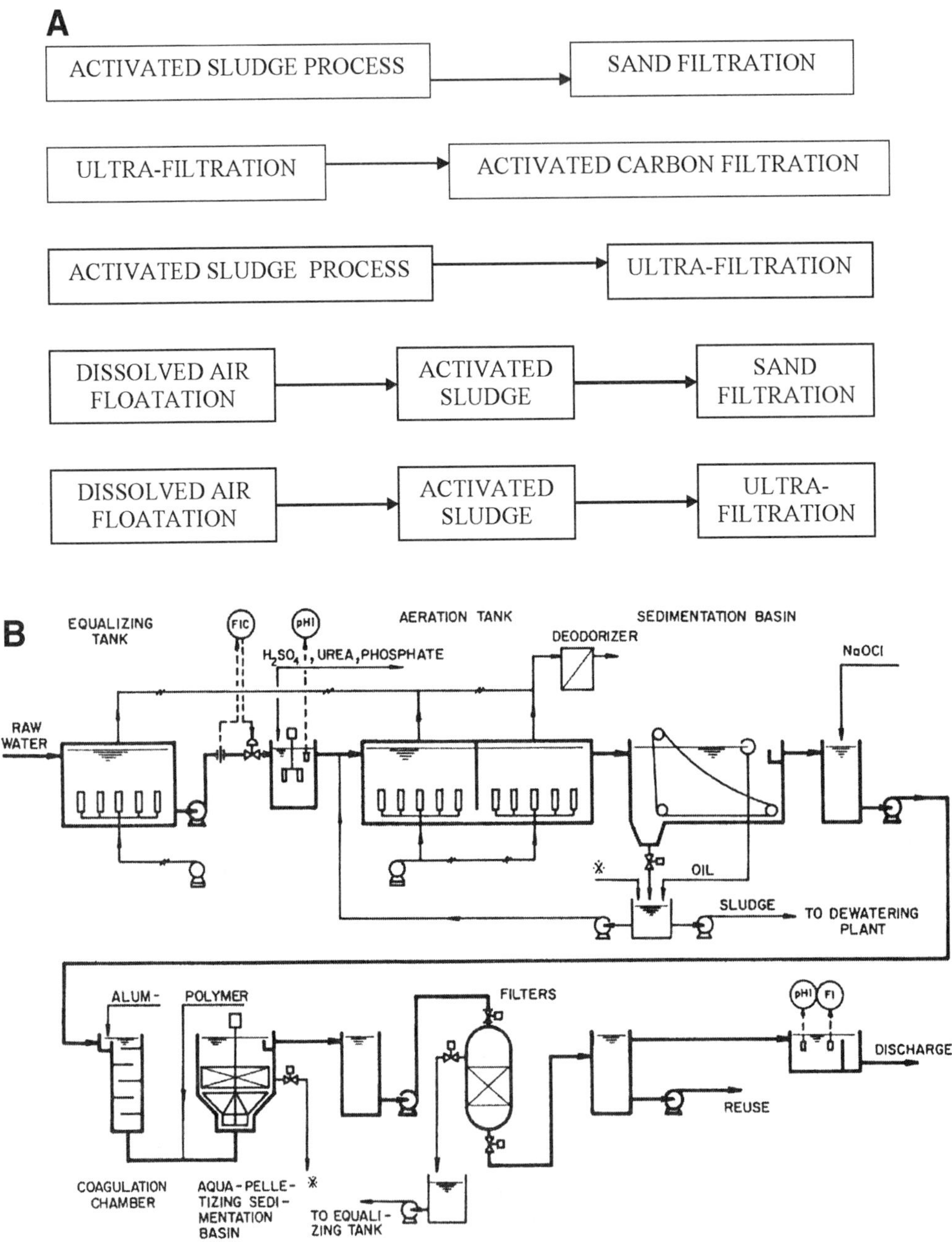

Fig. 14. (A) Typical combinations of treatment processes used in gray water reuse. **(B).** Flowchart of a wastewater recycling plant for an office building (11).

For the ions of the same valency, the extent of exchange reaction increases with decreasing hydrated radius and increasing atomic number:

$$Ca^{2+} > Mg^{2+} > Be^{2+}; \qquad K^{+} > Na^{+} > Li^{+}$$

Table 3
Design Criteria of Different Units Used in the Treatment and Reuse of Wastewater from Office Buildings (11)

Unit	Design Criteria
Aeration tank	BOD loading: 0.45 kg/m^3/d; detention time: 11 h
Primary sedimentation tank	Surface loading: 10 mm/min; detention time: 5 h
Nitrification tank	NH_4^+–N loading: 0.25 kg/m^3/d; detention time: 7 h
Coagulation tank	Detention time: 0.5 h
Secondary settling tank	Surface loading: 15 mm/min; detention time: 4 h
Filter	Filtration rate: 150 m/d; backwash rate: 0.6 m/min; air-scour rate: 0.6 m/min
	Filter media: Anthracite; effective size (ES) = 1.2 mm, depth (L) = 200 mm
	Sand ES = 0.6 mm, L = 400 mm
	Gravel size = 2–40 mm
Chemical dosage	Aluminum sulfate: 200 mg/L; sodium hypochlorite: 20 mg/L
	sodium hydroxide: 80 mg/L for coagulation and 80 mg/L for nitrification

2.2. *Application of the Physicochemical Processes in Wastewater Treatment and Reuse*

The physicochemical processes have extensively been used in treating domestic and industrial wastewater, in combination with other biological treatment processes. In this section, the application of the physicochemical processes in wastewater treatment and reuse are discussed with some design data.

Different combinations of treatment processes that can be used in grey water recycle are presented in Fig. 14A. The schematic flow diagram of the treatment process used in the recycle of wastewater in office buildings is shown in Fig. 14B. The general design criteria of different units that can be used in treatment and reuse of wastewater from office building is shown in Table 3.

Wastewater from a brewery consists of 30% from the brewery process and 70% from the bottle washing process by volume. Normally, the activated-sludge process is used to treat this wastewater as it contains biodegradable organic matter. The effluent from the biological treatment can be reused for the bottle washing process when the physicochemical processes such as coagulation, sedimentation, and filtration are added to the activated-sludge process. The schematic flow diagram of the treatment process is shown in Fig. 15. The design parameters of these processes are shown in Table 4. These physicochemical processes remove practically all the organics (BOD_5) and suspended solids.

In Australia, a combination of physicochemical processes has been utilized in a number of municipal wastewater-treatment plants for wastewater treatment and reuse. Wastewater reuse from sewage treatment plants (STP) was 171 GL (i.e., 11% of STP effluent) in 2000, and it is growing by 28 GL/yr (12). Rouse Hill STP, Sydney is one example where water-reuse practice has been strictly followed by the authority and community residing in the catchments.

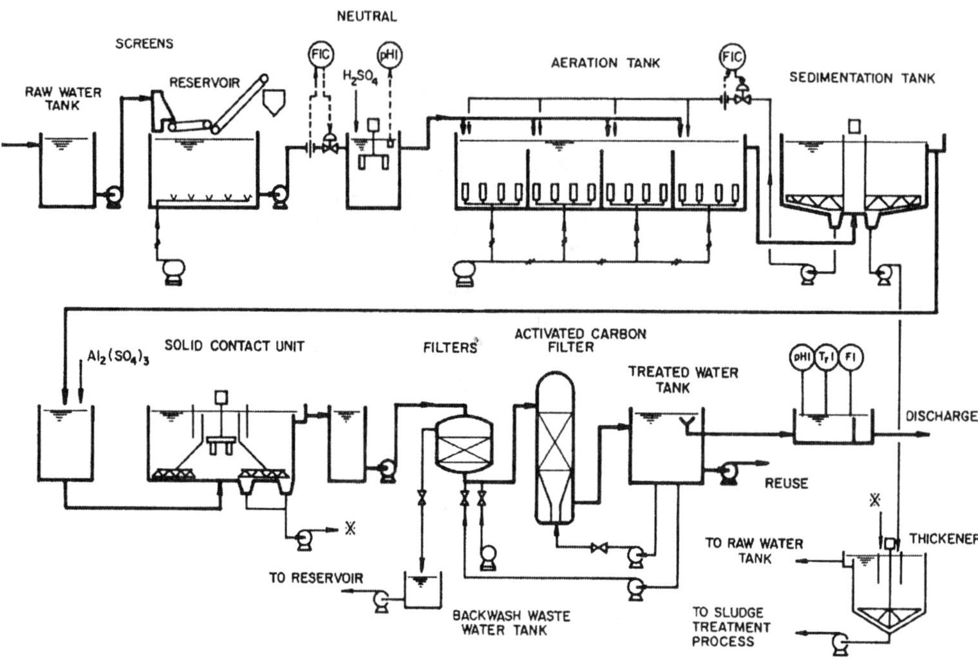

Fig. 15. Flowchart of a brewery wastewater treatment plant (11).

The Rouse Hill STP in Sydney, Australia, for example, uses multistage biological nutrient removal technology with chemical polishing for phosphorus removal. The capacity of the plant is nominally 25,000 equivalent population (EP) (6.75 MLD), but the capacity of individual processes varies from 25,000 to 100,000 EP. The STP has been designed to achieve the effluent standards shown in Table 5.

The sewage delivered to the plant first undergoes fine screening through fine continuous belt screens. Screenings are automatically transported, dewatered and discharged

Table 4
Design Criteria of Different Units Used in the Treatment of Beer Factory Wastewater (11)

Unit	Design Criteria
Neutralizing basin	Detention time: 8 min
Aeration tank	BOD loading: 1.2 kg/m^3/d; detention time: 22 h
Sedimentation basin	Surface loading: 10 m/d; detention time: 9 h
Coagulated Sedimentation basin	Surface loading: 50 m/d; detention time: 2.5 h
Activated carbon filter	Linear velocity: 5–20 m/h
Filter	Filtration rate: 200 m/d; backwash rate: 1.0 m/min; air-scour rate: 1.0 m/min; filter run: 12 h;
	Filter media
	Anthracite: effective size (ES) = 1.2 mm; depth (L) = 200 mm
	Sand: ES = 0.6 mm; L = 500 mm

Table 5
Treated Effluent Standards for Rouse Hill STP

Parameters	Standards
Biochemical oxygen demand (BOD_5)	5.0 mg/L
Non-filterable residue (NFR)/suspended solids	5.0 mg/L
Ammonia-nitrogen	1.0 mg/L
Phosphorus	0.3 mg/L
Total nitrogen	5.0 mg/L
Fecal coliforms	< 200 cfu / 100 mL

into 240 L garbage bins for land fill disposal. Screened sewage flows to a vortex-type grit trap. The grit from this trap is pumped to cyclone separators and then grit classifiers. Grit from this system is loaded into garbage bins for disposal. Sewage is then channeled via a flow-measuring flume to the solids separation tank where solids are concentrated via settling processes. Settled solids are recirculated in the tank, pumped away for disposal, or transferred to the biological reactor prefermenter. Flow less than three times average dry weather flow (DWF) goes from the solids separation tank to the biological reactor and the flow greater than three times average DWF is by-passed to chlorine contact tank.

Settled sewage is directed to a multistage biological reactor for organic and nutrient removal. The reactor includes a prefermenter, denitrification tank having anaerobic, anoxic, and aeration zones. Mixed liquor is recycled from the aeration zone back to the anoxic zones. Aeration is achieved using mechanical aerators. The mixed liquor from the biological reactor is settled in circular outward flow steel clarifiers. These are equipped with scum-withdrawal facilities. The return activated sludge is withdrawn from each clarifier individually by variable speed pumps at specified rates.

Secondary effluent gravitates to a rapid mix tank for chemical addition and a flocculation chamber for floc formation. Flow is then directed to a tertiary clarifier where the sludge is drawn off and recirculated or wasted to the digester. After final clarification, the effluent is filtered through shallow bed sand filters. Effluent is then chlorinated and the chlorinated effluent is then either directed to the wetlands for further polishing or to the recycle system. The treated effluent that meets the turbidity and pH requirements is pumped to another set of chlorine contact tanks. Here chlorine is dosed to give a residual of 0.5 mg/L after 1-h detention as required by the recycled-water specifications (Table 6).

The chlorinated recycle water is then checked for pH and adjusted if required. Recycled water that fully meets the specification is stored in holding tanks and is available for delivery to the distribution system. If it does not fully meet specification, it can be re-worked by directing it back for re-chlorination. Off-specification water is drained to a pumping station and passed through the treatment processes. To reduce the chlorine residual from that required for initial disinfection, each recycled-water reservoir is provided with a dechlorination facility using sodium bi-sulfite. To guarantee the integrity of the supply, the recycled-water reservoirs are cross connected to the potable water reservoirs. Connection to recycled water is mandatory for

Table 6
Recycled-Water Quality Criteria

Parameters	Limits
Microbial	
(a)At the STP:	
Fecal Coliforms	< 1 in 100 mL
Total Coliforms	< 10 in 100 mL
Viruses	< 2 in 100 mL
Parasites	< 1 in 50 mL
(b)At the user ends:	
Total Coliforms	< 2.5 in 100 mL (geometric mean over 5 days)
Physical	
Turbidity	< 2 NTU (geometric mean)
pH	< 5 NTU (95% of samples)
Color	6.5–8.5 (preferably 7.0–7.5)
Chemical	< 15 CTU
Free-chlorine residual	< 0.5 mg/L (at consumer end)

toilet flushing and outside (yard) uses throughout the new development areas in the catchments.

The recycled water, however, is not permitted for: (a) drinking, cooking, or kitchen purposes; (b) bath, showers, hand basins, or personal washing; (c) clothes washing; (d) swimming pools; (e) water-contact recreation; and (f) irrigation of crops for human consumption that are neither processed nor cooked.

2.2.1. *Semi-Pilot-Scale Application of High-Rate Floating-Medium Filter*

2.2.1.1. Floating Medium in Domestic Wastewater Treatment

The performance of the filter system was studied at a filtration velocity of 30 $m^3/m^2 \cdot hr$ using ferric chloride ($FeCl_3$) and polysilicato-iron (PSI) as flocculants. The effluent from the diversion chamber of a sewage treatment plant in Sydney, Australia after screening was sent to the filter system. The specific characteristics of wastewater used are summarized in Table 7. The optimum dose of $FeCl_3$ and PSI was found to be 15 mg/L and 2.5 mg/L, respectively. The filter was operated with frequent (once in 2 h) but very short backwash (30 s) during the filter runs.

The system was extremely effective in removing phosphorus and a majority of bacteria from the secondary sewage effluent. During the 16 h of filter run, the number of fecal streptococci (FS) and fecal coliform (FC) in the treated effluent were much lower than the effluent discharge standard of 200 No./100 mL. The chlorine dose required can thus be significantly reduced. The phosphorus (as orthophosphate) was also removed up to 90%. Head loss development was very low when $FeCl_3$ was used as flocculant (e.g., < 20 cm after 16 h of filtration time). In both cases, the average turbidity and COD of treated effluent were less than 1 NTU and 20 mg/L, respectively. The phosphorus removal efficiency was 93% at the filtration rate of 15 $m^3/m^2 \cdot h$ and 80% at 45 $m^3/m^2 \cdot hr$. The head loss development was lower than 24 cm for all cases. The results also indicated that the variation in filtration rates had no significant effect on

Table 7
Specific Characteristics of Wastewater

Parameter	Concentration
pH	6.4–6.8
Total suspended solids (TSS) (mg/L)	5.4–14.8
Turbidity (NTU)	3.7–13
Total phosphorus (T- P) (mg/L)	0.18–0.36
Chemical oxygen demand (COD, mg/L)	10–38
Fecal coliforms (cfu/100 mL)	3.2×10^3–7.9×10^4
Fecal streptococci (cfu/100 mL)	1.2×10^2-6.6×10^3

the fecal coliform and fecal streptococci removal (up to 30 $m^3/m^2{\cdot}h$) (Table 8). The filter system was just as effective in removing bacteria when operated at a filtration velocity of 15 $m^3/m^2{\cdot}h$, as it was when operated at 30 $m^3/m^2{\cdot}h$. However, a lot more numbers of FC and FS escaped through the filter bed when the filtration rate was increased to 45 $m^3/m^2{\cdot}h$. The total backwash water for the filter run was less than 1% of water production when using air and water backwash or combined mechanical and water backwash.

2.2.1.2. Floating Medium in Shrimp Farm Wastewater Treatment and Recycle

Pilot-scale studies were also conducted with the effluent from a shrimp farm in Queensland, Australia. In this study, the floating-medium filter was used as a flocculator/prefilter in the filter system together with sand filter as a subsequent polishing filter. The system was operated at 15 $m^3/m^2{\cdot}h$ with in-line polyaluminum chloride (WAC-HB) addition arrangement. The depth of floating medium (3.8 mm polypropylene beads) and sand (effective size, ES = 1.7 mm) was 60 cm each. The characteristics of prawn farm effluent used are summarized in Table 9.

The effluent quality from this combined filter system was acceptable for discharging in terms of turbidity (2.4 NTU), SS (5 mg/L), and $PO_{4\text{-}}P$ (0.04 mg/L). The head loss development was less than 20 cm after a 180 min filter run. A trial was also conducted with the same conditions but without addition of WAC-HB. The removal efficiency was about 50%. A comparative study conducted on backwash showed that 1 min wash per every 90 min could maintain high effluent quality for a long filter run. It is noted that the consumption of backwash water was about 1% of filtered water production. The quantity of backwash water was even smaller when a mechanical backwash method was

Table 8
Mean Percentage Bacteria Removal for Various Filtration Rates

Filtration velocity ($m^3/m^2{\cdot}h$)	Mean % removal (fecal coliforms)	Mean % removal (fecal streptococci)
15	96.4	97.1
30	95.4	95.3
45	57.8	61.3

Table 9
The Characteristics of Moreton Bay Prawn Pond Effluent (13)

Parameter	Concentration
pH	8.1–8.4
Turbidity (NTU)	18–49.9
Suspended solids (mg/L)	25–65
PO_4-P (mg/L)	0.08–0.22
Total organic carbon, TOC (mg/L)	5–6

applied. Figure 16 shows a system design for a high-rate floating-medium filter in prawn farm treatment and recycle.

The filter system is very compact compared to the traditional treatment systems such as settling tank, wetlands, and so forth, in prawn farm effluent treatment. For example, to treat 5 ha pond with 1 m depth and 10% exchange water per day, the space requirement for settling pond (overflow rate of 40 $m^3/m^2{\cdot}d$) is about 20 times larger than a high-rate floating-medium filter (filtration rate of 30 $m^3/m^2{\cdot}h$)

2.2.2. *Advanced Wastewater Treatment System at Dhahran, Saudi Aramco*

Wastewater from Dhahran, KFUPM, and local communities is treated at Dhahran North Sewage Treatment Plant (STP), Saudi Aramco, located on the Khobar Highway (14). The North STP has been designed to treat 8 MGD wastewater with 150 mg/L BOD

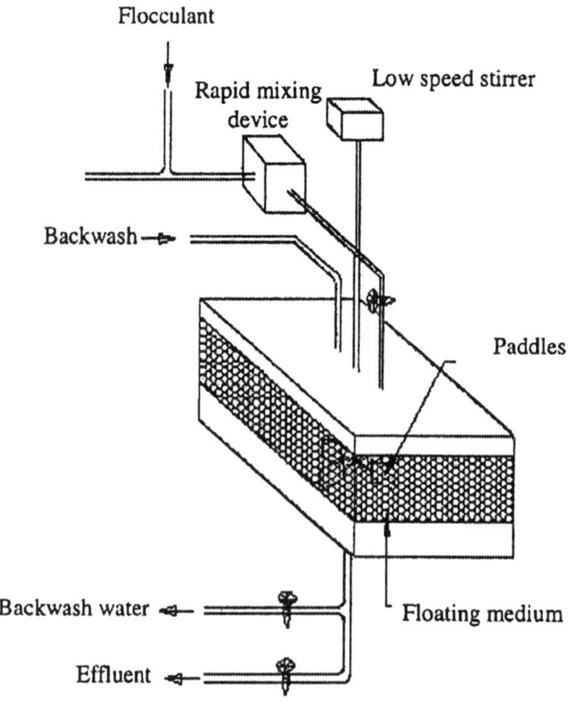

Fig. 16. A high rate filtration system for prawn farm effluent treatment and recycle.

(biochemical oxygen demand). The treatment processes include bar screen, grit chamber, Parshall flume, activated-sludge aeration tank with 13 h hydraulic detention time, final setting tank, effluent holding tank, and chlorine contact tank with residual chlorine of 2 mg/L. Sludge is treated at aerobic sludge digesters and digested sludge is pumped to sludge drying beds for drying by sun. Since Saudi Arabia is very hot country with a lot of sunshine, drying beds are ideal for sludge treatment. Dried sludge is disposed at a landfill. The effluent from treatment plant is pumped to percolation field by a 9 mile pipeline. The treatment processes at the percolation filed consist of percolation, evapotranspiration, and advanced wastewater-treatment system.

The flow to the advanced wastewater-treatment plant is 3.3 MGD. The plant is used to upgrade secondary effluent produced from Dhahran North Sewage Treatment Plant to meet unrestricted effluent standards for irrigation. The treatment processes include coagulation with alum and are followed by flocculation, sedimentation tank, rapid sand filters, and disinfection with chlorine with free chlorine residual of 1 mg/L. The treated effluent is reused for irrigation for watering lawns of homes, common greenbelt areas, recreational grounds, and roadside medians.

2.2.3. *Application of Ion-Exchange Process*

Ion-exchange process is commonly used to remove hardness (polyvalent cations), iron, and manganese from drinking water supply. It can be used to remove and/or recover many different types of ionic chemical species from the industrial wastewater (such as chromium and nickel from metal-plating wastewater). Ion-exchange process can also be used in municipal wastewater-treatment system to remove nutrients (nitrogen and phosphorus) and to demineralize the effluent for reuse purpose.

The ion-exchange system can be operated in one of the following modes: (a) batch, (b) fixed bed, (c) fluidized bed, and (d) continuous feed. Of these four modes, the fixed-bed system is the most commonly used.

A typical fixed-bed operating cycle consists of four steps: (a) service, (b) backwash, (c) regeneration, and (d) rinse. The service life of the fixed-bed ion-exchange system can be evaluated from the effluent curve or breakthrough curve. Backwash is provided to break up resin clumps and to remove finely divided suspended materials entrapped in the resin. It also eliminates gas pockets and restratifies the resin bed to ensure a uniform distribution of flow during service. Regeneration is provided to displace ion exchanged during the service run and to return the resin to its initial exchange capacity or to any other desired level depending on the amount of regenerant used. In general, mineral acids and alkalies are used to regenerate cation exchange resins and anion exchange resins, respectively. After the regeneration step, the ion-exchange resin must be rinsed free of excess regenerant before being put back into service

Ion exchange is preferable to chemical precipitation for the water-softening process when raw water contains low color and turbidity level, hardness is largely not associated with alkalinity (i.e., noncarbonate hardness is substantial), and hardness levels vary. Most widely used system for water softening is a continuous-flow fixed-bed column using a strong cationic resin in the sodium form, and the regeneration is done by NaCl (salt) solution.

Typical design criteria for a water softening resin system are:

Service flow rate: 12.5–20 $m^3/m^2 \cdot h$
Backwash rate: 50–70% expansion of resin bed
Regeneration: Regeneration solution concentration: 2–10 %; contact time: 30 min (2.5–5 $m^3/m^2 \cdot h$)
Rinsing: 1.5–3 m^3 of water/m^3 of resin volume
Bed depth: 75 cm minimum
Free board: 50–75% of bed depth

3. MEMBRANE PROCESSES

3.1. *Principle*

3.1.1. General

Membrane processes used in wastewater treatment can be categorized into four classes according to the size of particles that can be retained, namely, reverse osmosis (RO), ultrafiltration (UF), microfiltration (MF), and electrodialysis (ED). ED is a proven process for desalting brackish water. RO is also used extensively in desalting applications. It has an added advantage of being able to remove many organic compounds in addition to ionic species and microorganisms. The UF and MF techniques are useful in removing macromolecules, colloids, and suspended solids.

Filters and membranes differ in the mechanism by which the solute is retained. Filters, such as paper, rely on having the particles trapped within the fibrous network composing them, which eventually results in a decreasing flux rate due to plugging. Sand filters separate the solid particles through particle transport and attachment onto the filter grain. Membranes, on the other hand, rely on the discreteness of the pore size opening in contact with the feed solution and the porous substructure under the thin skin. It is unlikely that particles will become trapped within the membranes. The ability of membranes to reject macromolecules is based on the sieve mechanism. The size and shape of the macromolecule are important factors that determine whether the molecule will pass through the membrane.

The ability of membranes to separate ions from water depends not only on the pore size, but also on the surface adsorption phenomena. For instance, it is observed that above 97% of salt is rejected in RO membrane processes, although there is no major molecular size difference between water and common ions found in sea water. The high rejection occurs, in this case, due to the repulsion of ions away from the membranes, and the preferential adsorption of water molecules on the membrane surface. The pressure applied on the feed forces water through the membrane while retaining the solute in the bulk solution. Membrane processes are chosen according to the size range of solutes in the solution. Table 10 shows the operating conditions of membrane processes.

3.1.2. Reverse Osmosis and Nanofiltration

Reverse osmosis (RO) is based on the well-known phenomenon of osmosis, which occurs when two solutions of different concentrations are separated by a semipermeable membrane. In this process, pressure is applied on the side of the concentrated solution to reverse the natural osmotic flow. The thin RO membranes are essentially nonporous, and they preferentially pass water and retain most solutes including ions. The rejection

Table 10
Differences between RO, UF, and MF

Parameter	MF	UF	RO
Membrane	Porous, isotropic	Porous, asymmetric	Nonporous, asymmetric or composite
Pore size	50 nm–10 μm	5–20 nm	—
Transfer mechanism	Sieving mechanism (the solutes migrate by convection)	Sieving and preferential adsorption	Diffusion law (the solutes migrate by diffusion mechanism)
Law governing the transfer	Darcy's law	Darcy's law	Fick's law
Type of solution treated	Solution with solid particles	Solution with colloids and/or macromolecules	Ions, small molecules
Permeability range	10–100 $m^3/(m^2 \cdot bar \cdot d)$	1 $m^3/(m^2 \cdot bar \cdot d)$	0.01 $m^3/(m^2 \cdot bar \cdot d)$
Pressure applied	1 bar	1–5 bar	20–80 bar
pH (depending on membranetype)	—	1–10	4.5–7.5

(or retention) of ions is typically in the range 95–99.9% depending on the ions and the membrane. RO is characterized by high operating pressures (2,000–10,000 kPa). The RO unit was first used for desalination of sea water. At present, there are large installations with a capacity of 2×10^6 m^3/d to treat sea water for domestic and industrial purposes. In addition, there are small and medium-sized RO installations existing (0.4–95,000 m^3/d) to supply pure water for specific purposes like petroleum platforms, agricultural purposes, sterilized water for hospitals, laboratories, etc. In some countries, the surface water as well as water from aquifers contains nitrates in excess and therefore has to be treated.

The RO process is used for the production of pure water for industrial purposes. One of the main industrial uses of RO is to prepare ultrapure water for the electronic industry. RO membranes remove the contaminants except dissolved gases. RO units are also utilized to produce sterilized water for pharmaceutical industries and for medical purposes because they produce water absolutely free from bacteria and suspended solids. RO should be able to retain all species of concern, except for some organic species that may be partially transmitted by some types of RO membranes.

However, RO systems have some limitations. The operating range of RO falls in the order of a nanometer. The rate of permeate flux is very low although the pressure applied is as high as 2,000–10,000 kPa, whereas other membrane processes operate at comparatively lower pressure and a higher flux rate. Because of low flux, RO needs a larger membrane surface area. Furthermore, RO does not operate successfully for higher solute removal with solutions of high concentration. For optimum performance, a good pretreatment should be provided. In addition, owing to concentration polarization, gel-layer formation, fouling, and internal clogging, the permeate flux rate decreases as the membrane filtration proceeds.

Nanofiltration (NF) membranes have pores of size 2–5 nm and partially retain ions. Small and monovalent ions and low-molecular-weight organics tend to pass through the

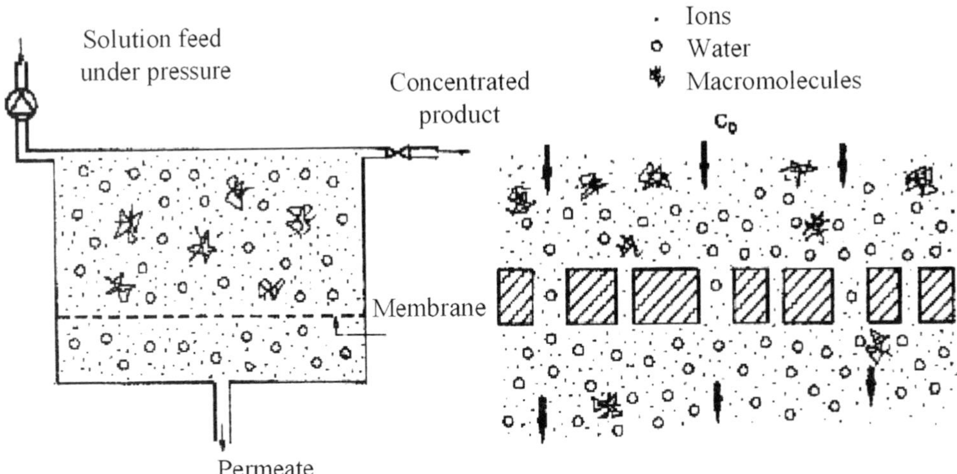

Fig. 17. Movement of molecules through membrane (17).

membrane. NF membranes usually have significantly higher water permeability than RO membranes and operate at lower pressures, typically 700–3,000 kPa (15).

2.1.3. *Ultrafiltration*

The ultrafiltration (UF) membrane allows the passage of water and low-molecular-weight solutes but retains macromolecules whose size is larger than the pore size of the membrane. UF utilizes permeable membranes to separate macromolecules and suspended solids from solution on the basis of size, separating compounds with molecular weights from 1000 to 100,000 (1 to 100 nm in size). UF enables concentration, purification, and fractionation of macromolecules in solution at ambient temperature and without phase change or addition of solvents. This protects the biochemical structure and activity of the product, giving increased yield over conventional technologies.

The application of high pressure to the feed side of the membranes enables the passage of water through the membrane. This makes the higher-molecular-weight compounds concentrate on the high-pressure side, while the concentration of lower-molecular-weight compounds remains same on both sides of the membrane (Fig. 17).

A variety of polymers such as cellulose acetate, polyvinyl chloride, polyacrylonitrile, polycarbonate, and polysulfone have been used to manufacture UF membranes. At present, inorganic membranes are also used which are more durable. Membranes must be compatible with the feed solution and, since membranes are subjected to surface fouling, they must be compatible with cleaning agents, too.

2.1.4. *Microfiltration*

Microfiltration (MF) is a pressure-driven membrane process for the separation of particles, microorganisms, large molecules, and emulsion droplets. The filter medium is a microporous membrane with a separation limit in the range of 0.02 to 10 μm. MF is a reliable separation process because separation of the abovementioned matter is difficult and not economical by the other separation methods (e.g., sedimentation, centrifugation, depth filtration). Hence, MF finds an important place in water and wastewater

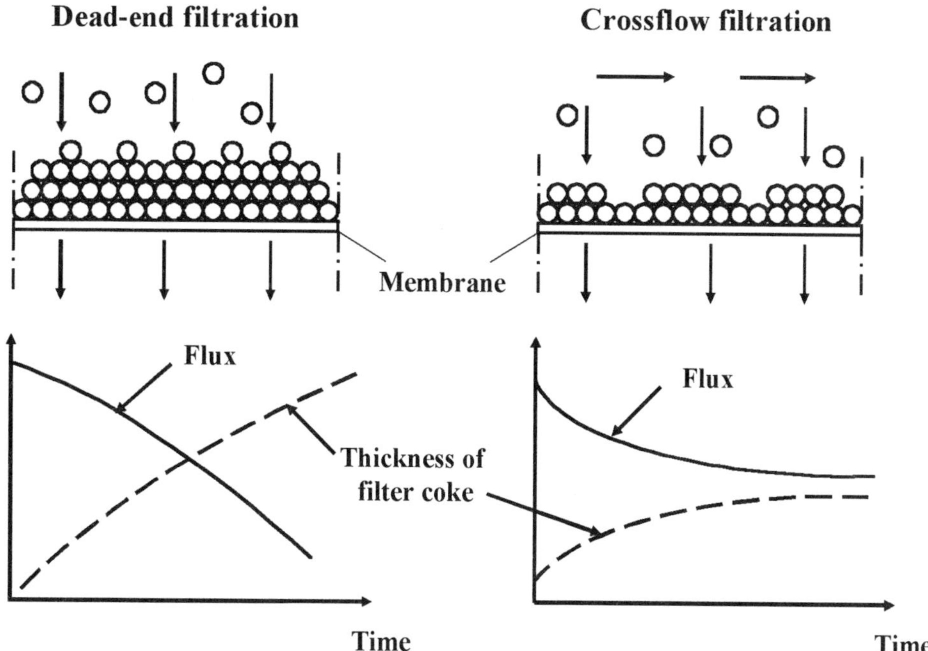

Fig. 18. Dead-end and crossflow microfiltration (16).

treatment for removal of dissolved materials and colloidal particles. In many instances, MF is used in combination with other separation processes to remove dissolved matter. In crossflow microfiltration (CFMF), the fluid to be filtered flows parallel to the membrane surface, i.e., crossflow with respect to the flow of the permeate (Fig. 18). The crossflow reduces the formation of a filter cake, thus ensuring constant filter conditions and, hence, permeate flow. This process is sometimes called tangential or dynamic filtration. CFMF is used for the production of pure liquids, the concentration of suspensions in order to recover valuable products, and the regeneration of process liquids. In some applications, CFMF would lead to considerable cost savings, because recycling of water and/or other valuable products are possible by this process. In addition, MF membranes operate at relatively low pressures (50–500 kPa), typically less than 100 kPa (17,18).

Microfilter membrane has a microporous structure and separates particles according to the size of pores, from a liquid or a gas phase. The separation is based on the sieve effect, i.e., the separation effect is limited to the outer surface of the membrane. The porosity of the inorganic (ceramic) membranes is much lower than that of polymer membranes, but their thermal stability enables them to be used at high temperatures.

3.2. *Application of Membrane Processes*

This section highlights application of membrane processes in water recycling and waste minimization both for industries and domestic purposes.

3.2.1. *Reverse Osmosis (RO) in Water Reclamation*

Eraring Power Station used to spend 4000 m^3/d of potable water from local potable water supply in the Hunter Region of New South Wales, Australia. The continued

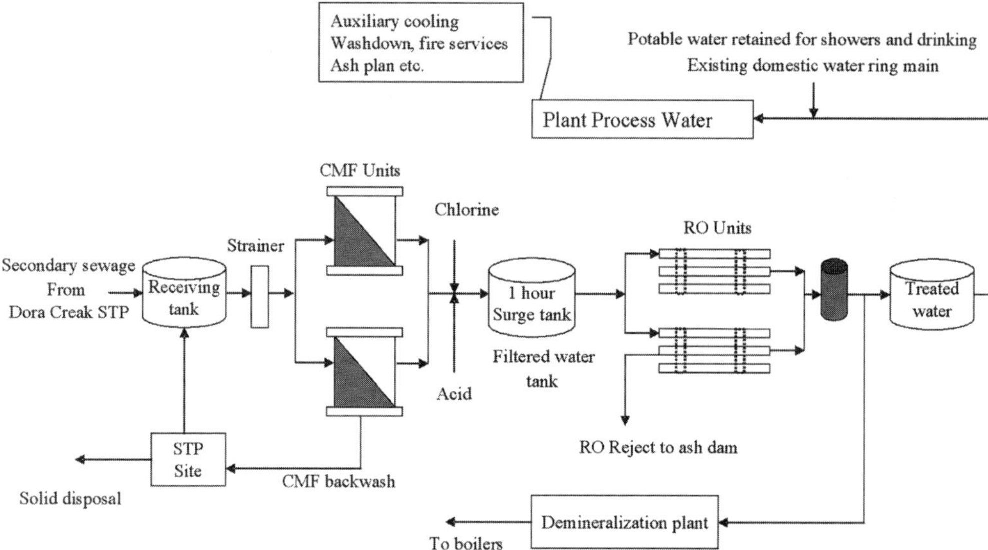

Fig. 19. Water reclamation scheme at Eraring power station (19).

residential growth required an expansion of potable water infrastructure. But, there were environmental issues, as construction of pipelines would have disturbed surrounding environmentally sensitive areas. Therefore, the expansion of potable water supply infrastructure was offset by replacing the potable water requirement of the Eraring Power Station with the reclaimed water from a nearby Sewage Treatment Plant located at Dora Creek.

This power plant now uses the treated-wastewater effluent. The treatment scheme by membrane processes of Eraring Power Station is presented in Fig. 19. Influent from the secondary treated Dora Creek STP flows under gravity from a 70,000 m^3 elevated tank through a 500 mm LDPE pipeline, to the suction of three centrifugal pumps, which deliver the feed to the two continuous microfiltration (CMF) units (19). Each CMF unit comprises 90 filtration modules containing hollow fiber membranes having an average pore size of 0.2 μm. Water enters the membrane modules at one end, flowing along the outside of the membrane and through the wall, removing all suspended solids, fecal coliforms, and *Giardia* cysts as well as significantly reducing human enteric viruses such as *Reovirus* and *Enterovirus*. Filtrate from the CMF unit is dosed with sodium hypochlorite to control biological growth and with sulfuric acid to reduce pH which helps to minimize hydrolysis of RO membranes.

The filtrate from CMF enters the reverse osmosis (RO) unit. The RO system comprises two-stage trains in a 6×3 array of seven-element membrane vessels. Water flows tangentially to the membrane surface. Two multistage centrifugal pumps of 150 kW power drive the RO plant. The membranes are cellulose acetate membranes rated at 98% rejection. Salts and organics are rejected allowing only water to pass through. Permeate is virtually free of all salts and virus and is piped to a degasser tower and stored in a 60 m^3 reclaimed water tank. Reclaimed water, because of its low total dissolved solids (TDS), is fed preferentially to the demineralization plant with the balance

Table 11
Performance of Membrane Processes at Eraring Water Reclamation Project

Parameter	Influent	Effluent
Nonfilterable residue (mg/L)	30–50	< 1
Turbidity (NTU)	approx. 50	< 0.1
Silt density index	not measurable	< 3
Particles (mg/L)	< 104	< 1
BOD (mg/L)	20–50	< 5
Fecal coliforms	< 106	< 1

of water used for non-potable purposes around the power plant. RO rejects are disposed to the station's ash dam. The performance of the membrane processes is presented in Table 11.

The total project cost approx US$ 2.4 million, which included detailed design, equipment supply, civil, electrical, and mechanical work plus some other equipment associated with transfer and pretreatment of wastewater to the facility. The annual cost savings are around US$400,000 from savings in potable water consumption and production of dematerialized water. The pay back period is estimated as 7–8 yr. The reclamation plant allowed construction of the 11.4-km-long sewer link between Eraring and the ocean outfall to be delayed for 15 yr at a saving of US$1.3 million. The augmentation of existing potable water infrastructure for the region was delayed for many years taking the combined saving to the Hunter Water Corporation to over US$ 3 million. Other major benefits of the scheme include conservation of the fragile environment, considerable reduced demand on the local water supply, and the opportunity to avoid further contamination of oceans, lakes, and rivers.

The simplicity and reliability of CMF and RO pretreatment greatly enhances the economic viability of wastewater reuse. The water-reclamation plant at Eraring Power Station demonstrates the potential to use this valuable resource in power and other industrial facilities located near municipal-wastewater-treatment plants.

3.2.2. Ultrafiltration (UF) in Domestic Wastewater Treatment

The combination of a membrane bioreactor with a high level of activated sludge and an ultrafiltration module allows a thorough treatment of wastewater, thus enabling it to be recycled and used in washing floors and flushing toilets. An industrial use of this system has been developed in Japan, where the compact nature of the system and local legislation has allowed it to be installed in office complexes and hotels (20).

Mitsui Petrochemical Industries (M.P.I.) has introduced the Rhone-Poulenc system, which combines ultrafiltration and activated-sludge bioreactors. This process has been tested in France in treating wastewater from hospitals and urban power stations. The M.P.I. has adapted a system called UBIS system for the treatment of wastewater from office buildings and hotels in Japanese cities.

The UBIS system enables the reprocessing of wastewater from kitchens, wash basins, and toilets. The wastewater is transferred to an activated-sludge bioreactor that has a high concentration of sludge (20 g/L) and a high level of agitation. These two factors reduce the residence time of the waste to about 1 h, compared with 24 h in a traditional system. Water passing through the membrane is free of suspended solids, viruses, and bacteria. This water is stored in a buffer tank where a small amount of sodium hypochlorite is added. It is then reused for flushing toilets.

The design of an ultrafiltration module using turbulence promoters allows very high fluxes over long periods. The flux varies between 100 and 120 L/m^2 of membrane over a period of 45 d. Normal chemical cleaning is performed every 45 d and this cleaning procedure requires approx 1–2 h. Electrical consumption is about 3 kwh/m^3 of treated water. For instance, to treat 100 m^3/d, a plant requires 45 m^2 of floor space, and a module with 34 m^2 of membrane area and a 6 m^3 bioreactor. Compared to the conventional process the UBIS system is very compact. The UBIS 20 system (capacity of 20 m^3/d) has a bioreactor of 2.2 m^3, buffer tank of 0.4 m^3, wash tank 0.3 m^3, and a membrane area of 7 m^2; the overall dimensions are 2.0 m (high) × 4.5 m (long) × 2.0 m. The small foot-print allows installation in the basement of commercial or residential buildings. For instance, in treating 100 m^3/d, the floor space saved is equivalent to 25 parking spaces. Operation and maintenance of the system is very simple and the system easily withstands the change in load levels. The treated water is of very high quality and can be reused. Energy consumption and running costs together are low compared with traditional processes because the treated water can be reused.

3.2.3. *Ultrafiltration (UF) in Industrial Wastewater Treatment*

Waste minimization at the source of waste generation point, is the correct approach and long-term solution to industrial pollution. It can be achieved by recycling wastes, process modification, product modification, substitution of raw and process materials, by-product recovery, etc. Membrane technology plays an important role in waste minimization. This section presents the membrane applications in industries such as paper and pulp, metal plating, cutting oil, textiles, and red meat abattoir industries. Many of these applications are examples of waste minimization with economic benefit.

3.2.3.1. Recovery of Lignosulfonates from Pulp Industry

Conventionally, an evaporator is used to recover lignosulfonates of all molecular weight fractions from black liquor. This material has little economical value. Ultrafiltration can be used to separate the high-molecular-weight lignosulfonates (which remain in the concentrate) from low-molecular-weight lignosulfonates (which escape into the permeate). The recovery of high-molecular-weight lignosulfonates from the concentrate is not only economical but also eliminates part of the pollution problem. This material can be used to produce industrial products like dispersing agents and spinning solvent for polyacrylonlitrile fibers. The low-molecular-weight lignosulfonates can easily be treated by conventional biological treatment. A pilot-scale study conducted using inorganic ultrafilters of carbon zircona indicated that this process is technologically and economically feasible. Operating conditions include temperatures in the range 90–140°C and pressure of 750 kPa. Cleaning with acid/alkali was done on monthly basis.

3.2.3.2. Recovery of Paints from Metal-Painting Wastewater

The ultrafiltration (UF) is used in the car painting industry especially where electrophoretic method of painting is adopted (Fig. 20). Around 25–45% of paint consumed goes into the wastewater stream. This can be completely recovered if a UF system is

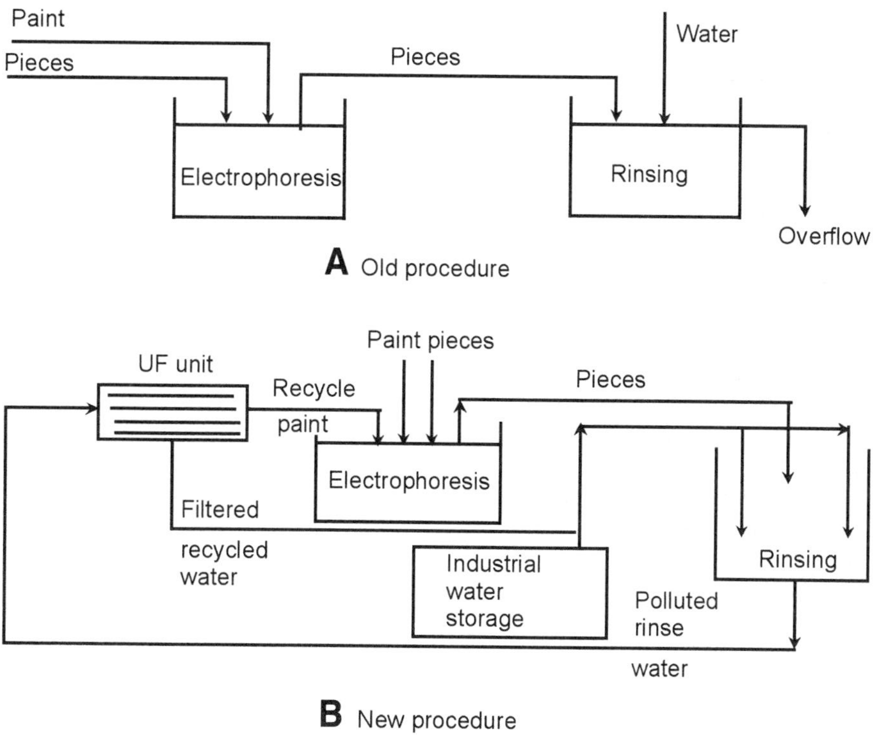

Fig. 20. Old and new processes used in car-painting industry (21).

incorporated in the process. This will also help in reusing the permeate and economizing the use of water.

Solvent/paint waste streams from automated paint-spraying operations may also be treated using UF. These types of wastes are typically generated when the paint color is changed and the lines are cleaned with paint solvent to flush out the old paint. An automobile manufacturing plant, which installed a UF system to recycle paint-cleaning solvent, reported a payback period of only 5.3 mo due to savings in waste disposal and fresh solvent costs. However, low fluxes were encountered because paint solutions typically contain high levels of dissolved polymers, which form a gummy, gel-like layer on the membrane surface. This layer is difficult to remove hydrodynamically and also has a low permeability.

In a particular car industry in France where they adopt cathodic painting, 70 cars are painted in an hour (250,000 cars/yr). The total surface painted is 4500 m^2/h (65 m^2/car). A good rinsing requires 1 L of water per square meter of surface; 95% of this rinse water can be recycled if UF is used in the process. In this industry, 135 m^2 of membrane area (three modules of 45 m^2 each) was used. Table 12 shows the cost estimation in implementing the recovery of raw material and related economic benefits. From the Table 12 it can be seen that the capital cost of UF can be recovered within 4 mo of operation.

3.2.3.3. Recovery and Recycling of Cutting Oil in the Metal Workshops

The cutting oil used in metal workshops can be reused after treating with ultrafiltration. This reduces the incineration of used oil and thus the associated pollution.

Table 12
Economic Benefit in Implementing the Recovery of Raw Materials in a Car Painting Industry (20)

(a) Capital cost modules, pumps, control units, etc.	\$ 450,000
(b) Operational cost energy cost (three pumps at 18 kW/h; 8000 h operation/yr at 6c / kW)	\$ 25,920 / yr
(c) Maintenance cost changing membranes once in 3 yr and regular maintenance	\$ 23,000 /yr
(d) Recovery paint recovery (9 g/m^2 × 4500 m^2/h × 3500 h/yr × \$ 0.01 / g = \$ 1,417,500) and Demineralized water recovery (\$135,000)	\$ 1,552,500 /yr
If the amortization period is 5 yr, amortization / yr [from (a)]	\$ 90,000/yr
Annual operation and maintenance cost [(b)+(c)]	\$ 48,920/yr
Annual saving due to the recovery of paints & demineralized water [(d)]	\$ 1,552,500/yr
Economy	\$ 1, 413, 580 /yr

Furthermore, energy savings can be made by the elimination of incineration. The reliability and longevity of the ultrafiltration membranes are satisfactory (they last more than 2 yr if they are maintained properly). Table 13 shows the benefits of using ultrafiltration to recover cutting oil in metal workshops. From the table, it is evident that, (a) pollutant discharge is decreased by 95% due to the incorporation of ultrafiltration and (b) the raw material requirement is reduced by 50%.

3.2.3.4. Recovery of Blue Colorant in Dyeing Industry

Recovery of blue colorant (indigo) from rinsing baths of jeans production units is another example for waste minimization using ultrafiltration. Figure 21 shows the processes used to recover blue colorant from rinsing baths. With rinsing water, 10–20% of the color input is taken away. This corresponds to a color concentration of 0.1–0.8 g/L. By using an ultrafilter at 60°C, this color can be concentrated up to 50 g/L and can be recycled directly.

As a first step, wastewater from the rinsing bath is mixed with hydrogen peroxide and passed through the first ultrafilter continuously. When this process is carried out for 60 h at the rate of 5 m^3/h (permeate flux of 0.2 m /m^2·h), 300 m^3 of dye is treated at 0.5 g/L concentration and 10 m^3 of concentrate can be obtained with color concentration of 15 g/L. In the second step, the 10 m^3 of concentrate with 15 g/L concentration is passed through an ultrafilter for 2 h (permeate flux of 3.5 m^3/h = 0.15 m /m^2·h) and 3 m^3 of concentrate is obtained at 50 g/L concentration. Thus every 60 hr, 150 kg of indigo is recovered. The payback period for the ultrafiltration system is about 1 yr and, after this, the savings every year will be around US \$ 400,000.

3.2.3.5. UF as a Pretreatment for RO in the Red Meat Abattoir Industry

In South Africa, there are more than 300 registered abattoirs. They consume a significant quantity (>7,000,000 m^3) of potable quality water every year and generate more than 6,000,000 m^3 of effluent per year, which generally goes into the municipal sewers. Therefore, an effluent treatment system was developed to:

Table 13
Benefits of Using Ultrafiltration to Recover Cutting Oil in Metal Workshops

Mass and pollution balance	Old process	New process
Raw materials required (m^3/yr)	993	473
Volume of waste (m^3/yr)	520	24
COD (kg/d)	250	12
Economic Balance (US$ in 1982)		
Investment	—	66,140
Annual costs	41,000	2,800
Payback period	—	21 mo
Savings after payback period	—	US $ 38,200/yr

(a) reduce the organic load in the effluent
(b) remove phosphate from the effluent
(c) recover re-usable water from the effluent
(d) recover organic concentrate from the effluent for by-product recovery

To achieve the aims described above, membranes manufactured in tubular form by Membratek Pty Limited in Paarl, South Africa were utilized. For posttreatment, the ultrafiltration (UF) polyethersulfone membranes with molecular mass cut-off of 40,000 were used in modules with each module having a membrane area of 2.0 m^2. This removed 90% of COD and 85% of phosphate from the abattoir effluent, and lead to a non-fouling feed for the posttreatment. In the posttreatment, RO is employed to produce a high-quality reusable water for abattoirs.

Twelve UF modules in a parallel-series configuration were housed in a skid-mounted frame, together with eight cellulose acetate RO modules of similar size. The rig is equipped with feed pumps, flow gauges, backpressure valves, and has automatic fail-safe

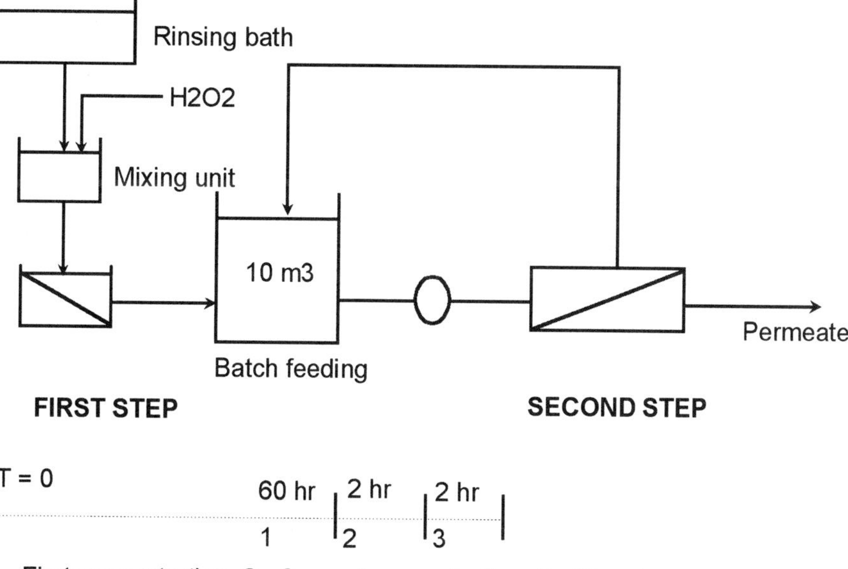

Fig. 21. Recovery of blue colorant from rinsing baths using ultrafilters.

Table 14
Performance of the Membrane System

Parameter (mg/L)	Influent	UF Effluent	RO Effluent
COD	6000	42–60	2–3
Soluble phosphorus	40	6	0.3

cutouts on temperature, pressure, and pH (for RO). The capacity of UF train was 1000 L filtrate/h and the capacity of RO train was 300–350 L of permeate/h. Table 14 summarizes the performance of the membrane system.

A cost comparison made based on a daily flow of 820 m^3 indicated that UF and RO treatment (capital cost of US$ 0.6 million and operating and maintenance cost of US$ 0.55 million) is more economical compared to the conventional treatment of anaerobic digestion (capital cost of US$ 0.75 million and operating and maintenance cost of US$ 0.85 million). Apart from the cost considerations, membrane processes lead to high quality effluent.

3.2.4. *Microfiltration (MF) in Domestic Wastewater Treatment*

Wastewater treatment and reuse are important parts of our aqueous environmental management, and they provide huge opportunities for the use of membranes. Table 15 presents water contaminants and typical removal range for membranes hybrid processes.

3.2.4.1. Water Mining Plant at Canberra, Australia

The first Water Mining plant was opened in 1995 at Southwell Park in Canberra, Australia (Fig. 22). The project was to demonstrate the feasibility of effluent reuse in urban public access areas and to establish community acceptance. Although non-potable treated wastewater is used widely for landscape irrigation and golf courses, this was the first time it would irrigate public access land within urban Canberra. Thus the plant design focused on health issues, noise and odor control, and preservation of neighborhood amenity.

Table 15
Water Contaminants and Typical Removal Range for Membranes and Hybrid Processes

Species	Size (μm, kD)	MF	UF	NF	RO	Chem + MF/UF	PAC + MF/UF
Protozoa	> 10	(a)	(a)	(a)	(a)	(a)	(a)
Coliform	>1	(a)	(a)	(a)	(a)	(a)	(a)
Turbidity	1–0.1	(a)	(a)	(a)	(a)	(a)	(a)
Cysts/oocysts	~0.1	(a)	(a)	(a)	(a)	(a)	(a)
Virus	0.01–0.1	(b)	(a)	(a)	(a)	(a)	(a)
THMP	< 10 kD		(b)	(a)	(a)	(b)	(b)
Color	< 10 kD			(a)	(a)	(b)	(b)
A1 species	< 1 kD			(a)	(a)	(b)	(b)
Ionic	< 0.1 kD			(b)	(a)		

[a]Near complete removal.
[b]Possible removal.

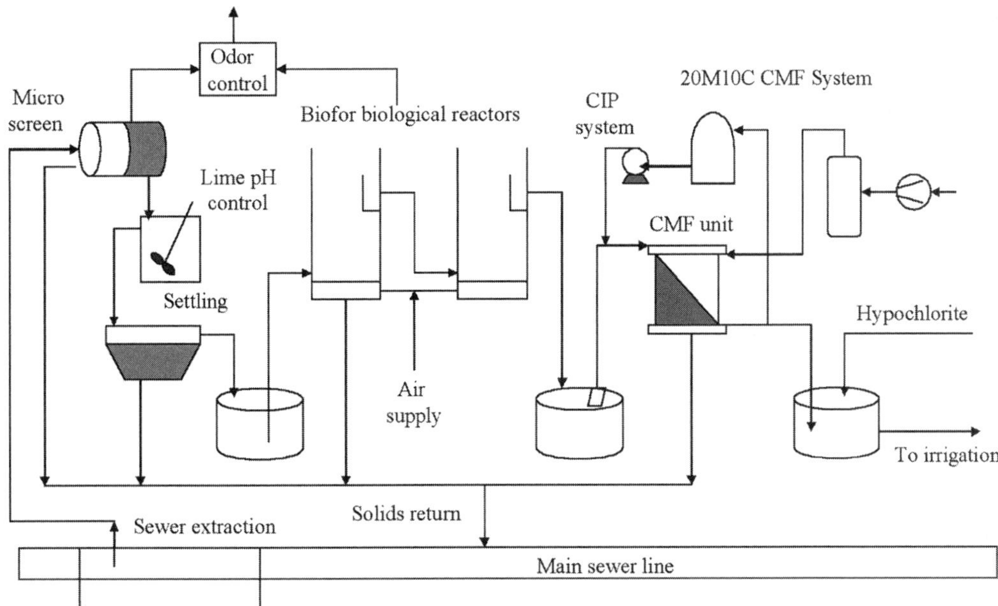

Fig. 22. Water Mining plant at Canberra.

Raw sewage is pumped from a pump well built into a 450 mm residential sewer passing close to a substantial recreational area incorporating Southwell Park, the Yowani Golf Course, Canberra Exhibition area, and the Race Course. Extraction is flow controlled to be no greater than 50% of sewer flow at any time. The flow passes through a 0.5 mm rotating contrashear screen (10 L/s) where debris, grit, and free oil (about 10% of solids and 5% of BOD) are removed and finally returned to the sewer. The screened sewage flows to a stirred tank (20 min retention) where lime is added to maintain alkalinity for the subsequent biological reactions. Hydrated lime is received in bags, made into a slurry, mixed in a flash mixer, and pumped to the reaction tank. The pH-adjusted wastewater flows to a solids separation tank (corrugated cross-flow interceptor plate design) before temporary storage in an underground 100 m^3 tank. Suspended solids are returned to the sewer.

Biological treatment takes place in a two-stage upflow biological aerated flooded filter (BAFF) containing granular media. Air is pumped up through the media to provide oxygen for biological activity. Stage 1 converts soluble carbonaceous BOD to biomass, which grows on and is captured by the media. Separate clarifiers are not required. The granular bed is backwashed periodically to remove captured solids using secondary effluent produced in the process. The backwash is returned to the sewer. Stage 2 converts ammonia to nitrate (nitrification). The reactors, both of which have a 5 L/s influent capacity, can also operate in parallel depending on the effluent quality required. The principal advantages of the process are its small foot prints, easy automation and low potential for odor production (19).

The secondary treated effluent is then pumped by submerged pumps to a CMF unit to remove fine solids and microorganisms. The CFMF unit comprises 20 filtration

Table 16
Performance of the Water Mining Plant

Parameter	Sewage ex-main	Settled sewage	Secondary treated	After CMF
BOD (mg/L)	250	220	16	1.9
Suspended solids (mg/L)	230	170	16	0.4
Fecal coliforms (cfu/100 mL)	—	—	$<10^6$	< 0.1
Turbidity (NTU)	—	130	8.2	0.4
Phosphorus (mg/L)	—	11	5.5	Soluble pollutants
Ammonia-Nitrogen (mg/L)	29	38	1.2	removed to a minimal
Total Nitrogen (mg/L)	—	—	32	degree

modules of hollow-fiber polypropylene membranes with an average pore size of 0.2 μm. Water enters the membrane modules at one end, flowing along the outside of the membrane and through the wall, removing virtually all suspended solids, fecal and other bacteria, parasites, and human viruses. A key element of the CMF system is the patented air backwash. When pressure across the fiber membrane reaches a pre-set level, filtration stops and high-pressure air is injected into the fiber tubes. The air blasts through the fiber wall dislodging accumulated sediments sitting on the outer surface. A sweep of feed flushes the dislodged sediment to sewer. When the driving pressure through the membranes climbs to around 150 kPa, which usually occurs over 3–7 d, the membranes are chemically cleaned using a 2% caustic solution and surfactant achieved by a fully automatic CIP (Clean-in-Pipe) system. The cleaning solution is reused and automatically topped up with cleaning concentrate and water. The CIP tank is desludged every 6–12 mo. The performance of the "water mining" plant is presented in Table 16.

To achieve double disinfection, the microfiltered wastewater is chlorinated using hypochlorite solution to maintain a chlorine residual of 0.5 mg/L. The treated effluent is stored in an underground tank from which it is pumped to irrigate the adjacent park and playing fields. The plant is unobtrusive, quiet, and odor free. It is contained in an attractive two-story building of 180 m^2 area. An equivalent area houses the underground storage. Air within the building and from the biological process is circulated through a hypochlorite solution wet scrubber before venting to atmosphere.

The plant cost US$1.4 million. It was designed to produce 300 m^3/d rising to 600 m^3/d with biological reactors operating in parallel. Being the first of its kind and built as a demonstration plant, it incorporates inevitable cost premium, for example, some over design and a high proportion of below-ground construction. Historically small sewage reclamation plants have been difficult to economically justify world over, and in Australia in particular, where the current potable water costs are US$ 0.35 per m^3.

3.2.3.3. MF as a Pretreatment to Reverse Osmosis at the Orange County Water District (OCWD)

The Orange County Water District (OCWD) has been treating clarified secondary effluent to potable water standards for groundwater injection since 1976. Water Factory 21, an advanced water-treatment plant, is considered to be the *de facto* industry standard

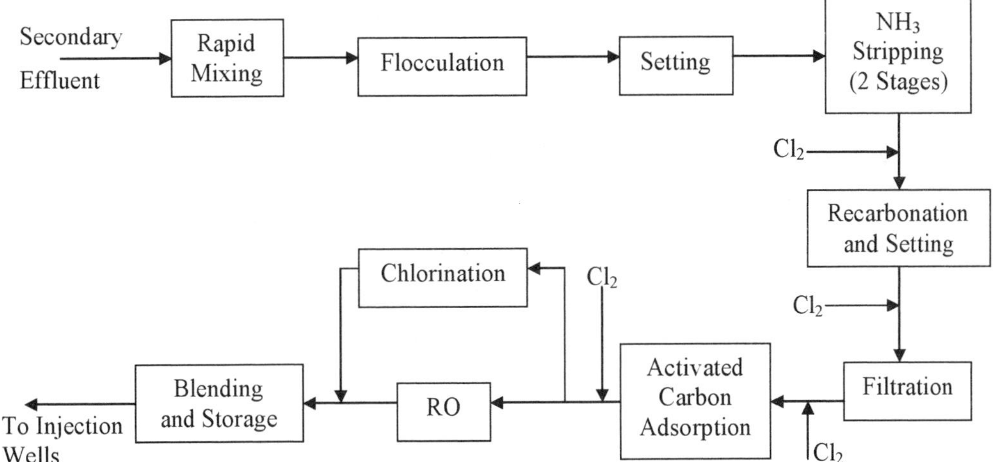

Fig. 23. Schematic flow diagram of Water Factory 21 treatment process.

for municipal wastewater reclamation using reverse osmosis. At present the pre-treatment process upstream of the cellulose acetate (CA) RO membranes consists of high-pH-lime clarification, recarbonation, and multimedia filtration. The high-pH–lime pretreatment process occupies approx 29,000 ft^2 (3000 m^2) and has a design capacity of 55,000 MLD. The CA membranes operate at flux of 17 $L/m^2 \cdot h$ and are cleaned on average once every 750 h (22) (Fig. 23).

In 1992, OCWD started the MF test program after the successful demonstration of a Memcor continuous microfiltration (CMF) system as pretreatment to thin film composite (TFC) RO membranes at the Reedy Creek Services District. In these trials OCWD investigated the use of MF as pretreatment for the cellulose acetate RO membranes, and MF and UF as a pretreatment for TFC RO membranes.

The MF membranes operated on clarified secondary effluent discharged from the Sanitation Districts of Orange County. The secondary effluent contained 1100 mg/L total dissolved solids, 10–11 mg/L total organic carbon, 5–15 mg/L biological oxygen demands, 5 mg/L suspended solids, 15–20 mg/L ammonia, and approx 10^6 coliforms per 100 mL. Secondary effluent turbidity is typically 2 NTU; however, it is not uncommon to have turbidity of more than 15 NTU.

CA RO membranes were operated downstream of the Memcor-60M10 (Memcor–US Membrane) for more than 2 yr. Microfiltered secondary effluent contained 1050–1150 mg/L total dissolved solids, less than 1 mg/L suspended solids, 7.5–9.5 mg/L total organic carbon (TOC), and a 3–5 mg/L combined chlorine residual at a pH of 5.5. The RO influent turbidity ranged from 0.07 to 0.2 NTU, and the average 15-min silt density index was less than 2. The CA membranes achieved an average TDS rejection of 96.7% on the microfiltered effluent. The average membrane run time, defined as the time to reach 75% of 1-h normalized flux decreased from 2140 h to 397 h as the instantaneous flux increased from 21 to 24 L/m^2h.

3.2.5. Membrane Bioreactor (MBR) in Wastewater Treatment and Reuse

The membrane bioreactor was initially developed for biotechnology applications. In the field of wastewater treatment, it can be considered as an emerging technology.

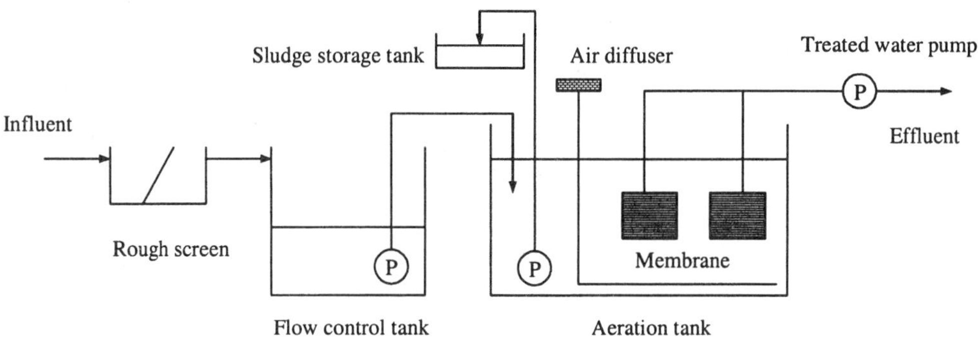

Fig. 24. Schematic flow diagram of submerged membrane process for industrial wastewater treatment.

However, its growth is significant due to the numerous advantages it offers, namely (a) possibility of very compact units: it is possible to operate the biological treatment at short residence time due to the high biomass concentration (20–30 g/L), (b) significant decrease in sludge production due to its operating conditions (high sludge age, low food to microorganisms ratio), and (c) high quality of the treated water that enables reuse: very low turbidity, high level of disinfection, lower BOD and COD than in conventional processes.

There are two main types of MBRs currently used at (small) industrial scale, submerged membranes and external loop membranes. Submerged membranes as initially proposed by Chiemchaisri et al. (23) are now used in industrial scale processes by Mitsubishi, Kubota, and OTV (BIOSEP process). An example of the submerged membrane system is illustrated in Fig. 24. This concept is attractive because (a) no recirculation, (b) air bubbles supplied for aeration of the missed liquor (biological oxidation) will also facilitate continuous declogging of membranes, and (c) membranes can be operated at very low transmembrane pressure (as low as 10 kPa) potentially below critical flux condition. This minimizes the fouling and thus the periodic chemical cleaning which leads to increased membrane life. Despite the relative large membrane requirement, the capital and replacement costs are reduced due to the simplification of membrane equipment (no cartridge) and increased membrane life, respectively.

External circulation of the sludge from the activated-sludge tank to the membrane unit is typified by the design proposed by Lyonnaise des Eaux. A number of units up to 150 m^3/d have been commissioned and a unit of 1500 m^3/d is proposed. In this application, tubular ceramic MF membranes are used with cross-flow velocities greater than 3 m/s.

3.2.5.1. Ceramic Membrane Bioreactor

The performance of a ceramic membrane bioreactor (MBR) at the Ville Franque (France) wastewater-treatment plant is shown in Table 17 (24,25). The permeate flux was kept at 150 L/m^2·h. Aeration and mixing in the activated-sludge tank were performed using a hydro-ejector. The plant was highly automated and thus the only labor cost was due to sludge treatment. The frequency of sludge withdrawal was very low and the system did not give rise to any odor problem. However, this type of MBR is still

Table 17
Performance of Ceramic MBR for Municipal Wastewater Treatment

Parameter	Influent	Effluent	Percentage removal
BOD_5 (mg/L)	154	2.5	98
COD (mg/L)	33	2.7	92
TOC (mg/L)	99	10	90
TKN (mg/L)	52	2.1	96
Total N (mg/L)	0	6.2*	84**
Total P# (mg/L)	8.25	0.45	95
Fecal colifom (in 100 mL)	1.3×10^6	26	99.9
Fecal strepto (in 100 mL)	6.9×10^5	26	99.9
Clostridium (in 100 mL)	9.5×10^3	26	99.7
Viruses (in 100 mL)	700	4	99.4
Cysts, *Giardia* (in 100 mL)	7.5×10^3	3	99.6
Cysts, *Cryptos* (in 100 mL)	700	3	99.6

* NO_3-N.

**

$$100\left[1-\frac{(\mathrm{TKN}+\mathrm{NO_3}-\mathrm{N})_{\mathrm{Effluent}}}{\mathrm{TKN}_{\mathrm{Influent}}}\right];$$

Ferric chloride addition.

more expensive than conventional processes such as activated sludge with settling and the trickling filter. Nevertheless, the MBR is attractive for small plants (<5000 people) as it produces a high quality of treated water suitable for reuse, and involves fewer processes than the conventional process. The conventional processes are more difficult to operate, potentially less reliable, and require lager land area.

3.2.5.2. Membio Process

The Membio process is a combination of an aerobic bioreactor and a microfiltration system developed by Memtec–US Filter Ltd. A simplified schematic diagram of the Membio treatment process is shown in Fig. 25. Wastewater is fed to the bioreactor and, after treatment, effluent goes to the continuous microfilter (CMF) where the solids, parasites, bacteria, and viruses are removed by filtration through a microporous membrane.

The key difference between the Membio bioreactor and other biologically aerated filters is the separation of the process into two distinct components. The bioreactor converts soluble material into insoluble biomass as this biomass is then removed by either backwashing the bioreactor or by capture on the membrane surface.

The Membio process has been operated successfully on a number of trade and sewage streams ranging from light domestic sewage at 100 mg/L BOD_5 to trade waste at up to 3700 mg/L BOD_5. Trials have been carried out with a loading rate of up to 8 kg BOD_5 per m^3 of bed per day although the feed stream had a low soluble percentage BOD_5. Plants are sized on 3 kg soluble BOD_5 per m^3 of bed per day.

The design of the Membio process minimizes the solids loading on the CMF unit, which has shell-side feed. In order to maintain throughput the membranes are regularly (approx 3 × per hour) backwashed with air and clean filtrate. On a daily basis the

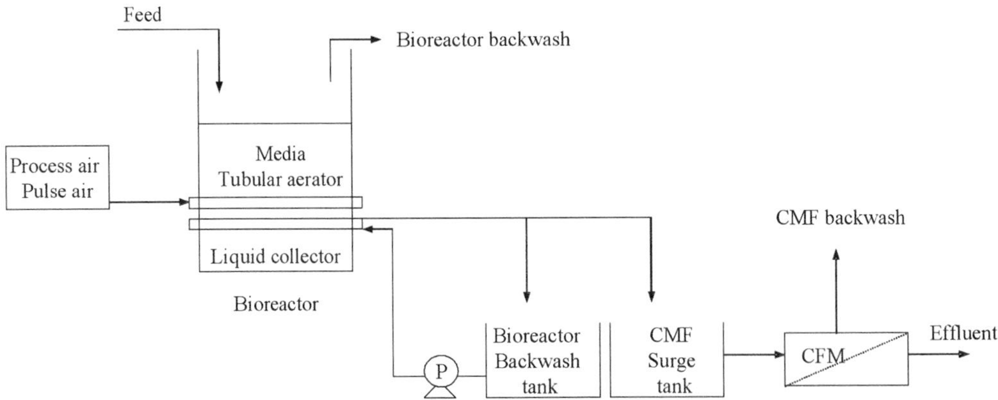

Fig. 25. Schematic flow diagram of Membio treatment process.

membranes are chemically cleaned. Results of a trial are summarized in Table 18. The data show that the Membio process achieves excellent removal of BOD_5 and suspended solids, and cell debris in the treated effluent could result in reduced rates of biofouling in the RO plant.

3.2.5.3. MBR in Tertiary Wastewater Treatment

Wastewater effluent is generally disinfected prior to discharge into a water body in order to protect the users downstream. Disinfection using chlorine would cause formation of trihalomethanes (THM) because of the organic matter left in the effluent. To avoid this, MF could be used as the treatment process.

In Australia, Memtec–US Filter MF membranes were added as the final stage of a sewage treatment plant that originally provided screening, grit removal, extended aeration, secondary sedimentation, dual-media filtration, and chlorination. In the retrofitted process the effluent from the secondary sedimentation stage was fed to the MF unit. A comprehensive microbiological testing program established that all indicator bacteria and viruses were essentially removed by this system operating with gas backwashing.

Table 18
Results of Membio Process at a Sewage Treatment Plant in Sydney, Australia

Parameter	Particular	Value
BOD_5	Input	101 mg/L
	Output	2–5 mg/L
	Removal	96%
	Loading	3.0 kg BOD/m^3·d
Suspended solids	Input	56 mg/L
	Output	0.44 mg/L
	Removal	99.2%
Fecal coliform	Input	76×10^5/100 mL
	Output	0.4/100 mL
	Log reduction	approx 7.0
Virus, coliphase	Log reduction	approx 3.0

Table 19
Comparison between Ceramic MBR and Conventional Activated Sludge in Dairy Wastewater Treatment

Characteristics	Conventional activated sludge	Ceramic MBR
Treated water (m^3/d)	500	500
Reused water (m^3/d)	0	400
Spend water (m^3/d)	500	100
Volume of aeration tank (m^3)	4300	600
Area requirement (m^2)	1300	260
COD of the treated water (g/m^3)	90	50
BOD_5 of the treated water (g/m^3)	30	<5
SS of the treated water (g/m^3)	30	0
Energy requirement (kWh/m^3)	5	8

Testing for chemical quality showed significant reduction in BOD, turbidity, and oil/grease as well as some reduction of heavy metals and phosphorus, and suspended solids being zero.

Consistent filtration rates over several months were achieved without significant biological membrane fouling. The study conclusively showed the capability of the microfiltration system to disinfect and clarify treated sewage. Table 18 shows the results of this application.

3.2.5.4. MBR in Industrial Wastewater Treatment

MF plays an important role in industrial-wastewater treatment. Its use in the Rennovex process for treatment of battery wastes in Newcastle, UK is a good example. This process uses tubular dynamic membranes formed in a curtain of woven fabric. The application recovers valuable solids as well as providing a water suitable for reuse in the process.

MF membranes are also incorporated into membrane bioreactors (MBR). For example, Lyonnaise des Eaux has used ceramic MBR in a number of industries, for example, dairy industry and cosmetic industry. A comparison between MBR and conventional activated sludge process in dairy industry is shown in Table 19.

REFERENCES

1. Ministry of Urban Development, *Manual on Water Supply and Treatment*, Ministry of Urban Development, New Delhi, 1991.
2. C. H. Lee and T. C. Chou, *Journal of the Chinese Institute of Chemical Engineers*, **25**, 239–245 (1994).
3. S. Vigneswaran and C. Visvanathan, *Water Treatment Processes: Simple Options*, CRC Press, Boca Raton, FL, 1995.
4. K. Sugaya, *6th World Filtration Congress*, Nagoya, 1993, pp. 729–733.
5. H. H. Ngo and S. Vigneswaran, *J. Water Science and Technology* **38**, 87–93 (1998).
6. H. H. Ngo and S. Vigneswaran, *J. Water Science and Technology*, **33**, 63–70 (1996).
7. V. L. Snoeyink, *Adsorption of Organic Compounds, Water Quality and Treatment—Handbook of Community Water Supplies*, AWWA, McGraw-Hill, New York, 1990.
8. C. Tien, *Adsorption Calculations and Modeling*, Butterworth-Heinemann Series in *Chemical Engineering*, 1994.

9. G. McKay, *Use of Adsorbents for the Removal of Pollutants from Wastewaters*, CRC Press, Boca Raton, FL, 1995.
10. D. Cooney, *Adsorption Design for Wastewater Treatment*, Lewis Publishers, Boca Raton, FL, 1998.
11. S. Vigneswaran and R. Ben Aim (eds.), *Water, Wastewater and Sludge Filtration*, CRC Press, Boca Raton, FL, 1989.
12. P. Dillon, *Water: Journal of Australian Water Association*, **28**, 18–21 (2001).
13. S. Vigneswaran, H. H. Ngo, J. Hu, and D. Y. Kwon, *AIT Civil and Environmental Engineering Conference, New Frontiers & Challenges*, Bangkok, 1999, Vol.1, pp. 1–7.
14. Y. T. Hung, *OCEESA J*, **19**, 48–53 (2002).
15. J. S. Taylor and E. P. Jacobs, *Water Treatment Membrane Processes* (Mallevialle, J. Odendaal P. E., and Wiesner M. R. (eds.)), McGraw-Hill, New York, 1996, pp. 9.1–9.70.
16. S. Rippenger, *Water, Wastewater, and Sludge Filtration*,(Vigneswaran, S. and Ben Aim, R. eds.), CRC Press, Boca Raton, FL, 1989, pp. 173–190.
17. S. Vigneswaran, B. Vigneswaran, and R. Ben Aim, *Environmental Sanitation Reviews* **31**, 5–46 (1991).
18. J. G. Jacangelo and C. A. Buckley, *Water Treatment Membrane Processes* (Mallevialle, J., Odendaal, P. E., and Weisner, M. R., eds.). Mc Graw-Hill., New York, 1996, pp. 11.1–11.39.
19. Bulletins A 2047, A 2051, A 2052, A 2056, and A 2069, *Memtec-US Filter Technical Bulletins*.
20. M. Roulet, *Selected Topics on Clean Technology*, (Vigneswaran, S., Mino, T., and Polprasert, C. eds.), Asian Institute of Technology, Bangkok, 1989, pp. 171–178
21. M. R. Overcash, *Techniques for Industrial Pollution Prevention—A Compendium for Hazardous and Non-hazardous Waste Minimization*, Lewis Publishers, Inc., Chelsea, MI, 1986.
22. G. L. Leslie, W. R. Dunivin, P. Gabillet, S. R. Conklin, W. R. Mills, and R. G. Sudar, Pilot testing of microfiltration and ultrafiltration upstream of reverse osmosis during reclamation of municipal wastewater, *1996 Technical Report* (unpublished paper), 1996.
23. C. Chiemchaisri, K. Yamamoto, and S. Vigneswaran, *J. Water Science and Technology* **27**, 171–178 (1993).
24. E. Trouve, L. Detons, G. Greaugey, and J. Manem, *Station d'epuration BRM de villegranque. Actes des 12emes Jounees Information Eaux - JIE 96*, Portiers, 60-1–60-2, 1996.
25. V. Urbain, E. Trouve, and J. Manem, *Proceedings of the 67thAnnual Conference and Exposition WEFTEC 94*, Chicago, IL, 1994, pp. 317–323.

17
Introduction to Sludge Treatment

Duu-Jong Lee, Joo-Hwa Tay, Yung-Tse Hung, and Pin Jing He

CONTENTS

1. THE ORIGIN OF SLUDGE

Sludge accumulates as a residue in all sorts of wastewater treatment. Sludge comprises the solids and colloids separated from wastewater as well as substances from biological and chemical operation units. Some sludges are produced during wastewater treatment, including primary sludge, which comprises settleable solids removed from the primary clarifier, and secondary sludge, which comprises biological solids generated in the secondary wastewater treatment plant. Figure 1 illustrates a waste activated sludge sample collected from a local sewage treatment works. Furthermore, drinking water producers produce coagulated solids–coagulant matrix, which contains all of the impurities in the raw water and the dosed coagulant. Additionally, numerous kinds of industrial sludges exist which are generated from treating different industrial wastewaters. The shared features of these sludges are high moisture content that is difficult to remove mechanically, relatively weak flocs that are easily torn off during shear, and concentrated pollutants that require further stabilization for safe disposal.

A typical sludge treatment/disposal system comprises four stages: (1) Pretreatment, during which sludge characteristics are altered to enhance subsequent process performance; (2) dewatering, for separating moisture from the sludge body; (3) post-treatment, for stabilizing or detoxicizing the sludge, and (4) final disposal, which aims to achieve safe and economically feasible disposal. Figure 2 displays a sludge management "network" that is adopted in practice. Sludge management system optimization treats the sludge in a manner that maximizes the benefits of recycling/recovery and is appropriate to local circumstances, including economy, geography, climate, etc.; links wastewater service to other waste management services via integrated planning; and

From: *Handbook of Environmental Engineering, Volume 3: Physicochemical Treatment Processes*
Edited by: L. K. Wang, Y.-T. Hung, and N. K. Shammas © The Humana Press Inc., Totowa, NJ

Fig. 1. The appearance of the collected activated sludge sample collected from a food manufacturing plant.

ensures a long-term service. However, to date, global process optimization for sludge management has been based on heuristics and experience.

This chapter introduces the typical chemical/physical treatment processes in sludge-management practice. The advantages and disadvantages of these processes are also addressed. Three biological treatment processes (aerobic and anaerobic digestion and composting) were also included in this chapter for the purpose of comparison. Useful references include Refs. 1–10. Detail of sludge treatment processes could be found in an accompanying book entitled "biosolids treatment processes."

2. CONDITIONING PROCESSES

Colloidal particles in sludge can remain stable owing to steric hindrance or charge repulsion. Coagulation or flocculation using chemicals is frequently used to enlarge the floc size or to compress the floc interior to facilitate solid–liquid separation efficiency. Through chemical addition and good dispersion through adequate mixing, the steric or charge repulsion by individual colloidal particles could be overcome and collision and agglomeration into large aggregates could occur.

Inorganic metal salts are the most widespread coagulant in sludge management practice. The use of polyelectrolyte flocculant recently has become popular owing to the rapid development of the polymer industry.

2.1. *Coagulation*

Ferric salts are the most popular inorganic coagulant in North America. Meanwhile, in the rest of the world, aluminum salts such as alum [$Al_2 (SO_4)_3.14–18H_2O$] and aluminum chloride ($AlCl_3$) are the most commonly used. The salts hydrolyze in water to form hydroxo complex ion, like $Al(OH)^{2+}$, $Al(OH)_2^+$, $Al(OH)_4^-$, $Fe(OH)^{2+}$, $Fe(OH)_2^+$,

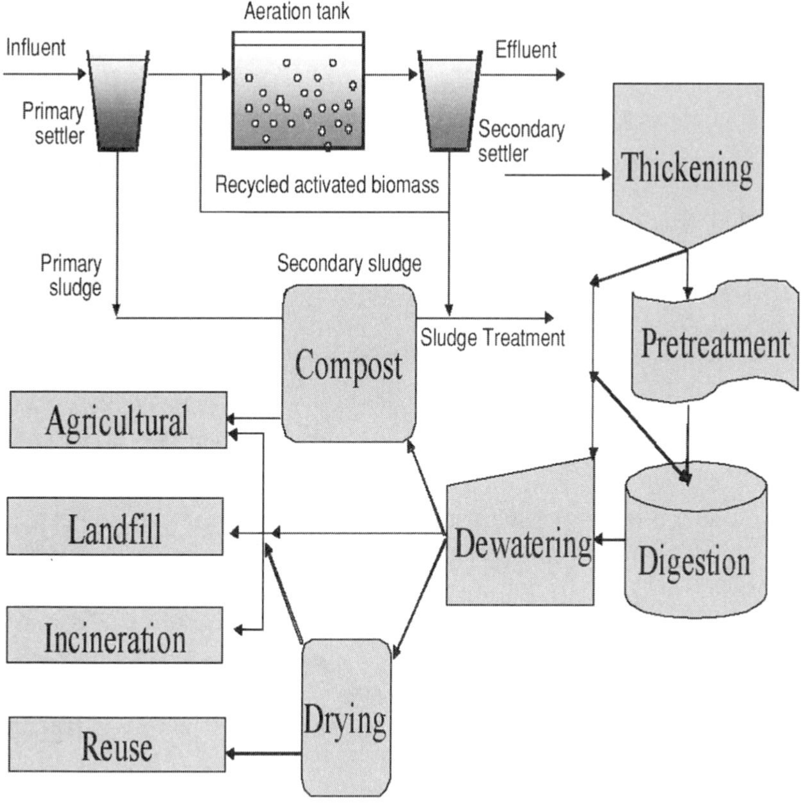

Fig. 2. A sludge treatment network.

$Fe(OH)_4^-$, $Fe_2(OH)_2^{4+}$, reducing the pH value of the suspension. The positively charged ions could be absorbed onto the negatively charged particles that could neutralize particle surface charge. The absorbed ions can form a participate given high ion absorption. This adsorption and charge neutralization mechanism facilitates particle aggregation. Absorption of too many ions can reverse the surface charge of the colloidal particles.

Precipitation occurs when the coagulant dose exceeds total metal hydroxide solubility. The precipitated floc can sweep over the suspension and trap and remove the fine particles during settling. This mechanism is termed sweeping enmeshment.

Figure 3A shows a magnified photograph of the fine particles in the raw water collected from one drinking water works in Taiwan. Figure 3B displays the entrapped floc by dosing the raw water with alum.

The aqueous chemistry of dosing inorganic coagulant could be interpreted using the "coagulation zone" concept. Figure 4 illustrates the correspondence of coagulant dosage with suspension pH value in the following four zones. (1) Zone I: the dosage is lower than the solubility and the whole sludge is in a stable state. (2) Zone II: Under weakly acidic or nearly neutral suspension the dissolved, positively charged ions are absorbed onto the colloidal particles thus destabilizing the suspension. (3) Zone III: With pH of over 5–6, if too many positive ions were absorbed onto the surface, the

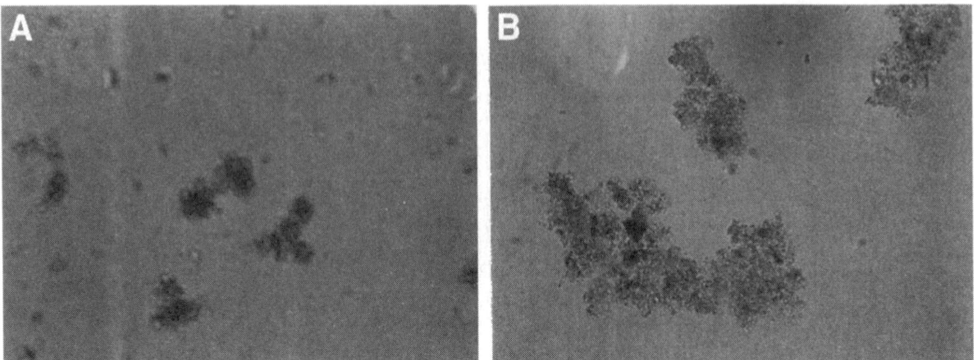

Fig. 3. (a) Particles in the raw water collected at the water intake port of PingTsan Water Works of Taiwan. (b) The coagulated floc by adding polyaluminum chloride (PACl) into raw water.

repulsion between colloidal particles may be built up again owing to charge reversal. (4) Zone IV: Under neutral or basic environment metal hydroxides could be produced when the dosage exceeds the solubility of total metal hydroxides. The precipitate could sweep over the suspension and entrap the particles. This occurrence is known as the zone for coagulation sweeping. Zone II and zone IV could co-exist under pH 7–8.

Lime is also commonly used in sludge conditioning, both as a coagulant and a pH regulator in the suspension. However, analysis of optimal lime dose is lacking.

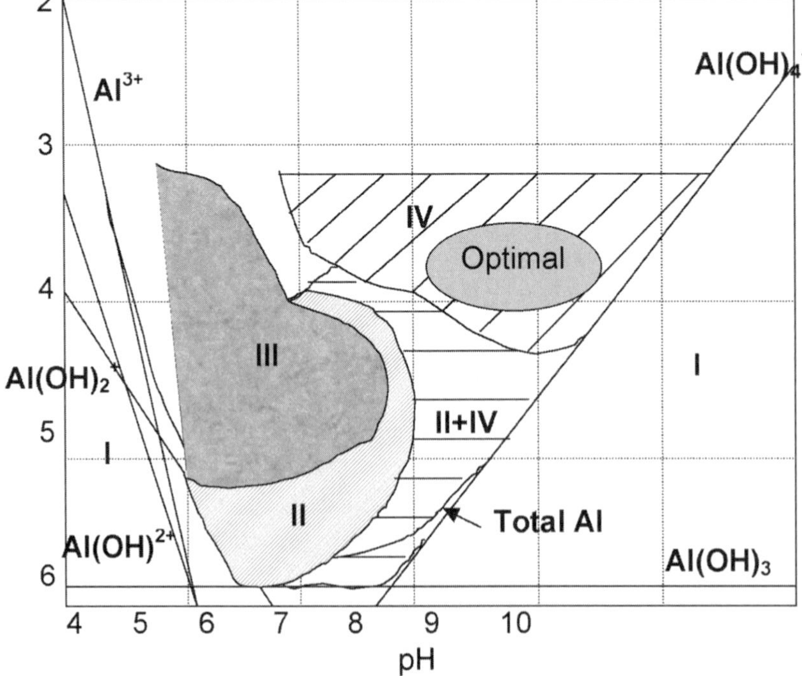

Fig. 4. Stability diagram for the alum salt in water.

2.2. *Flocculation*

Currently available synthetic organic flocculants include the following:

(1) Anionic polyelectrolytes carry negative charges on the functional groups. Commonly used flocculants include carboxylic acid and its derivatives (for example, hydrolyzed polyacrylic esters, amides, and nitriles), sulfonic salt (such as polystyrenesulfonate and polyethylenesulfonate), and anionic polyacrylamide
(2) Cationic polyelectrolytes carry positive charges on the functional groups. Commonly used flocculants include primary, secondary, ternary ammonium (for example, polyethylenimine hydrochloride), quarter ammonium (poly-2-methacryloyloxythrimethylammonium chloride), sulfonium (poly-2-acryloxyethyldimethsulfonium chloride), and cationic polyacrylamide.
(3) Nonionic polyelectrolyte carries no charge on the functional group (but generally is slightly charged in the aqueous solution). Common flocculants include polyester [poly(ethylene oxide)], polyols [poly(vinyl alcohol)], and polyacrylamides.

Two mechanisms, charge neutralization and bridging, mainly correspond to the observed flocculation behavior between the polyelectrolyte and the suspended particles. In the sludge suspension with negative-charged colloids, as the polymer chain is much smaller than the particle diameter, it adsorbs on the particle surface as a "patch." This adsorption partially neutralizes the surface charge and forms the opposite charged zone. As the particles collide, coagulation occurs and further improves the flocculation. On the other hand, when the length of "tails" or "trains" formed by polymer chains absorbed on the particle surface exceeds the repulsion distance between two colloids, flocculant molecules can interact with other flocculant molecules and particles simultaneously and create a "bridge" to connect two-particle aggregates into even bigger aggregates. For the nonionic polyelectrolyte or the anionic polyelectrolyte, the bridging becomes the predominant mechanism. Generally, the flocs produced through charge neutralization are relatively small and weak, while those produced through bridging are large and strong.

2.3. *Conditioner Choice*

Table 1 compares sludge conditioning performance using metal salts or polyelectrolyte. Apparently, the required dosage for metal salt is much higher than that for organic polymer. Consequently, coagulation by metal salt significantly increases the volume of the sludge that requires disposal. Additionally, the corresponding heat value and the digestibility of the coagulated sludge will be decreased. However, the price for the polymer is much higher than that for the inorganic coagulant. Figure 5 shows the magnified photographs of a flocculated sludge, the mean size of which increases from approx 5 μm for the original sludge to over 200 μm following flocculation. Additionally, unlike inorganic salt, the polymer cannot reduce the odor potential following conditioning.

Coagulant choice depends on sludge characteristics and subsequent treatment process. The "optimum" floc characteristics differ with treatment process. Table 2 lists the required floc characteristics for several solid–liquid separation processes. Notably, big floc size is essential for good clarification and filtration process performance, but is relatively insignificant for flotation performance. The inorganic coagulant generally produces porous and weak flocs, which worked efficiently during the clarification or low-shear filtration stages (such as in the pressure filter). While floc strength is important

Table 1
Comparison between Coagulation and Flocculation Using Inorganic Salts or Organic Polyelectrolyte

	Inorganic salt	Organic polyelectrolyte
Coagulation efficiency	Fair	Good
Dosage	High	Low
Floc size	Small	Large
Compressbility	Low	High
Floc strength	Weak	Strong
Corrosion potential	High[1]	Low
Suspension pH	Reduced	Unchanged
Cost	Low	High
Cake volume	Increased	Unchanged
Sludge digestibility	Decreased	Unchanged
Heat value	Decreased	Unchanged
Temperature/pH sensitivity	Sensitive	Insensitive
Dosing range	Wide	Narrow
Odor potential	Reduced[2]	Unchanged

[1]Particularly for ferric salts.
[2]Particularly for lime application.

for high-shear application, for example, in the belt filter press or the centrifuge, high molecular weight organic polyelectrolyte must be used to prevent floc breakage during dewatering. Table 3 lists some general guidelines for coagulant selection.

Owing to the high cost of flocculant, dual conditioning, namely, one flocculant combined with other conditioner (inorganic flocculant or other flocculant with different molecular weight or charge density), could help improve the dewaterability. Addition sequence may influence floc formation.

2.4. *Optimal Dose*

The so-called "optimal dose" of conditioner is that which can produce the effluent with the lowest solid content, the fastest filtration operation, and/or the dewatered cake with the lowest moisture content. Apparently, the optimal dose is an operationally

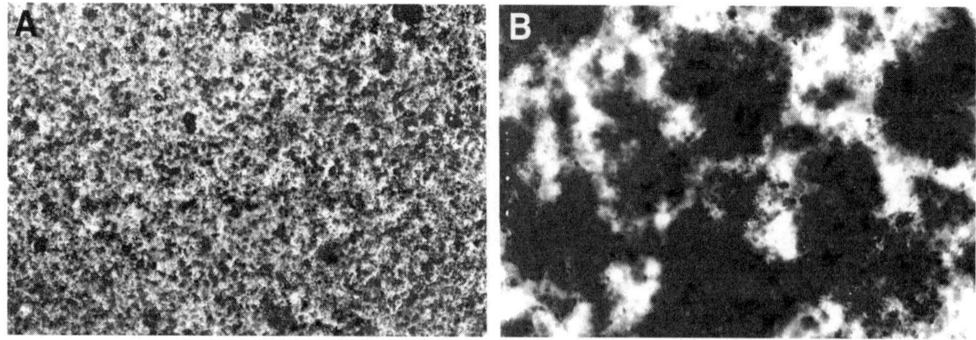

Fig. 5. Microphotographs of mixed sludge collected at a petrochemical plant. **(A)** original sludge; **(B)** Sludge flocculated with a high-MW cationic polyeletrolyte.

Table 2
The "Best" Floc Characteristics for Various Solid–Liquid Separation Processes

Solid–liquid separation	Best floc characteristics
Clarification	Large floc, loose floc structure, high permeability, with charge reversed flocs
Sedimentation	Large floc size, dense floc structure, regular floc shape, compressible, with no surface charge
Filtration	Large floc size, loose floc structure, high floc strength, high permeability, incompressible, with surface charge
Centrifugation	Large floc size, dense floc structure, incompressible cake, high floc strength
Flotation	Low floc density, high floc strength, uniform floc size, hydrophobic floc surface
Consolidation	High elasticity, dense floc structure, low bound water content, no surface charge
Electroosmosis	Low compensation voltage, with surface charge, large floc size, loose structure, high permeability

defined value that depends on the intended application. Restated, the dose that produces the driest dewatered cake does not necessarily yield the highest flow rate at filtration. An appropriate index should be selected to characterize the floc characteristics to determine the corresponding optimal dose.

For instance, Fig. 6 illustrates the zeta potentials of a biological sludge gathered at a food manufacturing plant subjected to flocculation using a cationic polyelectrolyte. As this figure reveals, the surface charge of the flocs is neutralized at a flocculant dose of 30 g/kg dried solids (DS). Moreover, the sludge dewaterability is also improved at this specific dose. The surface charge becomes positive at a high dose, indicating charge reversal. Meanwhile, the sludge dewaterability also worsens. This observation indicates that this specific sludge–flocculant system is surface-charge-controlled and the zeta potential offers an appropriate index for monitoring the flocculation performance.

Figure 6 also displayed the zeta potentials for the flocculated flocs subjected to ultrasonication. The newly exposed surface following ultrasonication means that the negatively charged surface requires more flocculant to achieve charge neutralization.

Laboratory tests could determine the optimal dose used in full-scale plants. As noted above, the appropriate laboratory tests should be selected to properly reflect the sludge

Table 3
Some General Guidelines for Conditioner Selection

Solid–liquid separation process	Recommended conditioner
Thickening	Inorganic salt or high-MW cationic polyelectrolyte
Clarification	Inorganic salt or low-MW polyelectrolytes
Filtration	Cationic polyelectrolyte with high MW, or dual conditioning with low-MW cationic polyelectrolyte followed by high-MW anionic polyelectrolyte
Centrifugation	Polyelectrolyte with very high MW

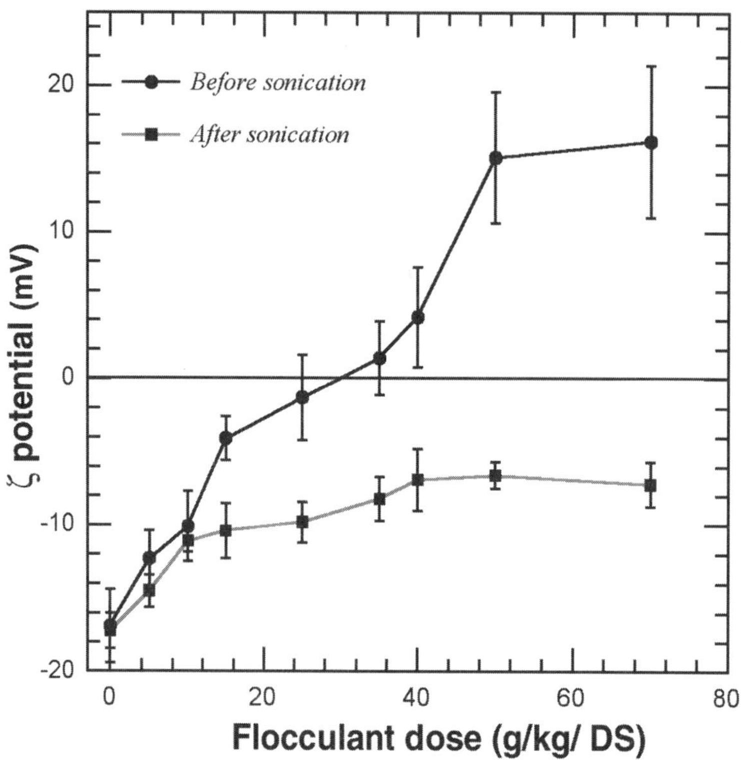

Fig. 6. The zeta potentials of biological sludge collected at a food manufacturing plant using high-MW cationic polyelectrolyte.

characteristics that control the process performance. Figure 7 shows the specific resistances of filtration measured for a biological sludge subjected to flocculation, using three cationic flocculants with similar molecular weight but different charge densities.

Apparently the KP201C (highest charge density) could yield the filter cake with the least filtration resistance at a dose of 18 g/kg-DS. Meanwhile, the flocculant carrying the lowest charge density (KP-108) requires the highest dose to reach a minimum resistance 2.5 times that achievable by dosing KP-156T.

The dosed conditioners would inevitably accumulate in the sludge. The possible effects of the polyelectrolyte flocculant remaining in the dewatered cake remain unclear. Particularly, the cationic polyelectrolytes most widely applied in sludge conditioning generally exhibited high toxicity to aquatic animals.

3. DEWATERING PROCESSES

3.1. *Dewatering Processes*

The solid-liquid separation could be classified into the following four categories: (1) pretreatment; (2) thickening; (3) filtration, and (4) post-treatment. Sludge dewatering reduces the sludge volume to facilitate the subsequent treatment/disposal processes. Inefficient sludge dewatering could significantly increase transportation, handling and final disposal costs. Table 4 lists the required solid fraction of selected treatment or disposal processes.

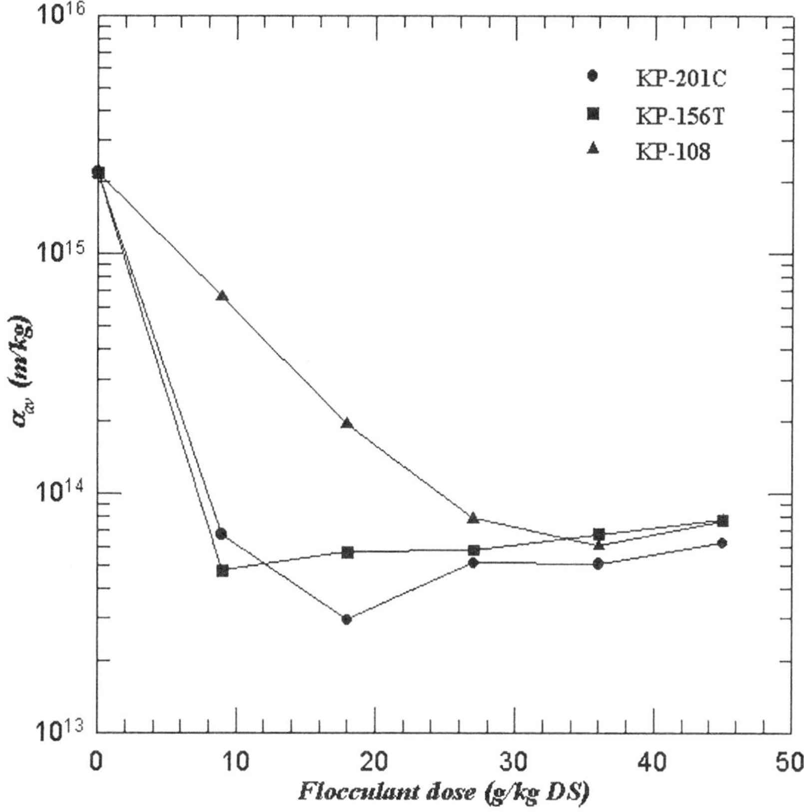

Fig. 7. The specific resistance of filtration for an activated sludge from a pulp and paper plant. The optimal dose of the three coagulants could be identified at the minimum of $\alpha_{\alpha\omega}$ versus dose curves.

3.2. *Sludge Thickening*

Sludge thickening/dewatering processes could be classified into "mechanical" and "non-mechanical" processes. Figure 8 illustrates the relative moisture removal ratio in the various dewatering stages. Over 75–80% of moisture removal occurs during the thickening stage. Proper conditioning can yield a drier cake following the filtration stage. The surface loading for waste activated sludge ranges between 18–22 kg/m^2-d,

Table 4
Required Solid Content for Various Treatment Processes

Subsequent treatment	Required moisture content
Pipeline transportation	>85%
Lime stabilization	<85%
Composting	<70%
Anaerobic digestion	3–6%
Transportation/storage	As low as possible
Applying liquid sludge on land	>94%
Incineration	<60% (self-ignition)
Landfill	As low as possible

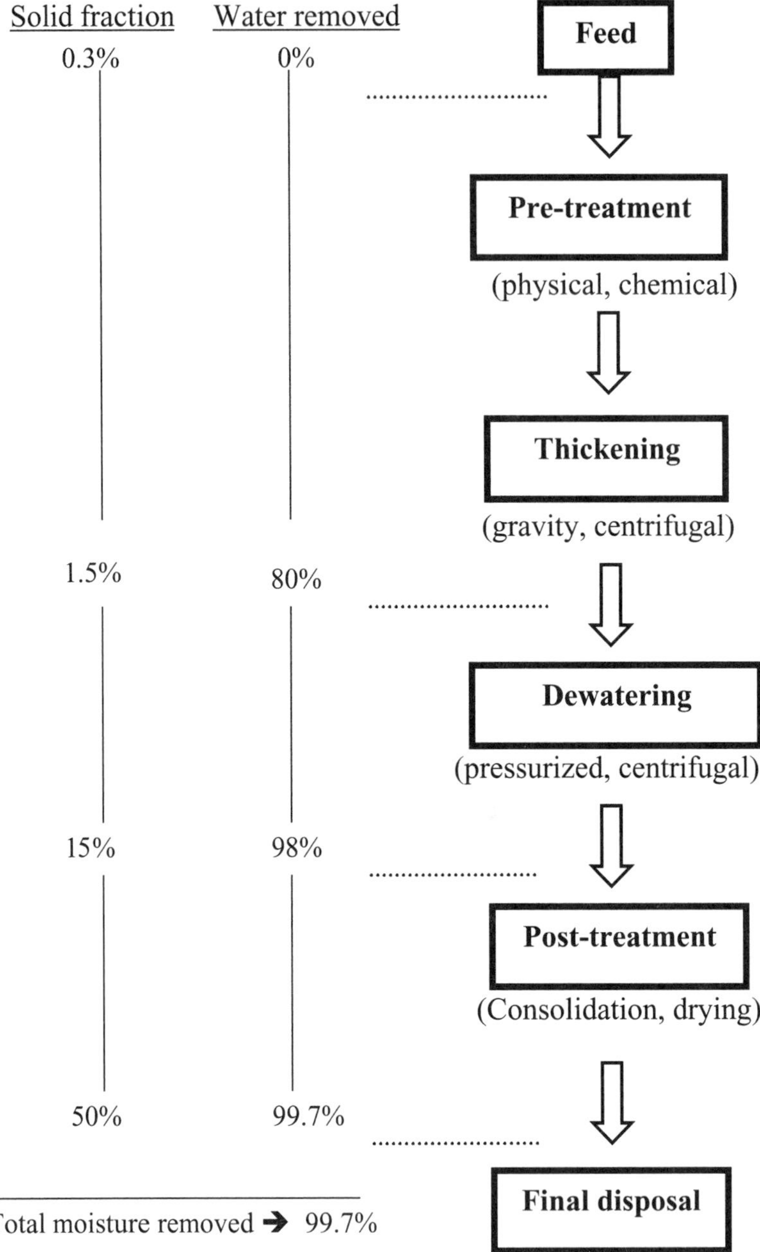

Fig. 8. Relative moisture removal ratio in four stages in sludge dewatering.

that of raw primary sludge ranges between 90 and 110 kg/m²·d, and that of mixed sludge ranges between 35 and 100 kg/m²·d.

Sludge thickeners include the gravity thickener, centrifugal thickener, and flotation thickener. The thickener design is mainly based on total flux method obtained in laboratory tests. The major assumptions in conventional thickener design include:

(1) bench-scale tests simulate the full-scale processes, (2) settling is a steady-state process, and (3) local floc settling velocity depends on concentration only. Table 5 lists the typical underflow concentration for gravity settled or centrifugal settled sludges. For waste activated sludge, the gravity thickener yields a concentrated solids sediment fraction of 5–7%, while the centrifugal settling yields a solid fraction of 9–11%. Meanwhile, for the chemical sludge the corresponding solid fractions are higher, at 15–17% for gravity settling and 19–24% for centrifugated sediment.

The steady-state operation of gravity thickener depends not only on the solids flux of the settling suspension, but also on the proper design and tuning of underdrain control. Figure 9 presents a record of sediment height in a full-scale thickener. The poorly controlled underdrain pump caused major fluctuations in sediment height, causing failure of the steady-state assumption used in flux theory. Operational performance of this full-scale thickener was poor.

3.3. *Sludge Dewatering*

The drying bed dewaters sludge through gravity draining and air evaporation. The sludge, normally digested, was spread on a sand bed or other medium at a thickness of 20–30 cm. Figure 10A schematically presents the configuration of a sand bed with underdrains. In 1 d, the sludge solid fraction could reach 12–15%, generally by water drainage. Air drying then could enable the solid fraction of the sludge to reach 20–25% within a few days. Owing to rapid evaporation, thermal stress could develop on the cake surface and cause cracking. Moisture permeability is also markedly reduced because a skin layer develops on the dried cake. Finally, the sludge can attain 60–70% solids in 6–12 mo. The dried sludge then can be removed and the process repeated.

Table 6 lists the advantages and disadvantages of the use of the drying bed. The relatively long drying time means that a large land area is needed. Odor is another essential problem in operation. However, the operation does not require skillful workers. In the absence of neighbors, the operational costs also are low compared to mechanical dewatering.

The belt filter press is a widespread sludge dewatering device, and contains an endless filter belt and a press belt. The belts have numerous "bends" for saving the footprint of the device. The sludge is fed to the gravity drainage zone to remove 60–75% moisture. The sludge then is squeezed between two belts to further remove moisture using rollers with decreasing radius to increase sludge shearing. Figure 10B displays the device schematics.

The operation is continuous and has a surface loading of 45-550 kg per hour per meter of belt width. The sludge can be dewatered to 10–20% solids for activated sludge, 25–30% for raw primary sludge, and 24–30% for digested sludge (Table 5).

Table 7 lists the advantages and disadvantages of using the belt filter press. The press has a small footprint and produces low noise and vibration. One major disadvantage of using the belt filter press is the odor problem. Additionally, appropriate conditioning with high-molecular-weight organic polyelectrolyte is a prerequisite to the success of the filter.

The plate and frame press consists of vertical plates covered with a filtering medium held side by side in a frame (Fig. 10C). On the cession of filtrate flow, the dewatered cake is discharged from the unit and carried away by a conveying system. The filtering

Table 5
Solid Content of the Dewatered Cake[1]

	Original solid fraction	Gravity settling	Centrifugal settling[2]	Vacuum filtration	Drying[3]	Centrifugal filtration[2]	Freeze/ thawed[4]	Consolidation[5]
Alum sludge[a]	4.0–6.0	10–16	18–23	22–27	35–40	38–43	40–45	57–62
Sewage sludge[b]	0.8–0.9	1–5	7–12	10–15	23–28	21–25	25–30	20–54
Activated sludge[c] (food industry)	0.7–1.4	5–7	9–11	10–18	28–33	24–28	25–35	20–35
Scum sludge[d] (fiber plant I)	2.0–4.0	7–10	9–14	17–21	24–30	21–25	30–40	30–35
Mixed sludge[e] (fiber plant II)	1.0–1.5	5–8	10–12	13–16	21–25	18–20	NA	27–32
Chemical sludge[f] (tannery plant)	8.0–9.0	15–17	19–24	24–28	24–30	NA	NA	29–34
Scum sludge[g] (refinery plant)	1.5–3.0	4–6	5–8	5–8	NA	NA	20–30	10–15
Heavy metal sludge[h]	2.0–3.5	NA	NA	NA	NA	NA	NA	40–50
Copper sludge[i] (electroplating plant)	4.5–5.0	NA	NA	NA	NA	NA	40–50	40–50
Clay suspension[j]	5.0	14–17	15–20	20–25	50–55	47–52	20–30	66–70
Kaolin suspension[j]	5.0	13–15	10–20	20–25	NA	44–50	20–30	62–67

[1]Data collected in a laboratory tester. Numbers are in units of wt%.
[2]Centrifugated at 1000 rpm for 1 h.
[3]Vacuum drying at 50–60°C for 1 h.
[4]Frozen at –17°C for 1 d, then vacuum filtering the thawed suspension.
[5]Consolidating at 3000 psi.
[a]Collected at PingTsan Water Works using PACl as coagulant.
[b]Collected at primary clarifier of a sewage treatment plant of Taipei.
[c]Collected at the recycling stream of activated sludge.
[d]Collected at the flotation basin of the wastewater-treatment plant.
[e]Mixed chemical and biological sludge.
[f]Collected at the primary clarifier, with 0.2% oil content.
[g]Collected at the scum storage basin with 0.5% oil content.
[h]Mainly ferric hydroxide.
[i]Mainly cuprous hydroxide.
[j]Synthetic raw waters.

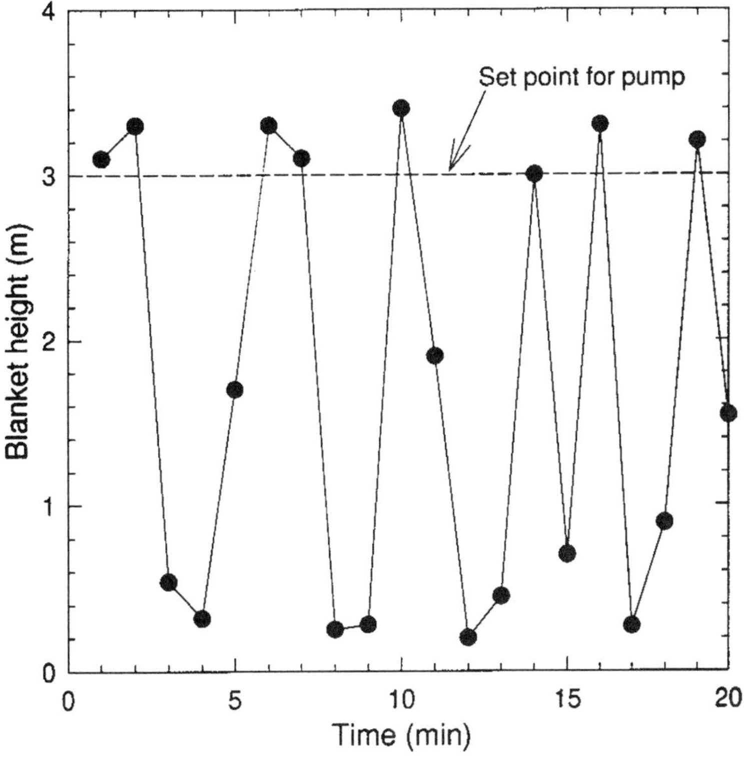

Fig. 9. Time variation of sediment height for a full-scale gravity thickener. The sediment height fluctuates considerably with time.

medium is then backwashed and the next filtration cycle begins. The sludge can be dewatered to 15–20% solids for activated sludge, 20–35% for raw primary sludge, and up to 35% for digested sludge. Diaphragm installation allows the activated sludge to be dewatered to achieve water content exceeding 30%.

Table 8 lists the advantages and disadvantages of using the plate and frame press. The press is a batch process. The low stress applied in the press enables the inorganic coagulant to be used for conditioning. Frequent cloth cleaning and device maintenance is one of the key disadvantages of the press.

The centrifuge uses centrifugal force (500–3000g) to increase the sedimentation rate of solids in the sludge (Fig. 10D). Continuous solid bowl conveyor type centrifuge has been used for over half a century to dewater sludge. Other centrifuges, such as disk type and basket type also are often used. The sludge is fed into the centrifuge together with the flocculants. The bowl is then rotated, moving the solid to the outer wall of the bowl and building up the cake. The cake is pushed by the conveyor toward the narrow end of the bowl, where further dewatering can be achieved. The centrifuge is cheaper and requires less space than other mechanical methods, and also causes less odor problems. Centrifuge is a more effective method of dewatering primary sludge than activated sludge. Bowl speed, sludge feed rate, and relative conveyor speed are the primary control

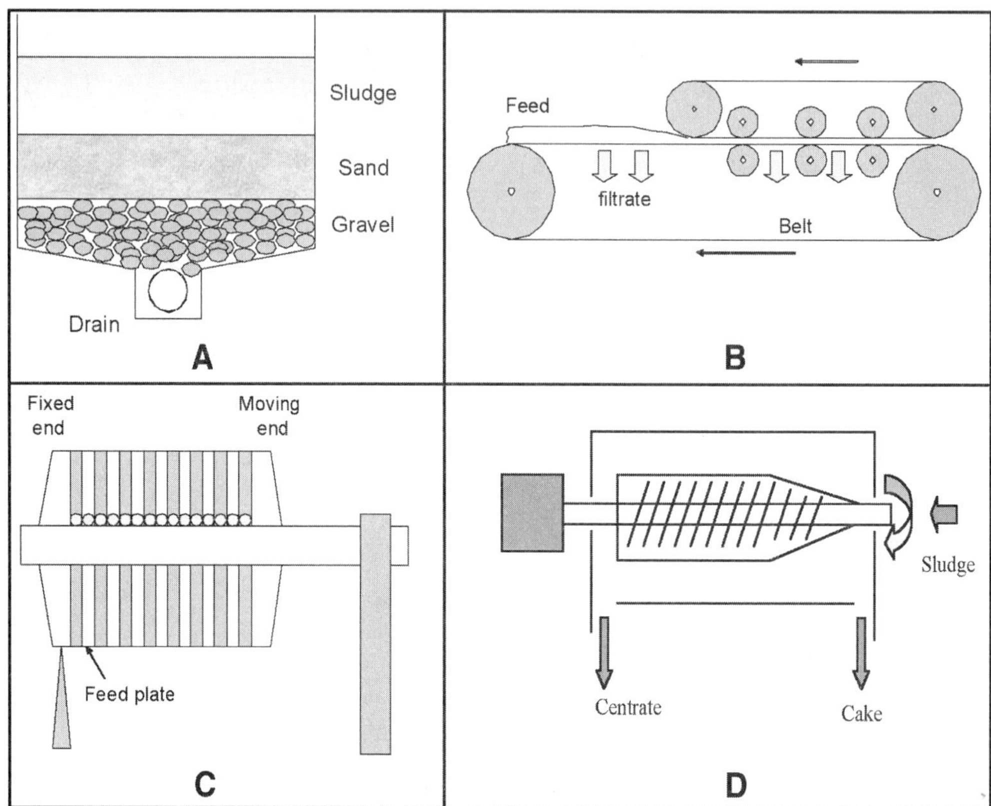

Fig. 10. Typical dewatering devices. **(A)** Sand bed, **(B)** belt filter press, **(C)** plate and frame press and, **(D)** centrifuge.

parameters. The residual moisture in the dewatered cake ranges from 8 to 10% for activated sludge, 25 to 35% for primary sludge, and 15 to 30% for digested sludge. High-solid centrifuge can reach a high solid content in dewatered cake.

Table 9 lists the advantages and disadvantages of the use of centrifuge. High-molecular-weight polyelectrolyte should be used before centrifugal dewatering. The press is a batch process.

Use of a centrifuge involves abrasion of the conveyor scroll and bowl. Another major problem with the use of the centrifuge is the scale formation that originates from the

Table 6
Advantages and Disadvantages of Sludge Dewatering Using Drying Bed.

Advantages	Disadvantages
No need for skilled personnel	Needs significant land area (300 m^2/1000 PE)
Low operation cost	Odor potential
	Vector control
	Suitable in dry or freezing climates

Table 7
Advantages and Disadvantages of Sludge Dewatering Using Belt Filter Press

Advantages	Disadvantages
Continuous operation	Odor potential
Low power consumption	Low consolidation pressure
Low noice and vibration	Sensitive to feed condition and conditioning efficiency, suitable to flocs of high strength
Small foot-print	Higher flocculant dose
Few cake pick-up problem	Needs skilled personnel

centrifugation of inorganic conditioner or lime biosolids. However, the newly developed centrifuge can minimize corrosion by using a durable material. Notably, the power and maintenance costs of centrifuge generally exceed the other mechanical dewatering processes.Normally, the sludge is best dewatered if it has porous and strong flocs and a low-viscosity liquid phase.

4. STABILIZATION PROCESSES

The sludge produced from treating domestic and industrial wastewater generally requires stabilization before final disposal or use. Treatment processes have been proposed either to improve efficiency of subsequent anaerobic stabilization and/or dewatering of the sludge or to boost its hygienic properties.

4.1. Hydrolysis Processes

These processes include the use of thermal energy, freezing and thawing, chemicals (such as ozone, acids, alkali, enzyme), mechanical energy (for example, high pressure, stirred ball mills, and ultrasound), and irradiation. Some investigations have considered combined treatment with alkaline addition and ultrasound. Aqueous-phase treatment processes have the advantages of eliminating the need for dewatering and producing few or no off-gases. Hydrolysis can increase subsequent anaerobic or aerobic digestion efficiency. Only a few full-scale applications currently exist (mostly using pasteurization). However, the potential to adopt hydrolysis processes to improve digestion efficiency recently has attracted increasing attention.

Changes in pH value yield a significant reduction in microbial density and the release of COD from the sludge body, particularly at pH>10. For example, a 2-h

Table 8
Advantages and Disadvantages of Sludge Dewatering Using Plate and Frame Press

Advantages	Disadvantages
High solids content achieveable	Batch process
High pressure applicable	Needs precoat material to prevent medium clogging
No need for strong flocs	Sensitive to conditioning performance
	Cake pick-up problem
	Labor intensitve process

Table 9
Advantages and Disadvantages of Sludge Dewatering Using Centrifuge

Advantages	Disadvantages
Continuous process	Needs skilled personnel
High solids content achieveable	Needs polyelectrolyte as flocculant to produce strong flocs
No odor problem	High power consumption rate
Suitable as pre-incineration stage	Noice and vibration
	Abrasion problem of bowl and conveyor scroll

treatment at pH 11 achieves 99% reduction ratio in total coliform bacteria compared to the original sludge. The filamentous bacteria were exposed outside the flocs given noticeable dissolution of the constituent components (Fig. 11B). However, at pH 3 the corresponding reduction ratio for total coliform bacteria was 87%. Acidification produces large floc size, thus improving filterability (Table 10). When using acids or alkali, the fact that sludge salinity will increase and cause possible disposal problems must be considered.

Thermal treatment has been recommended as the most appropriate technique for inactivating microorganisms in wastewater and sludge. As illustrated in Table 10, total coliform decays to very low levels ($<10^4$ MPN or CFU/mL). The SCOD also increases markedly after heating. However, because the floc global structure is not significantly disrupted, treated sludge still displays good dewaterability (Fig. 11C). Heating time is critical on the hydrolysis efficiency. Thermal treatment requires more energy than other mechanical processes, but cheaper thermal energy (such as waste steam, if available) can be used rather than the electrical energy that is required for mechanical processes.

Freezing and thawing treatment significantly enhance filterability (low CST and large d_f) by markedly adjusting floc structure (Fig. 11D) and reducing microbial density levels. In fact, the treated sludge can be classified as Class B sludge. The SCOD has increased to five to six times that of the original sludge. The ability of freeze/thaw treatment to condition sludge declines with increasing freezing speed. Slow freezing is required for sufficient sludge conditioning. In regions where natural freezing is feasible, such as in North America, running costs could be very low.

Ultrasound is a pressure wave that causes solution cavitation, causing local temperatures of over 1000°C and pressure exceeding 500 bars. Such cavitation can disrupt cell walls. In the present tests, at an intensity of 0.1 W/mL, the CST, particle size, and total coliform levels change only slightly following sonication. COD release into the supernatant is also limited. However, at 0.3 W/mL, the particle size and microbial density levels are substantially reduced. The SCOD increases up to 10 times following sonication, thus facilitating the subsequent digestion process. Meanwhile, sludge dewaterability reduces significantly by the significant disintegration of the floc structure following ultrasonication (Fig. 11E).

Alum conditioning reduces suspension pH value, thus influencing SCOD and microorganism levels (Table 10).

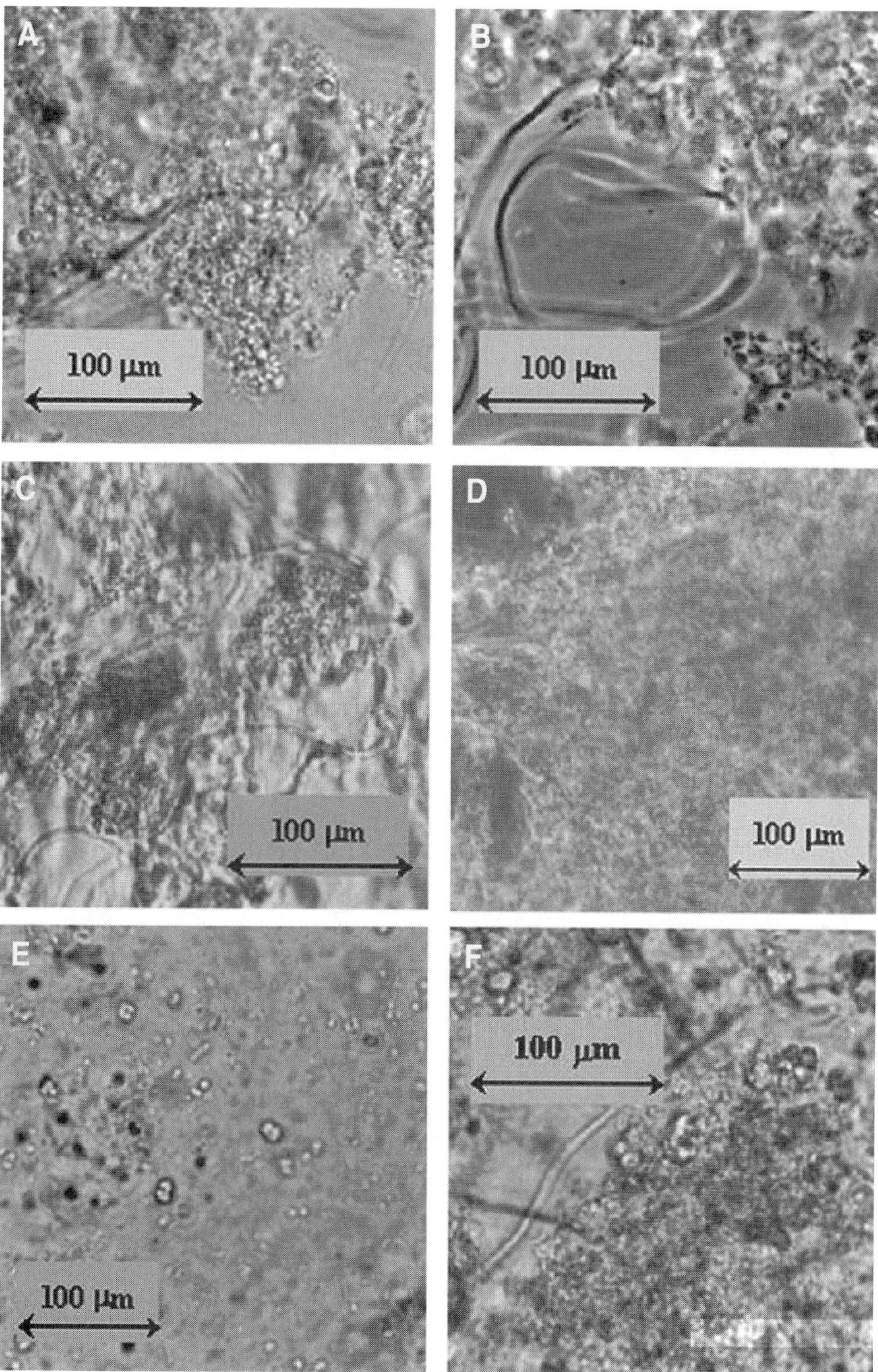

Fig. 11. The appearances of sludge floc subjected to pH change: **(A)** original sludge, **(B)** alkali treated at pH 11, **(C)** thermal treatment at 70°C for 30 min, **(D)** freezing and thawing at –17°C for 24 h **(E)** ultrasonidation at 0.33W/mL for 40 min; **(F)** alum coagulation at 40 ppm for 2 h.

Table 10
Effects of pH change on sludge floc characteristics

	Effects of pH change						
pH	4	5	6	6.9	8	9	10
ζpotential (mV)	–8.8	–13.4	–16.7	–16.7	–17.2	–17.5	–21.4
d_f (μm)	162	166	139	103	98	94	92
CST (s)	113	163	192	197	211	259	304
ZSV (μm/s)	20.4	5.4	0	0	0	0.1	0
$S_{Coliform}$	0.019	0.044	0.16	1.0	0.74	0.2	0.085
SCOD/TCOD	0.01	0.009	0.009	0.007	0.019	0.025	0.034

	Thermal Treatment (20 min)				
Temperature	40°C	50°C	60°C	70°C	80°C
ζpotential (mV)	–16.2	–18.1	–22.4	–23.5	–21.2
d_f (μm)	103	103	99	100	95
$S_{Coliform}$	0.99	0.47	0.27	0.22	0.25
SCOD/TCOD	0.029	0.074	0.130	0.185	0.263

	Freezing and Thawing					
		After freeze/thawed conditioning Freezing speed (μm/s)				
	Prior to freezing	14	4.7	2.3	0.72	0.51
ζpotential (mV)	–16.5	–11.2	–13.2	–11.1	–12.1	–12.4
d_f (μm) (Top)	103	145	202	268	315	321
CST (s)	197	41.8	42.8	39.9	38.5	41.2
ZSV (μm/s)	0	14	111	202	295	312
$S_{Coliform}$	1.0	0.98	0.49	0.36	0.28	0.27
SCOD/TCOD	0.007	0.007	0.008	0.012	0.018	0.017

	Ultrasonication						
	Original	0.11 W/mL 20 min	0.11 W/mL 40 min	0.11 W/mL 60 min	0.33 W/mL 20 min	0.33 W/mL 40 min	0.33 W/mL 60 min
ζpotential (mV)	–16.4	–15.9	–15.8	–16.7	–17.1	–17.5	–17.0
CST (s)	197	188	205	218	305	406	489
ZSV (μm/s)	0	0	0	0	0	0	0
$S_{Coliform}$	1.0	0.96	0.41	0.35	0.42	0.33	0.07
SCOD/TCOD	0.007	0.007	0.008	0.008	0.035	0.060	0.11

	Alum coagulation					
Coagulant dose (g/kg DS)	0	10	20	40	60	80
pH	6.88	5.19	4.64	4.24	4.06	3.91
ζpotential (mV)	–16.4	–16.5	–12.1	–10.1	–11.2	–9.4
d_f (μm)	103	105	107	105	109	110
CST (s)	197	188	193	180	175	179
ZSV (μm/s)	0	0	0.4	2.1	1.5	2.8
$S_{Coliform}$	1.0	0.99	0.99	0.98	0.96	0.93
SCOD/TCOD	0.007	0.006	0.006	0.005	0.003	0.003

Table 11
Comparisons between hydrolysis processes[a]

	Disinfection	Hydrolysis	Dewaterability
Adding acid	+	—	—
Adding alkali	++	++	X
Adding electrolyte	+	—	—
Alum coagulation	+	—	+
Polymer flocculation	—	—	++
Thermal treatment	+++	++	—
Freezing and thawing	+	+	++
Ultrasonication	++	++	XX

[a]++: considerably enhanced, + : enhanced, — : no effect, X : deteriorating, XX : very deteriorating.

Hydrolysis should be applied to the activated sludge rather than the primary sludge. Particularly, a high applied energy level is required to disrupt the cell walls and improve digestion. Strong oxidants could be useful for disintegrating the anaerobic sludge because they have low potential for further anaerobic degradation without disintegration. The application of enzymes to primary sludge with a high content of lignocellulosic material appears most appropriate. Anaerobically digested sludge that has been prehydrolyzed requires more flocculants than untreated sludge. Among the treatment processes discussed herein, only the freeze/thaw treatment significantly increased sludge filterability. All of the hydrolysis processes disinfected the microorganisms. In some cases, the treated sludge could be classified as Class A or Class B. Notably, thermal disintegration generated difficult to degrade organic compounds and strong odors. Treatment temperature exceeding 200°C could reduce organic matters into melanoidines, and, moreover, had a low bio-gas yield.

Capital and O&M costs vary considerably among treatment processes. Mechanical disintegration is extremely promising for enhancing digestion efficiency, because of its relatively low capital costs and energy consumption, and the release of no harmful off-gases. Comparing other pretreatment methods that could provide the same levels of COD-release efficiency, ultrasound and thermal pretreatment exhaust more energy than other processes.

Table 11 compares the effects of typical hydrolysis processes on sludge characteristics.

4.2. Digestion Processes

The digestion processes discussed here include lime stabilization, aerobic digestion, anaerobic digestion, and composting. Sludge lime stabilization is achieved by adding sufficient lime to sludge to raise the pH to 12 or more to disinfect microorganisms. The recent progress is to adopt the so-called lime post-treatment process (Fig. 12), by blending lime with the dewatered sludge instead of adding lime into undewatered suspension. Both hydrated lime and quicklime can be used, but quicklime is normally preferred owing to its high exothermic heat when reacting with water. Table 12 lists the advantages and disadvantages of lime stabilization processes. The stabilized sludge can serve as the Class B sludge based on US standards.

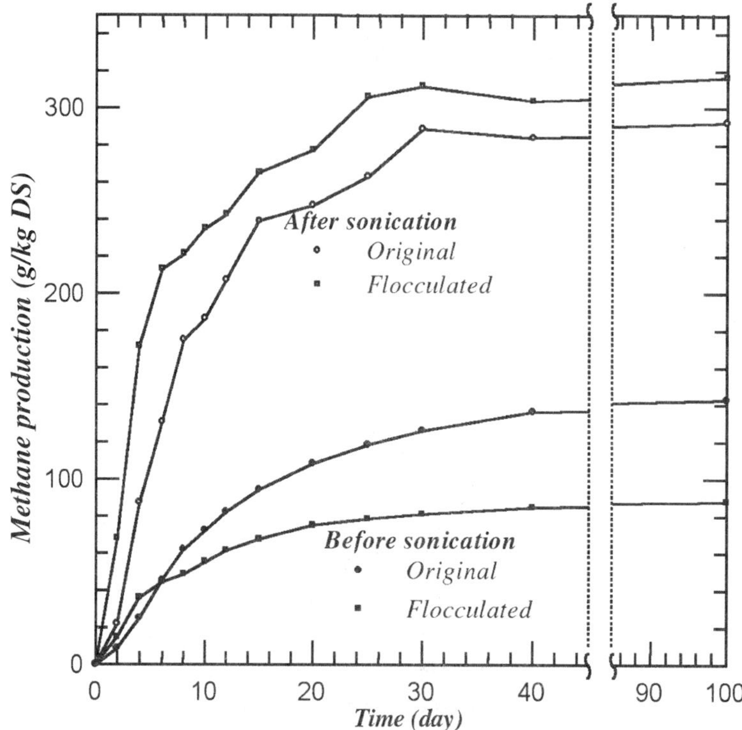

Fig. 12. Anaerobic digestion of sewage sludge. Treatment with original and with hydrolysed (ultrasonicated) sludges.

Lime treatment would not destroy the organic substances in the sludge, and hence an overdose of lime is required to prevent pH reduction during storage

Aerobic digestion simply aerates the sludge in an open basin until the biodegradable substance is oxidized sufficiently to become nuisance-free. The configuration resembles the activated-sludge process, but the operation is in the endogenous respiration phase. This process can be used to treat primary sludge, waste-activated sludge, and mixed sludge.

A typical aerobic digestion process comprises three phases: (1) mesophilic stage, dominated by fungi and acid-producing bacteria; (2) thermophilic stage, during which the temperature was increased to 70°C with the thermophilic bacteria and fungi, during which the maximum destruction and stabilization occur; (3) cooling stage. Most water

Table 12
Advantages and Disadvantages of Post-Lime Stabilization.

Advantages	Disadvantages
Dry lime can be used	High pathogen reduction rate
No requirement for subsequent dewatering	Increased sludge disposal volume
High pathogen reduction rate	Potential to pathogen revival during storage
Low cost	

Table 13
Advantages and Disadvantages of Aerobic Digestion of Sludge.

Advantages	Disadvantages
High volatile reduction ratio	High power cost (for aeration)
Low-BOD supernatant liquor	Poor dewaterability to digested sludge
Odorless, stable end product	sensitive to temperature change, tank material, etc.
High fertilizer value of end product	No valuable material recovered
Easy operation	pH would drop during digestion
Low capital cost	

can be released from the stabilized sludge during the aerobic digestion process, and, moreover, the pH value is stabilized. The following reaction dominates when digesting primary sludge:

$$\text{Organic matters} + O_2 \rightarrow \text{cellular cells} + CO_2 + H_2O$$
$$\text{Cells} + O_2 \rightarrow \text{digested sludge} + CO_2 + H_2O$$

During activated sludge digestion, the following reaction also occurs:

$$NH_4^+ + 2O_2 \rightarrow NO_3^- + 2H^+ + 2H_2O$$

Therefore, sludge pH and alkalinity both are reduced when nitrification occurs.

Table 13 lists the advantages and disadvantages of utilizing aerobic digestion. Recent developments in the aerobic digestion process have focused on heat recovery from the digestion system by utilizing higher digestion temperature and thicker sludge feed.

Anaerobic digestion is one of the oldest sludge treatment processes that solubilizes and ferments complex organic substances using microorganisms in the absence of oxygen. The products of anaerobic digestion include methane, carbon dioxide, other trace gases, and the stabilized sludge. The digestion process can effectively inactivate the pathogens in the sludge.

In digestion the long-chain organic matters are degraded to smaller particles via hydrolysis, which is normally conducted at a low reaction rate. Therefore, the digestion rate could be considerably increased if the hydrolysis processes in Section 4.1 were adopted. Following this step, the microbiological pathway is shown as follows:

$$C_6H_{12}O_6 \rightarrow 3CH_3COOH$$
$$3CH_3COOH + 3NH_4HCO_3 \rightarrow 3CH_3COONH_4 + 3H_2O + 3CO_2$$
$$3CH_3COONH_4 + 3H_2O \rightarrow 3CH_4 + 3NH_4HCO_3$$

The first equation produces acids, thus reducing pH value. The second equation then neutralizes the acid and buffers the suspension pH value. Meanwhile, if methanogenesis occurs in the third equation, the buffer is recovered and the methane is produced. If insufficient buffer exists, the pH decreases as determined by the first equation, and the conversion to methane then is inhibited as indicated by the third equation.

The most widespread anaerobic digester is the single-stage digester with heating and mixing. The digestion temperature ranges from 30 to 38°C for mesophilic digestion, and

Table 14
Advantages and Disadvantages of Anaerobic Digestion of Sludge.

Advantages	Disadvantages
Excess energy over that required by the process is produced	Easily upset and very slow to recover
Produced methane could be used to heat and mix the reactor	Heating and mixing equipments are required
Quantity of total solids for ultimate disposal is reduced	Large reactor volume is required
30–40% total solids (40–60% VS) may be destroyed	The resultant supernatant is a strong waste stream that adds loading to the WWTP
Pathogens are destroyed to a high degree	Cleaning operations are difficult
Thermophilic digestion enhances the degree of pathogen destruction	Possibility of explosion
Most organic substances in municipal sludge are readily digestible except lignins, tannins, rubber, and plastics	Gas line condensation/clogging

50 to 60°C for thermophilic digestion, with the latter providing a higher reaction rate and increased pathogen destruction ratio. Table 14 lists the advantages and disadvantages of using anaerobic digestion for sludge treatment. Owing to the low reaction rate for the anaerobic processes, the digesters have large volume, and the process is easily offset. However, anaerobic digestion is the only practical process in sludge management that can produce excess energy besides that required by itself. As a rule of thumb, each kilogram of organic matter destroyed during digestion can produce 1 m^3 of bio-gas containing 60% methane.

Figure 12 displays a typical anaerobic digestion process of sewage sludge, with and without flocculated polyelectrolyte. Evidently, the digestion takes about 30–40 d and the ultimate yield is approximately 140 g-CH_4/kg-DS for original sludge or 85 g-CH_4/kg-DS for flocculated sludge. Flocculation improves sludge thickening, but compromises digestion performance. On the other hand, the previously mentioned hydrolysis process (ultrasonication in this case) not only increases digestion rate, but also improves ultimate yield, up to 310 g-CH_4/kg-DS.

Sludge composting is a digestion process in which solid organic material undergoes biological degradation to produce a stable end product, with 20–30% of volatile solids converted to CO_2+H_2O. This process requires high oxygen concentration all over the composting pile to prevent creation of an anaerobic environment.

The sludge requires sufficiently high temperature for sufficient time to inactivate pathogens. Based on US EPA standards, the entire pile should be maintained above 55°C for at least 3 d in a vessel or static pile for the treated sludge to be classified as a Class A sludge. However, too high temperature in stage III prevents bacteria activity, causing sudden temperature drop. Moreover, excessive moisture (>70%) clogs pores and thus prevents aeration, and too little moisture prevents bacteria activity (<50%). The C:N ratio should range from 25:1 to 35:1 by weight, and volatile compound ratio > 50% is required to deliver sufficient heat for composting. Control of temperature

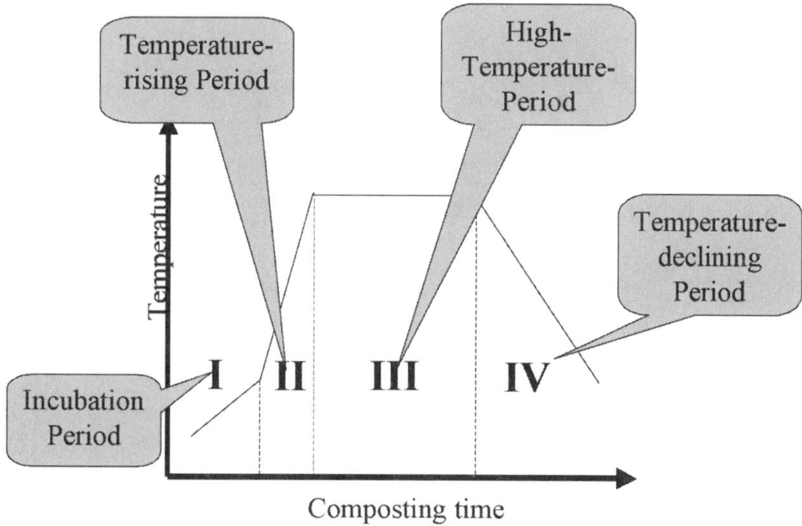

Fig. 13. The ideal temperature course of aerating composting.

course, moisture content, and sludge compositions is essential to the success of this process. Figure 13 reveals the "ideal" temperature course for the sludge compost, with the temperature level in regime III being between 55 and 65°C for at least 3 d. Bulking agent frequently is used to adjust the moisture content of the dewatered sludge. The aeration and the organic content of the sludge adjust the compost temperature. Table 15 lists the advantages and disadvantages of the various sludge composting processes.

5. THERMAL PROCESSES

5.1. *Sludge Incineration*

The first sludge incinerator was installed in Michigan in 1934. Since then, most incinerators have been multiple hearth furnace (MHF) types, which involve a vertically oriented and cylindrically shaped vessel containing 4–14 horizontal refractory hearths. Since the sludge melt is very sticky, each hearth is fitted with two or four rabble arms to rake the sludge across the hearth in a spiral fashion.

Table 15
Advantages and Disadvantages of Sludge Composting

Advantages	Disadvantages
Storable end products	Requires 18–30% DS
Saleable end products	Requires amendments
Low cost compared with incineration	Requires large land area
	High cost compared with direct land use
	Possible large amount of end products
	No market for end product
	Potential for aerosols
	Potential for odors

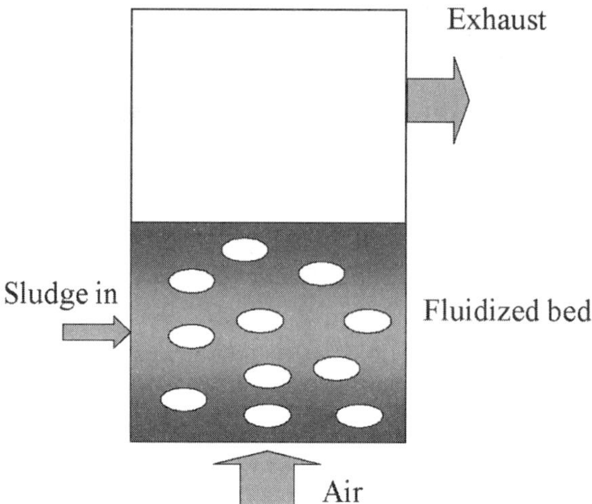

Fig. 14. The schematics of the fluidized bed furnace.

Before the 1970s incineration was a major sludge disposal method. However, significantly increased energy costs meant that incineration became the least preferred option by the 1980s. Nowadays, the energy consumption rate of modern incinerators has been minimized, landfill costs are increasing, and regulations governing land disposal of sludge are becoming increasingly stringent, meaning incineration remains a major route of sludge disposal, especially for densely populated, urban areas.

For incineration, the sludge temperature must be raised to 100°C to evaporate water from the sludge, then the water vapor temperature, air temperature, and the solid phase must be increased until ignition point is reached. This process is energy-intensive, consequently, sludge heat value is the key consideration in sludge incineration. In systems with no heat recovery, half of the available energy is required to heat fuel and air to the desired temperature.

Sludge organic fraction has a heat value of 25 MJ/kg-volatile matters. Given 40% inert fraction in solid phase and 80% residual moisture in the dewatered cake, the heat value of wet cake is reduced to 3 MJ/kg-dewatered cake, far less than that for the fossil fuel (40 MJ/kg). Generally, the moisture in the dewatered cake should be below 60% by weight to permit self-sustainable incineration. However, this value is significantly lower than that achievable by most mechanical dewatering devices. Auxiliary fuel thus is commonly needed for sludge incineration practice.

The fluidized bed furnace (FBF) has better thermal efficiency than the MHF, and thus has gained more attention during the past two decades. An FBF is a vertically oriented, cylindrically shaped, refractory-lined steel shell containing a sand bed and fluidizing air diffusers (Fig. 14). The air can fluidize the sand bed, together with the fed sludge, to a temperature of 760–820°C. The significantly better thermal contact between the violently mixed sand particles and the burned sludge provides greater thermal efficiency, and lower demand on the excess air for FBF than for MHF. Additionally, the high heat capacity of the sand bed can achieve a quick start-up. FBF presently is almost the only incinerator being installed for sludge incineration. The

Table 16
Advantages and Disadvantages of Sludge Incineration

Advantages	Disadvantages
Great volume reduction	Cause air pollution problem, requires control of heavy metals, dioxins etc
Phosphate recycling	Energy-intensive process
High contaminant destruction ratio	High capital cost
Minimisation of waste transport and disposal	Testing, operating, and controlling the process are labor-intensive and expensive
Most Suitable to densely populated countries	Ash must be disposed of Depleting natural resources
	Greenhouse gases and global warming effects

capital cost involved in installing a state-of-art FBF ranges from 800 to 1400 US$/dried tons/annum, and the operation and maintenance costs range from 70–140 US$/dried ton of sludge.

Table 16 lists the advantages and disadvantages of sludge incineration processes.

5.2. *Sludge Drying*

Drying recently has emerged as an important processing option for sludge management. Frequently used dryers include indirectly heated disc and paddle dryers, directly heated (flue gas) drum dryers, and fluidized bed driers. Most units comprise a hollow shaft with hollow discs or paddles attached to the shaft, through which steam or thermal oil is pumped to provide the thermal energy for drying. Dryer vapor is extracted via the vapor dome, after which the water is condensed and the non-condensable vapors are treated to eliminate odors.

Drying can considerably reduce sludge volume. During drying, the uneven thermal stress developed over the cake causes the formation of surface cracks. This occurrence facilitates the drying rate over the value predicted by conventional drying theory. Table 17 lists the advantages and disadvantages of the sludge drying processes.

The capital cost for installing a dryer with a capacity of 20–30 tons of sludge per day is around 5M US$, and the operation and maintenance cost ranges from 140 to 230

Table 17
Advantages and Disadvantages of Sludge Drying

Advantages	Disadvantages
Significant reduction in volume of sludge requiring disposal/reuse	Energy-intensive process
Production of a stable, pathogen-free product, normally in pelletised form	High capital cost
Minimisation of odors associated the sludge	High operational cost
Generation of a product with multiple end uses	Odor potential
	Testing, operating, and controlling the process are labor-intensive and expensive

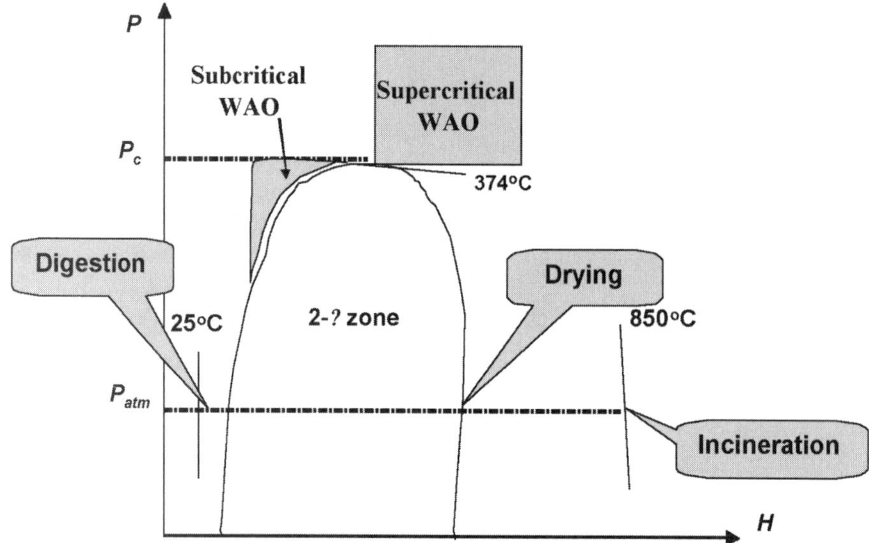

Fig. 15. Phase diagram of water and the thermal treatment processes.

US$/dried ton of sludge. O&M costs are more dependent on energy cost than is the incineration process, since the heat value of the volatile matters is not utilized in the drying. However, the end product can be used as a fertilizer, soil amendment, and fuel or raw material for other thermal processes, presenting the maximum flexibility to the final disposal options. Dozens of new drying plants are being installed globally.

5.3. *Other Thermal Processes*

Wet Air Oxidation oxidizes organic substances in an aqueous environment at temperatures between 120 and 400°C, much lower than those for conventional combustion (800 and 1500°C). The system pressure ranges between 70 and 100 atm. This process produces no ash, SO_x or NO_x. Moreover, no preliminary dewatering or drying is required; however, the feed concentration and heat value significantly influences operating costs.

Supercritical water has a solubility close to its liquid phase, and a diffusivity approaching its gas phase. Therefore, mass transfer is rapid with supercritical water. Supercritical water oxidation can completely destroy all organic material in wastewaters and sludges, including carcinogens and pathogens. Figure 15 illustrates the phase diagram of water and the regimes of typical thermal treatment processes. The abscissa is the enthalpy of water, and approximates the energy demand for conducting the indicated sludge treatment. The SCWO can operate more cheaply but with the same efficiency as the incineration process.

Thermochemical conversion produces liquid hydrocarbons from organic substrates with both thermal cracking and catalytic conversion. The first full-scale plant for sludge conversion was put into operation in Perth, and had a capacity of 25 dried tons per day. This plant is operated in an oxygen-free environment at 450°C and under atmospheric pressure.

REFERENCES

1. P. A. Vesilind, *Treatment and Disposal of Wastewater Sludge.* Ann Arbor Science, Ann Arbor, MI, 1979.
2. US EPA, *Process Design Manual for Sludge Treatment and Disposal.* EPA/625/1-74-006. US Environmental Protection Agency, Washington, DC, 1974.
3. US EPA, *Process Design Manual: Treatment and Disposal of Sludge.* EPA/625/1-79-011. US Environmental Protection Agency, Washington, DC, 1979.
4. US EPA, *Municipal Wastewater Sludge Combustion Technology.* EPA/625/4-85/015. US Environmental Protection Agency, Cincinnati, OH, 1985.
5. US EPA, *Dewatering Municipal Wastewater Sludges: Design Manual.* EPA/625/1-87-014. US Environmental Protection Agency, Washington, DC, 1987.
6. US EPA, *A Plain English Guide to EPA Biosolids Rule.* EPA-83Z/R-93/003. US Environmental Protection Agency, Washington, DC, 1993.
7. US EPA, *Process Design Manual Surface Disposal of Sewage Sludge and Domestic Septage.* EPA/625/R-95/002. US Environmental Protection Agency, Washington, DC, 1995.
8. US EPA, *Biosolids Technology Fact Sheets: Small Flow*, US Environmental Protection Agency, Washington, DC, 2004, VOL. 5, No. 1.
9. L. Spinosa and P. A. Vesilind, *Sludge into Biosolids: Processing, Disposal and Utilization.* International Water Association Press, London, UK, 2001.
10. P. Matthews, *A Global Atlas of Wastewater Sludge and Biosolids: Use and Disposal.* International Association of Water Quality Scientific and Technical Report No. 4, London, UK, 1996.

Index

D

P

Z